ISBN 978-0-364-62198-1
PIBN 10860849

THE

GEOLOGICAL

OBSERVER.

BY

SIR HENRY T. DE LA BECHE, C.B., F.R.S., &c

DIRECTOR-GENERAL OF THE GEOLOGICAL SURVEY
OF THE UNITED KINGDOM.

SECOND EDITION, REVISED.

LONDON:
LONGMAN, BROWN, GREEN, AND LONGMANS.
1853.

LONDON: PRINTED BY W. CLOWES AND SONS, STAMFORD STREET AND CHARING CROSS.

PREFACE.

It has been well remarked by Humboldt* that to behold is not necessarily to observe, that is, to compare and combine. The history of Geology, like that of all sciences depending for their effective advance on experiment or correct observation, amply proves the truth of this statement. We are not required to look far back to be fully aware of the many brilliant hypotheses which have given way before the advance of correct research. It was not that these brilliant hypotheses were intended as substitutes for sound geological knowledge, based on correct data, or that those who formed them were not as capable as any who may in after-times succeed in still farther systematically embodying the accumulated data of such times, but merely that correct observations were not then sufficiently abundant, and that powerful, and, sometimes, impatient minds supplied their place with conceptions more captivating than well founded. It is obvious that with a hundred well-established facts more can be accomplished than with ten, the deductions from which, however apparently correct, may even be fallacious as respects those derived from the consideration of the greater number. Let it not, nevertheless, be hastily concluded that the views which have passed away have not materially advanced geology, as those of a similar character have aided the progress of other sciences. Without them, though a few may have been impediments for the time, many a subject would have longer remained disregarded by its zealous investigator. Even the controversies which have from time to time appeared, many from differences of opinion arising the more readily as the subject was

* Kosmos.

less perfectly understood, gave a certain impulse to progress which the commencement of many inquiries so often demands.

The following work was undertaken in the hope that the experience of many years might assist, and, perhaps, abridge the labours of those who may be desirous of entering upon the study of geology, and especially in the field. Its object is, to afford a general view of the chief points of that science, such as existing observations would lead us to infer were established; to show how the correctness of such observations may be tested; and to sketch the directions in which they may apparently be extended. Having been, to a certain extent, founded upon a little treatise, entitled "How to Observe in Geology," long since out of print, a somewhat similar name has been retained for the present volume.

H. T. De la Beche.

CONTENTS.

Page

CHAPTER VII.

CHAPTER VIII.

CHAPTER IX.

CHAPTER X.

CHAPTER XI.

CHAPTER XII.

CHAPTER XIII.

Page

CHAPTER XIV.

CHAPTER XV.

CHAPTER XVI.

CHAPTER XVII.

CHAPTER XVIII.

Page

CHAPTER XXIV.

CHAPTER XXV.

CHAPTER XXVI.

CHAPTER XXVII.

CHAPTER XXVIII.

CHAPTER XXIX.

CHAPTER XXX.

Page

CHAPTER XXXI.

CHAPTER XXXII.

CHAPTER XXXIII.

CHAPTER XXXIV.

Page

CHAPTER XXXV.

CHAPTER XXXVI.

CHAPTER XXXVII.

APPENDIX.

INTRODUCTION.

Observations have now been sufficiently extended and multiplied to show that, during a long lapse of time, the surface of our planet has been undergoing modifications and changes. Of these the most marked have been produced by the uprise of mineral matter in a molten state from beneath that surface; by the wearing away and removal to other localities of this matter, either in its first state, after cooling, or in some secondary condition, by atmospheric influences and waters variously distributed for the time being; by the preservation of the remains of animal and vegetable life during at least a portion of this lapse of time amid deposits accumulated, for the most part, in horizontal layers beneath waters, and by the unquiet state of the earth's surface itself, from which, while considerable areas have been at different times raised slowly above, and depressed beneath the level of the ocean, whole masses of mineral matter of various kinds have occasionally been squeezed, bent, and plicated, sometimes ridged up into ranges of mountains.

To enable the geologist systematically to proceed with his researches, it became as needful for him as for other cultivators of science to have the power of classifying his observations. Of the various classifications proposed or modified at different times to satisfy the amount of knowledge of those times, it would be out of place here to make mention, further than to remark that at present a more mixed classification is often employed than seems desirable. For example, it is not unusual for the term *tertiary*, or *tertiaries*, to be applied to all accumulations posterior to the chalk of western Europe, while the other terms of secondary and primary or primitive, to which it has reference, are scarcely or seldom mentioned. We have, again, a mixed nomenclature for the groups of deposits, or the deposits themselves, for which it has been thought desirable to find distinctive names. While some groups are referred to localities, such as Cambrian, Silurian, Jurassic, and the like; others are named after some circumstance supposed characteristic, such as carboniferous, from containing the great coal deposits of Europe and North America; or oolitic, from many of the limestones in it being oolitic, that is, resembling the roe of a fish, or being composed of numerous small rounded grains.

It has been often considered that names derived from localities, where certain deposits have been taken as types, are preferable to those pointing to any mineral structure, inasmuch, as not only can the geologist readily make himself familiar with the kind of accumulations intended to be represented by the names, by visiting and studying the localities whence they are taken, but as also particular mineral structures having been repeated as often as the conditions for them arose, they form no guide for determining the relative age of rocks, whatever may have been the impression when names of that kind were given, and geological science less advanced than at present. The two structural names mentioned are thus liable to objection, carboniferous deposits extending from an earlier period than that supposed to be represented by the term, and up to the higher accumulations above the cretaceous series inclusive, and the oolitic character reaching from limestones amid the earlier fossiliferous rocks to the present day.* The mixed character of the present geological nomenclature arises, no doubt, from the manner in which, from time to time, various geologists have directed attention to different rocks or accumulations of them, those names having generally remained which have been found convenient and sufficient, up to the present time, for the purposes for which they have been employed.

The igneous products being those from which the chief part, if not the whole, of the detrital, and even chemical deposits have been directly or indirectly derived, it would appear desirable to consider them in the first place. Whatever the views entertained of the fluid condition of our planet, whence its form has resulted, such fluid condition produced by heat sufficient to keep all its com-

* One of the limestones of the lower Silurian series in North Wales, the Rhiwlas, near Bala, is oolitic.

ponent parts in that state, the present condition of the earth's surface in dispersed localities shows an abundance of points through which igneous products are now ejected, and the more extended the observat^on, the more certain does the inference appear correct, that the like has happened from the earliest times; at least since the seas wer nted by life. It has also been ascertained that molten matte. as risen from beneath in more massive forms, and in a manner with which we are not familiar, as now occurring, though such molten masses may, indeed, be formed at depths in the earth's crust, whence only future geological changes could bring them above the level of the sea. At all events, this massive form of intrusion is found amid comparatively recent geological accumulations, as well as among those of the most ancient date.

The mode of occurrence of the igneous rocks, which will be found treated of in its place in the following pages, would seem to point to their classification according to their chemical and mineralogical characters, so that any resemblance or difference that may exist between them, may be traced through the lapse of geological time, the relative dates of their appearance being obtained by means of the accumulations with which they may be associated, and to which relative geological dates can be assigned. Having entered upon these characters in the sequel, the following sketch of the more prominent of the igneous rocks may here suffice:—

Granitic Rocks.—Those composed of a granular mixture of quartz, felspar (whether orthoclase, albite, or labradorite), and mica, with, occasionally, the addition of schorl and some other minerals. As the aspect of these rocks varies considerably according to original chemical composition or the mode of cooling, a great variety of appearances are assumed, to which names have been assigned. It thus becomes desirable that these characters should be given whenever it can be accomplished, and that the mere term *granitic* be accompanied by mineralogical detail, and by a statement of the chemical composition, so that correct data may be collected for a proper appreciation of the real differences and resemblances of the rocks commonly thus named.

Felspathic Rocks.—The separation of these from the foregoing may often be regarded as somewhat imaginary, as indeed is the case with definite classifications of the great bulk of the igneous rocks, passing, as they sometimes do, into each other in masses of no very extraordinary volume. The variety known as compact felspar is most frequently a compound of the elements of some felspar, with a surplusage of silicic acid beyond that required for the silicates of that mineral, so that when opportunities have occurred for crystallization of the parts, the result has been a compound of felspar and quartz, or a *pegmatite*, as it has been sometimes termed, in that case a modification of the granitic rocks when the same minerals may alone constitute a portion of a general mass. The *trachytes* of active volcanos and those termed extinct, and of comparatively recent geological date, may represent the more pure felspathic rocks, when wholly formed of felspars, though it would appear that similar rocks are also found amid the igneous products of very ancient geological periods. Felspathic matter, that is, the various component substances in proportions which would form minerals of the felspar family (allowing for that substitution of one substance for another, termed *isomorphism*), if crystallized, should at least constitute the great bulk of these rocks, whatever others may be entangled among them.

Hornblendic Rocks.—These, including among them the rocks in which augite is substituted for hornblende, form a somewhat natural division, so far as the prevalence of these minerals may be sufficient to give a character to the mass of an igneous rock, inasmuch as silicate of lime is a marked ingredient, in addition to the silicate of magnesia, another essential substance, and protoxide of iron, generally present, sometimes replacing much of the lime and magnesia. In this division, therefore, are included the dolerites and basalts of active and extinct volcanic products, and the greenstones, generally of more ancient date. In dolerites, silicate of lime is also present in the labradorite, when that member of the felspar family is mingled with the augite of that rock. Taken as a whole, the hornblendic or augitic rocks are compounds of those minerals and some member of the felspar family, there being sometimes an excess of silica beyond the amount required for the various silicates in the hornblende or augite, and felspar; this excess, then, as it were, thrust aside as quartz.

Serpentinous Rocks.—To a certain extent these also appear a somewhat natural group of igneous products, especially when viewed with reference to a peculiar aspect, and to the presence of silicate of magnesia (constituting the bulk of the rock) and combined water. In the sequel we have endeavoured to show the correspondence between the varieties of serpentine, considered the most pure, and olivine, a common mineral in certain molten products of active and extinct volcanos. The rocks of this division vary, however, somewhat materially in their constituent substances, and in the proportions of them. Taking *bronzite* to be the mineral usually named diallage, it would appear little else than the silicate of magnesia of the matter of the purer serpentine mingled with a minor proportion of protoxide of iron, and a little alumina, crystallized, a small quantity of water also forming a part of it. The mineral now chiefly named *diallage* contains sufficient lime in addition, to make it essentially a silicate of lime and magnesia, with also a marked quantity of oxide of iron. In the compound, sometimes largely crystallized, termed diallage rock (gabbro), and not unfrequently associated with serpentine, the so-termed diallage has to be carefully examined. In all these rocks, whatever their variations, magnesia is a marked ingredient.

Porphyritic Rocks.—Though, no doubt, various kinds of mineral matter which have been in a molten state may be porphyritic, that is, have some mineral or minerals crystallized out and apart from the mass of the remainder of the rock, it seems nevertheless convenient, for the present, to notice these rocks as a group. Even amid vitreous matter, from comparatively quick cooling after fusion, definite chemical combinations may be crystallized, and dispersed through such matter. This can be artificially accomplished in our laboratories, and silicate of lime in crystals can be obtained dispersed through ordinary glass. In the arrangement of particles, beyond the vitreous condition, forming the compact and stony state, the porphyritic character is not rare among rocks; crystals, such as those of felspar, being dispersed amid a base of compact mineral matter. When the latter is chiefly felspathic, the rock is usually known as *felspar porphyry*. In like manner crystals of other minerals are also thus dispersed amid a similar base, such as those of quartz and mica. The base or general mass of the rock is occasionally granular, such as a compound of felspar and hornblende, constituting greenstone, with dispersed crystals of felspar or hornblende, such base having thus advanced to a state of confused crystallization. These are usually termed *greenstone porphyries*. In like manner certain granites become porphyritic, from separate crystals of felspar being scattered among the general compound, confusedly crystallized, and the rock is then called a porphyritic granite. Even serpentines become in a manner porphyritic when crystals of bronzite or diallage are dispersed through a base of that rock. The apparent conditions are, that the chemical composition and the mode of cooling of the general mass are such that certain constituent substances can combine and form separate and definite crystallized bodies, the remainder of the rock either not attaining the state when definite mineral compounds can be formed, or only doing so after the production of the first-formed minerals, and then in a confused manner, not interfering with the forms of the crystals first produced.

With regard to the mineral accumulations derived either directly or indirectly from the igneous rocks, and spread over areas of varied extent and form, by means of water, there is a large mass, more or less characterized by the presence among it of the remains of animals and plants which have existed at different periods, and so perishing, that portions of them, commonly only the harder parts, have been entombed in the mineral accumulations of such different times.

Observation has shown that these accumulations have succeeded one another, as the various detrital deposits in lakes and seas now succeed those which have preceded them, so that, when the ancient sea or lake bottoms, which, elevated into the atmosphere, now constitute so large a portion of dry land, can be studied in cliffs or other natural sections, or by artificial cuttings or perforations, their manner of succession can be ascertained. The more investigations have advanced, the more does it appear that these organic-remain bearing, or *fossiliferous rocks*, as they have been termed, have been deposited and arranged as similar accumulations now are in rivers, estuaries, lakes, and seas. Hence, the geologist, in endeavouring to ascertain the range of such fossiliferous deposits at any given time upon the earth's surface, has to consider the relative amount and position of the land and waters of that time, with all their modifying influences, as also the various conditions under which the life of the period may have been distributed, and its remains entombed amid the detrital and chemical deposits of the day. In fact, he has, from all the evidence he can collect, to suppose himself studying the state of the earth's surface, at such given time, as well with respect to its physical condition as to the existence and distribution of life upon it.

Viewing the fossiliferous rocks in this manner, it may be that some of those divisions among them, which it has been found convenient to make for their more ready description, and the tracing of certain states of a sea-bottom over minor areas, have been too minute, regarded as divisions applicable to the surface of the earth generally, since it is not to be supposed that particular mud or sand banks, however considerable locally, were more likely to have been formerly continued, even at intervals, over the earth's surface than they now are. At the same time such minor divisions, showing the constancy or modification of conditions, as the case may be, over the minor areas, are important, inasmuch as it is by a correct appreciation of this detail and the careful consideration of how much may be regarded in that light and how much as more general, that we learn the true value of the latter, and the restrictions which should be placed upon our views derived from the former.

Assuming the general condition of the earth's surface during the accumulation of the varied deposits in which the remains of animal and vegetable life have been entombed, to have been formerly much as at present, regarding the subject on the large scale, and without reference, for the moment, to the variable distribution of land and water, or to whether the heat in the earth itself may or may not, in remote times, have had a greater influence on the life of those times than at present, the sea would appear to have been the chief receptacle of the various mineral accumulations of all periods, so that classifications of the fossiliferous rocks, founded on a succession of deposits in it, would probably be alike the most useful and natural. The manner in which marine invertebrate animals now live, and the mode in which the remains of similar animals occur amid the fossiliferous rocks, are such, that this division of life seems now very generally admitted as the most appropriate on which to base classifications founded on the distribution of animals, the remains of which are discovered entombed in rocks. We must refer to succeeding pages for notices of the manner in which the remains of life are now preserved in mineral deposits, and for certain points connected with the occurrence of such remains in the accumulations of various geological dates which it appears desirable to bear in mind while studying the fossiliferous rocks. It will be sufficient here to mention that, after first duly ascertaining the actual relative superposition of the various mineral accumulations themselves for evidence of their real succession, and examining the remains of animal and vegetable life which have been found in them, it has been inferred that certain minor and major divisions may be effected in the general mass which shall represent the kinds of sea-bottoms marking given and succeeding geological times. Without, in the least, doubting that very great modification may be found needed in classifications based upon the examinations of even considerable areas, when an effective classification, representing the main facts connected with the accumulation and spread of fossiliferous rocks over large portions of the earth's surface, may be necessary, it still becomes desirable to have that which may satisfy the requirements for the time being. The following sketch, therefore, of the general divisions at present considered desirable for the area of Western Europe, and supposed, in part at least, to be found also convenient for the mode of viewing the fossiliferous deposits in many other parts of the world, may be useful, especially as respects the major divisions.

Stratified and Fossiliferous Rocks.

I. Tertiary, or Cainozoic.
II. Secondary, or Mesozoic.
III. Primary, or Palæozoic.

I. Tertiary, or Cainozoic.

A. Upper	*a* Mineral accumulations of the present time. *b* Pleistocene. *c* Pleiocene.
B. Middle	*a* Miocene.
C. Lower	*a* Eocene.

II. Secondary, or Mesozoic.

A. Cretaceous Group . .	*a* Chalk of Maestricht and Denmark. *b* Ordinary chalk, with and without flints. *c* Upper Green Sand. *d* Gault. *e* Shanklin Sands, Vecten, Neocomian, or Lower Green Sand.	
B. Marine equivalents of .	*a* Wealden clay . . *b* Hastings sands . *c* Purbeck series. .	Organic remains in these are of a fluviatile, lacustrine, or estuary character.*
C. Jurassic or Oolitic Group	*a* Portland oolite or limestone.† *b* Portland sands. *c* Kimmeridge clay. *d* Coral rag, and its accompanying grits. *e* Oxford clay, with Kelloways rock. *f* Cornbrash. *g* Forest marble, and Bath oolite. *h* Fuller's earth, clay, and limestone. *i* Inferior oolite, and its sands. *k* Lias, upper and lower, with its intermediate marlstone.	

* The recent researches of Professor Edward Forbes among the Purbeck series have fully illustrated the prudence of not trusting to fresh-water molluscs as characterizing particular divisions in deposits, at least those ranging downwards to that part of the fossiliferous series, he having ascertained that it required most careful critical examination to distinguish the fresh-water shells of that series, as it occurs at Purbeck, from those of certain existing fresh-water molluscs in England and part of Europe.

† The minor divisions of this group have been given with reference to those usually

D. Trias Group* . . .	*a* Variegated marls, Marnes Irisées, Keuper. *b* Muschelkalk.† *c* Red sandstone, Grès Bigarré, Bunter sandstein.

III. Primary, or Palæozoic.

A. Permian Group . . .	*a* Zechstein, Dolomitic or magnesian limestone. *b* Rothe todte liegende, lower new red conglomerate and sandstones, Grès Rouge.
B. Marine equivalents of ‡ .	*a* Coal measures, Terrain Houiller, Stein Kohlen Geberge.
C. Carboniferous limestone Group.	*a* Carboniferous and mountain limestone, with its coal, sandstone, and shale beds in some districts. Calcaire carbonifère, Bergkalk. *b* Carboniferous slates and yellow sandstone.
D. Devonian group . . .	*a* Various modifications of the old red sandstone series.
E. Silurian Group . . .	*a* Upper; Ludlow Rocks, Wenlock shale and limestone, Woolhope Limestone. *b* Middle; Caradoc sandstone and conglomerate. *c* Lower; Llandeilo, Bala and Snowdon beds.
F. Cambrian Group . .	*a* Barmouth sandstones, Penrhyn slates, Longmynd rocks, &c. Various rocks subjacent to the Silurian series in Wales and Ireland.

---

employed in England for the sake of English observers. Many modifications have been shown to be effected in other European countries. Of these divisions those of the Oxford clay and lias would appear much extended.

* The Trias and Permian groups afford an example, as regards the British islands, of a classification taken from organic remains in preference to the mode of occurrence of the rocks themselves, these groups here constituting parts of a general series of deposits with a somewhat marked general character, known as the new red sandstone. Certain general physical conditions were prevalent during the accumulation of these deposits in Great Britain, and certain portions of Western Europe, at the time that a modification in the life of the period was apparently effected in the same area and those adjacent to it on the north and east.

† In the collections lately brought to England by Captain Strachey, Bengal Engineers, after an examination of the Himalaya range, the forms of certain organic remains from the Thibet side of those mountains remind the geologist of those found marking the Muschelkalk of Germany; an interesting circumstance, considering the range of that rock in Europe.

‡ When the great thickness of these deposits in Europe and America is considered, it becomes very desirable to find their marine equivalents, inasmuch as the conditions under which the great mass of these coal measures has been accumulated, as has been noticed in the sequel, could scarcely constitute other than minor parts of those generally prevailing at the time. It is easy to conceive, as has indeed been done, that their marine equivalents might contain either the organic remains usually found in the deposit beneath them in parts of Western Europe, or those found in the group above them, or a mixture of both. In Northern England the alternations of conditions by which coal beds were included in the carboniferous limestone series, did not interrupt those for the existence of a marked kind of marine animal life in the same localities.

Altered or Metamorphic Rocks.

With the classification of detrital and chemical accumulations, effected hy the aid of water, and of earlier geological date than those last mentioned (the Cambrian), there are many difficulties. Indeed the limits which may be assigned to the latter, in descending order, are, in the present state of our knowledge, most uncertain. In the district of the Longmynd (Shropshire) the Cambrian group attains a thickness of 26,000 feet, almost entirely composed of detrital deposits. The same group, as exhibited in North Wales, not only presents a considerable depth of similar accumulations, but also shows, by pebbles in its conglomerates (vicinity of Bangor, Llanberis, &c.), that sands were firmly cemented into sandstones, and these ground into shingles, by water action, prior to the production of such conglomerates. These conglomerates, which also contain rounded fragments of hornblendic and felspathic igneous rocks, of a general character similar to those subsequently vomited forth in the same region amid the Silurian deposits, may only constitute portions of a series in the same manner that many other conglomerates are included in groups of rocks bearing given names; in fact, be the beaches of different portions of the time required for the whole deposit; yet they, with the muds, silts, and sands of the period, become important as pointing to causes in action at that time similar to those from which the like accumulations have been effected in after geological periods. Thus, as far as researches have yet gone, we do not arrive at physical conditions differing, as regards the production of detrital mineral accumulations, in any essential manner, that can be determined, from those which have afterwards influenced the accumulation of similar kinds of mud, silt, sand, and gravel, whether now constituting hard consolidated rocks, or found in a state more resembling that in which they were originally formed.

The aid to the classification of rocks, once supposed to be derived from certain of them having a crystalline or semi-crystalline aspect, yet still preserving a general stratified arrangement of parts, various modifications of them bearing distinct names, such as *gneiss*, *mica slate*, and others, is now well known to be unsatisfactory, inasmuch as such rocks have been ascertained to be of different geological dates. Whenever any heated mass of igneous rocks has been thrown into juxtaposition with detrital accumulations of various kinds (mud, silt, sand, or gravel), the conditions under which the latter are then placed become favourable for the modification of their component parts. Various circumstances, heat being regarded only as one of them, then so act that, besides a tendency of similar matter to gather itself together in irregular forms, particles can often so freely move and adjust themselves, that even minerals of distinct characters are formed, rarely, and sometimes not hitherto, discovered amid any other than these modified or altered rocks. As such modifications or alterations would be expected to depend upon the general chemical character and physical structure of the deposits acted upon, these minerals are found to be combinations of the substances which could readily move and unite in a definite manner. Thus, the altered rocks afford such minerals as *andalusite* (especially a silicate of alumina, the base of clays), *chiastolite* (another mineral, in which silicate of alumina is the chief ingredient), *cyanite* (another form in which the same substance is essential), *staurolite* (where silica, alumina, and peroxide of iron are required), and the *garnet*, with all its differences arising from isomorphism, and in which silica may be prominently combined with alumina or iron, magnesia or lime, as the case may be. While minerals of such kinds could be developed in these altered rocks, we should also anticipate that micaceous and siliceous sands or sandstones, as also those in which fragmentary portions of felspar were mingled, and especially when the latter were not decomposed, would be much modified by a free movement of certain substances. We might expect much obliteration of the grains of the sand, a disposal of the silica in planes, either of original deposit or of cleavage, should that have been effected in the rocks acted upon, as also that the micaceous and felspathic portions might be more gathered together in places, and even readjusted in crystals, since we know, as regards the solution of the component matter of such minerals, that in some veins evidently filling fissures, quartz, mica, and felspar are found either alone, or mingled with other minerals in a manner pointing to their production from solutions, their component parts derived from the adjacent rocks.

The like circumstances acting upon more simple substances, formed into beds of rock, such as limestones, and the combinations of ordinary calcareous matter with carbonate of magnesia in various proportions, would necessarily also produce modification in the arrangement of the component particles, a confused crystalline adjustment of them being effected, such as that seen in statuary marble, when conditions were most favourable. When these bodies were less pure, mingled with detrital matter of the ordinary kinds, the circumstances would be favourable to the development of different mineral substances, such as garnets and others,* amid the general mass. Looking at the varied modes in which detrital and chemical accumulations have been formed, and the different manner in which they can be acted upon by the influences noticed, either on the minor or large scale, the general result could scarcely be otherwise than of the most varied kind. To attempt, therefore, a classification of these modified or altered rocks relatively to geological dates, would be obviously useless.

In considering rocks of this kind it is needful also to bear in mind the general conditions under which beds of detrital or chemical deposits may be modified, or altered from their original state of accumulation, by other conditions than those of the contact or juxtaposition of mineral matter in a state of igneous fusion. Independently of chemical changes effected by the arrangement of the substances in different states of combination, adjusting themselves according to their affinities and the conditions under which they are then placed,† the circumstances which would arise when

---

* In this manner crystals of quartz have been sometimes produced in beds of statuary marble, as, for example, that of Carrara.

† We find quartz rocks (that is, grains of quartz, accumulated as sands, and firmly cemented together by silica, the separation of the old surfaces of the sand-grains from the siliceous cement sometimes obscure) as the continuation of ordinary beds of quartzose sandstone, the latter sometimes slightly consolidated, and have simply to infer, to account for the facts observed, silica infiltrated so as to consolidate the beds more in certain situations than in others. Such quartz rocks have often been supposed "altered or metamorphic" in the sense used for some of the same general aspect acted upon, with others, by juxtaposed igneous matter which had been in a fused state from heat; whereas they

such beds of rock were deeply buried beneath great accumulations of mineral matter, have to be carefully considered, if the temperature increases in the manner usually inferred as we descend beneath the surface of the earth, even to moderate distances. Huge masses, representing former wide-spread portions of the earth's surface, might thus be placed under conditions similar to those which produce modification and alteration when igneous matter rises from beneath and is forced amid or against detrital and chemical deposits. When again upheaved, as we know great and wide-spread masses of rock have often been during the lapse of geological time, it would be anticipated that similar matter, acted upon in a similar manner, would present like results, and there is much reason to consider that such influences have been the causes of the modification and alteration we sometimes find.

By carefully regarding altered or metamorphic rocks on the large scale, and with reference to all the conditions under which they may be produced, they are found to constitute a mass of mineral matter of much importance, showing us, in their most crystalline readjusted state, the extremes to which such matter may be modified without the mingling of parts inferred when in a state of igneous fusion. Igneous rocks themselves are often modified, their component particles having, as it were, striven to adjust themselves in a perfect manner as in the detrital and chemical deposits. Thus, ordinary greenstone can be sometimes observed to have its component minerals, hornblende and felspar, presenting the aspect of the rock known as *hornblende rock*, and beds of similar matter, either abraded from solid greenstones or vomited forth as ashes, and arranged in beds by the agency of water, to become the rock known as *hornblende slate*. Thus, then, without attempting to classify these modifications and alterations in the arrangement of the component parts of detrital, chemical, and

---

are merely more firmly cemented and purer quartzose or siliceous modifications of common hard grits, dispersed amid soft marls and shales in so many deposits. Again, the original crystalline accumulations of more chemically-formed beds have to be duly regarded, and separated from the "altered or metamorphic rocks" under notice, as we know that even confused crystalline deposits have been thus produced.

even igneous accumulations, geologically, as regards relative dates, they still, to a certain extent, constitute a class very convenient for investigation, it being always borne in mind that it is desirable only so to regard them, in the present state of our knowledge.

THE

GEOLOGICAL OBSERVER.

CHAPTER I.

DECOMPOSITION OF ROCKS.—FORMATION OF SOILS.—DECOMPOSITION OF GRANITIC ROCKS.—DECOMPOSITION OF SANDSTONES AND LIMESTONES.—INFLUENCE OF STRUCTURE AND ORGANIC REMAINS IN DECOMPOSITION.—DECOMPOSITION OF ROCKS CONTAINING IRON.

As geological knowledge advances, the more evident does it become that we should first ascertain the various modifications and changes which now take place on the surface of the earth, carefully considering their causes, and then proceed to employ this knowledge, so far as it can be made applicable, in explanation of the geological accumulations of prior date. This done, we should proceed to view the facts not thus explained, with reference to the conditions and arrangements of matter which the form of our planet, the known distribution of its heat, the temperature of the surrounding space, and other obvious circumstances, may lead us to infer would be probable during the lapse of geological time.

The geological observer cannot be long engaged in his researches before he will be struck with the tendency of rocks to decompose by the action of atmospheric influences upon them. He will soon perceive that this decomposition is both chemical and mechanical; that certain mineral bodies more readily give way before these influences than others; and that from altered conditions, as regards such influences, the same kinds of rock will more easily decompose in one situation than in another.

It is in consequence of this decomposition that we have a soil supporting that growth of vegetation upon which animal life depends; for soils are but the decomposed parts of more or less consolidated sea or lake bottoms and of igneous accumulations, with the remains of the vegetation which has grown on them, and of the animals which have lived upon the plants. From the varied configuration of surface the decomposed portions of rocks, forming soils, may not always cover those from whence they were derived, for they may and sometimes have been carried, mechanically suspended in water, to various distances, and there deposited, in such a manner as to be mingled with the decomposed portions of other rocks, or wholly cover over the latter. Be this, however, as it may, the decomposed parts of rocks form the base of the soils, affording soluble mineral matter to the plants requiring it, and presenting a physical structure capable of supporting their growth.

The decomposition of rocks, in its various stages, will require much attention, so that the observer may properly classify the facts coming within the range of his researches. Among rocks of igneous origin, such as granites, greenstones, and the like, he will find that the decomposition of felspar is among the chief causes of the disintegration of the igneous masses of which this mineral may form a part. It would be out of place here to enter upon the composition of the various minerals of the felspar family;* it will be sufficient to refer to those portions of them which are soluble, such as the silicates of potash or soda, as the case may be. These silicates, from the action of carbonic acid in the atmosphere, derived from the decay of vegetation, or brought into contact with them by waters containing it in sufficient abundance, are often readily decomposed. The particles once loosened by decomposition, and some of them carried off in solution, rains and changes of temperature, particularly in regions visited by frosts, act mechanically, and the surface of the rock, under favourable conditions, is removed. From a repetition of these causes the rock becomes decomposed to various depths, according to circumstances. In cases where the remaining portions are either too large or so situated as not to be readily carried away, a coating of the disintegrated insoluble part

* The four minerals of this family which chiefly enter into the composition of rocks, are orthoclase, albite, labradorite, and oligoclase, the general chemical composition of which may be regarded as follows:—

	Silica.	Alumina.	Potash.	Soda.	Lime.
Orthoclase	65·4	18	16·6*a*	–	–
Albite .	69·3	19·1	–	11·6*b*	–
Labradorite	53·7	29·7		4·5	12·1
Oligoclase	63	24·9	–	12·1*c*	–

a. Including a little soda and lime. *b*. In part often replaced by lime or potash. *c*. Commonly, also, containing potash and lime.

remains, and to a certain extent protects the solid rock beneath from that decomposition which it would otherwise have suffered.

In many granitic regions ample opportunities are afforded of observing the amount of decomposition thus produced; high tors or bosses of rock rising above a surface in a decomposed state

Fig. 1.

(fig. 1), while hard masses, having the fallacious appearance of boulders, rounded by attrition, are sometimes included in the loose decomposed granite, as represented beneath (fig. 2).

Fig. 2.

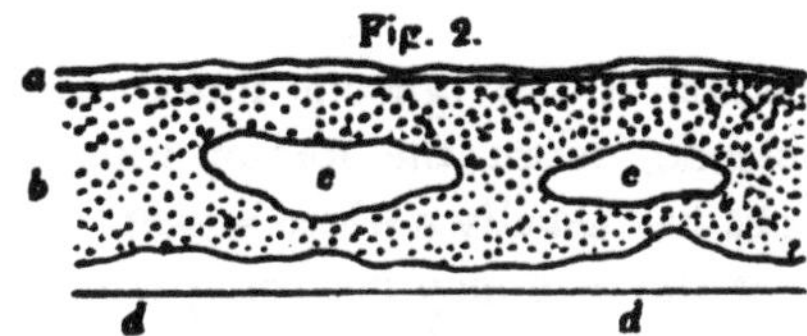

This illustration is taken from part of the road between Okehampton and Moreton Hampstead, Devon. *a* represents the vegetable soil; *b* decomposed granite; *c c* solid rounded masses of undecomposed granite, included in the decomposed part; and *d d* solid granite.

In such a section as this, great care should, however, be taken to ascertain that *c c* are not transported boulders of granite, included in smaller granitic gravel, as sometimes happens with granitic drift, near the sources whence it has been derived. Fortunately in this case the observer would be assisted by the presence of large crystals of felspar disseminated through all parts of the rock, both decomposed and undecomposed, and which are beautifully preserved, remaining uninjured in their forms and in their relative positions throughout the decomposed granite.

In granitic regions, sections such as that beneath (fig. 3), in

Fig. 3.

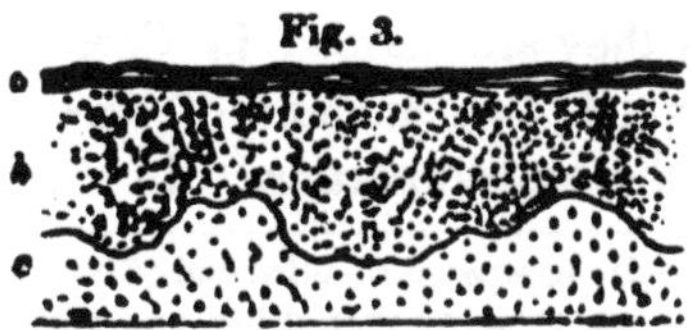

which *a* represents the vegetable soil, as it is commonly termed, *b* the decomposed, and *c* the solid granite, are not unfrequent. Sections of this appearance should, also, be carefully examined, and it be clearly ascertained that the granitic particles at *b* are of the same kind, and in the same general relative positions, as those at *c*, and that there can be no chance of their having been brought into their present position by moving water. The quantity of transported granitic matter around granite districts, as also among them, is sometimes so considerable that a superficial deposit of granitic particles, covering a very different kind of rock, may, without due care, be readily mistaken for a mass of decomposed granite. In the same way the remains of the rock of one part of a granitic region may be removed and cover the rock of another portion.

Among those igneous rocks in which hornblende* forms a marked component part, it sometimes happens that this mineral is also disintegrated.

In the decomposition of the igneous rocks chiefly composed of felspar and hornblende, when the former mineral prevails, the surface of these rocks has usually a white aspect, the soluble silicates of soda or potash, as the case may be, being removed, and a crust, principally formed of silicate of alumina remaining. Where hornblende much prevails, a brownish and reddish surface is common, the protoxide of iron of that mineral having been converted into a peroxide.

Rocks in which the felspar and hornblende have both been decomposed, are in some situations thickly coated with a loose covering. The variable manner in which a mass of igneous rock that has been placed under equal atmospheric conditions may have been unequally decomposed, will often afford an excellent illustration of the original differences in it, arising not only from a variation in the component parts, but also from the modified manner of their aggregation, in consequence of differences in cooling.

Some veins or dykes, as well of granitic as of other igneous rocks, afford excellent examples of their variable power of resisting the same order of decomposing influences according, chiefly, to the differences arising from modifications of cooling. Some granitic dykes, or *elvans*, as they are termed in Cornwall, show, as in the following section (fig. 4), an amount of decomposition gradually increasing towards the central portion. In this section *aa*, *bb*, and

* Hornblende is essentially composed of silica and of lime and magnesia, in variable proportions, these substances replacing each other, and sometimes also being partly replaced by protoxide of iron.

a represent the different parts of a granitic dyke, or elvan, traversing slate rocks, *d d*. Assuming, in this case, that the elementary com-

Fig. 4.

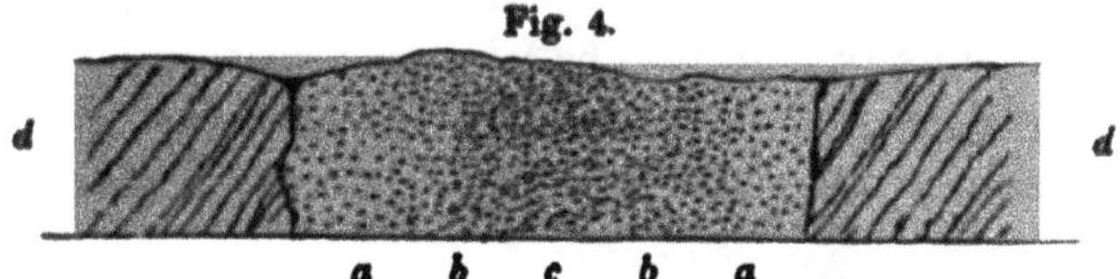

ponent parts of the dyke were originally the same, and that the differences found have arisen from variable cooling, as it is now well understood has frequently been the case, the decomposition has been effected according to the facility with which certain portions could be attacked by atmospheric influences and be subsequently removed. The two outward parts of the dyke (*a a*) are considered to be composed, as often happens, of a hard siliceous rock, the elements of the granitic matter having taken that form from comparatively quick cooling, so that this modification of it has resisted decomposition better than the rest. At *b b*, inside the hard rock, another modification, arising from more slow cooling, is supposed to exhibit a porphyry, some mineral, very frequently felspar, crystallizing out amid the base, itself less compact than the preceding variety. Not unfrequently in such cases the felspar is decomposed, and the insoluble portions even removed when directly exposed to the atmosphere. Still proceeding inwards, the rock becomes more and more granitic, until, finally, the central portions are well crystallized, and then exposed to the full action of the decomposing influences.

We often thus find, within a short distance, a good example of variable decomposition arising from differences in physical structure, the chemical composition of the mass remaining the same; a variation very instructive, since it enables the observer readily to appreciate the inequalities of surface which, in many regions composed of the same kind of igneous rocks, arise from changes in physical structure alone, some variations having better resisted decomposition or abrasion than others. At the same time he should carefully study the modifications in hardness, and the capability of resisting decomposition arising from changes in chemical composition, such, for instance, as those observable among the granites which occasionally graduate into schorl rock, in Devon and Cornwall.*

* The following may be taken as an estimated general view of the chemical difference between common granite, composed of two-fifths of quartz, two-fifths of

Of the curious forms assumed by granitic rocks from variable resistances to decomposition, those named the Kettle and Pans, at St. Mary's, Scilly (fig. 5), may be taken as a good example.*

Fig. 5.

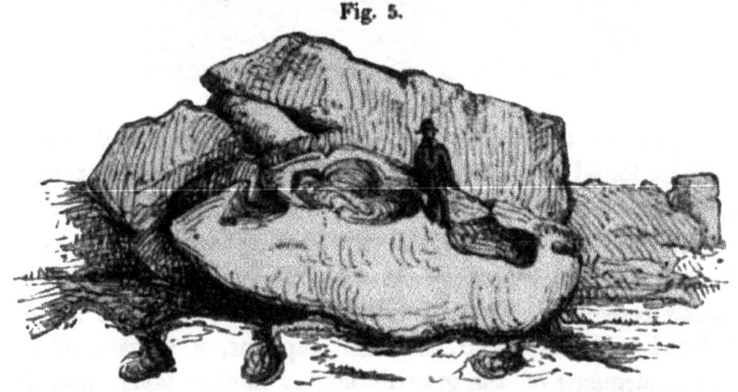

While rocks of a generally similar chemical composition, such as those above noticed, are found to decompose in a variable manner, according to the different aggregation of their component parts, it would readily be anticipated that any rocks formed of different materials, brought together as sands and gravel, and subsequently consolidated by some cementing substance, would be found to decompose irregularly and according to the different powers of their component parts to resist the chemical and mechanical influences to which they may be exposed. It will soon be perceived that, taken generally, the cementing matter of sandstones and conglomerates decomposes first, liberating the grains of sand and the pebbles, that have originally remained such from their hardness, and which are thus ready to be again carried by moving waters to other situations, there to form the parts of new accumulations. The rapidity of

---

orthoclase, and one fifth of mica, and schorl rock, supposed, for illustration, the proportions varying materially, to be formed of equal parts of schorl and quartz:—

	Granite.	Schorl Rock.
Silica	74·84	68·01
Alumina	12·80	17·91
Potash	7·48	0·35
Soda	–	0·98
Lime	0·37	0·14
Magnesia	0·99	2·22
Oxide of iron	1·93	6·85
Oxide of manganese . .	0·12	0·81
Fluoric acid	0·21	–
Boracic acid	–	1·79

* Though the true origin of the "Rock Basins," as they have been termed, is in general sufficiently clear, it may often have happened that, owing to a convenient situation, the Druids may have employed them for their purposes, either as they naturally occurred, or were artificially modified.

decomposition in such cases necessarily varies according to the nature of the cementing substance. A calcareous cement, though hard, will more readily give way before the chemical influences acting on limestones than ordinary siliceous matter, though the latter may be less compact; while a siliceous cement, if porous, may be more easily removable by the combined action of frost and thaw.

The hardest limestones, even those termed marbles when crystalline, will be observed to decompose on the surface.* The action is necessarily variable and dependent on the different resisting powers of the rock, on the one hand, and the exposure to the needful decomposing influences on the other. A crystalline and calcareous vein, running through an ordinary limestone, will often be seen standing out in salient relief, the arrangement of particles in the crystalline form, notwithstanding that the carbonate of lime is then generally more pure than in the body of the rock, being better able to resist atmospheric influences than in a less definitely arranged position.

Upon further examination it is perceived that not only the crystalline veins thus protrude upon the surface of the limestone rocks, but that many an organic remain does the same, and, in some instances, a limestone is only clearly distinguished as fossiliferous by this kind of decomposition, the common internal fracture ill exhibiting the fact. That this harmonises with the comparatively undecomposed condition of the crystalline vein becomes apparent when we examine the structure of these organic remains. The shells either retain to great extent the original crystalline or other definite arrangement of their parts, so essential to their well being when the animals of which they once constituted the hard portions were alive, or having been decomposed in the body of the rock during the lapse of time, the empty spaces, (or casts, as they are commonly termed) have been filled with crystalline carbonate of lime, which has percolated in solution through the pores of the rock into the cavities.†

By this kind of decomposition we often learn that many a limestone is really little else than a mass of organic remains cemented by a minor quantity of chemically deposited carbonate of lime. Some of the hardest limestones afford excellent examples of this

* This is often well shown in collections of antique marble statues.

† It was considered useless further here to remark on the composition of organic remains. It may, however, be noticed that the bones and teeth of fish, reptiles, birds, and mammalia have been often secured from removal by their composition, and that into the cavities left after the original decomposition of shells other less soluble substances than carbonate of lime have been infiltrated, such for example as into the cavities of the *Gryphæa incurva* and other shells, in the lias of Glamorganshire, where silica has replaced the original matter of the shells.

fact. The beds of carboniferous limestone of England, for hundreds of feet in depth, are occasionally found composed of little else than the disintegrated joints of encrinites, mingled with shells and a few corals.*

By the aid of decomposition we not only learn that many limestones are little else than such accumulations of the harder parts of molluscs and other creatures, many of which have lived and died in the places where we discover their remains; but we also find revealed the arrangements of the component parts of rocks, as well igneous as accumulated by means of water, which do not otherwise appear, arrangements of parts exceedingly important when we study the original manner in which rocks have been accumulated, or the modifications and changes to which, during the lapse of geological time, they have been subjected.† Many a sandstone, well *weathered*, as it is termed, will exhibit as beneath (fig. 6), a

Fig. 6.

honeycombed and irregular appearance, arising from the different character at parts of the cementing substance, either original or subsequent to the accumulation of the rock, as the case may be,

* This fact may be well studied, among other localities, on the southern coast of Pembrokeshire, where the cliffs afford excellent opportunities of observing the mode in which the materials of its carboniferous limestone have been accumulated.

† As regards the weathering of calcareous rocks, it can be seen to great advantage on part of the shores and amid the islands of the Lake of Killarney, where the carboniferous limestone is hollowed into most fantastic forms. A well-known and strangely-formed rock, standing out into the Great Lake, known as O'Donaghue's Horse, and which so well illustrated this decomposition, was unfortunately thrown down by the action of the fresh-water breakers upon it in 1851. Carbonic acid in the waters of the Lake, near its surface, has acted very conspicuously in this locality.

and many another structure, also of importance, such as the concretionary structure of some igneous rocks, then alone becomes apparent. We should, for example, probably be ignorant, without *weathering*, of this arrangement of parts in the granitic or elvan dyke,* *a a*, cutting through slates, *b b*, at Watergate Bay, Cornwall, and figured beneath (fig. 7).

Fig. 7.

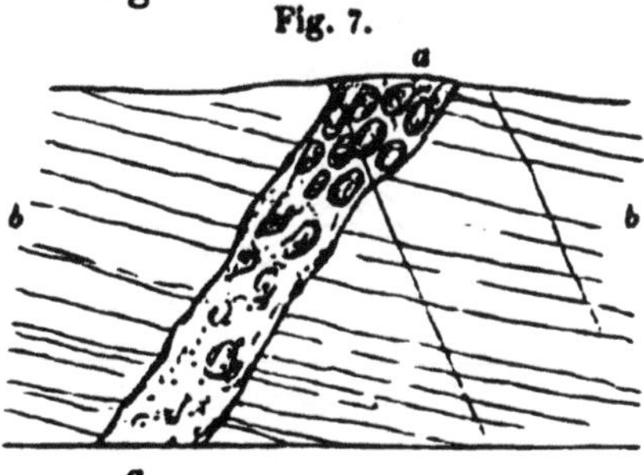

The division of rock masses by *cleavage* greatly aids their decomposition, since it renders them slaty, when this would not happen from any original accumulation of sand or mud in thin layers, one above the other, and the like arises from those separations in planes, named *joints*, more distant from each other, and which with the cleavage planes will be further considered in the sequel. By these means water more readily percolates through many rocks than it would otherwise do, and thus a greater amount of soluble matter may be attacked than would otherwise have happened in the same time.

Many hard rocks break up superficially in a manner showing little symmetry of form in the fragments, so much so that their shape seems more due to the irregular action of decomposing influences, than to differences of resistance from original structure. A compact limestone or hard sandstone may often be seen broken up beneath the soil, in the manner exhibited in the accompanying section (fig. 8), in which *a* represents the vegetable soil, *c c* a hard

Fig. 8.

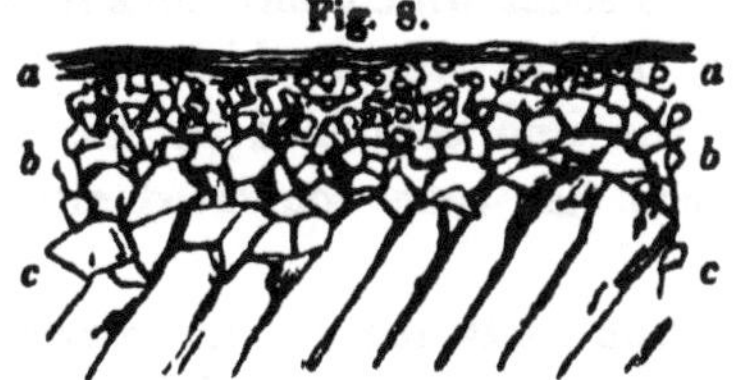

limestone or sandstone, and *b b* fragments of the same rock, largest

* This dyke is a compound of quartz, felspar, and mica, containing disseminated crystals of felspar.

towards *c c*, and evidently having constituted portions of the subjacent highly inclined beds, while the upper fragments are smaller and more confusedly mixed, though still angular. It sometimes happens when the rock, so broken up, is a sandstone that the chemical change of iron in the cementing matter subsequently to the formation of the fragments, is well seen. Upon breaking these fragments, sections, as beneath (fig 9), will often present themselves.

Fig. 9.

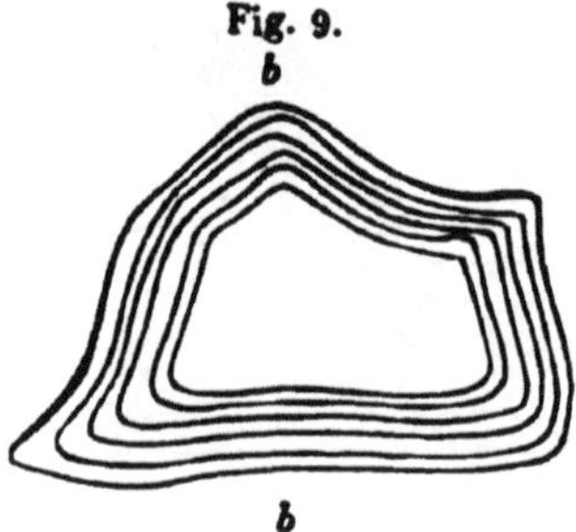

A central portion remains unchanged, surrounded by irregular zones (*b b*), commonly of a brownish red, arising from chemical action, by which the protoxide of iron has been converted into a peroxide. Similar changes of the protoxide of iron into the peroxide, are observable among the argillaceous limestones, such as the lias, and are indeed sufficiently common.

In some very earthy limestones, which may rather be considered to have been once silt, highly impregnated with calcareous matter, the disappearance of the latter in the higher parts of the rock, even to many feet in depth, has been so complete, and the peroxidation of the iron so extensive, that a rusty looking porous substance alone remains. Among some of the older accumulations such a rock may often be seen, and be found the only means by which beds, here and there containing a larger per-centage of carbonate of lime, can be traced or connected. Among the older rocks also, many a layer of a rusty colour shows a total disappearance of the carbonate of lime of the numerous shells which once constituted the bulk of the layer, their casts, or the spaces which they once filled, alone remaining, while the iron contained in the mud or silt which first enveloped them, has been converted altogether into a peroxide.

While thus the iron contained in many rocks exhibits a gradual change to a peroxide, many red marls and sandstones show an alteration from the peroxide of iron, giving a general red tint to these deposits, to a protoxide. Beneath the vegetation, and by the sides of natural joints the red colour will be seen converted into a green or bluish green, the change being due to the effects of decomposing vegetation, which has robbed the adjacent peroxide of iron of a portion of its oxygen. This is a point of much interest when we study the cause of the streaks of green or bluish green amid the red marls and sandstones of different geological ages, and which have probably arisen from causes in operation at the period when the whole has been accumulated. When we examine into the variations and modifications of colour arising from the present effects of decomposing vegetation, the old changes have to be carefully separated from the modern, since both are sometimes exhibited in the same sections.

The observer must be careful, in his estimate of the amount of decomposition which rocks may sustain from atmospheric influences, duly to consider the power of vegetation to prevent, assist, or otherwise modify it according to circumstances. Vegetation may prevent decomposition, by presenting a certain barrier to the effects of sudden frosts and thaws; assist the action of rains by keeping the higher parts of rocks more permanently wet than they would otherwise be; or greatly modify it by the various effects produced by the kind of plants which may cover the land at given times; for a portion of country covered by forest trees would be differently circumstanced, as regards the probable decomposition of the rocks of which it is formed, than when the same portion was either broken for tillage or spread over with pastures.

As a whole, the study of the decomposition of rocks is one of much importance, since by it we learn a variety of facts connected with the original accumulation of mineral masses, with which otherwise we should be unacquainted, and at the same time it often teaches us properly to appreciate the changes and modifications which have occurred since such original accumulation. It enables us to form a correct judgment of the amount of matter which may thus be prepared for removal and for accumulation elsewhere. We see causes and effects that have been in operation whenever land arose from beneath water into the atmosphere, however modified these may have been by alterations of conditions, such as those now found between the tropics, and in the arctic or antarctical regions, or which may have taken place in the atmosphere of our planet from its earliest state.

CHAPTER II.

REMOVAL OF SOLUBLE PARTS OF ROCKS BY WATER.—TRAVERTINE AND CALCAREOUS BRECCIA.—CHLORIDE OF SODIUM IN SPRING WATERS.—SILICA IN WATER.—HOT SPRINGS.—SPRINGS ON THE OUTCROP OF BEDS.—SPRINGS FROM FAULTS.—CAUSES OF LANDSLIPS.

As spring waters are not pure waters, but hold different substances in solution according to circumstances, and as it is evident that at least, the bulk of such waters are only rains which have percolated through rocks, and variably pour out again according to conditions, the substances so in solution must have been removed from the rocks. However small the soluble matter found in any single spring may be, on the average, collectively its amount is considerable, particularly when we regard the changes which rocks must have undergone from this cause alone during the lapse of any geological time, when circumstances may have thus permitted the removal of soluble matter from any given mass of them.

With the removal of lime as a bicarbonate we are commonly familiar, since by the loss of the excess of carbonic acid required to retain it in solution, this substance is thrown down in different forms varying from a simple incrustation upon vegetable matter, or upon stones or rocks, amid or over which water containing it may flow, to hard and compact limestones, some taking a crystalline form, as is frequently so well shown in the beautiful stalactites and stalagmites of many caverns in limestone countries. It is no uncommon thing in calcareous districts to find the fragments of limestones which have been detached from faces of rock by atmospheric influence, firmly cemented together, as a breccia, by carbonate of lime, left by the waters which have percolated through them.

In the calcareous countries of the tropics, where evaporation is more rapid than in temperate climates, the deposit of carbonate of lime may often be studied with much advantage. Heavy rains falling amid a mass of vegetation, the decaying parts of which furnish the needful carbonic acid, carry this with them amid the beds, joints, and caverns of the limestones; carbonate of lime is thus removed, and when the waters again emerge charged with bicarbonate of lime, and are exposed to the heats of a tropical sun, incrustations are formed in the shallow and slow-moving portions of the streams. Trees may even thus become imbedded by the shifting course of the waters, as is well seen at the Roaring River on the north side of Jamaica, where waters containing much bicarbonate of lime, after leaping over a cliff, run roaring amid a forest, the lower portions of the trees of which they encase with carbonate of lime, and shift their channels as new accumulations compel them to follow a new direction.

In shallow sheltered bays also of tropical coasts, to which water containing calcareous matter may slowly find its way, the solution becoming thus highly concentrated by evaporation as it flows onward, opportunities are occasionally afforded for observing the formation of the little rounded grains of calcareous matter in concentric coatings, termed oolites, a slight ripple being sufficient to produce a to-and-fro motion on the beach on which the calcareous matter is being deposited. Upon breaking these calcareous grains, sometimes a fine particle of common sand, or broken shell forms the nucleus, at others it would appear that a simple particle of the calcareous matter itself, before it became attached to any other solid substance, was sufficient for the purpose.

Though many countries show deposits of carbonate of lime from waters flowing over them, parts of Italy have so long been remarked on this account, that the name *travertino* has not unfrequently been given to such accumulations.* This deposit has also a peculiar interest in that land, inasmuch as we there sometimes find ancient architectural works, as for example, the remains of the temples at Pæstum, constructed of travertine, containing the remains of the same kinds of terrestrial and lacustrine shells which now exist in the vicinity, and become entombed in the travertine now forming. Of large accumulations of calcareous matter depositing under the atmosphere and not beneath bodies of water, the plains of Pamphylia would appear to afford a very striking example. The coasts of Karamania have long been known to present good instances of beaches

* Not only have we excellent opportunities of there studying the calcareous deposits thrown down from waters of ordinary temperatures, but those also from thermal springs, in which other substances are mingled in a manner to produce very interesting results. Of this kind is the intermingling of silica with the other deposits at the baths of San Filippo, where the waters have a temperature of 122° Fah[t]. (one spring being about a degree higher), and contain in solution, silica, sulphate of lime, bicarbonate of lime, and sulphate of magnesia. The ground around is composed of travertine deposited by the springs.

consolidated by the percolation of carbonate of lime amid the pebbles, thus forming a conglomerate. We may thus obtain not only breccias and conglomerates upon the land, by the evaporation of water, charged with bicarbonate of lime, without the aid of lakes, but also sheets of limestone, the overflow of rivers and the shifting of their courses causing the necessary deposits. It would be desirable, where fitting opportunities for studying the latter kind of accumulations may be found, carefully to examine the differences between them and those deposits effected in tranquil bodies of water, such as lakes. We should expect, while the gradual rise and overflow of the rivers may here and there bury, by means of the calcareous deposits from them, the fluviatile or lacustrine molluscs living previously in favourable situations, that there would be much showing the drift of animal and vegetable substances borne onwards to localities where their further progress was arrested, and where they became entombed beneath the limestone afterwards formed over them.

Although limestone may thus silently and unperceived be transported from one locality to another, since the clearest waters may contain the bicarbonate of lime in abundance, many other substances are also, in a similar manner, borne onwards in solution; and it becomes desirable, in the present state of geological science, that the mass of this matter, and the proportions of the substances commonly composing it, should be examined. Something is done by every analysis made of spring and river waters; and the desire to obtain good waters for domestic purposes, has lately led observers to connect the rocks from which springs issue and afford the supply to rivers with the quality of waters; but it would be well more systematically to study the soluble matter conveyed away in this manner by moving water.*

It should be recollected that when rivers are swollen by rains, though substances in solution amid the rocks may be then forced more abundantly out of some than at other times, the amount of soluble matter is not increased in proportion to the water, since much rain or melted snow then runs off the ground without penetrating amid the rocks. Common salt (chloride of sodium) will be found more frequent than may usually be supposed in spring and

* Much may be accomplished by taking up the water in clean bottles, well-corking, sealing, and securing them; noting the state of the springs, streams, and rivers at the time as regards the quantity of the water in them, and by obtaining a section of the rivers at some convenient situation, and a proper insight into their velocities at the time of taking the water, so that a fair estimate may be obtained of the amount of soluble matter transported.

river waters. When we consider the number of rocks which, from their organic contents, we have reason to suppose were formed beneath the sea, and which have been deposits of mud, silt, sand, or gravel, now elevated into the atmosphere, so that rain waters percolate through them, we shall not be surprised at the presence of chloride of sodium, since it is to be expected that this and other salts in solution in sea waters would, formerly as now, be disseminated amid mechanical deposits effected in the sea.

Silica is well known as in solution in some waters; chiefly, however, found in appreciable quantities in those which are thermal. The geysers of Iceland have been long celebrated for their abundant siliceous deposits.* Silica has borne such a part in the consolidation of rocks, that wherever opportunities occur of observing the effects arising from the action of silica-bearing waters, they should receive careful attention. The manner in which silica may be taken up in its nascent state, and in which it is discovered in heated waters, are circumstances of much importance when we have to consider its mode of occurrence in veins, or its agency in agglutinating the particles of mud, silt, and sands in beds of rock. It is now known not only that certain plants require this substance, but that it is essential to some animals; so that the study of the mode in which silica may be taken up in solution, distributed, and used not only by plants and animals, but also for the consolidation and filling up of the fractures of rocks, is one of much interest.

Springs are presented to our attention chiefly under two forms. First, from the combination of porous and less permeable rocks in such a manner that the water passing readily through the former, and with difficulty through the latter, lines of springs may form at any sides of hills or other exposures, where its outpouring is more easily effected than in other directions; and, secondly, from out of

* Sir George Mackenzie (*Travels in Iceland*) mentions that deposits from the Geysers extend to about half a mile in various directions, with a thickness of more than twelve feet. The leaves of birch and willow are fossilized, every fibre being discernible. Grasses, rushes, and peat are in every state of petrifaction. Very elaborate analyses of the Great Geyser waters by Dr. Sandberger and M. Damour, will be found in the sequel. From these it would appear that the silica constitutes about 0·55 of 1000 parts, including the water.

The siliceous deposits from hot springs (temperature 73° to 207° Faht.) in the volcanic districts of Fumas, St. Michael's, Azores, are important. Dr. Webster (*Edinburgh Phil. Journal*, vol. vi.) gives an interesting account of them. The siliceous deposits are noticed as most abundant in layers from a quarter to half an inch in thickness, accumulated to the depth of a foot and upwards. Compact masses of siliceous deposits are mentioned as having been broken up and re-cemented by silica, and the compound is represented as beautiful. The height of some of this breccia is estimated at thirty feet, and the general accumulation, including a clay, also deposited from the waters of the hot springs, as considerable, forming low hills.

those of breaks and dislocations of rocks which have been termed *faults*, and which become channels into which waters are either drained laterally, or forced up from beneath. Let the following section (fig. 10) represent one of a country composed of different rock deposits, somewhat similar to those in our oolitic districts, for example, *a a* being portions of a porous and calcareous rock,

Fig. 10.

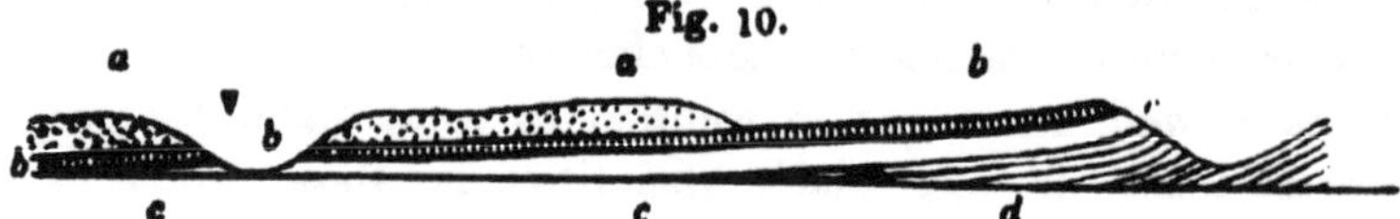

such as some of those oolites are, based upon a clay, *b b b*, itself reposing upon a sand, *c c c*, chiefly composed of siliceous grains, and this again resting upon a clay, *d*.

We should here have the conditions for a marked example of the springs of the first class. The rain falling upon *a a* would percolate through it, taking up calcareous matter by aid of the carbonic acid in the rain water, or obtained in its passage through the vegetable covering and soil. Not being able to permeate readily through the subjacent clay, *b b b*, it would be thrown out as spring water at the junction of the two rocks. This water would probably contain much bicarbonate of lime. The subjacent clay might furnish some water in the valley v, a slight portion of the rains finding its way amid the particles of clay, already moist. We will suppose that, as often happens, the spring water thus afforded would contain iron (from the decomposition of iron pyrites), and sulphate of lime (iron pyrites and selenite being often common in such clays). Beneath, in the two hills to the left of the section, the rain falling would not readily find its way from above to *c c*, though laterally this bed may be exposed to it, as a part is on the right of our figure. This bed has been considered as principally composed of siliceous grains, and to be based on a comparatively impervious bed, *d*, which may be a clay. Springs would find their way out of this bed in the valley v, and we should expect that, though they might contain certain matters in solution, these would not be the same, at least not in such abundance, as from the beds *a* and *b*.

A stream, therefore, flowing down the valley v, would collect waters differently charged with the substances which rains on their passage through the rocks had brought out in solution; and though the waters of such a stream would present us with a kind of mean of all the substances abstracted in solution from the various rocks, they would not show those obtained from any kind of rock taken by itself. These, consequently, would have to be studied where the springs flowed from each bed. The streams, moreover, contain the top waters which, during rains, flow over the surface, carrying off, independently of the matters mechanically transported, those which can be taken away in solution, and which had not formed component parts of any of the solid rocks passed over in their course, such matters being commonly derived immediately from animal and vegetable sources.

The observer would readily expect this simple mode of occurrence of dissimilar rocks, furnishing water holding different substances in solution, to be variously modified, so that while studying the kind of matter thus abstracted from rocks, he should carefully direct his attention to the connection of springs of this order with the kind of rocks traversed by rain waters.

The joints and cleavage among certain rocks greatly complicate the subject in some districts, and in others contorted and crumpled strata so occur, that long troughs and irregularly formed basins of water are held up amid the beds and rocks, pervious to water, in some localities, while dome-shaped masses tend to throw these reservoirs off in others. In the cases of such basins and troughs, the water remaining during the drier times may perfect many solutions, which, when the rainy seasons come to act, are borne away in springs, at that season only of importance.

Springs of the second class are commonly more constant as to the quantity and quality of the waters they deliver, and in this manner, when they traverse many dissimilar beds, furnishing the solutions of different substances, they are like the streams above noticed, as regards such substances. We do not, therefore, learn from them the kind of loss any particular rock may sustain from this cause, though they may be useful in showing the solutions delivered from the fissures. Let *f* in the accompanying section (fig. 11) be a

Fig. 11.

dislocation traversing various dissimilar beds, so that the bed *a* is thrown down, as it is termed, on the left, and that we find other and upper beds, *g h* and *i*, occupying the same general levels, as *a b c d* and *e*, on the other side of the fault. In such a case the various waters percolating through the latter would find their way into the dislocation with those of *g*, on the opposite side, and the solutions derived from all these beds would be mingled in the waters of the fault, flowing out at *f* in greater or less abundance, according to circumstances. We have here merely regarded the solutions derivable from the waters percolating through the upper beds; but as in the greater proportion of faults we possess no means of judging of the depths to which the dislocation may descend, we cannot form a correct opinion of the kind of rocks traversed by them, and affording solutions beneath.

Thermal springs, not in volcanic countries, have been traced either immediately to these dislocations, or the evidence has been such as to lead us to suppose that they may be merely covered over by beds, through which a sufficient passage has been found for the discharge of the waters rising among dislocated rocks beneath. The case of the Bath springs is not improbably one of the latter kind, the heated waters rising through some of those dislocations or faults which traverse the older rocks of the district (coal measures, carboniferous limestone, and old red sandstone), covered over unconformably by the new red sandstone series and lias (as these beds are known to do many dislocations of such older rocks in that country), the waters thus finding their way through cracks or passages in the superincumbent beds.

Fig. 12.

Connecting the heat of thermal fault waters with the increase of temperature of the crust of the globe inwards, as inferred from the increase of heat as we bore artesian wells, or descend in mining operations, the temperature of such waters would always be considerable, were it not that such temperature may be much modified by the conditions under which the waters are borne upwards and discharged. Let *f g*, in fig. 12, represent a fault traversing various rocks to a depth at which the water in it obtains a high temperature. These waters could only be discharged at that temperature, if the rate of outflow were so considerable, and the volume of water so large, as to be uninfluenced by the cooling conditions which would exist in the rocks through which they had to pass. Towards the surface, these rocks would take the temperature of the part of the world in which they may be situate, variable near such surface, but at a certain depth, according to latitude and local conditions influencing surface temperature, assuming a constant temperature unaltered by the climatal changes or modifications above. Between this fixed situation, which in fig. 12 we will for illustration assume to be at *a*, and that beneath, at *g*, where a very high temperature may exist, such as 212° Fahrenheit (the boiling point of water under a pressure of atmosphere equal to about 30 inches of mercury on the surface of the earth), the water in the cleft or fault, would be at intermediate temperatures. Some waters, supposing a ready discharge of them to exist laterally, might have a tendency to percolate through the adjacent rocks, and enter the main fissure at depths not far beneath that of the lowest constant temperature, thus assisting to cool the upflowing waters, independently of the decrease of temperature effected by that of the rocks themselves. No doubt, under the conditions supposed, the sides of the fissure would be heated at given depths beyond that temperature which, if the heated waters did not rise through them, they would possess, but the discharge of waters, as a whole constant, and other conditions the same, there would be a final adjustment of the order supposed. This would be a state of things conducive to the entrance of many substances in solution into the main fissure, which might not be introduced into spring waters, either at all or so readily and abundantly in the first class of springs. The greater heat, as the rocks increase in depth, and the permeation of waters through them, at high temperatures, would be favourable to the removal of silica, often perhaps, only to short distances, one kind of rock being modified by its gain in this manner, and another by its loss. Any thrown out in solution would be so much removed from them, to be employed elsewhere in the modifications now effecting on the surface, always assuming, for illustration, that the rocks traversed by the fissures furnished the matters held in solution by the waters flowing upwards through them. A supposition which will require to be modified if we consider that some substances or portions of them may be borne up into the cracks which had not previously formed parts of solid rocks. Under any view, the solutions contained in these fault waters, are conveyed away from the mouths of the fissures, and so much of them as have been added to waters percolating downwards from the atmosphere, or in any manner through or from the adjacent rocks, has caused a loss to such rocks,* and afforded matter, capable of ready transport, to be employed, as circumstances may permit, elsewhere in the formation of solid matter, or as an addition to solutions in the waters of lakes and seas.

Deep mines afford opportunities for observing the rate at which rain waters may percolate through the body or fissures of rocks downwards, and analyses of these waters so obtained, give the substances they have, during the time of their passage, taken up in solution. In mineral veins, the waters which would remain in them, or flow out as surplus, being in some mines pumped out to depths of even 1800 or 2000 feet, we no doubt have surface waters descending further than they would otherwise do in the same time, the check to their progress, interposed by the water disseminated amid the adjoining rocks, or in the fissure, being thus removed, but at the same time the evidence as to the power of the surface waters to descend in the time that may be observed, and as to the kind of solutions effected by them in that time is valuable.

Great care is required to give due importance to local conditions in such investigations, such as the comparative readiness with which the waters may be conducted downwards by means of an unworked continuation of the mineral veins—having easy water communications with the workings in the mines, the absence or relative abundance of great joints or other fissures in the adjoining rocks, the chance of any rivulet or stream passing over, when swollen by rains, fissures or cracks communicating with the main vein, and the like.

In some coal districts, the beds of under-clay (as those are often termed which are found supporting, or intermingled with the coal beds) are usually so impervious to water, that where faults or fractures of beds are rare, the collieries are little troubled with water. This impervious character, employing the term in a general manner, is well marked in coal measure districts where, as in parts of South Wales and Monmouthshire, the beds having a slight inclination, and being cut through by mountain valleys, springs of the class first noticed are thrown out in lines, marking those of the coal beds, the waters percolating through them being stopped downwards by the under-clays. A system of deposits in which such beds and others of tough shale occur, would present difficulties to the ready percolation of the water downwards. At the same time, slight

* Dr. Daubeny points to the very common presence of nitrogen in thermal waters as a proof that the water in them has been originally derived from the surface of the earth, that it there contained atmospheric air, and that, descending, this air was deprived of its oxygen by some process of combustion.

observation will soon show, that though water may not find its way in a sufficiently rapid manner in some collieries to be important, it is still most frequently there disseminated among the particles and joints of the rocks. Indeed, the manner in which water is disseminated among rocks is deserving of all attention, particularly when we regard it as a means by which a change and modification of chemical composition may be effected.*

The springs of the first class noticed as outflowing on the sides of hills and mountains, and on sea cliffs, are frequently productive of landslips, as they are often termed, the percolation of water in particular planes or directions so softening, or chemically removing the rocks, that a superincumbent weight not being held up by sufficient cohesion of the mass, is launched into the valleys or seawards as the case may be, thus producing a degradation of the land, throwing it into conditions fitted for more ready removal by rivers and the sea. Small landslips are very common, and are well seen in our oolitic districts, where the intermingled clays slipping into the valleys bring down the more consolidated superincumbent beds with them. In the coal district of South Wales good examples of a larger kind are to be found, and in many mountainous regions they are sufficiently common.

The slide or fall of the Rossberg or Ruffiberg on the 2nd September, 1806, afforded a memorable instance of the destruction produced by the percolation of water through bedded rocks in such a manner that, the needful cohesion of parts being destroyed, a great mass slid over an inclined plane of subjacent rocks. The following section (fig. 13) will serve to illustrate this fall, and some others

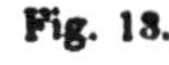

Fig. 13.

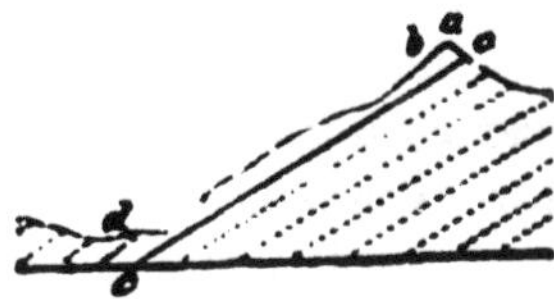

of the like kind. If in the mountain, *a*, water percolate through the porous strata *b* to the clay bed *c c*, the surface of the latter would become slippery, and the cohesion being insufficient to counteract the action of gravity, and no proper support be found

* The simple experiment of accurately weighing a piece of rock immediately after it is struck off in a metal mine or colliery, drying it thoroughly in a sand-bath, and then reweighing it, will often show more moisture to have been removed than might have been expected, the result being necessarily very variable from differences in the porosity of the substance.

below, the mass would be launched in the valley *d*. In the case of the Rossberg (a mountain 5196 feet above the sea), the upper beds were composed of conglomerates resting upon matter, which being partially removed by the percolation of water, and the beds at a high angle (about 45°), a launch of the upper beds took place, and a beautiful valley was covered with rocks and mud.*

The undercliffs between Lyme Regis and Axmouth, as well as those on the back of the Isle of Wight, illustrate the destruction of cliffs by means of springs. The following section (fig. 14) will show the conditions under which the undercliffs are produced at

Fig. 14.

Pinhay, near Lyme Regis. *a* is gravel; *b*, chalk; *c*, upper green sand, porous substances through which the rain waters percolate to the clay bed *d*, composed of the lower part of the green-sand beds *c*, and the upper part of the lias bed *e*, the upper green sands having over lapped the intermediate rocks observable in the south-east of England, and here resting upon the lias. The water being thus arrested in its progress downwards, escapes where it finds the least resistance; in this case towards the face of a cliff, originally formed by the action of the sea on the coast. The clay is gradually removed; the superincumbent green sand, chalk, and gravel lose their support, give way, and fall towards the sea. The lias *e* is not removed by the action of the coast-breakers so fast at the cliff *g*, as the rocks above are by the effect of the land springs, therefore the upper cliff retreats, leaving a mass of fragments confusedly intermingled at *f*, which has a constant tendency to move seawards, both from the destruction of the lias cliff *g*, by the breakers, and from the water percolating through the mass and loosening its base, so that it gradually moves towards the shore. The chalk and green-sand fragments are often sufficiently large and hard to afford, by their overfall, protection to the lias cliff, and thus a very confused but instructive coast section is exposed to the observer.

* The villages of Goldau and Busingen, the hamlet of Huelloch, a large part of the village of Lowertz, the farms of Unter- and Ober-Rothen, and many scattered houses in the valley, were overwhelmed by the ruin. Goldau was crushed by masses of rocks, and Lowertz invaded by a stream of mud. The lives lost were estimated at from 800 to 900.

CHAPTER III.

SUBSTANCES MECHANICALLY SUSPENDED IN WATER.—TRANSPORT OF DETRITUS BY RIVERS.—DEPOSIT OF DETRITUS IN VALLEYS.—ACTION OF RIVERS ON THEIR BEDS.—REMOVAL OF LAKES BY RIVER ACTION.—FORMATION AND DISCHARGE OF LAKES.—LACUSTRINE DEPOSITS.

THE rain waters not absorbed by the rocks, act mechanically on the surface of the land, removing to lower levels such decomposed portions of the rocks as their volume and velocity can transport. The mixed effects of decomposition from atmospheric causes, and of soaking of the surface on hill sides, are often well shown in slate countries, a certain depth beneath the soil exhibiting the turning over of the edges of the slates towards the valleys;—as it were the tendency of the moistened matter of the surface to slide by its gravity to the lower ground.

The accompanying figure will illustrate this fact, one of much importance to the observer, for without attention to it he might commit grave errors as to the true dip of strata, when only a

Fig. 15.

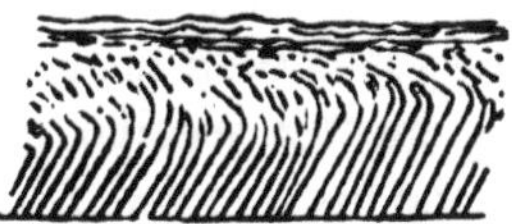

slight depth of section may be exposed on a hill side. In the above figure the real dip of beds is represented as the very reverse of that which might be inferred from a hasty glance at the surface. Although it may be supposed that the difference between this sliding down of the surface towards the lower grounds and the true dip was always so apparent as not to be mistaken, the depth to which this action has occasionally extended is sufficient to justify great caution in many districts.

Upon a hill side and among the rills, hollows, and little plains which may sometimes be there found, an observer may often have good opportunities of studying the power of water mechanically to transport the decomposed portions of rock brought within its influence. He will soon perceive, that not only according to the specific gravity, but to the form also of these portions is their removal effected, and that the manner of removal is of two kinds. In one case they are bodily carried in mechanical suspension in the water, while in the other they are swept onwards by its friction on the bottom. Small hollows will occasionally show the mode in which the matter so mechanically suspended or pushed onwards is brought to rest, and well illustrate the manner in which accumulations on the great scale may be and are effected.

If we suppose the observer placed in a granitic district where there is much decomposition of the felspar, such for example, as much of that near St. Austle, in Cornwall, he will soon find that while the fine decomposed remains of the felspar readily mingle with the waters which a heavy fall of rain may produce, the particles of quartz and mica are more commonly swept along the bottom, except where, from the slopes being considerable, the water may have sufficient rapidity to gather them up in mechanical suspension. While the volume of the particles of quartz may be larger, they are often more round, so that they are commonly more readily pushed along the bottom than the grains of mica, not only flatter but possessing greater specific gravity.* The milky-looking water containing the decomposed felspar is borne onwards, slight deposits taking place where an expansion of the bed of the rill or rivulet may permit comparatively still water, until sufficient quiet is found for the general deposit, while the quartz or mica are strewed in little ridges, or thrust into holes, remaining there if the force of the stream will permit.

Much information may be derived as to the manner in which detritus is pushed forwards by rivers into bodies of still, or comparatively still, water, by observing sand brought down by a rivulet into a small pool of stagnant water, where the sand ceases to be forced forwards, and consequently accumulates. It will be seen that little delta-form heaps of sand accumulate where the rivulet enters the pool, on the fan-shaped tops of which the channels, over which the moving water pushes the grains of sand, are continually shifting. Let *a* in the following sketch (fig. 16) represent a pool

Fig. 16.

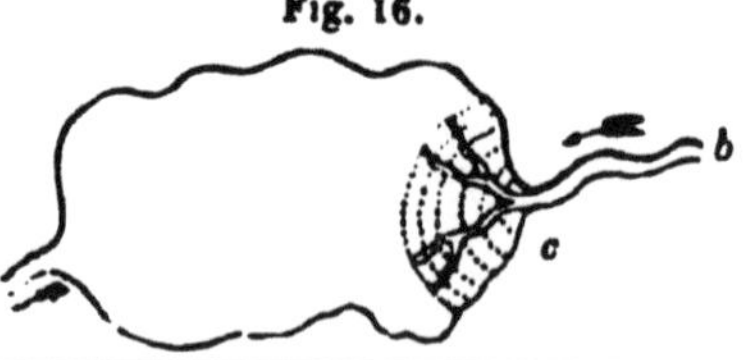

* The specific gravity of quartz is about 2·68, while that of common mica is 2·94.

of still water, into which a rivulet *b* pushes forward sand, then such sand will be found to accumulate at *c*, falling down into the pool *a*, in such a manner that a truncated heap of sand is produced, which increases superficially, as shown by the concentric lines at *c*. If now, attention be directed to the manner in which the grains of sand have been accumulated vertically, it will be found that they have been arranged as in the annexed section (fig. 17) in which *a*

Fig. 17.

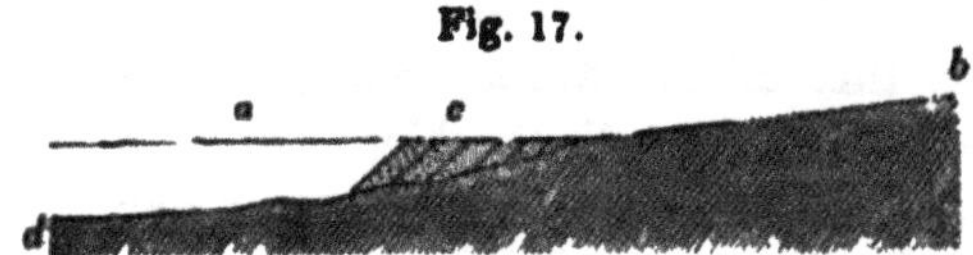

represents the surface of the pool, *d* its bottom, *b* the slope of the rivulet pushing forward the grains of sand, and *c* successive coats of sand formed by the grains falling over into still water, such grains supporting themselves in the same manner as in any rubbish heap, from the top of which rubbish is continually thrown over. By diverting from their courses the small streams of water which run down sandy sea beaches on many coasts, very valuable information may be obtained as to the manner in which grains of sand are forced forward, and arranged by the pushing action of running water. When brought into the deeper pools among the sands, the deltas produced are extremely instructive, and in such cases the angle formed by the layers or coatings above each other, as the sands accumulate, is commonly found to be about 28° or 30°.

Having examined the mode in which decomposed portions of rocks, as well as those worn off by the friction of the streams, can be transported by moving water on the small scale, an observer will more readily appreciate the transport and deposit of detritus on the great scale in the course of rivers, with or without the intervention of lakes, as the case may be, and its removal towards lower levels and the sea. The manner in which it is either taken up in mechanical suspension, or merely shoved along the bottoms of rivers, is precisely the same in principle as in the little rivulets, though the effects, from their greater magnitude, are more striking in the one case than in the other. Larger masses may be shoved forwards, because the volume of water may be larger, sufficient to move those onwards, the resistance of which the minor streams could not overpower, yet the cause of their removal is of the same kind.*

* The following list of the specific gravities of some rocks which we have elsewhere given (*Researches in Theoretical Geology*, 1834), may be useful in showing their power of removal, in fragments or pebbles, by running water, all other conditions as

It will soon be perceived, that while at one time detritus only of a given magnitude, form, or specific gravity can be either pushed onwards by, or be mechanically suspended in, the rivers, at another the detrius, previously at rest, is readily borne onwards, and effects produced which, without the needful evidence, would scarcely have been considered probable from examining those produced during the ordinary condition of the same river. From the details given of the effects of great floods, as, for example, that of the Moray, much valuable insight may often be obtained as to the effects which, during a long lapse of time, may be produced along the line of a river course by repeated action of this kind.

The minor floods, commonly known as *freshets*, more or less common in all rivers, are geologically important, not only as respects the greater movement outwards of detrital matter at such times by the mechanical action of the water, but also as they often surprise terrestrial animals in low localities, and transport them with plants to still lower situations, or into the sea, in the latter case covering up these as well as estuary and marine animals in a common deposit of mud and silt.

In some countries the freshets, or rises of river, are periodical, produced from periodical causes inland, as, for example, those of

---

to velocity and volume of the water, and volume and form of the fragments or pebbles, being the same:—

Rock	Sp. gr.
Calcaire grossier (Paris)	2·62
Chalk (Sussex)	2·49
Upper green sand (Wilts)	2·57
Lower green sand (Wilts)	2·61
Portland oolite (Portland)	2·55
Forest marble (Pickwall)	2·72
Bath oolite (Bath)	2·47
Stonesfield slate (near Stow-on-the-Wold)	2·66
Lias limestone (Lyme Regis)	2·64
Red marl of the new red sandstone (Devon)	2·61
Muschelkalk, fossiliferous (Göttingen)	2·62
Coal sandstone, Pennant (Bristol)	2·60
Coal shale, with impressions of ferns (Newcastle)	2·59
Millstone grit (Bristol)	2·58
Carboniferous limestone (Bristol)	2·75
Carboniferous limestone (Belgium)	2·72
Old red sandstone, micaceous (Herefordshire)	2·69
Old red sandstone (Worcestershire)	2·65
Silurian sandstone (Hartz)	2·64
Devonian sandstone (Ilfracombe)	2·69
Devonian sandstone, calcareous (Ilfracombe)	2·77
Silurian sandstone (Snowdon)	2·76
Argillaceous slate (Devon)	2·77
Carrara marble	2·70
Mica slate (Scotland)	2·69
Gneiss (Freyburg)	2·72
Domite (Puys de Dôme)	2·37
Trachyte (Auvergne)	2·42
Basalt (Scotland)	2·78
Basalt (Auvergne)	2·88
Basalt (Giant's Causeway)	2·91
Greenstones, various (different countries)	2·69 to 2·95
Sienite (Dresden)	2·74
Porphyry (Saxony)	2·62
Serpentine (Lizard, Cornwall)	2·58
Diallage rock (Lizard, Cornwall)	3·03
Hypersthene rock (Cock's Tor, Dartmoor)	2·88
Sienitic granite (Vosges)	2·85
Granite, gray (Brittany)	2·74
Granite (Normandy)	2·66
Granite, mica, scarce (Scotland)	2·62
Granite (Heytor, Devon)	2·66

the Nile, and deposits are then effected which do not receive additions until the annual time of rise again comes round. From this state of things to frequent alternations of floods and low states of rivers, there is every modification, so that the results of the deposits may be expected to be as modified as the causes of their production.*

When it is intended to ascertain the volume of water descending a river at a given time, and the amount of matter which may be then held in mechanical suspension by it, in order by a fair average to estimate the volume of water and the amount of matter, in mechanical suspension, borne seaward or into lakes during a year, or any amount of time thought desirable, much care is required so that the estimate may approximate toward the truth.

The section of a river presents us with waters moving with different velocities, and consequent transporting powers. Where the greatest weight of water occurs with equal velocities, there is the greatest pushing or forcing onwards of the bottom. If in the accompanying section (fig. 18) *g f g* represent that of a river

Fig. 18.

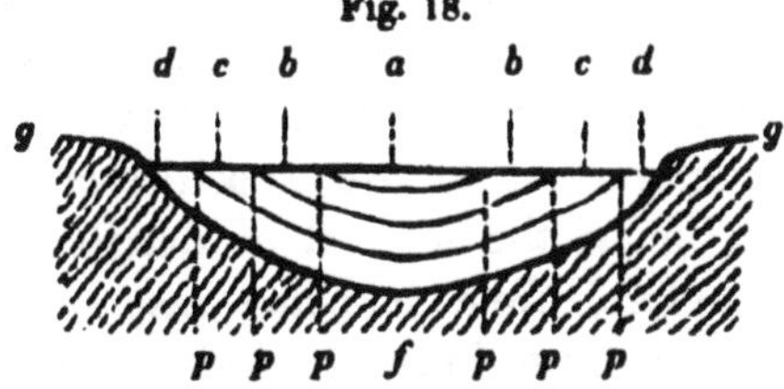

course, the greatest velocity of the water would be at *a*; and this will decrease towards the sides and bottom, where the friction would be greatest, as may be represented by the layers of water *b b*, *c c*, *d d*.

Let fig. 19 represent a longitudinal section of the layers of water corresponding with those in the cross section (fig. 18).

---

* As we have elsewhere observed (*Geological Manual*, 3rd Edition, 1833), there are few rivers more instructive than the Mississippi, man as yet not having effected many important changes on its banks, and we contemplate great natural operations, such as cannot be so well observed in those which have been more or less under his dominion for a series of ages. Its course is so long, and through such various climates, that the freshets produced in one tributary are over before they commence in another; and hence arise those frequent deposits of detritus at the mouths of the tributaries. These latter have their waters ponded back, and, to a certain distance, stagnant, by the rush of the floods in the great river across their embouchures, and in consequence a deposit is effected, which remains until a subsequent flood in the tributary removes it. (*Hall's Travels in North America.*) Captain Hall states, that when the Ohio is in flood it stagnates the waters of the Mississippi for many leagues, and that, when the Mississippi is in flood, it dams up the waters of the Ohio for seventy miles.

Then assuming that the motion of the particles of water in the layer *a* is sufficient to keep some of the matter mechanically suspended, and some not quite so suspended, the latter will sink by the action of gravity; not, however, at once falling to the bottom, but entering the second supposed layer of water, *b*, where the velocity being less, it descends in less time through it, and so on through the other layers *c* and *d*, describing a curve *i n*. As regards the amount of mechanically suspended detritus, in such a section, we should anticipate that it would be very unequally dispersed.

Fig. 19.

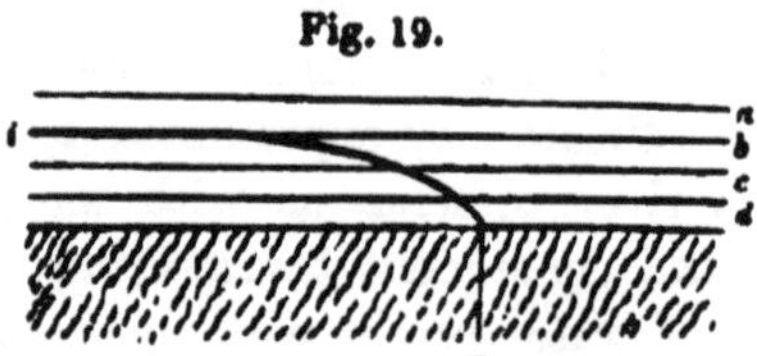

Considering the section, fig. 19, to be one taken through the centre of the stream, and that we add other longitudinal sections taken through the lines, *p p p p p p*, fig. 18, we should have two series, one on each side of the central section, the terms of which could rarely agree, either in respect to the velocities of the water, the power of transport, or in the amount of detritus contained in them. So far, therefore, from it being easy to estimate the amount of detritus borne down in mechanical suspension, or forced along its bottom from friction by a river, it is a subject requiring very great caution and skill, even to obtain an approximate rough estimate of the fact.

When the water has been obtained from which it is intended to separate the matter borne down by rivers, and by a sufficient number of trials, in different parts of the river, to estimate the amount of such matter passing a given locality, it is needful not to evaporate the water, as has often been done, for by this proceeding the matter in solution is obtained as well as that in mechanical suspension. A measured volume of water should be passed through a filter, and the weight of the matter thus collected should be carefully ascertained.

Fully to appreciate the distance to which the various kinds of detritus may be borne by moving water until they be deposited, attention should be directed to the quantity and kind which can merely be pushed forward by a given velocity of such water, acting by friction on the bottom or sides against which it may pass, and to the quantity and kind the same velocity may keep mechanically suspended at the same time.

As rivers are enabled to transport in mechanical suspension, or sweep forward detritus on the bottom, according in a great measure to their velocities, and as the latter, other things being equal, increase with the slope of the river channels, duly to estimate the power of a river to carry forwards to the sea or lakes the detritus thrown into the higher grounds, all the changes of slopes should be properly appreciated. Thus, if *a b* (fig. 20) represent the slope of a river in one place, and *b c*

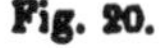

Fig. 20.

the slope of the same river in another, and the amount of water be neither increased nor diminished by tributary streams or diverging branches, the river will have greater velocity at *a b* than at *b c*, and consequently smaller pebbles and finer sand can remain at the bottom at *b c* than at *a b*.

The checks which a river may sustain in its course, such as by lakes, patches of level land, and the like, should be duly noted. Without this precaution it might be, and indeed has been, inferred that all the pebbles found far down a river course had been there swept by the river in its present state. While this is often true, care should be taken to ascertain that the needful conditions present themselves. Frequently, when a river takes its rise among high mountains, its onward course is, though often rapid, interrupted by tracts of level country, or even lakes, where the pebbles and heavier detritus are arrested; and yet pebbles derived from the rocks of the high mountains may be abundantly found in the river-bed further down than these obstacles, such pebbles having been brought to the channel in which the river now takes its course by previous geological conditions of the area. Thus, Alpine pebbles in some of the river courses of Northern Italy could not have been brought from the Alps into the plains of Lombardy, by existing rivers, since the Lago Maggiore, the Lago di Como, and others necessarily stop the progress of those borne from the high Alps by the torrents which now feed these lakes.

By attending to the kinds of rock traversing a valley, we

often find good opportunities afforded of studying the manner in which detritus derived from them, may become mingled by the action of the river waters. Care must, however, be taken to avoid considering as such those pebbles which may have been formed by the action of breakers while the land has been emerging from the sea, and which may have been at that time gathered into the lower parts of the valleys, or those which have subsequently been brought into them from the sides of hills or mountains by the long-continued action of rains and minor streams of water. Let *a b* in the annexed plan (fig. 21) represent the course of a river

Fig. 21.

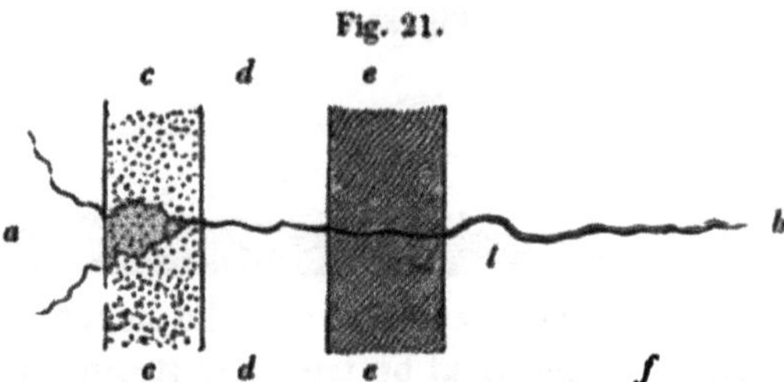

through a district composed of marked, but different, rocks *c c*, *d d*, and *e e*, into a low country, where its movement becomes sluggish, and let the fall of the river-bed be such as to give sufficient velocity to a needful body of water to push or sweep forward pebbles of the size of an egg, where the full force of the water can be directed upon them. The river being capable of forcing forward pebbles of this size on the bottom, those of minor size, other things being equal, would be driven onwards, and there would finally be a size, weight, and form of detritus held up in mechanical suspension by the movement of the water. Under such conditions there would necessarily be a deposit of the detritus pushed forward by the water, wherever sufficient obstacles produced a less velocity in the river; and, as the river varied in velocity according to the quantity of water in it, the accumulations thus formed would possess an irregular character somewhat as in the annexed section, one through several minor deposits, depending upon small shifts in the direction and force of the propelling current.

Fig. 22.

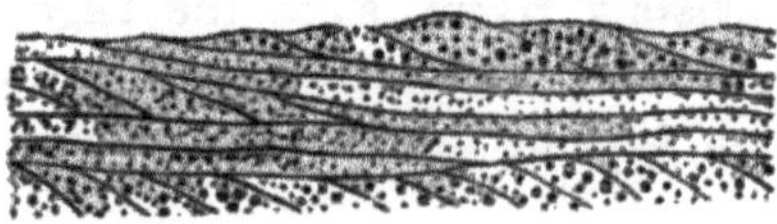

As the river in the plan (fig. 21) is supposed capable of shoving pebbles onwards to the commencement of the low ground *ff*, irregular accumulations of pebbles would be expected at *l*, where the force of the river could no longer drive them forwards. It would not, however, be anticipated that the finer silt or mud could be there accumulated, except in very minor quantities in still places; since the power to keep such detrital matter mechanically suspended would be gradually lost by the river. Indeed the time required for the settlement of the finer parts, might be such that the whole body of water could continue to move through the lowlands in a turbid and discoloured condition, slowly parting with such detrital matter disseminated through it.

It would be expected under the conditions noticed, that accumulations would take place along the line of the river course; and that, unless these deposits were cut up by floods and so carried further onwards,. the river-bed would be raised. The power of a river to keep its channel clear, and even to work it deeper, is commonly obvious where the river runs with rapidity; but it is not always so obvious, without careful investigation, that its bed has been raised, more particularly by the pebbles and sands shoved forward at the bottom.

In many plains, modified by rivers, the shoving forward of detritus is shown by the mode of its accumulation. Other accumulations so thin and wide spread as obviously to have been deposited from mechanical suspension are often, however, intermixed, so that both modes of deposit have contributed to the formation of these plains. Although we might feel certain that the beds of rivers must shift in great plains as such beds get raised, the waters taking the course of the lower levels, when such are presented, yet it is interesting to observe in some countries,—in Italy for example,—where artificial embankments have been formed to keep rivers flowing through fertile plains in their channels, that the beds of rivers become thus raised above the plains; and that roads rise up these banks on either side from the latter. In the little plain of Nice, the river ridges, formed by this cause, are striking, a loose conglomerate behind furnishing an abundance of pebbles to the river-bed. The following section (fig. 23) will serve to illustrate

Fig. 23.

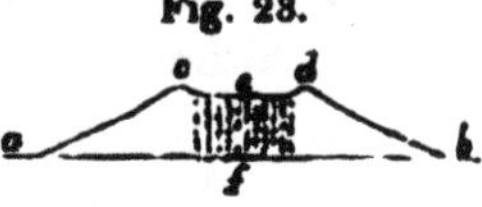

this fact, *a b* being the level of the country, in cultivation for many centuries, upon which artificial banks have been gradually raised to *c d*, to protect the cultivated lands from invasion by the detritus forced forward by the river *e*. Thus the detritus which would have naturally escaped upon the plain has been raised artificially from *f* to *e*, notwithstanding the somewhat general plan of throwing the detritus thus accumulated over the sides upon the protecting banks *c d*, thus deepening the channel when the waters in the river may be sufficiently low for the purpose. The Po presents on the larger scale a well-known example of the rise of its bed, so that it is higher than the houses in Ferrara, and the like may always be expected under similar conditions.

A river may so raise its bed as for some time not to find a new main channel amid the adjoining plain, its turbid waters when in flood escaping over the banks without actually causing a breach, as is shown in the annexed section (fig. 24), where *b* represents a

Fig. 24.

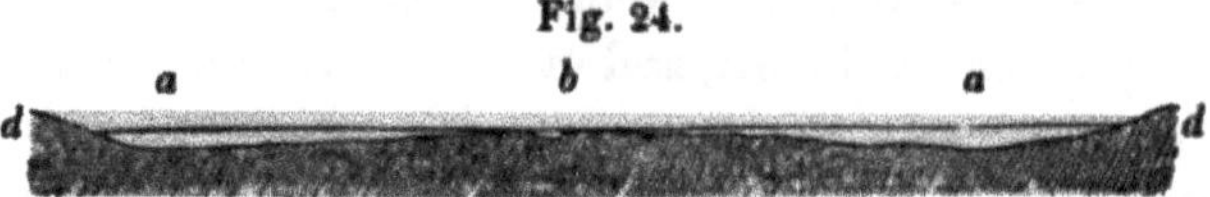

river which has so raised its bed that there are tracts of country on either side at a slightly lower level. In floods such a river, spreading over the adjacent land, would leave all the detritus mechanically suspended in its waters, *a a*, and which did not retire with the water until its level was that of the banks of the river, upon the ground beneath up to the rising slopes *d d*, thus eventually filling up the depressions. The more common action of a flood is as represented in the section beneath (fig. 25), where a river (*b*)

Fig. 25.

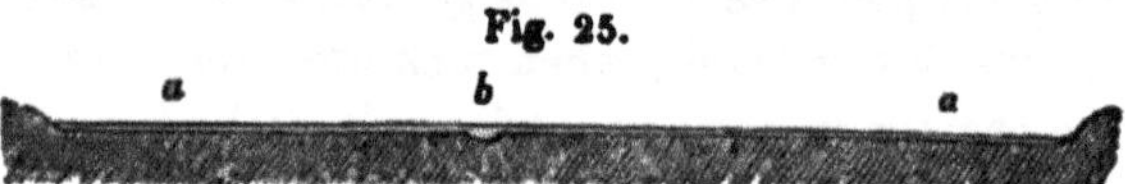

not raising its bed (the flood waters merely removing mud from the bottom, the only sediment there collected), the overflow of turbid water (*a a*) returns to the river bed, depositing only such matter in mechanical suspension as the time of repose may have permitted. In these ways much sedimentary matter is distributed over plains during floods.

The matter pushed forward by rivers, or held in mechanical suspension in their waters, has hitherto been regarded only with reference to the removal of that arising from the decomposition of rocks by atmospheric influences. We have now to consider the erosion of clays, sands, and gravels, and of hard rocks by means of the rivers themselves.

In many a river course it may readily be observed, that incoherent sands and gravels are cut into by the mere friction of the water, even when clear. That such a moving body should so act would be expected, and no doubt we should also anticipate that amid incoherent, or easily-removed substances, any modification in the course of a river would speedily produce change in other parts; but it is, nevertheless, extremely interesting to experiment on the course of streamlets passing among sands: as, for instance, on some extended shores at low tides, and trace the effects of even slight alterations in the stream courses. The cutting into one bank throws the water upon another, not previously worn away, and the whole bed of the stream gets modified. Such experiments tend to make us more readily appreciate those modifications of rivers, from the actual cutting powers of their waters, which are seen on the great scale in some parts of the world. They also show the distances to which the fall of a cliff, the filling up of a cavity, by which, as forming a lake, the force of a flood may have been previously stayed in its full course, and other obvious circumstances have produced modification and change.

There are few persons who have not noticed the manner in which rivers are disposed to take serpentine courses in level countries, a fact as easily observed amid the meadows of the flat portions of many valleys, of very limited dimensions, as among the vast bends of the Mississippi, or any other of the great rivers flowing under similar conditions. The rivers, by their friction, cut into the ground presented to their course, and by working away the earth, clay, sands, or gravel, of bend against bend, modify their channels. The waters necessarily cut away such banks at the bottom of each bend. Hence, if two bends be opposite to each other, as those in the next sketch (fig. 26), are at *a*, *b*, and *c*, the

Fig. 26.

river will tend, by continued erosion, to approximate them to each other, so that they finally meet, and the river course becomes shortened by the amount of the bends previously passed over.

Although some effects must follow the action of clear water upon bodies, the parts of which have not sufficient cohesion to resist removal, it is by the assistance of matter either mechanically suspended in, or forced onward by the water, that rivers most readily cut into their channels and erode their banks. By this assistance they wear even into hard rocks, removing the obstacles impeding their courses, and which prevent the formation of a convenient general slope. As among the simplest forms in which water acts by aid of mineral matter upon rocks, we may take the vertical holes drilled in even some of the hardest by means of pebbles so situated, that a rotatory action is given them, each in one place, by moving water. These are well known in many situations, where bars of rock stretch across river beds, and falls of water are thus produced. A pebble borne down by floods gets so established in an eddy that it remains there, and by constant friction, works a vertical hole downwards, sometimes to the depth of several feet. In some situations, where the obstacle has been much lowered by the erosive action of a stream, sections of the annexed kind may be seen. In rare instances the pebble, as at *a* (fig. 27), may still be seen, the section having been such as not to

Fig. 27.

have allowed it to fall out. In some situations this drilling into bars of rocks must have tended considerably to their ultimate removal.

It is, however, when a river is in flood, large pebbles grinding and driving against rocks which may be exposed to the fury of the torrent, and minor detritus, either hurried onwards on the bottom, or in mechanical suspension, grating against and rasping, as it were, such obstacles, that the erosive power is most effective. Huge blocks are forced onwards, leaving the furrows which have marked their course to attest that course in some situations, while the finer friction of small pebbles and sand produces a smooth surface in others.

When endeavouring to ascertain the abrasion which may be due to rivers, the amount of decomposition which any rocks in their course may have suffered, prior to the supposed abrading action, should be carefully estimated, so that too much importance should not be given to such action. It being known that the decomposition of many rocks is greatly assisted by such rocks being kept alternately in a wet and dry condition, the observer should notice if the water in any river course he may study, rises and falls, and in a manner sufficient to have an appreciable influence on the rocks washed by it.

Much care is required when we seek to refer the formation of a ravine through which a river may find its way to the cutting power of the river itself. There is no want of evidence that even minor streams, more particularly when swollen by rains, cut channels for themselves in various directions. In many a mountain region this is a fact of common occurrence. A little study will show the observer that some ravines are cut back very readily when, as beneath (fig. 28), beds, horizontal, or not far removed from that

Fig. 28.

position, and composed of comparatively hard rocks, such as sandstones, are based upon softer substances, such as clays or shales. From the combined action of atmospheric influences, and of the falling water, with sometimes also the aid of water percolating between the hard and soft rocks, the lower beds give way, and being composed of easily-comminuted substances, are soon removed in mechanical suspension by the torrent, while the hard rocks, losing their support, are precipitated to the base of the fall. This mode of cutting back a channel, with vertical or nearly vertical walls in the first instance, however they may be afterwards modified by subsequent falls, or erosion by small streams, may be as well seen in hundreds of little brooks, where the needful conditions of hard and soft and nearly horizontal strata are to be found, as in the valley of the Niagara, where the production of a ravine of this kind is exhibited on so large a scale.

If a barrier, such as a lava current, be suddenly thrown across a valley, the waters behind it, upwards, are necessarily sustained to the height of the lowest part of the new obstacle opposed to their further progress downwards. Should a section be presented to the attention of an observer, such as that beneath (fig. 29),

where a lava current, *a*, crosses a pre-existing valley in granite. *b b*, *d e* being a ravine, with *c* a river running through it, he should see if the stream of lava, *a*, has been actually cut through, or if it has never completely filled the valley, so that a space may have been left between the high part of the lava, *e*, and the bank of granite *d*, through which the waters readily found their way, the modifying action of the atmosphere and the river giving the fallacious appearance of a ravine wholly cut by the latter.

The observer will have carefully to distinguish between ravines which the rivers may have cut and those which are mere cracks or rents through which the drainage waters of any district may happen to find their way. Therefore he must carefully search for evidence sufficient to prove that the ravine may belong to either the one or the other of these classes. Let A and B (fig. 30) represent sections of

Fig. 30.

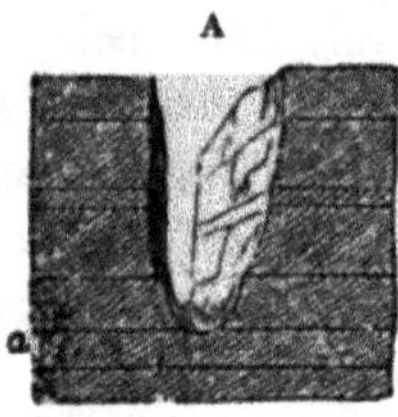

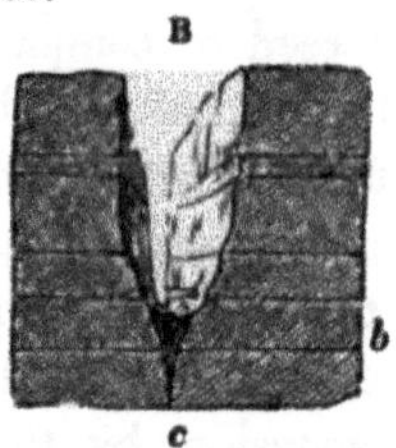

two ravines. In general appearance they might correspond; and even supposing a crack or rent, it may have been such as so slightly to move the opposite masses of rock as to be inappreciable. The geologist should endeavour to trace some bed of rock, such as *a*, unbroken from one side to the other, across the course of the river. Should he discover such a bed thus fairly connecting the sides of the ravine together (no twist in the crack or rent presenting a fallacious appearance of an unbroken bed), the ravine may still not be due to the cutting action of the river itself, for it may have been a channel of communication from one body of water to another at a time when the land may have been sufficiently submerged for the purpose. Hence fair evidence would still be required to show that the river really cut the channel.

If the observer should be unable to trace the rocks unbroken across the ravine, the evidence would remain uncertain, for under the supposition that the sides so correspond as to render a dislocation doubtful, blocks of rock, pebbles, and sand, may as well cover a crack, such as *c* in B, as a continuous mass of rock. Should, however, the beds on either side of the ravine, if prolonged, not meet, that is, if, as in the following section (fig. 31), a horizontal

Fig. 31.

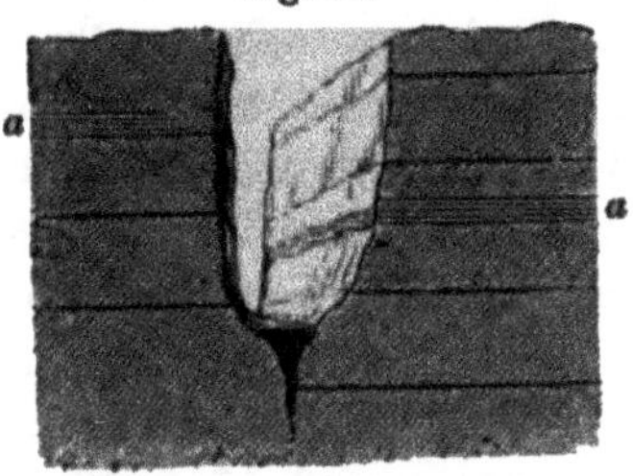

and marked bed *a*, be higher on one side than on the other, he will see that the line of ravine corresponds with a line of dislocation where this want of correspondence of sides is apparent, and by further search he should ascertain if this dislocation can be traced in the same line. Should this be so, it still remains to be ascertained if the river has really done more than modify the effects of an action, along the line of dislocation, by which the ravine may have been originally worked out. If, instead of horizontal, we find vertical beds of rock, as in the annexed map-sketch (fig. 32),

Fig. 32.

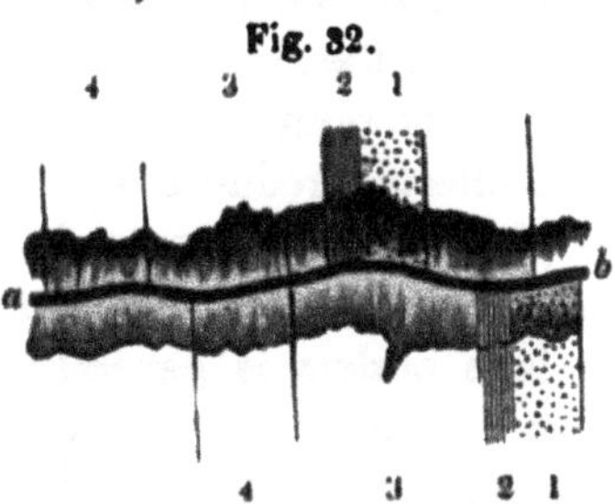

in which *a b* represents the course of a river through a ravine, and that a marked series of beds, 1, 2, 3, and 4, do not correspond if prolonged across the river, then also it would be evident that the latter flowed in a line of a dislocation.

Should the rise of the river-bed be such that a series of falls be found at the higher part of the ravine, so that eventually the level of the river-bed be equal to its most elevated portion, it will be evident that no strait with water, in the manner of a sea channel, was the cause of the excavation, since by submerging the land, the ravine would merely form an arm of the sea, and be liable to be filled up by the detritus borne by the river from higher levels into it.

Upon tracing up lines of valley for the purpose of studying any modifications they may have sustained from the action of rivers and other running waters upon them, it will often be seen, particularly in mountainous regions, that level spaces present themselves, having the appearance of lake bottoms, the river meandering through these plains, and not unfrequently finding its way to lower levels through gorges or ravines of various magnitudes. It is generally supposed that by lowering the level of the lake outlet, the barrier ponding back the water has been removed sufficiently for its passage under ordinary circumstances onwards, it being merely during very heavy floods, that any water is spread over these plains. On the small as well as the large scale, this explanation would often appear probable. If, as in the following section (fig. 33), supposed to represent three lakes, *a*, *b*, and *c*, on the line of a mountain valley, the erosive action of the river could lower the barriers *d*, *e*, and *f*, the cavities *a*, *b*, and *c*, would cease

Fig. 33.

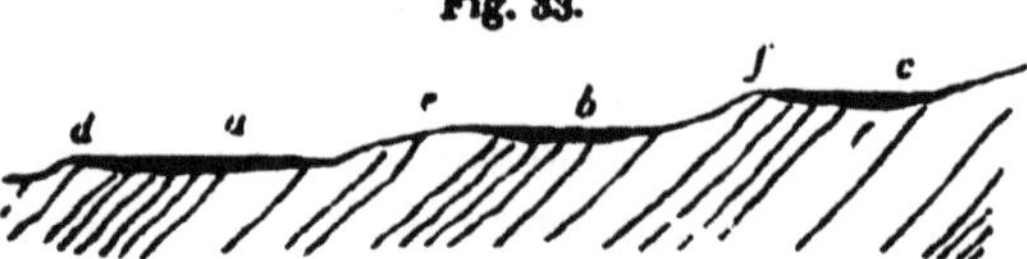

to be filled by water, and we should have plains in their stead, the old bottoms of the lakes, with the river meandering through them, and rushing through gorges or ravines at *d*, *e*, and *f*.

With respect to the effects produced by the cutting back of ravines to such bodies of water, once supposed capable of causing overwhelming floods, at lower levels, it should be observed that the depth of water at lake outlets is generally inconsiderable, so that the letting out and lowering of the lake waters would be gradual. To illustrate this, let the subjoined section (fig. 34) represent the case of a river cutting back its channel, in the manner of the Niagara (assuming that conditions were favourable for so doing), towards Lake Erie, so that the latter became drained by the operation. Let *h* s represent the slope, exaggerated, of the lake bed from *h*, where the surplus waters are delivered over the barrier ground, and *f′ o* the level of the river below the falls cutting back the channel. Supposing *f f′* to represent the place of the falls, at

Fig. 34.

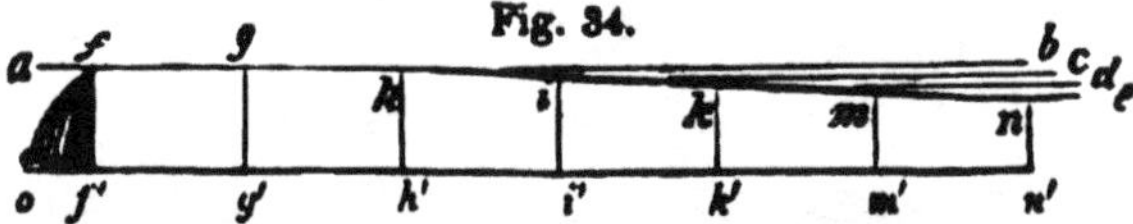

any given time, it is clear, the same effects continuing, that they may be further cut back to *g g′* and even to *h h′*, without diminishing the quantity of water in the lake. Once, however, at *h h′*, every succeeding cutting will occasion more water to pass over them, by draining the waters of the lake to the level of the top of the new falls, so that when these have retreated to *i i′*, the surface of the lake will sink to *i c*, and the mass of water, over the whole lake, and above the new level, will have passed over the falls in addition to the ordinary drainage discharge. This addition would add to the velocity and cutting power of the falls, which would be expected, all other conditions being the same, to retreat more rapidly to *k k′*, reducing the general level of the lake to *k d* in less time than it reduced it from *h b* to *i c*. In like manner, the level of the lake would be reduced to *n e*, which we may assume, for illustration, as its greatest depth; but every succeeding retreat of the falls lowering the general level so that the lake presented a minor area, the lake waters discharged would gradually become less until, finally, nothing more than the river would meander through the drained bottom of the lake. In considering the mode in which a lake may be drained by the cutting back of the outlet river channel, it should not be forgotten that, when large, the average loss from evaporation becomes less as the surface is diminished, so that the supply by the tributary rivers and streams is not much diminished by this cause, and more water finds its way through the outlet to the lower levels.

In volcanic regions we may expect a modification in the drainage of valleys by the flow of lava currents across them, and lakes may be formed in Alpine regions by the fall of masses of mountain into narrow valleys. From the former cause many permanent alterations in the drainage may be effected, the dammed-up waters finding a new outlet, more particularly amid accumulations of ashes and cinders. In the case of a lava current traversing a valley, the deepest part of a lake thus formed might be at the lower part, as in the annexed section (fig. 35), where the previous slope of a river-bed has been interrupted by the flow of a lava current *b* across a

Fig. 35.

valley, so that the river waters are ponded back, and form a lake at *a*. Supposing that a lava current fairly stopped the river course, even rising somewhat on the opposite side of such a valley, and thus preventing the conditions noticed above (p. 36), such a barrier might long remain, the stoppage of the river waters preventing any kind of detritus, which previously had been forced onwards along the bottom, from further progress, at the same time causing much of the mechanically suspended matter to fall. Both conditions would be favourable to the filling up of the lake, such deposits again to be cut through, should the barrier of the lava current be eventually removed. And it is to be observed that the cutting away of the barrier would be more easily effected when the lake was filled up, and gravel and sand could be brought to scour and wear away the channel of the rapids or waterfalls from *b* to *c*.

When mountain masses have fallen across narrow valleys, as they are known to have done, and have ponded back the waters, it may readily happen that debacles may be formed, producing very grea effects at lower levels, and causing the removal of masses of rock under such conditions, which the ordinary condition of the waters in the valley, with every regard to floods, would appear to render improbable. The observer may learn to appreciate the effects c such falls by throwing a dam of loose sand and gravel across an small stream, so that the waters be ponded back. At first the removal of the barrier will be slight, but after a time the waters rush out, sweeping a part of the dam before them, and removing, in their course downwards, stones and blocks which their vegetable coating show have for years well resisted all ordinary floods.

Sometimes also in mountain regions, a cross valley may, from thunder storm falling upon the area which it drains, thrust forwar such a mass of rubbish across a main channel as to pond back it waters, which finally clearing away the barrier thus formed, rusl suddenly onwards to lower levels. At other times the effects of .

tributary, delivering itself at right angles, or nearly so, to the main river, are more gradual; and in parts of a chief valley, where the fall of the latter is not so considerable as to produce a rapid current, more permanent changes are produced. The annexed sketch represents one of those cases, not uncommon in some regions,

Fig. 36.

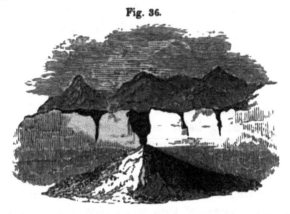

where a tributary comes through a lateral gorge, high above the main valley, thrusting forward the detritus borne along it, so as to form a sort of half cone. The increase of such a mass will modify the line of the main river, if the latter be unable to remove the detritus thus borne across its course. In favourable situations, such as in some parts of the Alps, cottages and cultivation will be seen on those parts of the mound where the more or less divided streams of the tributary do not rush furiously onwards to lower levels.

Among the causes of debacle and change in drainage depressions, we should not omit the consideration of glaciers falling across valleys from adjacent heights, since the great debacle down the valley of the Rhone in 1818, is still fresh in the memory of many who witnessed its transporting power, and who would scarcely otherwise have been disposed to credit the effects produced. After successive falls from the glacier of Getroz, during several years, into a narrow part of the Val de Bagnes, in the Vallais, the accumulation finally became such that the waters of the Dranse, which previously found their way amid the fallen blocks of ice, were ponded back. A lake was thus formed about half a league in length, and it was estimated to contain 800,000,000 cubic feet of water. By driving a gallery at a lower level in the icy barrier, this quantity was supposed to be reduced to 530,000,000 cubic feet, a mass of water which, effecting a passage between the ice and the rock on one side, was let off in about half-an-hour down the Val de Bagnes into the valley of the Rhone, and thus into the lake of Geneva, where fortunately, by the spread of the waters, their destructive force was lost. Huge blocks of rock were moved by this debacle, and a great mass of matter swept away to lower levels.

Mention has been already made of the deposits effected in the still portions of stream courses, and of the inclined angle which the layers of sand and gravel take, after being forced along the bottom of the stream bed, and thrown over little delta protrusions into the pools of water. The mode of detrital deposit to be observed in lakes is the same as in little pools, the difference is chiefly in the magnitude of the accumulations. The little pools differ principally from lakes from being liable to be swept by floods, and the deposited detritus to be thus once more lifted and borne onwards, which does not happen in lakes of fair magnitude. Moreover, discoloured flood waters spread over the pools, and not over pieces of water deserving the name of lakes. Lakes necessarily vary much as to the repose of their waters according to their depths. In the deeper parts of such a body of fresh water as that of the lake of Geneva,* there is no cause for movement from altered temperature of the water, for experiments would appear to show that this temperature always remains the same at the great depths, that of the greatest density of fresh water being found at all seasons of the year. In such situations also waves raised by winds on the surface are not felt, and whatever chemical or mechanical accumulations there take place would remain undisturbed, so long as the present conditions are continued.

In the shallow parts of the same lake, and necessarily also in shallow lakes generally, the waves (sooner raised in fresh water lakes than in the sea by the same force of wind, because the fluid put into motion is of less density) stir up the finer mud and silt, while the breakers act upon the shore, and for the time keep heavier matter in motion and mechanical suspension. As, therefore, the deep cavities holding lakes become filled up, there may be an irregularity in part of the accumulations of the higher portions not observable beneath.

If attention be directed to the mode in which detrital matter is protruded into great lakes, such as those of North America, Switzer-

* In a series of soundings of the lake of Geneva, made in 1819, and chiefly undertaken for the purpose of seeing how far the temperature of the water in it corresponded with that assigned to the greatest density of fresh water, an account of which was published, with a chart, in the 'Bibliothèque Universelle,' for 1819, we found the greatest depth of the lake to be 164 fathoms, or 984 feet, opposite Evian.

land, or Northern Italy, it will rarely happen that the contributing streams or rivers are not found to pour in detritus of various kinds and in different ways. Let us consider that the accompanying plan (fig. 37) represents that of a lake divided into two unequal portions, and that it is supplied with water, in addition to the rain which may fall upon it, by the rivers *c*, *d* and *e*; that *c* is a chief river, draining a large district, and *d* and *e* two torrents, descending occasionally from adjacent mountain heights with great force, while, at other times, they contain little water.

Fig. 37.

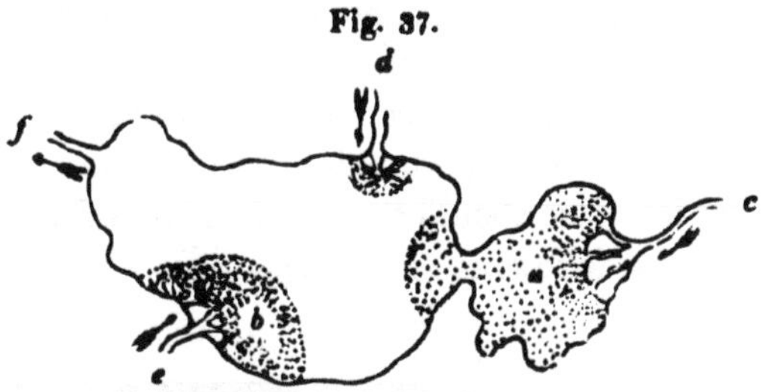

Let us further suppose that the waters of the river, *c*, are generally turbid, like those of the glacier rivers of the Alps, and that they vary in quantity at different times, so that the river both forces forward and holds mechanically in suspension variable amounts of matter. From such conditions as these we may assume that, though variable, the accumulations, brought down into the lake by the river *c*, would still be more uniformly spread than those resulting from the sudden rushes of water down the torrents *e* and *d*, the stones or pebbles, borne forwards by the latter, being larger than the detritus forced onwards by the main feeding river *c*.

In order to appreciate the difference of accumulation arising from these conditions, it may be desirable to assume that the depth of the lake is uniform, or nearly so, throughout, though of course the original form of the lake basin would influence the products. The river *c* would accumulate the detritus it can force along its channel, in the manner previously noticed, while at the same time it would discharge a body of turbid water into the still waters of the lake. The force of the former is checked by the latter; and the turbid water, being heavier than that of fresh-water lakes, would sink in clouds toward the bottom, as may be seen where the Rhone enters the lake of Geneva, and in various other similar situations. The velocity with which the turbid water would enter the lake would carry it to various proportionate distances, until its motion became finally checked. It is, however, interesting to observe that, from the difference in specific gravities, when turbid waters fall to the bottom, these steal quietly upon that bottom for considerabe distances, it being long before they part with the fine matter which they hold in mechanical suspension. The fine matter brought down by the Rhone is found in mud beneath the still deep waters of the lake of Geneva, many miles beyond the discharge of the turbid waters of the river into that lake.*

Assuming the depth of the lake to have been such that turbid could so creep beneath the clear waters as to form a deposit of mud or clay, we should have the bottom of the minor division of the lake coated with this finely-comminuted matter, while a delta-like protrusion of the sand and pebbles was formed over it. Supposing the commencement of such accumulations to be in a rock cavity, the basin of the lake, we should expect them to take somewhat of the form seen in the following section (fig. 38), where *a* represents

Fig. 38.

the first gravel and sand deposits, forced over at *c*, *b* mud, gradually accumulated over the rock basin, *d* the advance of the delta over the mud, and *g* the surface of the lake beyond the delta. Under such conditions we should have irregular beds of sand and gravel, with occasional patches of clay, the result of deposits in local stagnant places, based upon a clay which here and there, in its upper portion, might contain sand or sandy clay, the effects of floods carrying such matter in mechanical suspension beyond the delta into deeper water, and there depositing it upon the mud.

Still referring to the plan, fig. 37, we should expect the accumulations at the junction of the torrents, *d* and *e*, with the lake, to be much modified in character. To render the case more illustrative, we may consider that, from the nature of the rocks traversed by the respective torrents, little else than fragments of hard substances are shoved forward by *d*, while much earthy matter and soft rocks, easily comminuted by friction, are mingled with the harder fragments thrust into the lake by *e*. If a small amount of earthy matter be carried forward by *d*, the accumulation where the torrent enters the lake would form little else than a protruding mass of fragments,

* If a long trough be filled with clean water, and turbid water be very quietly poured into it at one end, the mode in which the latter finds its way beneath the former will be at once seen.

composed of beds different in position, but dipping at angles varying probably from 20° to 30° around the general curve of the protrusion; while such finely-comminuted matter as was held in mechanical suspension would descend to the bottom, and steal along beneath, as previously mentioned, adding to the mud derived from the chief stream *c*. The accumulations formed at *b* by the torrent *c*, would be of a mixed character between those produced by *c* and *d*. These causes continuing, the lake would be eventually filled up by clays, sands, and gravels brought into it by the rivers and torrents, the surface waves acting upon much of the higher accumulations as the general depth decreased. Finally, the outfalling river *f*, clear as that of the Rhone, where it quits the lake of Geneva, while the lake lasted, would be joined to the river *c*; *d* and *e*, as two tributary streams adding their waters to it, and the whole would traverse a plain, much as represented beneath (fig. 39), muddy sediment being added to the surface of the plain from time to time by floods, and the torrents still thrusting forward fragments of rock and pebbles where they joined it.

Fig. 39.

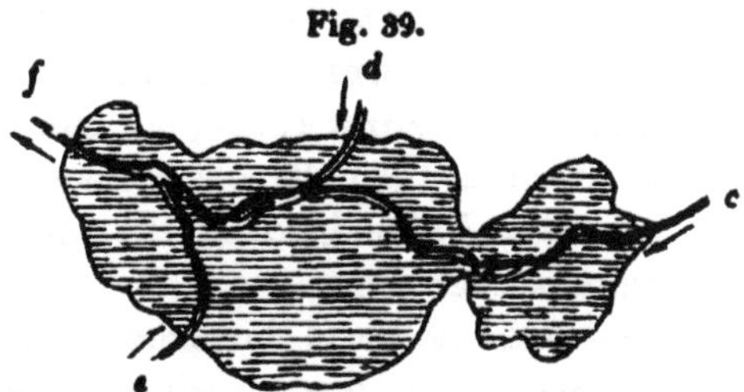

Great modifications of the mechanical accumulations here noticed will readily present themselves to the attention of an observer; aud, if he will combine some of them with the chemical deposits previously noticed, and add the harder parts of the animals which have either lived in, or been drifted into, the lakes, as also the leaves of trees and other plants, and the branches and trunks of trees which may eventually fall to the bottom after having been borne onwards sometimes quietly, at others confusedly and rapidly, he may better appreciate the still greater modifications to which lacustrine accumulations may be subject.

CHAPTER IV.

ACTION OF THE SEA ON COASTS.—DIFFERENCE IN TIDAL AND TIDELESS SEAS.—UNEQUAL ABRASION OF COASTS.—SHINGLE BEACHES.—CHESIL BANK.—COAST SAND-HILLS.

BEFORE we consider the accumulations effected in the sea, it is desirable to call attention to the action of the sea on coasts, since that action often contributes, in no small degree, to the matter of which such deposits are formed.

The sound produced by the grating and grinding of the pebbles of a shingle beach, even when the breakers on shore are comparatively unimportant, can scarcely have escaped the attention of those who have, even for a short time, visited coasts where such beaches, and they are common, are to be found. It will soon be apparent, that this friction, if continued for ages, must not only wear down the pebbles to sand, but grind away and smooth off even the hard rocks exposed to such powerful action. It is, however, when the observer sees the huge masses of rock moved by breakers arising from a heavy gale of wind, blowing on shore from over a wide spread of open sea, or from the long lines of wave known as a *ground swell*, that he not only learns to value the force of the water taken by itself, thus projected against a coast, but also the additional power it possesses of abrading the cliffs which may be opposed to the breakers by the size and abundance of the shingles they can then hold in mechanical suspension.

Properly to appreciate the power of breakers, a geologist should be present on an exposed ocean coast, such as that of Western Ireland, the Land's End (Cornwall), or among the Western Islands of Scotland, during a heavy and long-continued gale of wind from the westward, and mark the effects of the great Atlantic waves as they break and crash upon the shore. He will generally find in such situations that, though the rocks are scooped and hollowed into the most fantastic forms, they are still hard rocks; for no others could long resist the breakers, which, with little intermission, act upon them. Not only blocks of rock resting on the shore are driven forward by the repeated blows of such breakers, but those also firmly bolted down on piers are often thrown off and driven aside in far more sheltered situations. The history of many a pier harbour is that of the destructive power of breakers, and those who have witnessed a breach made in such a harbour during a heavy gale of wind, are not likely to remain unimpressed with the importance of breakers in the removal of land.*

Slight attention to the manner in which waves break on a coast will soon show that, upon the prevalent winds and the proportion of those which force the greatest waves, or *seas*, as they are generally termed, on shore, will depend, other things being equal, the greatest amount of destructive action. Thus, on a coast on which western winds prevail, and there is sufficient extent of open sea before it, we should expect to discover the greatest loss of land, the force of the breakers being there the greatest and most incessant. As a whole, the coasts of the British Islands are exposed to the heaviest and most incessant breakers from winds ranging from the N.W. to the S.W., and but slight acquaintance with our coasts will soon satisfy the geologist, that if the other coasts of our islands were exposed to an equal amount of abrading force, a large portion of them would soon be cut away at a far more rapid rate than at present.

With regard to the force of breakers on the coasts of the British Islands, Mr. Stevenson has found by experiments at the Bell Rock and Skerryvore lighthouses,† that while the force of the breakers on the side of the German Ocean may be taken at about a ton and a half upon every square foot of surface exposed to them the Atlantic breakers fall with about double that weight, or three tons to the square foot. Thus a surface of only two square yards would sustain a blow from a heavy Atlantic breaker equal to about fifty-four tons.

Taking an equal amount of prevalent winds and of open sea over which they may range, it will soon be observable that the abrasion

---

* During a heavy gale in November, 1824, and also in another at the commencement of 1829, blocks of limestone and granite, from two to five tons in weight, were washed about at the breakwater, Plymouth, like pebbles. About 300 tons of such blocks were borne a distance of 200 feet, and up the inclined plane of the breakwater. They were thrown over it, and scattered in various directions. In one place a block of limestone, seven tons in weight, was washed a distance of 150 feet. We have seen blocks of two or three tons, torn away by a single blow of a breaker and hurled over into a harbour, and one of one and a half or two tons, strongly trenailed down upon a jetty, torn away and tossed upwards by the force of another.

† Proceedings of the British Association for the Advancement of Science, Edinburgh, 1850.

of rocks, of equal hardness and similar position, is modified according as the adjoining seas are tidal or tideless. In the latter case, though no doubt the pressure of the wind upon water raises it to levels above those which it commonly occupies, the difference is not so considerable as to bring any large faces of cliff exposed to the action of the breakers. A beach, moreover, piled in front of a cliff is, in such seas, as rarely passed and the cliff attacked. In tidal seas, on the contrary, many feet are vertically exposed to the fury of the breakers as the tide rises and falls; and beaches piled up in moderate weather are, in fitting situations, removed by the return action of the breakers, so that the cliffs are again open to abrasion. Moreover, the rocks are exposed to greater decomposition from being alternately wet and dry, a consideration of some importance in many climates, particularly in those where the temperature falls below the freezing point of water during certain seasons of the year. It should not, nevertheless, be forgotten that coasts, where breakers reach the cliffs at high water, are frequently protected by beaches at low water; and that, therefore, they are moved from the abrading power of the waves during all the time that they fall on the protecting beaches—a time which changes with the varying state of the tides and of the weather generally.

Attention will not long have been given to the abrading action of breakers on coasts before it will be seen that there are many circumstances modifying the effects which would be otherwise produced. It will be observed that the wearing away of coasts is, among the softer rocks more especially, often much accelerated by land-springs, which, as it were, shove portions of the cliffs into the power of the breakers by so moistening particular beds or portions of them, that much of the cliff loses its cohesion, and is launched seaward. The loss thus sustained in some coasts is very considerable.

Fig. 40.

So far from being thus brought by, so to speak, inland influences within the reach of the sea, in other situations we find the higher parts of cliffs protruding over the sea beneath, as in the previous sketch (fig. 40), when we suppose the parts of the rock to be so coherent that the breakers have been enabled to excavate the lower part of the cliff in the manner here represented. The same action continuing, a time must come when the weight of the overhanging portion will outbalance the cohesion of the rock, and the mass above will fall. Breakwater, as it then becomes to a part of the cliff, much will depend as to the length of time it may so act, according to the manner in which it has fallen, particularly if stratified. If composed of beds of rock, and the slope of these beds face the sea, as in the following sketch (fig. 41), the breakers will have less power

Fig. 41.

to act upon them, than if the edges of the strata were presented to the sea, as represented beneath (fig. 42), in which position they offer the least resistance to the destructive action of the sea.

Fig. 42.

It will be sometimes found, that a hard rock constitutes the high part of a cliff, while the lower portion is composed of a softer substance, such as a clay or marl, and that masses of the harder rock falling from above afford protection, for a time, to the lower part of the cliff. Thus, let *a* in the annexed section (fig. 43) re-

Fig. 43.

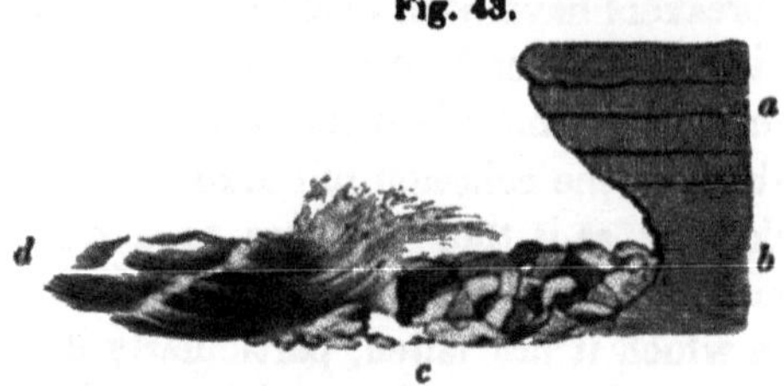

present the upper portion of a cliff formed of hard beds of rock, such as sandstone, while *b* is a marl or clay, then the action of the sea, *d*, upon the cliff would undermine it, and cause the fall of masses of the hard rock, *c*, which, accumulating at its base, would tend to protect it according to the quantity of fallen rock, the size of the masses, and their hardness. It will be found that cliffs composed as a whole of somewhat soft rocks, and clays, marls, or slightly indurated sandstones, are often protected at their bases by an accumulation of indurated portions of these rocks. Thus let the accompanying section (fig. 44) represent a clay in which there are

Fig. 44.

nodules of argillaceous limestones, as *a a* (and those of septaria in clays are often large), which, when washed out by removal of the clay, accumulate on the beach *b*. These then tend to protect the base of the cliff from the destructive action of the breakers. The study of any extended line of coast composed of horizontal or slightly-inclined beds of rocks of unequal hardness will present abundant examples of the modified protection afforded to the base of cliffs from the accumulation of masses derived from them.

Striking examples are often to be found on our shores of the wearing away of the land by the action of the breakers, so that rocks stand out in the sea detached from the main body of the land, but which once evidently formed part of it. Perhaps the accompanying sketch (fig. 45) of the cliffs near Bedruthan, Cornwall, may afford an idea of the manner in which some of our coasts are thus cut back by breakers. The islets here represented have been formed by such an abrasion of the rocks to the present cliffs

Fig. 45.

of the mainland, that portions, somewhat harder, and better resisting the action of the breakers than the rest, have remained. The breakers not unfrequently work round portions of the cliffs, forming a cave through a projecting point or headland. This, from the continuance of the same destructive action, becoming gradually enlarged, the roof, from the want of support, falls, and the point becomes an island, round which the breakers work their way, gradually increasing the distance between it and the mainland.

As might be expected, amid the wearing away of coasts by breakers, innumerable instances present themselves of unequal action on the harder and softer substances, according to their exposure to the destructive power employed upon them, so that long channels and creeks, and coves of every variety of form, are worked away in some situations, while hard rocks protrude in others.

Fig 46.

Coves afford shelter to the fisherman, from being hollowed out in some localities, while the hard ledges act as natural piers in others. The previous sketch (fig. 46), of Polventon Cove, on the east of Trevose Head, Cornwall, may be taken as a fair illustration of a harbour scooped out by the action of the breakers, which have so worn away the slate *a*, from a line of hard greenstone, *b*, that the latter forms a natural pier, named the Merope Rocks, affording shelter from the north-west winds, which, when strong, are much to be dreaded on this coast.*

It is not often, however, we should expect, though it must sometimes occur, that a mere trace of beds, superincumbent upon dissimilar rocks, can be found on coasts, showing how such may be entirely removed from the subjacent rocks by the action of the breakers. In this respect, the annexed sketch (fig. 47), may be

Fig. 47.

useful. It represents a small patch *a*, of a conglomerate of the new red sandstone series, named the Thurlestone Rock (in Bigbury Bay, South Devon), reposing, with a moderate dip seaward, unconformably upon the edges of Devonian slates *b*. Here the breakers have almost entirely removed the red conglomerate which was deposited upon the slates, and, no doubt, once covered them far more extensively than is now observable.

In estimating the abrading power of breakers on an extensive line of coast, it is desirable not only to direct attention to the relative hardness of the rocks of which it is composed, but also to the position of the beds (if the rocks be stratified), and to the planes of slaty cleavage and of joints. It will soon be apparent that among stratified rocks, lines of coast, under otherwise equal circumstances, depend on the directions and dip of the beds. Their position relatively to the force of the breakers is necessarily important; for if a series of beds, such as those in the accompanying

* Polventon Cove was at one time well known as a smuggling station, and is now often visited by vessels waiting for the tide into Padstow Harbour, a few miles distant.

sketch (fig. 48), dip seaward, the action of breakers falling on them in the manner represented would be comparatively trifling, since

Fig. 48.

the return of one breaker down the seaward slope of the beds, diminishes the force of the next falling upon it, and the power of the remainder, rushing up the slope, is gradually expended, and meets with no direct obstacle upon which it can destructively act. The positions in which the edges of the beds of any given rock are exposed to the action of the sea, are those where the abrading power of the breakers is most successfully exerted. Let us suppose that the annexed plan (fig 49) represents a line of coast exposed to the

Fig. 49.

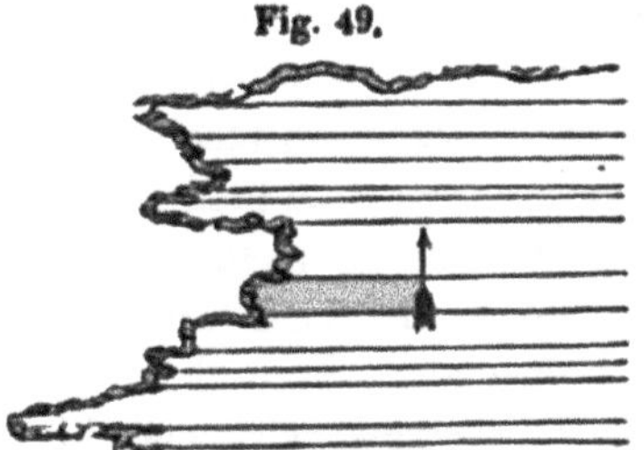

north and west, and that the abrading action of the breakers is equal from both points; then the effects produced will depend upon the resisting powers of the rocks themselves. Taking the country to be composed of beds of slates and sandstones, having a strike or direction from east to west, and a dip about 45° to the north; then, supposing no cleavage planes, and the slates to be parallel with the sandstone beds, the resisting powers of the rocks would be greatest on the northern coast, since the beds would there all slope seaward, while the same rocks would be liable to much abrasion on the west, the edges of the beds being exposed in that direction. Numerous indentations would be the result, similar to those represented in the plan, the softest beds being worn into the deepest coves, and the harder constituting the most prominent headlands.

In all investigations as to the loss of land from the action of the sea upon it, dependence can rarely be placed on old maps of coasts, which are for most part very inaccurate; indeed, there would be no difficulty in producing those which would, when compared with a good modern survey, apparently show an increase of half or three-quarters of a mile on a cliff coast, where, in fact, there had been considerable loss.

We have seen that cliffs become abraded by the action of the breakers, sometimes alone, at others combined with that of the atmosphere and of land-springs. The mineral matter so brought within the influence of the sea has to be removed, and observation soon shows, that while one part of it is caught up in mechanical suspension, and is then liable to be carried away by the movements of tides or currents, another portion remains and is exposed to the grinding action of the breakers on the coast. This latter portion necessarily varies in size from the block, which can only be shaken by the blows of heavy breakers discharged upon it, acting with their greatest power, to the small pebble temporarily caught up in mechanical suspension, even by minor breakers, but which again sinks to the bottom when not exposed to their influence.

It will be observed, respecting shingle beaches, that during a heavy on-shore gale, every breaker is more or less charged with the materials composing the beach, and that the shingles are forced forward as far as the broken wave can reach, their shock against the beach driving others before them, not held in temporary mechanical suspension. Shingles are thus projected on the land beyond the reach of the retiring waves, and there accumulate in long ridges parallel to the coast, especially where the land is low behind the shingle beach. Heavy on-shore gales and high tides combined necessarily produce the greatest accumulation of shingle in such localities, and although occasionally a breach may now and then be formed at such times, it becomes speedily filled up by the piling action of the breakers.

Attention to a shingle beach will soon show, notwithstanding the minor removal of portions from one place to another, backwards and forwards, and the modifications arising from the obliteration of the little lines of beach, not unfrequently produced during moderate weather, that as a whole it travels in the direction of the prevalent breakers until arrested against some projecting portion of the coast. This must happen, if any force act upon the shingles more in one direction than another, since they would be compelled to travel in conformity with it; and observation proves that such is the fact, for not only do we find pebbles of known rocks thus moved from the particular portion of cliff whence they have been derived, but, also, though breakers appear to adjust themselves to the tortuous character or outline of a coast, that there is always a slight oblique action in consequence of the main direction of the wind at the time.

One of the simplest forms in which the shingles of a beach are seen to have travelled is where, as in the annexed plan (fig. 50), we

Fig. 50.

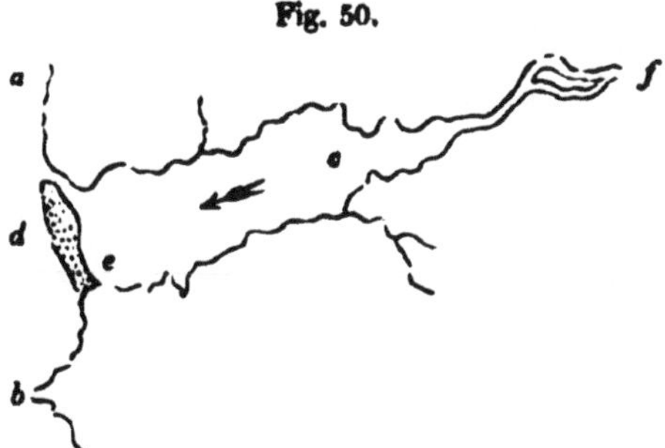

find a spit of shingle beach, *d*, composed of pebbles evidently derived from a coast, *b*, stretching in the direction to which the prevalent winds blow, the shingle beach being unable to cross over to the opposite coast (*a*) in consequence of the flow and ebb of the tide in and out of an estuary (*e*, *c*), into which a river (*f*) discharges itself at the higher end. In such cases, and they are to be seen in many situations, the rush of water is able to keep the channel open between the spit of beach *d* and the coast, *a*, not on the side of the prevalent winds, the ebb tide, especially when the river is in flood, effectually keeping the passage clear, and throwing off the shingle, which strives to cross over and block up the estuary.

There are good examples on the coast of Devonshire, at Teignmouth and Exmouth, of tongues of beach thus formed, but trending in different directions, exposure to the prevalent breakers being clearly seen to be the cause of the opposite directions taken by the beaches. At Teignmouth, a small portion only of the beach is derived from the rocks on the southward, and the river mouth is protected from the southerly and south-west winds, but exposed to the eastward and north-east. Hence, the beach is driven to the southward, and the river keeps its channel open by escaping against the hard cliffs of the Ness Point. The reverse of this action is observed at Exmouth.

We have various examples on our coasts (the Looe Pool, near Helston, Cornwall, and Slapton Pool, in Start Bay, Devon, are illustrative instances), where the river waters being insufficient to contend with the beach-piling action of the breakers, the outlet for the fresh waters is completely crossed by beaches, and lakes are formed behind them, the surplus waters percolating through the shingles. From this state of things to the escape of a river, by passing close to a hard cliff, there is every modification. In many localities exposed to open sea, the minor streams will be found dammed up by, or cutting through beaches, according to the state of the weather. A heavy on-shore gale throws up a bar of beach, which a flood from the land removes, and so the conditions alternate, with every kind of modification. The following (fig. 51) is a sec-

Fig. 51.

tion through the beach and lake at Slapton Sands, Start Bay, *a*, being the sea, which throws up the beach *b*; *c*, the freshwater lake behind the beach; *d*, the weathered and decomposed portion of the slate rocks *e*. This section is interesting also from showing that, at the present relative levels of sea and land in that locality, the sea has not acted on the hill *d e*, since the loose incoherent substance of *d* would have been readily removed by the breakers.

The Chesil Bank, on the coast of Dorsetshire, affords a good example of the driving forwards of shingle in a particular direction by breakers, produced by the action of prevalent winds. It is about 16 miles long, connecting the island of Portland with the mainland, and for about eight miles from that island, is backed by a narrow belt of tidal water, known as the Fleet. From its position, the heavy swells and seas from the Atlantic, often break furiously on this bank, which protects land that would otherwise soon be removed by them. The following (fig. 52) is a section

Fig. 52.

across the Chesil Bank, *a* being the bank; *b*, the water termed the Fleet; *c*, small cliffs formed by the waves of the Fleet, and by falls from the effects of land springs; *d*, various rocks of the oolite group, protected from removal by the Chesil Bank, and *e*, the sea, open to the Atlantic. In this case also we seem to have an example of the Atlantic breakers not having reached the land behind since the relative levels of the sea and land were such as we now find them.

A gradual sinking of the coast would appear to afford an explanation of the phenomena observed, and is a supposition harmonizing with the facts previously noticed at Slapton Sands.

The general travelling of shingles on a coast, much modified by conditions, may be illustrated by the following plan (fig. 53), in

Fig. 53.

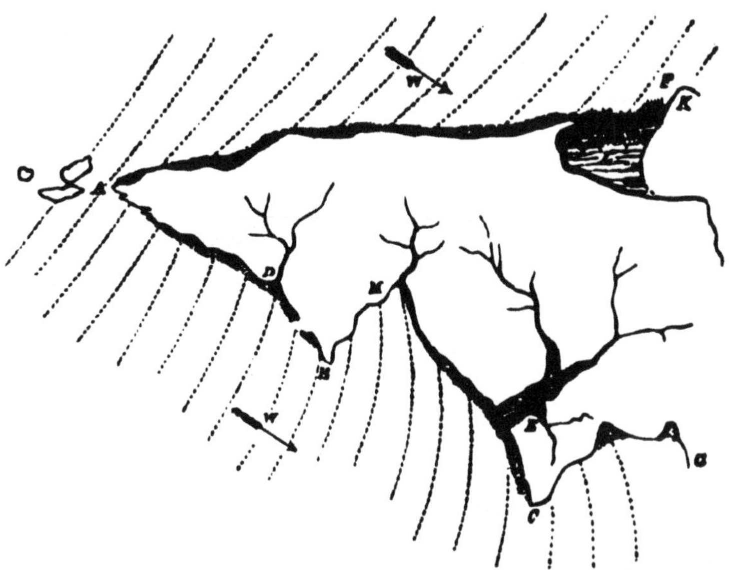

which G, C, B, A, and F, represent a line of coast exposed to the prevalent winds W. W. The lines of waves are shown by dotted lines, made to curve inwards behind protecting headlands. In consequence of the configuration of the coast, and its chief exposure to the action of breakers, the shingle would tend to travel from A to F on the one side, and from A to G on the other. There would be little impediment to their course along the line A F, until the river, on the right, presented itself, where K represents a cliff of hard rock, and F, the tongue of drifted beach, arising from the conditions previously noticed (p. 55). Between A and G the effects would be different, particularly if it be assumed that the point of land B projects into deep water. Considering the river at D as small, the beach would traverse its mouth and be only removed during heavy floods, so that the mass of shingle would tend to travel towards the point B, and there descend and accumulate in deep water. Supposing C another point of land jutting into deep water, it would bar the further progress of the shingle travelling from M to it, a beach closing the outlet of the lake at E, assumed to be shallow, and under the conditions previously mentioned as existing at the Looe Pool and Slapton, the back fresh-waters being unable to force outwards the beach accumulated by the breakers.

At L (fig. 53), we have shown a marsh accumulation behind the protecting influence of the shingle beach F, this accumulation being a deposit from the checked waters of the river, by the action of the flood-tide, when rains had caused detritus to be borne down in mechanical suspension by the river. The annexed plan (fig. 54) may aid in showing the modification often observable where the tongue of beach is composed of sand, backed by sand-hills: *a* repre-

Fig. 54.

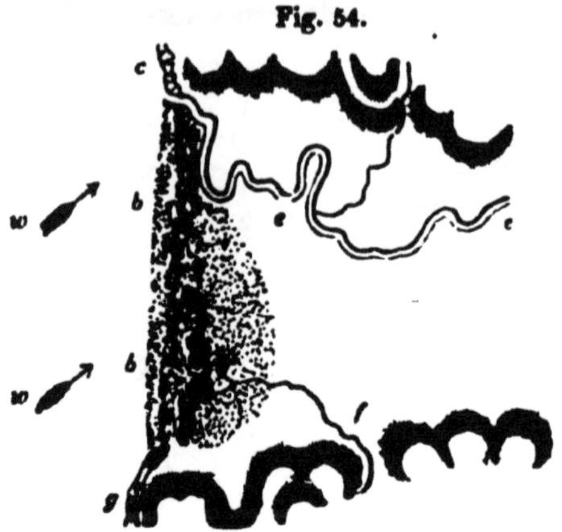

sents a tract of low level land, which may either have been formed by the filling up of an estuary under existing conditions, or be the bottom of an estuary of a previous time, now raised; *b*, *b* a sandy beach and sand-hills, protecting the low land from the ravages of the sea; and *e*, *e*, a river which makes good its course to the sea, by keeping close to the hard cliff *c*. We have also assumed that a small stream, such as *f*, occurs, so that it does not find its way to the main stream, but loses itself in pools amid the sand-hills, the mud from it tending to consolidate and cement the blown sands, binding them together, and hence supporting a vegetation which would not otherwise have found the conditions for its growth.

In these situations there is often a severe struggle between the action of the sea (swept by prevalent winds *w*, *w* (fig. 54), piling sand upon the beach *b*, *b*), assisted by that of the wind on the sand-hills, and the waters of the river. The effect of such a little stream as *f* is not unfrequently to give much firmness to the end of the beach and sand-hills towards *g*, while the sand blown over towards the main river is caught up by it and again carried out to sea, particularly during floods.

Let us now consider sandy beaches and sand-hills, bordering coasts generally. The sand on sea-shores is derived from the rivers bearing it down in mechanical suspension, or forcing it forward on the bottom to the sea; from the wearing away of cliffs of sand and sandstone by breakers, or from the attrition of the pebbles or shingles on beaches, so that finally they become mere sand. To these causes must, in certain localities, be added the trituration of shells and corals, ejected from the sea and piled up as beaches, in some places by themselves, at others variously mingled with ordinary sand.

Regarding the common occurrence of sea-shore sand of a certain average degree of fineness, it should be observed, that as detritus approaches that size it becomes more and more difficult to reduce it further, since it is then more and more easily caught up in mechanical suspension by breakers, and therefore grain cannot so readily be ground against grain.

The accumulation of sand-hills can as readily be studied on various portions of our own coasts, as in those parts of the world where the shores present little else than sandy dunes for hundreds of miles. A low line of coast with a shallow sea outside, and presenting a fair exposure to breakers, is usually sufficient for their production. The greater amount of shore dry at low water in tidal seas, and the greater the exposure to prevalent winds, the larger is commonly the accumulation of the sand-hills, other conditions being equal. The cause is sufficiently obvious. A large tract of sand, exposed between high and low water mark, and under the influence of a strong on-shore wind, is soon partially dried on its surface, and the dried sand is swept inland beyond the reach of the breakers of the rising tide, which could have again caught up this sand in mechanical suspension and have distributed it.

It is desirable, that the observer should select some day, when a strong on-shore wind blows over a tract of sand, and the drier the state of the atmosphere the better, to see the manner in which the grains of sand are transported inland, and to mark the various modifications of surface which arise from the deposit of the sand among the sea-weeds, or pebbles, should any occur. He will find that, while some grains of sand may be held in mechanical suspension by the wind at a height of an inch or so from the sandy surface beneath, the friction of the air on the latter produces such retardation of the wind current, that similar grains of sand are merely swept along the bottom. In such respects this perfectly accords with the movements of detritus in river channels, and above noticed. The difference is merely that the transporting power is air in the one case, and water in the other. Indeed, this action is so completely of the same kind, that the furrows and ridges produced by the friction of water currents over arenaceous accumulations, may be advantageously studied where wind currents drive over sand.

To observe the manner in which the sands furrow and ridge, and move onwards, a time should be chosen when the wind is not sufficiently powerful to hold the sand in mechanical suspension, but merely to drive or push it onwards. The ridging, as shown in the annexed section (fig. 55), is accomplished by the driving of

Fig. 55.

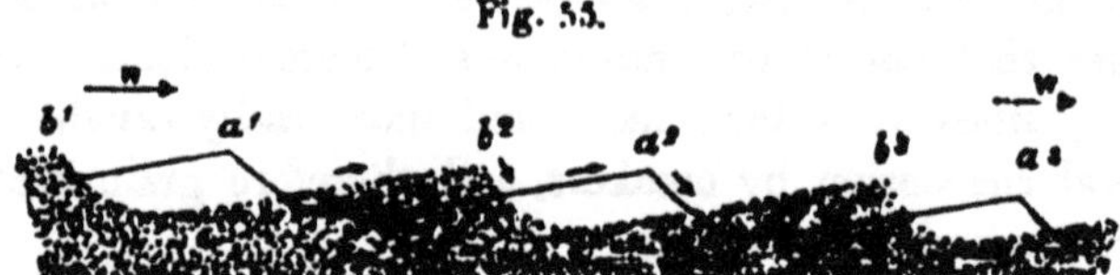

the grains with sufficient force by the wind acting in the direction *w*, *w*, merely to carry onwards those on the surface, the retardation of which by friction on those beneath so acts that the grains at b^1 are driven on to the ridge a^1, and by accumulation (the power of the wind being sufficient to cut down the ridges to a kind of general level, or curve, as the case may happen to be) fall over into the furrow b^2, and so on with the ridges a^2 and a^3. As the friction is continued, the crests of the ridges advance, and their places are occupied by furrows, to be replaced by ridges. When the velocity of the wind is favourable for researches of this kind, an observer will best see the advance of the ridges, by placing himself amid the moving surface, and directing his attention to the ridges nearest him, at the same time making due allowance for the obstacles presented by his feet, which will produce modifying influences, readily appreciated.

Arrived at the margin of the shore line, the sands pushed forward in the manner noticed, or caught up in mechanical suspension, when the winds are sufficiently powerful, accumulate, forming ranges of sand-hills, in some countries characteristic of long lines of coast. By their accumulation and tendency to move inland, in the direction of the prevalent and more powerful winds, they produce changes upon the adjoining low lands, and even upon considerable slopes of adjoining hills. The sands accumulated in the Bay of Biscay, may be considered as affording an illustrated instance of this encroachment on the land, and the modifications thence produced, inasmuch as great changes are known to have been there effected during the historical period.

The advance of these dunes is described as irresistible, and at a rate of 60 and 72 feet per annum. They force before them lakes of fresh water, formed by the rains, which cannot find a passage into the sea in the shape of streams. Forests, cultivated lands, and houses disappear beneath them. Many villages noticed in the middle ages have been covered, and a few years since it was stated, that in the department of the Landes alone, ten villages were threatened with destruction. "One of these villages, named Mimisan, has been," said Cuvier, "striving for 20 years against them; and one sand-hill, more than 60 feet high, may be said to be seen advancing. In 1802, the lakes invaded five fine farms belonging to St. Julien; they have since covered a Roman causeway, which led from Bordeaux to Bayonne, and which was seen about 40 years since, when the waters were low. The Adour, which was once known to flow by Vieux Boucaut, and to fall into the sea at Cape Breton, is now turned aside more than a thousand toises."*

There are few extended lines of coast which will not afford opportunities for the observation of sand-hills, and their mode of accumulation and change, for strong winds acting upon even a comparatively exposed surface, soon produce a marked alteration of their form. Successive accumulations, shown by the remains of surface vegetation grown during times where it could partially establish itself, are cut away and heaped up into other hillocks, new matter derived from the sea being added to the general mass. At times, a strong off-shore wind forces sand back to the sea, acting not only on the sand-hills over which it blows, but also on the dried surface of the sands bared between high and low tide, these still more easily carried seaward when left dry for a longer time, between the highest lines of neap and spring tides.

As the sand commonly found in sand-hills is not usually borne high in mechanical suspension by the winds, such districts will not long have engaged attention before the power of running water, even of small streams, if their courses be unobstructed and fairly rapid, will be seen to prevent the extension of blown sands. The sand drifted, falling into the streams, is carried onwards by these waters, and is thus prevented from traversing them.† Sand-

* Cuvier, *Dis. sur les Revolutions du Globe.* A thousand toises is about 6,400 English feet, or somewhat less than a mile and one quarter.

† Good examples of this fact may be observed on the coast of Cornwall. The Perran Sands are thus bounded for nearly two miles between Treamble and Holy

drifts are sometimes also found stopped by the flow of tidal waters in and out of lagoons. Of this kind, the accumulation of sand at the northern side of a spit of land, terminated by sand-hills, near Tramore, on the eastern coast of Ireland, may be considered as a good example.

As having a geological bearing, the observer would do well to direct his attention to the manner in which the remains of vegetable and animal life, both terrestrial and marine, become mingled in sand-hills. Portions of seaweeds will frequently be found blown, when dry, amid the terrestrial vegetation of the sand-hills; and the shells of the helices, which are often found in multitudes in such situations get mingled with marine shells, or their fragments.

In some situations, the sand-hills are largely composed of comminuted shells, ground to that state by the breakers; and in such cases, consolidation of parts of them may be observable, having the hardness of many sandstones. The carbonate of lime of the shells becomes acted upon by the carbonic acid in the rain waters, with additions from decomposing vegetation, when plants have established themselves on the surface of the sand, and a final deposit of the carbonate of lime, thus held in solution, agglutinates the grains of sand together. Indurated sands of this kind are sufficiently hard, occasionally, to be employed for building purposes.*

Well Bay. Much land is stated to have been covered by drifts from the Perran Sands, in consequence of a small stream having been covered by mining operations near Gear.

* The consolidated calcareous sand of New Quay, Cornwall, has been long used as a building stone. Not only is the neighbouring church of Crantoch built of this modern sandstone, but very ancient stone coffins have also been discovered, composed of the same consolidated sand, in the adjoining churchyard. The grains are so firmly cemented in this New Qnay sandstone, that where it graduates into a kind of conglomerate, pebbles of quartz and hard sandstone are generally broken through by a blow on the compound rock.

CHAPTER V.

DISTRIBUTION AND DEPOSIT OF SEDIMENT IN TIDELESS SEAS.—DEPOSITS OF THE NILE.—OF THE PO AND RHONE.—CONTEMPORANEOUS DEPOSITS OF GRAVEL, SAND, AND MUD.—DEPOSIT OF VOLCANIC ASHES AND LAPILLI.—DEPOSITS IN THE BLACK SEA AND THE BALTIC.—GULF OF MEXICO AND MISSISSIPPI.

As tideless seas might be considered as mere salt-water lakes, the distribution and deposit of detritus in them would, as a whole, resemble that of fresh-water lakes, particularly of those attaining the magnitude of the great North American lakes, but for the difference in the relative specific gravities of their waters. Slight attention to the overflow of rivers swollen by rains, and charged with mechanically-suspended matter, into the sea, will show that the discoloured waters of the rivers, instead of falling beneath the waters into which they flow, as is seen at the higher part of the lake of Geneva, and numerous other lakes, proceed seawards on the surface of the sea waters, and often to considerable distances. The cause is simply that, though discoloured by the detrital matter held in mechanical suspension, these river waters are still specifically lighter than the sea waters into which they flow.

The distances to which the river waters sometimes flow seaward, transporting fine detrital matter, parting with it gradually, must, when the great rivers of the world become full and turbid, be often very considerable. Colonel Sabine has stated, that at three hundred miles distant from the mouth of the Amazons, discoloured water, supposed to come from that river, was found, with a specific gravity of 1·0204, floating above the sea water, of which the specific gravity was 1·0262, the depth of the lighter water being estimated at 126 feet. It would be well that observers should direct their attention to such facts, for their accumulation would tend much to show us the extent to which fine sedimentary matter may be thus borne beyond the action of tides and coast currents.* As much matter may be thus distributed in chemical solution, valuable information might also be collected as to the kind and quantity of substances so held in solution.

From the varied depths near its shores, the Mediterranean affords us a good example of the deposits effected in seas which are commonly termed tideless. The great rivers which discharge themselves into it, such as the Nile, Po, and Rhone, now transport little sedimentary matter that is not finely comminuted, and of easy mechanical suspension. The Nile, which has been estimated to deliver a body of water annually into the Mediterranean about 350 times that which flows out of the Thames, beginning to rise in June, attaining its maximum height in August, and then falling until the next May, must thrust forward, from its periodical rise and fall, fine sedimentary matter with great regularity, tending thus to produce consecutive layers or beds of mud and clay of considerable uniform thickness and character, in those situations where modifying conditions do not interfere. Part of the fine matter brought down from the interior in mechanical suspension is deposited on the lower grounds traversed by the Nile; and it has been calculated that the surface of Upper Egypt has, in this manner, been raised more than six feet since the commencement of the Christian era. The fine matter not so deposited, passing with the river waters seaward, is necessarily borne furthest outwards when the greatest force of the river water prevails, namely, in August of each year.

The matter thus borne seaward may be kept a greater or less time mechanically suspended, according to the agitation of the surface by winds, but, as a whole, there must be an average area over which it is thrown down; the greatest distance of the deposit from the mouths of the Nile being attained in August, though the greatest thickness of a year's deposit will be nearer the land. As the river mouths advance, these sheets of fine sediment would be expected to extend further seaward, overlapping each other.

Where the surface of the sea cuts the slightly-inclined plane of sedimentary matter, partly in the sea, and partly on the land, the

* Very little practice would enable those who may have opportunities of making such observations to ascertain the amount of matter mechanically suspended in waters of this kind. If the scales be not very delicate, by pouring a large volume of the water through a filter, previously weighed, such an approximation to the truth may be obtained as might be useful. As previously observed (p. 28), mere evaporation of the water would give not only the matter in mechanical suspension, but that also in chemical solution.

breakers separate the finer from the coarser substances, keeping the former easily in mechanical suspension, and removing them from the shore outwards. The result is, an arenaceous boundary, with banks so formed as to include lagoons, such as are seen in the accompanying sketch of the delta of the Nile (fig. 56), at Lakes Mareotis, Bourlos, and Menzaleh.

Fig. 56.

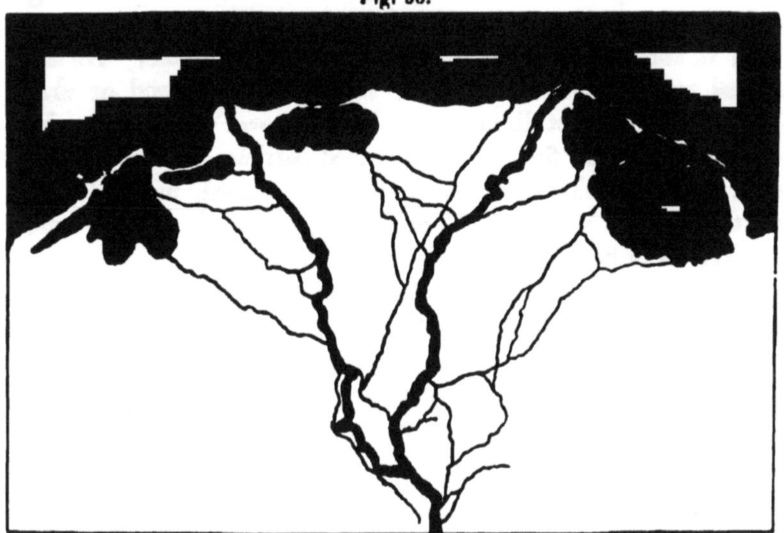

These lakes gradually fill up, the shore advances, and so, even supposing the same relative level of sea and land not to be altered through a long succession of ages, the bed of the Mediterranean becomes more shallow in that region, and a mass of matter, such, for the most part, as would eventually form clay, is accumulated; the upper portion sandy from the action of the breakers upon the level of the sea, and from the sifting action, so to speak, of the waves further seaward, at depths where that influence could be felt.

From the periodical character of the rise of water in the Nile, the equivalent periodical deposits might even be marked by bands or layers extending to distances bearing a relation to the amount of transporting power of the river waters, so that coarser particles could be carried further and over more extended areas at one time than at another. The general deposit, however, gradually advancing seaward, successive annual accumulations would, as a whole, overlap each other.

When we regard the Po and Rhone, we have not the same very marked periodical rise of their waters; though no doubt, taken as a whole, there may annually be times when more matter is borne outwards than at others. With the exception of the regularity of effects likely to be produced by the rise and fall of their waters, the accumulations formed by deposit from the detrital matter borne seaward by the Po and Rhone would, however, be similar to those of the Nile;* the same discharge of fresh waters holding matter in mechanical suspension over the surface of the sea, the same sifting of the detritus so borne seaward, where the action of the waves can reach it, and the same general order of accumulations.† As the general mass of matter advanced, there would be mud or clay formed at the greatest distance from the land, over which the sands, separated from the finer or mud-formed particles in the shallow water and along shore, would gradually be spread, mingled here and there with a patch of clay, or silt clay, deposited in the lagoons, behind lines of beach thrown up by the breakers.

These rivers are merely mentioned as marked examples. An inspection of a good chart of the Mediterranean will show that there are many others, the floods in which only bear mud and sands into it, the heavier detritus not reaching the shores, the fall of the river beds, and the force of their waters, being insufficient. In all such cases the accumulations would be mud or clay for a base, with an arenaceous top, so far as the causes we have noticed could prevail. It will be obvious, that clay may be accumulated in the depths seaward, while sands are advancing from the shore towards them, so that, if at any future geological period, the whole became uplifted above the level of the sea, we might have a sheet of arenaceous matter covering another of clay, the parts of each,

* As respects the Po, M. Prony considered himself authorised to conclude, from the examination of a large amount of evidence, "First, that at some ancient period, the precise date of which cannot now be ascertained, the waves of the Adriatic washed the shores of Adria. Secondly, that in the twelfth century, before a passage had been opened for the Po at Ficarrolo, on its left or northern bank, the shore had already been removed to the distance of 9,000 or 10,000 metres (5½ to 6 miles) from Adria. Thirdly, that the extremities of the promontories formed by the two principal branches of the Po, before the excavation of the Taglio di Porto Viro, had extended by the year 1600, or in 400 years, to a medium distance of 18,500 metres (about 11½ miles) beyond Adria; giving from the year 1200 an average yearly increase of the alluvial land of 25 metres (82 feet). Fourthly, that the extreme point of the present single promontory, formed by the alluvions of the existing branches, is advanced to between 32,000 to 33,000 metres (about 19¾ to 20½ miles) beyond Adria; whence the average yearly progress is about 70 metres (229½ feet) during the last 200 years, being a greatly more rapid proportion than in former times."—*Cuvier*, *Dis. sur les Rev. du Globe.* M. Morlot infers that the land round the head of the Adriatic is gradually sinking, but that the deposits of the rivers are still sufficient to effect a general gain upon the shores of that sea.

† It should be remarked that there are also calcareous accumulations at the mouth of the Rhone.

though continuous, formed at different times, and portions of the clay equivalent to parts of the sand. There would be zones, so to speak, of arenaceous matter corresponding with the advance of the coast, and not separated from the common sheet of that of which it constitutes a part, being formed at the same time with a layer of clay, which a prolongation of the sandy coating would cover at a subsequent period.

The same sea fortunately furnishes numerous examples of short rivers, with rapid falls of their beds, and occasional abundant supplies of water, thrusting pebbles into it. The effects produced are the same as when torrents discharge themselves into lakes, with the difference that the muddy part of the waters flows over the surface of the sea, the sand separating from it. According to the depth of water, and this is sometimes considerable, is the sand accumulated; if fairly deep, the sand falls not far distant from the coast, while the pebbles accumulate on the shore, and the embouchure of the river is extended. Though the general bed of shingles (the upper part acted upon by the breakers, as upon any other shingle beach) would advance as a whole, with an even upper surface, the accumulation of gravel or shingles would be formed by many irregular protrusions produced by changes in the direction of the river's mouth. The depth being favourable, we should expect, under such conditions, an accumulation of the following kind (fig. 57), *a a* being a section of the land, formed of beds of rocks

Fig. 57.

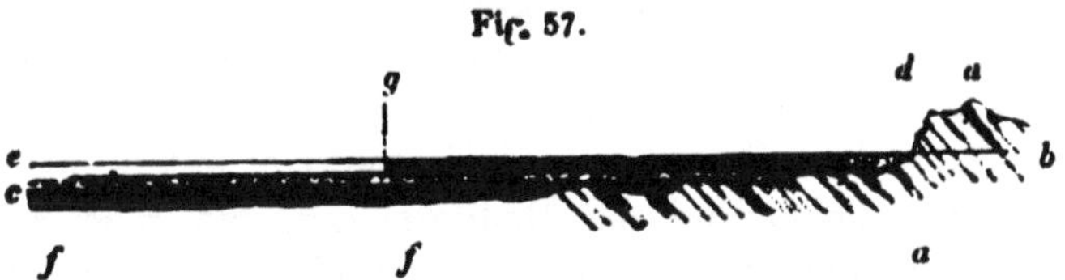

(represented as dipping inland, merely to separate them clearly from the other deposits), *b* the bed of the river, bearing down pebbles, sand, and mud into the sea, the level of which is shown by the horizontal line *e*; *d* exhibits the first accumulation of pebbles thrown over the steep shore, the pebbles falling to the bottom, and the sand only being deposited in a regular layer, more outwards; *c* the continuation of the sandy layer seaward; *f f* the mud deposited beyond the sand, and also continued; and *g* the extension of the pebbles over the sand, at some given time. In such an accumulation we should expect, after both sand and gravel had overspread the clay, a lower deposit of clay, above this anotehr of sand, and over the sand, gravel; parts of the gravel, sand, and clay, notwithstanding the extension of each layer continuously in the manner stated, being equivalent to each other in age, for the reasons before assigned.

Where depths were less considerable, we should expect an intermixture of the gravel and sand in a more irregular manner, and with an arrangement depending on the action of the breakers upon them; this action tending to pile back the shingles, as a whole, while it permitted the sandy sediment to be caught up in mechanical suspension, and thus it might be carried outwards by the river waters, in places where the stream of these waters could be felt. As previously observed, the finer and mechanically suspended particles would be borne over the surface of the sea, according to the volume and velocity of the outpouring river waters, eventually forming a layer of mud or clay where deposited. It will be obvious, that as the volume and velocity of the river waters varied, so would be their power to carry outwards, beyond the influence of the breakers, mechanically-suspended matter of different volume and weight, and hence that, within a certain range, there might be mixed layers of sand, silt, or mud, according to circumstances.

Not only do the rivers thus contribute matter, borne down by them to the shores, to be there arranged by the breakers, or thrust out into the sea and deposited in it, but every river also bears down some matter in chemical solution, to be added to the solutions present in the seas. In tideless seas, each river sends down its solutions into water which may, to a great extent, be considered stagnant, notwithstanding certain movements or currents sometimes in it, so that at the embouchures of the rivers the substances so borne down prevail within distances to which the river waters may act. In many localities around the Mediterranean, the river waters transport large quantities of bicarbonate of lime in solution. While we may consider that much of this substance is consumed by fish, crustaceans, and molluscs, for their harder parts, there is probably a large surplus which eventually takes the form of calcareous accumulations beneath the sea. The rivers which transport bicarbonate of lime abundantly would, when in flood, probably also carry forward sedimentary matter, so that at the mouths of such rivers we might have alternate times, variable probably in duration, when the rivers were clear, and carried forward, as compared with the volume of water, a large proportion of bicarbonate of lime, and when this substance bore a far less proportion to the volume of water, while fine detritus was abundant. Under such conditions we should have alternately layers of mud and calcareous matter, or mud more calcareous at one time than at another, so that eventually the calcareous matter might tend to separate into nodules, and in planes corresponding to the times when it was most abundantly thrown out of the rivers. In like manner we might have sulphate of lime, commonly enough in solution in some rivers, mingled with the mud, and eventually crystallizing out as selenite, a mineral so frequently discovered in various clay beds. Many other combinations of different substances, some in solution, others mechanically suspended, and borne down by the same rivers, will readily present themselves to the mind of the observer, and suggest attention to the conditions under which both are carried out into tideless seas.

When considering deposits in tideless seas, we must not forget those resulting from the fall of ashes and lapilli, thrown out from volcanos. The Mediterranean may fortunately be considered with reference to this kind of accumulation also, as there are volcanos in action in it, and on its shores. The great eruption in 79, which not only overwhelmed Herculaneum, but showered ashes in such profusion upon Pompeii as also to bury that town, could not fail to have thrown a large amount of ashes and lapilli into the sea; and considering the distances to which ashes are known to have travelled from volcanic vents, the ashes at least may have been widely spread. It will be obvious that whatever kinds of sedimentary accumulations they subsided upon through the sea, the ashes would mingle with them, coating over such deposits where tranquillity reigned, either from the depth of water or other causes, with a layer of ash. Where the action of waves on the bottom, or of breakers on the coasts, could be felt, in whatever tranquil state the ashes may have fallen originally to the bottom, they would be mixed up with the mud, sand, or pebbles, as the case might be, when thus acted upon, so that the particles of the ash would be disseminated among them. All rivers upon which the ashes fell would probably bear much of them outwards in mechanical suspension, for the fine matter which can be upborne and be carried by the winds to great distances would not readily subside through the river waters.

Under this view, the deeper parts of the Mediterranean, and especially those to which other sedimentary matter could not be carried by the movement of sea currents, or the drift of river waters outwards, would be those where the layers of ash would be most unmixed with other matter, excepting as regards the deposit of any substances from chemical solution in the sea, and to which its great tranquillity may be favourable. We do not know the depths at which calcareous accumulations may now be forming in the Mediterranean, but whether in shallow or deep situations, any ashes falling upon them would either accumulate in layers, or be mingled up with the limestone, according to the rapidity with which the one may subside, or the calcareous matter be deposited.

Not only are there volcanos on the borders of this sea, of the magnitude of Etna and Vesuvius, throwing out ashes and lapilli, but we have had evidence in our times, so lately as 1831, of the uprise of a volcano through the sea,* between Pantellaria and the coast of Sicily, and from deep water.† Columns of black matter are described as being thrown out of the crater, to the height of three or four thousand feet, spreading out widely even to windward. The upper part, above the sea at least, seemed to have been solely composed of ashes, cinders, and fragments of stone, commonly small. Among these, fragments of limestone and dolomite, with one, several pounds in weight, of sandstone, were observed, appearing to show that the volcanic forces had broken them off beds of these kinds of rock, when the igneous matter had been propelled through them.

An island so constituted, could not long resist the destructive action of the breakers, and thus, as soon as the supply of ashes, cinders, and fragments of rock ceased, it was cut away by them, and reduced to a shoal. During the time that this volcanic mass was accumulating, a large amount of ashes and cinders must have been mingled with the adjacent sea before it reached its surface, and no slight amount would be distributed around, when ashes and cinders could be vomited into the air. Add to this the quantity caught up in mechanical suspension by the breakers, and there would be no small amount to be accumulated over any deposits forming, or formed on the bottom around this locality, and out of the reach of any lava currents which might have flowed beneath the level of the sea. The breakers while they removed the lighter substances would, as it were, so sift the whole, that the heavier fragments would gradually subside to lower levels, and eventually beneath the action of seas breaking above, or simply moving the bottom during very heavy weather. Finally, there would be a

* To the island thus formed the various names of Sciacca, Julia, Hotham, Graham, and Corrao were given. Dr. Davy, who visited this volcanic island on the 5th August, 1831, has given a detailed account of it in the Phil. Trans. for 1832. M. C. Prévost was charged by the Academy of Sciences of Paris to visit and report upon it. He reached the island on the 28th September of the same year. It was then about 2300 feet in circumference, with two elevations, from 100 to 200 feet high, on different sides of the crater, the latter filled with boiling water.

† Captain Smyth proved (Phil. Trans. 1832) that the volcano did not rise from the Adventure Bank, as was first supposed, but to the westward of it, and from deep water.

collection of fragments, cemented by ash and cinders, in which there would not only be pieces of igneous rocks, but of limestone, dolomite, and sandstone also, for we are not to suppose that the pieces found accidently on the surface were those alone thrown out of the crater.

Thus, then, in the Mediterranean a very complicated series of contemporaneous accumulations is now in progress, its uneven bottom* being variably covered, according to conditions, by the matter brought into it either in solution or mechanical suspension by rivers; eroded from its shores by the action of the breakers, or ejected by volcanos, the whole, excepting lava currents or large sudden accumulations of ashes and cinders, more or less mingled with the remains of organic life, these remains themselves sometimes sufficient to form long-continued layers or beds.

Though, for convenience, the Mediterranean has been treated as a tideless sea and without motion, this is not strictly correct, inasmuch as small tides are felt in it, and currents are found. Indeed, as respects the latter, when powerful winds, by their friction, force the surface waters in some given direction for the time, well seen when driven against any part of the boundary coasts,† the movement is then sufficient to carry any substances, mechanically suspended, to distances proportionate to the power and continuance of the winds. When these waters again come to a state of repose, the return action will be similar. There are also currents in the Mediterranean, such as that out of the Black Sea into it through the Sea of Marmora, and the current at the Straits of Gibraltar, which sets in from the Atlantic,‡ the latter modified, however, by the

* In considering the deposits now taking place in this sea, we should bear in mind that it is divided into chief basins (see Captain Smyth's charts) by a winding shoal, the Skerki, connecting Sicily with the coast of Africa. The run of soundings upon this shoal, proceeding from the African to the Sicilian coast, gives 34, 48, 50, 38, 74, 20, 70, 52, 91, 16, 15, 32, 7, 32, 48, 34, 54, 70, 72, 38, 55, and 13 fathoms, whence its inequalities may be seen. There are soundings in 140, 157, and 260 fathoms on either side, and places where bottom has not been reached with 190 and 230 fathoms of line.

† An observer may often have opportunities in the ports of the Mediterranean of seeing the rise or depression, as the case may be, of the sea, according as the winds at the time may be blowing with strength off or on shore. Canals frequently afford good opportunities of observing this kind of action of wind on water; for the canal levels, in still weather, being accurately known, it becomes easy to see how much these waters are raised or depressed as the winds may press them in one direction or another. Mr. Smeaton found that in a canal, four miles in length, the water was kept up four inches higher at one end than at the other, by the action of the wind along the canal. The Caspian Sea is several feet higher, at either end, according as a strong northerly or southerly wind may prevail.

‡ Both these currents have been attributed to the evaporation of the surface waters of the Mediterranean, that sea not receiving a sufficient equivalent from the discharge

tides as respects the African and European shores of the Straits.* The current from the Atlantic is described as setting eastward into the Mediterranean at the rate of about 11 miles in 24 hours, passing along the African shore, and being felt at Tripoli and the island of Galitta.† An eastern current flows between Egypt and Candia, and at Alexandria. Arrived at the coast of Syria it turns northwards, and then advances between Cyprus and the coast of Karamania. Such currents would necessarily aid in transporting matter both in solution and mechanical suspension, the last-mentioned current especially acting on that brought down by the Nile.

From the lower specific gravity of the water in the Black Sea,‡ the fine detritus, borne into it by the waters of the Don, Dnieper, Dniester, and Danube, would be carried less distances, comparatively, over its saline waters than those of the Nile, Po, and Rhone over the Mediterranean, while from the same cause, supposing an equal force of wind to act upon both seas, any continued suspension of that matter which might be due to the agitation of waves, would be greater in the Black Sea than in the Mediterranean, the waters of the former offering less resistance to the wind from their inferior specific gravity. In the Baltic also, from its specific gravity, the deposit of detritus borne down the rivers discharging themselves into it, would approximate towards that observable in fresh-water lakes. Like most lakes, also, the Black and Baltic seas have out-

---

of rivers into it, or the fall of rain upon it, so that the Black Sea furnishes waters on the one side and the Atlantic on the other, in order to keep it at the height required.

* "On the European side, west of the island of Tarifa, it is high water at 11^h, but the stream without continues to run 2^h. On the opposite shore of Africa, it is high water at 10^h, and the stream without continues to run until 1^h; after which periods it changes on either side, and runs eastward with the general current. Near the shore are many changes, counter currents, and whirlpools, caused by and varying with the winds. Near Malaga the stream runs along shore about eight hours each way. The flood sets to the westward."—*Purdy, Atlantic Memoir.* The tide rises three feet at Malaga.

† An under and counter current has been considered to set westward, but of late this has been doubted. However this may be, Admiral Beaufort has shown, while noting the current which flows westward from Syria to the Archipelago, that "counter currents, or those which return beneath the surface of the water, are also very remarkable. In some parts of the Archipelago they are sometimes so strong as to prevent the steering of the ship; and in one instance, on sinking the lead, when the sea was calm and clear, with shreds of bunting of various colours attached at every yard of the line, they pointed in different directions all round the compass."—*Beaufort's Karamania.*

‡ According to the researches of Dr. Marcet (Phil. Trans. 1819), the specific gravities of the under-mentioned seas are as follows:—

Mediterranean . . .	1·02930	Baltic	1·01523
Black Sea	1·01418	Yellow Sea	1·02291

flowing currents,* so that the evaporation on their surface is not equal to the fresh water discharged into them.†

Supposing no counter and constant currents bringing in salt water from the Mediterranean to the Black Sea, and from the German Ocean to the Baltic, and that the discharged waters from both seas carry off the average saline waters of each, these seas would gradually become less saline in proportion to the different amount of salts in solution carried out to the adjoining seas, and those brought in by the rivers discharged into them.‡ Upon this view, therefore, both the Baltic and Black Seas may at previous periods have been more saline than at present. Considering, as geological evidence would lead us to infer, that the area now covered by the Caspian and that occupied by the Black Sea, were once beneath a common sea, changes subsequently effected have separated them as now found. In the Caspian we should have evaporation sufficient to overpower the influence of the fresh water poured in hy the Volga, Ural, and the minor rivers, while in the Black Sea the supply of fresh water is beyond the evaporation. Hence the Caspian remains a salt lake, while the Black Sea may be gradually becoming more and more a fresh-water lake, the Caspian not only retaining its original saline contents, but becoming more saline if either the salts brought down hy the rivers are beyond any deposit which may dispose of them, or the evaporation be greater than the supply of water from the Volga, Ural or Iaik, and minor streams.§ Upon such an hypothesis, though at first the deposits in each would be under the same conditions, these would gradually change as regards effects arising from the increasing difference in the specific gravities of the respective waters.‖

* The velocity of the current, in the narrowest part of the Sound (Baltic), is about three miles per hour; but the ordinary general rate, in fine weather, is about a mile and a half or two miles. The current flowing from the Black Sea runs commonly, in the Thracian Bosphorus, from three to five miles per hour, according to the direction and force of the winds.

† Strong opposing winds force back the current out of the Baltic, and, if sufficiently long continued, will raise the level of that sea.

‡ In equal weights (3 lbs.) of water taken from the East Friesland coast, and from Rostock in the Baltic, the following proportional differences in saline contents were found:—

	German Ocean.	Baltic.
Chloride of sodium	522	263
Muriate of magnesia	198·5	111
Sulphate of lime	23	12
Sulphate of soda	1·3	1
Residue	1·5	1

§ There is considered to be good evidence of the Caspian having stood at higher levels than at present, those more corresponding with the actual level of the Black Sea, beneath which the surface of the Caspian is now 81·4 feet.

‖ A peculiar bitter taste observable in the Caspian waters is attributed to the

Although ice may form in the shallow bays of the Black Sea, and the branch known as the Sea of Azof be often frozen over in the winter months, so that ice, floating away from the coasts, may be the means of conveying fragments of rock and pebbles into situations to which they would not be otherwise transported, the ice in the Baltic, from the geographical position of that sea, is a means of adding to deposits in it of a more important kind. In particularly severe seasons, extensive sheets of ice over parts of this sea occur, and cases are recorded where great distances could be, and were traversed by travellers. Large areas are commonly frozen for nearly three months in the year, the ice on the south commonly breaking up in April, while in the Gulfs of Bothnia and Finland it may continue until the middle of May. Though the Baltic may be, as regards the ordinary acceptation of the term, tideless, it is nevertheless liable to those local changes of level which are due to the pressure of powerful winds blowing for a time from particular points, and it is described as often vexed by such winds. Ice, therefore, around the shores of its numerous islets and uneven coasts, may often be broken up, particularly towards the warmer weather, with shingles from the shore, and fallen fragments from the cliffs in and upon it, and be transported seaward, the shingles and pieces of rock being there deposited, and thus adding gravels and distributed angular fragments to and among the more common accumulations formed in this sea, the depth of which varies from shallows, backed by marshes, to two localities on the south-east where the line gives respectively 110 and 115 fathoms of water.*

The Gulf of Mexico, its waters forced up by the pressure acting from the Atlantic through the Caribbean Sea, may, for geological purposes, be considered as a tideless sea, with, among others, a great river, the Mississippi, delivering matter in solution and mechanically suspended into it. The great movement of water coming round the Cape of Good Hope from the Indian Ocean, and considered as a constant current produced by the trade winds, assisted by the motion of the earth, sets from the Ethiopic Sea, united with an equatorial current of the Atlantic, across that ocean, against the West Indian islands. This pressure forces a constant stream of water into the Mexican Gulf, by the western side of the

presence of naphtha, which abounds in some localities on its shores. The basin of the Caspian appears of very unequal depth, this varying from the steep coast extending from the Balkan Bay to that of Mertroï Kultyuk—off which a line of 450 fathoms does not reach the bottom in some places—to long-continued, very shallow shores in others.

* The general depth has been estimated at 60 fathoms.

Yucatan Channel, with commonly a reflow close to Cape Antonio, at the west extremity of Cuba. Thus pressed up, the waters escape between Cuba and the Florida reefs in the current known as the Gulf Stream;* so that the waters in the Gulf of Mexico form a kind of comparatively tideless sea, in which deposits are effected much as in the Mediterranean. Though other rivers throw detritus into this area, collectively of much importance, the Mississippi is that, by its additions to the land, and by the discharge of matter mechanically suspended in its waters, which is the most important. The following is a plan (fig. 59) of the very characteristic advance of deposits from this river into the waters of the gulf.

The manner in which the main channel is bounded by lines of bank, rising above the sea, towards its final outlet, well marks the retardation produced by the friction of the banks as they arise. The various lakes, with the cross channels, are also highly illustrative of this order of accumulation.

As might be anticipated, when the fall of the Mississippi, during its greatest floods, is estimated at only one inch and a half in a mile between New Orleans and the sea, a distance of about 100 miles—while, when its waters are low, the fall is scarcely perceptible for the same distance—little mineral matter can be carried seaward in mechanical suspension, beyond that which, when deposited, would form silt, mud, or clay. This great river, therefore, now throws little other than fine mineral matter into the Gulf of Mexico, that which rises by accumulation above the surface of the sea being liable to be sifted by the shore waves, as at the mouths of the Nile. A vast mass of this fine sediment must have been thrown down, and is now accumulating in the Mexican Sea; the chief addition to such mass of mud or clay, independently of the hard remains of fish, crustaceans, and molluscs, being wood, the transport of which down the Mississippi and its tributaries is most abundant. Not only is this wood arrested in its progress in various places, or entangled among the channels of the delta, but much of it passes out seaward. Millions of logs and trunks of trees are transported several miles outwards during floods, so that it becomes difficult to navigate among them.†

* The breadth, length, and velocity of this long-celebrated current would appear, to vary. Winds often affect it, diminishing its breadth and augmenting its velocity, or augmenting its breadth and diminishing its velocity.

† Captain Basil Hall, *Travels in North America*, vol. iii.

Fig. 59.

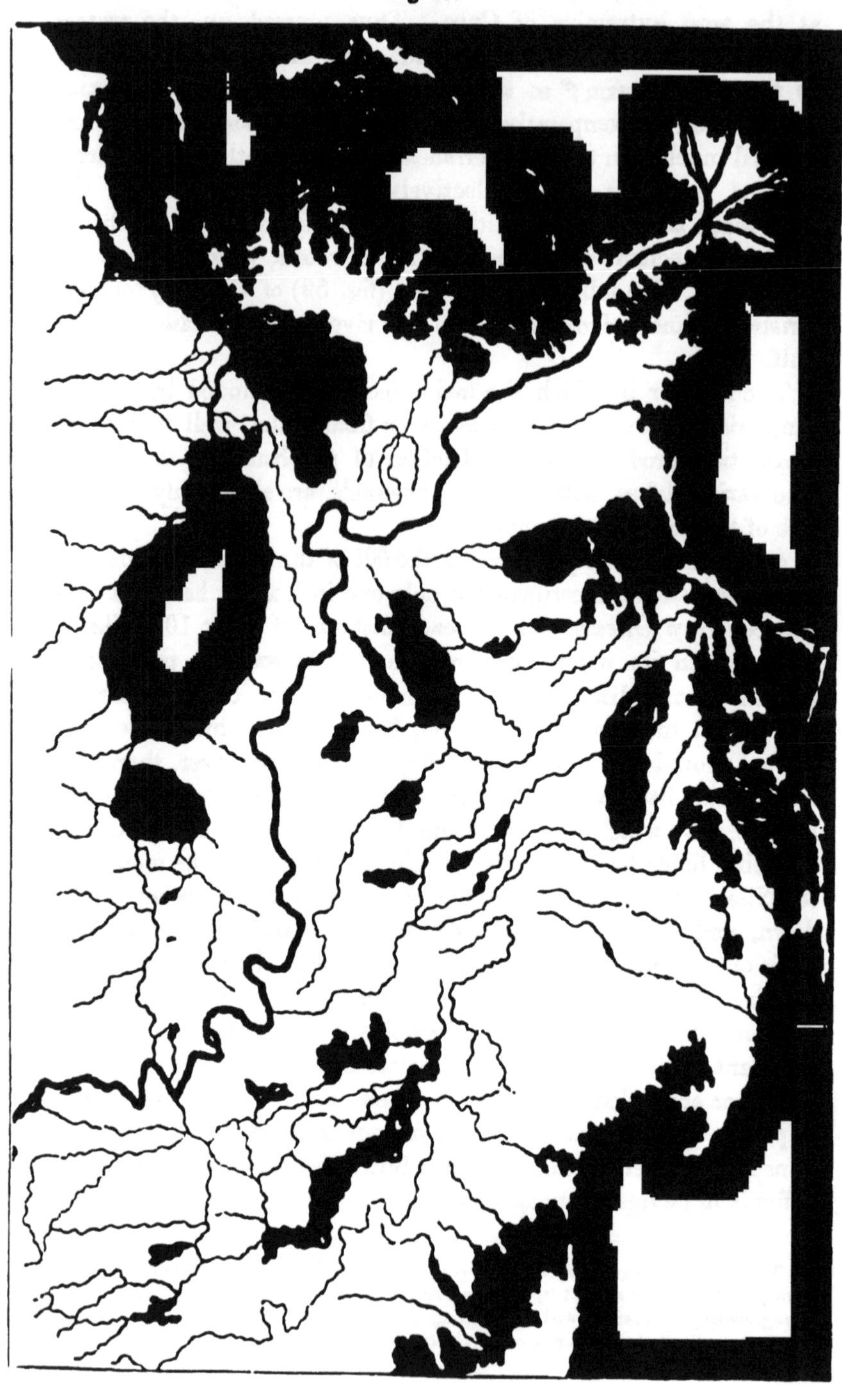

CHAPTER VI.

DISTRIBUTION AND DEPOSIT OF SEDIMENT IN TIDAL SEAS.—BARS AT RIVER MOUTHS.—RISE AND INFLUENCE OF THE TIDES.—DEPOSITS IN ESTUARIES.—DELTA OF THE GANGES—OF THE QUORRA.—DEPOSITS ON THE COAST NEAR SWANSEA.—INFLUENCE OF WAVES.—FORM OF THE SEA-BED ROUND THE BRITISH ISLANDS.—INFLUENCE OF CURRENTS.—SPECIFIC GRAVITY OF SEA WATER.—DISTRIBUTION OF SEDIMENT OVER THE FLOOR OF THE OCEAN.

UPON the coasts of the continents and of islands amid the ocean waters, not only is there a rise and fall of the sea-level twice in each day, but the river waters discharged into the ocean are, for the most part, ponded back by each rise of the tide, to be let loose at its fall with so much of the sea water as had been forced up the river channels during the flood tide. Here we have a very material modification of the discharge of the matter, either in solution or mechanically suspended in the rivers, as compared with its delivery into tideless seas by them. According to the varied character of the rivers where they discharge themselves into tidal seas; as regards the greater or less amount of water in them at different times; the kind of coast at their embouchures; depth of water, exposure to prevalent winds, and other conditions; so, no doubt, is the delivery of these waters modified; but in all they are exposed to checks from the rise of tide at their mouths. The opposition of the sea to the rivers at the height of the tide necessarily varies with the change from neap to spring tides; the amount of check which the sea gives to the outflow of the fresh water, thus alternating, on the minor scale; though, as a whole, a very constant effect is produced, the greatest resistance being offered at the heights of equinoxial spring-tides.

From the check thus given to the discharge of waters containing matter in mechanical suspension, or pushed forward by rivers in their channels, there is a tendency to form accumulations across the course of rivers, commonly known as *bars*. These will be found to occur variably, according as the real mouth of the river may be high up a deep branch of the sea (or in other words, where the sea level may cut high up a valley or depression, which thus becomes partly subaërial, partly submarine), or be situated on the general unbroken line of a coast, even, perhaps, protruding beyond it, into shallow water. It will soon be perceived that the breakers become important aids in the accumulation of bars according to such conditions; having little influence high up an arm of the sea, particularly where the channel is narrow, but assisting most materially in their formation when acting upon an exposed coast, more especially if the mouths of the rivers be open to strong and prevalent winds. This combination of checks given to waters pushing forward and carrying detrital matter in mechanical suspension, and by breakers striving again to thrust back that matter, produces bars at the mouths of many rivers, alike important as regards the subject under consideration, and the intercourse of nations.

The effects of tidal action in, for the time, arresting the outflow of rivers, will much depend upon the heights which the tides, on the average, attain; and it will readily be seen that, acording to the obstacles opposed to the tidal wave, and the form of the shores against which it moves, will be the change of sea level between high and low water. In the open ocean, where the tidal wave meets with, comparatively, little opposition, we find the difference of the sea level at high and low water far less than among funnel-shaped channels, and other favourable combinations of coast. Thus, while among the eastern Polynesian islands in the Pacific Ocean the tides rise and fall about 2 or 3 feet,* and in the Atlantic from 3 feet at St. Helena, and 4 to 6 feet at the Cape de Verde Islands, to 8 or 9 feet at Madeira, the equinoxial spring-tides in the Bay of Fundy rise from 60 to 70 feet.†

An observer need not travel from the shores of the British Islands to study the dependence of the rise and fall of tide upon local conditions: many situations will afford him the requisite opportunities. The Bristol Channel, since it fairly faces the tidal

* According to Mr. Dana (Geology of the United States' Exploring Expedition, 1838-42, p. 26), the tides rise only 2 or 3 feet through the eastern part of Polynesia; at Samoa 4 feet; at the Feejee Islands 6 feet; and at New Zealand 8 feet.

† A glance at the map will show how favourably this bay is situated for receiving a body of flood tide driven up between Cape Cod (Massachusetts), and Cape Sable (Nova Scotia), and forced onwards into Chignecto and Mines Bay. Though there is a very considerable bay between Gaspé Bay (Canada) and the North Point (Breton Island), on the north of the narrow isthmus separating Nova Scotia from New Brunswick, neither its form, nor the set of tide into it, cause a rise of water beyond about eight feet. There is, therefore, from local causes, a difference of high water, on either side of this narrow isthmus.

wave coming from the Atlantic, may be taken as a good example of a considerable rise of tide produced by the narrowing of an arm of the sea. Though strictly not an unmodified ocean tide, since the wave has to pass over nearly 300 miles of soundings, within the edge of the 100 fathoms line, before it strikes the Land's End, the change from the rise of 18 or 20 feet at St. Ives, Cornwall, to that of 46 to 50 feet at King Road (Bristol) and Chepstow, is striking; more particularly as the tides of 30 feet at Lundy Island, and 36 feet at Minehead, show this rise to be gradual. From the increasing elevation of channel, and friction, beyond Chepstow and King Road, and the withdrawing of the tidal pressure from behind when the ebb begins seaward, the height of tide soon decreases up the Severn. The tidal waters, however, so suddenly check the discharge of the river waters, that the latter are as suddenly forced back, the flood-tide rushing forwards in a great wave commonly termed the *bore*, and causing an instant rise of several feet in the lower part of the river, gradually fining off to the termination of all tidal action in the Severn.*

The annexed plan (fig. 60) will illustrate the example here

Fig. 60.

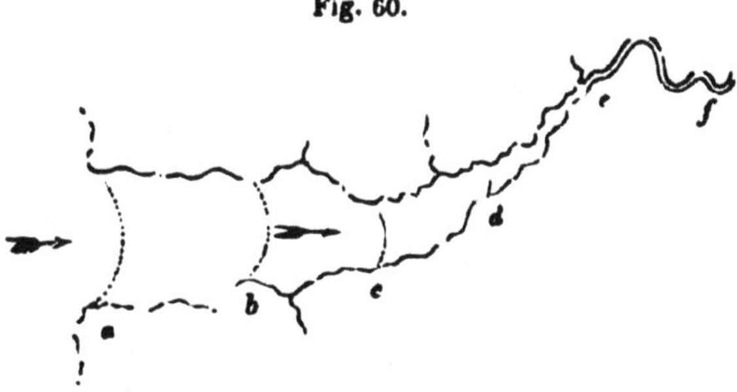

given. At *a* the tidal wave begins to be higher than in the open sea. At *b* its elevation is increased from the decrease of the depth and breadth of the channel; and at *c*, from similar causes, the height of tide is still greater. We may assume, for illustration, that at *d* the tidal wave becomes most elevated, and that afterwards, towards *e*, from the absence of propelling power behind, from the actual fall

* The same sudden rush of the flood, overpowering the ebb in tidal rivers, is observed in many other localities. The *bore-wave* up the Ganges is described as so rapid, that it scarcely takes four hours passing up a distance of nearly seventy miles, sometimes causing an instantaneous rise of five feet of tide at Calcutta. The boats on the shore on which it breaks take to the middle of the river for safety on its approach. A considerable bore-wave is stated to be observed at the mouth of the Maranon, or Amazons, during the equinoxes. The chief wave is from twelve to fifteen feet high, followed by three or four others. Its advance is very rapid, and its course is stated to be heard at the distance of two leagues.

of water on the ebb towards *c*, *b*, and *a*, and from the general rise of the channel, the tidal wave becomes less and less felt, until at *f*, its effect entirely ceases. The *bore* will depend upon local causes; but under the conditions noticed, the sudden check to the outflowing river, and corresponding sudden rise from the inflowing flood-tide, are not unfrequent, though the bore may not always be sufficiently important to arrest attention.

The English Channel affords us another good example of a considerable rise of tide produced by local obstacles, and the more instructive, as this rise does not extend across to the oppose coast, as is the case in the Bristol Channel. On the French side, the land of the Cotentin, terminating with Cape La Hogue, and the islands of Alderney, Guernsey, and Jersey, with the multitude of isles, islets, and rock, in the Bay of St. Malo, oppose a direct obstacle to the progress of the tidal wave coming from the Atlantic, while the English coast presents no such obstacle. In consequence, the sea level at high water is raised higher on the one side than on the other; and while the tides only rise 13 feet at Lyme Regis, 7 feet in Portland Road, 15 feet at Cowes, and 18 fect at Beachy Head, the difference of high and low water is 45 feet between Jersey and St. Malo, and 35 feet at Guernsey.

Not only are there these differences in the rise of tide from local causes, but the relative direction of the flood and ebb, with their consequent currents, also vary materially in some situations. Thus, at the Land's End the flood-tide runs 9 hours to the north, and the ebb 3 hours to the south; and numerous other modifications of the same kind, where the times of flood and ebb are different, are to be found on the coasts of the British Islands.

As regards the distribution of detritus by tidal streams, the direction of the latter will not only be found to change considerably during the progress of theflood or ebb, as the case may be, off many parts of coasts, but the ebb very frequently commences on shore, while a flood-tide is continued in the offing.*

As so much, not due to the friction of tidal streams on coasts has been attributed to it, instead of to the action of breakers—a destructive action more particularly felt when strong on-shore winds and high tides are combined—it would he well for an observer to study the velocity and transporting power of tidal waters on the

* It has been held that "the length of time between the changes of tide on shore and the stream in the offing is in proportion to the strength of the current and the distance from land; that is, the stronger the current, and the greater distance that current is from the land, the longer it will run after the change on the shore."—*Purdy, Atlantic Memoir.* 1829.

sea-shore. Those who dwell on, or visit, the coasts of the British Islands, where, fortunately, so many modifications in tidal streams may be more or less easily studied, will soon learn properly to estimate the value of tidal friction on land.

With respect to the tides around the British Islands, those flowing amid the Orkney and Shetland Islands, and through the Pentland Frith, between the mainland of Scotland and the former, would appear to be among the strongest. They vary considerably in force, according as they are neap or spring tides. While in Stronsa Frith and North Ronaldsha Frith the former only run at the rate of 1½ mile in the hour, the latter make a stream of 5 miles an hour. In the Pentland Frith, the spring-tides are stated to have a velocity of 9 nautical miles an hour, while at neap-tides they do not exceed 3 miles.*

Round the more prominent headlands, the tides, as we might expect, run with greater velocity than in the bays on each side of which they project, or in the offing outside. The tidal wave striking the headlands, and rising locally from this opposition, escapes round to the next bay, thus causing an accelerated stream of tide for a short distance. The friction of the water on the land is, however, commonly sufficient very materially to diminish the strength of the stream in immediate contact with it; so that, in calm weather, when the force of the tide is neither impeded nor accelerated by the force of opposing or favouring winds, chaff or other light bodies thrown into the sea will be seen to pass in a comparatively slow course along shore, while a strong stream of tide is running outside.

How little friction takes place in such situations may often be well seen by the presence of a coating of barnacles, or of sea-weeds, even upon steep headlands, though exposed to the action of breakers, these being, off such deep-water headlands, commonly unaided in their action by sand or gravel in mechanical suspension. It is desirable that the observer should carefully watch the shores of any district he may be examining, with respect to tidal friction, during calm weather, and from neap to spring tides. Except in the most exposed situations, he may perceive how rarely even grains of sand, much less small loose shingle, can be moved by any stream of tide in contact with the coast.

* The flood-tide there comes from the north-west, and is not of unusual strength until it meets with the obstacles of these islands and the mainland. The change of tide sooner on shore than at a distance from it, varies according to situation, amounting in some places to two or three hours.

The retarding effect of friction on the headlands is often well exhibited near the strong streams of tide off them, known as *races*, so dangerous, frequently, when opposed to powerful winds. Though the tides run in such situations with the greatest force of the locality, and the waters are thrown about in various directions, it often happens that, between the race and the headland, there is more quiet water, sufficiently broad for the passage of a boat in moderate weather.

Tidal waters rush with great force through channels formed between the horns of great bays and islands at a short distance from them; such is the case with the horns of Cardigan Bay and of St. Bride's Bay on the south of it, as shown in the following plan (fig. 61), where *a* represents Cardigan Bay, and *b* St. Bride's Bay.

Fig. 61.

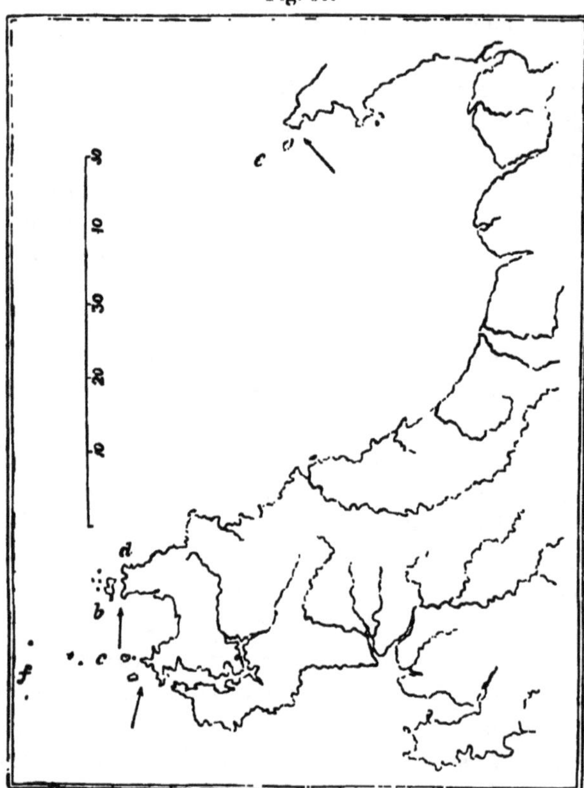

Foul rocky ground extends from the Smalls Light, *f*, to Skomer Island, *c*; between which and the mainland there is an exceedingly strong tide sweeping close to the cliffs. Supposing this to be a flood-tide, its force is diminished and almost lost in St. Bride's Bay, *b*.

This bay receives the flood-tide, not only through this channel, but also directly from the Atlantic; its flow over the foul ground between the Smalls Light and Grasholm, and thence to Skomer, being marked by broken water. Part of the tide driven into St. Bride's Bay escapes with much force between the mainland and Ramsay Island, and round the latter and the rocks and islets known as the Bishop and his Clerks, *d*, into Cardigan Bay. The latter also receives an abundant supply of the tidal wave direct from the Atlantic; and the flood passes with great strength between its northern horn and Bardsey Island, *e*.

In the chief channels noticed, no doubt little comparatively fine sedimentary matter could rest in the run of such tides, and any that might be thrown down by the eddies of one tide would probably be removed by the reverse action of the other; but these effects would be very local. That hard rocks readily resist such friction is well shown in the localities mentioned, barnacles and sea-weeds being commonly discovered on the sides of the channels at low water.

It will be at once perceived that the flood-tide passing up rivers would act very differently, according as the channels were continued deep outwards, or crossed by bars accumulated at their mouths. In the former case, the sea waters being specifically heavier than the river waters, as it were, wedge up the latter, discharging outwards, until the levels are so changed that the whole body of tidal water is driven inland, forcing and ponding back the fresh water.* In the more favourable situations of this kind, therefore, where great floods are running down a river, the heavier waters of the first of the flood-tide may be passing up the river while the lighter waters above are running outwards. In bar rivers the sea waters pour over the bars, and, if the channels be afterwards shallow, drive the river waters at once before them, while, if behind the bar there be water of much greater depth, as sometimes happens, the heavier sea waters first flow into the basin and raise the waters in it, so that when sufficiently elevated with the increasing tide, the whole passes up the river with the flood-tide, forcing back the fresh water. Between the action of the tide in such rivers as the St. Lawrence,† with its open estuary or arm of the sea, and the Ganges

* The passage of river waters outwards during freshets, from heavy rains in the interior, while the flood-tide waters are flowing beneath in a contrary direction, may occasionally be seen well shown when large vessels are at anchor in an estuary, as, for instance, in the Hamoaze, Plymouth, riding with their heads to the flood-tide, being sufficiently deep in the water to be influenced by it, while small boats, secured alongside, ride with their heads in the contrary direction, the outflow of the higher and fresh water alone acting upon them.

† The St. Lawrence affords a good example of the greater velocity of an ebb over a

and Quorra, the deltas of which protrude into the ocean, the one in the Bengal Sea and the other in the Gulf of Guinea, every modification will be found in the tidal rivers of the world.

While checked by the flood-tide, the waters of estuaries will deposit such of the matter, which they may hold in mechanical suspension as the time will permit, and according as the estuary waters may or may not be agitated by the friction of the winds. Slight observation is sufficient to show that highly-discoloured water is commonly found in estuaries, and that this is borne upwards and downwards by the tides, escaping seawards during the ebb in some estuaries in one direction, while the rivers add detrital matter to these bodies of water in others. In estuaries like the Severn, at the head of the Bristol Channel, the muddy water is carried backwards and forwards with such rapidity that it is only in the sheltered nooks and situations that it can find rest sufficient to deposit fine sediment, including among them the shores where retardation by friction also produces a sufficient state of repose during the tides.*

Many minor estuaries round the coasts of the British Islands show the filling up, not only of the sheltered places on their sides, but also of their upper parts, where detrital matter is gradually accumulated. If the course of the river has not been long through a level country, the deposits at the heads of estuaries may even be gravelly, while mud only is accumulated in the sheltered localities. If the annexed plan (fig. 62) represent one of these estuaries, then it will

Fig. 62.

usually be observed that the accumulation at the head *c* is more gravelly or sandy, particularly in its lowest parts, than in the sheltered situations, *a* and *b*. At *c* we have not only the heavier matter

flood-tide in an estuary. Where the ebb from the Saguenay unites with that of the St. Lawrence, it passes outwards with considerable strength, and is stated to run seven nautical miles per hour between Apple and Basque Isles. While the ebb is thus strong, the stream of flood-tide is scarcely perceptible.

* The difference of the friction on the sides of these estuaries, where mud is deposited, and more outwards in the stream of tide, is commonly well shown by the sandy bottom under the latter, the friction of the water being too great to permit finer sediment to remain in such situations.

thrown down by the check of the tide there felt, but also all the detritus which can be pushed along the bottom by the river *d*, during the ebb of the tidal waters, and during the common discharge of the river water when the tide has fallen, a combined time in some localities equal to nine and ten hours in each twelve. At *a* and *b* the fine sediment is commonly accumulated to the level of the highest ordinary spring-tides.

In estuaries of this class we should anticipate that there would be much gain of land where the discharging rivers entered them, and, accordingly, in such situations we often find extensive marshes and flats, which would justify this expectation, even if historical evidence could not be adduced. Of such evidence, however, there is commonly no want, and the heads of many estuaries around the British Islands, and along the ocean coasts of Europe, are known to have become more shallow and even to have moved further outwards, dry land supplying the place of marshes, and mud banks, within historical times.*

The mouths of the Ganges, extending across a distance of about two hundred miles (fig. 63), furnish us with the discharge of de-

Fig 63.

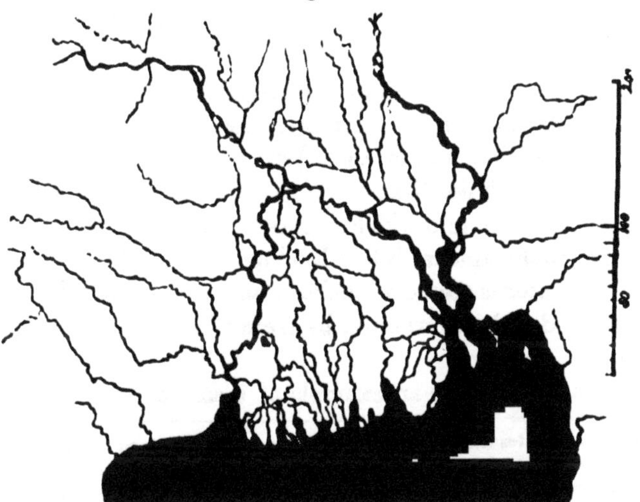

trital matter into a tidal sea of a different character. Here the abundance of the outflowing waters, particularly during floods, is sufficient to carry out a delta, more resembling those observed in tideless seas. In times long since passed, the Ganges may have discharged itself into an estuary, as far northerly as the com-

* These changes produced, independently of sea banks raised to keep out the tides.

mencement of its delta, now more than two hundred miles from the sea, and into the same estuary, the Brahmaputra may have delivered its waters, these two great drains of land extending to the Himalaya mountains having gradually filled up such an estuary, by depositing the matter transported mechanically in, or swept onwards by them. Coarse gravel is not forced forward by the Ganges within four hundred miles of its mouth, so that the sedimentary matter discharged into the sea is of a finer character. To this discharge, both by friction on the bottom and in mechanical suspension, checks are offered by the tides; but the body of fresh water is so considerable, as compared with such checks, that the sedimentary deposits rapidly gain upon the sea, notwithstanding that the general depth beyond the mouths of the Ganges is by no means inconsiderable. Innumerable changes in the direction of the various streams into which the delta is divided are produced inland.* The course of this river is described as affording good examples, on the great scale, of the alterations of channel, from the accumulation of banks upon small obstacles, to be equally well studied, as regards general principles, in hundreds of little streams. Thus a tree arrested in its course will produce an accumulation, gradually rising into an island, to be again swept away by another change of channel.

The great body of fresh water discharged by the Ganges in floods seems, to a great extent, to overpower the influence of the tidal wave, so that detrital matter then becomes accumulated more in the manner of the Nile, Rhone, Volga, and other great rivers, discharging themselves into tideless seas.† At the junction of the Ganges and Brahmaputra, below Luckipoor, there is a large gulf in which the water is scarcely brackish, and during the rainy season the sea is stated to be overflowed by fresh water for many leagues outwards.

In the Quorra we have an example of a similar kind, and a vast

* Major Rennel states that during the eleven years he remained in India, the head of the Jellinghy river was gradually removed three-quarters of a mile further down. He observed also, that "there are not wanting instances of a total change of course in some of the Bengal rivers. The Cosa (equal to the Rhine) once ran by Purneah and joined the Ganges opposite Rajemal. Its junction is now nearly forty-five miles further up. Gour, the ancient capital of Bengal, once stood on the Ganges."—*Phil. Trans.* 1781.

† The amount of detrital matter borne outwards by the Ganges has been estimated at about 2¼ per cent., and the average discharge of water at 500,000 cubic feet per second.—(*Gleanings of Science*, vol. iii. Calcutta, 1831.) If we take the quantity at 2 per cent., and consider the transported matter to give 15 cubic feet to the ton, we should obtain 57,600,000 tons per day, equal to a mass of ordinary granite, having a base of 1,000,000 square feet, rising to the height of 864 feet.

body of fresh water thrusts out a delta into the ocean. The great stream of water is checked, not overcome, in mid-channel, though felt between 30 and 40 miles up the river. In this river, and in many other tropical rivers, mangrove-trees add materially to the power of forming new land.* Wherever sufficient shelter can be obtained, they establish themselves in abundance; their stilt-like roots entangling any floating substances washed near them; producing a repose fit for the deposit of the finest sediment, and affording shelter to an abundance of reptiles, fish, crustaceans, and molluscs, which seek and enjoy the protection they afford.

When we regard the sea-shores of the world exposed to tides, we see a great destructive power in the breakers, as a whole in ceaseless action, grinding back and levelling off the land, and throwing a mass of matter into the tides sweeping round such shores, which mass, added to that thrust out of the rivers, has to be distributed by the streams of tide and such ocean currents as can receive any portion of it. Great rivers, as we have seen, may transport matter in mechanical suspension far outwards, particularly when swollen by floods, and thus place it within the distributing influence of the ocean currents. Through these it may take a long time to descend into those quiet depths where it can find a rest, one that may continue undisturbed until, perhaps, after a long lapse of geological time, the resulting deposit may be upraised, and placed within the destructive influences of the atmosphere and surface waters.

When detrital matter is thrown into the tides, it is borne to and fro by them, according to their flow and ebb, and the observer will have abundant opportunities of seeing on the coasts of the British Islands, and on the ocean shores of Europe, that the river waters when swollen by rains, bear outwards with the ebb, and in the direction that it takes along shore, much mechanically-suspended detritus, which does not again enter the rivers unless under very favourable circumstances. As a whole, much fine detritus, thus derived, is carried coastwise by the ebb, and accumulations are formed of it, if there be sufficient continued repose in that direction. So that should a sheltering headland run out, and a bay be formed between it and the embouchure of the river, there is a

* Alluvial land is described as forming into flat islands, covered by mangrove-trees and papyrus. These are sometimes so acted upon by floods as to be partially or wholly swept into the ocean. Professor Smith noticed a floating mass, probably washed out of the Congo, about 120 feet long, consisting of reeds resembling the *Doaar* and a species of *Agrostis*, among which branches of *Justicia* were still growing, further north off the coast of Africa.—*Tuckey's Expedition to the Zaire or Congo.*

tendency to deposit the finer sediment in the locality so sheltered. We may take the coast of Swansea as affording an easily-observed instance of the separation so affected.

Two rivers, *a* and *b* (fig 64), the Towey and the Nedd (Neath),

Fig. 64.

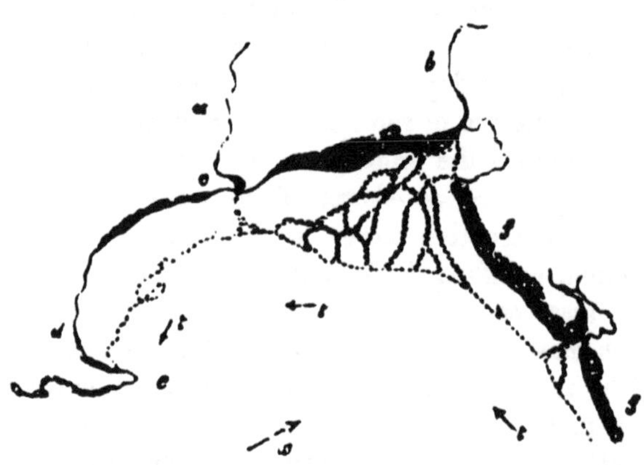

when in flood, bring down much sedimentary matter, the finer parts of which are carried by the ebb tide (*t*, *t*, *t*) towards the bay formed between Swansea (*c*) and the Mumbles (*d*). Here finding the necessary repose, the prevalent winds (*w*) blowing from the west and south-west, a part is deposited and mud is accumulated, the remainder of the detrital-bearing waters, escaping round the Mumble Rocks (*e*) into the general ebb tide passing westwards down the Bristol Channel. While this happens with the finer sediment, the arenaceous part of the detritus thrust out of the river is more quickly thrown down, and a large part of it becomes acted upon by the breakers, raised by the prevalent winds, and is forced partly into mechanical suspension during heavy gales, and then borne in the flood-tides, and partly brushed onwards by the waves, breaking upon much flat ground exposed at low water, towards the coast to the eastward (*f*, *f*). Here the conditions for the accumulation of sand-hills obtain, and the overplus of arenaceous sediment, borne outwards by the Towey and Nedd, and not retained by the sea, is blown by the winds upon the dry land. In this locality, therefore, the river-borne detritus, thrown into the tides, becomes in a great measure separated, mud being chiefly accumulated in one direction, sand in another, a surplus of the latter being restored to the land.

Though there is a tendency to accumulate the finer river-borne detritus in the direction of the ebb tides, this is often met by conditions so unfavourable to such a deposit that the finer matter does not there come to rest, but is gradually transported outwards to sea, and may thus be brought by tidal streams even within the influence of ocean currents. On a shallow coast the breakers alone, when they can act equally in the direction of the ebb and of the flood-tide, prevent the accumulation of the finer sediment, which, in consequence, can only find rest by being carried outwards into water of the needful depth and tranquillity.

The abrasion of coasts by breakers being the same, whether the tide be setting in one direction or another, as flood or ebb, the finer matter is carried in mechanical suspension equally by the stream of either along the coasts, finding rest in the situations where conditions are favourable, even entering estuaries by the flood-tide, when such estuaries occur in the line of its course, the indraught, on the flood, carrying it in with the tide. As we have seen (p. 54), the heavier parts, such as shingles and small pebbles, are distributed along shore, and the arenaceous portions, sometimes on the coast, sometimes more seaward, according to circumstances.

The agitation of the sea is felt at different depths in proportion to the magnitude of the waves raised by the friction of the wind. During heavy gales of wind, the depth at which this agitation has been observed, sufficient, as it were, to shake up fine sediment enough to discolour the water, is about 90 feet.* The disturbing effects of waves in minor depths is often well shown on shallow sandy coasts by the throwing on shore of many molluscs in a living state, known to inhabit the sands at moderate depths. By the agitation of the sea their sandy covering is removed, and they are swept onwards beyond their powers to retain their position at the bottom, and thus become finally thrown out upon the coast.

Besides the waves seen to arise on the spot from the action of the winds, the great indulations which are known as *swells* and *rollers* (so common on ocean shores, and due to the friction of winds out at sea which do not reach the land) disturb the sea bottom to a considerable extent, so that, both heavy seas and swells combined, the finer sediment becomes removed from all but favourable situations outwards, and is distributed off the coasts, outside the accumulations fringing them, and due to the action of the breakers.

The flow and ebb of the tides produce a motion tending to smooth out and flatten the accumulation of detrital matter deposited on the sea bottom within their influence. The smoothing action no doubt

* The depth at which the disturbing action of a sea-wave can be felt has been estimated even so high as 500 feet on the Banks of Newfoundland. *Emy, Mouvement des Ondes*, 1831, p. 11.

varies with the strength of the tides, as these may be springs or neaps, so that matter can be brought to rest during the latter, which becomes removed by the superior velocity and volume of water of the former; but, as a whole, there arises an adjustment, producing a sea bottom of a marked character. The friction of the tidal wave on the bottom forms ridges and furrows of the same kind with those previously noticed as produced by the winds on loose sand (p. 59). Where clear waters prevail, and the ridges and furrows are formed by this kind of friction alone, the resemblance is very striking, allowance being made for the relative weight of the particles of sand in the air and in the water. Where waves act on the bottom, it would be expected that such ridges and furrows would be modified by the to-and-fro action set up, although the on-shore might be greater than that of the counter movement, in proportion as the wave takes the onward force of a breaker, the higher part acquiring gradually a greater forward motion as the water becomes shallower, and the friction on the bottom becomes increased.

Almost every extensive sandy flat left by the tide, and of such the coasts of the British Islands afford abundant examples, shows the effects of friction on the sand. An observer should well study the various modifications to be seen in such situations, for among arenaceous accumulations of all geological ages, the effects of friction on sand and silt, by water in motion, is often very evident. In many situations peculiar arrangements of the surface sand will be observed to have arisen from the draining off of the tidal water, which has quitted a large tract of sand suddenly. We have thus friction on the bottom from the rise and fall of tidal waters on coasts, from the to-and-fro action of waves produced by winds (where the depths are favourable), and from streams of tide, variable in strength, usually acting in two directions, and often in more, from local causes.

From friction of all kinds much sedimentary matter is so shoved and pushed along the bottom in various directions, that from this cause alone a great flattening of the surface would be effected. If to this we add the deposit of matter borne in mechanical suspension, and derived either from rivers or the action of breakers, we should expect a distribution of detritus which, if raised above the level of the sea, would offer the appearance of a great plain. The accompanying map (fig. 65) will show the extent of area around the British Islands within a line of depth equal to 100 oms (600 feet), and which, if raised above the level of the sea, would present to the eye little else than a vast plain. To form this great tract of smoothed ground, no doubt the levelling action of breakers, cutting back the coasts, must be duly regarded; so that to this action, to that of the seas rolling in various directions, according to the winds stirring up the bottom in sufficiently shallow places, and to the distributing power of streams of tide, is mainly

Fig. 65.

due the present surface of this area,* the extent of which may be estimated by the annexed figure (fig. 66), representing 1000 square miles, on the same scale as the map (fig. 65).

* Always bearing in mind that there is a base beneath of tertiary and other rocks, over which the sands and mud are at present strewed, and which may here and there be still uncovered. In many a situation, a minor area, planed down by the action of

Fig. 66.

□

It is worthy of remark, that if, instead of the line of 100 fathoms beneath the sea, that of 200 fathoms had been selected, the second line would not have extended far beyond the first, the slope increasing far more rapidly outside the 100 fathom line than within it, so that, after preserving a very gentle slope, as a whole, outwards for the great area represented above (fig. 65), the bottom of the sea descends much more suddenly beyond it towards the Atlantic.*

Slight attention on the coasts will show that the water moving past them in a stream from tidal action travels backwards and forwards a somewhat limited distance, so that any detritus held by it in mechanical suspension, and eventually thrown down from such suspension, could only be deposited within a limited area, when no disturbing causes interfered. The water of a tidal stream, passing a coast at the average rate of three miles per hour, will only travel 18 miles, regarding the subject generally, before it is swept back again over the same ground for the like distance. The pressure of high winds, both on and off a coast, particularly if they be long continued, forces water against or away from the land, and so with any other direction a surplus of the ordinary body of water may take from the friction of the wind. Hence the mere backward and forward motion of the same body of water is somewhat modified, as also by the great additions made to the usual volume of tidal water by the discharge of great floods from rivers, striving to force their way over coast streams of tide.

Making, however, all reasonable allowance for these modifying influences, there remains enough of continued local action to procure local accumulations of detritus, more diversified in character near the coasts than at a distance from them, on account of the increased velocity of tides immediately off chief headlands, and their diminished strength of stream in sheltered bays, not forgetting estuaries, with and without bars of different kinds.

The observer has now to consider the distribution of fine matter in mechanical suspension by means of ocean currents. Some of

---

the breakers, may yet be kept clean from deposits by local causes. We may probably regard the whole area as the result of the cutting back of coasts by breakers, and of deposits from the causes pointed out, continued through a long lapse of geological time, movements of land, as regards its relative level with the sea, and on the large scale, having contributed to its present condition.

* Here and there, there are minor depressions in this area, and among them the trough-like cavities in the North Seas, known as the Silver Pits. The bottom around the chief pits is described as rising gradually to it, when suddenly the interior sides descend from a few to 40 or 50 fathoms, forming steep interior escarpments.

these are known to be very constant in their courses, others periodical, and many temporary. We have seen that the pressure of strong and long-continued winds forces up water by their friction on its surface in tideless seas, and consequently would expect that in the open ocean similar winds would force water before them, though the absence of land would produce a modification in the result. When the area so acted upon was bounded by a single range of coast, the modification would be less; and when two lines of coast presented themselves, between which the water could be forced, and lateral fall prevented, there would be an approximation to the effects observable at the north and south extremities of the Caspian, or on the east and west shores of the Black Seas, where the waters are pressed forward by the needful winds.

Independently of the pressure on the surface of the sea by winds either constant or nearly so, periodical or temporary, it has been supposed that the motion of the earth gives a certain movement to the waters of the ocean from east to west, thus increasing the power of some currents, due to the surface action of winds, and interfering with the movement of others. To the motion of water from this cause, the continent of America, with South Georgia, South Orkney, South Shetland, and the icy regions extending to Victoria Land, would interpose between the Atlantic and the Pacific, and the continent of Asia, with the Philippines, Borneo, Moluccas, New Guinea, and Australia, would oppose the westward movement of the Pacific, not forgetting New Zealand, and the multitude of islands and islets of Polynesia in that ocean.

The more open space for this supposed movement would be from the Indian and Southern Oceans into the Atlantic, the coast of Africa not offering it opposition beyond the latitude of 35° south. A constant current does run out of the Indian into the Atlantic Ocean, flowing up the west coast of Africa, to the equatorial regions, whence it strikes over to America, ponding up the water in the Gulf of Mexico. It has been inferred that this current is partly due to the motion of the earth, and partly to prevalent winds, those known as the Trade Winds especially driving the waters in the same direction.

The current into the Atlantic sweeps round the southern extremity of Africa by the Agulhas or Lagullas Bank, the soundings on which give mud to the westward of Cape Agulhas, and sand, containing numerous small shells, to the eastward. It might hence be assumed that this current acted upon the bank at a depth of 360 or 420 feet, sweeping off the finer sediment from the side exposed to its force, and parting with it in the more still water behind it. A mass of water is inferred to run up the west coast of Africa, from the Cape of Good Hope (between the coast and the waters of the adjacent ocean), 60 miles wide, 1200 feet deep, and of the mean temperature of the ocean, at an average rate of one mile per hour.* There are counter currents,† and the main current is considered to extend, as regards surface, to a comparatively moderate distance from the land. As a whole, this current reminds us of a body of water in movement westward, acquiring additional velocity against the southern extremity of Africa, as any minor mass of water in movement would against a common projecting cape or headland. We may regard another great Atlantic current, the Gulf Stream, as consequent on this main current, after it has traversed the Atlantic to the West Indies. Escaping from the Gulf of Mexico, as previously noticed (p. 74), the Gulf Stream waters flow northerly, a part passing off to the eastward, after passing the Straits of Florida, probably to equalize the general levels in that direction. As to the extent and velocity of the Gulf Stream, the contradictory evidence is sufficient to show that both are occasionally much modified. The winds, by their friction, necessarily affect the course of the stream, according to their duration, strength, and direction. In mid-channel, in the meridian of Havanna, the velocity is estimated at 2½ miles per hour; off the most southern parts of Florida, and about one-third over from the Florida Reefs, at 4 miles an hour. The stream is considered to range, in the meridian of 57° W. to 42° 45′ N. in summer, and to 42° N. in winter. A reflow, or counter current, sets down by the Florida Reefs or Keys to the S.W. and W.‡

Other currents are known in the Atlantic, such as that coming out of Baffin's Bay, through Davis's Strait,§ considered to join the Gulf Stream, the united body of water crossing over to the

* Sir James Ross. Voyage in the Southern and Antarctic Regions, vol. i. p. 35.

† Close to the shore there is an eastern current. The survey of the coast of Africa, to the east of the Cape of Good Hope, was made by Captain Owen, with the assistance of this current, against the force of the trade wind. Captain Horsburgh mentions having been carried by the eastern current, on the south of the main western current, at the rate of 20 to 30 miles in the 24 hours, and, in two instances, at the rate of 60 miles in the same time.

‡ Many small vessels are stated to make their passage from the northward by the aid of this counter current.

§ This current, commonly known as the Greenland Current, sets southerly down the coast of America to Newfoundland, bringing down large icebergs beyond the Great Bank. The velocity was found, by Captains Ross and Parry, to be 3 to 4 miles per hour in Davis's Strait. Off the coast of Newfoundland, it sometimes flows at the rate of 2 miles an hour; but is much modified by winds.

coasts of Europe and Africa. A southerly flow of water takes place from the coast of Portugal towards the Canary Islands, modified by the indraught of sea into the Mediterranean. Beyond these islands a S.W. current is noticed as probably due to the influence of the N. E. trade wind.

Constant currents are also mentioned in the Pacific. Currents are described as setting off the Galapagos to the N.N.W., and at Juan Fernandez, and 300 leagues to the westward of it to the W.S.W. (16 miles per day). Great quantities of wood are drifted from the continent of America to Easter Island by a stream of water passing in that direction. Between the Sandwich Islands and the Marquesas, currents have been found flowing westward at the rate of 30 miles per day. Among the Philippine Islands a current comes from the north-east, and runs with considerable force among the passages, dividing them from each other. Various other currents in the Pacific have been noticed. There are two, however, deserving of attention, inasmuch as one, flowing northerly through Behring's Straits, is thought to proceed eastward along the north coast of America,* and the other, passes round Cape Horn to the eastward for the greater part of the year.†

In the China and India Seas we find good examples of periodical currents. The water moves from the ocean into the Red Sea from October to May, and out of that sea from May to October.‡ In the Gulf of Manar, between Ceylon and Cape Cormorin, the current flows northward from May to October, setting the remaining six months to the S.W. and S.S.W. In the S.W. monsoon, the current between the coast of Malabar and the Lakdivas sets to the S.S.E with a velocity varying from 20 to 26 miles in the 24 hours. The currents in the China Seas, at a distance from shore, commonly flow, more or less, towards the N. E. from the middle of May to the

---

* Kotzebue describes this current as setting through Behring's Straits with a velocity of 3 miles an hour, to the N.E.

† This current has been doubted; but as there is a prevalence of strong westerly winds round Cape Horn, during the greater part of the year, the statement that there is such a current may be considered probable. A bottle, thrown overboard by Sir James Ross, near Cape Horn, was afterwards found near Port Phillip, Australia, having passed eastward about 9000 miles in 3½ years. Allowing 1000 miles for detours, this would be a rate of about 8 miles per day. It was Sir James Ross's practice, upon throwing bottles overboard, to load all but those intended for the surface, so that they took different depths. As sand was not stated to be found in this bottle, it was inferred that it was a surface bottle; hence the winds alone had much influence on its course.

‡ A current commonly flows from the Persian Gulf towards the ocean, during the whole time that the water runs into the Red Sea, and flows into the Gulf from May to October.

middle of August, taking a contrary direction from the middle of October to March or April. Their strength is most felt, as might be anticipated, among the islands and shoals.*

With respect to temporary currents, they are found to be innumerable; severe gales of wind, of long duration, readily forcing the surface water before them. Among channels and along coasts these are chiefly felt, the two boundary shores or the single coast opposing the further rise of water, and throwing them off in the manner of tidal waves.

While considering the movement of the ocean waters, the observer should not neglect any change in their position which may be due to their relative specific gravities. Experiments upon fresh water in lakes long since showed that a body of the heaviest water, that approaching towards a temperature of about 39°·5 or 40°, remained at the bottom undisturbed,† except by the influx of river waters, charged with detritus, which forced their way, spreading mud beneath them (p. 43). The researches of Sir James Ross in the Southern Seas have shown that in a similar manner water of a certain temperature, namely, of about 39°·5 Fahr., remains at the bottom, either colder or warmer water, as the case may be, floating above it. From many observations made, it was inferred that a belt of this water of a given temperature rose to the surface in southern latitudes, of which the mean is estimated at about 56° 26′, the whole body of ocean water in that circle being of this uniform temperature from the surface to the bottom, while on the north, towards the tropics and equator, water of a higher temperature floated above it, and on the south, that of a lower temperature.‡ Thus, considering the like belt of uniform temperature to appear in such parts of

* The strongest currents in these seas are experienced along the coast of Cambodia, during the end of November. They run with a velocity of 50 to 70 miles to the southward, in the 24 hours, between Avarilla and Poolo Cecir da Terra. Some parts of the stream setting into the Straits of Malacca, cause the tide to run nine hours one way and three hours the other.

† In 1819 and 1820, the author made experiments on the Lakes of Geneva, Neuchatel, Thun, and Zug, with a view of investigating this subject. An account of these experiments was published in the "Bibliothèque Universelle" for 1819 and 1820. It was found that, in the Lake of Geneva, the water, in September and October, 1819, had a temperature of 64° to 67° Fahr., from the surface to the depth of 1 or 5 fathoms, and that there was a general diminution of temperature downwards to 40 fathoms. From 40 to 90 fathoms, the temperature was always 44°, with one exception, when it was 45° at 40 fathoms. From 90 fathoms to the greatest depths, which amounted to 164 fathoms, between Evian and Ouchy, the temperature was invariably 43°·5. After the severe winter of 1819-20, the same temperature continued beneath. Experiments on the Lakes of Neuchatel, Thun, and Zug, alike pointed to water of a temperature approaching to the greatest density of water, between 39° and 40°, being at the bottom.

‡ The following were the observations on which Sir James Ross founded his view of

the northern hemisphere as is covered by the ocean,* there would be, where land did not occur, three great thermic basins, two towards each pole of the earth, and a middle trough, or belt through the central part of which the equator would pass. Sir James Ross points out that in lat. 45° S., the temperature of 39°·5 has descended to 600 fathoms, increasing in depth in the equatorial and tropical regions to about 1200 fathoms, the temperature of the surface in the latter being about 78°.† On the south of the belt of uniform temperature, the line of 39°·5 is considered to descend to 750 fathoms in lat. 70°, the surface being there at 30° Fahr.

To estimate a movement which might be produced by the settlement of any water of the density of 39°·5, striving to occupy an equal depth beneath those of inferior weight, either of greater or less temperature, as the case might be, to the north and south of these belts of uniform temperature, supposing that some approximation to such a belt was to be found in the northern hemisphere, we should compare the distance from these belts with the depths at which given temperatures have been observed. This done, we obtain for the slope on either side of the southern belt (assuming a plane for more ready illustration) of about 1 in 1723 to the 1200 fathoms of 39°·5 beneath the equator, and of about 1 in 1136 to the same temperature beneath 750 fathoms in 70° south latitude. So small an angle, with a change of temperature so gradual, could scarcely be expected to produce a lateral movement in the mass of ocean waters of geological importance.‡

---

the position of this circle, the water being ascertained in the localities noticed to have the same temperature from the surface downwards:—

Latitude.	Longitude.
57° 52′ S.	170° 30′ E.
55 09	132 20
55 18	149 20 W.
58 36	104 40
54 41	55 12
55 48	54 40

Voyage to Southern and Antarctic Regions, vol. ii.

* Allowing the same causes to be in operation in the northern hemisphere, we should expect similar effects, however modified by local circumstances. Scoresby obtained, in lat. 79° 4′ N., long. 5° 4′ E., 36° at 400 fathoms, the temperature increasing from 29° at the surface. Another observation by the same author, in lat. 79° 4′ N., gave 37° at 730 fathoms, the surface being 2[illegible]°. Again, in lat. 78° 2′ N. and long. 0° 10′ W., he found 38° at 761 fathoms, the surface being 32°.

† With regard to observations in the tropics, Colonel Sabine found, in lat. 20° 30′ N., and long. 83° 30′ W., a temperature of 45°·5 at 1000 fathoms, the surface water being at 83°. Captain Wauchope obtained in lat. 10° N., and long. 25° W., 51° at 966 fathoms, the surface water being at 80°; and he also found in lat. 3° 20′ S., and long. 7° 39′ E., a temperature of 42° at 1300 fathoms, the surface water being at 73°.

‡ It should be remarked that the temperature of 39°·5, found by Sir James Ross in situations leading to the inference that such temperature is that of the greatest

The agency of ocean currents in the transport of matter mechanically suspended in their waters, and derived from the decomposition or abrasion of land, will necessarily depend upon local conditions. Here and there streams of tide may deliver such matter to them, to be borne in the direction in which they may move, and great rivers, such as the Yang-tse-kiang, the Ganges, the Indus, the Quorra, and the Amazons, may thrust out bodies of water, flowing beyond the return of the tidal streams off coasts, and carrying detritus to ocean currents, through which it would have gradually to descend. It might thus be transported long distances, particularly if the depths it might have to descend, before stagnation of the lower waters would prevent any than a vertical fall of the matter, were considerable.* The matter obtained from the land seems chiefly to be thrown down as a fringe of various shapes and composition, skirting the shores; sometimes, from local conditions, extending to far greater distances than at others.

Although the great floor of the ocean may not be very materially

density of sea water, containing the ordinary amount and kinds of salts in solution, does not well accord with experiments in the laboratory. According to Dr. Marcet, those made by him show that the maximum density of sea water is not at 40° Fahr. In four experiments, Dr. Marcet cooled sea water down to between 18° and 19°, and found that it decreased in bulk till it reached 22°, after which it expanded a little, and continued to do so until the water was reduced to 19° and 18°, when it suddenly expanded and became ice at 28°. According to M. Erman, also, salt water of the specific gravity of 1·027 diminishes in volume down to 25°, not reaching its maximum density until congelation.

These results would seem to point either to some modifying influence acting upon the waters of the ocean, to faults in the instruments, to the mode of employing them, or to sources of error in the laboratory experiments not suspected. At considerable depths, the heavy pressure upon the bulbs of the thermometers, if used naked, might be supposed to produce an error as to the mass of water of uniform temperature from the surface downwards. If pressure, however, upon the bulb caused a higher apparent temperature, this should vary with such pressure; but the results do not bear out this view, unless it be assumed that the gradual increase of pressure exactly counterbalanced a decrease of temperature. It is worthy of remark, that the temperature of 39°·5 is about that assigned, from experiments, to pure water. It may be here observed that the water beneath 90 fathoms in the Lake of Geneva was found, both after a warm summer and a severe winter, to remain at 43°·5, not 39°·5 or 40°, as experiments in the laboratory would lead us to expect. From observations on the temperature of the western Mediterranean waters, at various depths, it is inferred that all beneath 200 fathoms remains at a constant temperature of about 55°. (D'Urville, Bul. de la Soc. de Geographie, t. xvii. p. 82.)

If we take 39°·5 for the temperature of the greatest density of sea water, we shall have to consider that the salts in solution produce no influence upon such density, the water alone having to be regarded. It would be very desirable that experiments respecting the density of sea water at different temperatures should be repeated in the laboratory, and that observations should be made at different seasons upon the temperature of deep fresh-water lakes, in order to see if we are in any way to regard the temperature obtained in the sea of 39°·5, observed by Sir James Ross, as a result to which some modifying influence may be attributed.

* Some very interesting observations respecting the surface density of the sea off the coast of British Guiana were made by Dr. Davy (Jameson's "Edinburgh

covered by deposits from ocean currents, conveying detritus from the great continents, Australia, and the larger islands of the world, the oceanic islands may collectively furnish matter of importance. The observer will find that many of these islands rise from comparatively considerable depths, so that detrital matter derived from them by the action of breakers (and they are very commonly exposed to a nearly-constant abrasion by the surf), moved by the tidal waves sweeping by the islands, and thence delivered into any ocean currents passing near, may be carried by the latter to considerable distances. These oceanic islands are found to be chiefly of two kinds, the one of igneous, the other of animal origin. With respect to the former, we have not only to consider the detritus they may now furnish by the action of breakers upon them, but also the transportable matter which may have been ejected from the igneous vents while they rose, by the accumulation of molten rock, cinders, and ashes.

Instead of simply accumulating around the igneous vent, as would happen, with certain modifications from the distribution of wind-borne ashes and small local movements of water in tideless seas, not only might there be a to-and-fro distribution of the volcanic matter carried various distances in mechanical suspension from the tidal wave acting against the new obstacle to its movement, but the finer substances could also be borne away by any ocean current passing near, and thus such substances be carried far onward in the direction of its course. As soon as any igneous matter is raised above the sea level, so soon is it attacked by the breakers, and only

Journal," vol. xliv, p. 43, 1848). He found that where the Demerara river meets the sea, near George Town, the density of the water was 1·0036, and subsequently as follows:—

1. 11 miles off shore	=	1·0210
2. 19 " "	=	1·0236
3. 27 " "	=	1·0250
4. 35 " "	=	1·0236
5. 43 " "	=	1·0250
6. 51 " "	=	1·0258
7. 80 " "	=	1·0266

The specific gravities of Nos. 4 and 5 were considered to have been influenced by heavy showers of rain which fell while the steamer on which Dr. Davy was on board passed. This modification in the density of the surface waters, by tropical rains, is well shown by the observations of the same author, off Antigua and Barbadoes. Towards the end of a very dry season, the specific gravity of the surface water, off the former, was found to be 1·0273, while, after three months of heavy rains, off Barbadoes, the specific gravity was reduced to 1·0260. The positions of these two islands give such observations considerable value. With respect to the matter mechanically held in suspension in the waters off British Guiana, Dr. Davy states that, for many miles near the land, it was sufficient to give a light-brown tint to the sea, like the Thames at London-bridge. It was only at about the distance of 80 miles from shore that the waters presented the blue colour of the ocean.

in proportion to its solidity and mass can the portion above water, and removed from the destructive action of the surf, remain to be more slowly wasted by atmospheric influences, and to be clothed with vegetation, if within climates fitted for its growth. Many an island in the ocean can be regarded as little else than the higher part or parts of a volcano, or some more extended system of volcanic vents, rising above its level, the mass and kind of matter ejected being sufficient to keep it there. As might be expected in a great volcanic region like that of Iceland, igneous vents have opened in the sea near its shores, as well as upon the dry land. A volcanic eruption is recorded as having taken place in 1783, about 30 miles from Cape Reikianes, and another off the same island about 1830.* In 1811 a volcanic eruption was effected through the sea off St. Michaels, Azores, and eventually, after the ejection of much matter, columns of black cinders being thrown to the height of 700 and 800 feet, an island was formed, about 300 feet high, and about one mile in circumference.

Fortunately the formation of this island was observed and recorded. It was first discovered rising above the sea on the 13th June, 1811, and on the 17th was observed by Captain Tillard, commanding the "Sabrina" frigate, from the nearest cliff of St. Michaels. The volcanic bursts were described as resembling a mixed discharge of cannon and musketry, and were accompanied by a great abundance of lightning. The following (fig. 67) was a sketch made at the time, and will well illustrate the manner in which ashes and lapilli may be thrown into any ocean current or tidal stream passing along, and be borne away by it.

This island, to which the name of Captain Tillard's frigate was assigned, subsequently disappeared, but whether simply by the action of the breakers alone, or from the subsidence of the main mass beneath, or from both causes, accounts do not enable us to judge.†

* In 1783, the eruptions of several islands were observed as if raised from beneath, and, during some months, vast quantities of pumice and light slags were washed on shore. "In the beginning of June, earthquakes shook the whole of Iceland; the flames in the sea disappeared, and a dreadful eruption commenced from the Shaptar Yokul, which is nearly 200 miles distant from the spot where the marine eruption took place."—(Sir George Mackenzie's Travels in Iceland.)

† This is not the only instance of a volcanic eruption forming a temporary island above the sea-level among the Western Islands. It is recorded in the MS. Journals of the Royal Society (a collection containing a mass of curious information respecting the progress of science after the foundation of the Royal Society), that Sir H. Sheres informed a meeting, of January 7th, 1690-91, "That his father, passing by the Western Islands, went on shore on an island that had been newly thrown up by a volcano, but that in a month or less it dissolved, and sunk into the sea, and is now no more to be found.

Fig. 67.

No doubt very many of the supposed banks in the ocean upon which the surf is stated to have been seen breaking, and never afterwards found, may be very imaginary, but it is still possible, that here and there statements of this kind may be founded upon more positive evidence; and that, making all allowance for incorrect views as to the latitude and longitude of the supposed banks, some due to the upraising of volcanic cinders and ashes have been observed, these finally so cut away that the sea no longer broke over them. However this may be, we can scarcely suppose that over the floor of the ocean all the eruptions from every volcanic vent upon it have reached above the surface of the water and remained there as islands, or that some, which have accumulated matter to depths not far beneath the surface waters, may not occasionally so vomit forth cinders and ashes, that these substances remain for a time above water until removed by the influence of breakers.

CHAPTER VII.

CHEMICAL DEPOSITS IN SEAS.—DEPOSITS IN THE CASPIAN AND INLAND SEAS.—CALCAREOUS DEPOSITS.—FORMATION OF OOLITIC ROCKS.—SALTS IN SEA WATER.—CHEMICAL DEPOSITS NOT NECESSARILY HORIZONTAL.

WE have previously adverted to the mixed deposits of calcareous and sedimentary matter in tideless or nearly tideless seas, from which alternate layers of argillaceous limestones and clays, or lines of argillaceous limestone nodules in the latter might result. According to the specific gravities of the waters of such seas, arising from the different amount of matter in solution in them, will, as we have seen, depend the distances over which river waters can flow outwards, supposing such rivers, for illustration, to be equal in volume and velocity, and as respects the amount of matter in solution or mechanically suspended. In this respect, the Caspian, the Black, and the Baltic Seas would all differ, the latter most approaching in the character of its waters to a fresh-water lake. Comparatively, these bodies of water would appear to afford greater tranquillity than tidal seas for the production of chemical deposits, always allowing for the depths to which their waters may be disturbed by surface causes, such as winds and changes in atmospheric temperature.

In tideless seas, such as the Caspian, where the substances brought down in solution by the rivers accumulate in comparatively still water, we should expect deposits which could not be effected with equal facility in the ocean, even in those parts which adjoin coasts. In the one case, evaporation keeps down the body of the water, probably even diminishing its volume during a long lapse of time; while, in the other, these solutions enter the great mass of ocean waters, and become so lost in it, that certain of them may only, under very favourable conditions, be able to accumulate as a coating or bed upon any previously-formed portion of the ocean floor. The way in which the tidal wave thrusts back river waters twice in each day (taking the subject in its generality), mingling the common sea waters with those of rivers, up the estuaries, is alone a marked difference from the outpouring of the rivers, with their contained solutions unmixed until the river waters flow over the sea. Instead of comparative quiet along-shore, except where disturbed by the action of surface waves, the whole body of water along tidal coasts is kept in motion, moving alternately one way or the reverse, and not unfrequently in various directions, in consequence of the modification of the bottom, and the mode in which the tidal wave may strike variously-formed or combined masses of dry land.

We have above called attention to the differences in tideless or nearly tideless seas, arising from differences between the evaporation of their surfaces, and their average supply of water from rivers or rains. Not only should we thence expect the modification of sedimentary deposits previously mentioned, but modifications also in the chemical coatings. An isolated area, like the Caspian, if the evaporation of its waters be greater than its supply, may, during such decrease, present us with conditions favourable to a deposit of some of its salts, while the main mass of the waters may yet be well able to hold much saline matter in solution. Any shallow parts adjoining the shores becoming isolated, and therefore cut off from the river supplies afforded to the main body, may readily be deprived of all their water by evaporation, and a sheet of saline matter be the result. Indeed, in this manner, any substances in solution would become deposited, and how far they might remain exposed without being removed by atmospheric influences, would depend upon the climate of the locality. That any such beds, the result of the evaporation supposed, may be covered by ordinary sedimentary deposits, due to geological changes of the locality, will be obvious.

Around such bodies of water as the Caspian, the observer possesses good opportunities for studying subjects of this kind, which are of considerable interest geologically, when we consider the mode of occurrence of gypsum and rock-salt in many situations, the not unfrequent connexion of these substances, and the kinds of sedimentary matter with which they are often associated. It may be also deserving of attention to consider in such parts of the world the probable annual evaporation of the surface of seas like the Caspian, and the annual supply of waters from rivers and rain.*

* It is interesting to consider, in any given land where such bodies of water may be found, even though of much less size, and where it seems certain, from geological evidence, that the present area occupied by such waters is less than formerly, how far

It may have happened from geological changes, such as might readily convert the Persian Gulf into an isolated sea, by raising the bottom between Cape Mussendom and the opposite coasts at Grou and Sereek, or the Red Sea, into another, by raising the bottom at Bab-el Mandeb, that these masses of water no longer communicated with the main ocean. Looking at the climatal conditions, and the absence of any great drainage from adjoining land flowing into it, the Red Sea would lose its waters from evaporation, while with respect to those of the Persian Gulf, it would depend upon the difference between the evaporation and supply of water chiefly obtained from the Euphrates, Tigris, and their tributaries. From existing information, we should anticipate that this supply would not equal the evaporation, so that both bodies of water might become Caspians.

It would be well if observers, when among such parts of the world, would gather information sufficient to show us the probable results of such alteration of conditions, especially as respects the deposits of substances now in solution in these seas, and their intermixture with common detrital matter. Observations directed to such points can scarcely fail to be valuable with respect to geological theory. Under the supposition of the conversion of the Red Sea into a Caspian, not only might there be a mixture, under favourable conditions, of chemical deposits and detrital accumulations, but coral banks and reefs would be also included in them.

By a glance at a map of Asia, it will be seen that a very large area, extending along 70 degrees of longitude from the Black Sea into China, with a varied breadth of 15 to 20 degrees of latitude, does not drain directly or indirectly into the ocean. There is reason to believe that it is a mass of land which, from geological changes, has been cut off from such drainage, the Caspian, the Sea of Aral, with numerous smaller bodies of water, now receiving such drainage waters as evaporation from the surface of this great area will permit, when gathered together in different positions.

the climatal conditions may so influence the evaporation and supply of water that a kind of balance is established. We may, for illustration, suppose that, in the first place, the climatal conditions are such, after the separation of a mass of sea waters from connexion with the ocean, that a considerable diminution of the volume of the separated water, and consequently, in all probability, of the area occupied by it, takes place. Then will arise the local conditions, under which this diminution may either continue or a balance of evaporation and supply become established. Evaporation, all other things being equal, will depend upon the area of water exposed. If large rivers, such as the Volga, for example, entering the Caspian, bring much sediment into the sea or lake, they tend to make it shallow, and also, by their deltas, to diminish the area, so that the conditions, as to general area, depth, and consequent volume of the water, alter. This alone might destroy any balanced conditions.

The evaporation may completely overpower the supply of water in certain parts of such an area, the salts in solution in the pre-existing waters forming sheets of matter corresponding with the minor areas or lakes when such solutions became in a condition to permit deposits, the least soluble substances being the first thrown down. A deposit of a particular substance once effected, similar matter would be more readily withdrawn from the solution by the attraction of the first deposits of such substance. In a dry climate, such portion of the common detritus, as did not become consolidated, would be swept about hy the winds, forming deserts, such as we find in the region noticed, the great Chinese desert of Kobi, or Shamo, being the largest of them. In all such lands the explorer will not lose his time by carefully examining the shores of these various inland seas and lakes, observing the physical conditions which may produce the isolation of shallow parts. It would he well also to study deposits of saline matter with reference to their origin from conditions, which may have readily obtained, in consequence of geological changes, by the separation of shallow-water indentations fringing the ocean, particularly in warm and dry climates,* as well as by the partial or total evaporation of salt lakes.

Amid the great flats which here and there occur on the shores of tidal seas, and which may become dry at certain times, so that patches of sea-water irregularly scattered over them are evaporated, leaving the salt, we have no doubt conditions, particularly in dry and warm climates, for the accumulation of thin sheets of salt, or other substances in solution, which, under favourable circumstances, might be covered up, and, to a certain extent, be preserved by detrital mud; but these deposits would scarcely have the importance of those previously noticed. At the same time, such situations should be examined with reference to the chemical accumulations which may be thus intermingled with detrital matter.

---

* In all cases, where practicable, it is desirable to obtain information as to the matters in solution in the various inland seas and lakes. They are known to differ in this respect, as might be anticipated. Thus, according to M. Eichwald, the waters of the Caspian contain much sulphate of magnesia, in addition to the other salts held in solution. Those who are possessed of sufficient chemical knowledge, if they have with them any of the little portable chests of the needful substances and apparatus, will have a local means of a qualitative analysis. It would be well if they could perform a quantitative one on the spot, seeing the difficulty of conveying bottles of water, to be kept, perhaps, a long time, and amid high temperatures. When the observer may not be a chemist, he may still assist, under favourable conditions as to transport, by obtaining the waters and putting a sufficient quantity into a clean bottle, immediately sealing it up carefully and tight, and forwarding it, as soon as circumstances may permit, to some experienced chemist for examination.

With respect to deposits from chemical solution, the calcareous may be considered as the most important geologically. We have previously adverted to their production in the air, and in fresh-water lakes. The cases of consolidated beaches on some coasts, like those noticed in Asia Minor, may be regarded as in a great measure due to the evaporation of the water containing the bicarbonate of lime in solution, as it percolates through these beaches. In the same manner, we seem to obtain their consolidation in some places by the oxides of iron and manganese, and by other substances. Respecting the actual formation of beds of limestone in the deeper sea by chemical deposit alone, though we feel assured that it is effected, the exact manner is scarcely yet well determined. The rivers flowing into both tideless and tidal seas alike transport calcareous matter in solution into them, though very variably; in scarcely appreciable proportions in some, abundantly in others. So long as the carbonic acid needful for the solution of the carbonate of lime remains, the latter will continue in the waters, but should it be withdrawn, either by evaporation of the sea waters in shallow places, or by separation in any other way, the carbonate of lime, if the lime be not taken up in any other combination, will be deposited.

With regard to shallow situations in tidal seas, particularly in warm climates, and where pools of water are left for sufficient time at neap tides, we should expect an evaporation of the water, at least in part, and a loss of the carbonic acid, enabling any carbonate of lime present to be held in solution, so that there was a consequent deposit of calcareous matter. This may be well seen where waters highly charged with bicarbonate of lime flow slowly into some nook or bay, on tropical coasts, and even in localities where the rise and fall of tide is small, as, for instance, around Jamaica. It is in such situations, under favourable conditions, that the little grains termed oölites, formed of concentric coatings of calcareous matter, may be sometimes observed to form. A slight to-and-fro motion, produced by gentle ripples of water, may occasionally be seen to keep the carbonate of lime depositing in movement and divided into minute portions, so that instead of a continuous coating of calcareous matter upon any solid substances beneath, a multitude of these little grains is produced. As might readily be anticipated, a small fragment of shell and even a minute crystal of carbonate of lime is sufficient to form a nucleus for the concentric coatings of these oolitic grains. An observer would do well, when an opportunity of this kind may present itself, to watch the mode in which the grains may be mechanically accumulated, like any other grains of matter, by the wash of the sea, or the drift caused by tidal streams, as he will thereby be the better enabled to judge of the differences or resemblances he may find between these accumulations and the beds formed of oolitic grains in the calcareous deposits of various geological ages.

While the mode in which calcareous matter may be deposited on the shores of seas may thus be advantageously studied, that in which it is effected in deep water must necessarily be matter of inference. By the means previously noticed, a large collective amount of carbonate of lime, held in solution by the needful addition of carbonic acid, is discharged by rivers into the sea; more, no doubt, in some localities than in others, but still as a whole, somewhat widely. Although we might expect solutions of a great variety of substances in the sea, the drainage of the land supplying them constantly, our knowledge on this subject would be more advanced than it is at present, if waters were more collected in different parts of the world, and off a variety of coasts, than they have been.

According to Professor Forchhammer, the greatest amount of saline matter in the Atlantic Ocean is found in the tropics far from land, in such places the sea-water containing 3·66 parts of saline matter in 100. He states, that the quantity diminishes in approaching the coasts, on account of the rivers pouring their waters into the sea, and that it also diminishes on the most western part of the Gulf Stream, where the proportion is 3·59 per cent. Professor Forchhammer proceeds to observe, that by the evaporation of the Gulf Stream waters, the quantity of saline matter increases towards the east, and reaches 3·65 per cent., in N. lat. 39° 39′ and W. long. 55° 16′. Thence it decreases slowly towards the N.E.; and at a distance of 60 to 80 miles from the western shores of England, the Atlantic contains 3·57 per cent. of solid substances in solution. The same proportion of salts is found all over the north-eastern part of the Atlantic, as far north as Iceland, at distances from the land not effected by the outflow of rivers.*

* It is desirable that in all researches as to the amount of the saline contents of the ocean, the depth from which waters for examination may be taken, be regarded. With respect to the specific gravity of sea water at different depths, Sir James Ross mentions (Voyage of Discovery and Research in the Southern and Antarctic Regions), that in lat. 39° 16′ S. and long. 177° 2′ W. (there being no bottom at 3,600 feet), the specific gravity at the surface was 1·0274; at 900 feet, 1·0272; and at 2,700 feet, 1·0268, all ascertained at 60° Fahrenheit. He further states that his daily experience gave this diminished kind of specific gravity in the depths. As evaporation would tend to render the surface waters more saline, it may be deserving of attention how far this cause may operate downwards in the sea.

With respect to the chemical character of the saline substances in the waters of the Atlantic, it would appear that they do not differ so much as might be supposed. At the same time, Professor Forchhammer's researches lead him to consider that lime is rather rare around the West India Islands, where myriads of polyps employ it for their solid coral structures; the proportion of lime to chlorine being there as 247 to 10,000, while the same substance is more common in the Kattegat, where part of the lime brought by numerous rivers into the Baltic is carried to the ocean. In the Kattegat the proportion of lime to chlorine is as 371 to 10,000. In the Atlantic Ocean 17 analyses gave 297 to 10,000; and between Faroe and Greenland 18 analyses afforded 300 to 10,000.*

Researches of this kind, limited as they are at present, are still sufficient to point out the modifying influences of proximity to land, of the heat of the tropics, of the melting of ice in the polar regions, and of oceanic currents flowing from one region, where certain conditions prevail, to another where these may be modified.

As geologists, we have to inquire if the salts in solution, and derived by means of rivers from the land, are thrown down on the sea-floor, either within a moderate distance from the land, or further removed in deeper oceanic waters. If we take the calcareous matter, we find that it can be transported, by means of rivers flowing outwards, for various distances over the heavier sea waters, to be still further carried outwards and into greater depths of water, probably, if an ocean current seizes on the river waters thus situated. No small aid would be afforded if, when fitting opportunities presented themselves, waters from the streams which might thus be traversed were carefully examined with reference to their chemical character. In warm climates there might be much evaporation from the upper part of river waters thus slowly passing along the surface of the seas, productive of results, as regards matter in solution, of appreciable value.

When we consult analyses of sea waters, to ascertain the condition in which lime may be present in them, we find enough to show that much is to be learnt by experiments made with the aid which the present methods of analysis can afford. We can readily understand that while lime may be pouring into some parts of the ocean, as a bicarbonate kept in solution by the proper amount of carbonic acid, it might be converted into solid matter by animal life in another, in regions where a balance of supply is not kept up, so

* Forchhammer. Memoirs of the British Association for the Advancement of Science, vol. xv. p. 90.

that eventually very unequal quantities are distributed in solution. But it would be well to ascertain such facts carefully, and especially with reference to the combination in which the lime may be found in the different regions of the ocean.*

With respect to the deposit of carbonate of lime from sea waters, Dr. Lyon Playfair suggests that, as river waters generally contain in solution a small quantity of silicate of potash, the carbonic acid, dissolved in sea water, enabling the carbonate of lime to be therein held in solution, would act on this silicate, decomposing it, and forming a carbonate of potash. The solvent being thus removed from the carbonate of lime, the latter would be precipitated, and a new portion would be formed from the double decomposition of the newly-formed carbonate of potash on the sulphate of lime and chloride of calcium when present. He suggests that this process of decomposition may account for the silica so frequently found in limestones. It is, however, to the action of vegetation, where this can flourish, on sea waters, that Dr. Lyon Playfair attributes a more general deposit of any carbonate of lime from them. He remarks, that marine, like terrestrial plants, constantly require and take away carbonic acid from the waters around them, so that the quantity necessary to keep any carbonate of lime in solution, and which may find its way into the sea waters, being removed, the carbonate of lime is thrown down.

Independently of the soluble matter thrown into the sea by rivers returning to it frequently that which in anterior geological times was accumulated in it, we have to reflect that the volcanic action which we know has been set up upon the ocean-floor, sometimes throwing up matter above the surface of the sea, forming islands, must as a whole have caused no small amount of soluble matter to be vomited forth. Looking at the gases evolved and substances sublimed from sub-aërial volcanos, we should expect many combi-

---

* We are indebted to Schweitzer for a very careful analysis of the waters of the English Channel. No doubt it is only good for the locality, one not favourable for a knowledge of the composition of oceanic waters, being too much shut in by land, from which river waters, differently charged with saline matter, are discharged. His analysis is as follows:—

Water - - - - - -	964·74372
Chloride of Sodium - -	27·05948
,, Potassium - -	0·76552
,, Magnesium - -	3·66658
Bromide of Magnesium - -	0·2929
Sulphate of Magnesia - -	2·29578
,, Lime - - -	1·40662
Carbonate of Lime - - -	0·03301

With, in addition to these constituents, distinct traces of iodine and ammonia.

nations to be formed and decompositions to arise. Seeing also the soundings around certain oceanic and volcanic islands, no slight pressure would have been exerted upon the earlier volcanic action beneath the seas, a modifying influence alone of no slight importance. Surrounded by seas of inferior temperature, closing in upon the volcanic vent as the heated waters rose upwards, there would be a tendency to have certain substances, only soluble at a high temperature, thrown down where the cooling influences could be felt; as also, when these substances may be borne upwards by the heated waters, to have them distributed by any oceanic currents acting over the locality, supposing that the heated waters either rose to, or were produced at distances beneath the surface of the sea where these currents could be felt. Without entering further upon this subject, we would merely desire to point out that, in volcanic regions, the sea may not only receive saline solutions marked by the presence of certain substances not so commonly thrown into it by rivers elsewhere, but that also submarine volcanic action may be effective in producing chemical deposits, either directly, or indirectly, which, under ordinary conditions, would either not be formed, or not so abundantly.*

With regard to the mode in which chemical deposits may be accumulated, it is very needful to consider that horizontality is not essential to them. They may be formed at considerable angles, against any previously-existing surface offering the needful conditions. Numerous deposits from solutions are effected as well on the sides as on the bottoms of vessels containing them.† Hence we may have deposits on the large scale, giving rise to deceptive appearances. Let *a*, for example, in the annexed section (fig. 68)

Fig. 68.

be the surface of a fluid, such as the sea, from which the beds, *b*, have been deposited from chemical solution (limestones for instance)

* It would be very desirable to ascertain points of this kind, so far as examining the sea waters around volcanic regions may enable the observer to do so; and more especially when, by any fortunate chance, opportunities are afforded after any submarine volcanic action may be evident or supposed.

† Pipes conveying waters containing much bicarbonate of lime, or many other substances in solution, are well known to be often coated all round.

upon the pre-existing surface, *c d*, of a stratified rock, *c c*, and it might, if only a portion of such a section was subsequently exposed, be concluded that there had been movements of the land tilting up these beds at *e*, when in reality there has been perfect repose as regards their relative position, since the time of their deposit. Even when, as a whole, somewhat horizontal accumulations of this kind might be expected, they are often found to have moulded themselves upon the irregularities of ground upon which they were thrown down.

CHAPTER VIII.

PRESERVATION OF REMAINS OF EXISTING LIFE AMID MINERAL MATTER.—OF PLANTS AND VEGETABLE MATTER.—BOGS.—DISMAL SWAMP.—RAFTS IN THE MISSISSIPPI.—ANIMAL REMAINS—ON THE LAND.—VERTEBRATA.—OSSIFEROUS CAVERNS AND LAKE DEPOSITS.—INSECTS.—LAND MOLLUSCS.—EFFECTS OF SHOWERS OF VOLCANIC ASHES.—ESTUARY DEPOSITS.—FOOTPRINTS ON MUD.

THIS is a subject of much importance to the geologist desirous of reasoning correctly upon the mode in which the fossiliferous rocks may have been accumulated. The habits of plants and animals engage the attention of the naturalist, and by his aid most important benefits are conferred upon the geologist. He is thus enabled to infer how plants or animals, found existing under certain conditions, may contribute by their remains to the mass of mineral accumulations now taking place, these occasionally even forming thick beds, spread over considerable areas, without the admixture of mud, and sometimes of any sediment derived from the decomposition or mechanical destruction of previously-existing rocks.

The observer should, in the first place, direct his attention to the manner in which the remains of terrestrial life may be entombed. Though when terrestrial plants die, the substances of which they are composed are, as a mass, returned to the atmosphere and soil whence they have been derived, the movements of animals which may feed upon them being regarded as so far local, that keeping to the grounds where their food is presented to them, their droppings restore to the soil what the plants had removed from it, the carnivorous animals which consume the graminivorous, returning that which the latter did not prior to death,—there are still conditions under which parts of existing vegetation may become permanently preserved.

Exposed to atmospheric influences after death, vegetation decays according to the structure of the different plants and the climate of the locality. The rapidity with which decomposition is effected in certain tropical regions is well worthy of attention. We not unfrequently find the outside of a large and prostrate tree retaining its form, and while the whole of the inside is hollow, filled with leaves that have fallen into it, and teems with animal life. This kind of decay is still more instructive when upright stems of plants, in tropical low grounds, liable to floods, retain their outside portions sufficiently long to have their inside hollows partially or wholly filled with leaves and mud or sand, the whole low ground silting up, so that sands, silt, and mud accumulate around these stems, entombing them in upright positions, without tops, though their roots retain their original extension. The study of the sedimentary accumulations of river deltas, amid the rank vegetation of some tropical countries, is very valuable as respects certain deposits in which the remains of vegetation form a conspicuous and important portion. Behind mangrove swamps much that has a geological bearing may be frequently seen; and indeed amid them, the observer not forgetting to direct his attention to the mode in which animal as well as vegetable remains become mingled with, and finally covered over by, sedimentary matter.

Not only in the tropics, but in other regions, large tracts of marsh land, interspersed with shallow lakes, are highly favourable to the accumulation of vegetable substances. The leaves of trees, growing in such situations, falling upon the patches of water, take a horizontal position, spreading in a layer in certain climates and seasons over their surfaces. These leaves gradually soak up water, and sink to the bottom. If, from time to time, flood waters bring fine mineral matter in mechanical suspension into such situations, it settles, and thus the leaves become preserved in thin layers alternating with the clayey sediment. Should it so happen that waters, charged with calcareous matter in solution, find their way either gradually and constantly, or by sudden rushes in floods, we may have the leaves or other remains of plants preserved in a deposit of carbonate of lime, more or less pure, according to the presence of any other matter brought into the lakes in mechanical suspension or chemical solution.

The manner in which bogs are formed should also be studied. Many no longer exhibit their progress over shallow lakes, while others will show it. In the latter case we find aquatic plants, like the large rushes and water lilies, accumulating mud about their roots, as also decaying vegetation, upon which finally the bog plants advance, the chief of which, in our climate, is the *Sphagnum palustre.* As these decay beneath, a new g owth continues above, up to levels where the requisite moisture can be obtained.* Trees are very frequently seen in these bogs (some of which are very extensive), in a manner showing that the conditions favourable for the growth of various trees have from time to time obtained, so that distinct levels of them have been found occasionally in the same bog.

The extent of bogs is very variable, as also the bottoms on which they repose. Sometimes the latter are formed of shell marls, accumulated at the bottoms of the shallow lakes, anterior to the advance of the aquatic vegetation over them. The thickness of bogs necessarily varies: in some 10 to 30 or 40 feet is not uncommon. Of the pauses in the accumulation of bogs, sufficient to permit a growth of trees upon them, as also a surface upon which habitations may be constructed, perhaps as good an example as any is that of the ancient wooden house discovered in June, 1833, in Drumkelin Bog, on the north-east of Donegal. It was 16 feet below the surface of the bog before the upper part was taken off, and 4 feet beneath the cuttings of the time, standing itself upon 15 feet more of bog, so that the total thickness at that place had been 31 feet. The house itself was a square of 12 feet sides, and 9 feet high, and was formed of two floors, the roof constructed with thick planks of oak, the wood employed for the whole dwelling, upon which no iron had been used. Upon clearing away the bog from the level of the house, a paved pathway was discovered extending several yards from it to a hearthstone, covered with ashes, some bushels of half-burned charcoal, some nut-shells, and blocks of wood partly burned. Near the house there were stumps of oak trees, which grew at the time it was inhabited. A layer of sand had been spread over the ground before the erection of the house. All seems to have marked a state of repose in the growth of this part of the bog; so that a change of conditions affecting the drainage would seem needful to account for the accumulation of 16 feet more above the surface, after the time when this wooden house was constructed. It may have been that one of those burstings of parts of a bog,

* Those travelling in North Wales will find, opposite Cwm-y-glo, below the bridge crossing the outlet of Llyn-Padarn (the lower Llanberis lake), a good example of a lake filling up, with the advance of water lilies and other aquatic plants upon a still remaining portion, while bog plants and bog creep on behind them. At the proper season, the locality is brilliant with thousands of water lilies thus advancing. It is easy to see that this was once a third Llanberis lake, but, being shallow, was the first to be nearly filled up.

some of which are recorded, had overwhelmed this locality, soft boggy matter having gradually accumulated to a higher level under favourable circumstances in some place adjacent.

Bogs are very irregularly dispersed, forming unequal patches as to area and thickness. The surface occupied by the bogs of Ireland alone, has been estimated at 2,800,000 acres. From the humic acid in them, animal and vegetable substances are often found well preserved, and, in consequence, numerous relics of ancient times have been handed down to us, which, unless entombed in bogs, would have remained unknown. Other things have evidently been lost in them, and have been brought to light by the progress of the turf-cutter. Many of the beautiful bronze swords, spear-heads, and other ornaments and weapons of its ancient inhabitants, have been thus preserved in Ireland. As might be expected, also, the remains of animals are found which have perished in the bogs.

Of bog-like accumulations in a warm climate, the "Dismal Swamp," as it is called—40 miles long, from north to south, and 25 miles in its greatest breadth, from east to west—partly in the State of Virginia and partly in North Carolina, seems an excellent example. Sir Charles Lyell describes this swamp as "one vast quagmire, soft and muddy, except where the surface is rendered partially firm by a covering of vegetables and their matted roots." * From the nature of the mass, which appears to be chiefly formed of vegetable matter, spongy for the most part, logs and branches of trees intermingled in it, water is so disseminated that the central portions of the swamp are the highest, rising on all sides above the surrounding firm and dry land, except for about 12 or 15 miles on the western side, where rivers flow into it from more elevated ground. The greatest height of the central part above the sides is estimated at about 12 feet, and in such central portion there is a lake, 7 miles long and 5 miles wide. The greatest depth of this lake is 15 feet; the sides are composed of steep banks of the vegetable mass, and the bottom is chiefly formed of the same matter in a highly-comminuted state, with sometimes a white sand, about a foot thick. Rivers flow out of the swamp from all other parts of its margin except that mentioned.

It is a highly-interesting fact as connected with this swamp, one having many geological bearings, pointed out by Sir Charles Lyell, that the surface supports a growth even of trees. He mentions

* Lyell's Travels in North America, vol. i., p. 143.

the juniper trees (*Cupressus thyoides*) as standing firmly in the softest places, supported by their long tap-roots. With other evergreens these trees form a shade, under which grows a multitude of ferns, reeds, and shrubs. The great cedar (*Cupressus disticha*) also flourishes under favourable conditions. Trunks of large and tall trees lie buried in the swamp. They are easily upset by extraordinary winds and covered in the mire, where, with the exception of the sap-wood, they are preserved. Much of this timber is found a foot or two from the surface, and is sawn into planks half under water. Bears inhabit the swamp, climbing the trees in search of acorns from the oaks, and gum berries. There are wild cats also, and occasionally a wolf is seen; so that there must often be conditions for the loss of these animals in the mire, and for the preservation of their bones. Indeed, in such a region as this, occupying an area of several hundred square miles, the amount and mixture of animal and vegetable matter, which may be collected in one great extended sheet, is not a little remarkable.

Rivers, in some regions, carry forward not only the small plants with the leaves and branches of the larger, but multitudes also of trees are thus sometimes transported, part of them retained within the sedimentary deposits of the rivers themselves, part swept out seawards. It is not among the long-cultivated lands that the amount of plants, great and small, carried downwards by rivers, is best observed, though during floods in them large trees are occasionally borne down their courses. It is in regions where man has not by his labours modified the growth of vegetation, or the course of rivers, that the transport of plants by running waters can be well studied. We then have conditions resembling those under which vegetable remains may in this way have been mingled with the sedimentary deposits of previous geological periods. On this account, the courses of rivers, such as those of the Mississippi and its tributaries, are still highly instructive, though in various ways other rivers, pursuing their courses through lands not yet cultivated in any part by man, may be still more so. The *snags* of the Mississippi, or great trees carried away from its banks, or those of its tributaries, and which are anchored, so to speak, by their roots upon the bottom of the stream, their heads bending with its strength, are well-known examples of the partial stoppage of trees on their course downwards. The same river, or rather one of its delta streams, named the Achafalaya, furnishes us with a good instance of a large accumulation of some of these drift trees within the last 80 years. About that time since numbers of these drift trees got entangled in the channel, so that they no longer passed freely down it. Eventually they formed a mass, termed the Raft, distributed irregularly, and rising and falling with the waters, for a distance of twenty miles, closely matted together in some localities. In 1808 the cubic contents of this collection of drifted trees was estimated at 286,784,000 cubic feet.* If by any change of conditions the channel of the Achafalaya became little supplied with water, and the raft consequently fell in the channel and was covered over with fine sediment derived from muddy waters quietly working their way into the old river course, a long line of lignite, corresponding with twenty miles of the old channel of this river, might be the consequence.

When we regard the great rivers of the world, we can scarcely avoid considering that a large amount of plants and trees, differing in kinds and structure according to climates, must be annually entombed, in a manner to prevent that decay they would have suffered if left, after death, solely to atmospheric influences. No doubt much of this vegetation is still decomposed after transport by the rivers to their deltas, yet much also must be entombed in deposits excluding ordinary atmospheric influences, and leaving the plants under conditions favourable for their gradual alteration into lignite, or to the more advanced state of coal, should geological changes so permit. In deltas, also, we have, in the pools and lakes formed by the advance of the sediments thrust forward by the rivers, circumstances in many regions favourable to the growth of aquatic and swamp vegetation. In such situations, as they fill up by the occasional inflow of the muddy waters of the rivers in flood, and by the growth and partial decay of the vegetation, we have also conditions suited to the preservation of some of the plants, or their parts, often in the positions in which they grew, mingled with carbonaceous matter and beds of sediment. It may so happen, in rivers where sands as well as mud are forced forward, that by the occasional shifting of a stream, or the breaking away of a bank, previously barring the entrance of any portion of a main stream, sands may be thrust forward over accumulations of

---

* The 20 miles of length were estimated at 10 miles, this distance being considered as representing a close packing of the trees. The average breadth was taken at 220 yards, and the depth at 8 feet.—(Darby, Geographical Description of the State of Louisiana.) Rafts of this description, but of less size, are, as might be expected, found in other divisions of the Mississippi and its tributaries. Captain Hall (Travels in North America, vol. iii., p. 370) mentions being a witness of one of those falls of the banks of the Missouri, covered with trees, which throw so much drift wood into the Mississippi, the banks of the latter also contributing largely to the general mass.

this kind, their deposit marked by successive lateral and sloping additions, such as have been previously mentioned.

With regard to the preservation of animal remains on dry land, or in fresh water, we have to recollect that the rapacious animals very frequently devour the bones of the vertebrata which they destroy, and that the scavenger animals eat up those which the former may have left unconsumed, so that few bones generally remain exposed on dry land to be decomposed by atmospheric influences. It is very probable that in deserts, the bones of animals which have perished in them may be often buried beneath great sand-drifts, there to remain, perhaps, if decomposing causes be slight in such situations, until geological changes may again bring such deserts beneath waters, and consolidation or removal of the sands be effected, as the case might be. We have seen the bones of rabbits and birds exposed by a shift of some of our coast sand-hills, by which portions of old accumulations, marked by successive growths of vegetation, have been carried off by the winds.

Vertebrate animals are, in some countries, overwhelmed by the fall of parts of mountain sides or cliffs, so as to become buried deeply in situations where their bones are under conditions favourable for preservation. Occasionally, they are destroyed by the partial fall of sea cliffs on tidal coasts, while wandering beneath them when the tide may be out, their harder parts, perhaps, washed out to sea when the breakers may have subsequently removed the fallen mass. Such harder parts may thus become mingled with any sedimentary accumulations then forming, should they not be ground to pieces on the coast by the breakers.

While studying the mode in which the remains of vertebrate animals may be preserved without the aid of streams, pools, or lakes of fresh water, it will be observed that the clefts of rocks, in countries where such occur, are places into which more animals fall than might at first sight be thought probable. In some of our limestone districts, where caverns are found open to the surface, many an animal is lost, notwithstanding the precautions usually taken, so that we are prepared to expect that, in uncultivated regions, animals chased by others, coming suddenly upon the brink of a fissure and unable to clear it at a bound, often get precipitated into it. How far their remains may be preserved will necessarily depend upon circumstances. While even inaccessible to scavenger quadrupeds, many of these fissures are open to scavenger birds who descend and devour the flesh, leaving the bones. Scavenger insects can readily also consume the softer parts. The ultimate preservation of the bones from the decomposing effects of atmospheric influences would depend upon their exclusion from them. The accumulation of clayey matter in the fissures, washed in from the tops or sides during rains, mingled often with fallen portions of rocks, forming the sides of the fissure, will tend to this end. Still better, however, would be their entombment by calcareous stalagmites and stalagtites, where these were formed in the fissures of limestones. In the latter case, we might have an ossiferous limestone breccia rising to the surface irregularly, the width varying with the form of the walls of the original fissure.

Caves, inhabited for a length of time by the same kinds of animals, during which they brought in their prey, so that such parts of themselves or of this prey which may have remained unconsumed accumulated, also afford opportunities for the preservation of vertebrate animal remains, according to circumstances. If these remains, even teeth, continued long under the decomposing conditions likely to obtain in such situations, without some protection afforded by clay in some caves, by stalagmites in limestone caverns, or by numerous fallen fragments, few traces would be expected, while, if these protecting influences existed, such remains might often be preserved.

It is, however, to the aid of water we have to look for the entombment of vertebrate remains in the largest quantities, though, no doubt, the labours of Buckland and others have taught us how much may be preserved in fissures and caverns. We have already noticed the loss of animals in bogs and swamps. In some regions, the collective amount of those which perish in this manner must be considerable. We have reason to believe that many mammals perish in lakes, sometimes sinking into soft ground on their borders, at others while endeavouring to cross them. In the former case they may be preserved, as in bogs and common swamps, in a nearly vertical position, their bones occurring relatively to each other as in life. In the latter, their bones may often be scattered. After decomposition had sufficiently advanced, so that the dead body floated, it may be either drifted to a shallow or deep side of the lake, supposing, for illustration, that both existed. If to the latter, and decomposition had still further advanced, and probably also the scavenger animals, both of the air and water, had consumed no small portion of it, the body might descend into deep water, with the bones still, as a whole, in their relative positions, so that if detrital or chemical deposits were there taking place, they would be in the condition to be so preserved. If drifted and stranded on a shallow part of a lake, the body would be liable to be attacked with facility by scavenger land quadrupeds, which might not have ventured into the water of the deep parts of the lake for this purpose. In many instances, as those who may have seen the dead bodies of animals under such circumstances are aware, the bones would be eventually much scattered, part of them pulled upon the dry land and decomposed, if not eaten, while another part may, under favourable circumstances, again enter the lake, and be there enveloped by deposits in the progress of formation.

Whether land animals floated or not after being drowned in lakes must often depend upon the consumption of their flesh while submerged. The various regions of the world furnish us with different creatures inhabiting such pieces of water. In many warm climates, the bodies would soon be attacked by reptiles, capable of easily destroying their softer parts. In some countries, the crocodilian family would speedily proceed to devour them, and not the less greedily that some decomposition had taken place. By their aid some animals might get dismembered in such a way that the bones became finally much scattered, and the parts of the same animal be somewhat spread among lacustrine deposits. The crocodilians themselves add not a little to the remains of terrestrial vertebrata entombed in lake accumulations, by seizing animals on the shores and dragging them into the water.*

With respect to the remains of aquatic reptiles and fish in lakes, the voracity of many of these creatures is commonly so great, and the system of mutual prey so incessantly kept up among them, that entire skeletons would have to be preserved under very favourable conditions. The deltas of the great rivers, especially those in tropical regions, will afford opportunities for the study of the manner in which the remains of aquatic reptiles may become embedded in detrital matter. We have seen the caiman of Jamaica, when pursued, so bury himself in the mud of the lagoons, in which he delights to live, that occasionally there must be some difficulty of withdrawal from it.

Floods in rivers, particularly those of large size, flowing amid great plains, where the sudden rise of water covers a large area in a short time, concealing the more shallow portions, would appear

* The caiman of the great West India Islands in this way frequently obtains dogs, and sometimes goats, incautiously approaching a place where he may be lurking, perhaps half depressed in mud, with the tip of his snout at the surface of the water. The caiman is considered by the negroes so fond of dogs' flesh, that when a bent mangrove tree, with a running noose, may be placed to catch one, a dog in a stout stockade, in the line traced out for the caiman, is thought one of the best baits.

the means by which many mammals are swept off their feeding-grounds, drowned, and their dead bodies buried amid the detritus borne down at the same time. At such times, also, bones of mammals which remain strewed about in the more exposed situations, not consumed or decomposed, get mingled with the mud, silt, or sands, carried forwards, and finally deposited. To delta accumulations, whether in lakes or seas, such floods must, in certain climates, often bring down terrestrial mammals, mingling their remains with those of many reptiles.

Though, from their powers of flight and consequent escape, we should not expect to find birds caught by floods so as to be carried away, drowned, and, under favourable circumstances, their harder parts entombed, yet, as we do occasionally, though rarely, find the body of a land bird borne down a stream in countries and at times of the year when we have no reason to suppose that it has been shot or otherwise destroyed by man, perhaps we may look to this cause as one, however occasional and rare, by which remains of birds may be preserved. It is in districts where great floods suddenly rise over very extensive flat lands, particularly at times when the young of many birds inhabiting and breeding upon them are unable to fly far or at all, that we anticipate the more frequent surprises of this kind. Land birds occasionally fall into lakes and perish. We have seen instances in which land birds chased by hawks have fallen into lakes. Accidents causing death also now and then happen to the waders frequenting the margins of lakes, as also to birds which live habitually on their waters, either supporting themselves by fishing in the shallow parts, like the swans, or by the aid also of diving, like the duck tribe. The preservation of their bones, once at the bottom, in lacustrine accumulations, would be the same as with other animal remains.

Under all circumstances, perhaps, to floods passing over extensive flats, raising to the surface of the water the bodies of birds which have perished by natural deaths, and which may be capable of floating, or sweeping forwards the bones of others, not yet consumed by scavenger animals, we may look for the chief causes of the transport by water and entombment of the remains of birds in the resulting deposits.*

* Neither should we forget, when considering the manner in which birds' bones may be preserved within the boundaries of land, that they may get entangled among travertines, and thus may be entombed in lines and patches corresponding with such calcareous deposits as they form in streams or pools, as under favourable circumstances in Italy.

In the great deserts of the world, birds, such as ostriches, perishing, their remains

During floods also conditions are very favourable to the sweeping off of numerous insects, even those having the power of flight being caught up in the waters before they could escape. Multitudes of these insects are no doubt consumed by fish, yet the remains of others may readily be so mingled up with the sediment of the flood waters where it can be deposited, as to remain permanently encased by mud, silt, or sand. Seeing the avidity with which, in general, insects cast by myriads, as they sometimes are, on the surface of lakes or pools of water, are devoured by fish, when we discover their remains embedded in calcareous matter, as they have been, we should expect circumstances ill-suited to the habits of insectivorous fish and aquatic reptiles. It may be that in waters in certain pools or lakes charged with large quantities of carbonate of lime in solution by means of the needful carbonic acid, the latter may be so abundant as to drive off the insectivorous fish, and insect-eating aquatic reptiles.

We find the remains of land molluscs mingled with soils in many localities in sufficient abundance to show how capable the shells of these animals are of preservation when circumstances will permit. Though light as regards the absolute weight of each shell, the specific gravity of land shells is considerable, more approaching that of arragonite than of common calcareous spar.* In soils, the shells are ill placed for resisting decomposition beyond a certain amount of time, the waters containing carbonic acid readily percolating to them, so that in such situations they are, if not lately embedded, usually brittle, and not unfrequently broken. Among blown sands land shells are often abundant, some land molluscs especially delighting in such habitats.

In volcanic countries, or those over which, from their proximity to such countries, volcanic ashes may be scattered, and sometimes abundantly, land shells, and, indeed, various other land animals, may be completely covered over with coatings sufficient not only

may be often covered over by great sand drifts, and remain so long beneath, even supposing some change of drift to expose them, as to be no longer available as food to the animals which would otherwise consume them. Some may remain permanently covered, until, as previously mentioned, by a change of geological conditions, these deserts may be again submerged, and their sands be either removed or consolidated into rocks.

* When experimenting some years since upon the specific gravity of shells, we found those of the following land molluscs to be :—

Helix Pomatia	2·82
Bulimus decollatus	2·85
——— undatus	2·85
Auricula bovina	2·84
Helix citrina	2·87

to kill them, but to aid in the preservation of their hard parts. The fall of large quantities of ashes and cinders, discharged in some volcanic eruption, would appear to cause a greater sudden entombment of terrestrial animals, with the probability of preserving their more solid parts entire, than can be obtained without the aid of water, even including the moving sands of deserts. Volcanic districts are, in temperate and tropical regions, often fertile, abounding in vegetable and animal life, so that in regions, such as Sumbawa and Java, for example, land animals, including an abundance of molluscs, may be readily buried beneath discharges of lapilli and ash, such as were vomited forth from the volcano of Tomboro, in Sumbawa, in April, 1815.*

---

* The eruptions commenced on the 5th April, and continued more or less until the 10th, when they became more violent. A Malay prahu was on the 11th, though distant from Sumbawa, enveloped in utter darkness from the ashes in the air. Upon landing afterwards on the island, the commander found the country covered to the depth of three feet by ashes and cinders; and difficulty was experienced in sailing through the cinders floating on the sea. At Macassar, 217 nautical miles from Tomboro, the volcanic discharges were heard to such an extent that, supposing there was an engagement with pirates near at hand, the East India Company's cruiser "Benares," was despatched with troops on board to look after them. The following account, by the commander of the "Benares," obtained by Sir Stamford Raffles, will show the amount of ashes and cinders vomited forth:—

Proceeding south to ascertain the cause of the explosions heard, at 8 o'clock on the morning of the 12th, "the face of the heavens to the southward and westward had assumed a dark aspect, and it was much darker than when the sun rose; as it came nearer it assumed a dusky-red appearance, and spread over every part of the heavens; by ten it was so dark that a ship could hardly be seen a mile distant; by eleven the whole of the heavens was obscured, except a small space towards the horizon to the eastward, the quarter from which the wind came. The ashes now began to fall in showers, and the appearance was altogether truly awful and alarming. By noon the light that remained in the eastern part of the horizon disappeared, and complete darkness covered the face of day. This continued so profound during the remainder of the day that I," continues the commander of the "Benares," "never saw anything to equal it in the darkest night; it was impossible to see the hand when held close to the eyes. The ashes fell without intermission throughout the night, and were so light and subtile that, notwithstanding the precaution of spreading awnings fore and aft as much as possible, they pervaded every part of the ship."

"At six o'clock the next morning it continued as dark as ever, but began to clear about half-past seven, and about eight o'clock objects could be faintly observed on deck. From this time it began to clear very fast. . . . . . The appearance of the ship when daylight returned was most singular; every part being covered with falling matter. It had the appearance of calcined pumice-stone, nearly the colour of wood ashes; it lay in heaps of a foot in depth on many parts of the deck, and several tons of it must have been thrown overboard; for though an impalpable powder or dust when it fell, it was, when compressed, of considerable weight. A pint measure of it weighed twelve ounces and three-quarters; it was perfectly tasteless, and did not affect the eyes with a painful sensation; had a faint smell, but nothing like sulphur; when mixed with water it formed a tenacious mud difficult to be washed off."

Approaching Sumbawa on the 18th, the "Benares" encountered an immense quantity of pumice, mixed with numerous trees and logs with a burnt and shivered appearance. The fall of ashes at Bima, 40 miles from the volcano, was so great as to break in the Resident's house in many places. The Rajah of Saugar described some of the stones which fell there to have been as large as two fists, though not generally

The great eruption of Vesuvius in 79 furnishes us with an excellent example of the manner in which the surface of a country may be covered up by the discharge of volcanic ashes and lapilli, so that various works of art and use are preserved for our instruction. Pompeii not only shows us paintings still remaining on the walls of the houses, but also a great variety of delicate articles, extending to those of the women's dressing-cases. At Herculaneum we have even the writings of the time on papyri, in part still legible. We see an abundance of men's works as they were overwhelmed by the discharge of the ashes and cinders upon them, and often in a condition, after being thus buried beneath mineral matter, permeable to water, for 1800 years, which might not at first be expected. So little general injury seems to have been sustained by the town, even by the shocks of explosions so near, or earthquake movements, that the crushing in of house-tops by means of the weight of ashes and cinders, and the filling up of all corners by the finer dust, appear to have been the chief effects produced. Walking in the street of tombs at Pompeii it seems to require little else than the presence of persons clothed in the costume of the place when overwhelmed by cinders and ashes, to have that street presented to us as it appeared 1800 years since. As showing that not only bones may be preserved under such conditions, but the form of the flesh itself which clothed them, two remarkable instances have occurred at Pompeii, where parts of the human form retained their external shape, the enveloping ash having been sufficiently consolidated, before the decomposition of the fleshy parts. The thickness of the ashes and lapilli which covered up Stabiæ, Pompeii, and Herculaneum in 79, has been estimated as varying from 60 to 112 feet in depth.*

There are few things we can consider more suddenly destructive of terrestrial animal and vegetable life than these great volcanic eruptions, particularly within areas where several feet of lapilli and ashes can be accumulated over a considerable area within a few days. The whole surface previously clothed with vegetation,

---

above the size of walnuts. A great whirlwind is mentioned by the Rajah, "which blew down nearly every house in the village of Sangar, carrying the tops and light parts along with it. In the part of Saugar adjoining Tomboro, its effects were much more violent, tearing up by the roots the largest trees, and carrying them into the air, together with men, houses, cattle, and whatever else came within its influence." Many thousands of lives were lost, and the vegetation of the north and west sides of the peninsula was completely destroyed, with the exception of a high point of land where the village of Tomboro previously stood, and where a few trees still remained.—*Life of Sir Stamford Raffles.*

* Daubeny, "Description of Active and Extinct Volcanoes," 2nd edit., 1848, p. 221.

with a multitude of land molluscs and insects, with many birds and mammals, may be all covered with a thick coating of these volcanic products; many of the molluscs and insects close to the plants on which they may have been feeding. In regions where bogs prevail, large tracts of these vegetable accumulations may be buried, with any birds, insects, or molluscs frequenting them, by a thick layer of ashes and lapilli, the subsequent consolidation of which, by geological causes, might produce the deceptive appearance of a molten rock having flowed over them without producing those effects which would, under the latter supposition, have been anticipated. Indeed, when we have to study the fossil vegetation of some regions, a reference to the conditions under which trees and even bogs may be covered by volcanic ashes is one by no means to be neglected.*

In tideless seas, terrestrial animal and vegetable substances, borne down floating on the rivers, necessarily pass out over the dense waters of the sea to various distances, according to circumstances, and may be transported still further than the force of the river waters have carried them by favouring currents, should there be such, or by winds, the latter capable of driving them about in various directions, should they change. The body of a drowned animal, the decomposition of which is sufficiently advanced to give it the specific gravity capable of floating (and it should be recollected that it would float easier in sea than in fresh water, as regards its own specific gravity), may be thus drifted a considerable distance until eaten, or too much decomposed to float. Small animals may be readily consumed, bones as well as flesh, by the larger voracious fish; but the bones of the larger mammals might, under favouring circumstances, find their way to the bottom, even in deep tideless seas, like parts of the Mediterranean, to be there mingled with the remains of molluscs or other creatures inhabiting the same depths.

The observer has, in like manner, to consider the various land plants and trees which can be carried long distances, sometimes with live creatures still upon them, parts of the latter subsequently, at least those which may escape the voracity of marine animals,

* It is stated that in consequence of the great eruption of Skaptar-jökul in 1783, the atmosphere over Iceland was impregnated with dust for a long time. Traces of this dust were observed in Holland. It is evident that bogs in Iceland may readily become buried beneath volcanic ashes and cinders under such conditions. We may take the great explosion of the Souffrière, in Guadaloupe, in 1812, as an example of the destruction of vegetable and animal life, and of a considerable covering of both in many places in a tropical region. It was during this eruption that ashes were conveyed to Barbadoes by an upper current of wind, opposite to the trade wind.

scattered over various depths of the sea bottom. It will require little attention to see how often the dead shells of land molluscs get thrust out seawards, their modes of floatation at first being such as to keep them above water. The positions necessary for this purpose will depend upon the state of the sea surface at the time. If, notwithstanding the state of weather which may have caused floods in the interior of adjoining lands, lifting off the dead shells from the low grounds in multitudes, the sea be moderately calm, the land shells will be carried on with the river waters, but if there be a breaking sea they soon get upset and sink.

In such situations we have also to regard the mingling of detrital with organic matter, which may be effected by the pushing forward of the sands and gravel on the bottom of the rivers. Many a drowned animal may thus become mixed up with a delta advance, and many a river and land mollusc be included amid a general subaqueous drift. Trees often get entangled and buried on the coast, as well as floated off seaward.

Thus in tideless seas we have the ready means of transporting terrestrial and fluviatile vegetable and animal remains to various distances seaward, some under favourable circumstances, capable of being embedded in marine deposits at various depths, while others are included amid the detrital accumulations formed by the action of the rivers, thrusting out silt, sand, and gravel from the shores, not forgetting any calcareous deposits which may sometimes be added.

In estuaries we obtain a state of things somewhat different. In them a check is afforded at each flood tide, to all borne floating out by rivers, so that when great freshets prevail in the rivers, all caught up by the floods in the interior and floated off low grounds, or borne to the main streams by tributaries, are arrested in their progress. The floating bodies of animals, trees, and smaller plants, are thus not permitted to escape directly seaward, but are lifted by the height of the tide over any low grounds bordering the estuary, these flats, at such times, being more than commonly covered with water. When the ebb tide lowers the waters, the various substances floated over the estuary lowlands not unfrequently remain upon them, more particularly if any wind prevailing at the time forces them on the edges of the flooded lands. There is often a curious mixture of terrestrial, fluviatile, estuary, and more marine animal and vegetable remains, scattered over the estuary flats after such floods, more particularly should it happen, as it sometimes does on the western parts of the British Islands, that a heavy gale, accompanied by much rain, occurs at a time of spring tides, so that the high tides combined with an on-shore wind, rising the sea waters still higher, are met by strong freshets from the land. Under ordinary conditions, fringes of estuary fuci, mingled with land plants, estuary crustaceans and molluscs and land shells, with here and there the remains of some creature, more strictly marine, are familiar to all visiting estuaries.

Although amid the deltas of rivers delivering their waters into tideless seas, among the lagoons formed and the coasts adjoining, there may be variable mixtures of fresh and sea waters, affording proper places for the growth and increase of vegetables and animals fitted for living in brackish water, the conditions are different from those of an estuary. In the one case the waters are stationary, except so far as floods from the interior may force forward an extra amount of fresh water, or a prevailing on-shore wind may drive in a greater volume of sea water; while in the other, large tracts are sometimes bare at one time and covered by water at another, the amount of the saline mixture being variable also, depending on the state of the tide and the volume of fresh water falling for the time into the estuary. And here it is necessary to remark that the observer should not consider as an estuary one of those great indentations of a coast, commonly termed an "arm of the sea," and which is but the consequence of the sea level cutting a previously-formed inequality of the land surface, not unfrequently the prolongation of some valley. No doubt the one kind of coast may sometimes shade into the other, but as regards the kind of life inhabiting estuaries, we should consider brackish water as essential to the latter; at all events to such an extent that at low tide a river, the waters of which become fresh or brackish, should occupy the channel left.

Under the conditions of an estuary silting up in the manner previously noticed, it must necessarily happen that the molluscs and other creatures inhabiting different surfaces, or small depths beneath them, died, such harder parts of them as might be preserved remaining at levels corresponding with such surfaces, here and there mingled according to circumstances, with vegetable and animal remains, drifted as above mentioned. It will he well to examine the manner in which the different parts of an estuary surface may vary at the same time as to the animal life existing upon it, from the creatures inhabiting the little rills of water which only get checked at spring tides, otherwise meandering amid the higher estuary mud or clay-flats, to those in or upon the sands in the more exposed situations, covered by every tide.

The manner in which terrestrial animals may become caught in the softer places should also receive attention, especially where springs, readily finding their way beneath silt and sand, form quaking or quick sands which engulf them, their bones remaining after the flesh has been consumed by the scavenger animals. An observer should by no means neglect the foot-prints of terrestrial animals, nor indeed of any leaving marks or trails, such having lately, and very deservedly, become of geological importance. These foot-prints are often excellently well preserved upon the mud or clay flats, or gently-sloping grounds of estuaries. Very many estuaries around the British Islands afford abundant opportunities for the study of the mixed foot-prints of birds and mammals upon the mud or clay, more especially during the heats of summer, and at neap tides, when extensive surfaces, covered at spring tides, may be bare and exposed to the drying influence of the sun. We have often seen the foot-prints of common gulls, where these birds have been busy around some mollusc, crustacean, or fish drifted on shore, and sufficiently in a fresh state for their food, most beautifully impressed upon clay or mud, hard dried by the sun, the courses of the birds, sometimes single, at others in pairs or more numerous, well preserved. In the same way the tracts of other birds are common, crossed here and there by those of rabbits, hares, stoats, and weazles, and occasionally of dogs. In some localities, after an area of mud or clay, thus trod upon during the difference of time between the spring and the neap tides, has been well dried by the heats of the summer sun, with deep cracks formed from loss of moisture, pieces of the most instructive kind may with care be taken away, further dried and preserved, and even baked into a brick substance, if the composition of the clay be well suited to the purpose. Mingled with these marks we have often also the trails of molluscs, as also those of estuary crustaceans, striving to regain the water, after finding themselves left by the tide.

It might at first be supposed that the rise of the tides over this, for the time, somewhat hard surface, marked by the foot-prints and trails of different animals, would entirely obliterate all traces of them. How far this may be effected will, however, depend upon circumstances. If the rise of the tides from neaps to springs were accompanied by much ripple or waves from winds, it would be anticipated that the fine detritus constituting the mud or clay would, when remoistened, be readily caught up in mechanical suspension, so that all traces of foot-prints and trails would be removed. In all situations where such ripple or waves could be felt this would be expected. All parts of estuaries are, however, rarely exposed to such influences at the same time: many a nook remains tranquil; and in those where the accumulation of detritus is in progress, and films or fine layers of mud succeed each other, if one becomes hardened before another is deposited, a line of separation more or less permanent is usually established between them. We are sometimes able to separate these layers from each other, after careful drying, so that foot-prints are seen upon many surfaces, beneath each other. We have been fortunate in this respect with some portion of sun-dried mud of the Severn estuary, and Sir Charles Lyell has pointed out the manner in which the foot-prints of the sandpiper (*Tringa minuta*) are not only preserved in the red mud of the Bay of Fundy (a locality so favourable from its tides, for the exposure of much ground at the neaps), but also repeated upon the different layers of accumulation.

In some estuaries, long necks of sands and sandhills so, in part, cross their mouths, that bays of still or comparatively still water, occasionally of considerable area, occur behind them, the main streams of tide flowing elsewhere. Let us assume, for illustration, that fig. 50 (p. 55) represents some estuary of this kind, and that, instead of a shingle beach, *d* is a tract of sandhills, perhaps extending several miles in length, then *e* would be the kind of bay noticed, left in comparative quiet, as regards the stream of tide, flowing chiefly on the opposite coast. Much would of course depend upon conditions as to the kind of deposits effected at *e*, but under the supposition that the set of the tides was such as not to cause a sweep of the stream round this bay, it would be favourable for the occasional deposit of the finer sediment or mud borne down the river, *f*, by floods. At the same time it would be exposed to the drift of sand from the sandhills, *d*. In such localities, we have seen the foot-prints of mammals and birds, hardened in the sun, well strewed over by the drift sand from the sandhills; and it should be observed, that the same winds which were powerful enough to disturb the sandhills and cause the drift, would be prevented by the shelter afforded behind the same hills from disturbing the bay waters near the shore, these waters being under the lee of the sandhills, so that even in the shore and shallow waters the sand may be drifted over the mud or clay, filling up the hollows of the foot-prints.

Should the general surface of the land be subsiding gradually, as regards the sea level, it will be obvious that great estuaries may present conditions highly favourable to the preservation of the foot-prints of animals, the actual remains of which, amid the detrital accumulations, may be most rare. Many aquatic birds frequenting estuaries at particular times, often when driven to seek their food in such situations, from tempestuous weather in their more common sea haunts, may thus leave their foot-prints, the conditions for the preservation of whose bones in the estuary deposits themselves would be of the most rare kind, indeed not to be expected, except under the accident of some individual being killed when up the estuary. With the most truly estuary birds, those which build and commonly live on estuary shores, the case might be different. Upon the supposition of a gradual change in the level of the sea, the land descending, we might have sands abundantly thrust forward over clay with foot-prints and trails. A lowering of a mass of sandhills, partly barring the mouth of an estuary, would at once place much arenaceous matter within the transporting influence of the tidal waters, to be drifted over mud flats, formed previously behind them. In some regions the mass of sand, either accumulated as partial and sub-aërial bars, or more gathered together by the sides of estuary mouths, to be again thrown into tides, however eventually other sandhills and tracts might arise (conditions continuing favourable), would be considerable.

That the remains of cetaceans should be found amid estuary accumulations, as also those of numerous fish, some of them more known as purely marine than estuary, will not surprise those who may have seen the porpoises dashing up the estuaries of our coasts in chase of fish which they have driven before them, and their occasional entanglement in shoal waters, when left by a quick-falling tide. Other cetaceans also get sometimes stranded. It is more common to find the chased fish, especially the smaller fry, driven on shore. The birds, no doubt, then pick up the fish abundantly, so that only a minor portion may leave their hard remains for entombment, and doubtless, also, the cetaceans often escape in the pools where they may be caught upon the rise of the tide, but there are still many chances for the preservation of the harder parts of these animals amid estuary accumulations which should not be neglected.

CHAPTER IX.

ORGANIC REMAINS IN MARINE DEPOSITS.—MODIFICATION OF CONDITIONS ON COASTS OF AMERICA.—OF PACIFIC OCEAN.—OF THE INDIAN OCEAN.—OF COASTS OF AFRICA AND EUROPE.—OF ARCTIC SEA.—DISTRIBUTION OF MARINE LIFE.—MODIFICATIONS FROM TEMPERATURE AND PRESSURE.—FROM LIGHT AND SUPPLY OF AIR.—RESEARCHES OF PROF. E. FORBES IN THE ÆGEAN SEA.—ZONES OF DEPTH.—PROF. LÖVEN ON THE MOLLUSCS OF NORWAY.—ZONES OF DEPTH IN THE BRITISH SEAS.—ORGANIC REMAINS DEPOSITED IN THE DEEP OCEAN.—ON COASTS.

It is in connexion with the sea, looking at the evidence afforded us by the various fossiliferous rocks of different geological ages, that we should look for the preservation of the great mass of animal remains amid the detrital and chemical deposits of the time. We have seen that, by means of rivers and winds, various plants and animals, or their parts, may be borne into the sea, and that in estuaries we may have a mixture of terrestrial and marine remains, and of others suited especially to such situations. In respect to estuaries, some so gradually change into arms of the sea, to be seen on the large scale in the Gulf and River of St. Lawrence, and other situations, and equally well in numerous localities of far less area, in various parts of the world, as for instance, in the Bristol Channel and the Severn estuary, that no marked distinctions can be drawn between the one and the other.

Viewing the coasts of the world generally, we not only have to regard all the modifications for the existence of marine animal life, arising from the more or less exposed or sheltered situations of headlands, bays, and other forms of shore, but also the mingling of fresh waters with the sea under the various circumstances connected with the drainage of the land into the sea. Let us consider the modifications of condition for the existence and entombment of marine animal life from Cape Horn to Baffin's Bay. First, there is the difference of climate, producing modifications of no slight order, more especially in moderate depths. From Cape Horn to the

West India Islands, with the exception of the Straits of Magellan, there is an unbroken oceanic coast, subject to the action of the tides, upon which bodies of fresh water are thrown by drainage channels in different places, the chief of which are the Rio de la Plata, the Rio de San Francisco, the Tocantins, the Amazons, and the Orinoco rivers, delivering the portion of rains and melted snows not taken up by the animal and vegetable life, or required for the adjustment of springs or other interior conditions of a large part of South America. After a line of coast little broken by rivers, we find extensive estuary conditions at the mouth of the Plata, and not far beyond Lake Mirim, about 100 miles long, a body of water apparently cut off from the ocean by coast action, and draining into another lake or lagoon, Lago de los Patos, having a channel still open to the main sea, and about 150 miles long, with an extreme breadth of about 50 miles. In these two bodies of water, receiving the drainage of the adjoining land, there are necessarily modifications of the ocean conditions for life, and for the entombment of its remains outside in the main sea. A range of coast succeeds, to which comparatively small rivers discharge themselves, until the San Francisco presents itself, and so on afterwards until the mouths of the Para and Amazons join in forming (including between them the Island of Marajo) great estuary conditions, the tides being felt up the latter river, it is stated, 600 miles, so that there are several in the river at the same time.

The mouths of the Orinoco present us with delta-form accumulations, and then comes the Carribbean Sea influenced by the ponded-back waters of the Gulf of Mexico, so that a kind of tideless sea shades into one where the tides are more felt. More northerly the Gulf stream is seen, transporting warmer waters to colder regions, and skirted by a shore, marked by a line of lagoons for above 200 miles on the coast of Florida, one of them named the Indian river, about 110 miles in length, with an extreme breadth of 6 miles; another, the Mosquito lagoon, being about 60 miles long, with the like extreme breadth. Thence a much-indented shore, on the minor scale, continues until we come to Cape Fear (Carolina), where the lagoon conditions obtain, a kind of barrier, broken by passages termed *inlets*, permitting the ingress and egress of sea waters. In Core, Pamlico, Albemarle, and Currituck Sounds, we find a great body of water of an irregular shape, measuring along the line of barrier separating them, except where broken by inlets from the ocean, about 160 miles in length. Rivers drain into this body of water in various directions, so that estuary conditions obtain in different places, while the great barrier banks, a point of one of which forms Cape Hatteras, place it under a modification of the conditions outside in the main sea. More northward, we obtain the great indentation of Chesapeak Bay, with its minor breaks into the land, the chief of which is the Potomac; and then the Delaware Bay, with its river extending inland, the lagoon coast and its inlets continuing from Cape Charles (north entrance of Chesapeak Bay) towards the Delaware, and from near Cape Mary (Delaware Bay), about 85 miles to the northward. Next follows the mouth of the Hudson, and the modifications arising from the shelter of Long Island up the sound at its back, the lagoon character still apparent on part of its ocean coast. After shores variously indented, we reach the Bay of Fundy with all the modifications due to the great rise of tide (p. 78) at its northern extremities. This is succeeded by the great estuary conditions of the St. Lawrence, and finally the large indentions of Baffin's Bay and Strait and Hudson's Bay and Strait, and all the other channels of the cold regions of North America communicating with the Atlantic Ocean.

It is impossible, when directing our attention to this long line of coast, so variously modified in character, and necessarily so different in climate, not to see how very modified must also be the conditions for the existence of life and the preservation of any of its harder parts. One contemporaneous coating of sedimentary or chemically deposited matter must include the remains of very different creatures, either living upon or in the surface accumulations, as well as the vegetable and animal remains drifted into it from the land. The molluscs inhabiting the coasts of the cold regions would be expected to differ materially from those in the tropics, and the plants and terrestrial animals and amphibious creatures of the latter would vary from those in the former. The organic remains buried in the deposits of the Gulf of Mexico, though entombed at the same time as those in Baffin's Bay, could scarcely be expected to offer the same characters.

If, instead of the eastern coast of America, we look to the western, the first marked difference which presents itself is the absence of great rivers up the whole of the southern Continent and the land connecting it with the wide-spread northern part. Numerous sheltered situations are to be found amid the islands and inlets extending from Cape Horn to, and including the island of, Chiloe; after which, for about 6000 miles of coast, to the Gulf of California, the shores are little broken by indentations, except at Guayaquil and Panama, and do not present a single estuary of importance as on the eastern side of the continent. The mixture of fresh water with the oceans on either side is very different, as are also the conditions for estuary life and the transport of terrestrial and fluviatile organic remains for entombment in the coast sedimentary accumulations. Even after we have passed the Gulf of California, and the Colorado delivering its waters at its head, there is, for about 2000 miles, from Cape S. Lucas to Vancouver's Island, a slightly-indented coast and a minor discharge of drainage waters, with the exception of those delivered by the Columbia or Oregon. Subsequently more northward, for about 800 miles, islands and inlets are common, offering modifications for the existence of marine life, as regards shelter and exposure to waves produced by winds, to Sitka Island and Cross Sound. After which comes the variously-indented coast extending to the Aleutian Islands, and so on to Behring Straits.

Though we have the same range through climates, the character of the two coasts of the American continent varies so materially that we can scarcely but expect very important modifications, as well in the life as in the physical conditions under which it is placed. We have not only to regard the very great difference in the amount of fresh waters discharged on the east and on the west, with its consequences, but also the ponded waters of the Mexican Gulf and their continuation into the Carribbean Sea, with the result, the Gulf Stream, on the one side and not on the other, not neglecting the important difference presented by the great Mediterranean Sea, of Hudson's Bay and Baffin's Bay on the east, and the kind of coast found on the west. To this also should be added the great barrier offered by America to the passage of tropical marine animals from one ocean to the other.*

It may be useful to glance at the great modification of conditions on the western side of the Pacific. Though a great portion of the drainage of Asia is disposed of in other directions, the surplus waters of a large area still find their way to the east coast. The

* According to M. Alcide d'Orbigny, of 362 species of molluscs in the Atlantic and Great Oceans, there is only one common to both, *Siphonaria Lessoni*. Of these 362 species, omitting the last, 156 belong to the Atlantic, and 205 to the Great Ocean. He also remarks that, if the two sides of the American continent be compared, the proportion, in the Atlantic, of gasteropod to lamellibranchiate mollusca, is 85 to 71, while in the Pacific it is 129 to 76. Of 95 genera considered to be proper to the shores of South America, 45 only are common to the two seas. This M. D'Orbigny attributes to the steep slopes of the west side, the Cordilleras rising near the coast, and rocks being more numerous than sandy shores, so that gasteropods would be expected to be more common, while the Atlantic coasts present mud, silt, and sand in great abundance, with gently-sloping shores for a large proportion of their length.—*Recherches sur les lois qui Président à la Distribution des Mollusques Côtiers Marins*. *Comptes Rendues*, vol. xix. (Nov. 1844). *Ann. des Sciences Naturelles*, Third Series, vol. iii., p. 193 (1845).

Saghalian river throws its waters, derived from a considerable area, behind the island of the same name, to be driven into the Okhotsk Sea on the north, or the Japan Sea on the south, as the case may be. Both these seas are, to a certain extent, separated from the main ocean by the range of islands, composed of the Kourile and Japanese islands, extending from Kamschatka to Corea, the Japan Sea especially, from the great mass of island land interposing between it and the Pacific, offering the character of a Mediterranean Sea.

Proceeding southerly we arrive at the Yellow Sea, which receives the abundant drainage effected by the Hoang Ho and its tributaries, and more southerly still we find the body of fresh water discharged into the sea by the Yang-tse-kiang. Thence, to the south, until the Si-kiang with its tributaries presents itself in the Canton estuary, comparatively minor rivers flow into the ocean, the coast being much indented, smaller rivers and streams often discharging in the upper parts of the indentations.

The Island of Hainan, with the great promontory stretching to meet it from the main Chinese land, forms the Gulf of Tonquin, into which the San-koi and other rivers discharge their waters. The amount of fresh water poured into the sea on the eastern coast of Cochin China is subsequently of no great importance, and it is not until we arrive at the delta of the Maikiang or Camboja that the sea is much influenced hy the influx of fresh waters, an influence again, however, to be repeated at the head of the Gulf of Siam, by the outpouring of the Meinam, a river remarkably parallel with the Maikiang for about 700 miles, the latter holding a singularly straight course, as a whole, to the N.N.E., for about 1750 miles.* The remaining portion of the Asiatic continent, formed by the Malayan promontory, throws no important body of fresh waters into the sea in the form of a main river.

From Kamschatka nearly to the equator we thus have a continental barrier, for the most part not wanting in the outflow of bodies of fresh water, sufficient to produce marked influences on parts of the coasts, and consequently upon the conditions under which animal life may exist along it, and the remains of terrestrial and fluviatile plants and animals be drifted outwards into any sedimentary or chemical deposits now forming adjoining it. Minor parts of the ocean are also, to a certain extent, separated off by islands, the range of the Philippines and Borneo, in addition to those mentioned, tending to portion off the ocean down to the

* Considering the inference to be correct, as it appears to be, that the Latchou is the upper part of the Maikiang.

equator, so that, as a whole, a marked modification of physical conditions is observable on the east and west coasts of the Pacific Ocean.

From the equator southward we have no longer a mass of unbroken land on the west to compare with the continuous continent of America on the east. A barrier to the free passage of tropical animal life, supposing other conditions equal, is not presented on the west. Although much land rises above the surface of the sea, the mass of Australia not so very materially of less area than that of Europe, and Borneo and New Guinea exposing no inconsiderable surfaces, there are channels of water amid them permitting tropical marine creatures to extend themselves under fitting circumstances. Though, with the exception of Australia, the various islands may not offer areas sufficient for the accumulation and discharge of fresh waters equal in one locality to some of the great rivers of the world, collectively they embody conditions for the outflow of much fresh water around many of them, so that estuary and brackish water conditions obtain, and consequently physical circumstances fitted for the modification of life. So far as the eastern coast of Australia is concerned, it presents about 2000 miles of shore not more broken or affording more fresh water than the opposite coast of South America. The western part of the Pacific differs from the eastern portion in the multitude of points and small areas through which the floor of the ocean reaches the atmosphere, productive of a combination of influences affecting animal life and the accumulation of its harder remains.

While on this subject, it may be well to call attention to the material changes which would be effected if, by any of those alterations of the level of sea and land which the study of geology teaches may be reckoned by differences very far exceeding the depths required, channels of communication were established between the Atlantic and Pacific Oceans by a sufficient subsidence of the Isthmus of Panama, or the communication cut off between the Pacific and Indian oceans by an uprise of the land and sea bottom between Australia and the Malayan Peninsula, one stretching through Timor, Floris, Java, and Sumatra. If the multitude of oceanic islands in the Western Pacific did not too much break up currents, we may suppose a certain amount of ponding up of waters inside the Moluccas, Borneo, and the Philippine Islands somewhat resembling that now effected behind the West India Islands, with perhaps also a modification of the Gulf Stream, escaping along the coast of China. Startling as, at first sight, such changes may appear, the geological student has to accustom himself to consider modifications in the distribution of land and water, and elevations and depressions of a far more extended kind when he comes to reason upon facts connected with the accumulation and distribution of mineral and organic matter constituting rocks, formed at various geological periods.

In the Indian Ocean we have shores confined to the tropical and temperate regions. For nearly 2000 miles the coast of Australia, from Cape Leeuwin to Cape Bougainville, presents us with no known great river pouring out a volume of water sufficient to influence an extended area. The same with the island range of Timor, Floris, Java, and Sumatra, and up the Malay Peninsula, to the head of the Gulf of Martaban, where the Irawaddy thrusts out its delta and discharges a volume of fresh water, the drainage of a large area. From thence to the mouths of the Ganges no important amount of fresh water is carried out into the sea. The great volume thrown into the sea by this river has been already mentinned (p. 85). Hence to Cape Comorin we find rivers of varied magnitude, the most important of which are, proceeding southwards, the Mahanuddy, Godavery, Kistna, and Coleroon, draining, with minor streams, the great area of Southern India. As a whole, the Bengal Sea and Martaban Gulf receive a considerable quantity of fresh water, the discharge of which conveys a mass of detritus into the sea, and produces conditions in the waters and the sea bottom, which, beyond Cape Comorin, are not found for about 1000 miles, until we reach the Gulf of Cambay, into which the Nerbudda and other rivers discharge themselves. We find another volume of fresh water thrown into the sea by the Indus, still more northerly, after which we obtain the moderate outflow of fresh water of the coast of Beloochistan, the great indentation of the Gulf of Oman, and its continuation the Persian Gulf, the nearly-dry coast of Arabia, to the Arabian Gulf and its long-continued indentation, the Red Sea. From Cape Guardafui to the Cape of Good Hope, for about 4400 miles, the sea seems little influenced by any considerable discharge of fresh water on the coast, excepting in such places as at the mouths of the Zambesi and two or three other localities.

Looking at the Indian Ocean as a whole, any influences upon marine animal life from fresh waters poured into the sea, with the greater amount of terrestrial and fluviatile plants and animals drifted into the ocean by rivers, would be chiefly found in the Bengal Sea (including the Martaban Gulf) and upon the north-east shores of the Arabian Sea, with one or two places on the east coast of Africa. Excepting Madagascar and Ceylon, the area occupied by islands is inconsiderable. The coasts bounding it on the east are those chiefly of considerable islands (the mass of Australia better deserving the name of a continent), so that in the tropical regions there is a free communication by means of sea channels with the Pacific. On the west, Africa bars all direct communication with the Atlantic, though at the same time the region terminated by the Cape of Good Hope and Cape Agulhas, trends southward, so comparatively little southward of the tropics, and currents (p. 93) so set from the Indian Ocean, round Cape Agulhas and up the south-western coast of Africa, that there is no great land boundary between tropical marine life in the one ocean and the other.* The Indian Ocean is now cut off from marine communication with northern regions (however this may have been effected in former geological times, even as late as the tertiary period, by means of waters uniting the Red and Mediterranean Seas), while it is well open to all marine life which may enter it, under fitting conditions, from the south. Herein it differs from the Atlantic and Pacific Oceans, which range from the Northern to the Southern Polar regions.

In the run of the African coast which bounds the Atlantic for so long a distance on the east, fresh waters flowing outwards through great drainage channels seem chiefly to occur at the Orange River, the Nourse, the Coanza, and the Congo, or Zaire, on the south of the equator, and at the Quorra, Gambia, and Senegal, on the north. The coast northward of the Senegal bounds for about 1000 miles the Atlantic on the one side, and the great African Desert on the other. From the Desert to Cape Spartel minor streams only fall into the sea. The great indentation of the Mediterranean then succeeds.

The European rivers discharged into the Atlantic, or the tidal seas and channels communicating with it, are inconsiderable streams as compared with the great rivers of the world; indeed a large portion of the European drainage finds its way into the Mediterranean, Black, Caspian, Baltic, and Arctic Seas. Such drainage as falls into the Caspian is evaporated in that sea, and that not so treated in the Black Sea is evaporated in the Mediterranean; with all which directly finds its way into the latter. So that from the

* Due regard has, however, to be paid to the temperature of the current, considered to be that of the mean of the ocean, which flows for some distance up the west coast of Africa, from the Cape of Good Hope, as also to that stated to run from the south end of Africa some way up the eastern coast.

Baltic alone the drainage waters of Europe find their way into the Atlantic, in addition to those which flow directly into it, or the tidal channels and seas communicating with it. Enough, however, escapes in this way to give a varied character to the coast conditions, as regards the mingling of fresh with sea waters, under which aquatic life may be found and, in part, entombed, and the remains of terrestrial and fluviatile plants and animals be also accumulated.

In the Arctic Ocean, the coasts present us with much mingling of fresh water and sea, the drainage of a large portion of Asia and of a minor portion of Europe falling into it; part of the fresh water discharged into great indentations or arms of the sea, such as the White Sea and the Gulfs of Obi, Ieniseïsk, Khatangskii, and Kolima; part through deltas, as the Petchora and Lena; and part in a more ordinary form. The fresh water so supplied to the coasts of these regions is interrupted or lessened during many months of the year by the climate; much of it being arrested in the form of ice, to be let loose in the warmer months. The ice, also, in the seas of these high latitudes, necessarily modifies the coast conditions for life as it exists in the temperate and tropical shores of the world. The drainage delivered into the same ocean from North America is less important than from Europe and Asia. Of the North American rivers flowing into this ocean, the Mackenzie would appear the most important, succeeded by the Back and Slave Rivers. The land and sea are so mingled on the north coast of America, and the ice and snows so abundant, that the shore waters become much influenced thereby.

Looking to the Southern Ocean, we find the ice and snow of the Antarctic land most important, as regards the shore conditions. A great barrier of ice, indeed, there occupies the position of the coast for a great extent, so that both in the Arctic and Antarctic regions we have to regard ice accumulated round the land, or formed in the sea, as most materially influencing the existence of marine life and the preservation of its remains amid sedimentary and chemical deposits.

In such regions, also, we see the extension of marine life (vegetable and animal), and of air-breathing creatures (birds and mammals) feeding upon it beyond the range of terrestrial vegetation, and of animals directly consuming it or the creatures which first feed upon it.

Though such is the general fact, the conditions for the entombment of the remains of terrestrial animal and vegetable life in the

Arctic and Antarctic regions are, as respects the present distribution of land and sea, different. In the former, we have the delivery of important rivers into the sea, an abundance of water being discharged during the warm season when the ice is broken up at their mouths, and the interior ice and snows are melting. The Obi and its tributaries alone drain a large Asiatic area, extending from lat. 47° to 67°. The Jeniseï, rising from the Tangnou and Little Altai Mountains, likewise flows through 20° of latitude to 70° N., while the Lena and its tributaries, considered to drain 785,565 square (English) miles, rises (in lat. 57°) from the Jablonnoï or Stannovoï Mountains (the eastern portion of which looks upon the sea of Okhotsk), delivering itself into the Arctic Ocean, in about lat. 73° 38′ N. Other rivers, also, flow northerly for considerable distances from the south, such as the Dvina, Petchora, Khatanga, Anabara, Olia, Olenek, Iana, and Kolima. In Northern America, also, the rivers, though not numerous, flowing northerly, still show a drainage extending to the south for several degrees of latitude, though much interrupted by lakes.* Thus the Mackenzie, delivering itself into the Arctic Ocean in about 69° N., flows from the Slave Lake by an outlet in about 61° N., giving 8 degrees of latitude for this course, during which the river receives the drainage from the Great Bear Lake. Regarding the Slave Lake as a mere interruption, by which the waters are spread over a wider space in a depression, the waters discharging themselves by the Mackenzie are derived from a drainage extending over a considerable area (estimated at about 510,000 square miles), and reaching down to lat. 52° 30′ N., by means of the Slave River (running out of the western end of Athabasca Lake), and the Athabasca (flowing into the same lake also at its western extremity).

In the northern parts of Europe and Asia, 3,000,000 square miles of which have been estimated as draining into the Arctic Ocean, and in some portion of North America, there are, therefore, conditions, particularly during the floods caused by the melting of the ice and snows, for thrusting forward the remains of terrestrial and fresh-water life into the northern seas, there to be mingled with detritus, upon the transport and accumulation of which ice has an important influence. We should expect that amid the intermixed land and sea, terrestrial animals often perish while crossing among the ice, at times when the latter is breaking up in the channels

* Collectively the lakes of North America constitute a marked feature in the physical geography of that part of the world. The volume of water in the chief lakes has been estimated at 11,300 cubic miles.

and gulfs, so that their bones are, under favourable conditions, preserved in any sedimentary matter accumulating beneath. No such conditions prevail in the southern continent, which navigators have lately made known to us. No great rivers there flow outwards, and neither terrestrial plants nor animals, directly or indirectly living upon them, furnish their remains for mixture with any sedimentary deposits which may be forming. All aid which great river drainages afford to the latter is cut off, and the little detritus that can be obtained from the land seems only capable of being so derived directly in the few localities exposed to the breakers during the short period of the year when the shores are not bounded entirely by ice. For the finer matter, not ice-borne, entombing the remains of life, we may probably look, as affording the chief supply, to ashes and lapilli vomited from volcanos, and scattered over adjacent seas.

Enough has been stated to show the unequal conditions as to climate and the mingling of fresh with sea waters along coasts. The observer has next to consider the varied character of the shores themselves as regards the shallowness or depth of the adjacent seas, and the modifications of temperature, pressure, access of light, and conditions of intermixed air thence arising. It has been above seen (p. 96), that the volume of ocean is so arranged as to the specific gravities of its waters, that an equal temperature, considered to be 39° 5′, reigns in the sufficiently deep parts from pole to pole, water of higher temperature rising above these more dense waters in tropical regions, and of a lower temperature towards the poles. Though this even temperature may prevail at the proper depths, it is necessarily modified as the seas become shallow, or currents may transport warmer or colder waters, as the case may be, from one oceanic area to another. As the coasts are approached in the parts of the world where warmer waters float above those of 39° 5′, we have conditions under which the temperature decreases downwards below the level of the sea to 7200 feet, and upwards in the air to the greatest heights of land. Viewing the subject generally, therefore, and as far as temperature is concerned, marine animals which could support a decrease of temperature equal to about 39° 5′ (the surface temperature being taken at 78°), could live from the level of the sea to the greatest depths in the equatorial ocean.* A higher temperature may be found under favourable conditions of shallow water and small tides in some

* It will be at once obvious that this difference of temperature is easily sustained by many land animals in different parts of the world.

tropical regions. In the polar areas, included within the belts of equal temperatures from the surface to the bottom of the sea, and within which colder waters, as a whole, float above those of 39° 5′, there is a different state of things. Still regarding the subject merely with respect to temperature, the animals capable of living in the tropical regions, and unable to support lower temperatures than 39° 5′, could not occupy the higher waters.

While in the equatorial parts of the world the temperature of the ocean may not be very materially altered on its surface, it is different with those portions of its higher waters exposed to the changes of winter and summer, so that the temperature of the surface waters is there more considerably modified, especially upon coasts. The animals living in the shallow waters of such regions are, therefore, liable to an amount of difference in temperature not experienced by those inhabiting the seas of the tropics. This is more particularly the case on the shores of tidal seas, with their estuaries, where, even at high water, large tracts of coasts may only be covered by shallow waters, becoming dry at low tide.

As regards mere temperature, it will be apparent that a vast volume of the southern ocean might be tenanted by similar life, extended over its floor at any depths from about 7200 feet at the equator, 3600 in lat. 45° S., the surface in 54° to 58° S., and 4500 feet in lat. 70° S.; and, probably, under the needful modifications, considering the different distribution of land and water on the south and on the north, in a similar manner towards the northern regions. Modifications, also, arising in the various seas communicating with the main ocean, and more or less separated from it, such as the Mediterranean, in the western part of which the waters beneath 200 fathoms have been supposed to remain at about a temperature of 55°,* must also be borne in mind.

With differences of depth, the observer has to consider the differences of pressure to which any animal or vegetable living in the sea would have to be subjected, so that such life would be very differently circumstanced, though under equal temperatures, at the depth of 7200 feet at the equator, and in the shallow waters of the

* M. Berard found, at a depth of 1200 fathoms (without reaching bottom), between the Balearic Islands, a temperature of 53°·4, the surface water being at 69°·8, and the air at 75°·2. From other observations in the western part of the Mediterranean, at the respective depths of 600 and 750 fathoms, and another not stated, it was found that the water was still at 55°·4, though the temperature of the surface water varied materially. M. D'Urville remarks that these experiments accord with some made by himself, also in the Mediterranean, at 300, 200, 250, 600, and 300 fathoms, when he obtained the respective temperatures of 54°·5, 54°·1, 57°·3, 54°·6, and 54°·8 —*Geological Manual*, 3rd edit., p. 25.

oceanic regions where that of 39° 5′ rises to the surface. We cannot suppose an animal so constructed as to sustain a pressure of more than 200 atmospheres at one time, and of 2 or 3 atmospheres at another. A creature inhabiting a depth of 100 feet would sustain a pressure, including that of the atmosphere, of about 60 pounds to the square inch, while one at 4000 feet, no very important depth, would have to support a pressure of about 1830 pounds to the square inch.

Animals, among other conditions for their existence, are adapted to a given pressure, or certain ranges of pressure, so adjusted that they can move freely in the medium, either gaseous or aqueous, in which they live. All their delicate vessels, and the powers of their muscles are adjusted to it. When the pressure becomes either too little or too great, the creature perishes; and, therefore, when acting freely in such a medium as the sea, an animal will not readily quit the depths in which it experiences ease. All are aware of the adjustment of an abundance of fish to the depths, to or from which they may frequently descend, by means of the apparatus of swimming bladders. This arrangement, however, only changes their specific gravities as a whole, the relative volume occupied by the air or gases in the swimming bladders, being the chief cause of difference, though, no doubt, also the squeezing process at great depths would diminish the volume of such other parts of their bodies, as were in any manner compressible, the reverse happening with a rise from deep waters to near the surface. So adjusted to given depths do these swimming bladders appear for each kind of fish, that it has been observed that the gas or air in the swimming bladders of fish brought up from a depth of about 3300 feet (under a pressure of about 100 atmospheres), increased so considerably in volume, as to force the swimming bladder, stomach, and other adjoining parts outside the throat in a balloon-formed mass.*

* Pouillet, Elémens de Physique Expérimentale, tom. i., p. 188. Seconde Edition.

As regards the pressure and different depths in the sea, Dr. Buckland mentions (Bridgewater Treatise, vol. i., p. 345) that "Captain Smyth, R.N., found, on two trials, that the cylindrical copper air-tube, under the vane attached to Massey's patent log, collapsed, and was crushed quite flat, under the pressure of about 300 fathoms (1800 feet). A claret bottle, filled with air, and well corked, was burst before it descended 400 fathoms (2400 feet). He also found that a bottle filled with fresh water, and corked, had the cork forced in at about 180 fathoms (1080 feet). In such cases the fluid sent down was replaced by salt water, and the cork which had been forced in was sometimes reversed." Dr. Buckland adds that Sir Francis Beaufort had informed him that he had frequently sunk corked bottles in the sea more than 600 feet deep, some of them empty, others containing some fluid. "The empty bottles were sometimes crushed, at others the cork was forced in, and the fluid exchanged for sea water. The cork was always returned to the neck of the bottle, sometimes, but not always, in an inverted position." Dr. Scoresby (Arctic Regions, vol. ii., p. 193) gives an account

While thus some kinds of marine animals have the power to adjust their specific gravities to the medium in which they may be placed, some molluscs, such as the nautilus, possessing it, others appear unable, under ordinary conditions, to raise themselves much above the sea bottom. It will be evident that the more their component parts are incompressible, and the fluids in them agree with the specific gravity of the sea in which they live (and the specific gravity of the sea does not appear to vary from any increase of saline matters in it to great depths, though water being slightly compressible, it will become more dense according to depth), the less they would experience the difficulties of a change of depth. On the contrary, the more any parts may be compressible, and air or gaseous matter be included in their bodies, the less would they suffer changes of depth with impunity.

That light should have its effect upon marine as upon terrestrial vegetation we should expect, the light of day being important as well to one as the other, viewing the subject as a whole. It would evidently, also, be important to all marine creatures possessing the organs of vision, so that we should anticipate that the great mass of fish, crustaceans and molluscs which possessed eyes, would occupy situations and levels in the sea where they could obtain the light needful to them. The *Pomatomus Telescopium*, caught at considerable depths in the Mediterranean (near Nice), is considered to afford an example of adjustment to the minor amount of light reaching its ordinary abode, its eyes being remarkable for their magnitude, and apparently constructed to take advantage of all the rays which can penetrate the depths at which it lives. While, however, light may be absolutely needed for the existence of some marine life, it is not obviously necessary to others, those not possessing eyes. Many marine creatures seem to flourish under conditions in which it can be of little value, at the same time the influence of light may often be of importance where it is not suspected.

It is not improbable that to the power of obtaining a proper amount of disseminated atmospheric air in waters, we may look for a very important element in the existence of animal and, indeed, of vegetable life in the sea. To the one and the other oxygen seems essential. At the junction of the sea and atmosphere, we have the best conditions for the absorption of the air by water, the agitation of the surface from the friction of the atmosphere on the sea, particularly during heavy gales of wind, being especially

of a boat pulled down to a considerable depth by a whale, after which the wood became too heavy to float, the sea water having forced itself into its pores.

favourable for a mechanical mixture of the two, assisting the absorption of the air.* Of the amount of air, or rather of the apparently needful element, oxygen, at various depths in the sea, we seem to possess no very definite information, so that researches on this head are very desirable. From observations by M. Biot, on the gaseous contents of the swimming-bladders of fish, it has been inferred that such contents probably vary according to the depths at which such fish live. He found these swimming bladders nearly filled with pure nitrogen when they were those of fish inhabiting shallow waters, and with oxygen and nitrogen, in the proportion of $\frac{9}{10}$ of the former to $\frac{1}{10}$ of the latter, when those of fish living at depths of from 3000 to 3500 feet.

According to M. Aimé, the amount of air disseminated in the waters of the Mediterranean, opposite Algiers, is nearly constant from the surface to the depth of 5250 feet.†

We might assume that, from its immediate contact with the air, surface waters would more readily obtain any needful dissemination of it than those situated at greater depths, so that the mode of consuming oxygen would be adjusted to such conditions, animal life inhabiting great depths being so formed as to require it at considerable intervals. In tidal seas we find certain molluscs adjusted to live in situations exposed to the atmosphere during the fall of every tide, while others inhabit places always covered by sea, except, perhaps, at equinoxial spring tides. From inhabiting shores some molluscs are commonly considered as littoral species, while others are well known as rarely obtained except in deep waters.

Although general views may have been some time entertained with respect to the modification of marine life, depending upon the temperature, pressure, light, and ability to procure oxygen under which it may be placed,‡ it could scarcely be said that we had sufficient data for the philosophical consideration of this subject until the labours of Professor Edward Forbes in the British and Ægean Seas supplied the necessary information.

Professor E. Forbes pointed out that with regard to primary

* The friction of air upon fresh-water lakes produces the same result, intermingling the air and water. Cascades and waterfalls intermix them in many rivers, those especially in which fish swimming high, or inhabiting minor depths, most flourish.

† Comptes Rendus de l'Académie des Sciences, 1843, vol. xvi., p. 743.

‡ The author entered somewhat at length on this subject, in 1834, in his Researches in Theoretical Geology, chapters xi., xii., and xiii. To this work a table was appended by Mr. Broderip, containing all the information then known respecting the depths and kind of bottom at which recent genera of marine and estuary shells had been observed.

influences, the climate of the Eastern Mediterranean was uniform, and that the absence of certain species in the Ægean Sea, characteristic of the Western Mediterranean, was rather due to a modification in the composition of the sea water, from the impouring of the less saline waters of the Black Sea, than to climate.* The influx of river water produces its consequences; and it is remarked that, among 46 species of testacea collected on the shore at Alexandria, there were 4 land and fresh-water molluscs, 3 of which are of truly subtropical forms,† so that while in one part of the Mediterranean forms of this character are mingled with the ordinary marine testacea, in another, as at Smyrna or Toulon, the *Melanopsis* is mixed with them near the former, and characteristic European pulmonifera near the latter. It is also shown by Professor E. Forbes, that while vegetables of a subtropical character may be borne down by the Nile, into the Mediterranean on the one side, accompanying the remains of crocodiles and ichneumons, the Danube may transport parts of the vegetation of the Austrian Alps, with the relics of marmots and mountain salamanders, the marine remains mingled with these contemporaneous deposits retaining a common character.

With respect to modifications in conditions arising from depth, Professor E. Forbes divides the Eastern Mediterranean into eight regions, each considered to be characterised by its fauna, and also by its plants, where they exist. Certain species were found confined to one region, and several were ascertained not to range into the next above, whilst they extended into that beneath. "Certain species," he adds, "have their maximum of development in each zone, being most prolific of individuals in that zone in which is their maximum, and of which they may be regarded as especially characteristic. Mingled with the true natives of every zone are stragglers, owing their presence to the action of the secondary influences which modify distribution. Every zone has also a more or less general mineral character, the sea bottom not being equally variable in each, and becoming more and more uniform as we descend. The deeper zones are greatest in extent; so that whilst the

* He attributes to this cause the dwarfish character of the molluscs, with few exceptions, when compared with their analogues in the Western Mediterranean. "This is seen most remarkably in some of the more abundant species, such as *Pecten opercularis*, *Venerupis irus*, *Venus fasciata*, *Cardita trapezia*, *Modiola barbata*, and the various kinds of *Bulla*, *Rissoa*, *Fusus*, and *Pleurotoma*, all of which seemed as if they were but miniature representations of their more western brethren. To the same cause may probably be attributed the paucity of *Medusæ*, and of corals and corallines. Sponges only seem to gain by it."—Report of the British Association, vol. xii., p. 152, (Meeting of 1843).

† *Ampullaria ovata*, *Paludina unicolor*, and *Cyrena orientalis*.

first, or most superficial, is but 12, the eighth, or lowest, is above 700 feet in perpendicular range."*

While tracing the first region or littoral zone, which is thus limited to 12 feet, all the modifications arising from kind of bottom, rock, sand, or mud, are shown to have their influences, and the effects of wave action, bringing up the exuviæ of animals inhabiting the next region beneath, are pointed out. The second region is estimated at 48 feet, ranging from 2 to 10 fathoms; the third at 60 feet, between the levels of 10 to 20 fathoms; the fourth at 90 feet (20 to 35 fathoms); the fifth at 120 feet (35 to 55 fathoms); the sixth at 144 feet (55 to 79 fathoms); the seventh at 150 feet (80 to 105 fathoms); and the eighth, all explored below 105 fathoms, amounting to 750 feet, more than twice the depth of all the other regions taken together, the total depth amounting to 1380 feet.

So complete are the modifications in invertebrate life, produced by the conditions in these various zones or regions, that only two species of molluscs were found common to the whole eight—viz., *Arca lactea* and *Cerithium lima*, "the former a true native from first to last, the latter probably only a straggler in the lowest." Three species were found common to seven regions;† nine to six regions:‡ and seventeen to five regions.§ With regard to geographic distribution and vertical range in depth, Professor E. Forbes remarks that those species which possess the one exhibit the other, more than one-half of those having an extensive range in depth, extending to distant localities, in nearly every case to the British seas, some still further north, and some in the Atlantic, far south of the Straits of Gibraltar. He concludes "that the extent

* British Association Reports, vol. xii., p. 154.

† *Nucula margaritacea*, *Marginella clandestina*, and *Dentalium 9-costatum*; the second considered to have possibly dropped into the lower zones from floating sea-weeds.

‡ *Corbula nucleus*.
Neæra cuspidata.
Pandora obtusa.
Venus apicalis.
Turritella 3-plicata.
Triforis adversum.
Columbella linnæi.
Cardita trapezia.
Modiola barbata.

§ *Neæra costellata*.
Tellina pulchella.
Venus ovata.
Cardita squamosa.
Arca tetragona.
Pecten polymorpha.
—— *hyalinus*.
—— *varius*.
Crania ringens.
Natica pulchella.
Rissoa ventricosa.
—— *cimicoides*.
—— *reticulata*.
Trochus exiguus.
Columbella rustica.
Conus mediterraneus.
Terebratula detruncata.

of the range of a species in depth is correspondent with its geographical distribution."*

As regards the influence of light, Professor E. Forbes presents us with facts connected with the molluscs and other animals, deserving much attention and extended research, due allowances being made for the modifications produced, as he points out, and to be attributed in many cases to an abundance of nullipores, and to a beautiful pea-green sea-weed, *Caulerpa prolifera*. The majority of shells in the lower zone were found to be white, or, when tinted, of a rose colour, few exhibiting any other hues. In the higher zones, the shells, in a great many instances, exhibited bright combinations of colour. The animals also of the testacea and radiata, in the higher zones, were much more brilliantly coloured than iu the lower, where they are usually white, whatever the colour of the shell may be.†

The researches of Professor E. Forbes have led him not only to attach great value to the bottom in or on which marine animals may live (and it will be obvious that creatures whose habits may be suited to mud would find themselves ill at ease upon rocky ground alone), but also to point out the effects produced by the accumulation of the harder parts of successive generations of marine animals

* Reports of British Association, vol. xii., p. 171.

"If," observes the Professor, "we inquire into the species of Mollusca which are common to four out of the eight Ægean regions in depth, we find that there are 38 such, 21 of which are either British or Biscayan, and 2 are doubtfully British; whilst of the remaining 15, 6 are distinctly represented by corresponding species in the north. Thus among the Testacea having the widest range in depth, one-third are Celtic or northern forms: whilst out of the remainder of Ægean Testacea, those ranging through less than four regions, only a little above a fifth are common to the British seas. One half of the Celtic forms in the Ægean, which are not common to four or more zones in depth, are among the cosmopolitan Testacea, inhabiting the uppermost part of the littoral zone."

† Professor E. Forbes adds, "In the seventh region, white species (of Testacea) are also very abundant, though by no means forming a proportion so great as in the eighth. Brownish-red, the prevalent hue of the Brachiopoda, also gives a character of colour to the fauna of this zone; the crustacea found in it are red. In the sixth zone, the colours become brighter, reds and yellows prevailing, generally, however, uniformly colouring the shell. In the fifth region, many species are banded or clouded with various combinations of colours, and the number of white species has greatly diminished. In the fourth, purple hues are frequent, and contrasts of colour common. In the second and third, green and blue tints are met with, sometimes very vivid, but the gayest combinations of colour are seen in the littoral zone, as well as the most brilliant whites."

Respecting the colour of the animals of Testacea, the genus *Trochus* is selected as "an example of a group of forms mostly presenting the most brilliant hues both of shell and animal; but whilst the animals of such species as inhabit the littoral zone are gaily chequered with many vivid hues, those of the greater depth, though their shells are almost as brightly coloured as the covering of their allies nearer the surface, have their animals for the most part of an uniform yellow or reddish hue or else entirely white." Reports Brit. Assoc., vol. xii., p. 173.

upon the same bottom, thus, in fact, altering its condition, so that they may die out from this increase.* He considers that until the old conditions be restored by a new accumulation of detrital matter different from that presented by the animal exuviæ, the same animals would not find the kind of bottom suited to them; and the geological bearing of this view is shown to be illustrated by the bands or layers of fossils so frequently found interstratified with common sedimentary matter.† In conclusion, Professor E. Forbes adverts to the evidences of the existing fauna of the Ægean which would be presented if its bottom were elevated into dry land, or the sea filled up by sedimentary deposits. While the remains of some animals would afford the needful evidence of their existence, and occur under circumstances whence the probable depths at which they lived might be inferred: of other animals, very abundant in the present seas, no trace would be found.‡

While Professor E. Forbes was thus investigating the conditions under which marine life existed in the Eastern Mediterranean, it fortunately so happened that Professor Löven was engaged in researches leading to general and similar conclusions respecting the modifications in marine life on the coast of Norway. Though both localities are so far similar that the shores are for the most part rocky, and deep water to be often obtained near the coast, they differ as to climate, and as to the sea being tideless in the Eastern

* He illustrates this point by beds of scallops (*Pecten opercularis*), or of oysters, which, when considerably increased, give rise to a change of ground, by the accumulation of their shells, so that the race dies out, and the shelly bottom becomes covered over by sedimentary matter.—*Edinburgh New Phil. Journal*, vol. xxxvi., p. 324.

† In his paper on the light thrown on Geology by Submarine Researches, being the substance of a communication made to the Royal Institution of Great Britain, on the 23rd February, 1844 (Edinburgh New Phil. Journal, vol. xxxvi., p. 318, 1844), Professor E. Forbes, while remarking on all varieties of sea bottom not being equally capable of sustaining animal and vegetable life, observes, "In all the zones of depth, there are occasionally more or less desert tracks, usually of sand or mud. The few animals which frequent such tracks are mostly soft and unpreservable. In some muddy and sandy districts, however, worms are very numerous; and to such places many fishes resort for food. The scarcity of remains of testacea in sandstones, the tracks of worms on ripple-marked sandstones, which have evidently been deposited in a shallow sea, and the fish remains often found in such rocks, are explained in a great measure by these facts."

‡ The following are the inferences on this head, inferences extremely valuable respecting the animal life existing at different geological periods:—

"1. Of the higher animals, the marine Vertebrata, the remains would be scanty and widely scattered."

"2. Of the highest tribe of Mollusca, the Cephalopoda, which though poor in species is rich in individuals, there would be but few traces, saving of the Sepia, the shell of which would be found in the sandy strata forming parts of the coast lines of the elevated sea-bed."

"3. Of the Nudibranchous Mollusca there would not, in all probability, be a trace to assure us of their having been; and thus, though we have every reason to suppose

Mediterranean and oceanic off Norway. While adverting to the modifications of life at different depths, Professor Löven attributes much of the character of the submarine life off the coasts of Norway

---

from analogy that those beautiful and highly-characteristic animals lived in the tertiary periods of the earth's history, if not in older ages, as well as now, there is not the slightest remain to tell of their former existence."

"4. Of the Pteropoda and Nucleobranchiata, the shell-less tribes would be equally lost with the Nudibranchiata, whilst of the shelled species we should find their remains in immense quantity, characteristic of the soft chalky deposits derived from the lowest of our regions of depth."

"5. The Brachiopoda we should find in deeply-buried beds of nullipore and gravel, and from their abundance we could at once predict the depth in which those beds were formed."

"6. The Lamellibranchiate mollusca we should find most abundant in the soft clays and muds, in such deposits generally presenting both valves in their natural positions, whilst such species as live on gravelly and open bottoms would be found mostly in the state of single valves."

"7. The testaceous Gasteropoda would be found in all formations, but more abundant in gravelly than in muddy deposits. In any inferences we might wish to draw regarding the northern or southern character of the fauna, or on the climate under which it existed, whether from univalves or bivalves, our conclusions would vary according to the depth in which the particular stratum examined was found, and on the class of mollusca which prevailed in the locality explored."

"8. The Chitons would be found only in the state of single valves, and probably but rarely, for such species as are abundant, living among disjointed masses of rock and rolled pebbles, which would afterwards go to form conglomerate, would in all probability be destroyed, as would also be the case with the greater number of sublittoral Mollusca."

"9. The *Mollusca tunicata* would disappear altogether, though now forming an important link between the Mediterranean and more northern seas."

"10. Of the Arachnodermatous Radiata, there would not be found a trace, unless the membranous skeleton of the *Velella* should, under some peculiarly favourable circumstances, be preserved in sand."

"11. Of the Echinodermata, certain species of *Echinus* would be found entire; species of *Cidaris*, on account of the depth at which that animal lives, would not be unfrequent in certain strata, as the region in which it is found bounds the great lowermost region of chalky mud; the spines would be found occasionally in that deposit, far removed from the bodies to which they belonged. Starfishes, saving such as live on mud or sand, would be only evidenced by the occasional preservation of their ossicula. Of the extent of their distribution and number of species no correct idea could be formed. Of the numerous *Holothuriadæ* and *Sipunculidæ*, it is to be feared there would be no traces. The single Crinoidal animal would be easily preserved entire, but its ossicula and cup-like base would be found in the more shelly deposits."

"12. Of the Zoophyta, the corneous species might leave impressions resembling those of Graptolites in the shales formed from the dark muds on which they live. The *Corals* would be few, but perhaps plentiful in the shelly beds, mostly, however, fragmentary. The *Cladocora cæspitosa*, where present, would infallibly mark the bounds of the sea, and, from the size of its masses, might be preserved in conglomerates where the testacea would have perished. The *Actinia* would have disappeared altogether."

"13. Of the sponges, traces might be found of the more silicious species when buried under favourable circumstances."

"14. The Articulata, except the shelled annelides, would be for the most part in a fragmentary state."

"15. Foraminifera would be found in all deposits, their minuteness being their protection; but they would occur most abundantly in the highest and lowest beds, distinct species being characteristic of each."

to variations in the sea-bottom, always, however, making allowances for the depth,* thus agreeing with the general views of Professor Forbes.

While marine life is thus found adjusted to different zones of depth on the ocean shores of Norway and the east part of the Mediterranean, always carefully considering the local and physical conditions, it becomes the more interesting to have direct evidence of the adjustments which may be effected on the great and gentle slopes bounding some coasts, such as those so important on the eastern coasts of America. Respecting these great detrital fringes off coasts, among which may be classed, though very small, comparatively, the shallow seas around the British Islands, the area of

---

"16. Tracts would be found almost entirely deficient in fossils, some, such as the mud of the Gulf of Smyrna, containing but few and scattered; whilst similar muds in other localities would abound in organic contents. On sandy deposits, formed at any considerable depth, they would be very scarce and often altogether absent. Fossiliferous strata would generally alternate with such as contain few or no organic remains. Whilst at present the littoral zone presents the greatest number and variety of animal and vegetable inhabitants, including those most characteristic of the Mediterranean Sea, when upheaved and consolidated, their remains would probably be imperfect as compared with those of the natives of deeper regions, in consequence of the vicissitudes to which they are exposed, and the rocky and conglomeratic strata in which the greater number would be embedded. A great part of the conglomerates and sandstones found would present no traces of animal life, which would be most abundant in the shales and calcareous consolidated muds."—Prof. E. Forbes' Reports, Brit. Association, vol. xii., p. 176.

* Professor Löven observes, "As to the regions, the littoral and laminarian are very well defined everywhere, and their characteristic species do not spread very far out of them. The same is the case with the floridеous Algæ, which is most developed nearer to the open sea. But it is not so with the regions from 15 to 100 fathoms (90 to 600 feet). Here there is at the same time the greatest number of species, and the greatest variety of their local assemblages; and it appears to me, that their distribution is regulated, not only by depths, currents, &c., but by the nature of the bottom itself, the mixture of clay, mud, pebbles, &c. Thus, for instance, the many species of Amphidesma, Nucula, Natica, Eulima, Dentalium, &c., which are characteristic of a certain muddy ground at 15 to 20 fathoms, are found together at 80 to 100 fathoms. Hence it appears, that the species in this region have generally a wider vertical range than the littoral, laminarian, and perhaps as great as the deep-sea coral. The last-named region is with us characterised, in the south, by *Oculina ramea* and Terebratula, and in the north, by Astrophyton, Cidaris, *Spatangus purpureus* of an immense size, all living, besides Gorgoniæ and the gigantic *Alcyonium arboreum*, which continues as far down as any fisherman's line can be sunk. As to the point where animal life ceases, it must be somewhere; but with us it is unknown. As the vegetation ceases, at a line far above the deepest regions of animal life, of course the zoophagous mollusca are altogether predominant in these parts, while the phytophagous are more peculiar to the upper regions. The observation of Professor E. Forbes, that British species are found in the Mediterranean, but only at greater depths, corresponds exactly with what has occurred to me. In Bohuslan (between Gottenburg and Norway), we found, at 80 fathoms, species which, in Finmark (on the north), may be readily collected at 20, and on the last-named coast, some species even ascend into the littoral region, which, with us here on the south, keep within 10 to 11 fathoms."—*On the Bathymetrical distribution of submarine life on the northern shores of Scandinavia.*—British Association Reports, Notices, and Abstracts, vol. xiii., p. 50.

which inside depths, not exceeding 600 feet, will be seen by reference to fig. 65 (p. 91), we should anticipate disturbing conditions much affecting the distribution of some portion of the marine life upon them. With regard to a knowledge of the distribution of marine life in the British seas, we are indebted to the researches of Professor E. Forbes, commenced anterior to those undertaken in the Ægean Sea.* It was while prosecuting these researches that he ascertained the value in these investigations of the power of mollusca to migrate.† He has pointed out that they do so in their larva state, ceasing "to exist at a certain period of metamorphosis, if they do not meet with favourable conditions for their development, *i. e.*, if they do not reach the particular zone of depth in which they are adapted to live as perfect animals."‡

Professor E. Forbes divides the British seas into four zones of depth: 1, the Littoral; 2, the Laminarian; 3, the Coralline; and, 4, the Coral.§ The littoral zone lies between high and low water mark, varying in extent according to the rise and fall of tide, and the shallowness or depth of the shore. "Throughout Europe, wherever it consists of *rock*, it is characterized zoologically, by species of *Littorina*; botanically, by *Corallina*; where *sandy*, by the presence of certain species of *Cardium*, *Tellina*, and *Solen*; where *gravelly*, by *Mytilus*; where *muddy*, by *Lutraria* and *Pullastra*." The littoral is divisible into minor zones.‖ The Laminarian zone is the belt commencing at low-water mark, and extending to the depth of 7 to 15 fathoms (42 to 90 feet). Algæ are common, and numerous animals inhabit the forests composed of them. "Among

---

* The first notice of them was published in the Edinburgh Academic Annual for 1840.

† In 1840, he gave a summary of seven years' observations at a particular season of the year.—*Annals of Natural History*, vol. iv.

‡ Edinburgh New Phil. Journal, vol. xxxvi., p. 325, 1844. Speaking of the manner in which the larvæ and eggs may be transported, it is observed that "if they (the larvæ) reach the region and ground of which the perfect animal is a member, then they develop and flourish; but if the period of their development arrives before they have reached their destination, they perish, and their fragile shells sink into the depths of the sea. Millions and millions must thus perish, and every handful of the fine mud brought up from the eighth zone depth in the Mediterranean, is literally filled with hundreds of these curious exuviæ of the larvæ of mollusca."

§ These zones, originally pointed out in 1840, are considered to have been established by subsequent researches (Memoirs of the Geological Survey of Great Britain, vol. i., p. 371, 1846). Professor Forbes remarks that the two first regions had been previously noticed by Lamouroux, in his account of the vertical distribution of sea weeds; by Audouin and Milne Edwards in their observations on the natural history of the coast of France; and by Sars, in the preface to his Bagtivelser og Jagtivelser.

‖ A table of the characteristic animals and plants, of four sub-zones, is given in Professor Forbes' Memoir on the Geological Relations of the existing Fauna and Flora of the British Isles.—*Memoirs of the Geological Survey of Great Britain*, vol. i., p. 373.

the mollusca, the genera *Lacuna* and *Rissoa*, the *Patella pellucida* and *lævis*, *Pullastra perforans* and *vulgaris*, and various *Modiolæ* are especially characteristic of this zone, and numerous zoophytes and *Radiata*, especially *Echinus sphæra*, *Tubularia*, *Actinea senilis*, though ranging both higher and lower, are more prolific here than in any of the other regions." Lastly comes the *Nullipora*, bounding the marine vegetation in depth, and rarely ranging down to more than 120 feet in our seas.*

The region of corallines is so termed from the greatest abundance of corneous zoophytes, which appear to take the place of plants, being found in it. The carnivorous mollusca are abundant, species of *Fusus*, *Pleurotoma*, and *Buccinum* are common, and many species of *Trochus* are found; *Naticæ*, *Fissurellæ*, *Emarginulæ*, *Velutinæ*, *Capulus*, *Eulimæ*, and *Chemnitziæ* are abundant; and among bivalves, *Artemis*, *Venus*, *Astarte*, *Pecten*, *Lima*, *Arca*, and *Nucula*. "Numerous and peculiar *Radiata*, including the largest and most remarkable species, abound, and for number, variety, and interest of the forms of animal life in the British seas, this region transcends all others."† This zone extends from about 90 to about 300 feet, its greatest development being between 150 and 210 feet.

The fourth region is that of deep-sea corals, and is local. The greater part of the area of the British seas does not attain the depth at which this zone commences. Professor E. Forbes considers this region as hitherto but very partially explored. "As far as we know," he observes, "it is well characterized by the abundance of the stronger corals, the presence in quantity of species of the *Dentalium*-like genus of *Annelides*, called *Ditrupa*, by a few peculiar *Mollusca*, and by peculiar *Echinodermata*, as *Astrophyton* and *Cidaris*, and *Amorphozoa*, as *Tethya cranium*. All our British *Brachiopoda* inhabit this zone, and probably range throughout it."‡

* Professor E. Forbes points out, that the Nullipora likewise bounds marine vegetable life in the Mediterranean, where it descends to 420 and 480 feet. With respect to the depths to which marine vegetable life extends, he remarks, that it does so further than is commonly supposed, stating that in the Eastern Mediterranean, *Codium flabelliforme* is found at 30 fathoms, *Microdictyon* at 30 fathoms, *Rityphlæa tinctorea* at 50 fathoms, *Chrysymenia uvaria* at 50 fathoms, *Dictyomenia volubilis* at 50 fathoms, *Constantinea reniformis* at 50 fathoms, and *Nullipora polymorpha* at 95 fathoms, (570 feet).

† Forbes, Mem. Geol. Survey of Great Britain, vol. i., p. 374.

‡ Professor E. Forbes remarks respecting the Brachiopoda, that when found, in certain localities, in more shallow water among the corallines, there are reasons for believing that their occurrence there may be explained by geological changes affecting the conditions of the sea bottom.

[We

The advance thus made will be sufficient to stimulate other observers, so that at no very distant period a valuable mass of evidence may be anticipated.* Probably the general views, based on the local investigations above noticed, may be found capable of extensive application. However this may be, it can scarcely but happen that an accumulation of additional data would most materially aid the progress of the geological inferences to be deduced from the mode of occurrence of organic remains in rocks.

With respect to the littoral zone, that most influenced by climate, while in tideless seas or those where tides are of little consequence, the marine animals inhabiting it are under conditions of slight change, as regards the vertical rise and fall of water; in tidal seas the case is different. In tidal seas many littoral molluscs are exposed to atmospheric influences for different periods, those near high-water mark the longest; so that while the latter may remain uncovered by water six or eight hours at a time, those nearer low-water mark may be so for only an hour or two, some merely for a short time at spring tides. Neap tides also leave a belt surrounding land, the higher part of which is only covered by water for a few days at a time, and then only at spring tides. It thus happens that while in the tropics the littoral zone may not be under very material changes of temperature during the year, this condition, looking at the subject as a whole, gradually shades off on either side towards the polar regions, where the water becomes solid along shore for part of the year, and the coasts are often only partially clear in the summer, portions being still

We would refer for a valuable view of the characteristic plants and animals inhabiting the four zones into which the area of the British seas has been divided, to the table given by Professor E. Forbes, in his Memoir on the Geological Relations of the existing Fauna and Flora of the British Isles, Memoirs of the Geological Survey of Great Britain, vol. i., p. 375.

* When we recollect that under favourable circumstances the officers of our Navy and of our Merchant Service, may render great assistance to this inquiry, when properly conducted, it is to be hoped that we may eventually obtain, through their exertions alone, more extended facts connected with the subject. Under their care the dredge might often be applied with advantage; and if specimens of the animals obtained were stowed away safely, properly ticketed, for the examination of competent naturalists, far more would be known in the next half-century touching the distribution of marine life, particularly at depths where surface waves could not so act as to drive its remains on shore, than could be accomplished by naturalists alone, however zealous.

During the surveying voyage of H.M.S. Rattlesnake, commanded by the late Captain Owen Stanley, R.N., on the coast of Australia and New Guinea, numerous valuable observations were made upon the distribution of marine animals in depth, and an account of the zones of life, in the regions explored, is contained in the "Narrative" of the voyage by Mr. Macgillivray.

subject to the occasional pressure of fices and masses of ice. In certain intermediate regions all animals and plants inhabiting the distance between high and low water mark, with its modifications according to the state of the tide, must be adjusted to sustain the extremes of a long range of temperature, in order to support the atmospheric changes to which they are exposed. The differences of temperature observable round the British Islands, notwithstanding the advantage of their position, are sufficiently considerable to produce a movement among many marine animals, as is well known, so that certain of them are only seen close in shore, among the pools left by the tide, in the warmer season. Others again appear organized to sustain a considerable change of temperature. We have seen the common limpet (*Patella vulgaris*) apparently doing well on our coasts, at temperatures of 92° (close to the rock), and of 24°, a range of 68°. As far, therefore, as temperature is concerned, such a mollusc could live in the ocean waters, and at any depths, in all parts, except in the higher portions of the sea during the winter months in the icy regions. Its organization is no doubt adjusted to a littoral life, and to changes of temperature, as part of the littoral conditions in such climates as that of the British Islands, but the amount of change which it can in this manner support, may make us careful at giving too much importance to temperature alone in the distribution of marine animal life. Once beneath a moderate depth of sea, the mass of water is less acted upon by atmospheric influences, and the adjustment to the specific gravities of water at different temperatures is such as to produce much uniformity in the temperature of the deeper zones, and minor modifications in those above them; in the warmer regions tending to keep the sea temperature beneath that of the atmosphere, and in the colder to raise that of the water above it. As, therefore, the sea level is approached, so as a whole must the animals inhabiting the higher zones be adjusted to support changes in the temperature of the sea in those regions where the summer differs materially from the winter as regards heat.

Quitting coast conditions, and regarding the ocean beyond the depths of 200 or 300 fathoms, we have a large area, on the bottom of which we have no reason to suppose any vegetation exists, seeing that observations on coasts would lead us to conclude that the needful conditions for its growth terminate at comparatively very minor depths. All phytophagous marine creatures would not be expected beyond their ability to obtain the food fitted for them, while the carnivorous animals have necessarily the power to extend vertically and horizontally, far beyond the growth of marine vegetation, however this vegetation may support the mass of life upon which the carnivorous animals have, as a beginning, to feed. In the region of the Sargasso weed, we have an example of a floating vegetation, tending to support animal life, and forming the abode of multitudes of marine creatures in the open sea. This, however, is an exception to the general fact of the absence of marine vegetation in the open ocean, except so far as stray portions of sea-weed, borne by currents from coasts, may be concerned.

In the oceanic depths there exist, apparently, conditions under which some portions of the remains of the fish, crustaceans and molluscs, to be found on the surface above, may be preserved. Although much may be consumed and continued in the mass of life so inhabiting the surface, from time to time some part of the harder portions of animals may descend to the bottom, assuming that the specific gravity of such remains be such as to permit their fall through the water.* Shells of the *Ianthina communis*, having a specific gravity of 2·66, and of the *Nautilus umbilicatus* with that of 2·64, would, after the fleshy matter of these molluscs was decomposed or consumed, and no air entangled in the interior of the shells, be capable of descending to any depths which we may consider at all probable in the ocean, supposing its saline contents not to materially differ in depth, and the compressibility of sea water such as experiments upon fresh water would lead us to infer. We may thus have the remains of marine animals scattered over the bottom of the ocean floor, in certain localities especially, as also those of stray animals drifted from coasts, attached to seaweeds or pieces of wood, both of which decomposing, the harder portions of these animals may fall to the bottom at great depths. It can scarcely be supposed that all the logs of wood bored by the Teredo, or covered over by the common barnacle, *Anatifa striata*, are drifted on shore, and that they do not often become so decomposed as to permit the descent of the harder parts of these animals to the bottom. Indeed we might anticipate a somewhat singular

* With respect to the compressibility of the ocean waters; according to Poisson (Nouvelle Théorie de l'Action Capillaire, p. 277) it would require a pressure equal to 1100 atmospheres to reduce water six-hundredths of its volume. In the experiments of MM. Colladon and Sturm, water not deprived of air, was compressed equal to 47·86 millionths for each atmosphere, and deprived of air, equal to 49·65 millionths. The experiments of M. Oersted gave a compression of 46·65 millionths for each atmosphere. Water containing salts in solution was found, as might be expected, somewhat less compressible.

mixture of the harder parts of some marine animals in different parts of the ocean, especially in the vicinity of islands rising out of considerable depths, such, for example, as near Hawaii, Mani, and other islands of the Sandwich group.

Returning to the coast, we find with the animal life the vegetation on which it feeds, from that exposed to the atmosphere at low water, on tidal shores, to that only known by dredging and fishing. Those accustomed to examine the rocky shores of tidal seas well know how much sea-weed may be cast on shore in the little bays and creeks, or be drifted to the larger bays, during and after heavy gales of wind, producing breakers on such coasts, and which tear up the marine plants, especially towards low-water mark, where during calmer times they may have been abundantly produced. A sandy bay beyond a long line of steep rocky coast, the latter exposed to some heavy gale during the rise and fall of several successive tides, so that sea-weeds, detached by the breakers from it, are driven by wind and tide into the bay, will be often seen hy the observer to have its beach covered in various places with matted masses of these plants. Frequently, as might be expected from the forces employed, these lines of sea-weed are cast up bigb on the beach, beyond the reach of calmer seas to float them off. They will there be disposed of according to the climate. In warm countries, or in the summer months of the temperate regions, they soon decompose, and their remains, not borne off in a gaseous form, become intermingled with the beach. An observer, by studying the sections of sandy beaches exposed by rills or small streams of water, may occasionally find irregular layers of black carbonaceous matter, the result of the decomposition of masses of sea-weeds cast on shore, intermingled with the ordinary sand, and in some localities, parts of a shingle beach will be seen with an abundance of intermixed sea-weed in a decomposed or decomposing state. He may also find the light matter of decomposed sea-weeds borne to deeper waters in sheltered situations, its entombment in such places depending upon the disturbance to which it may be subsequently exposed, and the amount of ordinary sedimentary substances which may collect permanently over it. In some localities much mud, black with carbonaceous matter thus derived, may be accumulated.

Molluscs, living among the sea-weeds thus detached and cast on shore, are occasionally observed to be entangled amid the plants, their harder parts remaining intermingled with the sands or shingles after the decomposition of the plants, so that the shells of rock-frequenting molluscs become embedded with those of others living in and upon sands. The little *Patella pellucida* is very commonly thrown on shore on our coasts, adhering to the cavity it has made for itself at the root of some large fucus, and which, indeed, has weakened the power of the plant to keep its place, when acted on by the sea in heavy gales. It is also very common to find drifted marine creatures of other kinds entangled in these masses of detached sea-weeds; on some coasts the remains of crustaceans being abundant.

With regard to steep coasts, vertical or nearly vertical cliffs plunging suddenly into deep water, it may happen that molluscs, feeding upon marine plants growing at various depths, and themselves inhabiting different depths, according to their kinds, get knocked off by the sea. While those uninjured may again recover their positions, a few perish, and their shells be preserved in sand, silt, or mud, with the remains of other molluscs living on such bottoms; so that the remains of littoral, shallow, and deep-water molluscs become preserved together in the same deposit. Molluscs as they die must have their shells washed away by the sea on such coasts, and thrown into deep waters. Some account has also to be taken of birds picking the animals out of shells which they may have obtained upon the rocks at low tide, or have brought from adjacent bays where they may have been cast alive or recently killed on shore. We have seen the common oyster-catchers busy knocking off and eating limpets upon projecting portions of steep coasts, leaving the shells, all of which, when there is breaker action, must have been washed into deep water as the tide rose. Such circumstances have to be considered upon the steep coasts of the world, of which there is no want, many fathoms of depth being found, with occasionally a few projections in different places, close along shore, various marine vegetables and animals occupying zones of the depths best suited to them. The sea adjoining some of the ocean islands, where great depths are obtained all round, may, perhaps, afford some of the best conditions for collecting together the remains of marine life which had inhabited different zones of depth.

While the remains of marine animals which have existed in different zones of depth, with the modifications due to sheltered and exposed situations and other variations of conditions, may be collected either in the immediate vicinity of, or at no great distance from, steep coasts, it is in tidal seas, to the fringes of detrital or chemically-formed matter around the chief masses of land, rising above the sea, that we look for the preservation of the great amount of organic remains. Indeed, the modifications of the actual coasts as to depth, are commonly but variations of the manner in which these submarine fringes join the subaërial portions of the solid land. Such fringes may be regarded as forming extensive plains on the margins of tidal seas (here and there a projecting mass of rock rising above them), with usually a somewhat gentle slope to the depth of 600 to 1000 or 1200 feet, after which they often appear, as a whole, to descend more abruptly. Gentle as the slope may be, the differences of depth appear sufficient, as above stated, for the modification of life upon it, so that while some animals live near the coast, others keep far out in the deeper water. While some portion may be enabled to live at varied depths, there exists a mass of life, the remains of which would be entombed far from shore in one case, and near it in the other, and not commingled, as in the case of steep coasts, and adjoining deep seas. A glance at the charts of a large portion of the eastern side of the American continent will show how far separated, horizontally, such masses of remains may be.

Let it, for illustration, be supposed that the following map (fig. 69) represents an extended line of coast, so that 1, 1 ; 2, 2 ;

Fig. 69.

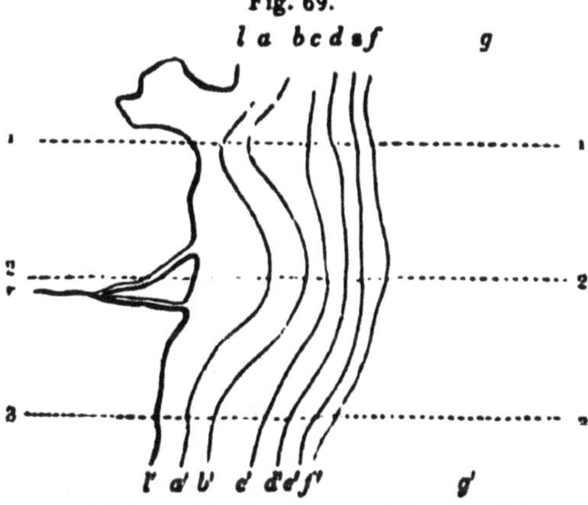

and 3, 3, are parallels of latitude sufficiently distant from each other to render surface temperature different enough to be important as regards marine life. Let *l l'* be a line of coast extending from north to south, and *f f'* the outer verge of about 100 fathoms off the same coast, a more sudden increase of depth taking place at this depth into the area *g g'*; equal depths, or zones being represented by the lines *a a'*, *b b'*, *c c'*, *d d'*, and *e e'*.

For still further illustration, we have supposed a large river (*r*) to deliver itself upon the coast. Upon such a subaqueous area, we have the conditions for the entombment of the remains of the life distributed over it in certain bands, coinciding with depths ranging in lines with the coast, and with the power of tidal and wave action upon the detritus thrust forward by, or carried in mechanical suspension out of, the river, in addition to any sedimentary matter directly obtained from the coast. The effect of the river waters in rendering the shore water brackish would vary in depth, according to circumstances, the tendency of such waters, from their relative specific gravity, being to float above the sea water, and not to be much mingled with the latter to any great amount of depth, though, upon the ebb tide, brackish water might be carried along shore if the tide took that course. Different states of the weather would modify the conditions for the mixture of fresh and sea waters. Thus during heavy on-shore gales of wind, and freshets in the rivers, as are often combined on the western portions of the British Islands, the conditions for mingling sea and river waters would be more favourable than during calm weather.

Let us suppose the following section (fig. 70) to represent (though upon a very exaggerated scale) that of the map (fig. 69),

Fig. 70.

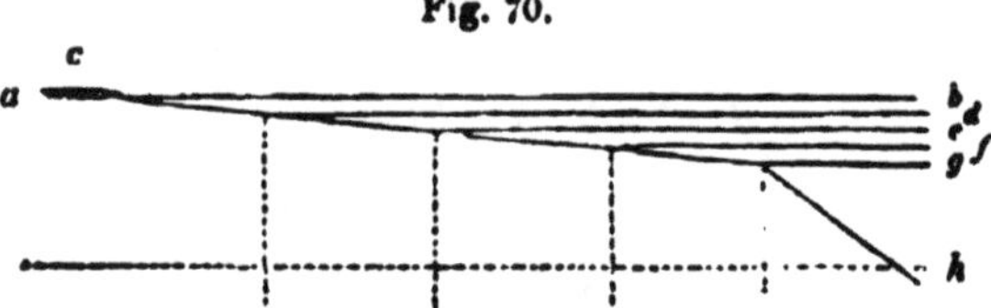

a b being the sea-level, *c*, coast, *d*, *e*, *f*, *g*, different depths of sea, and *h*, the more sudden descent into deep water. In tideless seas these various depths would remain undisturbed, except by movements arising from the waves produced by the winds above, unless, indeed, there be currents acting in such seas. In tidal seas the case would be so far different, that the level of the sea itself would be altered during every tide; on some coasts making a change of many feet. With this change of level, any motion in the waters produced by waves above would also descend more or less deep, supposing equal wave action on the surface. In addition, the sweep of the tidal stream will extend to the depths it may, according to conditions, reach, and occasionally an ocean current may range sufficiently near a coast to act on the bottom, the shores of ocean islands sometimes offering conditions for the latter. We have now to consider that while the shells of molluscs would often remain in the mud, silt, or sand which the animals may frequent, penetrating into them to various depths, according to their habits, so that such remains are preserved after their death in the position usually occupied by the molluscs, numerous other shells remain on the surface to be acted upon in the manner of any inorganic substance.

That shells are so scattered about, multitudes brought up by the arming of the sounding-lead abundantly attest. Moreover, collections of certain species are found to mark particular portions of soundings off given coasts. Thus off the shores of the British Islands, charts give localities as marked by *Hake's teeth*, as they are termed; commonly nothing else than a multitude of the shells of *Dentalium* scattered over particular areas. Other collections of shells are equally well known. While these shells, scattered over the sea bottom, are often either the entire hard parts of univalves, or single and uninjured valves of the bivalves, at other times they are crushed or broken. Whether in the one state or the other, according to their specific gravities, volume, and form, they will be acted upon by streams of tide, by ocean currents sweeping within sufficient depths, or by surface wind-wave action transmitted to the bottom. With respect to specific gravities, though there is apparently much variation in this respect, the floating molluscs being, some of them at least, provided with shells of comparatively minor specific gravity, the range seems something between 2·67 and 2·85*. With equal forms and volumes, fragments or rounded grains of a great proportion of marine shells would apparently be specifically heavier than grains of quartz and rock crystal (2·63—2·65), of common felspar (2 53—2·60), of albite (2·61—2·68),

* The author obtained the following specific gravities of a few marine shells some years since.—Researches in Theoretical Geology, 1834, p. 76.

Argonauta tuberculosus	2·43	Chiton	2·
Nautilus umbilicatus	2·64	Pholas crispata	2·
Ianthina communis	2·66	Cytherea maculata	2·
Lithodomus dactylus	2·67	Bulla	2·
Teredo (great, East Indies)	2·68	Voluta musica	2·
Haliotis tuberculatus	2·70	Cassis testiculus	2·
Cyprina vulgaris	2·77	Strombus gibberulus	2·
Mytilus bilocularis	2·77	Pyrula melongena	2·79
Strombus gigas	2·77	Tellina radiata	2·88

It is not improbable, that if experiments on this head were much multiplied, individual differences would often be found in the same species. While the shell of the *Argonauta tuberculosus* is lighter than pure Sussex chalk (2·40), and that of *Haliotis tuberculatus* is equal in specific gravity to Carrara marble (2·70), the greater numbers exhibit a packing of particles more approaching Arragonite.

and of chlorite (2·71), while they would be lighter than mica (2·94).

The forms of shells or their fragments, except they have been ground down to rounded grains by breaker action on beaches, commonly agree little with those of the sedimentary matter among which they lie superficially mixed. When, therefore, we have to regard any movement of water around whole shells or their fragments, their forms become important, as also the mode in which they may be exposed to any moving force employed. Thus the same shell, if a conical univalve, would offer a different resistance, according as it might be placed with its apex or its base to the moving water, when acted upon, though we might expect that the moving water would soon turn such body, so that its apex would be presented to the line of action. With the valve of a bivalve, its hold on a bottom of sand or silt would be very different, whether it were turned with the margin of the valve downwards, or merely rested upon some part of its bombed surface. How far the valve of a shell could be transported along the bottom without being upset, will depend on very obvious conditions. In all cases we have to consider that shells, or their fragments, having a specific gravity rarely, perhaps, exceeding 2·85, and often presenting forms readily moved, are not difficult of transport in a medium of the specific gravity of 1·027—1·028.

Referring to the plan and section above given (figs. 69 and 70), the observer will have to distinguish between the remains of those molluscs which may die amid the mud, silt, or sand, and so have their harder parts preserved in the situations where they live, and the remains on the surface of the sea bottom. How far these may retain their positions relatively to the zones of depth suited to their animals, will depend upon the circumstances above noticed. Looking at the subject generally, they would be liable to be moved at the depth at which surface-wave action could reach, and therefore to be moved shorewards in shallow waters; so that the remains of molluscs accumulated near the coast in the zones *b a l*, *b' a' l'* (fig. 69), varying in the depths *b d*, *d e* (fig. 70), would, at the proper depths, have surface-wave action added to tidal streams able to transport the shells or their fragments, tending to move them on-shore. In the outer zone *e f* (fig 69), and at the depths *f g* (fig. 70), the effects of the tidal movement may not only be little felt, but also any action upon the bottom from surface-waters be inappreciable. Still further out, in the deep waters *g* (fig. 69), or *h* (fig. 70), there may be no movement sufficient to produce transport of loose matters on the bottom. There might, therefore, be movements in the water producing considerable mixtures of the remains of molluscs in shallow situations, extending even to the casting of shells or their fragments upon the shore, from depths depending upon various local modifications of the causes of transport above noticed.

On many exposed ocean-coasts we have even the accumulation of sandy dunes, composed, for the most part, of fragments of mollusc shells ground down to sand, these cast on shore and dealt with by winds in the manner of common sand. The western coasts of Ireland, Scotland, and of part of England, afford many examples of this fact.*

The induration of sands formed of comminuted shells has been previously mentioned (p. 62), and, as may be expected, such indurated sands occasionally include remarkable mixtures of organic remains. The rock in which the human bones were discovered at Guadaloupe would appear to be of this character. Not only corals and shells from the neighbouring sea, but terrestrial shells also, including the *Bulimus guadaloupensis* (Ferussac), are preserved in it. Teeth of the caiman, with stone hatchets and other remains of human art, are mentioned as having been found in this consolidated sand.

The study of the manner in which the shells of molluscs, and the harder parts of marine animals generally, are thrown on shore, of the depths from which they may be borne by the action of on-shore waves and breakers, and of the various arrangements of whole, broken, or comminuted shells in layers, from their accumulation like ordinary detrital matter at various depths in the sea to their rejection upon the land, is one which will amply reward the observer anxious to compare the manner in which these remains are now distributed and arranged with that of the organic remains found in fossiliferous rocks. He may at times see the

* This shell sand is often employed as manure; it is known to have been so employed in Cornwall in the reign of Henry III. A charter of Richard, King of the Romans, granting the liberty of taking this sand for manure, was confirmed by Henry III. (Lysons, "Mag. Brit.," Cornwall, p. ccclii, who cites Rot. Chart., 45 Hen. III.) Carew notices the use of it in his Survey of Cornwall (1602), and it is largely employed for agricultural purposes to the present day. Mr. Worgan, in 1811, estimated the cost of the land carriage of this sand in Cornwall at more than 30,000*l.* per annum. Large quantities are obtained at the Dunbar Sands, in Padstow Harbour, the annual amount estimated at 100,000 tons. It has been calculated that 5,600,000 cubic feet of sand, chiefly composed of comminuted sea-shells, are annually taken from the coasts of Cornwall and Devon, and spread over the land in the interior as a mineral manure.—Report on the Geology of Cornwall, Devon, and West Somerset (1839), p. 479.

pushing action of the small wash of the sea driving the larger shells and their fragments before it into convenient localities, there accumulating in a mass those which may have been distributed by breaker action along a line of coast, while at others he will find the shells jammed in amid the joints and crevices of rocks so firmly that they become difficult to remove.

CHAPTER X.

CORAL REEFS AND ISLANDS.—DISTRIBUTION OF CORAL ANIMALS.—CHEMICAL COMPOSITION OF CORALS.—KEELING ATOLL.—FORM OF CORAL ISLANDS.—BARRIER REEFS.—LAGOON ISLANDS.—ISLE OF BOURBON.

GREAT as the accumulations of the harder remains of molluscs may be in the sea or on its shores (and regarding the amount of matter, chiefly calcareous, abstracted from the sea or contained in their food the volume of these harder remains added annually to common detrital and chemical deposits must be very considerable), the coral accumulations of tropical regions present us with the most striking additions, by means of animal life, to the mineral deposits now in progress. They have for many years attracted the attention of navigators and naturalists, so that much information has been obtained respecting them.*

With regard to the distribution of corals, Mr. Dana states, that the *Astræacea*, *Madreporacea*, and *Gemmiporidæ* among the *Caryophyllacea*, are, with few exceptions, confined to the coral-reef seas, a region included between the parellels of 28° north and south of the equator,† these corals forming the principal portion of the reefs, and being confined to depths within 120 feet from the surface. Other corals, as is well known, extend to far greater depths, and into colder regions. Sir James Ross, in his voyage to the South Polar Regions, obtained live corals from a

* We would more especially call attention to the labours of Mr. Darwin, who has not only been personally engaged in the investigation of coral reefs and islands, but has also carefully studied the works of navigators and naturalists relating to the subject. The results of his investigations are contained in his work, entitled, "Structure and Distribution of Coral Islands," London, 1842. We would also refer to the labours of Mr. Dana, contained in his "Structure and Classification of Zoophytes," Philadelphia, 1846. Mr. Dana's views are also founded on the personal examination of coral reefs and islands.

† Locally, coral reefs are found further north and south than 28°. They extend in the Bermuda Islands to lat. 32° 15′ N., the greatest distance from the equator, as Mr. Darwin observes, at which they are known to exist, and to lat. 30° N. in the Red Sea. Houtman's Abrolhos, on the western shores of Australia, in lat. 29° S., are of coral formation.

depth of 1620 to 1800 feet off Victoria Land. Mr. Charles Stokes notices a species of *Primnoa (lepadifera)*, as found from 900 to 1800 feet off the coast of Norway; and Professor E. Forbes a species of the same genus from a depth of 1668 feet off Staten Land.* As respects the range of corals, Mr. Dana observes that "*Caryophyllidæ* extend from the equator to the frigid zone, and some species occur at the depth of 200 fathoms or more. The *Alcyonaria* have an equally wide range with the *Caryophyllidæ* and probably reach still higher towards the poles. The *Hydroidea* range from the equator to the polar regions, but are most abundant in the waters of the temperate zone."† Regarding the distribution of species, Mr. Dana states, that of 306 species, 27 only are common to the East Indies and Pacific Ocean, while only one species, and that with doubt (*Meandrina labyrinthica*), is considered to be common to the East and West Indies.‡

Mr. Darwin remarks on the entire absence of coral reefs in certain large areas in the tropical seas. No coral reefs have been found on the west coast of South America, south of the equator, or round the Galapagos Islands; neither have any been yet noticed on the west coast of America, north of the equator. In the central parts of the Pacific there are islands free from coral reefs, and there do not appear to be any coral reefs on the west coast of Africa, or round the islands of the Gulf of Guinea. St. Helena, the Cape Verde Islands, St. Paul's, and Fernando Noronha are also without such reefs.§

Regarding the occurrence of corals as a whole, we thus see that they may be more or less strewed over a very large submarine area, one extending from the polar to the equatorial regions, some of them keeping to small depths within a portion of the general area comprised between the parallels of 28° north and south of the equator, and even rising to the surface of the sea in certain parts of that minor area. However great occasional accumulations of their harder parts may be, under favourable conditions elsewhere, concealed beneath the ocean waters, we have

* Sir James Ross, "Voyage to the Antarctic Regions."

† "Structure and Classification of Zoophytes."

‡ Referring to the causes of distribution and original sites, or centres of distribution, Mr. Dana observes:—"There is sufficient evidence that such centres of distribution, as regards zoophytes, are to be recognized. The species of corals in the West Indies are, in many respects, peculiar, and not one can with certainty be identified with any of the East Indies. The central parts of the Pacific Ocean appear to be almost as peculiar in the corals they afford. But few from the Feejees have been found to be identical with those of the Indian Ocean."—"Structure and Classification of Zoophytes."

§ Darwin, "Structure and Distribution of Coral Reefs," pp. 61, 62.

in the masses of dead and living corals which constitute islands and reefs, enough to show the geological importance of these animals, which thus, from their food and the surrounding waters, secrete a mass of matter constituting rocks, so acted upon, under fitting condition, by the breakers and by atmospheric and chemical influences, that dry land rises sufficiently above the sea to support terrestrial vegetation and animal life.

It would appear that the calcareous secretions* of corals only begin to be formed after the last metamorphosis of the young animal, one effected when it quits the swimming state and attaches itself to some support. Until that time the young can move, by their own powers and the transporting action of tidal streams or oceanic currents, to situations where, under the needful conditions, they can settle and flourish as perfect corals. No doubt myriads of the young animals perish, or are consumed as food, so that a part only is available for supplying the loss by death of the old stock, for increasing the amount of coral life in localities where it previously existed, or for the formation of new colonies. Under all such circumstances, if there be no cause producing a removal of the harder parts of the corals after the death of the polyps which secreted them, they will accumulate. Portions of the harder parts would appear to be destroyed by the animals which feed upon or bury themselves among the corals while living, others are broken off by the action of the sea, and some would appear to be taken up in solution. In the first case, the portion not required for the harder parts of the animals feeding upon the corals appears to be thrown down with their fæces amid the coral reefs; in the second, the fragments torn off by the breakers are distributed, like any

* Dana, "Structure and Classification of Zoophytes," p. 52. Speaking of the mode in which the secretions are formed, Mr. Dana observes:—"In a *Madrepora* the surface between the cells becomes covered by minute points by the continued secretions, and then a layer forms, connected with the preceding, by these points or columns. The interior usually becomes, afterwards, nearly solid by additional secretions. This variety of structure may be observed also in the *Dendrophyllia*; and even the compact species, in which there are no traces of cellules, will often show evidence of having been deposited in layers. I have seen it brought out with singular distinctness, in a specimen half fossilized. In many corals, however, we fail to detect this deposition in layers. This is the case in the *Astræa* tribe. The *Pocillopora*, and some allied corals, have transverse plates crossing the cells internally, which are intermitted secretions from the lower part of the polyp; but no appearance of layers has been detected in the spaces between the cells. The *Favosites*, and many *Cyathophyllidæ*, are examples of similar interrupted secretions across the cells," (p. 53.)

Respecting the foot secretions, he remarks:—"The foot secretions appear to be entirely independent of the tissue secretions. The former are often horny when the latter are calcareous, and when they occur together they constitute separable layers, one enveloping the other. The united polyps of a branch have their mouths opening other detritus, also among the reefs; and in the third, the part not appropriated by the living corals, or by other animals, for their harder parts, appears to be deposited amid the matter of the reefs, tending to bind them together, and adding to their solidity.

From the chemical researches of Mr. B. Silliman, who analysed numerous specimens of calcareous corals sent him by Mr. Dana, it would appear that, after the animal matter had been separated, carbonate of lime formed from 97 to 99 per cent. of the inorganic matter which remained; 1 to 3 per cent. being composed of silica, lime (probably united with the silica), carbonate of magnesia, fluoride of calcium, fluoride of magnesium, phosphate of magnesia, alumina, and iron.* The animal matter varied from 2·11 to 9·43 per cent. From many analyses of corals made at the Museum of Practical Geology, London, carbonate of lime was found to range from 82 to 95·5 per cent., carbonate of magnesia from a mere trace to 7·24 per cent., sulphate of lime from a trace to 2·76 per cent., and organic matter from 3 to 8·27 per cent. Silica, alumina, iron, phosphates, and fluorides were also obtained as in the analyses of Mr. Silliman. As a mass, therefore, we may regard the hard matter secreted by the coral polyps of a reef as chiefly formed of carbonate of lime, mingled with animal matter, of occasionally a notable quantity of carbonate of magnesia, with a minor per centage of other substances, among which are found fluorine and phosphoric acid.

The young of the reef-making coral polyps attaching themselves where the needful conditions obtain,† and according to the habits and requirements of each species, it becomes important to learn how far these may differ, and yet each species aid in building up the general mass of a reef. Mr. Darwin's detailed description of Keeling, or Cocos atoll, situated in the Indian Ocean in lat. 12° 5' S., affords a valuable view of the manner in which the

---

outwards on every side, while the bases are directed inward towards a common central, or axial line. The simultaneous secretions of the bases, therefore, must necessarily produce a solid axis to the branch," p. 54.

* Dana, "Structure and Classification of Zoophytes," pp. 124—131.

† Respecting the needful conditions for the establishment and distribution of reef-making corals, Mr. Couthouy (Boston Journal of Natural History, vol. iv., 1842, and American Journal of Science, vol. xlvii., 1844,) and Mr. Dana (American Journal of Science, vol. xlv, 1843), independently of the views of each other, refer to the temperature of the sea rather than to its depth, as limiting the range of the reef-making corals, and attribute the absence of coral reefs in the inter-tropical and eastern portions of the Atlantic and Pacific to the influence of the cool and extra-tropical currents which there set in. Mr. Dana limits the distribution of the reef-forming corals to a temperature of, and above, 60° Fahr.; and Mr. Couthouy considers that they thrive best in water, at a temperature of between 76° and 80° Fahr.

various corals forming a reef in those seas are adjusted. Having, under favourable circumstances, reached the outer edge of the reef, where the coral was alive, he found that it was almost entirely composed of living porites, forming great irregular rounded masses from four to eight feet broad, and little less in thickness. On the furthest mounds which he reached, and over which the sea broke with some violence, the polyps in the upper cells were dead, but three or four inches lower down they were living. In consequence of the check given to their growth upwards, the corals extended laterally. Further outwards the porites were all seen to be alive. Next in importance is the *Millepora complanata*, growing in thick vertical plates, and forming a strong honeycomb mass, generally of a circular form, the marginal plates being alone alive. Between these plates, and in the crevices of the reef, a multitude of zoophytes and other productions flourish, protected by the porites and millepora from the breakers. Masses of living coral, apparently similar to those of the margin, descend very gradually outwards to the depth of 60 or 70 feet. The arming of the sounding lead brought up fragments of *Millepora alcicornis* within these depths, and there was an impression of an astræa, apparently alive. Examining the fragments thrown on shore by the breakers, the porites and a madrepore, apparently *M. corymbosa*, were the most common; and as this coral was not found alive in the hollows of the reef, Mr. Darwin concludes that it must occur abundantly in a submerged zone outside. Between the depth of 72 and 120 feet the arming of the lead came up an equal number of times marked by sand and coral. Beneath 120 feet sand was obtained. After the depth of 150 feet the outward sides of the reef plunged, at an angle of 45°, into the sea, the depth of which was not found at 2200 yards from the breakers, with a line of 7200 feet in length.*

Close within the outer margin of the reef, where the coral life ceases, three species of nullipora flourish, either separately or mingled together, forming by their successive growth a layer two or three feet in thickness, of a reddish colour. This layer fringes the reef for about 20 yards in width, constituting a continuous

* Darwin, "Structure and Distribution of Coral Reefs," pp. 6—8. "Out of 25 soundings," observes Mr. Darwin, "taken at a greater depth than 20 fathoms, every one showed that the bottom was covered with sand; whereas, at a less depth than 12 fathoms, every sounding showed that it was exceedingly rugged, and free from all extraneous particles. Two soundings were obtained at the depth of 360 fathoms, and several between 200 and 300 fathoms. The sand brought up from these depths consisted of finely-triturated fragments of stony zoophytes, but not, as far as I could distinguish, of a particle of any lamelliform genus: fragments of shells were rare."

"At a distance of 2200 yards from the breakers, Captain Fitzroy found no bottom with a line of 7200 feet in length; hence the submarine slope of this coral formation smooth convex surface, when the corals are united into a solid margin, and forming a protecting breakwater.*

The form of this atoll will be seen by the subjoined plan, (fig. 71.)† The reef is broken by two open spaces, through

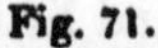
Fig. 71.

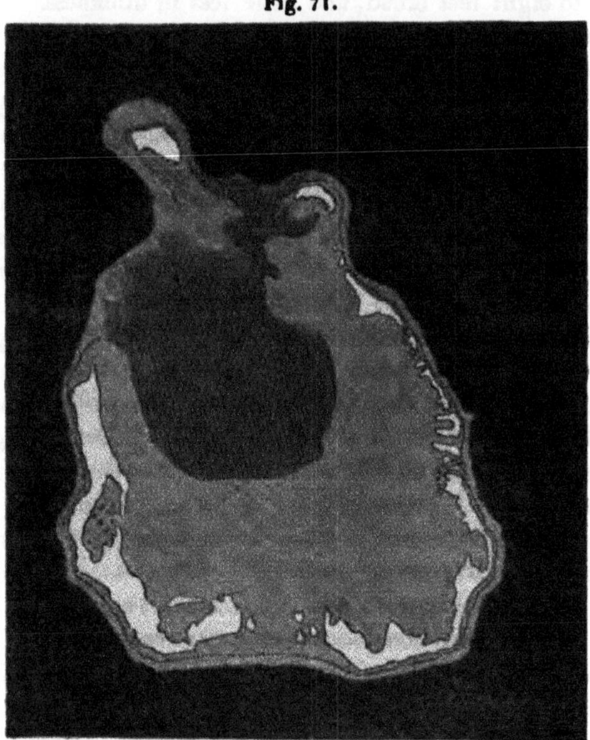

one of which ships can enter; it varies from 250 to 500 yards in breadth, with a level surface, or one very slightly inclined towards the interior lagoon, and at high tide the sea breaks entirely over

is steeper than that of any volcanic cone. Off the mouth of the lagoon, and likewise off the northern point of the atoll, where the currents act violently, the inclination, owing to the accumulation of sediment, is less. As the arming of the sounding-lead from all the greater depths showed a smooth sandy bottom, I at first concluded that the whole consisted of a vast conical pile of calcareous sand; but the sudden increase of depth at some points, and the circumstance of the line having been cut, as if rubbed, when between 500 and 600 fathoms were out, indicate the probable existence of submarine cliffs," pp. 8—9.

* "These nulliporæ," observes Mr. Darwin, "although able to exist above the limit of true corals, seem to require to be bathed during the greater part of each tide by breaking water, for they are not found in any abundance in the protected hollows on the back part of the reef, where they might be immersed, either during the whole or an equal proportional time of each tide. It is remarkable that organic productions of such extreme simplicity, for the nulliporæ undoubtedly belong to one of the lowest classes of the vegetable kingdom, should be limited to a zone so peculiarly circumstanced," p. 9.

† An interesting selection of plans of coral reefs, either surrounding mountainous islands or forming atolls or lagoon islands, among which that of Cocos or Keeling

those parts which do not rise into islets on its surface. *Pocillopora verrucosa* is a common coral in the hollows, as also a madrepore, closely allied or identical with *M. pocillifera*. When the breakers are, by the formation of an islet, prevented from washing entirely over the reef, and channels and hollows are filled up, a hard smooth floor is formed, uncovered only at low water, and strewed with a few fragments torn off during heavy gales. The islets which are formed by an accumulation of fragments, about 200 or 300 yards from the outer edge of the reef, vary in length from a few yards to several miles, with an ordinary breadth of less than a quarter of a mile. On the windward side of the atoll the increase of the islets is by the addition of fragments thrown on their outer sides by the breakers, the highest part thus formed rising from six to ten feet above ordinary high-water mark, and upon this there may be hillocks of blown sand, some of which rise to an elevation of 30 feet. On the leeward side of the atoll, from the sweep of the wind across the lagoon, the little breakers thus formed cast up sand and fragments of thinly-branched coral from the lagoon on the inner sides of the islets in that part of the atoll, thus adding to them inwards. These islands are lower than those to windward, though broader. The fragments beneath the surface are cemented into a solid mass, so as to form a ledge from two to four feet high, from being worked by the breakers acting beyond ordinary high water. Chemical changes take place occasionally among the calcareous fragments thus cemented together, so that the altered coral passes gradually into spathose limestone.*

The lagoon within is necessarily a sheltered situation, and is

---

Island is one, will be found in Plates 1 and 2 of Mr. Darwin's work on the Structure and Distribution of Coral Reefs; and a most valuable map in the same work, showing the distribution of the different kinds of coral reefs, with the position of active volcanos in the Indian and Pacific Oceans.

* "The fragments of coral which are occasionally cast on the 'flat,' are, during gales of unusual violence, swept together on the beach, where the waves each day at high water tend to remove and gradually wear them down; but the lower fragments having become firmly cemented together by the percolation of calcareous matter, resist the daily tides longer, and hence project as a ledge. The cemented mass is generally of a white colour, but in some few parts reddish from ferruginous matter: it is very hard, and is sonorous under the hammer; it is obscurely divided by seams, dipping at a small angle seaward; it consists of fragments of the corals which grow on the outer margin, some quite and others partially rounded, some small and others two or three feet across; and of masses of previously formed conglomerate, torn up, rounded, and re-cemented; or it consists of a calcareous sandstone, entirely composed of rounded particles, generally almost blended together, of shells, corals, the spines of echini, and other such organic bodies." "The structure of the coral in the conglomerate has generally been much obscured by the infiltration of spathose calcareous matter; and I collected a very interesting series, beginning with fragments of unaltered coral, and ending with others where it was impossible to discover with the naked eye any trace of organic structure. In some specimens I was unable, even with the aid of a lens and by wetting them, to distinguish the boundaries of the

described as much more shallow than those of atolls of considerable size. About half the area consists of sediment, including mud, and half of coral reefs, the corals composing the latter having a very different aspect from those on the outside, and being very numerous in kind.* The sediment from the deepest part of the lagoon was like a very fine sand when dry, though it appeared chalky when wet. Mr. Darwin points out that much fine sediment may be supplied by means of the excrements of the scarus and holuthuriæ, which feed on the coral; large shoals of two species of the former, one of which inhabits the lagoon while the other keeps outside, feeding entirely on the corals, while swarms of various species of holuthuria browse upon the lagoon corals. "The amount of coral yearly consumed and ground down into the finest mud by these several creatures, and probably by many other kinds, must be immense."† The tide flows in and out of the lagoon through the channels, and the latter also carry out the water thrown over the reefs by the breakers.

Thirty-two coral islands in the Pacific Ocean were examined by Captain Beechey;‡ they were of various shapes, and 29 had lagoons in their centres. The dry coral forming islets on the reefs is rarely elevated more than two feet above the sea when divested of any sandy materials heaped upon it, and but for the abrupt character of the outer margin would be inundated by the breakers. Captain Beechey found in the islands seen by him no instance in

---

altered coral and spathose limestone. Many even of the blocks of coral lying loose on the beach had their central parts altered and infiltrated." Darwin, "Structure of Coral Reefs," p. 12.

Mr. Beete Jukes mentions masses of meandrina, six or eight feet in diameter, turned upside down, and much worn, as torn by the force of the breakers from their places of growth on the weather edge of a coral reef, and driven 200 to 300 yards inwards. "Narrative of the Voyage of the 'Fly,' 1847."

* "Meandrina, however, lives in the lagoon, and great rounded masses of this coral are numerous, lying quite or almost loose on the bottom. The other commonest kinds consist of three closely-allied species of true Madrepora in thin branches; of *Seriatopora subulata*; two species of Porites, with cylindrical branches, one of which forms circular clumps, with the exterior branches only alive; and, lastly, a coral, something like an Explanaria, but with stars on both surfaces, growing in thin, brittle, stony, foliaceous expansions, especially in the deeper basins of the lagoon. The reefs on which these corals grow are very irregular in form, are full of cavities, and have not a solid flat surface of dead rock, like that surrounding the lagoon; nor can they be nearly so hard, for the inhabitants made with crowbars a channel of considerable length through these reefs, in which a schooner, built on the south-east islet, was floated out. It is a very interesting circumstance, pointed out to us by Mr. Leisk, that this channel, although made less than ten years before our visit, was then, as we saw, almost choked up with living coral, so that fresh excavations would be absolutely necessary to allow another vessel to pass through it."—Darwin, "Structure," &c., p. 13.

† Darwin, "Structure of Coral Reefs," p. 14.

‡ "Narrative of a Voyage to the Pacific and Behring's Straits, &c., in the years 1825, 26, 27, and 28." London, 1831.

which the strip of dead coral exceeded half-a-mile from the usual wash of the sea to the internal lagoon. In general it was only 300 or 400 yards. "Beyond these limits, on the lagoon side in particular, when the coral was less mutilated by the waves, there was frequently a ledge, two or three feet under water at high tide, 30 to 50 yards in width; after which the sides of the island descended rapidly, apparently by a succession of inclined ledges formed by numerous columns united at their capitals, with spaces between them, in which the sounding-lead descended several fathoms."* The windward sides of the reefs and islets upon them are higher than the others, the islets not unfrequently well wooded,† while, on the opposite sides, the reefs are "half drowned" or wholly under water. The breaks or entrances into the lagoons generally occur on the leeward side, though they are sometimes found in a side that runs in the direction of the wind, as at Bow Island. The points or angles of the islands were found to descend less abruptly than the sides. The lagoons vary in depth, from 20 to 28 fathoms being found in those which were entered, though the appearance of the water in others would lead to the inference that they were very shallow. The accompanying figures are the sections given by Captain Beechey as affording a general view of these coral islands. Fig. 72 is a section across one about five miles wide; *a a* being dry islets on the reef; *b b*, lagoon; and *c c*, open ocean; and fig. 73 a section across an islet and part of a lagoon, with the slope towards the sea, AB being the habitable part of the island; *a b*, water line; *a h*, general descent seawards towards the points; *a i*, general descent at the points; CC, part of the lagoon; DD, coral knolls in the lagoon; Z, the ocean; *s s s*, soundings on coral.‡

While the coral reefs above mentioned exhibit no traces of rocks

* "Narrative of a Voyage to the Pacific and Behring's Straits, &c., in the years 1825, 26, 27, and 28." Vol. i. p. 256. London, 1831.

† With respect to the vegetation on Bow Island, Mr. Collie mentions that the pandanus and pemphis grow in the sheltered parts of the plain between the ridges; that the loose dry stones of the first ridge are penetrated by the roots of the tefano, which rises into a tall spreading tree, accompanied by the suriana and tournefortia, under the shelter of which the achyranthus and lepidium thrive best. Beyond the first ridge the scævola flourishes. "Beechey's Voyage," vol. i., p. 248. At Ducie's Island the trees are stated to rise 14 feet, making, with the island, 12 feet above the sea, 26 feet from its level. Ibid. vol. i. p. 59.

‡ Captain Beechey gives a more detailed account of Matilda and Bow Islands than of the others. The windward side of the former "is covered by tall trees, while that to leeward is nearly all under water. The dry part of the chain enclosing the lagoon is about a sixth of a mile in width, but varies considerably in its dimensions; the broad parts are furnished with low mounds of sand, which have been raised by the action of the waves, but are now out of their reach, and mostly covered by vegetation.

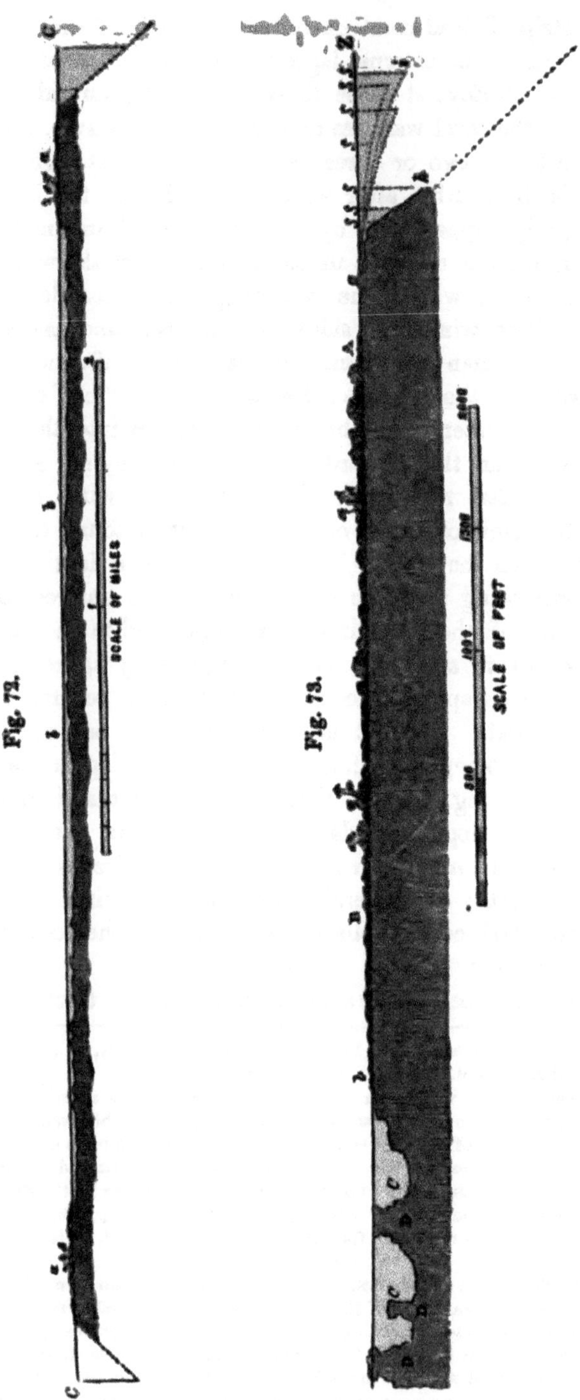

Fig. 72.

Fig. 73.

further than those formed by the consolidation of the matter, chiefly calcareous, secreted by the polyps or derived from it, and distributed chemically and mechanically, other reefs surround islands or groups of islands formed of different rocks, or range along the shores of seas, such as the Red Sea, or those of large masses of land, as on the east coast of Australia. While the reefs

---

The violence of the waves upon the shore, except at low water, forces the sea into the lake at many points, and occasions a constant outset through the channel to leeward.

"On both sides of the chain the coral descends rapidly; on the outer part there is from 6 to 10 fathoms, close to the breakers; the next cast is 30 to 40; and, at a little distance, there is no bottom with 250 fathoms. On the lagoon side there are two ledges; the first is covered by about three feet at high-water: at its edge the lead descends three fathoms to the next ledge, which is about 40 yards in width; it then slopes to about 5 fathoms at its extremity, and again descends perpendicularly to 10; after which there is a gradual descent to 20 fathoms, which is the general depth of the centre of the lagoon. The lake is dotted with knolls or columns of corals, which rise to all intermediate heights between the bottom and the surface." "Voyage," &c., vol. i., p. 218.

"Bow Island is 30 miles long by an average of 5 miles broad. It is similar to the other coral islands already described, confining within a narrow band of coral a spacious lagoon, and having its windward side higher and more wooded than the other, which, indeed, with a few clusters of trees and heaps of sand, is little better than a reef. The sea in several places washes into the lagoon, but there is no passage, even for a boat, except that by which the ship entered, which is sometimes dangerous to boats, in consequence of the overfalls from the lagoon, especially a little after the time of high-water.

"The bottom of the lagoon is in parts covered with a fine white sand, and is thickly strewed with coral knolls, the upper parts of which overhang the lower, though they do not at once rise in this form from the bottom, but from small hillocks. We found comparatively few beneath the surface, though there are some: at the edge of such as are exposed there is usually six or seven fathoms of water; receding from it, the lead gradually descends to the general level of about 20 fathoms. The height of water in the lagoon is subject to the variations of the tides of the ocean; but it suffers so many disturbances from the waves, which occasionally inundate the low parts of the surrounding land, that neither the rise of the tide nor the time of high-water can be estimated with any degree of certainty. The strip of low land enclosing the lagoon is nearly 70 miles in extent, and the part that is dry is about a quarter of a mile in width. On the inner side, a few yards from the margin of the lake, there is a low bank formed of finely-broken coral; and at the outer edge a much higher bank of large blocks of the same material, long since removed from the reach of the waves, and gradually preparing for the reception of vegetation. Beyond this high bank there is a third ridge, similar to that skirting the lagoon; and outside it again, as well as in the lagoon, there is a wide shelf, three or four feet under water, the outer one bearing upon its surface huge masses of broken coral, the materials for an outer bank, similar to the large one just described." "Voyage," &c., vol. i., p. 245, 246.

Mr. Beete Jukes ("Narrative of the Voyage of the 'Fly' to Torres Straits, &c., 1847") presents us with the following description of one kind of small coral island as seen from the mast-head:—"A small island, with a white sand beach and a tuft of trees, is surrounded by a symmetrically open space of shallow water of a bright grass-green colour, inclosed by a ring of glistening surf, as white as snow, immediately outside which is the rich dark blue of deep water. All the sea is perfectly clear of sand and mud; even where it breaks on a sand beach it retains its perfect purity." "It is this perfect clearness of the water which makes navigation among coral reefs at all practicable, as a shoal with even five fathoms water on it can be discovered at a mile distant from a ship's mast-head, in consequence of its greenish hue contrasting with the blue of deep water."

touch the land in some places, they are removed from it in others; and many present the appearance of the lagoon islands—land either in one mass or in several masses rising through the interior lagoon. Mr. Darwin has classed these various modes of occurrence into atolls, or lagoon islands, barrier reefs, and fringing or shore reefs.*

The following map (fig. 74) of the Gambier's group,† may be taken in illustration of the barrier reefs, and as also showing

Fig. 74.

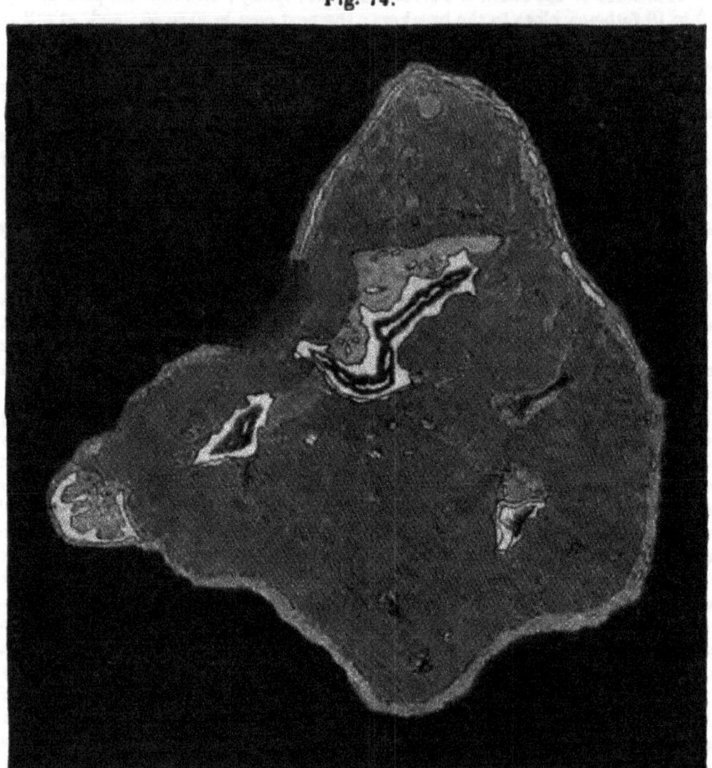

coral reefs fringing the contained islands. All the interior islands are steep and rugged, Mount Duff, on the largest, rising to the height of 1248 feet, and they would appear to be of igneous origin.‡ The outside reef on the north-east, the windward side,

* We would refer to Mr. Darwin's work, "Structure and Distribution of Coral Reefs," for great detail respecting the different kinds of reefs.

† Reduced from that given by Captain Beechy, "Voyage to the Pacific and Behring's Strait," vol i.

‡ They are described as composed of porous basaltic lava, sometimes passing into tufaceous slate, at others into a columnar basalt. Dikes cutting the mass were observed.

has portions raised above the sea, bearing trees and other plants, while in the opposite direction it dips 30 or 40 feet beneath the sea. The outer sides plunge, as usual, into deep water, while the inner descend gradually to 120 or 150 feet. Patches of coral are scattered over the lagoon, and adhere as fringing reefs to the steep islands rising out of it. On the larger island the coral reef rendered the water so shallow, that the larger boats could not come within 200 yards of the landing-place.

The annexed plan (fig. 75) of Ari Atoll, one of the Maldiva Islands,* exhibits a modification of those coral islands which have a general reef surrounding a lagoon. Here a number of reefs form an outer line, and the interior is occupied by a number of others. Many of these are ring-formed, so that the general group reminds us of many minor atolls, rising above an area of a tabular character, round which the sides plunge rapidly into deep water. The common depth between these reefs and islets, some rising above the level of the sea, varies from 150 to 200 feet, and in the basins of the ring-form detached reefs from 24 to 60 feet. According to Captain Moresby, the central and deepest part of the lagoons in the Maldiva Islands is formed of stiff clay, of sand near the border, and of hard sand-banks, sandstone, conglomerate, rubble, and a little live coral in the channels of the reef.† The other large islands, or rather groups of islets and reefs, of the Maldives, present the same general characters, while the smaller, one of which (Ross Atoll) is represented in the following page (fig. 75, *a*), offers the usual atoll character.

From the observations of Mr. Darwin on part of the coral reefs of the Mauritius, it would appear that the edge of the reef is formed of great masses of branching madrepores, chiefly *M. corymbosa* and *M. pocillifera*, mingled with a few other kinds of coral. To the depth of 48 feet, the coral ground appeared free from sand; but from that depth to 90 feet a little calcareous sand was brought up by the arming of the sounding-lead; more frequently, however, it came up clean. The two madrepores above mentioned, and two species of astræa, with large stars, seemed the commonest corals for the whole of this depth. Some fragments of *Millepora alcicornis* were brought up, and in the deeper parts were large beds of a seriatopora, allied to, but differing from, *S. subulata*. From

* Reduced from the chart of the Maldives, by Captain Moresby and Lieutenant Powell.

† As stated by Captain Moresby to Mr. Darwin. "Structure," &c., p. 34. Captain Moresby informed Mr. Darwin that *Millepora complanata* was one of the commonest kinds of corals on the outer margin of reefs of the Maldives. Ibid. p. 33.

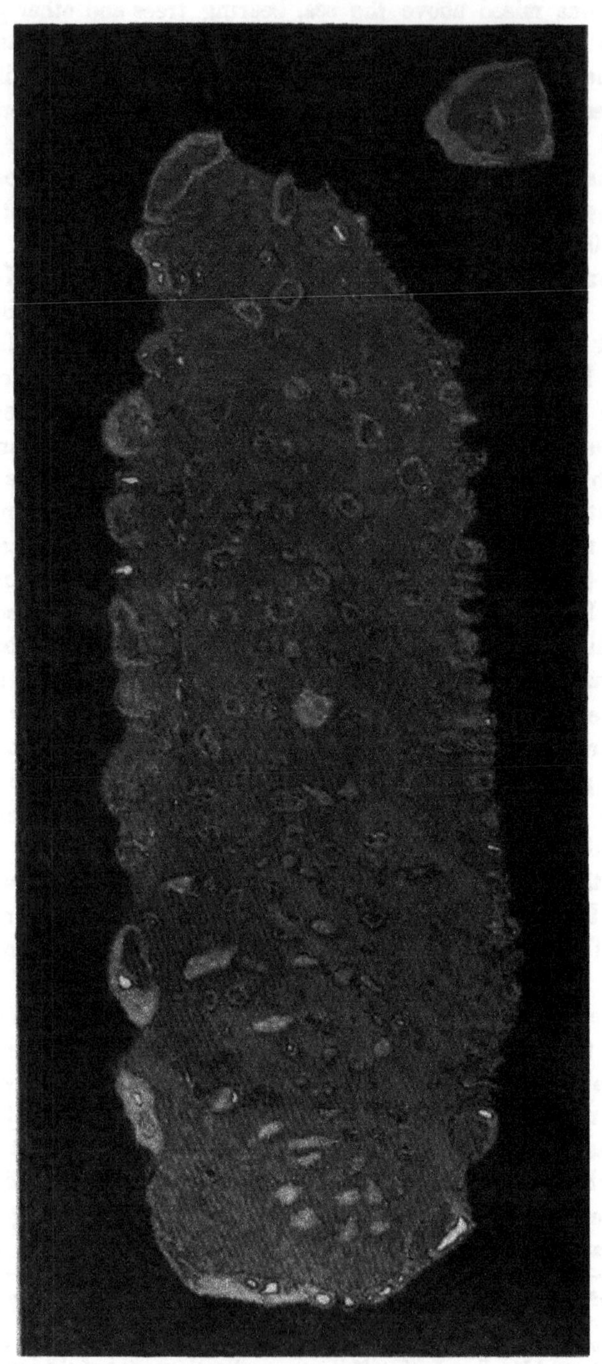

90 to 120 feet the bottom, with few exceptions, was covered by sand, or strewed with seriatopora. From 120 to 198 feet, the soundings showed a sandy bottom, with one exception, at 180 feet, when the arming came up as if cut by the margin of a large caryophyllia. On the beach, the rolled fragments consisted chiefly of madrepores and astræa of the smaller depths, of a massive porites, like that at Keeling atoll, of a meandrina, *Pocillopora verrucosa*, and of numerous fragments of nullipora.*

The reef surrounding the Mauritius, excepting in two or three parts where the coast is almost precipitous,† generally ranges at a distance of one, two, or even three miles from the shore. Opposite every river and streamlet the reef is, as is common, breached, and the slope outside the reef seems generally to be moderate, bearing a relation to the slope of the adjoining land.

The Isle of Bourbon is also surrounded by coral reefs, only broken through at the embouchures of the rivers, and opposite the chief ravines. M. Siau, who had excellent opportunities of observing these reefs in 1839 and 1840, has stated ‡ that the channels or passages through the reefs are kept open by the streams of fresh water passing outwards through them, and that they would be otherwise soon filled up. As it is, they are considered to have decreased in size, in consequence of a diminished quantity of rain having, of late years, fallen upon the Isle of Bourbon. These channels being, as usual in such situations, the passages to roadsteads behind the reefs, their condition is a constant subject of attention, and, as illustrative of the quick growth of certain at least of the reef-making corals of that locality, M. Siau mentions that, in one of the channels (that of the Rivière d'Abord), a coral rock has risen from the bottom, and in the middle of it, to the height of 29 feet (English) in 12 years. M. Siau§ presents us

* Darwin, "Structure of Coral Reefs." † Darwin, Ibid.

‡ Comptes Rendues, tom. xii., 1841.

§ M. Siau observes, that "the labours of the coral polyps are as varied as the species. Some (and these are the most widely spread) establish themselves by families at the bottom of the sea, on a volcanic or any other rock, unattackable by the action of the waves. Each family constructs a detached boss (*mamelon*) which may rise to the height of two or three yards by the labours of many generations. These bosses are known in the country by the name of *pâtés de coraux*. The bottom is thus covered by bosses, which most frequently join, touch, or approximate to each other, sometimes leaving open spaces between them, into which (coral) sand and shingle are washed by the sea. Such spaces are known as *rigoles de sables*.

"Upon this fresh bed new families establish themselves, constructing another bed. The latter are independent of the former. Sometimes they entirely repose with a very interesting account of the mode of growth of the reef-making corals of this island, showing the establishment of a series of coral bosses upon each other, with the admixture of coral sand, and shingle, in the interstices between them, up to the level of the sea, where the labours of the reef-making coral polyps terminate.

---

on the first *pâtés*, sometimes on the *rigoles*, so as to conceal them; sometimes an isolated *pâté* completely covers a primitive *rigole*. The spaces between this second bed are also converted into *rigoles*, the sea throwing in sand and small shingles. Above this second bed other generations raise a fourth and a fifth, and thus the mass is formed of those immense reefs so common in intertropical seas.

"It would be wrong to conclude, from the description given, that the beds thus formed have a uniform thickness. It should be understood that very great differences exist in the height of the *pâtés*, and that the entire reef would present a shapeless and divided assemblage of superimposed monteoules, the interstices between them filled with sand and shingles, their contiguous portions joined together by a coral cement.

"The corals of which we have spoken are the most common, forming the mass of the reefs. The coral produced is grey, very compact, of a very close grain, and often harder than marble. This coral is not worked away by the waves, and is not entirely soluble in acids. Upon the firm base of the bosses, above described, a variety of small and delicate corals, of different kinds, establish themselves. It is these fragile corals which alone furnish the white sand and shingles to the shore and the *rigoles*, and they are entirely soluble in acids."

The same author remarks upon the depths agitated by the waves, and infers (Comptes Rendues, tom. xii., p. 775) that he had evidence of that action at the depth of 578 French feet (616 English feet), on the north-west of the roadstead of St. Paul, Isle of Bourbon. It will be obvious that, in such researches, the friction on the bottom, by tidal streams, and ocean currents, has carefully to be distinguished from the movement produced among the particles of water beneath by the action of surface wind-friction waves above. Whatever the cause of motion in the superficial parts of the sea bottom, either from surface-wave action, or the friction of tidal streams, or ocean currents, the observation of M. Élie de Beaumont (appended to M. Siau's Paper), respecting an inquiry as to the depths at which *fixed* animals are found upon bottoms liable to this motion, such animals depending for their food upon the prey which may pass them, is equally important.

GREAT BARRIER REEF OF AUSTRALIA.—CORAL REEFS OF THE RED SEA.—CONDITIONS FOR THE OCCURRENCE OF CORAL REEFS.—INFLUENCE OF VOLCANIC ACTION ON CORAL REEFS.—COMPOSITION OF CORAL-REEF ACCUMULATIONS.—INFLUENCE OF CHANGES IN THE LEVEL OF SEA AND LAND.—REEFS NEAR BERMUDA.

THE Great Barrier Reef, extending off the east coast of Australia for about 1100 miles, with a mean breadth of about 30 miles, from Breaksea Spit, in lat. 24° 30′ S., and long. 153° 20′ E., to Bristow Island in lat. 9° 15′ S., and long. 143° 20′ E. off the coast of New Guinea, presents an area of about 33,000 square miles, chiefly covered with organic, mechanical, and chemical accumulations resulting from the secretions of the coral polyps. This great mass is broken northwards by the influence of river waters discharged from the south-eastern portion of New Guinea, carrying detritus with them, and covering the bottom of the adjacent seas with a muddy sediment. These conditions ceasing, we find the great coral accumulations continued to Louisiade, thus extending the surface, allowing for the great break above mentioned, over many more thousands of square miles.

The survey of Torres Straits, between Australia and New Guinea, by Captain Blackwood, has added materially to our knowledge of the Great Barrier Reef, and Mr. Beete Jukes, naturalist to the expedition, has afforded us very valuable information respecting it.* He divides the coral accumulation into—1st, linear reefs, forming the outer edge, or actual barrier; 2nd, detached reefs, lying outside the barrier; and 3rd, inner reefs, or those which lie between the barrier and the shore. With respect to the linear reefs, they are described as generally long and narrow, more or less parallel to the coast of Australia, and separated by narrow

* "Narrative of the Surveying Voyage of H.M.S. 'Fly' commanded by Captain Blackwood, R.N., in Torres Strait, New Guinea, &c." By J. Beete Jukes, M.A., &c. London, 1847.

breaks or passages, varying from 200 yards to a mile in width, and from half a mile to 15 miles in length. They have commonly great depths of water on the ocean side, lines of 100 or 200 fathoms rarely finding bottom close to the reefs, while the depths inside generally vary from 60 to 120 feet. The detached reefs occur only in one locality—somewhat in front of Cape Grenville, Australia (if we except the reefs eastward of the Great Barrier, eastward of Torres Strait), rise from deep water all round, and have more or less of a circular form, with lagoons inside. The inner reefs are very numerous, scattered over the platform beneath the more shallow water between the outer reefs and the coast of Australia, sometimes leaving an open channel between them and the land on the one side, or the barrier on the other. They are of different forms, have sometimes gradual slopes around, and at others are steep-sided.*

Mr. Beete Jukes observes, that up to about lat. 21° 10′, at Swain's Reefs, it can scarcely be said that any true barrier exists, there being merely a bank of soundings off the shore, "with large masses of coral reef settled upon it, and within its outer boundary, —almost equally large clear spaces intervening between the different groups of reefs. In Swain's Reefs, the individual reefs on the outer edge of the group can scarcely be distinguished in form from those inside them, although they may have a little more linear shape, and their greatest length runs more invariably along the line of the boundary of the group. It is only at their northern extremity that they assume one of the characteristics of a true barrier, that of rising like a wall from a deep and almost fathomless

---

* Beete Jukes, "Surveying Voyage of the 'Fly,'" vol. i., pp. 317-18. Mr. Beete Jukes gives a detailed account of the range of coral accumulations from Breaksea Spit (vol. i., pp. 318-332). Respecting the most southern portion, it is stated, that "from Sandy Cape (Australia), a sandy shoal runs out, partially covered by coral, as it proceeds outwards. It is formed of siliceous sand, with 10 or 20 fathoms of water upon it, sloping to 30 fathoms, after which it plunges into deep water. At the Capricorn Group, about 50 miles more northward, all—even the smallest grains of sand—was calcareous, and so it seemed to continue to the sedimentary matter brought down by the New Guinea rivers, eastward of Torres Strait.

"North of the parallel of 23° 10′, there is an open space of sea, in which no reefs occur, about 50 miles wide from north to south; and the bank of soundings instead of being a steep, well-defined edge, slopes out very gradually far to the eastward. The flat of about 20 fathoms, extends out as usual from the mainland for about 30 or 40 miles, and then gradually deepens, till 70, 80, 90, and 100 fathoms are successively attained, 20 or 30 miles eastward of the boundary of the line of soundings, as it exists to the southward. The character of the bottom likewise changes from a coarse coral to the finest possible mud, of a light olive-green colour, in which the lead often wholly buried itself on reaching the bottom. This, when dried, was also entirely calcareous, and wholly soluble in muriatic acid."—Ibid. vol. i., p. 320.

sea."* To the northward the reefs become more numerous. Where the detached reefs occur opposite Cape Grenville is a great bay in the barrier, with very deep water, a line of 1710 feet having failed to strike the bottom on its southern side, four miles inside the reefs forming the bay. Near this bay Yules' Detached Reef rises from an unknown depth, greater than 600 feet, and appears to have a lagoon in its centre. The Great Detached Reef rises on the northward of this reef, also from a great depth, containing a lagoon with 180 feet of water in it.

Raine's Islet is also another detached reef rising with steep sides (in one place at an angle of 55°) from deep water. Bottom was found at 960 feet one mile north of the islet, and at 1080 feet two miles and a half north-east of it. On the southern side, bottom was not found until close to the breakers of the Great Barrier Reef, when fine coral sand was brought up from 1050 and 1200 feet.† Pandora's

* "Between Swain's Reefs and the mainland there is a space of 50 to 60 miles wide, clear of reefs, with a depth of 30 to 50 fathoms."—Beete Jukes, "Surveying Voyage of the 'Fly,'" vol. I., p. 321.

† Raine's Island is described as about 1000 yards long and 500 wide, rising in no part more than 20 feet above high-water mark. "It is formed of a plateau of calcareous sandstone, which has a little cliff all round, 4 or 5 feet high, outside of which is a belt of loose sand, forming a low ridge between it and the sea. Some mounds of loose sand also rest upon the stone, especially at its western end. The length of the island runs in about a N.N.W. and S.S.E. direction. It is surrounded by a coral reef, that is narrow on the lee side, but to windward, or towards the east, stretches out for nearly two miles. The surface of this reef is nearly all dry at low water, and its sides slope rapidly down to a depth of 150 or 200 fathoms." "The island is covered with a low scrubby vegetation," and "the central part of the island had a rich black soil several inches deep." The stone forming the base of the island is described as "made up of small round grains, some of them apparently rolled bits of coral and shell, but many of them evidently concretionary, having concentric coats. It was not unlike some varieties of oolite in texture and appearance. It contained large fragments of corals and shells and some pebbles of pumice, and it pielded occasionally a fine sand that was not calcareous, and which was probably derived from the pumice. Some parts of it made a fair building stone, but it got softer below, till it passed downwards into a coarse coral sand, unconsolidated, and falling to pieces on being touched. In the quarries opened next year for the beacon (constructed for the purposes of navigation), many recent shells, more or less perfect, were found compacted in the stone, and one or two nests of turtle's eggs, of which, in some cases, only the internal cast had been preserved, but in others the shell remained in the form of white carbonate of lime. Some drusy cavities were also found in the stone, containing crystals of gypsum." "It is evident from the fossil turtle eggs that the consolidation of the stone had taken place after it was raised above the sea. It was due, probably, to the infiltration of the rain-water percolating through the calcareous sand, that had been gradually piled above high-water mark by the combined action of the winds and the waves. The thickness of the vegetable soil in its centre shows that it has been above water for a great length of time."—Beete Jukes, "Voyage of the 'Fly,'" vol. I. pp. 126-128. The whole surface of the island was covered with birds—all but one kind, a land-rail—sea birds, such as frigate birds, boobies, gannets, &c. "On walking rapidly into the centre of the island, countless myriads of birds rose, shrieking on every side, so that the clangour was absolutely deafening, like the roar of some great cataract." There were turtle tracks on the beach, and the shells and skeletons of dead turtles were scattered about the island.

Entrance, through the Barrier Reef, occurs in 11° 10′ S., northward of the deep bay, and the detached reefs, after which the great reef is made up of long and closely-connected masses, with few and small gaps for 40 miles. From 10° 40′ to Flinders' Entrance, in lat. 9° 41′, the reefs consist of numerous spots and patches (too close to afford good entrance for vessels), forming submarine pinnacles or towers, rising from a depth of 90 or 120 feet, still, however, preserving the line of the barrier, with deep water outside, in which the bottom was not found with a line of 960 feet.

From Cape Weymouth and Restoration Island, in consequence of the altered run of the Australian coast and of the barrier reefs, the difference between the outer reef and the mainland in the parallel of Cape York (the N.E. point of Australia), has increased to 80 and 90 miles. The whole of the intermediate distance has not been surveyed, but Mr. Beete Jukes states, that there appear to be many inner reefs at a short distance from the land. Between these and the great eastern barrier, the sea is comparatively free from them, many sunken patches being, however, scattered about, and the bottom irregular in places. "The general depth varies from 12 to 20 fathoms, the bottom being coarse sand (with many foraminifera and detached corals and corallines), gradually passing as we approach the land into finer sand and detritus, and from that into the finest possible mud, wholly calcareous and lying close to the shore."*

The outer barrier terminates at Anchor Key, in lat. 9° 20′ S., and no coral reef is found further towards the coast of New Guinea, in this direction, except the Bramble Reef, described as fringed round other rocks. The chart shows coral sand and fragments on the bottom in 38 fathoms, increasing to 54 fathoms, and stretching out 50 miles to the eastward of the Bramble's Key, while all the soundings on the north, in front of a low coast, with a large discharge of fresh water from various channels in New Guinea, are of mud and sand. In front eastward of Flinders' Entrance, Portlock's reefs rise from a depth of 360 to 400 feet, so that on the north, as on the south, as is observed by Mr. Beete Jukes, the corals rise from the ocean in shallow water as compared with the central portions. Between Cape York and the opposite coast of New Guinea, extensive reefs seem to prevail adjoining the latter, rising out of 30 to 70 feet of water, and a considerable reef connects Warrior Island with the mainland of New Guinea. All the central parts of Torres Strait,

* Beete Jukes, "Voyage of the 'Fly,'" vol. i., p. 330.

from north to south, between Cape York and Turtle-Back Island, are remarkable for a nearly uniform bottom, 9 to 11 fathoms, formed of sand and mud. No coral reefs were found in this central band, except narrow fringing reefs round islands, formed of other materials,—porphyries, granites, and quartz rocks.*

Coral reefs are abundant in the Red Sea, fringing the coasts to a great extent. Numerous localities have been examined for a distance of about 200 miles by MM. Ehrenberg and Hemprich, and 150 species of corals were observed. According to the former,† these reefs form shallow incrustations on the rocks of the coasts, from 3 to 12 feet beneath the surface of the sea, often sloping outwards. They do not always adjoin the coast, but often form narrow parallel bands at various distances from it. The reefs are composed of Madrepora, Retepora, Millepora, Astræa, Favia, Caryophyllia, Mæandrina, Pocillopora and Stephanocora, mixed with

---

* Beete Jukes, ("Voyage of the 'Fly,'" vol. i., p. 331,) from his experience among the great coral accumulations of Eastern Australia, has given the following account of an individual coral reef.—"A submarine mound of rock, composed of the fragments and detritus of corals and shells, compacted together into a soft spongy stone. The greater part of the surface of this mound is quite flat and near the level of low water. At its edges it is commonly a little rounded off, or slopes gradually down to a depth of 2, 3, and 4 fathoms, and then pitches suddenly down with a very rapid slope into deep water, 20 or 200 fathoms, as the case may be. The surface of this reef, when exposed, looks like a great flat of sandstone with a few loose slabs lying about, or here and there an accumulation of dead broken coral branches, or a bank of dazzling white sand. It is, however, chequered with holes and hollows more or less deep, in which small living corals are growing; or has, perhaps, a large portion that is always covered by two or three feet of water at the lowest tides, and here are fields of corals, either clumps of branching Madrepores, or round stools and blocks of Mæandrina and Astræa, both dead and living. Proceeding from this central flat towards the edge, living corals become more and more abundant. As we get towards the windward side, we of course encounter the surf of breakers long before we can reach the extreme verge of the reef, and among these breakers we see immense blocks, often two or three yards (and sometimes much more), in diameter, lying loose upon the reef. These are sometimes within reach by a little wading; and though in some instances they are found to consist of several kinds of corals matted together, they are more often found to be large individual masses of species, which are either not found elsewhere and consequently never seen alive (Mr. Beete Jukes saw an irregular block of Mæandrina, of irregular shape, 12 to 15 feet in diameter), or which greatly surpass their brethren on other parts of the reef in size and importance. If we approach the lee edge of the reef, either by walking or in a boat, we find it covered with living corals, commonly Mæandrina, Astræa, and Madrepora, in about equal abundance, all glowing with rich colours, bristling with branches, or studded with great knobs and blocks. When the edge of the reef is very steep, it has sometimes overhanging ledges, and is generally indented by narrow winding channels and deep holes, leading into dark hollows and cavities where nothing can be seen. When the slope is more gentle, the great groups of living corals and intervening spaces of white sand can be still discerned through the clear water to a depth of 40 or 50 feet, beyond which the water recovers its usual deep blue. A coral reef, therefore, is a mass of brute matter living only at its outer surface, and chiefly on its lateral slopes."—Ibid. vol. i., pp. 314–316.

† "Uber die Natur und Bildung der Corallen-Bänken des Rothen Meeres," Berlin, 1834.

the shells of molluscs, the remains of fish, &c. According to M. Ehrenberg, the height, resulting from the accumulation of the same corals, is small. With respect to the banks and reefs lying some distance from the shore, Captain Moresby states that they appear more elongated than they really are when correct plans are constructed of them. Though many of these reefs rise to the surface, the greater number are found at depths from 30 to 180 feet, and consist of sand and living coral, the latter covering the largest part of their surfaces. They run parallel with the shore, sometimes connected with the mainland by transverse banks. Deep water occurs close to them.*

With respect to the varied conditions under which coral reefs are found, probably the observer may conveniently first consider the manner in which different species of coral have hitherto been known to occur. As Mr. Beete Jukes has remarked, though the reef-making coral polyps are only known to us as living at depths not extending beyond 120 to 180 feet, there may be others forming masses of calcareous matter at greater depths with which we are unacquainted.† The evidence respecting corals of various kinds would lead us to infer that, like the molluscs above mentioned (p. 146), while some prefer, or are adjusted to particular bottoms, whether solid rock, sand or mud, at various depths, moderate or considerable, others are only to be found in shallow water. Viewing the subject in this light, the corals living at the surface of the sea may be compared with littoral molluscs keeping situations peculiar to them. While some appear adjusted to the nearly constant movement of ocean breakers, others, even at small depths, require tranquil water; so that at nearly equal depths the corals, forming the hard mass of the reef, or finding shelter amid its cavities, in the lee of lagoons, when there are such, divide themselves into two classes.

Referring to the early and swimming state of the reef-making coral polyps, we may assume that, wherever fitting conditions presented themselves, they could settle, adhere to a sufficiently hard substance, and commence the foundation of a reef. If we take coasts as they are variously presented to us, we find that, as regards depth, we may have the 120 or 180 feet, for the reef-making corals either close to the shore, or removed to various distances

* Darwin ("Structure and Distribution of Coral Reefs," p. 192), from information communicated to him by Captain Moresby.

† It would be well carefully to examine the coral reefs, which have been undoubtedly raised above the sea by geological movements, for the species contained in their lower parts.

from it. So that, assuming the swimming germs to meet with the requisite bottom, they can commence their reef-rearing labours at various distances from the land, and, raising the reefs, form very different lines around or adjoining it. Let, in the annexed diagram (fig. 76), *a a* be the surface of the sea round an island, *b b* the

Fig. 76.

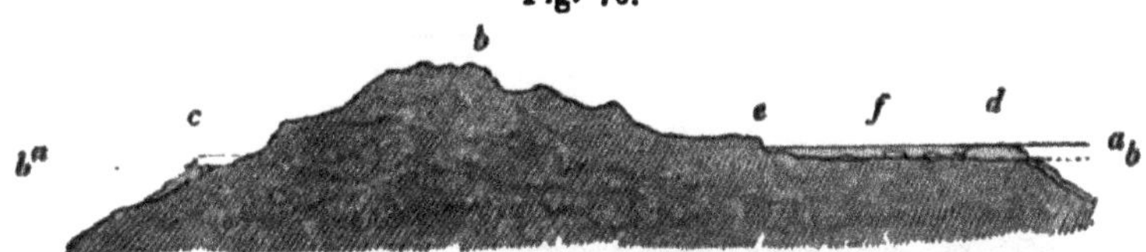

level beneath that surface at which the swimming coral germs can attach themselves and begin their labours, then at *c* the reef would be fringing and adjoining the coast; while at *d*, a bank might be raised up, forming a barrier reef to the coast *e*. Such a bank once established, the space *f*, between the coast, *e*, and the barrier, *d*, becomes fitted for those corals which require the shelter afforded by the latter. Whether from being best adapted for procuring food, or as affording conditions ill-suited to the coral-eating animals, the surface reef-making corals flourish in the surf of breakers, so that they grow, as a mass, outwards. With respect to original bottom, if there be sufficient tranquillity at the depth of 120 feet from wind-wave action, either directly produced on the spot by winds, or transmitted, as a ground or ocean-swell, from a distance, there appears no reason why the corals found at that depth, in lagoons and other sheltered situations inside barrier reefs, should not live and die under such circumstances, besides other corals, not yet known. These would form a base on which the more shallow water and littoral corals, among them those able to resist the breaker-surf itself, would begin their work. So long as these keep at sufficient depths, the mechanical action of the breakers will little affect them, but as they rise with the reef they gradually come within its influence, so that finally the coral masses are dealt with as the rocks of any other coast would be under similar conditions.

While corals thus forming a coast, may be, to a certain extent, adjusted to the powers of ordinary breakers, any increase in the force of the breakers over the resisting powers of the corals would break off portions of the latter, so that, during heavy gales of wind, the resistance becoming very unequal to the force employed, large masses of the coral are torn off and hurled over the reef inwards. This can scarcely happen without minor portions being also thrown over, or broken off from the detached masses, and the general action such that fragments of the coral mass fall outside the steep slope of the outward growth, a steep slope which we should expect to have been gradually formed as the coral reef rose within the mechanical action of the breakers. Let *a b* (fig. 77) be the surface

Fig 77.

of the sea in calm weather (for the moment considered without reference to changes of level produced by tides), *c d*, a depth at which reef-making corals can, other conditions being favourable, establish themselves, and *e*, *e*, the commencement of a reef, not raised so high as materially to feel any of the mechanical effects arising from any wave action, W, W, W, though every successive addition to the reef would bring it more and more within that influence. When, by vertical increase in the coral mass, a breaker could be formed by sufficient proximity to the wave, W, W, W, abrasion would commence as the coral resistance became unequal to the force employed, and the detritus would be scattered on each side, the inside probably, from the direction in which the force was applied, receiving the chief portion, while some fell outwards towards *b d*. As the coral growth rose to the surface, under ordinary weather, the increase more than meeting the loss by abrasion, the interior would be filling up also by corals, some of which required the shelter there afforded them. Outside, the breaker action would remove the smaller fragments in mechanical suspension, leaving the larger blocks, so that hollows amid the latter would get filled with a portion of the finer matter, the greater part of which would be carried out at the base of the reef, more or less ground into sand by the friction to which it may have been exposed.

If we suppose the reef to have so risen that it touches the surface of the sea, the growth of the coral still increasing the mass beyond the power of the surf to break off portions of the reef, a time would come when, from the usual breaker action upon coasts previously mentioned, fragments of various sizes, with coral pebbles and sand, from continued friction of the fragments in the surf, would be thrown up in a bank upon the reef, with sand added by the winds. The decomposition of the animal matter in the more freshly broken pieces of coral, added to any animal matter entangled amid the reef, would assist in its consolidation by, among other things, the production of carbonic acid, for combination with the water to act on the carbonate of lime of the corals, so that sufficient would be taken up in solution to cement them together, by subsequent deposit among the coral fragments, thus forming conglomerates and sandstones. A dry portion once above water, the often-described vegetation succeeds, the decomposition of which also affords free carbonic acid for the further solution of carbonate of lime, and an additional consolidation of the mass beneath by chemical means. Considering the mixture of animal matter in the coral mass itself, entangled among it in various ways, and by its decomposition affording carbonic acid, and the conditions under which this carbonic acid could be brought to aid in the solution of the carbonate of lime of the coral polyps, it will be seen that circumstances may often arise for the obliteration of the organic texture, and the substitution of calcareous matter, presenting an inorganic character, such as has been often remarked.

The nearer the surface the greater would be the power of the breaker action to peel off the upper coating of the reef, during heavy gales of wind, and cast the fragments inwards as well as the rounded pebbles which may have been formed in fitting situations at ordinary times by the common force of the breakers. We should expect this to be effected to distances beyond the margin of the reef, dependant upon circumstances, among which the rise and fall of the tide, both during ordinary weather and at the time of any heavy gale of wind, have to be regarded. Assuming *c*, *t* (fig. 78) to be the difference of the tide level, it will be obvious

Fig. 78.

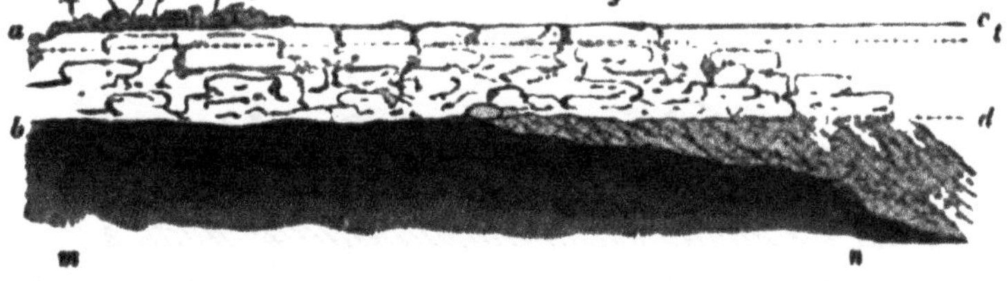

that any power which the breakers may have at the level *a*, *c*, will be changed during the rise and fall of the tide, ranging up and down all the portions of the reef exposed within the depth *t*, *c*. We have assumed as in the accompanying section (fig. 77), that the coral animals in their free swimming state met with a bank *m*, *n*, so that at the level *b*, *d*, they found the conditions, as to depth and other things suited to them. Assuming that they would not work beneath this level, as the reef rose to the island *f*, *g*, and the detritus outside accumulated, the latter would cover over the deeper part *n*, of the original bank, by successive coatings, over which the coral polyps would advance their work laterally, thus covering horizontally a detrital mass, laminated at an angle according to the slope on which it may accumulate. Taking the outside detrital increase at any given amount of cubic contents, it would follow that, according to the small slope of the original bank would be the rapidity of the lateral advance over which the corals might be disposed to work, steep slopes affording the least ground at a given level for such increase. For the sake of easy illustration we have assumed a bank such as that near Breaksea Spit, on the coast of Australia, and above mentioned (p. 182), which after retaining a certain general depth, and presenting a rounded margin, plunges into deep water.

It may now be desirable to consider the effects which would result, in the regions of coral reefs, from volcanic action. We have seen that within our own times, volcanic action has brought ashes and cinders to, and above the sea level in the Mediterranean (p. 70) and in the Atlantic (p. 100), that the islands so produced have been temporary, and that very probably the incoherent matter of which they were composed has been cut down to the depths at which breaker, or wave action could disturb and remove such matter. At least this could scarcely but happen, supposing no subsidence from the pressure of the water into the crater in such a manner as to lower the volcanic mass, independently of any subsidence from volcanic causes themselves, carrying down the mass of ashes and cinders beyond the influences of breakers and waves.

If in the annexed diagram (fig. 79) we consider *a*, *b*, *c*, *d*, to

Fig 79.

be a section of a volcanic cone, the top of which was forced during some eruption above the level of the sea, *e*, *f*, that this condition ceasing, breaker and wave action cut down the loose materials to the level *g*, *h*, one to which their influence could extend, even probably sifting the ashes and cinders, as in the case of the Island of Sciacca (Mediterranean), so that a somewhat stony bottom might be the result, we should have conditions fitted, in the coral-reef seas, for the settlement of the germs of the reef-making polyps at *i* and *k*. At those points of the section the reefs would increase as above noticed, when they rose sufficiently high to be acted upon by the breakers, fragments broken off, and partly thrown down on the outsides, towards *a* and *d*, and the corals spreading over them as previously noticed. Inside there would be a lagoon, which, as soon as a general barrier of coral was established outside, would be filled in the usual manner with corals and other marine creatures suited to the sheltered conditions there found. The overflow of the breaker-waters, and the rise and fall of tides combined, would tend to keep open a channel or channels between the lagoon and the sea outside, and finally, terrestrial vegetation would establish itself upon a coral bank chiefly raised into the atmosphere by the piling influence of the breakers upon the coral ridge. The forms of such islands would necessarily depend much on the horizontal section of the volcanic accumulation, when cut down by breaker and wave action. We should expect the submarine and steep flanks of the mass to be encrusted by coral sands and fragments in proportion to the time during which the reef-corals may have been increasing outwards in any particular locality, so that the sounding-lead could bring up little else around the coral reefs and island except coral detritus, and the marine animals which could exist under the needful conditions at various depths around the main mass.

As we have abundant proofs that, not only ashes and cinders have been vomited out of volcanic vents, reaching to and beyond the sea-level, but molten rock also, the whole even attaining considerable altitudes, such as the volcanic heights of Hawaii, and others of the Sandwich Islands, with deep water around them, it may not be undesirable to consider the conditions under which coral reefs might be gathered around such volcanic masses. Let, in the annexed section (fig. 80) *a*, *b*, *c*, represent the remains of a mixed volcanic mass of molten rock, and of ashes and cinders, cut away by atmospheric influences and breaker action, so that a portion of hard rock, *b*, perhaps once molten matter in the crater of a volcano, stands above the level of the sea, while at *g* and *f* incoherent ashes and cinders are cut back by breaker action (as in fig. 79) to this hard rock. We should now have conditions for the formation of reefs at *f* and *g* under the same circumstances as above noticed (fig. 79),

Fig. 80.

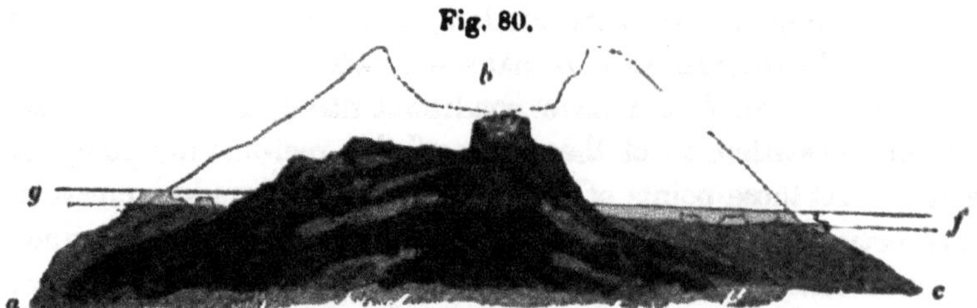

with this difference, that instead of an uninterrupted lagoon in the interior of the coral reefs, there would be land emerging from it, so that these reefs would be encircling. In addition to the usual causes of keeping channels open, the islands, if of good size, might contribute fresh-water streams, at times charged with detrital matter, preventing the increase of the coral reefs in the lines which they traversed.

It would appear desirable, in the first instance, that an observer should direct his attention to the conditions under which coral reefs and islands could be formed, either as lagoon islands, reefs touching or encircling land composed of ordinary detrital or igneous rocks, or upon shoals and banks ranging in front of considerable lines of coast. It is not a little interesting further to consider the mode in which a general mass of coral matter would be composed, after a lapse of time sufficient to complete the filling up of lagoons, with or without the protrusion of dry land formed of ordinary rocks through them, or the space between an outer line of coral reefs and a considerable range of coast, such as that of a portion of Eastern Australia.

As to the height to which corals may rise, Mr. Beete Jukes found coral polyps alive six or eight inches out of water, and so remaining for nearly an hour, until the return of the tide. He often observed the same fact, and believes that an exposure to the air and sun will not kill many of the polyps, so long as the coral remains in a position of growth, the cells retaining their moisture. He has seen blocks of living astræa, the tops of which were 18 inches above water. This shows that we may take the ordinary tide level for that to which the reef-making coral polyps can work under favourable conditions; and that there may be a mass of matter coinciding with the line of a main reef round a lagoon-encircling island, or in front of a long range of coast, which may, from the top to the other substances on which the reef reposes, be chiefly formed by the growth of corals upon each other, occasionally mixed with the hard remains of marine animals inhabiting the cavities amid the corals, and with detrital portions driven in amid the hollows of the rising mass.

To whatever extent the germs of coral polyps could settle upon any surface, beneath the sea-level, suited to their development, conditions would change, as the reef portion rose seaward, behind the shelter gradually afforded from the roll of the waves, and their action on the bottom beneath; so that while sands and fragments of corals, broken off by the breakers, arranged themselves, as above mentioned, outwards, (the reef-making polyps working over this detritus,) a very complicated series of deposits and coral growths would be formed inwards. In the case of coral lagoon-islands, there would finally be calcareous plateaux of very variable areas, some many square miles in extent, of equal levels, separated, in such regions as the coral island groups of the Pacific Ocean, by irregular intervals of deep water. These isolated sheets of matter of general similar character would be, to a certain extent, stratified, though coral growths may pierce the general mass in various directions; the strata composed of beds of coral sand and mud, as these gradually accumulated, mingled with the shells of molluscs, the spines and coverings of echinoderms, the hard remains of fish, with possibly also those of certain birds and turtles, even the eggs of the latter being preserved in the higher sand-banks. There would be deposits of calcareous mud outside also, at depths where it could accumulate in an undisturbed manner; this calcareous mud borne out of the outlet channels during ebb-tides, and when heavy gales drove an abundance of water over the weather-side of the encircling reefs, to escape out of the same channels. Such mud might be widely spread by tidal streams and ocean currents, and so far constitute a kind of connexion, enveloping uneven and submarine ground, between the coral plateaux.

In the case of the reefs, more or less encircling islands of varied magnitudes, and composed of ordinary sedimentary and igneous rocks, there would be a modification of the deposits inside the reefs, so far as a supply of decomposed mineral matter from atmospheric influences and ordinary detritus from such lands would be concerned. The remains of a larger and more varied amount of terrestrial vegetable and animal life would be there expected; as, also, under favourable conditions, the addition of the harder parts of fluviatile creatures. Where intermingled with the simple lagoon reefs, there would be corresponding modifications of the interior deposits at the same general level.

As respects the accumulations, for so many thousand square miles, inside the Great Barrier Reef, off the eastern coast of Australia, there would be a great sheet of matter, as a whole, having a certain general character. Viewing, generally, this range of coast, there is a great absence of fresh waters draining from the adjoining land; indeed, water is scarce along it under ordinary conditions. Hence no material influence is exercised on the growth of the coral polyps, and their associated life, by rivers and streams of fresh water, either clear or charged with detritus in mechanical suspension. Seaward we have the same conditions as the outward portions of the lagoon reefs, for about 1000 miles; the southern portion ranging beyond the circumstances fitted for the development of the reef-making coral germs, and resting on banks of ordinary silicious sand, while the northern portion is terminated by the influx of river waters, bringing down muddy matter from New Guinea; thus also preventing the same germs from properly establishing themselves, though conditions would otherwise appear to be fitted for their development, for passing the outflow of the river waters, coral reefs are again established to the northward.

Inside this long line of outer reefs, accumulations are effected as in the ordinary isolated lagoon reefs, until the main line of coast is approached, where modifications would be expected, though on a larger scale, of the kind found around the islands, composed of ordinary rocks, inside encircling reefs, and above noticed.* Such a small volume of fresh waters flowing outwards from the land, comparatively little detrital matter from the interior seems transported far seaward, so that the calcareous detritus derived directly from the reefs, and ground finer by friction from breaker action, or passed through the animals feeding on the coral polyps, readily becomes forced towards the land from the prevalent action of the waves in that direction. It there mingles near the coasts with such detritus as may be derived from the land by breakers, however modified these may be from the shelter afforded by the outer reefs, or be carried out into such tidal streams as prevail by the rivers in flood. Viewed as a whole, we should expect much continuity in some of the deposits, particularly the finest, in many parts of the great area comprised between the coasts and the outer great barrier reefs, in which an abundance of molluscs, radiata, and layers of certain corals, with the harder parts of fish and crustaceans, would be entombed. Near the land, and particularly where mangrove swamps prevail, there would be modifications of these continuous deposits. As a whole, it would constitute a great mass

* The green mud off Cape Direction, east coast of Australia, is wholly calcareous. —Beete Jukes, "Narrative," &c.

more or less stratified, intermingled here and there, especially towards the outer barrier reefs, with complicated mixtures of coral growth in reefs, the detrital matter derived from them, the harder parts of other marine animals living among them, and the alterations of structure produced by chemical means.

Stratification, or an approximation to it, is not confined to the coral sands and mud, and the layers of organic remains which may be intermingled with them, for a tendency to split into slabs is often noticed in the mass of the reefs. Indeed, Mr. Beete Jukes not only mentions such a mode of occurrence at Heron Island (part of the eastern Australian coral accumulation), but joints in that reef also, splitting the coral rock into blocks of from one foot to two feet in the sides. These joints or divisional planes are parallel to the dip and range of the beds respectively, and the coral beds dip seaward at an angle of from 8° to 10°.

The observer has next to turn his attention to the consequences, as regards coral reefs and islands, which would follow any of those changes of the relative levels of sea and land, both on the small and large scale, and to be subsequently further noticed, which the study of geology teaches us has so frequently occurred.

There can be little doubt of coral banks and reefs similar to those in the seas of our times, and in coral-reef regions, having been raised above the surface of the sea, like other marine accumulations, forming dry land. Such have been long known. MM. Quoy and Gaimard, who accompanied the expedition of M. Freycinet, and who remarked on the moderate depths to which the reef-making corals appeared to extend,* mention that on the coasts of Timor coral banks so occur above the sea level as to have induced M. Peron to consider the whole land formed of them.† At Oahu, and other places in the Sandwich Islands, coral banks have been long known to extend inland, and more modern researches have confirmed these observations. Mr. Couthouy, in particular, gives an account of ancient reefs, now raised above the sea level, at the islands of Maui, Morokai, Oahu, and Tauai.‡ We had occasion to remark also some years since on the raised coral reefs on part of the coast of Jamaica.§ Elizabeth Island, off the eastern side of the low archipelago,

* Quoy and Gaimard, Sur l'Accroissement des Polypes Lithophytes considéré géologiquement, Annales des Sciences Naturelles, tom. vi.

† Upon proceeding inland a short distance, MM. Quoy and Gaimard found these coral banks resting on vertical beds of slate.

‡ Remarks on Coral Formations.

§ Geological Manual, 3rd edit. 1833, p. 165.

(between Ducie and Pitcairn Islands,) has been considered, from the description of Captain Beechey,* to be a good case of a raised coral island, its flat summit 80 feet above the sea. Mr. Darwin has accumulated a mass of information† from the personal communications of the Rev. J. Williams, Mr. Martens, (of Sydney,) and Mr. G. Bennett, and from numerous voyagers, and other authors, showing coral banks elevated to various heights above the level of the sea in the Cook and Austral Islands, Savage Islands, the Friendly Islands, the Navigator Group, the New Hebrides, New Ireland, the Marianas, the East India Archipelago, the Loo-Choo Islands, Ceylon, Mauritius, Madagascar, part of the eastern coast of Africa, the Red Sea, and the West India Archipelago. In fact, this list comprises examples in all seas where coral reefs and islands have been noticed, and leaves little doubt that since coral reefs and islands were formed, as they now are, in the fitting regions, many throughout those regions have been raised above the level of the sea into the atmosphere.

Lafû Island, one of the Loyalty group, has also been noticed by the Rev. W. B. Clarke as a raised coral island. It is about 90 miles in circumference, and surrounded by a fringing reef, upon which the depth gradually increases outwards for a quarter of a mile, the reef then plunging into deep water. The whole island is composed of dead coral. Its average height above the sea is about 120 feet; and it attains, at points on the eastern side, an elevation of 250 feet. There is a ledge or shelf, like that now surrounding the island, at 70 or 80 feet above the sea. The surface is table land, with hollows and elevations, such as characterise a coral reef. Mr. Clarke infers that this island has been elevated at two distinct periods; at the first to the amount of 170 feet, at the second to 80 feet additional height.‡

In considering the elevation of coral reefs above the sea-level, the portions of a sea-bottom should also be taken into account, which may by the same means be brought within such a distance of the surface water, that the germs of the various coral polyps, which aid in the establishment of a reef, could find the needful conditions for establishing themselves. It has to be borne in mind that inequalities of the sea-bottom exist as much in coral regions

* Beechey, Voyage to the Pacific.

† Darwin, Structure and Distribution of Coral Reefs, pp. 132-137.

‡ Clarke, Quarterly Journal of the Geological Society of London, vol. iii., p. 61, 1847.

as in others; indeed, the igneous character of many islands, particularly in the Pacific and Indian Oceans, would lead us to expect no slight variations in this respect. Many a boss or extended collection of volcanic inequalities may be raised by the same movements as those which have elevated the coral banks to various altitudes above the sea. While some pierced the surface-level of the water, to be dealt with by the atmosphere and the breakers, as far as their varied resistances to the destructive action of the one or the other might permit, others would be differently circumstanced. Some formed of incoherent cinders and ashes would readily be cut down to the level to which breaker action could extend; while others would only just reach the needful depth beneath the surface-water for the establishment of coral reefs.

As respects inequalities of sea-bottom, if the Great Bank of Newfoundland were in coral regions, and were elevated from its present relative level, so that its broad platform with its common depth of from 40 to 50 fathoms, one small portion of the area being occupied by the Virgin Rocks, were raised about 20 fathoms, by which coral reef-making germs could fix and develop themselves under fitting conditions, we should have an area of between 35,000 to 40,000 square miles, around the irregular margin of which there would be conditions, as represented in fig. 81 (p. 198), for an extended border of coral reefs. The sea around the margin would often be suddenly deep. The new Admiralty Chart of the North Atlantic gives 106, 137, 147, 132, 107, and 149 fathoms, as now found close off the south-eastern side of the Great Bank. There would be an island of small size, now the Virgin Rocks, above water with still 20 fathoms close to it; and supposing a somewhat questionable shoal (with 3½ fathoms upon it), about 40 miles to the eastward of the Virgin Rocks to be really existing, there would be another small island in the same area. The Great Bank, with its continuation the Green Bank, would be separated by a channel, then 55 to 79 fathoms deep, from the St. Pierre Bank, round the edges of which there would be conditions for a fringe of coral reefs, enclosing a large area of water with no other land above its surface. Indeed, the St. Pierre Bank would then hear an external resemblance to a great atoll, about 140 miles across from south-east to north-west, with a maximum breadth from south-west to north-east of about 70 miles, and having a somewhat steep slope

Fig. 81.

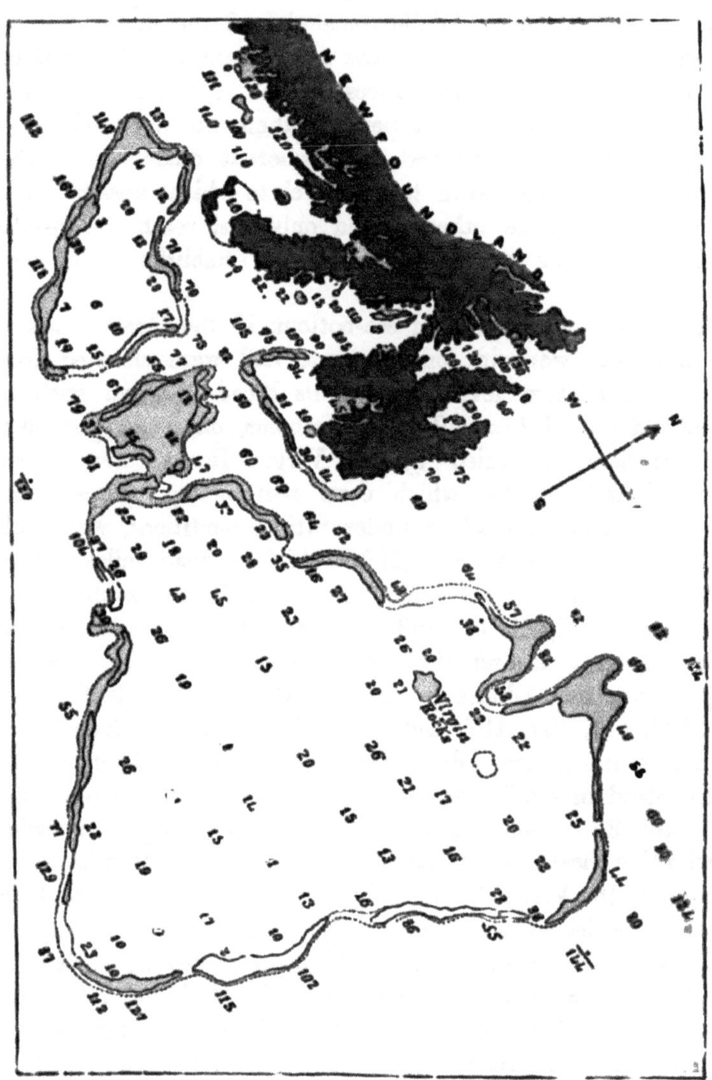

outside, on the south-west side, into 118, 160, and 149 fathoms, the change from the present depths being taken into account.*

In the Bermudas we seem to have an instance of an isolated bank in the Atlantic, far distant from land, and rising from deep water, upon the upper part of the crown of which coral reefs have established themselves, mingled with others which have been described as chiefly composed of serpulæ, nulliporæ incrusting the work of the marine animals as upon the coral reefs of the Pacific (p. 169).† The remarks of Captain Nelson, of the Royal Engineers, having chiefly reference to the geological structure of these islands, there is yet much to be accomplished by the experienced naturalist respecting the reefs themselves, which are especially interesting from their geographical position, and the marine life connected with it.

Although deep water is reported to surround the bank upon which the reefs and isles of Bermuda are situated, the reefs themselves are immediately bounded outwards by shallow water, only 6 and 7 fathoms being marked on the charts as a somewhat common depth immediately beyond the reef, deepening somewhat further distant to 12 and 15 fathoms. Captain Nelson describes ‡ the Bermudan group to consist of about 150 islets, lying in a north-east and south-west direction, within a space of 15 by 5 miles, and containing altogether an area of about 21 square miles. This group is situated very near, and conformably to, the south-east side of a belt of reefs, partly formed by corals, and partly by serpulæ, of a rude elliptical form, 25 miles long by 13 miles broad.

* The extensive submarine area of the Newfoundland banks is also highly interesting, as exhibiting a very slight difference in level. From 250 to 300 feet beneath the surface-water seems a very common depth, though there appear to be gradual swelling portions bringing the bottom more upwards. If these banks were elevated above the sea, they would present an irregularly bounded platform, divided by one main channel, many thousand square miles in extent, the chief height above which would be a rocky eminence, about 240 feet above the general surface, where the Virgin Rocks now occur; and, if the questionable shoal on the eastward really exists, a boss of ground of about the same altitude in that direction also. If these banks have been in previous geological times raised into the atmosphere, all traces of considerable hills and valleys, which may then have existed, have been obliterated. And this may readily have happened from the levelling effects of breaker action, combined with the distribution of the detritus by tidal streams and ocean currents, as the land may have slowly subsided. Be this as it may, detritus would not readily be now borne to these banks from the adjoining coasts of Newfoundland by any drifting action along the bottom, since, as previously mentioned, deep water occurs between the banks and that land.

† The occurrence of coral reefs so far northward in the Atlantic is referred to the influence of the heated water carried northerly by the Gulf Stream.

‡ Nelson, Transactions of the Geological Society of London, 2nd series, vol. v.

The channels amid the islets are shallow, and the depth of water within the boundary reefs rarely exceeds 12 to 14 fathoms. The highest land rises to about 260 feet above the sea at Sears Hill, and Gibbs Hill has an elevation of 245 feet.

Captain Nelson describes the islands as altogether calcareous, the beds varying from loose sand to limestone, so compact as to receive a good polish, the whole derived from animal secretions, chiefly marine, though the remains of land shells, and birds' bones are also mentioned. From the mode of occurrence of the calcareous beds, and especially from the saddle-shaped sections observable throughout the islets, Captain Nelson infers that the deposits have been effected by means of the wind, driving the calcareous sands, including fragments of, and whole shells, in the usual way before it, and heaping them up irregularly into sand-hills, the component parts of which have been variously consolidated. A foot of red earth, containing vegetable matter, commonly covers the calcareous accumulations. Fragments of coral and shells are noticed as common, and the remains of *Lucina* (*Venus*) *Pennsylvanica* are especially pointed at as frequent. *Turbo Pica* is also common; and Captain Nelson is inclined to refer its occurrence on the heights to the hermit crabs, which he has seen running about with these shells.* Coral reefs occur inside the main, or outside reefs, and do not rise above low water, except at spring tides. Over the bottom of this basin calcareous sand and chalky clay (the best anchoring ground) are distributed. The tides average a rise and fall of about 4½ feet, and at low water the main reefs stand about 2 feet above the sea.

Although there may be good evidence of much of the calcareous accumulations of these islets having been effected by means of the wind, piling up sand and fine calcareous particles driven, in the usual way, by breaker action at high tides, and by gales of wind,† still there would also appear evidence that there may have been some elevation of the general mass, since a bed containing shells of the *Lucina Pennsylvanica*, and now about 6 feet above water, has an even range from Phyllis Island to Harris Island. Under this

* *Mellita* (*Scutella*) *quinquefora* is noticed as found, the pores of the crusts filled with crystalline carbonate of lime, like the echinites in the European chalk. Turtles' bones have been discovered in the accumulations, as also the remains of cypræa and bulla.

† Sand drifts are now in progress; and Captain Nelson especially refers to one encroaching on the land,' and arising from works executed a few years since, by which a protecting vegetation was removed, and the wind acted on a sufficient area of free sand to work its destructive way into a more considerable mass.

hypothesis a mixed accumulation, by means of both wind and water, would not appear inconsistent with the sections given by Captain Nelson. The following (fig. 82) is one similar to many sections of sandstone deposits formed beneath water.

Fig. 82.

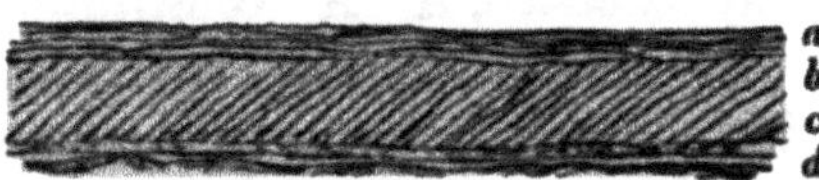

a, b, c, Ordinary friable calcareous rock.
d, Recent loose deposit in front of the cliff.

Having so much evidence of the elevation of coral reefs in various parts of the world, and seeing that depressions of ordinary rock-accumulations (to be hereafter noticed) have been effected, often on the large scale, also in various regions, it would be expected that coral reefs, and even extended areas in which they occur, may in like manner have been subjected to the like movements. Mr. Darwin has very ably sustained this view, both as respects single coral reefs, and extended regions in which they may be found.* To account for barrier reefs and atolls, which have been produced by the subsidence of land, around which fringing coral reefs only were first attached, he gives the following illustration:—†

Fig. 83.

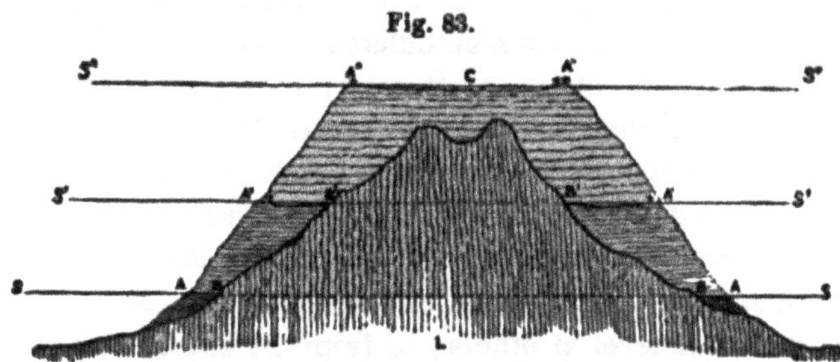

Let A, A, be the outer edges of fringing reefs, on two opposite sides of an island L, at a sea-level, S, S; B, B, shores of the island; A′, A′, outer edges of the fringing reef, after its upper growth, during the gradual subsidence of the island L, by which the relative sea-level became transferred to S′, S′; then A′ B′,

* Darwin, Structure and Distribution of Coral Reefs, chap. vi. On the distribution of coral reefs with reference to the theory of their formation.

† In this section the two diagrams given by Mr. Darwin (Structure of Coral Reefs, pp. 98 and 100), have, for convenience, been thrown into one. In other respects they are the same. As Mr. Darwin points out, the sections of the lagoons are exaggerated.

B′ A′, are sections of the lagoons inside barrier reefs on each side of the land, after this subsidence, and B′ B′ the new shores of the island. Should the gradual subsidence continue, so that the relative level of the sea, as regards the island L, be changed to S″ S″, then the original island round which the corals first formed a fringing belt, would be completely concealed, A″ A″ constituting the outer reefs of an atoll, and C its contained lagoon. In this manner the original mass of land, which might be a volcanic cone, or some modification of that form, would become encrusted by the remains of marine animals, or the detrital and chemical accumulations arising from such remains, including the fæces of the various reptiles, fish, crustaceans, molluscs, and other marine creatures which inhabited the reefs and lagoons. This would be contained within a general crust, due chiefly to the work of the outer reef-making polyps; this crust again covered, after a certain depth, by the débris of the reefs, broken away by breaker action, as the subsidence continued, and accumulated over the first formed reefs in the usual talus. With the exception of certain portions of this outside distribution of the débris from breaker action, there would be a general horizontal arrangement of the rest; even the crust of the outer reef exhibiting, to a certain extent, this mode of accumulation. A large volume of calcareous matter, obtained by marine animals from the sea and their food, mingled with some terrestrial vegetable and animal matter, would be thus accumulated;—and no small amount would be required to fill in, as it were, the space between the outer crust of the rising reefs and the original land.

Assuming the hypothesis good for the single case adduced, there would appear no difficulty in applying it to a variety of modifications, arising either from the form of the original land, which may be either of small or considerable extent, mountainous or hilly in one part, and more level at others; or from the altered and changing arrangements of the surface distribution of land and water, at different times as the subsidence continued, being more sudden at one time than at another, or being interrupted by pauses of greater or less duration.* The Maldives are considered as affording a good example of the effects of the submergence of the land, after the first incrustation of its shores by the reef-making corals, so that a considerable fringing reef round a large island, like that of New Caledonia, became divided up into numerous rings of coral

* The varied effects of submergence of coral reefs and islands will be found treated at length, and with reference to reefs and islands considered to bear out this view, in Mr. Darwin's Structure of Coral Reefs, chapter v., entitled Theory of the Formation of the different Classes of Coral Reefs.

reefs, crowning different heights of the original land, the general outline of the latter still preserved from the upward increase of the general mass of the corals. The Great Chagos Bank, on the south of the Maldives, presents peculiarities well worthy of attention;* and Mr. Darwin considers it as an instance, in which the corals of the main reef perished before or during a submergence, now such that the great outer reef, instead of being within the break of the sea, is sunk from 5 to 10 fathoms beneath the surface of the water, two or three spots only rising into islets. As the depth of the outer reef does not appear such as to prevent reef-making corals from developing themselves, and as interior knolls present themselves at the same depth with luxuriantly-growing corals, there would appear some other reason than mere depth of water, with its consequences, preventing the establishment of more than a slight amount of living coral polyps on these reefs. Supposing the outer reef to have once flourished in the common manner, at and near the surface (and there are still two or three islets above water), perhaps the somewhat sudden submergence of an extended range of islets, thickly studded over the outer reef, with their vegetation and sands, would, for a long time at least, be very unfavourable to conditions well suited for the re-establishment of the upper reef-making corals. Wave and tidal action would tend to distribute and move about the sands over the sunk reef in a manner scarcely fitted to the habits of the upper reef-making coral polyps, or the firm establishment of their germs.

With regard to the incrusting of islands by coral masses, including the accumulations mechanically and chemically obtained from the stony matter, chiefly calcareous, secreted by the

* "The longest axis is ninety nautical miles, and another line drawn at right angles to the first, across the broadest part, is seventy. The central part consists of a level, muddy flat, between forty and fifty fathoms deep, which is surrounded on all sides, with the exception of some breaches, by the steep edges of a set of banks, rudely arranged in a circle. These banks consist of sand, with a very little live coral; they vary in breadth from five to twelve miles, and on an average lie about sixteen fathoms beneath the surface; they are bordered by the steep edges of a third narrow and upper bank, which forms the rim of the whole. The rim is about a mile in width, and with the exception of two or three spots where islets have been formed, is submerged between five and ten fathoms. It consists of smooth, hard rock, covered with a thin layer of sand, but with scarcely any live coral; it is steep on both sides, and outwards slopes abruptly into unfathomable depths. At a distance of less than half a mile from one part, no bottom was found with 190 fathoms; and off another point, at a somewhat greater distance, there was none with 210 fathoms. Small steep-sided banks or knolls, covered with luxuriantly-growing coral, rise from the interior expanse to the same level with the external rim, which, as we have seen, is formed only of dead rock."—Darwin, Structure and Distribution of Coral Reefs, p. 39.

polyps and other marine creatures forming or inhabiting coral reefs, if we can have a growth and accumulation adjusted to the submergence,* the geological results would be such that upon again being elevated above the sea-level, (as has happened so often with many regions during the lapse of geological time after submergence), in areas such as that of the corallian portions of the Pacific, numerous masses of calcareous accumulations would present themselves, often, perhaps, corresponding in general character at equal, or nearly equal levels, and having the appearance of being the remains of limestone deposits, once continuous, though in reality they had never been united. If the Cape de Verde Islands, the Canaries, and the Azores, were to be incrusted with coral reefs, and be gradually depressed beneath the level of the sea, so that the reefs and the consequent accumulations inwards could be adjusted to the rate of submergence, and were again raised above the sea (as at least some of them would appear to have been, since notwithstanding their general igneous character, sea-bottoms and shores around them, of the more recent geological times, termed the tertiary, are uplifted), these fallacious appearances would be very marked.†

The observer would readily expect to find, in regions where coral reefs abound, and volcanos are now, or have been active during their formation, that there are occasional mixtures of igneous products with the coral accumulations. In Mr. Beete Jukes' account of the Great Barrier Reefs of Australia, he mentions mingled substances of this kind at Murray and Erroob

---

* If the submergence were so rapid that the growth, and consequent accumulations inside the reefs did not become adjusted to it, and supposing the reef-making corals only able to flourish in certain minor depths, it is obvious that the reefs could not increase upwards, but remain beneath like any mass of inorganic matter, the reef-making polyps perishing.

With respect to the rate of growth of reef-making corals, the evidence is at present somewhat uncertain and contradictory. Some contend that the growth is very slight, reefs having been known in their present state for a long time; while others consider their increase as more rapid. There is evidently a want of more information on this subject, especially as respects the conditions under which the appearances supporting these different views may have been caused.

† There are suddenly very considerable depths around the Cape de Verde Islands. Even in the channel between Sal and Bonavista, a line of 232 fathoms found no bottom. The same with the Canaries. There is no bottom at 309 fathoms close on the north of Palma. The like with Madeira, off the west end of which a line of 280 fathoms does not reach the bottom. Very deep water surrounds the various islands of the Azores. There is a depth of 300 fathoms close on the south of Santa Maria, and no bottom with 320 fathoms of line between that island and the Formigas, on the north-east. Around Pico, Fayal, San Jorge, and Terceira, there are depths of 200 and 300 fathoms near the land; and there is very deep water around Flores, the most western isle.

Islands, where beds are found containing variable portions of trachytic lava and calcareous rocks, some of the lumps of lava and limestone being apparently rounded by attrition. There are also beds of volcanic ash, or sand, in which calcareous grains are dispersed. In Erroob, igneous rocks cover the sandstones and conglomerates. In this region, also, pumice would appear to have been at one time much drifted about, arising probably from volcanic eruptions in directions whence a portion of this light substance could be driven by prevalent winds and currents. It seems to have become mingled with the coral deposits. Portions of it are embedded in the coral rock of Raine's Islet, and are frequent in the coral conglomerate on the north-east coast of Australia.* Captain Wilkes describes† portions of vesicular lava as found among blocks of coral conglomerate at Rose Island, a small and low coral island, forming the most eastern of the Samoan group. We should expect many mixtures of volcanic rocks with coral sands and pebbles on the beaches of volcanic islands, fringed by coral reefs, as appears often to be the case, and also an occasional overflow of lava on reefs adjoining land liable to volcanic eruptions.‡

* Beete Jukes, Narrative of the Voyage of the "Fly." He describes flats of coral conglomerate, half a mile wide, as frequent along shore on the N.E. coast of Australia. Upon all these flats, and about ten feet above high-water, there is an abundance of pumice pebbles. They occur on the east coast of Australia, under similar conditions, for 2000 miles; are rarely seen on the present beach, or found floating at sea; and Mr. Beete Jukes infers, that this proves either the stationary character of the coast, or that it has been equally affected, for this distance, by elevation or depression. He allows for the piling action of the breakers, and considers it as not improbable that the coast has been slightly elevated, or, at least, has not suffered any depression through a long lapse of time.

† United States' Exploring Expedition, vol. ii., p. 64.

‡ A coral bed, ten feet thick, is stated to occur between two lava streams at the Isle of France; the coral bed elevated, since its formation, above the level of the sea.

CHAPTER XII.

TRANSPORT OF MINERAL MATTER BY ICE.—HEIGHT OF SNOW-LINE.—GLACIERS.—CAUSE OF THE MOVEMENT OF GLACIERS.—GLACIER MORAINES. MOTION OF GLACIERS.—GROOVING OF ROCKS BY GLACIERS.—ADVANCE AND RETREAT OF GLACIERS.—GLACIERS OF THE HIMALAYA.

VERY considerable attention has, of late years, been directed to the influence of ice in the distribution of detritus, both upon dry land and over the bottom of the sea, and to the mechanical effects ice may produce on hard rocks, or loose accumulations, on or against which it may move or be thrown, upon the land or beneath the sea.

We observe the influence of the sun's heat to be now such (whatever view may be taken of any supposed heat in the body of the earth itself, sufficient in previous times, to prevent the formation of ice on its surface), that the cold of the planetary space, as it has been termed, so acts upon the earth, that it is, as it were, encased in a comparatively thin warmer space, outside which, water remains permanently solid; this space having a spheroidal form somewhat more oblate than the sea-level, so that, at the equator, there is a difference of from 16,000 to 17,000 feet between the two, and that it joins that level in the Arctic and Antarctic regions. Above this, it is inferred that the temperature continues to decrease in the atmosphere until, finally, that of the planetary space alone prevails.*

Taking thus the heat derived from the sun as so influencing the present surface temperature of the earth, that the cold of the planetary space does not render the waters solid over the whole face of the world, we should, from the conditions under which this heat could prevail, anticipate many minor modifications in its

* Fourier inferred that the temperature of the planetary space was −50° centigrade (58° Fahr.), and Svanberg held it was −49°·85 centigrade, employing another method. Observing this near approach to the result given by Fourier, the latter calculated the temperature according to Lambert's statements, and obtained −50°·35.

action.* These would arise from its different absorption and radiation according as it fell upon land or water, and in different latitudes; from the varied relief and character of the land, and its intermixture with surface waters; from the variation in the waters as to depths, and the motion of some portions of them from colder to hotter regions, or the reverse; from the movement of the atmosphere and its varied conditions; and from the periodical change in the position of portions of the earth's surface, according as one hemisphere or the other becomes most exposed to the influence of the sun.

Numerous observations have shown the exact regularity of the space, in which water commonly remains liquid, to be much disturbed by the modifications noticed; so that, for all the purposes required by the animals and vegetables of our planet, certain regions are rendered habitable which would otherwise scarcely support life. A very marked instance of this kind is found on the north flank of the Himalaya, where the perpetual snow-line, as it is termed, is, from a combination of physical conditions, more elevated by 1500 or 2000 feet than on the southern side of the same great range of mountains.† Minor modifications of the same kind are abundant, as also from the influence of great surfaces occupied by the sea, and from prevalent winds sweeping over it and reaching land; thus producing marked elevations of the general temperature above that at which ice would be common.

To obtain the snow reposing on the regions or elevated mountains, piercing through the space above noticed into those portions of our atmosphere where the temperature is such that snow more or less encrusts them during the whole of the various climatal changes of the year, we have to infer evaporation from the land and water, modified according to their various states, surfaces, and

---

* Respecting the temperature of our atmosphere, M. Arago has remarked, (Ann. de Phys. et de Chim., tom. 27,) that, "1st, in no part of earth on land will a thermometer, raised from two to three metres (6·5 to 10 English feet) above the ground, and protected from all reverberation, attain 46° centigrade (114°·8 Fahr.); 2ndly, in the open sea, the temperature of the air, whatever be the place and season, never attains 31° centigrade (87°·8 Fahr.); 3rdly, the greatest degree of cold which has ever been observed upon our globe, with the thermometer suspended in the air, does not descend 50° centigrade below zero (58° Fahr.)." To this he adds, "4thly, the temperature of the water of the sea, in no latitude, and in no season, rises above 30° centigrade (86° Fahr.)."

† With reference to the snow-line on the northern flank of the Himalaya, Dr. Hooker states (letter to Sir William Hooker, dated Tongu, N.E. Sikkim, altitude 13,500 feet, July 25, 1849), "that the snow-line, in Sikkim, lies on the Indian side of the Himalayan range at below 15,000 feet; on the Thibetan (northern) slope, at about 16,000 feet."

localities, sufficient to afford the needful falls of water in this form.*

From the polar regions, where we find such a great amount of climatal change, that the influence of the sun, as far as it can be there experienced, is uninterrupted, or nearly so, during one-half of the year, and unfelt during the remainder, to the tropical regions, where portions of mountain masses may rise so high into the atmosphere as to support a covering of snow, there are necessarily great variations of temperature, the latter becoming less changeable, as a whole, in the equatorial portions of the earth.

When attention is directed to the effects arising from these variations of temperature, it is found that the production of glaciers

---

* Experiments do not seem to give the temperature at which the evaporation of snow or water ceases, so that while a limit may be inferred for this evaporation at some height to which parts of a mountain-chain might be elevated, it might readily happen that there are none such on the face of our planet, the vapour of water always mixing with the gases of the atmosphere up to all the heights in it to which parts of the earth's surface have been protruded.

Humboldt (Fragmens Asiatiques, p. 549) has given the following table of the snow-line on certain mountain ranges:—

Mountains.	Latitude.	Height above the Sea.
		English Feet.
Cordillera of Quito . .	0° to 1½° S.	15,730
——— Bolivia .	16° ,, 17¾° S.	17,070
——— Mexico .	19° ,, 19¼° N.	15,090
Himalaya:—		
Northern Flank . .	30¾° ,, 31° N.	16,620
Southern Flank . .	. .	12,470
Pyrenees	42½° ,, 43° N.	8,950
Caucasus	42½° ,, 43° N.	10,870
Alps	45¾° ,, 46° N.	8,760
Carpathians	49° ,, 49¼° N.	8,500
Altai	49° ,, 51° N.	6,400
Norway		
Interior	61° ,, 62° N.	5,400
,, ...	67° ,, 67½° N.	3,800
,, ...	70° ,, 70¼° N.	3,500
Coast	71° ,, 71¼° N.	2,340

To these may be added the following observations:—

Locality.	Latitude.	Height above the Sea.	Authority.
		English Feet.	
Bolivia	16° to 18° S.	16,963	Pentland.
Cordillera of Chili . .	33° S.	14,500	Gillies.
Chiloe.	40° to 43° S.	6,000	Fitzroy.
Tierra del Fuego . .	54° S.	3,750	King.
Kamtschatka . . .	57° N.	5,308	Erman.
Bären Island. . . .	74° 30′ N.	590	Durocher.

M. Durocher (Mémoire sur la limite des neiges perpétuelles. Voyage de la Recherche, 1845) places the line of perpetual snow in the Arctic Ocean in 78° N.; so that, at Spitzbergen (N.W. coast), it descends to the level of the sea.

stands somewhat prominently forward among those which have a geological bearing. In the Alps, Europeans have been long familiar with the elongated masses of ice, so called, descending from the regions of snows, through ravines and rocky depressions of various forms, even into fertile valleys, where ripening crops and ice may be almost in contact under the heats of summer and autumn, in latitudes ranging from 44° to 47°. De Saussure, though not the first to examine them, by the charm of his writings, directed no little attention to glaciers, and to the effects produced by them. Other authors have, at various times, since described them; and, among those of late years, M. Charpentier* and M. Agassiz† have written much in support of a particular hypothesis as to the mode in which these masses of ice moved outwards from the mountain heights whence they originated, and as to their former more considerable range and extension than at present, pointing to many circumstances connected with this subject, and of geological value, though the hypothesis itself may not be adopted.

The progress of researches respecting glaciers and their geological effects, affords a fair example of the necessity of careful observation in a right direction; so many assertions connected with the mode of occurrence and advance of these masses of ice, upon which hypotheses were based, having been found, upon actual investigation, unsupported by facts. Though this has been the case, many observations have, from time to time, been recorded, which have borne the test of careful investigation; and no one would appear more desirous of admitting the value and importance of real additions to our information on this subject, than Professor James Forbes, to whom so much of our present knowledge of the Alpine glaciers is due.‡

A glacier commences near the line of perpetual snow, but lower somewhat than that on the adjacent ground. "There is often a passage, nearly insensible, from perfect snow to perfect ice; at other times, the level of the superficial snow is well marked, and the ice occurs beneath it. No doubt the transition is effected in this way: —the summer's thaw percolates the snow to a great depth with water; the frost of the succeeding winter penetrates far enough to

* "Essai sur les Glaciers et sur le Terrain Erratique du Bassin du Rhone," 1841. Lausanne.

† "Études sur les Glaciers," 1840. Neuchatel.

‡ See his Travels through the Alps of Savoy and parts of the Pennine Chain, with observations on the Phenomena of Glaciers, 2nd edit., Edinburgh, 1845; and his papers printed in the Philosophical Transactions, for 1846.

freeze it at least to the thickness of one year's fall, or, by being repeated in two or more years, consolidates it more effectually."* The part of a glacier, where the surface begins to be annually renewed by the unmelted accumulation of each winter, is commonly known as *névé*, and true stratification has been here recognized by De Saussure and other writers. Professor Forbes agrees with M. de Charpentier† in thinking that this stratification becomes entirely obliterated as the *névé* passes into complete ice.‡ The *crevasses*, or great fissures, in the *névé* are considered to differ from those lower down the glacier in their greater width and irregularity, and the caverns in it to be more extensive and singular in their forms, from the greater facility with which the *névé* is thawed and water-worn.

Further down the valley, or ravine, in which the glacier finds its way, much will necessarily depend, as to the form and appearance of the latter, upon the general character of the ground traversed. The ice changes its character: it is not like that produced by the freezing of still water in the lake, but "laminæ, or thin plates of compact transparent blue ice, alternate, in most parts of every glacier, with laminæ of ice not less hard and perfect, but filled with countless air-bubbles, which give it a frothy semi-opaque look." "The alternation of bands, then, is marked by blue and greenish-blue or white curves, which are seen to traverse the ice throughout its thickness whenever a section is made. It is, therefore, no external accident—it is the internal structure of a glacier, and the only one which it possesses, and may be expected to throw light upon the circumstances and formation of these masses."§

* Forbes' Travels through the Alps of Savoy, 2nd edit., p. 31.

† For his views respecting glaciers, consult M. de Charpentier's Essai sur les Glaciers et sur le Terrain Erratique du Bassin du Rhone. Lausanne, 1841.

‡ "The granulated structure of the névé is accompanied with the dull white of snow passing into a greenish tinge, but rarely, if ever, does it exhibit the transparency and hue of the proper glacier. The deeper parts are more perfectly congealed, and the bands of ice, which often alternate with the hardened snow, are probably due to the effect of thaw succeeding the winter coating, or any extraordinary, fall. On exposed summits, where the action of the sun and the elements is greater, the snow does not lie so long in a powdery state, and the exposed surface becomes completely frozen." Forbes' Travels through the Alps, &c., 2nd edit., p. 32.

§ Forbes' Travels through the Alps, &c., 2nd edit., p. 28. It is remarked, respecting this structure, that it is the consequence of the viscous condition of the mass and its movement. It is observed that it "has all the appearance of being due to the formation of fissures in the aërated ice or consolidated névé, which fissures having been filled with water drained from the glacier, and frozen during winter, have produced the compact blue bands" (p. 372). Professor Forbes considers, that, as the viscous mass moves onward, the central parts faster than the sides, these fissures, filled with ice, take a more horizontal position in the general mass, with such modifications as may be expected on the sides and bottom where the friction is greatest,

Below the *névé*, the glacier commonly finds its way, amid various depressions of different forms, to the lower ground, far beneath the line which marks the usually constant presence of snow throughout the year. The accompanying view (fig. 84, p. 212) of Mont Blanc, taken from the Bréven, a mountain rising high above the valley of Chamonix, which separates the Bréven from the Mont Blanc, will give a better idea of the passage of the glaciers downwards into the lower valley, than a verbal description, and more especially as the altitude and position of the Bréven itself prevents that foreshortening and less instructive view obtained from beneath.

As the great icy mass descends from the region of the *névé* to the lower ground, the *crevasses* vary much in length and breadth, sometimes extending across the whole glacier,* and this, as might be expected, according to the character of the surface on which it may repose. As it descends into warmer regions, the glacier is necessarily exposed to the influence of higher temperature, and if it did not obtain the needful supply from above, it would there diminish in bulk and disappear. As this supply varies, the extension of a glacier will correspond with the kind of seasons experienced, so that it may descend further into the lower valleys at one time than another;† and thus its actual amount of protrusion into a valley may depend, for the time, upon effects produced through many seasons, and be liable to frequent change.

---

accompanying his remarks by ideal sections of glaciers and real sections of viscous bodies experimented upon for illustration. He observes, "that this ribboned structure follows a very peculiar course in the interior of the ice, of which the general type is the appearance of a succession of oval waves on the surface, passing into hyperbolas, with the greater axis directed along the glacier. That this structure is also developed throughout the thickness of a glacier, as well as from the centre to the side, and that the structural surfaces are twisted round in such a manner, that the *frontal dip*, as we have called it, of the veins, as exhibited on a vertical plane cutting the axis of a glacier, occurs at a small angle at its lower extremity, and increases rapidly as we advance towards the origin of the glacier" (p. 372).

* In the account of his passage over the Col de Géant, Professor Forbes mentions an immense chasm or crevasse, extending wholly across a glacier in the descent on the Chamonix side, and at least 500 feet in width. "It terminated opposite to the precipices of the Aiguille Noire in one vast *enfoncement* of ice, bounded on the hither side by precipices not less terrible." Travels through the Alps &c., p. 238.

† Among the numerous examples of the varied extension and volume of glaciers known in modern times, there would appear none more illustrative than that of the Brenva, on the Italian side of Mont Blanc. In 1818 it attained a height different from that found by Professor Forbes, in 1842, of at least 300 feet, as proved by that of a rock, upon which a well-known chapel (Chapelle de Berriér) was placed, which, with the rock on which it stood, was heaved and fissured by the rise of the ice. This great increase of volume, and its decrease in the 24 years, is well attested. The Professor remarks, that the mean temperature for the five years preceding 1818, when the glacier was thus of such increased volume, presented no marked change, the mean temperature at Geneva being for that time 7°·61, Reaumur (49°·12 Fahr.); the mean for the last 40 years, in the same town, being 7°·75 (49°·44 Fahr.)—Travels, &c., p. 205.

The turbid waters, rushing out from beneath the glaciers of the Alps will be familiar to all who have visited those mountains, as also the caverns of ice through which these waters commonly find their way when they are most abundant. With respect to such waters, Professor Forbes has pointed out that they may not only be due to the ice melted by contact with the rocks on which it moves, to the fall of rain upon the ice drained by the glacier valley, in the season when rain falls, and to the waste of the glacier itself by the sun and rain, but also to the natural springs rising from beneath the ice, as in any other locality.*

With respect to the cause producing the motion of glaciers, different views have been taken:† Professor James Forbes considers it as "convincingly proved,‡ that the motion of a glacier varies not only from one season to another, but that it has definite (though continuous) changes of motion, simultaneous throughout the whole, or a great part of its extent, and therefore due to some general external change," and that "this change has been shown to be principally or solely the effect of the temperature of the air, and the conditions of wetness or dryness of the ice."§ With regard to the movement itself, the Professor has pointed out "that the ice does not move as a solid body,—that it does not slide down with uniformity in different parts of its section,—that the sides, which might be imagined to be most completely detached from their rocky walls during summer, move slowest, and are, as it were,

* With respect to the waste of the glacier by the sun and rain, Professor Forbes remarks, that it is "a most important item, and which constitutes the main volume of most glacier streams, except in the depth of winter. It is on this account that the Rhine and other great rivers, derived from Alpine sources, have their greatest floods in July, and not in spring or autumn, as would be the case if they were alimented by rain-water only. On the same account, the mountain torrents may be seen to swell visibly, and roar more loudly, as the hotter part of the day advances, to diminish towards evening, and in the morning to be smallest." "Winter is a long night amongst the glaciers. The sun's rays have scarcely power to melt a little of the snowy coating which defends the proper surface of the ice; the superficial waste is next to nothing, and the glacier torrent is reduced to its narrowest dimensions."—Travels through the Alps of Savoy, &c., 2nd edit., pp. 20, 21.

† The chief of these views will be found in the works of MM. de Saussure, De Charpentier, Agassiz, Élie de Beaumont, Mr. William Hopkins, and of Professor James Forbes. In the latter they will be seen discussed in much detail, and the Professor's own views advocated, especially in his Travels through the Alps of Savoy, &c., in chap. xxi. (entitled "An attempt to explain the leading Phenomena of Glaciers"), and in his papers entitled "Illustrations of the Viscous Theory," published in the Philosophical Transactions for 1846 A very detailed account of the works and views respecting glaciers will also be found in the Histoire des Progrès de la Géologie, de 1834-1845, by the Vicomte d'Archiac, Paris, 1847.

‡ Travels through the Alps of Savoy, &c., and alluding to chap. vii.

§ Ib., chap. xxi., p. 363.

dragged down by the central parts." * His theory is, that "a glacier is an imperfect fluid, or a viscous body, which is urged down slopes of a certain inclination by the mutual pressure of its parts." † He does not, however, "doubt that glaciers slide on their beds, as well as that the particles of ice rub over one another, and change their mutual positions." ‡

The movement of glaciers is important as regards the transport of mineral substances, inasmuch as by it they bear onwards upon their surfaces any fragments of rock that may fall upon them from the heights amid which they pass; thrust before them any loose accumulations of blocks, gravel, sand, and earth which may oppose their course, and even break off portions of rocks where the resistance of the latter is less than the force of the glacier, divisional planes, such as joints and cleavage, with the natural bedding of rocks, often rendering the mass of such rocks less resistant than it would otherwise be. The observer, accustomed often to see the steep cliffs of many a mountain region covered towards their bases

* Travels through the Alps of Savoy, &c. chap. xxi. p. 363. As to the different rates of motion of a glacier, it is observed that a glacier. "like a stream, has its still pools and its rapids. Where it is embayed by rocks it accumulates—*its declivity diminishes, and its velocity, at the same time.* When it passes down a steep, or issues by a narrow outlet, its velocity increases. The *central* velocities of lower, middle, and higher regions of the Mer de Glace are—1·398, ·574, and ·925; and if we divide the length of the glacier into three parts, we shall find these numbers for its declivity, 15°, 4½°, and 8°.—Forbes' Travels, &c , p. 371.

† Ib., p. 365. It would be somewhat out of place in this work to enter more fully into the theory of glacier movements. Respecting the theory of the viscous condition of a glacier, Professor Forbes alludes to its spreading as a viscous body would do when a glacier passes out of a narrow gorge into a wide valley, stating that this fact had been first brought prominently forward by M. Rendu, now Bishop of Annecy (Travels, p. 367.) M. Rendu (Théorie des Glaciers de la Savoie, Chambery, 1840) divides glaciers into *glaciers réservoirs* and *glaciers d'écoulement*, the former in the high regions, and the latter descending into the lower valleys. He estimates the height of the separation between the two in Savoy at 2,923 metres (9,590 English feet). He points to the accumulation of snow in the higher regions, the rain, when it falls there, freezing, and to the feeding of the lower glaciers by the descent of this snow and ice.

‡ "But," he adds, "I maintain that the former motion is caused by the latter, and that the motion impressed by gravity upon the superficial and central parts of a glacier (especially near its lower end), enables them to pull the lateral and inferior parts along with them. One proof, if I mistake not, of such an action is, that a deep current of water flows under a smaller declivity than a shallow one of the same fluid. And this consideration derives no slight confirmation, in its application to glaciers, from a circumstance mentioned by M. Élie de Beaumont, which is so true that one wonders that it has not been more insisted on—namely, that a glacier, where it descends into a valley, is like a body, pulled asunder or stretched, and not a body forced on by superior pressure alone" (p. 370). In a note to this passage, the Professor remarks, "that a state of universal distension, or a state of universal compression, is equally incompatible with the existing phenomena of most glaciers, and that compression in some parts and distension in others are plainly indicated by their natural features."

by the débris detached by atmospheric influences from above, and especially in climates or regions where frosts and thaws often alternate, would readily expect these fragments to move onwards with the glacier on the edges of which they may fall.* Instead, therefore, of accumulating in a talus of débris, as can be well studied in many mountainous regions, the mass of fragments moves slowly onwards, and the protection from atmospheric influences afforded by this talus to the solid rocks beneath it (in some mountain countries collectively very considerable), is removed precisely in localities where the vicissitudes of climate are often so great that it can be the least spared.

The blocks and smaller fragments (necessarily very variable in form and volume, according to the character of the overhanging rocks, and the amount of their decomposition anterior to their fall), thus strewed upon the glaciers, are well known as *moraines*.† These moraines also necessarily differ in general volume, according to the amount of matter which may be detached from the heights above the glacier, and the rate of movement of the glacier itself on the sides adjoining the sources of the detached fragments. Two lines of moraine will mark the edges of a glacier, should the heights on either side of it afford the needful supply, as also a mass of rock rising through a glacier, should it also afford fragments, as has been pointed out by Professor Forbes. When two or more streams of glaciers unite, each bearing its two, or even as we have seen, three lines of rock fragments, the union will so dispose of the lines as to form a less number for the remaining course of the glacier, as in fig. 85, where the glaciers coming down the valleys A and B, and uniting the four moraines, two on the sides of each glacier, become three (*a*), (*b*), (*c*), by the union of the lateral moraines 2 and 3 into a central moraine (*b*). Various other unions, easily imagined, are produced, as minor contribute to main glaciers. A great central moraine may be established by the junction of two long lines of glacier sides, unbroken for a considerable distance, and upon which

* The débris on mountain sides often completely masks their character as left anterior to such coverings. There are few mountainous regions which do not show this when carefully examined. Mining operations often prove it on the sides of hills. Ravines, where ravines may not be uncommon, are usually favourable for observations of this kind; as, for example, many instances are found in Derbyshire, where the faces of steep cliffs are often modified in this manner, the long-continued action of atmospheric influences having smoothed off many a precipitous hill side, where the same effects may be seen in daily progress. This action has greatly modified the face of most countries, especially when combined with landslips.

† This name has become common with us from the works of De Saussure and others writing in French. *Guffer* is the German term for them.

a great fall of fragments may take place, while the opposite sides may be marked with slighter lines of moraine, derived from tributaries receiving a less amount of fragments.

Fig. 85.

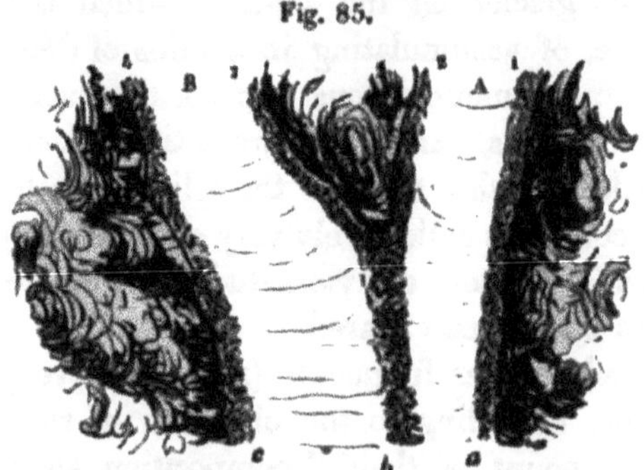

The following view (fig. 86), representing the upper part of the glacier of the Aar,* well shows the lines of moraine coming down from the glacier of the Finsteraarhorn, on the left, and from that

Fig. 86.

of the Lauteraarhörner, on the right. It also illustrates the formation of a single line of moraine in the centre, by the union of the

* Reduced from a view in the Études sur les Glaciers by Agassiz. Neuchatel, 1840.

two lateral moraines of the glaciers above noticed. Examples are also seen of the mushroom-like appearance produced by the unequal melting of the surface of a glacier, so that protection being afforded, as long since pointed out by De Saussure, by a block of rock (particularly if it has so fallen on the glacier as to rest in a tabular manner), the ice beneath has not disappeared so rapidly as around it, and thus the block is raised upon a stem of ice. Some of the blocks thus supported are very large.* It is, in the same manner, to the protection from the sun and rain afforded to the ice beneath by the mass of the moraine, that it often rises above the ice.†

The following view‡ (fig. 87, p. 218), of the upper part of the glacier of Zermatt, also shows the effects produced, as regards moraines, by the union of glaciers. On the left, the lines of moraine are derived from the glacier of Monte Rosa and of the Gornerhorn, with the lateral moraine of the foot of the Riffelhorn and the great moraine of the Breithorn. On the right are the glaciers of the Little Cervin and of the Furke-flue. The *crevasses* across the united glaciers are well exhibited in front, as also the rise of the ice above the side of the glacier, showing that the blocks and other rock fragments, there borne onwards, would have a tendency to fall over and accumulate in a lateral moraine, off the ice, and upon the adjoining lower ground.

If the rate of movement of a glacier depends upon the slope and

* A large one, observed by Professor Forbes in 1842, is represented in his Travels through the Alps, &c., pl. 1, and he gives the following instructive account of it:—"There lies on the ice a very remarkable flat block of granite, which particularly attracted my attention on my first visit in 1842 to that part of the glacier. It is a magnificent slab, of the dimensions of 23 feet by 17, and about 3½ in thickness. It was then easily accessible, and by climbing upon it and erecting my theodolite, I made observations on the movement of the ice; but as the season advanced it changed its appearance remarkably. In conformity with the known fact of the waste of the ice at its surface, the glacier sunk all round the stone, while the ice immediately beneath it was protected from the sun and rain. The stone thus appeared to rise above the level of the glacier, supported on an elegant pedestal of beautifully veined ice. Each time that I visited it, it was more difficult of ascent, and on the 6th August, the pillar of ice was *thirteen feet high*, and the broad stone so delicately poised on the summit of it (which measured but a few feet in any direction), that it was almost impossible to guess in what direction it would ultimately fall, although by the progress of the thaw, its fall in the course of the summer was certain. During my absence in the end of August, it slipped from its support, and in the month of September it was beginning to rise on a new one, whilst the unmelted base of the first was still very visible on the glacier." (p. 92.)

† The glacier cones, as they are called, are accounted for on the same principle of protection from the influence of the sun, sand washed by rain-water into cavities on the glacier finally so accumulating that it prevents the melting of the ice beneath at the rate experienced around, so that the sand still remaining on the ice, the latter takes the form of a cone with a sandy covering. They have been found 20 to 30 feet in height, and 80 to 100 feet in circumference.—Agassiz, Études sur les Glaciers, chap. x., and Forbes' Travels, &c., chap. ii., p. 26.

‡ Also taken from the Études sur les Glaciers, by Agassiz.

form of the subjacent and boundary rocks, all other conditions being equal,* we should expect it to vary very materially in the course of the same glacier, and in different glaciers. The very

Fig. 87.

careful investigations of Professor James Forbes have proved the correctness of the view taken by M. Rendu,† that the central portions move faster than the lateral,‡ so that the blocks and frag-

* When the rates of advance of different glaciers are compared with the slopes on which they move, it is very essential to take all other conditions into account, a precaution which does not appear to have been always adopted.

† Théorie des Glaciers de la Savoie, p. 63.

‡ Travels through the Alps of Savoy, &c, chap. vii., entitled "Account of Experiments on the Motion of the Ice of the Mer de Glace of Chamouni." The means adopted were of an order to insure success. The Professor selected a point on the ice, and determined its position with respect to three fixed co-ordinates, having reference to the fixed objects around. He found, after the observations of four days, that the ice on which his instrument was placed moved during each 24 hours at the rate of

15·2 : — 16·3 : — 17·5 : — 17·4 inches,

a variation which he considered due to the increasing heat of the weather. On trying the rate of nocturnal motion, as compared with the diurnal, the Professor found exactly one-half, the night having been cold. The general motion was not by fits of advance and halts, but orderly and continuously. By well-considered arrangements, he also found that the somewhat common opinion of the sides of a glacier moving ments of the medial lines of moraine descend further in less time than those on the sides, even when the latter are not thrown over, and left on the ground bounding the sides of a glacier. From the nature of the transport, any enormous mass of rock, detached from a height above a glacier, will move as readily onwards as a small fragment; indeed, from its protecting influence against the action of the sun and rain, it would tend to preserve the ice beneath far more effectually, considering the subject generally; not, however, forgetting that from the unequal melting of the ice around and partly beneath, it may be tilted off, not only into a crevasse, where it might advance with the general march of the glacier, but also into some situation where its progress may, for a time at least, be arrested. Some of the blocks observed on the glaciers are of very considerable dimensions. Professor Forbes mentions having seen one on the ice of the glacier of Viesch, in the Valais, nearly 100 feet long, and 40 or 50 feet high.*

---

faster than the centre was incorrect, and that, on the contrary, their motion was slower. From the 29th June to the 1st July (1842), while the sides of the part of the Mer de Glace experimented upon moved at a rate of 17·5 inches for each 24 hours, the centre advanced 27·1 inches. Other experiments on other parts of the glacier led to similar results. It was found that, "(1) The motion of the higher parts of the Mer de Glace is, as a whole, *slower* than that of its lower portion, but the motion of the middle region is slower than either. (2) The Glacier de Géant moves faster than the Glacier de Léchaud. (3) The centre of the glacier moves faster than the sides. (4) The difference of motion of the centre and sides of the glacier varies with the season of the year, and at different parts of the length of the glacier; and (5) The motion of the glacier generally varies with the season of the year and the state of the thermometer." Subsequent investigations enabled Professor Forbes to state (Philosophical Transactions, 1846, "Illustration of the Viscous Theory of Glacier Motion," parts 1, 2, and 3), that the movement of the Mer de Glace went on continuously for several days, and he gives a valuable table of the apparent and relative motion of 45 points, two feet apart, in a line traversing the axis of this glacier, in 1844 (p. 171).

The motion of a particular stone, named the Pierre Platte, on the Mer de Glace, was observed to be as follows:—

From the 17th September, 1842, to 12th September, 1843, the advance in 360 days was	256·8 feet.
Reduced to the year of 365 days	260·4 ,,
Mean daily motion	8·56 inches.
From 12th September, 1843, to 19th August, 1844, 342 days	270 feet.
Proportional motion for 365 days	288·3 ,,
Mean daily motion	9·47 inches.

There are also important tables of the motion observed at two stations on the Glacier des Bois (one observed from the 2nd October, 1844, to the 21st November, 1845, and the other from 4th December, 1844, to 21st November, 1845); and at two stations on the Glacier des Boissons (one from 20th November, 1844, to 22nd November, 1845, and the other from 2nd October, 1844, to 22nd November, 1845), showing the variable, but continued progress, of these glaciers during the intervals. Among the results, it appeared that "in both glaciers the summer motion exceeds the winter motion in a greater proportion as the station is lower, that is, exposed to more violent alternations of heat and cold."

* Travels, &c., p. 46. A very large granite block, also seen by the Professor upon the Mer de Glace, in 1842, is figured by him in the same work, p. 57.

While thus fragments of all dimensions, and in great abundance, find their way with an unequal rate of movement, according to their position on a glacier, to lower levels, numerous others are arrested in their progress, tilted off and left on the ground adjoining its sides, should circumstances permit. When a glacier so changes its volume as to occupy a higher relative level at one time than another, amid the mountain depressions and ravines over and through which it may move, and the conditions for leaving marginal accumulations of rock fragments on the outside of it obtain, such accumulations, should the ice afterwards decrease in volume, would remain to attest this previous state of the glacier.* No marks of this kind would be left where the sides of a ravine or cliff were so steep that the blocks could not find rest. The fragments would either rise or descend with the glacier, some probably falling into any space left between the ice and the wall of rock, and open either from a certain amount of melting of the glacier at its contact with the rock, or from the passage of the mass of ice along the uneven front of a cliff, cavities of different kinds thus presenting themselves.

From fragments of rock becoming jammed between the ice of a glacier and its rocky walls, as cannot fail often to be the case, and indeed is well known, the friction of these fragments, pressed by the great force of the glacier, grooves and furrows the adjoining rocks in lines corresponding with their motion. Professor Forbes gives the following interesting view (fig. 89) of the 'Angle,' Mer de Glace, where granite blocks are jammed in between the ice and the rock, wearing "furrows in the retaining wall, which is all freshly streaked, near the level of the ice, with distinct parallel lines, resulting from this abrasion. The juxtaposition of the power, the tool, and the matter operated on, is such as to leave not a moment's doubt that such striæ must result, even if their presence could not be directly proved."† This friction alone would tend to

* Professor Forbes (Travels, &c., p. 24) has given a very illustrative section (fig. 88), showing the manner in which fragments (c) of rock may be left by the de-

Fig. 88.

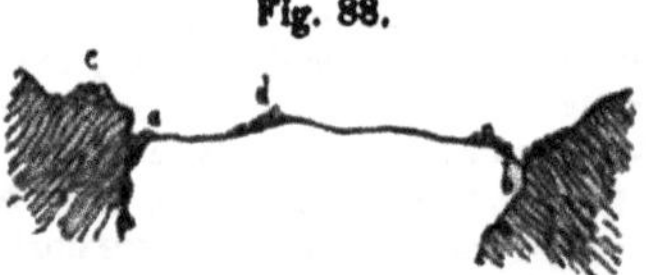

crease of a glacier, and in which a part of a lateral moraine may fall into a cavity, a, between the ice and the boundary rock, or be left stranded on an inclined shore, b. The section of a central moraine is seen at d.

† Travels, &c. The woodcut is a reduced sketch from pl. 3, p. 76.

reduce the fragments to less size, and even to fine powder. Looking at the general conditions of glacier movements, the kind of ground these masses of ice pass over, and to the introduction of fragments from the sides, and even through the crevasses to the bottom, we

Fig. 89.

should expect that the grooving and scratching would be considerable on the bottom and sides, mingled with an extensive smoothing of surface, as if, in the application of a huge polishing apparatus, acting, as a whole, with minor deviations, in one direction, harder grains were strewed about, so that scratching as well as polishing was effected.* This scratching and smoothing by glaciers has been chiefly observed, with reference to their geological value, in modern times, though the rounded and polished surfaces frequently seen have been long known by the name of *Roches Moutonnées*, that assigned them by De Saussure.

From the general grinding of glaciers on their beds, the friction

* Though the grooves are usually long, parallel, and polished, the minor scratches often cross each other.

of the fragments on each other, and the decomposition of many kinds of rock in regions where the alternations of frost and thaw are so common, particularly in the warmer parts of the year, much finely-comminuted mineral matter could scarcely fail to be exposed to the action of any running water, finding its way amid the glacier, and along its sides and bottom.* The streams and rivers derived from glaciers have commonly a marked character, as above noticed, from the quantity of fine mineral matter in mechanical suspension. These sometimes fall into lakes, and leave the fine sedimentary matter behind them, as is the case with many amid and on the skirts of the Alps, while some have a considerable course, as, for example, the Durance (bearing the glacier waters of Monte Viso), and many tributaries of the Po, fed by glacier streams from the southern side of Mont Blanc, and other Italian portions of the high Alps.

Independently of the mass of fragments which may be borne forward by a glacier, when it is on the increase outwards, from a fitting combination of conditions, it ploughs up the ground before it, thrusting forward the loose substances, no matter how accumulated, and with them, should they come in its course, fields, woods, and houses. We remember seeing the Glacier des Bois thus crushing and forcing all before it during its advance in 1819.† These accumulations, to which the transported blocks and minor fragments of rock are being added, as the ice melts, which once supported and carried them onwards,‡ are known as terminal moraines, and by their position a glacier is inferred to

---

* There is much finely-comminuted mineral matter distributed over some parts of many Alpine glaciers. It is sometimes so fine as to enter the interstices of the more porous ice, thus distinguishing the latter from the more compact bands. These "dirt bands," as Professor Forbes terms them, were of much service to him in his examination of the structure of glaciers. Alluding to the discoloration from this finely-comminuted detritus, the Professor observes, "The cause of the discoloration was the next point, and my examination satisfied me, that it was not, properly speaking, a diversion of the moraine, but that the particles of earth and sand, or disintegrated rock, which the winds and avalanches and water-runs spread over the entire breadth of the ice, *found a lodgement* in those portions of the glacier where the ice was most porous, and that consequently the 'dirt bands' were merely *indices of a peculiarly porous veined structure traversing the mass of the glacier in these directions*."—Travels, &c., p. 163. Upon careful examination these "dirt bands" were found to be quite superficial.

† In 1820 it attained its greatest known modern advance into the valley of Chamonix.

‡ Respecting the blocks and fragments of rocks thus carried outwards, M. Rendu has remarked that some of them can be occasionally traced to the very commencement of a glacier.—Théorie des Glaciers de la Savoie.

be, for the time, either retreating, advancing, or stationary.* That glaciers advance and decrease is well known, and this to considerable distances, so that many a terminal moraine left at one time, may be again forced forward at another, part of it so caught in the advance of the ice as to be employed in grooving and scratching the solid rocks beneath, then bared and passed over by the glacier. Enormous blocks† are often left by glaciers in their retreat; indeed, under such circumstances, they would not only leave the terminal moraines, marking their extension for the time, and during periods of increase, but also their whole load of blocks and fragments, up to the new limits of the decreased glaciers.

Supposing a glacier to advance and retreat from causes which, though variable on the minor scale, are constant for considerable intervals of time, there would be no small amount of blocks and fragments of rock, too considerable to be borne onwards by river action, left either perched on various parts of the mountain sides, or distributed over the valleys, within the range of increase and decrease of these masses of ice in glacier regions. This great and constant general action, continued through long time, would scarcely otherwise than very considerably modify the state of the area from that original condition, when the glaciers were first formed, even supposing no alteration in the relative level, as respects the sea, of the mountain masses amid which they occur. Avalanches aid in the general descent of fragments of rocks, carrying many, with their snows, to lower levels, sometimes falling on glaciers, sometimes into deep valleys, where the fragments are merely exposed to the ordinary action of rivers.

Taking the general causes and movements of glaciers in the Alps for his guides, the observer is enabled to infer how far glaciers would be found in other regions. M. Élie de Beaumont has pointed out,‡ that from the little variation of climatal con-

* Professor Forbes, after quoting M. Venetz (vol. i. of the Transactions of the Swiss Nat. Hist. Society), as pointing out "that passes the most inaccessible, traversed now, perhaps, but once in twenty years, were frequently passed on foot, sometimes on horseback, between the eleventh and fifteenth centuries," considers the evidence important, as "showing that a *very notable* enlargement of these boundaries (glacier boundaries), was consistent with the limits of atmospheric temperature, which we know that the European climate has not materially overpassed within historic times."—Travels, &c., pp. 43, 44.

† Professor Forbes mentions one of green slate, pushed forward by the glacier of Swartzberg, valley of Saas, and now left at a distance of about half a mile from the glacier by its retreat, estimated by M. Venetz to contain 244,000 cubic feet. This mass, if about 14 cubic feet be taken to the ton, would weigh no less than 17,428 tons.

‡ Remarques sur deux points de la Théorie des Glaciers, Annales des Sciences Géologiques, 1842. He observes that glaciers being due to annual and not merely to diurnal conditions, there could be only perpetual snows, and not glaciers, under the equator, where the variations of temperature are only diurnal.

ditions in tropical regions, glaciers would not be expected among the mountains there situated, and sufficiently high to be clothed with perpetual snow. Where the alternations of frost and thaw, snow and rain, would be insufficient to produce the needful amount of névé, assuming this to be the storehouse whence the glaciers are supplied, these would not be found. Looking, therefore, at the different known regions of the world, their varied relief, as regards the distribution of high and low land, the different amount of water supply from the atmosphere, either in the shape of snow, hail, or rain; changes of temperature during various times of the year, and their amount; prevalent or periodical winds —one set dry, the other bringing abundant moisture, and proximity or distance from the sea—the observer finds no want of modifying conditions for the presence or absence, and geological importance of glaciers. At one time glaciers were somewhat doubted among the great range of the Himalaya, but several are now known. The height of the lowest part of the Pinder glacier is estimated at about 11,300 feet above the sea, and that of the Kuplinee glacier at 12,000, which, the height of the perpetual snow line near them being considered at about 15,000 feet, would give a glacier descent of 3,700 feet for the former, and 3,000 feet for the latter.* The lowest part of the glacier of the Ganges is 12,914 feet above the sea, according to Captain Hodgson.

* Captain Strachey, Bengal Engineers, Jameson's Edinburgh New Phil. Journal, vol. xliv., p. 119, and Journal of the Asiatic Society of Bengal, No. viii., p. 794.

CHAPTER XIII.

ARCTIC GLACIERS REACHING THE SEA.—NORTHERN ICEBERGS AND THEIR EFFECTS.—ANTARCTIC GLACIERS AND GREAT ICE BARRIER.—GEOLOGICAL EFFECTS OF ANTARCTIC ICEBERGS.—GLACIERS OF SOUTH GEORGIA.—GLACIERS OF SOUTH AMERICA.—TRANSPORT OF DETRITUS BY RIVER ICE.—GEOLOGICAL EFFECTS OF COAST ICE.—EFFECTS OF GROUNDED ICEBERGS.—GENERAL GEOLOGICAL EFFECTS OF ICE.

PROCEEDING from the temperate parts of the world, where lands rise sufficiently high into the atmosphere to obtain a constant covering of snow, and the fitting conditions permit glaciers to descend amid the adjacent valleys at lower levels,* to the Arctic and Antarctic regions, we find glaciers not only covering various portions of land, but jutting into the sea, the line of perpetual snow having descended towards its level. If the observer will in imagination, and by reference to the view of part of it previously given (fig. 84), fill up the valley of Chamonix with sea to the height of about 4000 feet above the village of Chamonix (3,425 feet above the sea), and, therefore, so that the perpetual snow line descended (in round numbers) to within about 1,000 feet from the sea level,† it will readily be seen that numerous glaciers would jut into the sea, resting upon and grating along the rocks forming their bases and sides, until the emersion in the water became such that they floated at their extremities, the transport of fallen fragments being continued in the manner that it now is, until the glacier reached the sea. Here the conditions for their further transport would be modified. Instead of terminal moraines, the blocks would be thrown into deep water, and those which now fall off the lateral moraines would be distributed at greater or less

* In the Pyrenees, the conditions for the production of glaciers would appear to be such, that, where they occur, they are almost always found on the northern slopes of the mountains.

† Taking 8,500 feet above the sea as the snow line for the Alps, the altitude inferred by Professor Forbes.

distances from the new shores. Modifications would also arise from the increase or decrease of the mass of the glaciers, assuming the needful climatal changes. If we now add wave and breaker action, and tides, it will be seen that there would be a tendency to have the protruding portions of the glaciers, where they floated, broken away by the one and the other, more particularly when the glaciers were weakened by lines of crevasses, formed, as now, upon the land, before the protrusion seaward was effected. Great masses of ice would thus be borne away, supporting their moraines, gathered and transported outwards, as at present.

This imaginary case may be considered as realised, from the near approach of the perpetual snow line to the sea, with certain obvious modifications, in portions of the Arctic regions. Glaciers formed, necessarily, at minor altitudes above the sea, there descend to the shores in various situations, as, for example, in parts of Greenland and Spitzbergen,* even advancing beyond them, so that their extremities become separated and are borne away by tidal streams and sea-currents, the masses of ice often loaded with the fragments of rock detached from the cliffs and heights amid which the glaciers moved outwards, as in the Alps. Let *a*, *b*, *c*, *d*, and *e*, (fig. 90), represent the section of a portion of coast, partly beneath the level of the sea, and partly along a ravine or hollow, in which a glacier, *f*, *g*, *c*, *h*, finds its way outwards to the sea, *s*, so that at

Fig. 90.

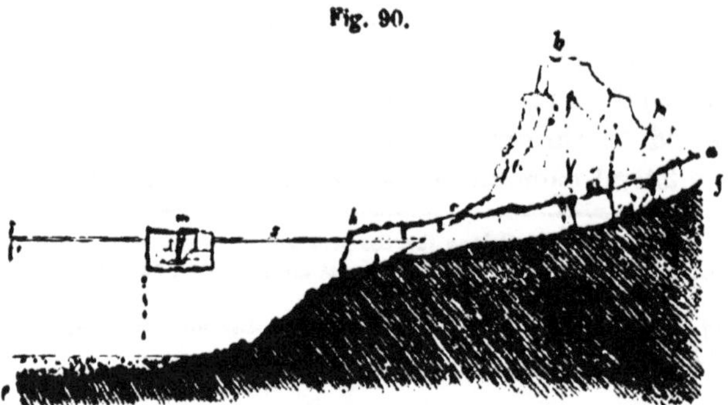

h, it has a tendency to float at its extremity, from its relative specific gravity, as regards the sea, and it should be recollected

* With respect to the alternations of temperature productive of glaciers, it would appear that in these regions there is no want of the needful alternations of frost and thaw. In Greenland the heat of the days in the summer months is considerable, thawing the snow and ice, while the nights are commonly cold, with frosts. Even during the winter at Spitzbergen, when strong southerly winds prevail, thaws are known. The temperature of the warmest months at Spitzbergen is estimated at 34°·3, and the longest day lasts four months, the northern portion of these islands being within 10° of the North Pole.

that glacier ice would sink less deeply in sea than in fresh water. And let *t* be the level of ordinary high water in a tidal sea, and *t t′* the difference of level between high and low water. The ordinary glacier movements and their consequences would go on uninterruptedly, as in the Alps, allowing for the modifications due to an Arctic climate, from *f* to *g*, where the sea level cuts the coast and glacier; while from *g* towards *d*, a change in the polishing, grooving, and scratching of the rocky sides and bases would gradually be effected as the final floating of the ice removed its pressure from them. Still much of the polishing, grooving, and scratching would take place beneath the sea level, and the fragments which may have fallen between the glacier and its sides, or through crevasses of sufficient depth, while above the level of the sea, would be squeezed out beneath the ice under that level, accompanied by the finer detrital matter, derived in the manner above mentioned (p. 221). and borne away in mechanical suspension by glacier rivers. As the ice moved seawards, instead of the terminal moraine of an inland glacier, the blocks and fragments of rock of the lateral and central moraines, should there be such, would fall over into the sea, accumulating in different ways beneath it, according to the depth of water and configuration of the coast. It is assumed, for illustration, that at *d* such blocks do accumulate. With respect to the finer detritus, instead of being removed, amid dry land, by running waters, as in the Alps, its outward movements by such means would be checked at the sea level, *t*, with the difference due to the fall of tide *t′*. Its further course outwards would depend upon the specific gravity of the water loaded with this matter in mechanical suspension, and the general motion of the glacier. We have seen that the turbid waters of the Rhone readily sink beneath the clear water of the Lake of Geneva, spreading over the bottom (p. 44), and we should anticipate, from the melting of the glacier in the sea, and the consequent less specific gravity of the latter, that the turbid waters under notice, finding their way beneath Arctic glaciers in the usual manner above the sea level, would also be discharged outwards beneath the glacier. Taking all the circumstances into consideration, there would appear much probability of the finer detritus finding its way beneath the glacier into the sea, to be distributed over its bottom according to conditions, tidal streams and sea currents producing their usual effects. Along steep coasts, such as those of Greenland, where glaciers are so common, much mud may be thus distributed under the deep water which usually adjoins them, and into this mud glacier-borne fragments of rock, sometimes of considerable volume, would from time to time be discharged, so that the resulting mixture would be a clay without apparent stratification, amid which fragments of rocks, of very varied form and volume were dispersed.

The transport of fragments by glacier ice, the latter jutting into the sea, does not cease in the cold regions of the globe with the extension of the glacier itself. Not only is the glacier subject, at its seaward extremity, to the breaker action, which observers inform us undermines its base, and finally brings down huge fragments into the water, but also to the pressure of tidal streams or sea currents, and to the fracturing influence of the up-and-down motion produced by the rise and fall of tides in tidal seas. Some of the masses of ice thus broken off and and floated away, as at *m* (fig. 90), with any load of blocks and minor fragments of rock, which in the ordinary inland glaciers of temperate climates might be carried towards the terminal moraines, would contribute, as at *e* (fig. 90), by their melting, and during a long lapse of time, no small amount of blocks to be dispersed amid the clay or mud, even of deep waters, such as those in Baffin's Bay.

Greenland has been considered as a mass of land nearly covered by perpetual snows and interlaced with glaciers, many of the latter protruding beyond the ordinary coasts into the sea. Their seaward extremities are well known, after having been detached from their main masses, to be floated away, often bearing fragments of rock in and upon them, even to and beyond Newfoundland.* In the western and mountainous part of Spitzbergen, glaciers reach and protrude into the sea, exposing ice-cliffs from 100 to 400 feet in height. A little northward of Horn Sound, a great glacier is noticed as occupying 11 miles of the sea-coast, the highest portion

* The current from the northward bears a mass of ice with it to the southward along the east coast of Greenland; sea ice, as well as the glacier ice noticed above. The ice is described as sometimes extending across from Greenland to Iceland; polar bears being occasionally ice-borne to the latter, where they commit great havoc until destroyéd. The accumulation of ice is stated to extend occasionally from 120 to 160 miles, seawards, around Cape Farewell. Its movement thence is described as northward to Queen Anne's Cape, passing afterwards to the western side of Davis's Strait, and from Cape Walsingham (Cumberland Island) along the American shore to Newfoundland.

Mr. Redfield (American Journal of Science, vol. xlviii., 1845) gives a valuable chart, illustrative of a paper, on the Drift Ice and Currents of the North Atlantic. Touching the general quantity of drift-ice, it is stated to vary considerably. "It is sometimes seen as early in the year as January, and seldom later than the month of August. From March to July is its most common season. It is found most frequently to the west of longitude 44°, and to the eastward of longitude 52°, but icebergs are sometimes met with as far eastward as longitude 40°, and in some rare cases even still further towards Europe."—p. 373.

rising in a cliff of 400 feet above the water. On the east coast of North-East Land great glaciers are also found.

M. Ch. Martins* mentions that the glaciers of Spitzbergen are commonly even and not much broken, and that the ice resembles that of the upper glaciers of Switzerland, pointing out that of Aletsch as a good illustration of the Spitzbergen glaciers. There are lateral, but no central moraines, the former proceeding with the glacier to the sea.† The cliffs of ice rising above the sea he estimates, as previous observers have done, as varying in height from 30 to 120 metres (98 to 393 English feet), and he states, that the seaward terminal portions of the glaciers rest on water. Respecting the height and slope of the Spitzbergen glaciers, he estimates the difference between the foot and the summit of a Bell Sound glacier at 1,150 feet, and its slope at 10°. The principal glacier of Bell Sound is also stated to be nearly horizontal, in consequence of its great length. M. Eugene Robert, who likewise visited Spitzbergen, remarks on the destruction of the ice by the breakers, and considers that where this is not effected, the masses of ice are very stationary. M. Durocher, who has also visited these lands, observes‡ that the glaciers do not there rise more than from 1,300 to 1,650 (English) feet above the sea; the snows above not taking the character of névé, being too much elevated above the needful conditions for its production.

The masses of ice detached from the land, floating about, and commonly known as icebergs, are sometimes of very considerable dimensions. The accompanying (fig. 91, p. 230) is a view of one seen by Sir Edward Parry§ in his first voyage, and is interesting, not only as showing the magnitude of this mass of ice (the far greater portion being concealed beneath the sea), but also as exhibiting something of its structure. It may be here observed,

* Observations sur les Glaciers du Spitzberg comparés à ceux de la Suisse et de la Norwége. Bull. de la Soc. Géol., vol. xi., 1840; Bibliothéque Universelle de Genève, 1840.

† Respecting the moraines of Spitzbergen, M. Martins observes that the bases of the nearly-vertical cliffs bounding the glaciers are covered with a mass of débris, fallen from the heights. Between these heights and the glacier there is sometimes a small valley or depression. The great glacier of Bell Sound is thus separated from its boundary heights. This glacier was merely stained with earth in its lateral portions. Those of Madalina Bay were covered with stones at their lower portions, occupying about an eighth part of their breadth. Not only were blocks seen in their upper surfaces, but also imbedded in the ice. M. Martins never saw them in the front of the glaciers bordering the sea.

‡ Mémoire sur la limite des neiges perpétuelles, sur les glaciers du Spitzberg comparés à ceux des Alpes—Partie de Géographie Physique du Voyage de la Récherche, 1845. Scoresby gives the height of the Horn Sound glacier as 1300 feet.

§ Reduced from a plate, in Parry's First Voyage, 4to edit.

that such masses of ice, remaining, as they are often known to do, stranded for a long time in some high latitude, might become covered with snows, marked by alternations of frosts and thaws, and even frozen rain, so that their upper parts may be in the condition of névé, thus covering over the remains of old moraines, resting on more ordinary glacier ice. Indeed, as respects the latter itself, in regions where the perpetual snow line closely approximates to and even cuts the level of the sea, we might expect the névé condition more and more to prevail, and it has been considered that icebergs are frequently of that character.

Fig. 91.

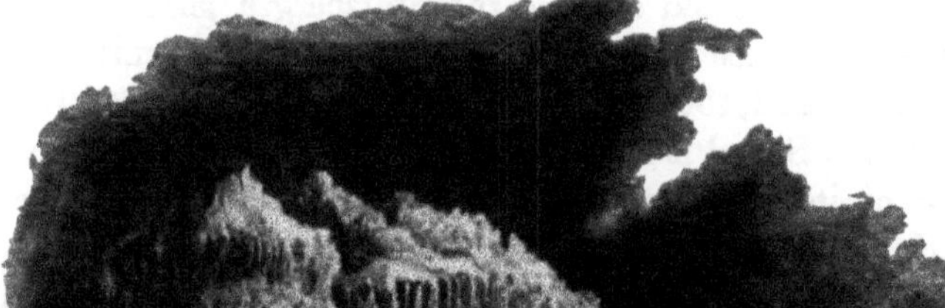

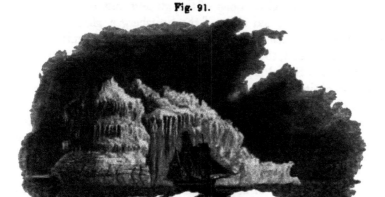

The northern icebergs may be regarded as the great carriers of rock fragments, often of a great size, from the lands where the bergs have been formed, as portions of glaciers, over a part of the Northern Atlantic, distributing them upon the bottom in various directions, and upon parts of it to which no other cause now contributes detritus.* Blocks and minor fragments may even be thus dropped upon bare rocks beneath, and upon every kind of inequality. Should a constant supply of block-bearing icebergs, regarding the subject generally, be thrown into any constant current, corresponding lines of deposit would result, assuming the melting of masses of ice, of various sizes, at different times and distances during their progress in such current; these lines having no reference to the form of the bottom, or to its modifications from any other deposits accumulated now, or at previous geological

* Mr. Couthouy mentions an iceberg, with apparently boulders upon it, as low down as latitude 36° 10′ N., and longitude 39° W. The same author states that he had often met with icebergs between the parallels of 36° and 42° N., in his voyages to and from America and Europe. American Journal of Science, vol. xliii., 1842.

times. Stranded near shores, or upon mud or sand-banks, these though somewhat deep in the sea, still catching their submerged portions, icebergs would tend much to disturb detrital deposits beneath them, particularly when moved by the waves produced during heavy gales of wind, as also by the rise and fall of tides. The heavy thumping of such huge masses, as some of these icebergs are, would cause great derangement of deposits effected tranquilly; and in many situations, blocks and fragments of rocks, with gravels, sands, and clays, would be irregularly mixed by the application of such force—singular intermixtures, and contortions of any previously bedded structure being produced. The icebergs which ground upon the Banks of Newfoundland can scarcely fail to produce much disturbance of the bottom, often adding to it great blocks and minor fragments of rocks, borne by them from more northern regions.*

As is well known, glaciers reaching the sea are not confined to the northern hemisphere,† they are also found in the antarctic regions. Sir James Ross mentions a great glacier, at Ætna Islet, South Shetland, as descending from a height of 1,200 feet into the ocean, where it presented a vertical cliff of 100 feet. Adjoining the termination of the glacier, Sir James found the largest aggregation of icebergs, evidently broken from it, he had ever seen collected together. Glaciers are also noticed by Sir James Ross as descending from the Admiralty Range (mountains 7,000 to 10,000 feet high) in Victoria Land, and projecting many miles into the sea, bare rocks in a few places inland breaking through the covering of ice. As in the arctic regions such glaciers may be expected to bring down with them those fragments of rock which

* Mr. Couthouy (American Journal of Science, vol. xliii., 1842, p. 155) mentions having seen (in September, 1822) a large iceberg aground on the eastern edge of the Great Bank of Newfoundland, and considered to be in about 720 feet water (120 fathoms), soundings three miles inside giving 630 feet (105 fathoms). A fresh wind from the eastward kept forcing it on the bank, the sea causing it to rock with a heavy grinding noise. On another occasion (August, 1827) he observed another iceberg aground upon the Great Bank, in between 480 and 540 feet water (80 to 90 fathoms). The huge mass rocked with the swell, going at the time, and even turned half over when struck by the breakers. The sea, for about a quarter of a mile around, was discoloured by mud worked up from beneath. Above water the iceberg was 50 to 70 feet high, and about 1,200 feet long. It suddenly fell over on its side, with much disturbance of the sea.

† Although glaciers are so common in Iceland, they do not appear actually to reach the sea. Those descending from the high Jökulls are noticed as separated from it by great moraines. Some of the glaciers are described as black in parts from the quantity of volcanic cinders and ashes with which they are covered. The sudden melting of snows and glaciers, from volcanic action in Iceland, is represented as producing great rushes of water, bearing large accumulations of volcani products outwards.

can fall upon them, to grate over the hard rocks on which they move, and to aid in contributing fine detritus to the adjacent sea-bottom, should the temperature be such that water could flow between the ice and the supporting rock. When we consider the volcanic character of so much of the great southern land as has been seen, we should expect that, as in Iceland, volcanic eruptions and the heating of the ground would occasionally produce the sudden melting of snows, and descent of the water, which could remain fluid sufficiently long to find its way to the sea. In this manner, not only the transport of ashes and cinders, and the larger volcanic substances vomited out of craters, may be moved to the lower ground, or into the sea, but also the fragments of rock which might have fallen upon snow or ice from any cliffs or steep places wherever atmospheric influences could detach them; not forgetting the effects of earthquakes (so common in great volcanic countries) upon the glaciers and snows, especially in localities where great avalanches could be produced.

Though from its general mode of occurrence, the great icy barrier of the antarctic regions might not, at first, appear any important agent in the transport of mineral matter, it has been found that, under certain conditions, portions of the ice, detached from it, may bear no inconsiderable amount of mud, sand, and rock fragments of various sizes into milder climates, depositing their loads over the bottom of the sea upon which they may be carried. This icy barrier presents a very singular appearance, stretching over a vast distance, with ice-cliffs rising from 150 to 200 feet above the sea, large fragments of them and minor pieces of ice floating in front of it, as shown in the annexed view* (fig. 92), representing a great detached mass in a long creek or bay in the barrier itself. From the relative specific gravity of the ice and sea-water, the former necessarily descends from beneath the level of the sea to a depth which might be estimated if the ice were of a uniform kind, with a known specific gravity. This is, however, far from being the case, for the layers of which it is composed, would appear to present somewhat the character of the névé of the higher parts of glaciers in temperate regions, being formed of alternations of snow, sleet, frozen mist and rain, with the refreezing of

* Taken from Captain Wilkes's "United States' Exploring Expedition," vol. ii. The vessel represented is the "Peacock," which had been driven against this great mass of ice. The view will at the same time afford an idea of the great barrier itself, which would be but an extension of a similar range of ice-cliffs. A long illustrative view of the great antarctic ice-barrier is given in Ross's "Voyage of Discovery and Research in the Antarctic Regions," vol. i. p. 232.

portions which in the summer months may be thawed at times by the influence of the sun.* As detached portions of this barrier were found by Sir James Ross aground, 60 miles from its main edge, and 200 miles from Victoria Land, in 1,560 feet of water, the ice was there at least of that thickness.

Fig. 92.

The depth of water obtained not far distant from the barrier† would show, as Sir James Ross has observed, that much of it must be upborne by the sea, and not rest on the sea-bottom, however the general mass may be held fast by adhering to land, or by reposing upon mud, sand, gravel, or solid rock, at minor depths. It will be obvious that the ice must be limited in depth by the temperature of the water to which it descends. We have seen (p. 96) that at the depth of 4,500 feet, the most dense water, with its temperature of 39°·5, appears to remain somewhat fixed in these regions, the waters of the upper parts of the sea necessarily varying in temperature according to the seasons. In January (1841), consequently in the summer of that portion of our globe, Sir James Ross found, about 12 or 14 miles from the barrier, a temperature of 33° at a depth of 900 feet, one which could not fail, widely spread beneath

* Sir James Ross describes gigantic icicles depending from the projecting parts of the ice-cliffs, proving that thaws sometimes took place. Notwithstanding that the time of the observation (February 9, 1841) corresponded, as respects season, with August in England, the temperature was at 12° (Fahr.) and did not rise above 14° at noon.

† Sir James Ross found (lat. 77° 56′ S., long. 190° 15′ E.) a depth of 1,980 feet (330 fathoms), within a quarter of a mile of the barrier, the bottom green mud. He also obtained 2,400 feet (400 fathoms) 12 or 14 miles off the icy barrier in another situation, about 100 miles from Victoria Land, the bottom being also a green mud, so soft that the sounding-lead descended into it 2 feet. "Voyage of Discovery and Research in the Southern and Antarctic Regions," vol. i.

as we might expect it to be, to act upon the lower part of the great mass of ice descending into the sea.*

Seeing that numerous and large masses of ice are annually detached from the great ice-barrier adjoining Victoria Land, and are floated off into milder regions, the question arises of whence the needful supply for this loss is obtained, assuming a certain general icy frontier to bound the barrier, and due allowance being made for the variation of seasons. The great thickness of the detached masses would lead us to consider that they were not portions formed on the outskirts of the main mass during certain seasons as additions to it, and were subsequently broken off, to be replaced by other additions; but rather that they were essential portions of the main mass, formed at the same time and in the same manner with it. Under this view there would be a motion outwards of this mass, sufficient to supply the annual waste of icebergs at the outer edge. Such a movement, though very slow, would yet produce a corresponding effect on the bottom of the sea over which this great mass of ice passed, grating over it, heavily pressing upon and scratching bare rocks and shingle beds, in the manner of a common glacier, though over a far wider area. Shingle beds, produced by some previous condition of land and sea, might thus, as well as any supporting rock, be scratched throughout, pebbles moved against pebbles, in lines of a general parallel character, over very extended areas.†

As the various layers of which the ice-barrier is formed indicate accumulations from atmospheric causes, unless the melting of the beds‡ beneath was equal to the deposit of snow, sleet, fog,§ and

* The temperature at 1,800 feet was 34°·2, at 900 feet 33°, at the surface 31°, and of the air 28°. In another situation (lat. 77° 49′ S. and long. 162° 36′ W.), and about one mile and a half from the barrier, Sir James Ross found the temperature of the bottom (green mud) at 1,740 feet (290 fathoms) to be 30°·8, only 2° lower, he observes, than would be obtained at a more considerable distance from the barrier, and showing the small influence of the mass of ice upon the sea adjoining it.

† Any outward motion of the great ice-barriers, however slow, by bringing portions of it forward which were based on rock, or shallow sea-bottoms, into depths where their bases could be melted, would also tend to keep those parts flattened which might otherwise have a large amount of snow or ice accumulated upon them, supposing such accumulation beyond the loss of evaporation and melting.

‡ As regards these layers, Captain Wilkes ("United States' Exploring Expedition," vol. ii.) observes, "that 80 different beds, on the average 2 feet thick, were counted in the large icebergs, detached from the main ice, and 30 in the smaller." Assuming similar beds beneath the sea level, the whole would constitute no small amount of ice and snow accumulated in horizontal layers and beds, in parts supported like beds of solid mineral matter by subjacent ground.

§ Respecting fog, Captain Wilkes remarks, "that it may make, when frozen, a marked addition to the ice accumulations, since he has known it frozen to the depth of a quarter of an inch upon the spars and rigging of the ships in a few hours."

rain (frozen upon its fall) above, there would be a continued increase of icy matter. The marked general uniformity in height of the ice-cliffs, and the tabular character of the surface of the barrier inwards,* would point to some cause having an extended and uniform action, so modifying any accumulation of the kind as to keep the mass at a general uniform thickness. The temperature of the sea at a fitting depth would appear sufficient to effect this, any addition from above to the general mass, so long as it plunged into water and did not rest on the sea-bottom, being compensated by the melting of the lower surface, pressed down by the increased accumulation above.

Captain Wilkes refers the formation of the ice in the first place to ordinary field ice, upon which layers from rain, snow, and even fog so accumulate, that the mass descending, takes the ground, part of it trending outwards into deeper water, and floating when conditions permit.†

Huge masses of this barrier, detached from it, float to more temperate regions, borne onwards by currents and prevalent winds. The accompanying sketch‡ (fig. 93) will afford an idea of the

Fig. 93.

tabular character of numerous icebergs before they have been much melted in more temperate climates, and also will show the stratified appearance noticed. Sir James Ross found many§ in about

* Where an opportunity occurred of seeing over the ice-cliff (about 50 feet high), Sir James Ross describes the mass as quite smooth in its upper part, and looking like "an immense plain of frosted silver."

† Wilkes, "United States' Exploring Expedition," vol. ii. Respecting that portion of the mass which reposes on the bottom beneath the level of the sea, we have also to consider the effect, for any value it may have, which may be due to terrestrial heat beneath, the ground protected from great atmospheric depressions of temperature by the mass of ice and snow above.

‡ Taken from Wilkes's "United States' Exploring Expedition," vol. ii.

§ 27th December, 1840. "Voyage of Discovery," &c. They extend often with a similar tabular character, according to particular seasons, more northerly. According to such seasons, also, the icebergs generally of the southern regions range to very different warmer latitudes. Upon returning from the antarctic regions in 1840, the different vessels of the United States' Exploring Expedition saw the last in 55° S.,

63° 30′ south, rising with tabular summits to the height of from 120 to 180 feet, several more than 2 miles in circumference. They were falling rapidly to pieces, and their course was marked by the portions of ice detached from them.

Respecting the mode in which icebergs are separated from the main mass of the ice barrier, and from the few he observed near it during the summer months, Sir James Ross infers that they are chiefly detached during the winter, the temperature of the sea and the air being then so different, whereas it more closely approximates during the summer. He points to the great cracks, some many miles in length, observed in the ice of arctic regions upon a sudden fall of 30° or 40° in the temperature, and more especially well seen in the great freshwater lakes, where the sudden rents are accompanied by loud reports. The unequal expansion of the ice exposed, on the surface, to 40° or 50° below zero (Fahrenheit), while beneath the temperature is 28° to 30° above it, could not, Sir James Ross infers, but produce the separation of large masses of ice. However little the action of the waves could affect a mass descending so low beneath the surface of the sea, we should expect that the influence of a rise and fall of tide would be felt, tending alternately to lift and depress much of it, especially at spring tides, so that supposing fissures formed, this very constant up and down movement would also tend to separate masses at the outer edge of the barrier.

While numerous icebergs are but the detached portions of the great ice barrier, which have not rested on a sea-bottom, and therefore transporting no mineral matter to milder regions, beyond any volcanic ashes or cinders discharged over the icy area, of which they may have formed a part, from such volcanic vents as Mount Erebus, and be interstratified with the layers of ice and snow,* others carry onwards no small amount of mud, sand, and rock fragments of different sizes. We have accounts of some covered with such detritus, blocks, so found, weighing several tons.† The detached portions of the glaciers, such as those de-

51° S., and 53° S. They were known to range so much northerly in 1832, that vessels bound round Cape Horn from the Pacific were obliged to put back to Chili for a time, in order to avoid them.

* Sir James Ross (Antarctic Voyage) mentions "that having observed new-formed ice off Victoria Land, covered with some colouring matter, a portion of the ice was melted and filtered, and an impalpable powder collected, considered as volcanic dust."

† Ross, "Voyage in the Antarctic Regions," vol. i. p. 173. Mr. Couthouy observed masses of rock embedded in an iceberg seen in lat. 53° 20′ S., long. 104° 50′ W., 1,450 miles from Tierra del Fuego, and 1,000 miles from St. Peter's and Alexander's Islands, whence he supposes the ice to have drifted. One of the rock masses seemed to show

scending from the Admiralty range, would be expected to transport the fragments which could fall upon them, as in the arctic regions. It would appear that, in addition to whatever may be thus carried, large icebergs which have rested upon the sea-bottom are often capsized, so that the mud, sand, and pieces of rock adhering to them beneath are suddenly upturned, a very great change in the relative position of such detritus being in this manner quickly produced. Sir James Ross mentions one suddenly capsized off Victoria Land, bringing up a portion of the bottom 100 feet above the surface of the sea, so that it was, for the moment, supposed to be an island not previously seen.* In this manner detritus may not only be transported directly from the land upon detached portions of glaciers, but also the mud, sand, and stones of a sea-bottom be uplifted several hundred feet, and carried great distances into milder climates.† A somewhat constant supply and a general course of the floating ice, from currents and prevalent winds, would cause a vast quantity of the detritus, thus obtained and floated away, to be distributed over the sea-bottom; mud, sand, and fragments of varied sizes mingled together. Though the finer matter would take longer to sink through the sea,‡ and so far become strewed over the bottom more widely and in a more even form, enveloping various inequalities that may occur (as well covering the tops as the sides, if not too steep, of submarine hills), the larger fragments would fall more irregularly upon and into the finer sediment. Submarine hill-tops would be as much covered by them as any depressions, and they would often be plunged into

---

a face of about 20 square feet. When within half a mile of this iceberg, the temperature of the air was 35°, and of the water 34°. The water to leeward of the ice was 7° colder than 4½ miles to windward of the berg.—"American Journal of Science," vol. xliii., 1842.

* "Antarctic Voyage," vol. i. p. 196.

† Captain Wilkes ("United States' Exploring Expedition") considered that he landed upon an upturned iceberg, part of the icy barrier weathered by storms, about eight miles from the main land, in latitude 65° 59′ 40″ S. Upon it were boulders, gravel, sand, and mud or clay. The larger specimens were of basalt and red sandstone. One piece of rock was estimated at 5 to 6 feet in diameter. The stones were cemented by very compact ice, thus forming an icy conglomerate.

As regards the distances to which the icebergs from the southern ice are carried, Captain Wilkes infers that they are conveyed westward the first season by the south-east winds, about 70 miles north of the barrier, being the second season driven northwards until they reach 60° S., after which they rapidly move more northward and disappear. Sir James Ross mentions a tabular iceberg, rising 130 feet above the sea, and three-quarters of a mile in circumference, in about 58° 36′ S.

‡ Sir James Ross ("Antarctic Voyage") considers the bottom as usually to be found in the Antarctic Ocean at 12,000 feet. Inequalities to a considerable amount also exist. No bottom was obtained by a line of 24,000 feet in latitude 68° 38′ S., and longitude 12° 49′ W.

mud, in the same manner as the sounding-lead above mentioned (p. 233), and which descended two feet into the fine green mud beneath 2,400 feet of sea, at a distance of 100 miles from Victoria Land. This fine mud would not appear an uncommon sea-bottom off Victoria Land,* and as icebergs discoloured by mud seem not unfrequent in these southern latitudes, such mud may be widely

---

* This mud seems, from the soundings obtained by Sir James Ross ("Antarctic Voyage"), to be common for about 400 miles along the great icy barrier near Victoria Land. It has been noticed previously (p. 233) that a detached portion of this barrier was found aground upon it, beneath 1,560 feet of water, 200 miles from that land. Respecting its composition, those minute bodies the *Diatomaceæ*, which were considered by Ehrenberg and many naturalists as infusorial animals, and by others as vegetables, and which seem now, especially from the researches of Mr. Thwaites, to be admitted by Dr. Hooker, Dr. Harvey, and other highly-qualified persons as the latter, would appear to form no inconsiderable portion of it. At the same time, as no rivers of Victoria Land bear out fine sediment, and great volcanos are there in activity, we may look to the distribution of ashes and cinders vomited forth from the latter as adding such products from time to time to this mud.

"The water and the ice of the South Polar ocean," observes Dr. Hooker ("Flora Antarctica," vol. ii. p. 505), "are alike found to abound with microscopic vegetables belonging to this order (Diatomaceæ). Though much too small to be discerned with the naked eye, they occurred in such countless myriads as to stain the berg and pack ice wherever they were washed by the swell of the sea; and when enclosed on the congealing surface of the water they imparted to the brash and pancake ice a pale ochreous colour. In the open ocean northward of the frozen zone, this order, though no doubt almost universally present, generally eludes the search of the naturalist, except when its species are congregated amongst that mucous scum which is sometimes seen floating on the waves, and of whose real nature we are ignorant, or when the coloured contents of the marine animals which feed on these *Algæ* are examined. To the south, however, of the belt of ice which encircles the globe, between the parallels of 50° and 70° S., and in the waters comprised between that belt and the highest latitude ever attained by man, this vegetable is very conspicuous, from the contrast between its colour and the white snow and ice in which it is embedded, insomuch that, in the eightieth degree, all the surface ice carried along by the currents, the sides of every berg, and the base of the great Victoria Barrier itself, within reach of the swells, are tinged brown as if the polar waters were charged with oxide of iron.

"As the majority of these plants consist of very simple vegetable cells, enclosed in indestructible silex (as other *Algæ* are in carbonate of lime), it is obvious that the death and decomposition of such multitudes must form sedimentary deposits, proportionate in their extent to the length and exposure of the coast against which they are washed, in thickness to the power of such agents as the winds, currents and sea, which sweep them more energetically to certain positions, and in purity to the depth of the water and nature of the bottom. Hence we detected their remains along every ice-bound shore, in the depths of the adjacent ocean, between 80 and 400 fathoms. Off Victoria Barrier the bottom of the ocean was covered with a stratum of pure white or green mud, composed principally of the siliceous cells of *Diatomaceæ*; these on being put into water rendered it cloudy, like milk, and took many hours to subside. In the very deep water off Victoria and Graham's Land this mud was particularly pure and fine; but towards the shallower shores there existed a greater or less admixture of disintegrated rocks and sand, so that the organic compounds of the bottom frequently bore but a small proportion to the inorganic."

Respecting the distribution of the *Diatomaceæ*, Dr. Hooker remarks (ibid. p. 505) that many species are found from pole to pole, "while these or others are preserved in a fossil state in strata of great antiquity. There is also probably no latitude between that of Spitzbergen and Victoria Land, where some of the species of either country do not exist: Iceland, Britain, the Mediterranean Sea, North and South distributed and be irregularly supplied with sand, stones, and large fragments of rock, as the icebergs melt away and drop their loads of mineral substances.*

Captain Cook long since (1777) made known the fact that, at the mountainous island of South Georgia, included between latitude 53° 57′ and 54° 57′ S., glaciers descended into the sea, detached masses from which floated outwards, to be distributed by ocean currents and prevalent winds, in given directions. The following view of Possession Bay† (latitude 54° 5′ S.), in that island presents us with a glacier reaching the sea, the depth of which was more considerable than that of an ordinary sounding line (204 feet) employed at the time. Captain Cook says, "The head of the bay, as well as two places on each side, was terminated

Fig. 94.

America, and the South Sea Islands, all possess Antarctic *Diatomaceæ*. The siliceous coats of species only known living in the waters of the South Polar ocean have, during past ages, contributed to the formation of rocks, and thus they outlive several successive creations of organized beings. The phonolite stones of the Rhine and the tripoli stone contain species identical with what are now contributing to form a sedimentary deposit (and perhaps at some future period a bed of rock), extending in one continuous stratum for 400 measured miles. I allude to the shores of the Victoria Barrier, along whose coast the soundings examined were invariably charged with diatomaceous remains, constituting a bank which stretches 200 miles north from the base of Victoria Barrier, while the average depth of water above it is 300 fathoms, or 1,800 feet."

* As respects sand intermingled with ice and carried away, Captain Wilkes mentions ("United States' Exploring Expedition") a floating mass, composed of alternate layers of snow and ice, the former mixed with sand. Upon this pieces of granite and red clay were also found.

† Taken from the plate, vol. ii., p. 213, of Cook's "Voyage to the South Pole," 4to, 1777.

by perpendicular ice-cliffs of considerable height. Pieces were continually breaking off, and floating out to sea; and a great fall happened while we were in the bay (January 17, 1775), which made a noise like cannon." He also calls attention to the bottom of the bays generally in this land being filled by glaciers, supplying an abundance of icebergs; and it is easy to infer that, from amid the mountain cliffs among which these glaciers find their way to the coast, many a fragment of rock may be ice-borne, and deposited at the bottom of the sea, remote from South Georgia. Not a stream or a river could be seen throughout the whole coast explored, though it was visited in the summer of that region. Captain Cook also mentions bays full of glaciers, descending from the heights of Sandwich Land, discovered by him upon leaving South Georgia, on the south-east of that island.*

Quitting the far southern land and remote islands, the climate is such in Tierra del Fuego, although comprised between latitude 52° 30′ and 56° S. (a range corresponding in the northern hemisphere with the position and distance between Birmingham and Edinburgh), that the line of perpetual snow occurs, according to Captain King, at between 3,500 and 4,000 feet above the sea in the Straits of Magellan, and that glaciers descend into the sea.† Mr. Darwin states that on the north side of the Beagle Channel (a remarkable strait, running east and west across the southern part of Tierra del Fuego) the mountains are covered with perpetual snow, whence, in many places, magnificent glaciers descend to the water's edge, fragments falling from them into the sea, and floating about as miniature icebergs.‡ He remarks that glaciers occur at the head of the sounds along the whole western coast of the southern

* Cook's "Voyage to the South Pole," vol. ii., p. 224. He remarks also upon the flat surfaces, and even heights, of the icebergs in that region, some two or three miles in circumference, reminding us of the character of those off the great ice barrier near Victoria Land.

† Mr. Darwin gives the following table of the climate of Port Famine, Straits of Magellan, and of Dublin:—

	Latitude.	Summer Temperature.	Winter Temperature.	Difference.	Mean of Summer and Winter.
	° ′	°	°	°	°
Dublin	53 21 N.	59·54	39·2	20·34	49·37
Port Famine	53 38 S.	50·	33·08	16·92	41·54
Difference . . .	0 17	9·54	6·12	3·42	7·83

‡ Darwin, "Voyage of Adventure and Beagle," vol. iii., p. 243.

part of South America.* It would appear that as far north as latitude 48° 30′ S. glaciers advance into the sea. Eyre's Sound is terminated by glaciers descending from the range of the Sierra Nevada on the east. Mr. Bynoe saw numerous detached masses of ice floating about, 20 miles from the head of the sound; and upon one, drifting outwards, found an angular block of granite, described as a cube of nearly two feet, partly imbedded in it, the ice thawed around.† Mr. Darwin directs attention to the occurrence of a glacier at the level of the sea, even in latitude 46° 40′ S., in the Gulf of Penas, reaching to the head of Kelly Harbour, pointing out that thus "glaciers here descend to the sea within less than nine degrees of latitude from where palms grow, less than two and a half from arborescent grasses; and, looking to the westward, in the same hemisphere, less than two from orchideous parasites, and within a single degree of tree ferns."‡

The transport of mineral matter by floating ice is not limited to portions of glaciers, broken off where they have protruded into the sea, or to masses detached from great continuous ranges of ice, such as the barrier off Victoria Land. Rivers, in regions where the temperature descends sufficiently low, remove no small portion of such matter by means of ice down their courses, and coast ice distributes no inconsiderable amount of it in various directions. As regards the mode in which detritus may be conveyed by rivers, it may often be studied in our brooks and streams, when a sudden thaw suddenly fills them with water, lifting away ice which may bind gravel, sand, or pieces of frozen mud together, by their sides or in shallow places. According to the relative specific gravities of the detached portions of ice, stones, sand, and mud, will they be seen to move, some larger pebble, perhaps deeply set in its support of ice, trailing along, and leaving the mark of its passage on the bottom. Other portions will float more freely onwards, some acquiring rotatory motion, and, by grinding against each other,

* Darwin, "Voyage of Adventure and Beagle," vol. iii., p. 282. Mr. Darwin observes (p. 283), "In the Canal of the Mountains no less than nine (glaciers) descend from a mountain, the whole side of which, according to the chart, is covered with a glacier of the extraordinary length of 21 miles, and with an average breadth of 1½ mile. It must not be supposed that the glacier merely ascends some valley for the 21 miles, but it extends apparently at the same height for that length, parallel to the sound, and here and there sends down an arm to the sea-coast. There are other glaciers having a similar structure and position, with a length of 10 or 15 miles (Tierra del Fuego).

† "Voyage of Beagle," vol. iii., p. 283. Mr. Darwin calls attention to this sound being in a latitude corresponding, in the north, with that of Paris, and also to an "Iceberg Sound," as given in the charts still further north.

‡ Ibid., p. 285.

parting with some parts of their load, especially the heaviest, while here and there they become jammed in the narrower parts of the stream, and stranded upon shoals, there remaining, in great part, until, the thaw proceeding, the ice melts, and the detrital matter is dealt with by the stream in the usual manner.*

The transport of mineral matter which may often and easily be seen in this minor manner, under the fitting conditions, is but carried out upon a larger scale in many great rivers, where the relative magnitude of the effects produced more engages our attention, especially when those objects to which we attach interest are endangered or sustain injury. In the regions where ice is common upon great rivers during part of the year, and that part of the year the time when the water supply is the least, and the river level the lowest, the fragments of rock, pebbles, sand, and mud of the sides, and shoal grounds become, as it were, a piece of the main sheet of ice, should it extend entirely over the river, or of such portions of one as may exist. These are ready to be broken off, lifted, and borne down the stream as the waters of the river rise before any general increase of temperature melts the ice upon the banks, shoals, or general surface of the river. It will be obvious that the transport of detritus will depend upon circumstances, as in the little brooks, and that while some portions are carried long distances, others will be left in various situations; sometimes fragments of rock being carried to, and accumulated in, situations where the ordinary force of the river cannot readily dislodge them, and indeed sometimes be altogether insufficient for the purpose. We have various accounts of detritus so borne downwards in rivers by means of ice. In the St. Lawrence there would appear to be good opportunities of studying the transport of mineral matter on the large scale. Captain Bayfield

* It is while studying the effects of ice in the brooks and minor streams that an observer may sometimes see the formation of ice at the bottom. M. Arago, whose attention this subject has engaged, remarks respecting it ("Annuaire du Bureau des Longitudes pour 1833," p. 244), that the movement of these running waters mixes those of different temperatures and densities, so that when the whole is at the freezing point, the pebbles and other substances at the bottom of the brook constitute so many projections, as in a saline solution, and thus ice is formed upon them. The ice thus produced is spongy, from the crossing and confused grouping of its crystals, the movement of the water preventing an uniform arrangement of parts. The ice accumulates, and gradually envelopes numerous pebbles and other substances, and will rise to the surface with its mineral load if the general specific gravity of the whole will permit. M. Leclercq has observed ("Mémoires Couronnés par l'Académie de Bruxelles," tom. xiii. 1845) that the ice is first formed upon the face of the pebbles or other objects opposed to the current of water, and that, although a rapid flow of water contributes to the first production of the ice, the increase of ice is in proportion as the movement of the water is moderate, the extreme cold considerable, and the sky clear.

has pointed out that there, where the temperature in winter sometimes descends 30° below zero (Fahr.), large boulders are entangled in the ice, and carried considerable distances upon the surface of the water in the spring. Shoals are thickly strewed with them.* Conditions being favourable for keeping blocks and fragments of rock in the lower part of the river ice, thus carried onwards, and indeed often driven forwards rapidly, wherever the general masses grated upon any bottom, over which they could be forced by the volume of water behind (and heavy piles of ice sometimes accumulate, obstructing the free flow of the waters), much scratching and furrowing would be expected, according to the relative hardness of the rocks passed over and of the ice-borne fragments, to the pressure of the mass of ice and detritus, and to the velocity with which that mass may be driven upon the rocky ledge or shoal. Fragments of rock, set in the ice, and grating against vertical cliffs rising from comparatively deep water, such as frequently occur on the bends of rivers, would also horizontally scratch and abrade the rocks, according to their relative hardness, the ordinary river action not removing these marks, though they may become obliterated by atmospheric influences at lower states of the river, especially where the cliff-rocks were composed of somewhat incoherent materials. Thus while some ice-supported boulders and fragments of rocks were grooving and furrowing the horizontal surface of a ledge of rocks at *b* (fig. 95), and others, encased in ice, were borne down the river at the same time, scratching and wearing away the vertical cliff at *c*, another collection might be leaving permanent traces of its passage upon previously ice-borne boulders, accumulated from local causes at *a*.

Fig. 95.

It is interesting to consider that by such means large rounded portions of rock, with minor pebbles, may thus be borne towards the Gulf of St. Lawrence, and be thrown down, after being scratched in their passage over hard ledges of rock, or over boulders in shallow water, in situations where such marks would not be removed by any attrition to which they would be exposed under existing circumstances, there accumulating with finer detritus, even mud deposited from water in which it had been held in ordinary

* Bayfield, "Proceedings of Geolog. Soc. of London" (1836), vol. ii., p. 223.

mechanical suspension. Thus the scratching of the ledges of solid rock and heavy stranded boulders in shallow situations might be accomplished, and the boulders and pebbles by which this was effected be themselves often also scratched, carried onwards under favourable circumstances, and be deposited, with these marks still upon them, amid fine sediment in depths beyond the reach of wave or breaker action for the attrition necessary to remove such scratches.

The great rivers of Northern Europe, Asia, and America delivering themselves into the Arctic Sea, flowing as they do from milder into colder climates, present us with the conditions for the formation of ice sooner, and its continuance later at their embouchures than towards their origin. The effects produced are especially interesting, inasmuch as when, from the melting of the snows and ice on the southward, floods are produced, these meet with the obstruction of the ice towards the mouths of rivers. In consequence, it not unfrequently occurs that the resistance of the ice being suddenly overcome, it is violently upheaved and broken, and in parts thrown aside, with any masses, or minor fragments of rocks attached to it. Sir Roderick Murchison has pointed out the banks of rock-fragments thus produced on the sides of rivers in Russia, and especially notices the fluviatile ridges of angular blocks towards the mouth of the Dwina. White carboniferous limestone there occurs (about 110 versts from Archangel), and the waters of the river entering amid its chinks and joints, separates them when frozen, so that subsequently they are entangled in the ice adjoining the banks, and are thus carried with it.* By the sudden rise of waters thus caused, many a block of rock must be borne over low ground, stranded on shoal water, or be occasionally carried seawards, and thrown down amid fine sediment, the conditions for the transport of which outwards would be increased during these sudden discharges of water. The crashing and jamming together of the broken masses of ice would be highly favourable to the scratching and scoring of blocks and fragments of rocks entangled among them, and such blocks and fragments may also be often transported to situations where, under existing circumstances, the markings thus produced would not be obliterated.

---

* Murchison, "Geology of Russia in Europe and the Ural Mountains," vol. i., p. 567. He quotes M. Böhtlingk as noticing large granitic boulders, weighing several tons, entangled in the branches of pine trees, 30 or 40 feet above the level of the streams. Speaking of blocks of rock ice-borne down rivers, Sir Roderick Murchison, after noticing their modes of transport and deposit, remarks, that old drift from the north may thus be brought back to the northward by the rivers, p. 565.

When we consider the state of sea-coasts in those regions where the temperature falls sufficiently low during a part of the year that ice is formed upon them, entering amid the substances of which they are composed, and binding blocks of rock, shingles, sand, and even mud, with the remains of any marine animals there occurring, into one solid mass, we see that when the warmer season in such regions comes round, mineral matter may be readily removed from one place to another upon the breaking up of the coast ice.

Upon the breaking up of this coast ice, which sometimes rests on shallow ground, and at others covers deep water, we should expect much grinding of the masses on the shore, scratching and grooving the sides of cliffs and shallow rocky bottoms, when shingles or other fragments of rock are frozen into the ice, so as to be brought into contact with the one or the other.* The force employed would appear to be often very considerable, great sheets of ice being set in motion, and being driven with tremendous crashes against the land, so as not only to act upon shore ice, in which rock fragments and shingles may be embedded, thus pressing them heavily against bare rocks, but also forcing beaches before them, grinding the pebbles and boulders against each other, and upon exposed rocks, by which both may be scored and marked. In this manner friction marks may be produced, which in some situations may not be very readily removed by the ordinary rounding and smoothing of breaker action.

When an observer studies the maps and charts which we as yet possess of the northern seas of America, Europe, and Asia, he will find enough to show him that portions of beaches may readily be removed upon the breaking up of ice from the coasts, and be transported to other situations, where, upon the melting of that ice, they may be thrown down in depths amid any fine detritus there

* M. Weibye, of Kragero, is quoted by M. Frapolli ("Bulletin de la Société Géologique de France," 1847), as inferring, respecting the marks left by the block-and-shingle-bearing ice of the Scandinavian coasts, that on those bordering the sea in the Bradsbergsamt, "the scratches and furrows on horizontal, or nearly horizontal surfaces, take a direction always perpendicular to the general line of coast in open bays, and always parallel to the range of the channels in narrow fiords, that the horizontality or the greater or less inclination of the scratches on the inclined or vertical surfaces depends on the relief of the coasts of the locality, and always corresponds with this relief and with the action of the different winds." M. Frapolli himself also calls attention to the effects of coast ice armed with blocks and pebbles of rock, driven about in numerous fragments by the storms of winter and spring, and grinding against the cliffs of Scandinavia, polishing and scratching the rocks according to their surfaces and position, the cliffs scratched in horizontal lines along the fiords and in other simillar positions.

accumulating. Should any of their component pebbles or fragments of rock have been so acted upon as to be scratched before they were thrown down, they would retain those marks amid the fine deposits in such depths. As ice adheres to coasts in many localities during winter, upon which, from the ordinary action of the sea on shores, breakers throw whole and broken shells of molluscs and other marine animal remains during the summer, these remains would be liable to be entangled in portions of beach removed by the ice, and be scattered over various depths of water, in the same manner as the transported mineral matter, and thus the remains of littoral molluscs, often in fragments, may be dispersed amid a mixture of mud, and ice-borne blocks, and fragments of rock accumulating in deep water.

In tidal seas account has to be taken of the movement of ice in estuaries, and in those long deep loughs or arms of the sea, in Norway termed *fiords*,* up and down which the flood and ebb tides are felt according to circumstances. Coast ice, borne backwards and forwards by the tide, and having pebbles and fragments of rock so set in it that they can grind upon or against bare rocks, spread horizontally or rising vertically, or nearly so, in the estuaries and fiords, could scarcely fail to become an instrument of importance in the scratching and grooving of such bare rocks, these markings being also, especially in the case of the cliffs, not easily removable. This action continuing through many successive ages, certain kinds of rocks might, in favourable localities, retain marked scratches and grooves thus produced, independently of the influence of winds driving the fractured coast ice about against lines of coast, upon the breaking up of such ice. Fragments of ice and any mineral matter they may sustain are thus piled up at the bottom of bays or in shoal water, a combination of a heavy on-shore gale of wind and a spring tide leaving many a fragment of rock in a situation whence it could not readily be removed under ordinary circumstances.

No small amount of rounded boulders and pebbles of various sizes may thus become strewed near coasts, or be mingled beneath deep water with the angular fragments which have either been transported by icebergs, broken off the terminal portions of glaciers,

* The channels which divide Tierra del Fuego into its many islands, and the Straits of Magellan separating it from the mainland of America, with the very numerous indentations and channels found between the east entrance of the Straits of Magellan and the Gulf of Penas, and into which glaciers often descend, and ice floats about, would appear to be frequently very deep and steep-sided. In mid-channel, eastward of Cape Forward, Captain King found no bottom in the Straits of Magellan with a line of 1,536 feet.

or which may have fallen from cliffs upon coast ice, with the addition even of the remains of littoral or shallow-water molluscs, or of other marine animals, such as the bones of fish, whales, and seals carried off by the coast ice. A good example of the removal of a block of rock by coast ice, so far from the polar regions as Denmark, is mentioned by Dr. Forchhammer, who states that one, about 4 to 5 tons in weight, and resting on the shore, was encased in coast ice during the winter of 1844, and carried out to sea with the ice in the following spring, leaving, as it moved seaward, a deep furrow in the sandy clay of the shore, not quite obliterated six months afterwards.*

As modifying the accumulations which may be formed on the bottoms of seas liable, from time to time, and, sometimes, as a whole, periodically, to sustain icebergs grounded upon them, the observer has to bear in mind that not only may the icebergs, by being forced against banks, jumble together, and singularly mingle beds of clay and sand, even occasionally adding transported fragments to the disturbed mass, but also act as rocks round and amid which streams of tide, or sea-currents, may become for the time modified. We should expect this to be most experienced in the regions where, from the general intensity of the cold, the icebergs could the longest remain. Sir James Ross mentions that the streams of tide were so strong amid grounded icebergs at the South Shetlands, that eddies were produced behind them,† so that, as far as such streams were concerned, they acted as rocks. Navigators have observed icebergs sufficiently long aground in some situations, that even mineral matter might be accumulated at their bases in favourable situations, while streams of tide may run so strongly between others, that channels might be cut by them in bottoms sufficiently yielding, and at depths where the friction of these streams could be experienced. Much modification of sea-bottoms might be thus produced by grounded icebergs, not forgetting those seasons of the year when many become joined together by ordinary sea-ice, con-

* Forchhammer, "Bulletin de la Société Géologique de France," 1848. He observes, respecting the transport of blocks and pebbles on the coast of Denmark by coast ice, that although the latter envelopes the blocks and pebbles on the shore, to enable these to be borne away, it is necessary that the thaw or rupture of the ice should coincide with a rise of the waters. Respecting blocks and fragments of rock borne out by the ice from the Baltic, by means of the current setting through the Kattegat in the spring, Dr. Forchhammer mentions that, in 1844, a diver found the remains of an English cutter, blown up during the bombardment of Copenhagen in 1807, covered by blocks, some of which measured from six to eight cubic feet. The same diver affirmed that all the wrecks he had visited in the roadstead of Copenhagen were more or less covered by rock fragments.

† Ross, "Antarctic Voyage."

stituting part of a mass to be dealt with on the large scale, when such ice is broken up. However firm the grounded icebergs may, like so many anchors, often tend to hold the main mass, it is not difficult to conceive that conditions arise by which many are dragged, cutting and ploughing up the sea-bottoms in their courses.

Ice thus transports portions of rocks, either in the shape of glaciers, descending under the needful conditions in various extra-tropical regions, as floating ice down rivers, as coast ice, as fragments of glaciers descending into the sea, or as masses which, having been aground, capsize, and bring up a portion of the bottom on which they previously rested. Huge fragments of rock are by these means moved to distances from their parent masses, of which no other known power, now in force on the surface of our globe, appears capable. It has been seen that glaciers increase and decrease according to the variations of the climates under which they are formed. What the amount of that increase and decrease may be under the conditions now existing, and where glaciers have been noticed, seems not well ascertained, though the differences in their volume and extent would appear to have been greater than was once supposed. Be that as it may, they distribute rock fragments outwards from mountain regions, these generally angular, unless ground between the glacier sides and bottom, the larger blocks and fragments remaining where the glaciers left them, while minor portions and finely-comminuted mineral matter are thrown into the torrents and rivers, to be disposed of by them according to their powers. River ice may carry detritus entangled in it, distributing the mineral matter over areas corresponding with their courses, and which may be sufficiently flooded by them, transporting many a block and fragment which the power of the stream could not otherwise have moved. With the exception of rock fragments, which may have fallen from cliffs overhanging the rivers and not afterwards have been rounded, which may have been broken up from the sides in the manner previously noticed (p. 244), or which may have been left by some prior geological condition of the area, we should expect much of the detritus borne down by river-ice to be composed of the ordinary pebbles, sand, and mud of river courses.

The sea deals with any ice-borne detritus received from rivers, or from the coasts, according as it is tideless or tidal, and as the portions into which these are carried may be in movement as sea and ocean currents, or the ice be acted on by the wind. Looking at the northern regions, where rivers of sufficient importance discharge themselves, carrying ice outwards, and coast ice is common, it may be anticipated that much coast shingle, with rounded river pebbles, lumps of the frozen mud, and sands of estuaries, the occasional remains of marine animals, and now and then those of terrestrial animals, suddenly swept outwards by the river floods, would be strewed about upon the sea-bottom. Many a bone of elephants, rhinoceroses, and other animals, imbedded in the mud, sand, and gravel, of these regions, may also, after having been washed out of the beds which contained them, be ice-borne into the sea, and be mingled with remains of existing animals. To these may be added angular fragments carried out by the ice of rivers, or borne by coast ice from beneath cliffs whence such fragments have fallen upon it, independently of those carried into parts of the same seas by icebergs detached from the terminal part of glaciers.

Although the arctic seas are so shut in by the lands of America and Asia, a comparatively small opening (Behring's Strait) only occurring between them, a space sufficiently wide exists between America and Europe, notwithstanding the interruption presented by Iceland, to permit the escape outwards of a certain portion of ice. We have seen that over the bottom of part of the North Atlantic blocks and fragments of rocks, with minor detritus, are now being strewed, without reference to its inequalities. In the antarctic seas very different conditions present themselves. Great rivers bearing ice-borne blocks and fragments of rocks, with minor detritus, are not found. The land, now commonly supposed to occupy so large an area in the South Polar regions, supports little else than water in its solid form, and the coast, for the most part, seems so encased by huge icy barriers, that common coast ice would there appear considerably limited, as compared with the arctic regions, in its power to carry off rounded boulders and shingles. Such glaciers as reach the sea, transporting fragments from the inland cliffs amid which they may move, would appear the principal agents in carrying mineral matter directly from the land, allowing for a portion transported by coast ice. The ice aground off Victoria Land would nevertheless appear to have the power of transporting much detritus when broken up into icebergs and upset, strewing blocks and minor fragments, sand and mud, over a part of the Southern Pacific. The South Shetlands, South Orkneys, South Georgia, Sandwich Land, and the lands more or less encased with ice between the South Shetlands and Victoria Land doubtless also contribute, by means of glaciers, coast ice, and probably also, as capsized grounded ice, blocks and fragments of rock (some rounded), sand and mud, to the bottom of the Southern Atlantic, and the ocean southward of Africa and Australia. The southern portion of America adds its glacier-borne fragments, and thus, both on the north and on the south, portions of rocks, formed in the colder, are ice-borne, and left beneath the seas of the more temperate, regions of the earth.

CHAPTER XIV.

INFLUENCE OF A GENERAL INCREASE OF COLD.—MODIFICATIONS OF TEMPERATURE FROM CHANGES IN THE DISTRIBUTION OF LAND AND SEA.—ERRATIC BLOCKS.—EFFECTS OF GRADUAL RISE OF THE SEA BOTTOM STREWED WITH ICE-TRANSPORTED DETRITUS.—EFFECTS OF A SUPPOSED DEPRESSION OF THE BRITISH ISLANDS.—INCREASE OF ALPINE GLACIERS.—TRANSPORT OF ERRATIC BLOCKS BY GLACIERS.—FORMER EXISTENCE OF GLACIERS IN BRITAIN.—ELEVATION OF BOULDERS BY COAST ICE DURING SUBMERGENCE OF LAND.—ERRATIC BLOCKS OF THE ALPS.—ERRATIC BLOCKS OF NORTHERN EUROPE.—ERRATIC BLOCKS OF AMERICA.

THE geological effects now due to ice being as previously represented, it becomes desirable to consider those which would probably arise either from a general diminution of temperature on the surface of the globe, or from partial changes of that temperature. With respect to the first we have to look to some general cause common to the whole globe. Whatever the conditions for the distribution of temperature may have formerly been, we see that the influence of the sun now causes the heat of the tropics, and the different exposure of the polar parts of the earth's surface to it, the great variations of seasons there experienced. Any changes of sufficient importance, therefore, in the influence of the sun, which should produce a corresponding change on the face of the earth, so that the line of perpetual snow, as it is termed, should descend lower towards the sea in the equatorial, and cut its level at less high latitudes in the polar regions, would materially alter the climates of many parts of the world. Geological effects due to ice would be more widely spread than they now are, and the equatorial space within which ice-transported masses of rock and other detritus cannot be borne, would be more limited. Glaciers, where they could be formed, would not only become more extended than they now are in certain mountainous regions, but ranges of mountains, amid which they do not at present occur, the line of perpetual snow not descending sufficiently low, would contain them; so that, in the one case, mineral matter would be distributed by them over a wider area; and in the other, over districts where no transport of the kind exists at the present time. Fragments angular, subangular, and rounded, would be distributed by river-ice and coast-ice, where none such are now formed, and sea-bottoms would then be strewed over by them, where, at present, nothing of the kind is carried. Animal and vegetable life would be adjusted to the new conditions (that adapted to the colder climates of the earth moving more towards the equator), its remains, at least such as were preserved, spreading over those of the animals and plants which flourished in the same regions under higher temperatures.

The like general effects would be expected if, without supposing a diminished influence of the sun, our whole solar system, moving through space, should pass from the temperature now inferred to be that of the portion amid which that system takes its course (p. 206) to one less high. And it may well deserve the attention of the geologist to consider the effects which would follow such a change, even to the amount of a few degrees, as commonly measured by thermometers. In his observations on the distribution of masses of rock, apparently ice-borne to their present positions, and about to be noticed, it is very desirable that he should regard the subject generally as well as locally, so that, whatever may eventually appear the right inference to be drawn from the facts recorded, such as may bear upon the former should not be omitted in the search for the latter. As regards the evidence of many climates having remained much the same, with certain modifications, during those comparatively few revolutions of our planet round the sun, of which we have any records, and from which we may infer that the climates generally of the surface of the globe have not suffered material alteration since the historical period, as it has been termed, the geological observer will soon perceive that he is forced to consider it as affording him very limited aid in his inquiries respecting the former climatal conditions of the earth.

The present different conditions as to the production of ice capable of transporting mineral matter, in the manner above noticed, in the northern and southern cold regions of the globe, are sufficient to prove that partial changes of great importance may arise from differences on the surface of the earth itself. Every-day experience in geological research will show the observer that he has to consider the surface of the earth to have been in an unquiet state from remote geological times to the present, and that while he so often stands, amid stratified deposits, on ancient sea-bottoms now elevated to various altitudes above the ocean level, many a region shows that its area has more than once been beneath that level and above it. Thus, although a mass of land may now rise above the sea-level at the South Pole, separated by a broad band of ocean from other great masses of land to the northward, producing certain effects as regards the climate of that part of the globe, and the northern polar regions are otherwise circumstanced, it by no means follows that such has always been the case, even in more recent geological times. If we change the conditions of the two polar regions, a difference of results is obtained of an important geological character. Mr. Darwin has skilfully touched upon the effects which would follow such a modification of conditions, and which require to be borne in mind in researches of this kind.*

In like manner any elevation or depression of a considerable area of dry land, which should raise parts of it above, or lower others, now above, beneath the line of perpetual snow, would produce modifications in the transport of mineral matter which could be effected

* He transports, in imagination, parts of the southern region to a corresponding latitude in the north. "On this supposition," he observes, "in the southern provinces of France, magnificent forests, intwined by arborescent grasses, and the trees loaded with parasitical plants, would cover the face of the country. In the latitude of Mont Blanc, but on an island as far eastward as Central Siberia, tree-ferns and parasitical orchideæ, would thrive amidst the thick woods. Even as far north as Central Denmark, humming birds might be seen fluttering about delicate flowers, and parrots feeding amidst the evergreen woods, with which the mountains would be clothed down to the water's edge. Nevertheless, the southern part of Scotland (only removed twice as far to the eastward) would present an island "almost wholly covered with everlasting snow, and having each bay terminated by ice-cliffs, from which great masses, yearly detached, would sometimes bear with them fragments of rock. This island would only boast of one land-bird, a little grass and moss; yet, in the same latitude, the sea might swarm with living creatures. A chain of mountains, which we will call the Cordillera, running north and south, through the Alps (but having an altitude much inferior to the latter), would connect them with the central part of Denmark. Along this whole line nearly every deep sound would end in 'bold and astonishing glaciers.' In the Alps themselves (with their altitude reduced by about half), we should find proofs of recent elevations, and occasionally terrible earthquakes would cause such masses of ice to be precipitated into the sea, that waves, tearing all before them, would heap together enormous fragments, and pile them up in the corner of the valleys. At other times, icebergs, charged with no inconsiderable blocks of granite, would be floated from the flanks of Mont Blanc, and then stranded in the outlying islands of the Jura. Who, then, will deny the possibility of these things having taken place in Europe during a former period, and under circumstances known to be different from the present, when, on merely looking to the other hemisphere, we see they are under the daily order of events?" Mr. Darwin then calls attention to the island groups, "situated in the latitude of the south part of Norway, and others in that of Ferroe. These, in the middle of summer, would be buried under snow, and surrounded by walls of ice, so that scarcely a living thing of any kind would be supported on the land."—Narrative of the Surveying Voyages of the Adventure and Beagle, vol. iii. p. 291.

by ice. If the region comprising the Alps were raised 3,000 feet above its present relative level, the area fitted for the formation of glaciers would be greatly extended, many a valley would be filled with ice, and many a mountain would contribute its glacier, not so filled or contributing at the present moment. Blocks and minor fragments of rocks would be ice-borne over, and left at distances from the main range not now attained; and, under the supposition of a gradual rise of land, many modifications would attend the change in the perpetual snow line, whence the glaciers for the time took their rise. Many a ravine and mountain side would be grooved and scratched not now touched by glaciers, and huge masses of rock be accumulated in heaps or lines, in localities where no ice now transports such masses. Assuming a depression of the same area, if we take the present relative levels only into consideration, the transport of glacier-horne blocks and fragments of rock, with the polishing, grooving, and scratching of valleys and their sides by the moving ice, would be limited to the areas now occupied by glaciers, duly allowing for their extension and contraction within the range of the present climatal conditions.

Thus, by the elevation and depression of large areas of dry land, very varied conditions for the existence, extension, or contraction of glaciers, with their geological consequences, may arise, without reference to those due to floating ice, excepting such as could be formed in great lakes, such as that of Geneva, for example, where effects similar to those observed in northern America would be produced. On the shores of such lakes coast ice would be formed, enclosing fragments of the rocks, and the shingles of beaches, to be borne away, should circumstances permit, if raised to an altitude permitting a depression of temperature sufficient for the production of such ice. There is also no difficulty in imagining conditions under which glaciers could protrude into large fresh-water lakes, carrying rock fragments with them, and having their extremities broken off and floated away with their detrital loads, under proper depths of water, as now takes place in the sea in the polar regions. Such masses of ice, though not moved onwards by streams of tide or ocean currents, would still be under the influence of the winds, to be driven to, and stranded in minor depths, where the ice could melt, and leave any blocks or fragments entangled in or resting upon them.

With respect to the distribution of ice-borne blocks of rock upon lakes, Sir Roderick Murchison has called attention to effects which would follow the lowering of lakes in regions where ice could be formed of sufficient thickness and importance for the transport of detritus.*

When the depression of an area of dry land, with the needful modifications of surface, in climates where glaciers had been formed, was such that the sea entered amid the valleys in which these streams of ice occurred, the change might or might not, according to the general climatal conditions produced, affect the glaciers. Should the change in the northern be of an order to introduce the climate of the southern hemisphere, it has been above seen (p. 240), the cold might be so increased, that Alpine glaciers became more extended, delivering icebergs into surrounding seas, so that, as Mr. Darwin has remarked (note, p. 253), they might float away, and be stranded on the Jura, then an island range.

Hitherto we have regarded these alterations of level as slowly produced, so that the changes, of whatever kind, were gradual, causing no sudden alteration of conditions. This, however, is far from necessary in geological reasoning, there being evidence connected not only with actual mountain ranges, but also with many a district wherein the rocks are broken and contorted, which would lead us to infer, with every allowance for the repeated effects resulting from the multiplied application of minor forces, that considerable forces had often been somewhat suddenly called into action. The waves produced during the disturbances of the land, known to us as earthquakes, and which will be noticed hereafter, are sufficient to show how, in that mode alone, glaciers, protruding into the sea, or great lakes of fresh water, may be lifted at their ends, and their fragments, with any load of detritus they may sustain, be whirled about and stranded in unusual situations. Greater waves would produce greater results, and when we unite them with land suddenly depressed beneath the sea-level, even only a few hundred feet, in such regions as those of Victoria Land and South Georgia, or of Greenland and Iceland, we have the means of removing ice and producing a complicated mixture of blocks and minor fragments of rock of great geological importance. In like manner, the sudden elevation of land, covered by snow and glaciers, if accompanied by the transmission of heat through fissures then formed, or by the increased temperature of the supporting mineral matter from the protrusion of igneous rocks among it, so that the snow and ice were suddenly and in part melted, would be productive of no slight geological effect, more especially if the glaciers of the land so acted upon protruded, or nearly so, into the sea.

* "Geology of Russia in Europe and the Ural Mountains," vol. i. p. 568.

Huge blocks of rock, often angular, are found scattered in such a manner over parts of the northern portions of Europe and America, and again in part of South America, and amid and around mountainous regions, such as the Alps, that, comparing their mode of distribution with that now known to be taking place by means of ice, attention has of late been very generally given to this explanation of their mode of occurrence. The masses of rock so found are commonly termed *Erratic Blocks*, and correct observations respecting the conditions under which they are found are material to a right understanding, particularly as respects the northern hemisphere, of the manner in which they have been accumulated.

As there are occasionally blocks of rocks scattered over a country, which are merely portions of some harder beds, interstratified with more yielding substances, or are the remains of dykes and veins of igneous rocks, the continuity and mode of occurrence of which may not be clear, the more readily disintegrated rocks having been removed by the effects of atmospheric influences, or breaker action at some prior geological time, the observer has in some districts to employ much caution as respects their origin. This is especially needed where the dykes or veins of the igneous rocks may have decomposed, as often happens, in an irregular manner, so that portions of the more unyielding, or harder parts, are scattered about, while traces of the softer are not easily found. From the liability of certain igneous rocks to decompose in spheroidal forms (fig. 2, p. 3, and fig. 7, p. 9), such blocks will sometimes present the false appearance of having been rounded by attrition, as if worn on some coast. Let, for illustration, *a*, *b*, be a dyke of greenstone, liable to

Fig. 96.

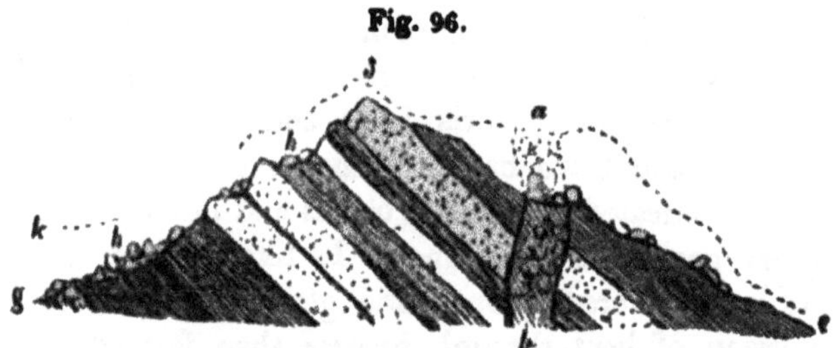

unequal decomposition in different parts, at *a* decomposed in spheroidal portions, then during the loss of general surface upon the hill side *e f*, the harder parts of the disintegrated portion, *a c*, might fall over towards *e*, and present the appearance of rounded boulders of greenstone resting upon some other rock. Again, on the other side of the hill, *f g*, there might also be angular fragments of rock, *h h*, detached from the harder beds above them, during a loss of matter from an old surface, *f k*. This kind of precaution has frequently to be taken in granitic regions, the blocks of granite often decomposing in a rounded form, so as, when scattered about amid bogs, and much disintegrated rock, to present the appearance of boulders rounded by attrition.

The tendency to decompose in spheroidal forms has also to be sometimes well considered when it is inferred that such rocks, even when they are true erratic blocks, have been rounded by attrition before they were ice-transported. A block of granite, for example, such as that represented beneath, *a* (fig. 97), though now rounded,

Fig. 97.

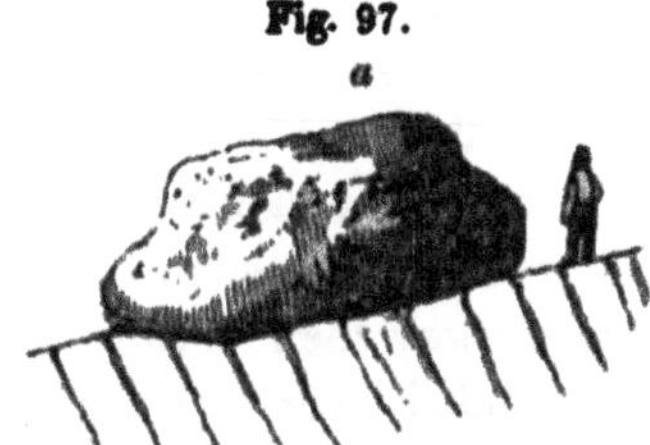

may have been transported in a more angular condition, the removal of the angular parts having been effected by decomposition, from atmospheric influences, since it occupied its present position. In this manner, rounded blocks of granite may be scattered down a mountain side, as in the following section (fig. 98), where granite,

Fig. 98.

c, rising in a tor, *d*, above certain stratified deposits, *b*, has fallen in blocks, down the slope, a large rounded block presenting itself at *a*. Although it may have so happened that such a state of things had been brought about by the motion of a glacier, leaving lateral moraines (other fitting conditions obtaining), or by coast ice carrying blocks of rock, it still becomes needful to ascertain that such are not blocks fallen from the heights, and simply rounded by decomposition, which a careful examination of the granite at *d*, would aid in showing.

As under the hypothesis of cold having once prevailed in the northern hemisphere, greater than at present, much of the land then submerged is now raised above the level of the sea, and consequently an upward movement of a large portion of northern Europe, Asia, and America inferred, it becomes of no slight interest to see how far ice, in its various modes of occurrence, could be the means of producing the distribution of the rock fragments, often of great magnitude, there found. Assuming the submergence, it becomes desirable to see if its amount can be ascertained. There is always the difficulty of knowing how much portions of rock, of various sizes, may have been rounded and left on coasts and in river courses over the older accumulations, anterior to this supposed ice or glacial period in the northern hemisphere. Giving this, however, its full value, we should expect, as the land rose and the temperature became gradually elevated to that which we now find, that, under certain favourable circumstances, glaciers which were previously cut off by the sea, floating away their terminal portions, might for a time become more extended over dry land, thrusting forward their moraines further than formerly. Thus the levels at which the remains of true terminal moraines could be found, might not give the amount of submergence sought, even supposing that they could be fairly separated from other accumulations of rocks which they may more or less resemble. Coast accumulations of the time, if they could be traced, would be more certain guides.

Still assuming a gradual disappearance of ice, up to the amount now found in the northern regions, and consequently the entire disappearance of many glaciers on lands, such, for example, as the British Islands, where they are supposed to have occurred at the glacial period, the various moraines, as also the polished surfaces, grooves, and scratches formed by the glaciers, would be gradually left to be dealt with by atmospheric influences, and the modifications and changes brought about by them, vegetation spreading over the land as the snow and ice disappeared.

The land rising, and the deeper parts becoming more shallow, mud, previously beyond the action of the wind-waves moving on the surface, would be caught up in mechanical suspension, to be carried to more quiet situations by streams of tide (in tidal seas), or sea-currents, where these began to act. The same with the other portions of the sea-bottom: fragments of rock, of various forms and sizes, thrown down from portions of glaciers, river ice, and coast ice, as they floated above and gradually parted with them, rising with the rest. While much fine sediment would be separated from the larger detritus, as the wind-wave action became more and more felt, so that much of this sediment might be removed from amid the larger detritus, bringing the portions of the latter gradually into closer contact, it would be when the sea-bottom came within the action of the breakers, that the chief modifications of such previous sea-bottom would be effected. The new coasts would be adjusted to the conditions arising from their exposure to the force of the breakers, and the rise and fall of tides, where these were felt, and the angular fragments which had reposed quietly at the bottom, in the manner above noticed (p. 230), would be brought within the action of the breakers, to be rounded by attrition, large blocks standing out as many rocks now do on the sea-coasts. While previously ice-borne and rounded blocks and shingles would again be more worn, the angular fragments would be more or less rounded by the same action, according to their exposure to the breakers. Lines of beach would be thrown up in the usual manner, sandy or shingly, according to circumstances, and be left and be modified by atmospheric influences as the land rose, and the drainage of the old sea-bottom became adjusted to its various levels and inequalities of surface.

Under such circumstances, very variable results would be produced as conditions changed, and the component portions of the old sea-bottom were partly removed and partly left; dispersed ice-borne fragments of rock, rounded or angular as the case may have been, brought together, the angles of the latter sometimes completely rounded by breaker action, at others not much injured; the shells of molluscs and the harder parts of other marine animals sometimes removed and redeposited in a nearly uninjured state, at others, broken into fragments, and variously arranged amid the new accumulations of mud, sand, shingles, and boulders. Should there have been a tendency, under the old conditions of the sea-bottom, to have glacier ice, loaded with rock fragments, or coast ice, bearing away shingles, boulders, and also angular blocks floated away in particular directions, dropping their mineral burdens in lines, upon that bottom, such lines, as it rose, would be preserved according to circumstances. However separated large blocks might be by any other deposits effected during their gradual accumulation, there would be a tendency to remove the finer sediment from among them, so that they would finally present the aspect of lines, often, when the blocks were very thickly thrown down from the ice, forming ridges. Such ridges would, however, be acted upon by breakers during the rise of the land, so that detritus might be strewed upon them in the manner of beaches, and thus a complicated arrangement of parts be produced.

During such changes, icebergs derived from glaciers would float about until the parent glaciers either disappeared or became separated from the sea, and the coast ice formed would become gradually limited in its production up to its present adjustment. Various new modifications would arise from the formation of coast ice, as also from the river ice, as the drainage of the old land found its way amid the new land, with the rain and spring waters of the latter, to the sea. Many blocks of rock would be caught up on the coast, and be transported elsewhere, as was the case with the block on the coast of Denmark, mentioned by Professor Forchhammer (p. 247), and rivers flowing in certain directions might carry back blocks of rock towards their parent masses, as noticed by Sir Roderick Murchison* in the manner that blocks are now moved northwards by the Volkof and Msta.

Under the hypothesis, therefore, of lower temperature accompanied by more sea, the bottom of much of which has since become dry land in the northern hemisphere, the observer has not only to study a wide range of country for evidence of the land supposed to be originally above the water, variously snow-clad, and furnishing glaciers, the terminal parts of which, from time to time, floated away, with the coast ice and extension probably of ice barriers, but also the modifications which the old sea-bottom has undergone in its rise above the sea. Thus he would often have to separate, and duly weigh, much evidence which might, at first, appear somewhat contradictory as to erratic blocks having been transported by land ice or sea ice—as to the polishing, grooving, and scratching of subjacent rocks by the one or the other, and as to the original arrangement and rearrangement of many detrital accumulations.

It may be instructive to consider the effects which would follow the submergence of the British Islands, and of an adjoining portion of France, to 1,000 feet beneath the level of the seas which now surround and adjoin them. And it should be noticed that of a submergence to this, and even a larger amount at a comparatively recent geological period, there would appear good evidence. A glance at the accompanying map (fig. 99), which represents the

---

* "Geology of Russia in Europe and the Ural Mountains," vol. i., p. 565.

land that would, under this hypothesis, be above water, will show numerous islands and islets variously distributed.* The largest amount of dry land would be found in Northern Scotland, and be divided into two main portions by a strait, now occupied by the low ground and lakes between the Murray Firth and Loch Linnhe. Off these principal islands there would be many minor islets, chiefly on the south and south-west. In Southern Scotland there would also be a patch of dry land, of some size, and in Cumberland and Westmoreland another; while a somewhat comparatively large island would extend, in a north and south direction from Westmoreland, by Yorkshire into Derbyshire. In Wales there would be much land above the level of the sea, with many detached islets there and in some parts of England; among them the tops of the Malvern hills, which now at a distance present so much the appearance of an island.† In Ireland there would be numerous islets, the chief island being formed by the Wicklow mountains and their continuation. From them, to the westward, many islets would rise above the sea. As a whole, the Irish islets would be principally gathered into two groups, one on the north, the other on the south.

Taking this submergence, with a climate resembling that of Tierra del Fuego and South Georgia, so that such islands as were sufficiently high were snow-clad, glaciers would descend into the valleys, even occasionally reaching the sea, their terminal portions loaded with blocks and fragments, these floated off by the ice, and strewed over the bottoms of the neighbouring seas according to circumstances. And respecting the heights of the islands, many would rise to sufficient altitudes for these effects to be produced, Lugnaquilla being still 2,039 feet above the sea, Ben Nevis 3,373 feet, Skiddaw 2,022 feet, and Snowdon 2,571 feet. If to this we add the coast ice, with its effects as above noticed (p. 245), there would be no want of conditions for the distribution, by means of ice, of blocks of rock of various sizes and kinds, and of fragments

* The light portions of the map represent the parts of the present land, which would appear above water if it were submerged 1,000 feet; the next shade will be readily recognized as the present outlines of the British Islands; the darker shade corresponds with the depth of 600 feet (100 fathoms) around these islands, and the black portion with the deeper oceanic waters.

† A study of the Malvern district is not only interesting as showing how long the Malvern hills retained their insular character during the emergence of the British Islands to their present relative level, but also as regards the island state of the same hills at a far more remote geological period, one anterior to the accumulation of the rocks commonly known to British geologists as the New Red Sandstone. A detailed account of this district is given by Professor John Phillips, "Memoirs of the Geological Survey of Great Britain," vol. ii., part 1.

of all forms over the area now presented by the British Islands, at various levels beneath that corresponding with an altitude of 1,000 feet above the present sea level. While this was being accomplished, the formation of moraines, and the polishing, grooving, and scratching of rocks, through the instrumentality of glaciers, would be effected above that level, up to altitudes where glacier action of that kind could be then felt. At the sea level, and at such depths beneath it as its influence could be felt, coast ice would be the means of polishing, grooving, and scratching rocks exposed to its action; icebergs would ground, producing their effects, and such rivers as moved rocks by means of ice would add their ice-transported detritus.

A submersion of the British Islands to 1,000 feet beneath the present level—a change in the relative level of sea and land which, however startling it may be to those unaccustomed to geological investigations, the observer will soon learn to consider as one of a minor kind,—could scarcely fail to be accompanied with a submersion of various adjoining portions of Europe. It is not needful to infer that the relative change of level was of equal amount through a very considerable area. It may have been greater in some regions, less in others; but let this have been as it may, such a change would probably bring about a very material difference in the distribution of land and sea, as we now find it. Among other modifications, the Scandinavian regions might be brought under conditions by which, should currents permit, blocks and fragments of rocks, and of various sizes and forms, could be borne by icebergs or coast ice, and be distributed over the bottoms of the seas then on the southward of them, some even being drifted to the area of the British Islands, mingling here and there with their own ice-distributed detritus.

In such changes, not only has the geologist to bear in mind the different distribution of sea and land, but also the modification of tidal action and sea currents effected, duly giving attention to the probable extension of coast-ice, even, perhaps, sometimes amounting to great icy barriers. Though some value would have to be attached to the influence of the outstanding group of islands and islets then rising above the area now more extensively occupied by the British Islands, the waves of the Atlantic would roll over a large tract now forming a portion of Northern France, with Belgium, Holland, Denmark, Northern Germany, and an extended area in Russia. The conditions producing the action of the tides surrounding the British Islands being changed, others would arise suited to the new arrangement of land and sea, and many a mass of ice in the Scandinavian regions, so long as it rested on sea-bottoms, would act as land in the modification of tidal streams and sea currents.

How far the outlines of the land may have generally resembled the present at the commencement of these changes it would be difficult to say, since many modifications have been produced while such changes were effected, and the submergence may have commenced when more land was above the sea level than at present, somewhat more corresponding with the line of 600 feet now beneath the sea, around the British Islands, as in the plans, fig. 65 (p. 91), and fig. 99 (p. 261). Taking, however, the present distribution of sea and land as a guide, and looking chiefly to the production of ice (other consequences of submerging and emerging land being reserved, in a great measure, for subsequent notice), we have to consider an increase of cold on the one side, and a decrease of dry land accompanied by a loss of height, on the part still above water, on the other. For convenience we may regard these changes as gradual, the modifications arising from more rapid change being readily appreciated.

The gradual increase of cold would tend to lower the line of perpetual snow over the dry land, while the rate of its descent down any mountain range would depend upon the rate of submergence of the land. They might balance each other. Should the rate of decrease of temperature be more rapid than would be compensated by the submergence, pre-existing glaciers would increase even during the descent of the land, and new glaciers would establish themselves elsewhere under the needful conditions. Assuming, however, the continued increase of cold, a time would come, even if the pre-existing glaciers did not much increase during the submergence of the land, when those formed in Scandinavia could reach the sea, as now in Greenland, distributing detritus by their detached portions bearing rock fragments to the adjacent seas.

Looking to other portions of Europe with reference to this submersion of 1,000 feet, or thereabouts, it may not be uninstructive to consider the effects of the cold inferred upon the glaciers of such regions as the Alps, and the establishment of new glaciers in other mountainous districts where the needful conditions may have been produced. In the Alps the glaciers would increase, as they now do, under the influence of certain seasons; but instead of that decrease which brings them back to a certain state from a modification of the seasons in another direction, the increase would continue, an extension of the sea from the Atlantic being not unfavourable for this purpose, independently of the greater cold produced. Under such circumstances, glacier-borne blocks and other rock fragments, which would have been left in many a locality, or carried forward to the terminal moraines, would continue to advance with the augmented length and volume of the glaciers, until they were finally arrested in their progress by the conditions affecting the extent of the glaciers themselves. If the observer will study the occurrence of existing glaciers upon maps or models of the Alps and adjoining districts,* he will perceive that the outward courses of existing glaciers would be greatly extended, while many a new glacier would contribute its ice to the general mass, sometimes carrying its own moraines, and at others modifying the courses of the main streams of ice into which it might merge. With a change of temperature and of relative level of sea and land, which should bring down the altitude of the present line of perpetual snow in the Alps to that of Chiloe (between 40° to 43' S., the Alps being between 42' and 47' N.), it would descend about 2,500 feet, and with it the névé of the glaciers. This descent of the snow-line being supposed gradual, the glaciers would advance as gradually, and the blocks derived from the present interior portions of the Alps would be moved onwards in front. Let, in the following section (fig. 100),

Fig. 100.

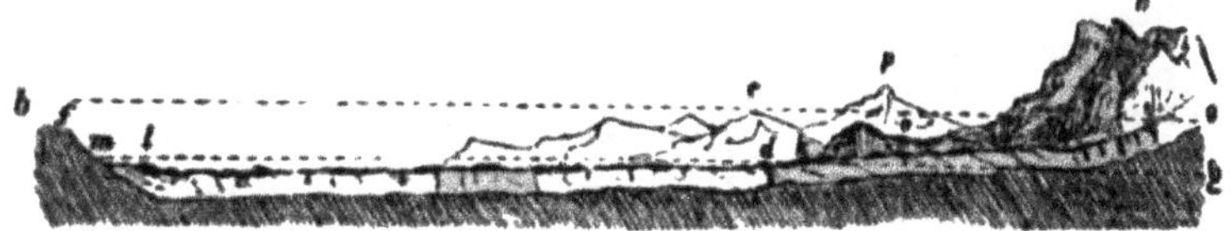

a, *b*, be the level of perpetual snow in a range of mountains amid which glaciers are formed, *d*, the extension of one of these glaciers under any given, yet needful, conditions; *c* and *f*, mountains, just beneath the line of perpetual snow. If now the conditions so change that *g h* becomes the perpetual snow line, those for the production of glaciers continuing, the supply of the original glacier will take place at a lower level, while the ice which only extended to *d*, would be forced onward, on the same principle as the ordinary, however temporary, increase of a glacier may be affected.

* The map accompanying "Travels in the Alps of Savoy," &c., by Professor James Forbes, upon which the glaciers of the districts visited are very carefully entered, will be found very useful for this purpose, and more especially with reference to the inferred extension of glaciers down the valley of the Rhone to the Jura.

With it any collection of blocks, thrust forward in the usual manner to *k*, would be moved onward, with the ice, to *l*, and possibly to *m*, the proper conditions prevailing.* With such increase a collateral glacier might come in from a valley *o*, between *n* and *c*,

* Regarding the extension of Alpine glaciers from increased cold, continued through a certain amount of geological time, the slopes over which they may be inferred to have passed require attention, due allowance being made for the effects which would arise from the supposed greatly-increased volume of many glaciers. As connected with this subject, M. Élie de Beaumont has given ("Note sur les pentes de la limite supérieure de la zone erratique," &c., Annales des Sciences Géologiques, 1842) the following table for the upper limit of the erratic block zone of the valley of the Rhone, &c.:—

Localities.	Distances.	Differences in the Height of the two Localities.	Inclination.		
	Metres.	Metres.	°	′	″
Grimsel . . . . Aernen	25,000	487	1	6	57
Aernen . . . . Brieg	10,000	293	1	2	56
Brieg . . . . Martigny	80,000	70	0	3	1
Great St. Bernard . . . . Plan-y-beuf	15,000	731	2	47	24
Plan-y-beuf . . . . . . Martigny	18,000	319	1	0	55
Martigny . . . . Monthey	18,000	293	0	55	57
Montigny . . . . Mimisse	44,000	425	0	33	12
Mimisse . . . . Geneva	49,000	535	0	41	2
Martigny . . . . Playau	44,000	228	0	17	48
Martigny . . . . Chasseron	92,000	400	0	14	55
Playau . . . . Chasseron	49,000	172	0	12	4
Plan-y-beuf . . . . . . Chasseron	110,000	719	0	22	28
Great St. Bernard . . . . Chasseron	125,000	1,450	0	33	52
Grimsel . . . . Martigny	121,000	850	0	24	9
Grimsel . . . . Playau	165,000	1,078	0	22	27
Grimsel . . . . Chasseron	213,000	1,250	0	20	10
Aernen . . . . Playau	140,000	591	0	14	3
Névé of the Ober Aer . . Grimsel . . . . Level of the Roches Moutonnées	13,500	624	2	38	45
Grimsel . . . . Brunig	29,000	1,037	2	2	52

M. Élie de Beaumont remarks that he does not know in the Alps any glacier which moves through any considerable extent, such as a league, with a slope much less than 3°.

perhaps the extension of a small glacier previously formed at *p*, or altogether new; and thus blocks and glaciers may descend against the extension of the main glacier to *m*. The face of the Alps, as regards snow and ice, would be most materially changed by a descent of the snow-line, so as to be of about the same altitude as that of Chiloe, and a further decrease of temperature would necessarily still further extend the glaciers.

Assuming a depression of this kind, the observer has to take into consideration the rise of the sea-bottom to the present European levels of sea and land, accompanied by an elevation of general temperature to that now found. As the land rose, beaches would be left in various situations, showing the different alterations of the relative levels of sea and land. Should considerable pauses in the elevation of the land have taken place, these would be marked by lines of cliff, where the rocks could be sufficiently worn by the breakers. The production of coast ice would gradually become less, so that its formation would cease in the southern lands, and the glaciers generally would decrease, leaving their lines of moraines, and many angular blocks of rock, perched on the sides of mountains, as in the following sketch (*a*, *b*, fig. 101), at altitudes corresponding

Fig. 101.

with the volumes of their transporting glaciers at the periods of their chief extension down valleys, where only a remnant of such glaciers may be now left at their higher extremities, or even, as in the British Islands, no portion of one may remain.

The land continuing to rise, not only would the previous sea-bottom, with its varied accumulations (in some of which the remains of animal life would be entombed, often in regular beds of sand, silt, and mud), be brought within the destructive influence of the breakers, as above noticed (p. 258), but rivers also would begin to flow amid the old sea-bottom. According to circumstances, such rivers would present varied characters, and some would carry forward ice-borne detritus to the sea, or leave it on their courses, as it might happen, until only certain of them, those now possessing the needful conditions, so transported mineral substances.

From the interest which has been excited respecting the transport of erratic blocks, many of great volume, by means of ice, a mass of information has been collected, rendering the submersion of a large portion of Northern Europe, Asia, and America, accompanied by a considerable depression of temperature, extremely probable. The effects of floating ice have for a long time engaged attention. Professor Wrede, of Berlin, would appear to have been among the first to account for the erratic blocks on the south of the Baltic, by means of floating ice, there having subsequently been a change of level in that region, by which the sea-bottom became dry land.* Sir James Hall also long since referred to floating ice, combined with earthquake waves, as a means of transporting erratic blocks;† and its aid, under various conditions, has been sought in explanation of the transport of large and often angular blocks of rock from their parent masses to considerable distances. Though Professor Playfair long since (1802) pointed out glaciers as having been the means of carrying erratic blocks,‡ even (in 1806) inferring that those on the Jura may have been transported by the extension of ancient Alpine glaciers to that range of mountains, the subject engaged no great attention for some time. M. Venetz appears to have been the first who, having had occasion to study glacier

* "Geognostical Researches relative to the Countries on the Baltic, and particularly to the Low Lands at the Mouth of the Oder, with Observations on the gradual change of the Level of the Sea in the Northern Hemisphere, and its physical causes, as quoted by De Luc, Geological Travels, 1810." Professor Wrede supposed a slow change in the centre of gravity of the earth, so that the waters retreated from the northern hemisphere, leaving the sea-bottom dry, with the ice-borne blocks of rock upon it. He calculated the ice needed to float an erratic block, estimated to weigh 490,000 lbs., occurring at the mouth of the Oder.

† "On the Revolutions of the Earth's surface" (1812), Transactions of the Royal Society of Edinburgh, vol. vii., p. 157. After noticing the removal of a block of rock four or five feet diameter, being a boundary mark between two estates on the shore of the Murray Frith, by the tide, while encased in ice, for 90 yards, and also the magnitude and effects of earthquakes, he asks, respecting the erratic blocks of Northern Europe, if both combined would not produce the effects required, "the natural place of these blocks being covered perfectly with ice, in the state best calculated for fulfilling the office here assigned it," p. 157. He inferred that in the Alps similar waves, assuming the fitting conditions, would wash off portions of glaciers with their load of blocks.

‡ Playfair, "Illustrations of the Huttonian Theory," § 349.

movements, subsequently (1821) took the same view;* one adopted afterwards (1835) by M. de Charpentier,† and further extended (in 1837) by M. Agassiz.‡ The subject then attracted more general interest, especially from the writings of M. de Charpentier § and M. Agassiz,‖ and the consideration of the effects produced by existing glaciers and floating ice, with the probability of a colder state than at present of the northern portions of Europe, Asia, and America, at a comparatively recent time, now form one of the usual objects of geological investigation.

Sir Charles Lyell long since called attention to the distribution of blocks and minor fragments of rock over the sea-bottom by means of icebergs, and to the manner in which such detritus would be found scattered over various levels, if this sea-bottom were upraised and formed dry land.¶ Subsequently (in 1840) after noticing the action of drift ice, charged with mud, and blocks of rocks, he pointed out the manner in which floating ice may, by grounding upon coasts or banks, so squeeze the upper layers of mud, sand, and gravel, that contorted masses of these layers may repose upon undisturbed and horizontal beds beneath.** It was, however, in consequence of a visit to this country by M. Agassiz, in 1840, and upon the extension of his views respecting glaciers to the British Islands,†† that the former existence of glaciers in them has attracted

* Venetz, "Bibliothèque Universelle de Genève," tom. xxi., p. 77, and "Denkschriften der Schweizerischen Gesellschaft;" 1 Band, Zurich, 1833.

† De Charpentier, "Notice sur le cause probable du Transport des Blocs Erratiques de la Suisse, "Annales des Mines," 3me Series, tom. viii., 1835

‡ Agassiz, "Address before the Helvetic Society of Natural Sciences, at Neufchatel," 1837.

§ "Essai sur les Glaciers et sur le Terrain Erratique du Bassin du Rhone," Lausanne, 1841.

‖ "Etudes sur les Glaciers," 1840.

¶ "Principles of Geology," 1832.

** In a communication on the Boulder Formation or Drift, and associated fresh-water deposits, composing the mud cliffs of Eastern Norfolk, "Proceedings of the Geological Society of London" (January, 1840), vol. iii., wherein the contortions observed on that coast are thus explained.

†† In the "Proceedings of the Geological Society of London," vol. iii., p. 328 (1840), M. Agassiz has given a summary respecting his views of the former existence of glaciers in the British Islands. Ben Nevis, in the north of Scotland, and the Grampians in Southern Scotland, are considered by him as the great centres of dispersion of erratic blocks by glacier ice in that part of Great Britain. He pointed out the mountains of Northumberland, Westmoreland, Cumberland, and Wales, as well as those of Ayrshire, Antrim, Wicklow, and the West of Ireland, as also centres of dispersion, "each district having its peculiar débris, traceable in many instances to the parent rock, at the head of the valleys. Hence," observes M. Agassiz, "it is plain the cause of the transport must be sought for in the centre of the mountain ranges, and not from a point without the district." The Swedish blocks on the coast of England do not, he conceives, contradict this position, as he adopts the opinion that they may have been transported by floating ice," p. 329. He considered that the best example of glacier striated rocks in Scotland is to be seen at Ballahulish.

attention. Numerous facts have since been adduced in support of this opinion by Dr. Buckland, Sir Charles Lyell, Professor James Forbes, Mr. Darwin, and others.* The amount of submergence at

* Dr. Buckland ("Proceedings of the Geological Society of London," vol., iii. p. 332, 1840), in his paper, "On the Evidences of Glaciers in Scotland and the North of England," points out localities which he infers show the remains of moraines near Dumfries, in Aberdeenshire, in Forfarshire, at Taymouth, Glen Cofield, and near Callender, with evidences of ancient glaciers on Schiehallion, in and near Strath Earn, and near Comrie; and of glacial action at Stirling and Edinburgh. He also mentions moraines in Northumberland, the evidence of ancient glaciers in Cumberland and Westmoreland, and the dispersion of Shap Fell granite by ice.

In his address to the Geological Society of London, as its President, in February, 1841, Dr. Buckland gave a condensed statement of the progress of investigations on this subject during the preceding year, one in which the "Glacial Theory," was so much considered.

Dr. Buckland subsequently, in his memoir on the Glacia-Diluvial Phœnomena in Snowdonia, and the adjacent parts of North Wales (December, 1841), "Proceedings of the Geological Society," vol. iii., p. 579, described the rounded and polished surfaces, often accompanied by grooves and scratches, attributed to glacier action, in the valleys of Conway, of the Llugwy, of the Ogwyn, of the Sciant, and of Llanberis, of Gwyrfain or Forrhyd, of the Nautel or Lyfni, and of the Gwynant.

Sir Charles Lyell, in his paper "On the Geological Evidence of the former existence of Glaciers in Forfarshire," stated that though, for several years he had attributed the transport of erratic blocks, and the curvature and contortions of the incoherent strata of gravel and clay, resting upon the unstratified till, to drifting ice, he had found difficulty in thus accounting for certain other facts connected with the subject, until Professor Agassiz extended his glacial theory to Scotland. After a description of various minor districts, Sir Charles Lyell observes, "that it is in South Georgia, Kerguelen's Land, and Sandwich Land, we must look for the nearest approach to the state of things which must have existed in Scotland during the glacial epoch."

Professor James Forbes, in his "Notes on the Topography and Geology of the Cuchullen Hills, in Skye, and the traces of ancient glaciers which they present," (Edinburgh New Philosophical Journal, 1846, vol. xl., p. 76), points out groovings and scratchings upon polished rocks of a marked kind. He observes, respecting the valley of Coruisk, that "the surfaces of hypersthene, thus planed or evened, present systems of grooves exactly similar to those so much insisted on in the action of glaciers on subjacent rocks, and as evidence of glaciers in parts of the Alps and Jura, where they are now awanting. These grooves or striæ are as well marked, as continuous, and as strictly parallel to what I have elsewhere shown to be the necessary course of a tenacious mass of ice urged by gravity down a valley, as anywhere in the Alps. They occur in high vertical cliffs, as near the Pissevache; they *rise* against opposing promontories, as in the valley of Hasli; they make deep channels or flutings in the trough of the valley, as at Pont Pelissier, near Chamouni; and as at Fee, in the Valley of Saas. At the same time these appearances have a *superior limit*, above which the craggy angular forms are almost exclusively seen, where the phenomena of wearing and grooving entirely disappear. In short," adds Professor Forbes, "it would be quite impossible to find in the Alps, or elsewhere, these phenomena (except only the high polish which the rocks here do not admit of) in greater perfection than in the Valley of Coruisk." Other evidence of the like kind is also adduced.

Mr. Darwin, in his "Notes on the effects produced by the ancient Glaciers of Caernarvonshire, and on the boulders transported by floating Ice" (Philosophical Magazine, 1842, vol. xxi., p. 180), after mentioning the labours of Dr. Buckland, on the same country, and that Mr. Trimmer had first noticed ("Proceedings of the Geological Society," vol. i., p. 332, 1831) the scoring and scratching of rocks in North Wales, adduces additional evidence of glacial action in that district. He observes this period has been variously estimated. Mr. Darwin infers, from a large greenstone boulder on Ashley Heath, Staffordshire, at 803 feet above the sea, and apparently derived from Wales, a considerable depression of England beneath the sea, and that Scotland, from other data, must have been submerged 1,300 feet.* Looking at the heights to which gravels extend in Wales, often apparently the remains of masses of coast shingles and sand, a like, if not a greater depression beneath the present sea level would be there required. In Ireland, we find large blocks of granite sometimes perched on the heights, amid grooves and furrows on the surface of the rocks beneath, at altitudes of 1,000 feet and more. In some cases we almost seem to have before us a portion of the very blocks which scratched and scored the subjacent rock-surfaces.†

Erratic blocks, occasionally of considerable magnitude, are found, in some localities, at various elevations above rocks of their kind, and from which they are considered to have been detached. Although it is obvious that each fragment, so detached, has deprived the mass of rock whence it has been derived, of so much of its volume, and perhaps also of its height, as regards elevation above

---

that, "within the central valleys of Snowdonia, the boulders appear to belong entirely to the rocks of the country. May we not conjecture," he continues, "that the icebergs, grating over the surface, and being lifted up and down with the tides, shattered and pounded the soft slate rocks, in the same manner as they seem to have contorted the sedimentary beds of the east coast of England (as shown by Mr. Lyell) and of Tierra del Fuego?" * * * The drifting to and fro and grinding of numerous icebergs during long periods near successive uprising coast lines, the bottom being often stirred up, and fragments of rocks dropped on it, will account for the sloping panes of unstratified till, occasionally associated with beds of sand and gravel, which fringes to the west and north the great Caernarvonshire mountains." Mr. Darwin further remarks (p. 186), as not "probable, from the low level of the chalk formation in Great Britain, that rounded chalk flints could often have fallen on the surface of glaciers, even in the coldest times, I infer, therefore," he continues, "that such pebbles were probably enclosed by the freezing of the water on the ancient sea-coasts. We have, however, the clearest proofs of the existence of glaciers in this country, and it appears that, when the land stood at a lower level, some of the glaciers, as in Nant Francon, reached the sea, where icebergs charged with fragments would occasionally be found. By this means we may suppose the great angular blocks of Welsh rocks, scattered over the central counties of England, were transported." The deposits of this date in Ireland have occupied the attention of several geologists, among whom may be mentioned, Mr. Weaver, Mr. Griffith, Colonel Portlock, Mr. Trimmer, Professor Oldham, Mr. Bryce, Dr. Lloyd, Mr. Hamilton, and Dr. Scouler.

* Philosophical Magazine, 1842, vol. xxi., p. 186.

† Although in several parts of Ireland the facts relating to the transport of erratic blocks can be well studied, and the altitudes at which they and the smoothing and scratching of surface rocks are found well observed, there are few places where the latter can be seen in greater perfection than the beautiful neighbourhood of Glengariff, county Cork. The scoring and rounding of the sides and bottom of the valley from the lower part of the demesne of Glengariff to Bantry Bay, and thence to the southward, in the direction of Cape Clear, are particularly worthy of attentive study.

the sea level, and consequently that if multitudes have been thus detached, previous heights, composed of such rocks, may have been much reduced by the loss thus sustained, there are instances where it would not appear a sufficient explanation to infer that a transport of erratic blocks had been effected by ice in such a manner, that, while higher portions of the parent rock floated away at the required levels, the remaining lower portions were denuded, in the usual manner, as the land emerged. To account for such instances, Mr. Darwin considers that we should regard the probable effects of submerging land, where coast ice could be formed, upon blocks of rock which may have been ice-transported to its shores. He points out that erratic blocks and other portions of the beaches of such shores might gradually be raised as the land became submerged, so that finally coast detritus, including the blocks of rocks ice-transported from various distances, would be elevated to heights above that at which it was accumulated or stranded. Blocks, with other coast fragments and shingle, would thus, when the land again emerged from beneath the sea, be found raised above the level at which the remains of their parent rocks are now found.*

Respecting the erratic blocks of the Alps, and of the adjoining countries, a large mass of information has been collected.† The main fact of the blocks and associated minor detritus having been transported from the higher Alpine mountains outwards on both sides the main ranges, showing that the cause of their dispersion had been in the Alps themselves, forms the base of the chief modern hypotheses connected with the subject, whether the sudden melting of snows and glaciers by the heat and vapours accompanying the last elevation experienced in these mountains,‡

* Darwin, "On the Transportal of Erratic Boulders from a Lower to a Higher Level."—Journal of the Geological Society, 1849, vol. v. Mr. Darwin remarks that the fragments of rock "from being repeatedly caught in the ice and stranded with violence, and from being every summer exposed to common littoral action, will generally be much worn; and from being driven over rocky shoals, probably often scored. From the ice not being thick, they will, if not drifted out to sea, be landed in shallow places, and from the packing of the ice, be sometimes driven high up the beach, or even left perched on ledges of rock."

† A valuable summary of the labours of geologists on this subject will be found in the "Histoire des Progrès de la Géologie, de 1834 à 1845," tom. ii., chap. 5, by the Vicomte d'Archiac. Appended to it is a list of the publications which may advantageously be consulted.

‡ As regards the transport of blocks of rock by the sudden melting of snow from the escape of gases rising through fissures during the elevation of mountain chains, the observer will find the subject carefully treated in the "Note relative a l'une des causes présumables des phénomènes erratiques,' by Elie de Beaumont (Bulletin de la Société Géologique de France, t. iv. p. 1334, 1817). On the supposed heat of the gases required for the melting of the snow, M. Elie de Beaumont remarks, after

the former great extension of Alpine glaciers, or the latter combined with a considerable submergence of land, so that the sea entered many of the valleys of the Alps, coast ice being possibly also produced.

Von Buch, De Luc, Escher, Élie de Beaumont, and other geologists, long since pointed out that, from the mode of occurrence of the Alpine erratic blocks, the great valleys of the Alps existed prior to their dispersion, and much observation has been directed to the sources whence particular kinds of blocks have been derived.* The magnitude of the blocks on both sides of the Alps, in connection with the distances they must have travelled from their parent rocks, has also long engaged attention. The *Pierre à Bot*, above Neuchâtel, and represented beneath (fig. 102),† affords a good example of an erratic block, perched on the side of

Fig. 102.

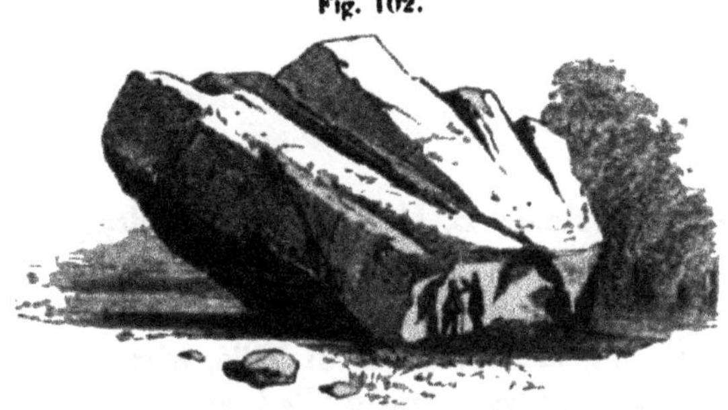

noticing many circumstances bearing on the subject, that "it is unnecessary to attribute to the gaseous current, considered to have been disengaged from fissures in the ground, a temperature higher than that needed to overcome the atmospheric pressure. Little would be gained by giving this current a very high temperature." . . . "The hypothesis which admits the *erratic thaw* to have been produced by vapours of moderate temperature, appears to me," he continues, "also that according to which nature would have worked with the minimum loss of heat."

* With reference to the mode of distribution of the erratic blocks in the basin of the Rhone, as also to the kinds of rocks so distributed, M. Guyot has remarked (Bulletin de la Soc. des Sciences de Neuchâtel, 1846, Archives de Genève, Sept., 1847):—

1. That a kind of rock which is abundant in one part of the basin, is rare, or absent, in another.
2. That the blocks of different kinds, commencing with the locality of their origin, form parallel series, preserved in the plain; blocks of the right side of the valley keeping to the right, of the left side to the left, while those of the centre preserve their central position.
3. That groups composed of a single kind of rock, to the exclusion of others, are here and there found in the midst of various rocks.

These views M. Guyot considers as borne out by numerous facts, and he infers that the blocks have been distributed by glaciers in the manner in which similar blocks now are by the moraines of actual Alpine glaciers. He states that similar facts are observable in the valleys of the Rheuss and Rhine.

† Taken from a view in the "Travels in the Alps of Savoy," &c., by Prof. James Forbes, 2nd edition.

the Jura, far distant from its source. This granite mass is estimated as containing about 40,000 cubic feet, and considered to have been transported 22 leagues from the crest of the Follaterres on the north of Martigny.* The blocks on the Jura have always attracted much attention from the circumstance that they must have been transported over the great valley of Switzerland, intervening between that range and the Alps. The blocks on the Chasseron are estimated as rising to the height of about 3,600 feet.† On the southern side of the Alps striking masses of erratic blocks are to be seen in the vicinity of the Lakes of Como and Lecco. They will be found high up the northern side of Monte San Primo, a mountain well separated from the high Alps by the intervening Lake of Como. The following (fig. 103) is a section of this mountain, showing the manner in which the erratic blocks rest upon it.

Fig. 103.

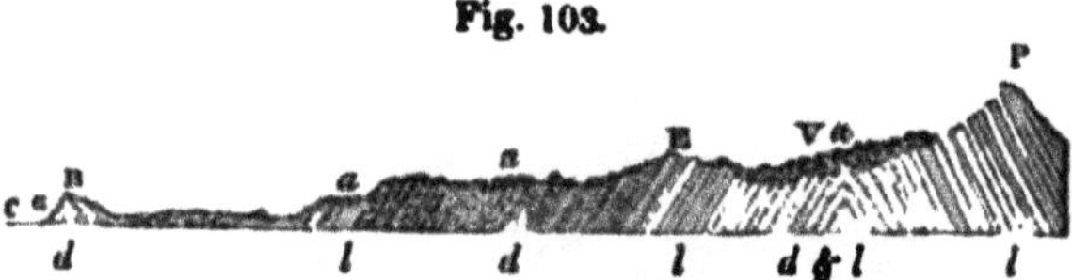

P, Monte San Primo; B, bluff point of Bellaggio, rising out of the Lake of Como, C; *a a a a*, blocks of granite, gneiss, &c., scattered over the surface of the limestone rocks, *l l l l*, and the dolomite *d d d*. V, the Commune di Villa, where a previously-existing depression has been nearly filled with transported blocks and minor detritus. On the north side of the Alpi di Pravolta, E, the block represented beneath, (fig. 104), is seen, one however not

Fig. 104.

* M. d'Archiac remarks ("Histoire des Progrès de la Géologie," t. ii., p. 249), that granite and gneiss generally form the blocks of the largest size. "A block of granite, on the calcareous mountain near Orsières, contains more than 100,000 cubic feet. Above Monthey, many blocks derived from the Val de Ferret, and which have thus travelled a distance not less than 11 leagues, contain from 8,000 to 50,000 and 60,000 cubic feet." . . . "The blocks of talcose granite of Steinhof, near Seeberg, one of which measures 61,000 cubic feet, has travelled about 60 leagues."

Considering the 40,000 cubic feet supposed to be contained in the *Pierre à Bot*, as French measure it would weigh about 3,000 tons.

† Necker, "Etudes Géologiques dans les Alps," vol. i. Paris, 1841.

so remarkable for size, as for showing the little attrition it could have suffered during its transport from the higher Alps to its present position.

A large amount of information has been obtained respecting the distribution of erratic blocks in Northern Europe, and the sources in Scandinavia whence they have been detached.* The area over which they have been so distributed has been shown in a map by Sir Roderick Murchison, M. de Verneuil, and Count Keyserling,† the boundary line exhibiting the southern and eastern limits of the erratic blocks extending from Prussia, to Voroneje, in Russia, and thence northwards to the Gulf of Tcheskaia, on the North Sea. It is remarked that from the German Ocean and Hamburg on the west, to the White Sea on the east, an area of 2000 miles long, varying in width from 400 to 800 miles (which may, perhaps, be roughly estimated at about 1,200,000 square miles), is more or less covered by loose detritus, amid which there are blocks of great size, the whole derived from the Scandinavian mountains.

While regarding the kind and extent of country thus more or less covered with erratic blocks, and the position which the Scandinavian mountains would occupy relatively to a large submerged area, the opinion that glaciers, icebergs (detached from them), and coast ice, may have been the chief means of dispersing the blocks and other detritus from a large isolated region, as that of Scandinavia would then be, appears far from improbable. Careful examination of the Scandinavian region itself shows that the whole land has been elevated above the present level of the adjoining seas in comparatively recent geological times, and there has been found a scoring of subjacent rocks, and dispersion of blocks outwards from it, according with this view.‡

---

* The observer would do well to consult the Rapport sur un Mémoire de M. Durocher, entituled "Observations sur le Phénomène Diluvien dans le Nord de l'Europe," by M. Élie de Beaumont (Comptes Rendus, tom. xiv., p. 78, 1842), wherein an excellent summary and general view of the subject, including the marking of subjacent rocks, up to the date of the observations, will be found. He should likewise consult the "Geology of Russia in Europe and the Ural Mountains," 1845, by Sir Roderick Murchison, M. de Verneuil, and Count Keyserling; chapter xx., Scandinavian Drift and Erratic Blocks in Russia; and chapter xxi., Drift and Erratic Blocks of Scandinavia, and Abrasion and Striation of Rocks; and also the "Histoire des Progrès de la Géologie de 1834 à 1845," tom. ii., prèmiére partie, Terrain Quaternaire ou Diluvien. Formation erratique du Nord de l'Europe. Paris, 1848. Notwithstanding the title, this valuable work contains information up to the date of publication. A most excellent and impartial summary of the labours relating to this subject, with original observations, will be found in this 'History.'

† "Geology of Russia in Europe and the Ural Mountains," 1845.

‡ M. Daubrée states (Comptes Rendus, vol. xvi., 1843), that the traces of transport of detritus and of friction diverge from the high regions precisely as in the Alps. This was observed up to an elevation of 3,800 feet (English). M. de Böhtlingk (Poggendorff's Annalen, 1841,) states that Scandinavian blocks have been transported

In the region occupied by these erratic blocks, ridges of them and other detrital matter have been observed to run in lines, often for considerable distances. These are commonly known as *skars*, or *ösars*.* Count Rasoumouski would appear (in 1819) to have been among the first to remark upon those in Russia and Germany, observing that they usually occurred in lines having a direction from N.E. to S.W. M. Brongniart pointed out (in 1828), that those of Sweden, though sometimes inosculating, took a general direction from north to south.† Much discussion has arisen respecting the origin of these lines of accumulation. Upon the supposition that lines of blocks may have been accumulated by glaciers, and the drift of iceberg and coast ice in particular directions, and that upon the uprise of such lines of deposits, breaker action had been brought to bear upon them for a time, we should expect very complicated evidence.

In Northern America erratic blocks are found to occupy a large area, some being strewed as far south as 40° N. latitude. Here, as in Northern Europe, the general drift of detritus appears to be from the northward to the southward, and blocks perched at various altitudes, scored and scratched surfaces of subjacent rocks, and ösars or lines of accumulation ‡ occur in the same manner. Such similar effects point to similar causes, and hence the explanations

---

from the coast of Kemi into the Bay of Onega, and from Russian Lapland into the Icy Sea, that is, in northerly, north-westerly, and north-easterly directions, as quoted also in the "Geology of Russia," vol. i., p. 528.

* It is worthy of remark that similar accumulations of this date, in Ireland, are known as *Escars*.

† "Annales des Sciences Naturelles," 1828. M. d'Archiac observes ("Histoire des Progrès de la Géologie," 1848, tom. ii., p. 36,) that "the form of the ösars, their disposition, and their parallelism with the furrows and scratches of erosion, naturally lead to the idea of a current which has swept the southern part of Sweden from N.N.E. to S.S.W. M. Durocher has found, with M. Sefström, that the ösars were heaped up on the southern side of the mountains which, in that direction, opposed their course. The ösars in Finland, though less marked, have a direction from N. 25° W. to S. 25° E., one which, with the preceding, represents the radii of the semicircle in which the great erratic block deposit of Central Europe occurs"

In the "Geology of Russia in Europe and the Ural Mountains" will be found the views of its authors respecting skärs or ösars. A figure is given of an iceberg aground, and the consequences of its melting stated, lines of angular and rounded blocks being strewed, as the ice dissolved, by a current acting constantly in one direction.

‡ An interesting account of two remarkable trains of angular erratic blocks in Berkshire, Massachusetts, is given by Professors Henry and William Rogers, in the "Boston Journal of Natural History," June, 1846. These two trains, one extending for 20 miles, both previously noticed by Dr. Reid and Professor Hitchcock, were traced to their sources. The blocks are generally large, the smaller being several feet in diameter. One weighs about 2,000 tons. The blocks gradually decrease in size to the S.E., those which have travelled farthest being the most worn. They are stated not to mingle with the general drift beneath them, the boulders and pebbles in which bear "the traces of a long-continued and violent rubbing." "Other long and narrow lines of huge erratic fragments are seen elsewhere in Berkshire, and abound,

offered have been of a similar general character.* A large amount of information has also been collected respecting the occurrence of these blocks, and of the polishing and scoring of subjacent rocks.† It is stated that the divergence of any blocks, such, for example, as those of the Alps, is not observed in the United States. Professor Henry Rogers points out that the scorings do not radiate from the high grounds; but that, amid the mountains of New England and in the great plains of the west, and in Pennsylvania, Vermont, and Massachusetts, they preserve a south-east direction at all their elevations; the lower parts of the great valleys being alone excepted. In the mountainous portions of the region, the heights and flanks exposed to the north and north-west are the most polished and scored. Blocks of large size have been found in New England, New York, and Pennsylvania, from 1,000 to 1,500 feet above the sea.

Erratic blocks are also found in South America. Mr. Darwin discovered them up the Santa Cruz river, Patagonia, in about 50° 10′ S. latitude, and at about 67 miles from the nearest Cordillera. Nearer the mountains (at 55 miles) they became "extraordinarily numerous." One square block of chloritic schist measured 5 yards on each side, and projected 5 feet above the ground; another, more rounded, measured 60 feet in circumference. "There were innumerable other fragments from 2 to 4 feet square."‡ The great plain on which they stood was 1,400 feet above the sea, sloping gradually to sea cliffs of about 800 feet in height. Other boulders were found upon a plain, above another, elevated 440 feet, through

---

we think, in nearly all the mountainous districts of New England. One such train, originating apparently in the Lennox ridge, about two miles on the south of Pittsfield, crosses the Housatonic Valley, south-easterly, as far at least as the foot of the broad chain of hills in Washington. Some very extensive ones are to be seen on the western side of the White Mountains.

* These will be found in the works and memoirs of Hitchcock, Mather, Emmons, Hall, Rogers, Hubbart, Redfield, Jackson, Christy, Ch. Martins, and other geologists.

† We are indebted to Dr. Bigsby for an early notice of the erratic blocks of North America.—(Trans. Geol. Soc., London, vol. i., second series.)

In 1833, Professor Hitchcock ("Report on the Geology of Massachusetts," art. Diluvium,) adduced abundant evidence of the northern origin of these blocks in the districts described by him. The like was also done at an early date for other portions of North America, by Messrs. Lapham, Jackson, Alger, and others. The observer will find an able summary of the facts known in 1846, on this subject, in Professor Hitchcock's Address to a meeting of the Association of American Geologists in that year. Professor Henry Rogers also treated in a general manner of the American erratic blocks in his Address to the same scientific body in 1844, (American Journal of Science, vol. xlvii.) Another general summary, up to 1848, is given by the Vicomte d'Archiac, ("Histoire des Progrès de la Géologie," tom. ii., chap. 9, Terrain Quaternaire de l'Amérique du Nord).

‡ Darwin, "On the Distribution of Erratic Boulders, and on the Contemporaneous Unstratified Deposits of South America."—Geol. Trans., second series, vol. vi. p. 415.

which the same river flows, and at 800 feet above the sea. In the valley of the Santa Cruz, and at 30 or 40 miles from the Cordillera, (the highest parts in this latitude rise to about 6,400 feet,) blocks of granite, syenite, and conglomerate, not found in the more elevated plains, were detected. Mr. Darwin infers that these are not the wreck of those observed on the higher plain, but that they have been subsequently transported from the Cordillera. He had not opportunities of observing other erratic blocks in Patagonia, but refers to the great fragments of rocks noticed by Captain King on the surface of Cape Gregory, a headland, about 800 feet high, on the northern shore of the Strait of Magellan. Mr. Darwin also describes rock fragments of various dimensions and kinds in Tierra del Fuego and the Strait of Magellan, amid stratified and unstratified accumulations of a similar general character to those of this geological date in Europe.* Many of the erratic blocks are large, one at St. Sebastian's Bay, east coast of Tierra del Fuego, was 47 feet in circumference, and projected 5 feet from the sand beach. The general drift of these deposits is considered to be from the westward, the manner in which the transported fragments of rock would be carried by a current similar to that which sweeps against the present land. On the north of Cape Virgins, close outside the Strait of Magellan, the imbedded fragments are considered to have been transported 120 geographical miles or more from the west and south-west. On the northern and eastern coasts of the Island of Chiloe, extending from 43° 26′, to 41° 46′ S. latitude, Mr. Darwin detected an abundance of granite and syenite boulders, from the beach to a height of 200 feet on the land. He infers that these boulders have travelled more than 40 miles from the Cordillera on the east.†

* At Elizabeth Island, Strait of Magellan, there occurs, "fine-grained, earthy or argillaceous sandstone, in very thin, horizontal, and sometimes inclined laminæ, and often associated with curved layers of gravel. On the borders, however, of the eastward part of the Strait of Magellan, this fine-grained formation often passes into, and alternates with, great unstratified beds, either of an earthy consistence and whitish colour, or of a dark colour and of a consistence like hardened coarse-grained mud, with the particles not separated according to their size. These beds contain angular and rounded fragments of various kinds of rock, together with great boulders."—Geol. Trans., second series, vol. vi., p. 418. Variations of these accumulations are noticed as occurring in other places, and two sections of contorted and confused beds at Gregory Bay are given, and Mr. Darwin infers that this disturbance may have been produced by grounded icebergs.

† "The larger boulders were quite angular." . . . "One mass of granite at Chacao was a rectangular oblong, measuring 15 feet by 11 feet, and 9 feet high. Another, on the north shore of Lemuy islet, was pentagonal, quite angular, and 11 feet on each side; it projected about 12 feet above the sand, with one point 16 feet high: this fragment of rock almost equals the larger blocks on the Jura."—Geol. Trans., second series, vol. vi., p. 425.

CHAPTER XV.

MOLLUSC REMAINS IN SUPERFICIAL DETRITUS.—ARCTIC SHELLS FOUND IN BRITISH DEPOSITS.—EVIDENCE OF A COLDER CLIMATE IN BRITAIN.—EXTINCT SIBERIAN ELEPHANT.—CHANGES OF LAND AND SEA IN NORTHERN EUROPE.—EXTINCTION OF THE GREAT NORTHERN MAMMALS.—RANGE OF THE MAMMOTH.—FROZEN SOIL OF SIBERIA.

UPON the supposition of the submergence of a large portion of the present dry land of Northern Europe, Asia, and America, beneath seas upon which ice was formed, and into which glaciers protruded in lower latitudes than at present, we should expect to discover in the marine deposits of these regions, and of the period now upraised into the atmosphere, evidences of the marine animal life of the time having corresponded with the low temperature to which it was then exposed. This evidence is considered to have been found.

As regards the British Islands, Mr. Trimmer pointed out, in 1831, that amid the detrital accumulation referred to this date, and at a considerable height above the sea (since ascertained to be 1,392 feet), upon Moel Trefan (one of the hills on the outskirts of the chief Caernarvonshire mountains), fragments of *Buccinum*, *Venus*, *Natica*, and *Turbo* of existing species were found. He also stated that on the flanks of the Snowdonian mountains, and between them and the adjoining sea, in the Menai Straits, there were large accumulations of boulders and fragments derived from a distance, (among them chalk flints,) mingled with others of a local kind. Mr. Trimmer subsequently (1838) published a more general statement on the same subject, noticing various localities where he and others had found shells, of a similar character, in deposits referred to this date.*

* The first communication was made to the Geological Society of London (Proceedings of that Society, vol. i.); the second to the Geological Society of Dublin, in a memoir, in two parts, entitled, "On the Diluvial or Northern Drift on the Eastern

Commenting on the facts observed by Mr. Trimmer on Moel Trefan, Sir Roderick Murchison (in 1832) inferred from the previous discovery of shells of existing species in the Lancashire gravels and sands by Mr. Gilbertson, one which he was enabled to confirm from actual observation, and from finding similar accumulations over a large tract of country, that the materials of the ancient shore of Lancashire and of the estuary of the Ribble, were deposited during a long protracted period, and "were elevated and laid dry after the creation of many of the existing species of molluscs."* Numerous facts of the like kind were noticed by different observers;† but the inference as to a temperature less at that geological time than at present, as shown by the remains of molluscs, does not appear to have taken a distinct form until Mr. Smith, of Jordan Hill, published his views on the subject in 1839.‡ He discovered shells in places where their animals had lived and died, in the counties of Lanark, Renfrew, and Dumbarton, and hence inferred their entombment by depression, a half-tide deposit being converted into one in a deeper sea. From these and other researches, Mr. Smith obtained a mass of evidence which led him to conclude, from the remains of the molluscs discovered in deposits of this date in different localities, that the climate of the British Islands had then been colder than it now is, more especially as Arctic molluscs, not

---

and Western side of the Cambrian Chain, and its Connexion with a similar Deposit on the Eastern side of Ireland, at Bray, Howth, and Glenismaule."—(Journal of the Geological Society of Dublin.) Mr. Trimmer mentions that, prior to his discovery of the shells on Moel Trefan, Mr. Gilbertson had found shells of existing species in gravel and sand near Preston, Lancashire, and that Mr. Underwood had observed furrows and scratches on the surface of rocks laid bare among the Snowdonian mountains, when the great road from Bangor to Shrewsbury was in progress.

* Address, as President, to the Geological Society of London, February, 1832.—Proceedings of that Society, vol. i, p. 366.

† Among the observations of the time, and as important for the locality noticed, should be mentioned those of Sir Philip Egerton, "On a Bed of Gravel containing Marine Shells, of recent Species, at Wellington, Cheshire" (Proceedings of the Geological Society, vol. ii., p. 189, April 1835). Sir Philip notices the remains of *Turritella terebra*, *Cardium edule*, and *Murex arenaceus*, and infers that there had been an alteration of 70 feet in the level of land and sea, as regards the locality, since the deposit was formed. In 1837, Mr. Strickland ("On the Nature and Origin of the various kinds of transported Gravel occurring in England," read at the British Association in that year) took a general view of the stratified and unstratified character of these deposits, and divided them into—1. *Marine drift*, formed when the central portions of England were under the sea; and, 2. *Fluviatile drift*, when they were above its level, forming dry land, the first composed of (*a*) erratic gravel, without chalk flints; (*b*) erratic gravel, with chalk flints; and (*c*) local, or non-erratic gravel.

‡ "On the late Changes of the relative Levels of the Land and Sea in the British Islands" (Memoirs of the Wernerian Natural History Society, Edinburgh, vol. viii., p. 49, &c.) In this memoir Mr. Smith most carefully cites all those who had previously discovered facts relating to the subject, giving an account of these facts.

now found round the British coasts, were obtained from these accumulations.*

Professor Edward Forbes, in 1846, availing himself of the information then existing, and of his own researches on the same subject, pointed out that the total number of species of molluscs discovered in the deposits of the British area, and referred to this geological time, was about 124, all, with a few exceptions, now existing in the seas around the British Islands, and yet indicating by their mode of assemblage a colder state of the area than at present.† While carefully noticing the error which might arise

---

* Alluding to the researches of M. Deshayes, to whom the unknown shells discovered were transmitted, and who stated that those still found recent, but not in the British seas, occur in northern latitudes, Mr. Smith remarks that this view confirmed that which he had previously entertained from finding many of the shells common with those obtained by Sir Charles Lyell, at Uddevalla, in Sweden, and figured by him (Phil. Trans., 1835); from having been informed by the same geologist that the *Fusus Peruvianus* still inhabited the Arctic seas; and from Mr. Gray (of the British Museum) having, from a cursory examination of the shells discovered, remarked that they had all the appearance of Arctic shells. Mr. Smith adds, "In the Clyde-raised deposits, shells common to Britain and the northern parts of Europe occur in much greater abundance than they do at present. The *Pecten Islandicus*, which has probably entirely disappeared, and the *Cyprina Islandica*, which, if found recent in the Clyde, is extremely rare, are amongst the most common of the fossil species." Most valuable catalogues are appended to the memoir of Mr. Smith, consisting of lists of recent shells in the basin of the Clyde and north coast of Ireland (including land and fresh-water shells); of shells from the newer Pliocene deposits of the British Islands (also including land and fresh-water shells); and of recent species (then new) from the Firth of Clyde.

† Professor E. Forbes, "On the Connexion between the distribution of the existing Fauna and Flora of the British Isles, and the Geological Changes which have affected their Area during the Period of the Northern Drift" (Memoirs of the Geological Survey of Great Britain, vol. i., p. 367, &c.). The Professor observes that, "as a whole, this fauna is very unprolific, both as to species and individuals, when compared with the preceding molluscan fauna of the red and coralline crags, or that now inhabiting our seas and shores. This comparative deficiency depends not on an imperfect state of our knowledge of the fossils in the glacial formations—on that point we now have ample evidence—but on some difference in the climatal conditions prevailing when those beds were deposited. Such a deficiency in species and individuals of the testaceous forms of mollusca, indicates to the marine zoologist the probability of a state of climate colder than that prevailing in the same area at present. Thus the existing fauna of the Arctic seas includes a much smaller number of testaceous molluscs than those of Mid-European seas, and the number of testacea in the latter is much less than in South-European and Mediterranean regions. It is not the latitude, but the temperature which determines these differences." "That the climate," he subsequently observes, "under which the glacial animals lived, was colder, is borne out by an examination of the species themselves. We find the entire assemblage made up, 1st, of species (25) now living throughout the Celtic region in common with the northern seas, and scarcely ranging south of the British Isles; 2nd, of species (24) which range far south into the Lusitanian and Mediterranean regions, but which are most prolific in the Celtic and northern seas; 3rd, of species (13) still existing in the British seas, but confined to the northern portion of them, and most increasing in abundance of individuals as they approach towards the Arctic circle; 4th, of species (16) now known living only in European seas, north of Britain, or in the seas of Greenland and Boreal America; 5th, of species (6) not now known existing,

from neglecting the occurrence of species at different depths in the sea, he observes, that among those found in these deposits, and in situations where they must have lived and died, there are shells, such as the *Littorinæ*, the *Purpura*, the *Patella*, and the *Lacunæ*, "genera and species definitely indicating, not merely shallow water, but, in the three first instances, a coast line."*

Taking a general view of the flora of the British Islands, and of the probable sources whence its parts have been derived, Professor Edward Forbes has inferred that a portion was obtained from northern regions when the higher parts of these islands were alone above the sea, at a time corresponding with that when the marine molluscs living in the seas around them were of the character above noticed, and when the climate was colder than it now is, the evidence of the land flora thus corroborating that afforded by the remains of the marine molluscs. Under such conditions he infers that "plants of a subarctic character would flourish to the water's edge." The whole area being subsequently upraised, in the manner above noticed, the previous islands would become mountain heights, and the plants, uplifted with them, not being deprived of the climatal conditions fitted for them, continued to flourish and be distributed as we now find them.†

---

and unknown fossil in previous deposits. Two other species, from southern deposits in Ireland, were, one the same as one (*Turritella incrassata*) still existing in the South-European, though not in the British seas, and the other (*Tornatella pyramidata*) extinct, but found fossil in the crag." Professor E. Forbes remarks, that it is "of consequence to note the fact that the species most abundant and generally diffused in the drift are essentially northern forms, such as *Astarte elliptica*, *compressa*, and *borealis*, *Cyprina communis*, *Leda rostrata* and *minuta*, *Tellina calcarea*, *Modiola vulgaris*, *Fusus bamfius* and *scalariformis*, *Littorinæ* and *Lacunæ*, *Natica clausa* and *Buccinum undatum*; and even *Saxicava rugosa* and *Turritella terebra*, though widely distributed, are much more characteristic of North-European than of Southern seas."

* "Memoirs of the Geological Survey of Great Britain," vol. i., p. 370. The Professor adds, "a most important fact, too, is that among the species of *Littorina*, a genus, all the forms of which live only at water-mark, or between tides, is the *Littorina expansa*, one of the forms now extinct in the British, but still surviving in the Arctic Seas."

† "Memoirs of the Geological Survey," vol. i. Professor E. Forbes divides the general flora into five parts, "four of which are restricted to definite provinces, whilst the fifth, besides exclusively claiming a great part of the area, overspreads and commingles with all the others." With regard to his general view, the Professor takes, as his main position, that "the specific identity, to any extent, of the flora and fauna of one area with those of another, depends on both areas *forming*, or having formed, part of the same specific centre, or on their having derived their animal and vegetable population by transmission, through migration, over continuous or closely-contiguous land, aided, in the case of Alpine floras, by transportation on floating masses of ice." As respects the vegetation to which reference is made in the text, Professor E. Forbes observes, "The summits of our British Alps have always yielded to the botanist a rich harvest of plants which he could not meet with elsewhere among these islands. The species of these mountain plants are most numerous on the Scotch mountains—com-

As confirming his views respecting the effect of great cold at this period upon the marine molluscs in the seas around the British Islands, Professor E. Forbes found, while dredging, that there were depressions off the coasts in which molluscs of Arctic character still remained, as if imprisoned in cavities during the general rise of the sea-bottom, so that while their germs still found the needful conditions for their development in such depressions, when they passed beyond them, they perished.

Quitting the minor area of the British Islands, and extending our views to the great region ranging from Scandinavia eastward along Northern Asia to Behring's Straits, we should, in the higher latitudes, expect no great aid, as regards evidences of a colder climate having more prevailed at that geological time than at present, from the remains of marine molluscs entombed amid detritus,* or from the existing flora there found. Under the hypothesis of a depression of land, accompanied by increased cold, it is not difficult to conceive that the marine fauna and terrestrial flora of the region became adjusted to the conditions obtaining at the different times, the one accommodating itself to the new shores, the other creeping to the proper grounds, as the sea-bottom changed and the general temperature became lowered or elevated. The discovery, however, of large animals entire in ice, or frozen mud or sand, with their flesh and hair preserved, in high northern latitudes, and of kinds not now existing there, has been considered as affording somewhat of the evidence required.

It is now about half a century since that the body of an elephant, of a species not now living, but the remains of which are widely

---

paratively few on more southern ridges, such as those of Cumberland and Wales. But the species found on the latter are all, with a single exception (*Lloydia serotina*), inhabitants also of the Highlands of Scotland; whilst the Alpine plants of the Scotch mountains are all, in like manner, identical with the plants of more northern ranges, as the Scandinavian Alps, where, however, there are species associated with them which have not appeared in our country."

* The well-known mass of shells at Uddevalla, in Sweden, raised to the height of 216 feet above the level of the sea, and beneath part of which M. Alexandre Brongniart long since found *Balani* still adhering to the supporting gneiss rocks on which they grew ("Tableau des Terrains que compose l'Ecorce du Globe," p. 89), is described as composed of species still existing in the neighbouring seas. A list of these shells was given by M. Hisinger, "Esquisse d'un Tableau des Petrifactions de la Svéde," ed. 2me, Stockholm, 1831. Professor E. Forbes has pointed out that this accumulation of shells was noticed by Linnæus in 1747, and that the species discovered by him are now known as *Balanus Scoticus*, *Saxicava rugosa* or *sulcata*, *Mya arenaria*, *Littorina littorea*, *Mytilus edulis*, *Fusus scalariformis*, *Pecten Islandicus*, *Fusus antiquus*, and *Balanus sulcatus*. In 1806, the Uddevalla shells, and others of existing species, raised above the present level of the sea in Norway, were observed by Von Buch. They were also described by Sir Charles Lyell, in his account of the rise of land in Sweden, "Philosophical Transactions," 1835.

dispersed amid the later geological accumulations of the northern portions of the northern hemisphere,—its flesh so fresh that bears and wolves devoured it,—was found frozen in 70° N. latitude, near the embouchure of the Lena in Siberia.* The body of a rhinoceros also, of a species now extinct, whose hard remains are also discovered in somewhat similar positions, had been obtained in the state of a mummy by Pallas thirty years previously, in latitude 64° N., from the banks of the Wiljue, which falls into the Lena, the carcase smelling like putrid flesh, the hair still partly on the body. These discoveries long since led to speculations respecting a change of climate in Siberia, one suddenly destroying the animals mentioned by cold, so that their carcases were preserved. Professor Playfair (in 1802) would appear to have been the first to infer that the elephants and rhinoceroses of Siberia, now extinct, may have been fitted for a cold climate, though the elephants of the present day inhabit regions of a higher temperature, and that "they may have migrated with the seasons, and by that means have avoided the rigorous winters of the high latitudes."† He also considered that they might have lived farther to the south than the localities where their remains are now found, and "among the valleys between the great ranges of mountains that bound Siberia on that side." Sir Charles Lyell, in 1835, took a similar but more extended view of the subject.‡ Adverting to the mode of occurrence of the abundant remains of elephants in the deposits of Siberia,—an abundance so great that a trade in their tusks for ivory has long been established,§—to the deposits themselves in which they are discovered having been formed beneath the sea, since they contain the remains of marine shells; and to a slow upheaval of the borders of the Icy Sea, as is now taking place, he considered that a considerable change in the physical geography of the

* Mr. Adams, who carefully preserved what remained of this animal, relates that it was first observed as a shapeless mass by Schumakof, a Tungusian chief, and owner of the peninsula of Tamset, in 1799; that this ice-covered mass fell upon the sand in 1803, and that, in the next year, the chief cut off the tusks, the fossil ivory, if it may from its comparative freshness be so termed, found in these regions, being an article of commerce. Mr. Adams, visiting the spot two years afterwards, obtained the skeleton, still in part covered by the fleshy remains, with portions of its hair, which, together with the tusks, subsequently purchased, is now preserved in the Museum at St. Petersburg; and a description is given of it in the "Memoirs of the Imperial Academy of Sciences," vol. v., of which a translation was published, with a figure, in London, in 1819.

† Playfair's "Illustrations of the Huttonian Theory," Edinburgh, 1802.

‡ "Principles of Geology," 4th edition, 1835.

§ This fossil ivory is still imported from Russia into Liverpool, where it finds "a ready sale to comb-makers and other workers in ivory."—Owen, "History of British Fossil Mammals," p. 249.

whole region had been effected, a great increase of land northwards being the result of a long-continued and slow uprise of land and sea-bottom. He inferred a general decrease of temperature, so that the elephants and rhinoceroses, though they may have been fitted to live in colder regions than any of the kinds now existing, gradually perished.

Sir Roderick Murchison and his colleagues, in the examination of the geology of Russia and the Ural Mountains, adopted similar general views, inferring that the Ural, Altai, and neighbouring regions of Siberia, were above the sea when these great mammals existed, and that they lived in herds adjacent to lakes and estuaries,* into and down which their remains were swept. It would appear, especially by the researches of M. Middendorf, that the shells found with these remains are of kinds now existing in the seas of the region, so that the molluscs of that time and the neighbouring seas have not been exposed to conditions effecting their destruction. M. Middendorf also mentions, that in 1843, the carcase of an elephant was found in the Tas, between the Oby and Yenesei, in about latitude 66° 30′ N., "with some parts of the flesh in so perfect a state, that the bulb of the eye is now preserved in the Museum of Moscow."† Sir Roderick Murchison, M. de Verneuil, and Count Keyserling also remark, when describing the range and boundaries of the erratic blocks of Russia, that the area of the districts of Perm, Viatka, and Orenburg, was probably "above the waters and inhabited by mammoths"‡ at this period.

With regard to the probable habits and food of the elephant (*Elephas primigenius*) and the rhinoceros (*R. tichorhinus*), the researches of Professor Owen have shown,§ that on physiological

* "Geology of Russia in Europe and the Ural Mountains," vol. i., p. 500.

† The discoveries of M. Middendorf, of 1843, were communicated to Sir Charles Lyell in 1846 ("Principles of Geology," 7th edition, 1847). "Another carcase, together with another individual of the same species, was met with in the same year (1843), in latitude 75° 15′ N., near the river Taimyr, with the flesh decayed. It was embedded in strata of clay and sand, with erratic blocks, at about 15 feet above the level of the sea. In the same deposit, M. Middendorf discovered the trunk of a larch tree (*Pinus larix*), the same wood as that now carried down in abundance by the Taimyr to the Arctic Sea. There were also associated fossil shells of *living northern* species, and which are moreover characteristic of the drift, or *glacial* deposits of Europe. Among these *Nucula pygmaea*, *Tellina calcarea*, *Mya truncata* and *Saxicava rugosa*, were conspicuous."—Lyell's Principles, 7th edition, p. 83.

‡ Alluding to their map, it is further observed that this probably happened, "when the erratic blocks were transported over the adjacent north-western line marked in the map, as the extreme boundary of the granitic erratics, which were, we believe, stranded on or near the shelving shore of this ancient land."—Geology of Russia, vol. i., p. 522.

§ "History of British Fossil Mammals and Birds," 1846. To the previous inference that the elephant, from its warm, woolly, and hairy coat, was an animal fitted to live grounds the *Elephas primigenius* "would have found the requisite means of subsistence at the present day, and at all seasons in the sixtieth parallel of latitude," so that by adopting, with Professor Playfair and Sir Charles Lyell, the inference that this animal migrated northwards during the warmer parts of the year, as many northern mammals now do, the mammoth, as that kind of extinct elephant has been termed, would have lived easily on the land considered to have been above water at this period. The Professor adds, "in making such excursions during the heat of that brief season (the northern summer), the mammoths would be arrested in their northern progress by a condition to which the rein-deer and musk-ox are not subject, viz., the limits of arboreal vegetation, which, however, as represented by the diminutive shrubs of Polar lands, would allow them to reach the seventieth degree of latitude." With regard to the habits and food of the two-horned rhinoceros,* found frozen in Siberia, the inferences do not appear so clear as for the mammoth. From the greater amount of hair found on the extinct and frozen rhinoceros, noticed by Pallas, than upon existing rhinoceroses, he seems to have concluded that it might have lived in the temperate regions of Asia. Professor Owen remarks that, "although the molar teeth of the *Rhinoceros tichorhinus* present a specific modification of structure, it is not such as to support the inference that it could

---

in a cold climate (the skin of the carcase from the Lena, and the ground on which it fell, affording many pounds weight of reddish wool and coarse long black hairs), Professor Owen showed that its teeth especially were adapted for the apparently cold climate in which its remains have been so abundantly detected. "The molar teeth of elephants possess," observes the Professor, "a highly-complicated, and a very peculiar structure, and there are no other quadrupeds that derive so great a proportion of their food from the woody fibre of the branches of trees. Many mammals browse the leaves; some small rodents gnaw the bark; the elephants alone tear down and crunch the branches, the vertical enamel-plates of their huge grinders enabling them to pound the tough vegetable tissue and fit it for deglutition. No doubt the foliage is the more tempting, as it is the most succulent part of the boughs devoured; but the relation of the complex molars to the comminution of the coarser vegetable substance is unmistakeable. Now, if we find in an extinct elephant the same peculiar principle of construction in the molar teeth, but with augmented complexity, arising from a greater number of triturating plates, and a greater proportion of the dense enamel, the inference is plain that the ligneous fibre must have entered in a larger proportion into the food of such extinct species. Forests of hardy trees and shrubs still grow upon the frozen soil of Siberia, and skirt the banks of the Lena as far north as latitude 60°. In Europe arboreal vegetation extends ten degrees nearer the pole; and the dental organization of the mammoth proves that it might have derived subsistence from the leafless branches of trees, in regions covered during a part of the year with snow."—p. 267.

* The horns of this rhinoceros have been ascertained to have been of large size. One of the horns of an individual, probably the front or nasal horn, in the Museum at Moscow, measures, according to Professor Owen, nearly three feet in length.

have better dispensed with succulent vegetable food than its existing congeners; and we must suppose, therefore, that the well-clothed individuals who might extend their wanderings northwards during a brief but hot Siberian summer, would be compelled to migrate southward to obtain their subsistence during winter."*

Considering the general evidence thus adduced as to the climate of Northern Europe at this geological time, we have to suppose a considerable depression of a large area beneath the level of the Atlantic; an increase of cold, causing glaciers to descend into the sea in Scandinavia, and even in the British Islands; a great increase, if not extension into the sea, of the glaciers of the Alps, icebergs and coast ice distributing masses and minor fragments of rocks over a considerable European area, as also the shingles of beaches, sand, and mud, accompanied by the transported remains of terrestrial and marine creatures, and a movement of land plants, with terrestrial and marine animals, in accordance with the low temperature then existing. The amount of land rising above the sea, prior to the inferred depression, is uncertain. It may have been more or less than that which we now find, though deposits of varied thickness were accumulated at this time, and now constitute a part of the dry land of Europe, and probably also a portion of the bottom of the adjoining seas.

Respecting the great mammals, the carcases of which have been so well preserved in Siberia, and admitting, with Professor Owen, their perfect fitness to have lived in a climate such as that at present found in Northern Europe and Asia, up to a high latitude, we have to consider that at the time of greater cold, their food being adjusted to it, their range, even in the summer season, would be more limited northward, not only by any coasts which might then be thrown back by the depression beneath the sea level, but also by the supposed decreased temperature. The great rivers, flowing northward, would, as Humboldt, Sir Charles Lyell, and Sir Roderick Murchison have pointed out, be then under similar conditions to the present, their embouchures exposed to lower temperatures than their courses in more temperate regions, such courses, though somewhat shorter, being still liable, as now, to be blocked up by ice at their mouths. In such a state of things there is little difficulty in inferring that the elephants and rhinoceroses lived, as they are supposed to have done, in a climate of low

* "History of British Fossil Mammals and Birds," 1846, p. 353.

temperature, and that their remains were buried in the detritus accumulated in lakes and at the embouchures of the northern rivers of the time, numerous carcases being washed out to sea and preserved amid ice, or frozen mud and sand, among deposits containing the remains of marine molluscs, such as are now living in the adjoining Arctic sea.

The cause of the extinction of the great mammals mentioned requires much consideration, and a careful observation of the facts connected with the entombment and preservation of their remains. Humboldt has remarked that the low temperature at present experienced across Poland and Russia to the Ural mountains, "is to be sought in the form of the continent being gradually less intersected, and becoming more compact and extended,—in the increasing distance from the sea,—and in the feebler influence of westerly winds. Beyond the Ural, westerly winds blowing over wide expanses of land, covered during several months with ice and snow, become cold land winds. It is to such circumstances of configuration and of atmospheric currents that the cold of Western Siberia is due."* By the immersion of the present dry land to the extent supposed,† unaccompanied by the general decrease of temperature inferred in Northern Europe, there might, no doubt, be reason to expect that such northern portions of European and Asiatic Russia as were above water would have a higher temperature than at present, but how far this would be met by such a decrease of the present temperature of Scandinavia, the British Isles, and a portion of Central Europe, that glaciers descended to the then sea level, it is more difficult to infer. Because icebergs may have floated from Scandinavia, and have become stranded on the shores of the districts of Perm, Viatka, and Orenburg, and thence along the line pointed out by Sir Roderick Murchison, M. de Verneuil, and Count Keyserling to the westward, it is not a necessary inference that the temperature of those regions, making every allowance for the influence of multitudes of icebergs at certain seasons, had been very low, more than that the temperature of Newfoundland should be that of Greenland and Baffin's Bay, whence the icebergs stranded near it are derived. Even supposing that as the land rose the temperature of Siberia became such as we now find it, it does not seem to follow, judging from the researches

* Cosmos, 7th Edit. (Sabine's Translation), vol. i., p. 323.

† The observer would do well to refer to the map given by the authors of the Geology of Russia in Europe and the Ural, for the area bounding the occurence of erratic blocks.

and reasoning of Professor Owen, that the mammoths necessarily perished from cold or the want of food.* Assuming that the great cold was unfavourable to their continuance in Siberia, that the country towards the mountains on the south was equally so to their habits, and that thus they may have been there extirpated, the same reasoning does not seem to apply so well to the districts on the west of the Ural.

It is now well known that the mammoths must once have existed widely spread over the northern portions of Europe, Asia, and America; whence the inference, on the hypothesis that they all proceeded from a common stock, or centre, that they spread themselves over continuous portions of land, dry for the time, however now separated they may be by seas. Their remains are not uncommon in Great Britain, though less so apparently in Ireland, and Professor Owen has pointed out the connection of these Islands with Europe when these and other contemporary animals passed into them.† The

* On this subject Professor Owen remarks, that "with regard to the geographical range of the *Elephas primigenius* into temperate latitudes, the distribution of its fossil remains teaches that it reached the fortieth degree north of the equator. History, in like manner, records that the rein-deer had formerly a more extensive distribution in the temperate latitudes of Europe than it now enjoys. The hairy covering of the mammoth concurs, however, with the localities of its most abundant remains, in showing that, like the rein-deer, the northern extreme of the temperate zone was its metropolis. Attempts have been made to account for the extinction of the race of northern elephants by alterations in the climate of their hemisphere, or by violent geological catastrophes, and the like extraneous causes. When we seek to apply the same hypothesis to explain the apparently contemporaneous extinction of the gigantic leaf-eating megatheria of South America, the geological phenomena of that continent appear to negative the occurrence of such destructive changes. Our comparatively brief experience of the progress and duration of species within the historical period, is surely insufficient to justify, in every case of extinction, the verdict of violent death. With regard to many of the larger mammalia, especially those which have passed away from the American and Australian continents, the absence of sufficient signs of extrinsic extirpating change or convulsion, makes it almost as reasonable to speculate with Brocchi, on the possibility that species, like individuals, may have had the cause of their death inherent in their original constitution, independently of changes in the external world, and that the term of their existence, or the period of exhaustion of the prolific force, may have been ordained from the commencement of each species."—History of British Fossil Mammals and Birds, p. 269.

† "History of British Fossil Mammals and Birds," 1846, Introduction, p. xxxvi. "If," Professor Owen observes, "we regard Great Britain in connection with the rest of Europe, and if we extend our view of the geographical distribution of extinct mammals beyond the limits of technical geography,—and it needs but a glance at the map to detect the artificial character of the line which divides Europe from Asia,—we shall there find a close and interesting correspondence between the extinct Europæo-Asiatic Mammalian Fauna of the pliocene period and that of the present day. The very fact of the pliocene fossil mammalia of England being almost as rich in generic and specific forms as those of Europe, leads, as already stated, to the inference that the intersecting branch of the ocean which now divides this island from the continent did not then exist as a barrier to the migration of the mastodons, mammoths, rhinoceroses, hippopotamuses, bisons, oxen, horses, tigers, hyænas, bears, &c., which have left such abundant traces of their former existence in the superficial deposits and caves of Great Britain."

depth of Behring's Straits is comparatively trifling, varying from 132 to 192 feet, so that we feel little surprise in finding at Eschscholtz Bay, in about 66° 20′ N. on the North American shores, inside the Straits, the remains of the *Elephas primigenius*, associated with the bones of the urus, deer, horse, and musk ox, in a cliff about 90 feet high, extending about 2½ miles in length. These remains were first noticed by Dr. Eschscholtz (during Kotzebue's voyage), in 1816, and the bones were supposed to be imbedded in ice; but the observations of Captain Beechey's party, in 1826, showed that the ice was merely superficial, arising from the freezing of water descending over the face of the cliff, and that the remains of these mammals were really imbedded in a deposit of clay and fine quartzose and micaceous sand. A smell, as of heated bones, was observed where the animal remains abounded.*

This facing of ice having been thus deceptive, Dr. Buckland was led to infer† that there also might have been some error respecting the elephant of the Lena having really been encased in ice, and not in mud, the face of which was covered by ice, as at Eschscholtz Bay. Correct observations respecting the mode of occurrence of the animals preserved in a comparatively fresh state, with their fleshy portions in part or wholly remaining, are somewhat important, inasmuch as, if found in ice, we have to infer either that such ice had always remained unthawed in the atmosphere (at least so far as the portions enveloping the animals were concerned), from the time when these mammals were encased in it to the present time, or that it became depressed beneath detrital accumulations of the period, and also remained unthawed, until the whole being elevated again into the atmosphere, it became, with the accumulations among which it had been buried, exposed to the climatal and denuding conditions of the present day. Though there would be difficulty in submerging ice, from its specific gravity, beneath water, and especially sea-water, unless sufficiently well loaded with detritus to render this of the proper kind, it may readily happen that, in very cold climates, coast-ice may be anchored, so to speak, in such a manner, by penetrating amid shingles, sand, or mud beneath, that it could be covered over in part, or in thickness, according to variations in seasons, by detrital matter, so as to be in the condition to descend, thus covered over, to those depths where it could remain unthawed, with any animals

* "Beechey's Voyage to the Pacific and Behring's Straits." The bones were examined, and the animals to which they belonged were determined, by Dr. Buckland.
† Ibid.

entombed in it. Indeed, certain facts noticed by travellers and voyagers in the Arctic regions would lead us to infer that this might be the case, and accounts are given of beds of actual ice being found beneath detrital deposits in those regions.* Descended to a proper depth beneath the surface, but not sufficient to bring it within the influence of the heat found to exist beneath certain depths in different parts of the globe, ice might remain there, only to be thawed by a great increase in the temperature of the general climate, or by being again elevated, with a sufficient denudation of protecting detritus, so that the heat of the atmosphere in summer would dissolve it, and disclose any animal remains which may have been therein preserved. At the same time, mud and silt, into which the bodies of such animals as the elephants and rhinoceroses, above noticed, may have been borne during floods, could readily have become frozen, and covered with other detritus, and thus descending, have retained, from what we learn of the depth to which the frozen ground extends in Siberia—a depth apparently very different from that found in North America, in the same latitudes—the remains of the animals in as fresh a state as when first embedded in them, to a level, beneath that of the sea, of 400 feet, if the cold approached that now experienced in northern Siberia.†

* M. Middendorf informed Sir Charles Lyell, that in 1843, "he had bored in Siberia to the depth of 70 feet, and, after passing through much frozen soil mixed with ice, had come down upon a solid mass of pure transparent ice, the thickness of which, after penetrating two or three yards, they did not ascertain."—Principles of Geology, 7th Edition, p. 86.

† The depth to which frozen mud and sand could descend in these regions, without being thawed by the influence of terrestrial heat beneath, would appear from the information of M. Helmersen ("Observations on a Pit sunk at Jakoutsk," Ann. des Mines de Russie, vol. v., 1838), to be between 300 and 400 feet. On the 25th April, 1837, the temperature of the bottom, 378 (English) feet deep, was 31°·1, the strata on the sides of the pit at 75 feet being 21°·2 Fahr. The accumulations passed through were composed of clay, sand, and lignite, mixed with ice.

Some experiments made by M. Middendorf, as reported to the Academy of Sciences of St. Petersburg in 1844, showed that, in a shaft and the galleries of some works near the Lena, and at a depth of 384 (English) feet, the frozen crust was still not passed through, though a marked gradual increase of temperature was observed in the descent. While, in one series of experiments, a thermometer, in the ground, 7 feet from the surface, gave on the 25th March,—1° Fahr., the temperature gradually advanced to 26°·6 Fahr. According to M. Erman ("Proceedings of the Academy of Sciences at St. Petersburg," 1838), the depth of ground thawed in September, 1838, in Northern Siberia, was 4 feet 8 inches in woody tracts, and 6 feet 8 inches in the marshy situations.

From Sir John Richardson having found the depth of the frozen ground not to exceed 26 feet at Fort Simpson, on the Mackenzie, a station in the same latitude as Jakoutsk (62° N.), M. d'Archiac has inferred ("Histoire des Progrès de le Geologie," vol. i, p. 88), that the cold must be far more intense in Northern Asia than in North America, at these high latitudes. Under this view, the bodies of animals could now be preserved in Northern Siberia, by descending and ascending land, which could not be so preserved in North America.

U 2

CHAPTER XVI.

OSSIFEROUS CAVERNS AND BRECCIA.—FORMER CONNEXION OF BRITAIN WITH THE CONTINENT.—MAMMOTH REMAINS FOUND IN BRITISH SEAS.—OSSIFEROUS CAVE OF KIRKDALE.—MUD IN OSSIFEROUS CAVES.—GENERAL STATE OF OSSIFEROUS CAVERNS.—HUMAN REMAINS IN PAVILAND CAVE.—CAVES FORMERLY DENS OF CERTAIN EXTINCT CARNIVORA.—HUMAN REMAINS IN OSSIFEROUS CAVERNS.—COMPLICATED ACCUMULATIONS IN CERTAIN CAVES.—PEBBLES IN OSSIFEROUS CAVES.—DEPOSITS IN SUBTERRANEAN RIVER CHANNELS.—OSSEOUS BRECCIA IN FISSURES.—CHANGES IN THE ENTRANCES OF CAVES.—OCCURRENCE OF MASTODON REMAINS.—ASSOCIATION WITH THOSE OF THE MAMMOTH.—EXTINCT MAMMALS OF CENTRAL FRANCE.

THE bodies of elephants and rhinoceroses being found so well preserved in Siberia,—and nowhere, as has often been remarked, are the remains of the *Elephas primigenius* more abundant than in the lowlands, adjoining the icy sea of Northern Asia,*—it is desirable to consider the remains of the same kinds of elephant and rhinoceros, with those of contemporary mammals, found embedded amid accumulations in caves and clefts of rock. The connection of the British Islands with the continent of Europe and

* Dr. Mantell states ("Wonders of Geology," vol. i., p. 148, 6th Edition, 1848) that, a company of merchants having been formed in 1844, to collect fossil ivory in Siberia, sixteen thousand pounds of jaws and tusks of mammoths were obtained during the year, and these were sold at St. Petersburg, under the denomination of Siberian ivory, at prices from 30 to 100 per cent. above those of recent elephantine ivory.

From the researches of M. Hedenström, multitudes of the remains of elephants, rhinoceroses, oxen, and other mammalia, occur in the frozen ground between the Lena and the Kolima, and he mentions that one of the islands of New Siberia, or the Liakhor Islands, in the Arctic Ocean, off the coast of Siberia, between the embouchures of the Lena and Indigirka, is composed of little else than a mass of mammoth bones, which has been worked for many years by the traders for the fossil ivory it yields.

This statement is confirmed by those of other travellers. The high preservation of fossil ivory is not confined to Siberia. Mr. Bald mentions (Wernerian Transactions, vol. iv.) that tusks found between Edinburgh and Falkirk were made into chessmen.

Asia has been above noticed (p. 289), as needed for the migration of the *Elephas primigenius* and *Rhinoceros tichorhinus* into the former, the remains of these mammals so occurring as to leave no room for doubting, that the animals themselves here found the conditions fitted for their existence and increase.* The observer has carefully to weigh the evidence afforded as to the precise geological period when these great mammals thus prospered upon lands now divided from the continent by sea, which it would

---

* Respecting the mammals existing at this time in the area of the British Islands, Professor Owen remarks, ("History of British Fossil Mammals"—*Introduction*,) after noticing the probable disappearance of the mastodon from it, that "gigantic elephants of nearly twice the bulk of the largest individuals that now exist in Ceylon and Africa, roamed here in herds, if we may judge from the abundance of their remains. Two-horned rhinoceroses, of at least two species, forced their way through the ancient forests, or wallowed in the swamps. The lakes and rivers were tenanted by hippopotamuses as bulky and with as formidable tusks as those of Africa. Three kinds of wild oxen, two of which were of colossal size and strength, and one of these maned and villous like the bonassus, found subsistence in the plains. Deer, as gigantic in proportion to existing species, were the contemporaries of the old *Uri* and *Bisontes*, and may have disputed with them the pasturage of that ancient land; one of these extinct deer is well-known under the name of the 'Irish Elk,' from the enormous expanse of its broad-palmed antlers [the Professor states elsewhere, Hist. Brit. Foss. Mammals, p. 467, that the remains of this animal have been found in the ossiferous cavern of Kent's Hole, Devon]; another had horns more like that of the wapiti, but surpassed that great Canadian deer in bulk; a third extinct species more resembled the Indian hippelaphus; and with these were associated the red-deer, the rein-deer, the roebuck, and the goat. A wild horse, a wild ass or quagga, and the wild boar, entered also into the series of British pliocene hoofed mammalia.

"The carnivora, organized to enjoy a life of rapine at the expense of the vegetable-feeders, to restrain their undue increase, and abridge the pangs of the maimed and sickly, were duly adjusted in numbers, size, and ferocity to the fell task assigned to them in the organic economy of the pre-Adamitic world. Besides a British tiger of larger size, and with proportionally larger paws than that of Bengal, there existed a stranger feline animal (*Machairodus*) of equal size, which, from the great length and sharpness of its sabre-shaped canines, was probably the most ferocious and destructive of its peculiarly carnivorous family. Of the smaller felines, we recognise the remains of a leopard, or large lynx, and of a wild cat.

"Troops of hyænas, larger than the fierce crocuta of South Africa, which they most resembled, crunched the bones of the carcases relinquished by the nobler beasts of prey; and, doubtless, often themselves waged the war of destruction on the feebler quadrupeds. A savage bear, surpassing in size the *Ursus ferox* of the Rocky Mountains, found its hiding-place, like the hyæna, in many of the existing limestone caverns of England. With the *Ursus spelæus* was associated another bear, more like the common European species, but larger than the present individuals of the *Ursus Arctos*. Wolves and foxes, the badger, the otter, the foumart, and the stoat, complete the category of the pliocene carnivora of Britain.

"Bats, moles, and shrews were then, as now, the forms that preyed upon the insect world in this island. Good evidence of a fossil hedgehog has not yet been obtained; but the remains of an extinct insectivore of equal size, and with closer affinities to the mole-tribe, have been discovered in a pliocene formation in Norfolk. Two kinds of beaver, hares and rabbits, water-voles, and field-voles, rats and mice, richly represented the Rodent order. The greater beaver (*Trogontherium*) and the tail-less hare (*Lagomys*) were the only sub-generic forms, perhaps the only species, of the pliocene *Glires* that have not been recognized as existing in Britain within the historic period. The newer tertiary seas were tenanted by cetacea, either generically or specifically identical with those that are now taken or cast upon our shores."

appear scarcely probable they safely crossed, either by will or accident. The geological time when the needful connection was formed between the British Islands and the continent of Europe, so that these and other contemporary mammals freely roamed from the one part of a general area to the other, is, therefore, a matter of no slight interest.

It has to be borne in mind that, during any modified distribution of land and sea formerly existing, by which deposits were accumulated, and the carcases of animals were floated out to sea, or swept into fresh-water lakes, so that their harder parts became embedded in calcareous matter, mud, silt, or gravel, the lighter portions of the accumulation, amid which they were entombed, would, as now in the German Ocean and some other parts of the sea adjacent to the British Islands, be liable to be washed off, either at the proper depths beneath the surface of the sea by the action of the wind-waves, or on the shores by the breakers, when changes of level of the sea and land so took place that this action could be experienced. Tusks, teeth, and the bones of the *Elephas primigenius* have thus been fished up by the trawlers and dredgers on the south-east of England, and in a state sometimes showing little marks of attrition, bearing more the appearance of having been merely relieved, by the wave action, of the mud, silt, or sand which once enveloped them.* Supposing the elephants and rhinoceroses, with other

* Professor Owen, in his "History of British Fossil Mammals," mentions (p. 246), that "most of the largest and best-preserved tusks of the British mammoth, have been dredged up from submarine drift near the coasts. In 1827, an enormous tusk was landed at Ramsgate; although the hollow-implanted base was wanting, it still measured nine feet in length, and its greatest diameter was eight inches; the outer crust was decomposed into thin layers, and the interior portion had been reduced to a soft substance resembling putty. A tusk, likewise much decayed, which was dredged up off Dungeness, measured 11 feet in length; and yielded some pieces of ivory fit for manufacture. Captain Byam Martin, who has recorded this and other discoveries of remains of the mammoth in the British Channel (Geological Transactions, second series, vol. vi., p. 161), procured a section of ivory near the alveolar cavity of the Dungeness tusk, of an oval form, measuring 19 inches in circumference. A tusk dredged up from the Goodwin Sands, which measured 6 feet 6 inches in length, probably belonged to a female mammoth." * * "This tusk was sent to a cutler at Canterbury, by whom it was sawed into five sections, but the interior was found to be fossilized and unfit for use." * * "The tusks of the extinct elephant, which have reposed for thousands of years in the bed of the ocean which washes the shore of Britain, are not always so altered by time and the action of surrounding influences, as to be unfit for the purposes to which recent ivory is applied." Mr. Charlesworth, after mentioning that a large lower jaw of a mammoth, of which he gives a figure ("Magazine of Natural History, new series," vol. iii., p. 348, 1839), had been dredged up off the Dogger Bank, in 1837, and quoting Mr. Woodward ("Geology of Norfolk"), as stating that more than 2,000 elephants' teeth had been dredged up off Hasbro', on the Norfolk coast, in 13 years, relates that a mammoth's tusk, dredged up by some Yarmouth fishermen off Scarborough, about 1836, was so slightly altered in character, that it was sawn up into as many pieces as there were men in the boat, each

contemporary animals, the remains of which are found with those of these mammals, to have been spread over the land prior to the great depression, accompanied by increased cold, as above noticed, and that they gradually retreated before the advance of the sea, diminishing the amount of low ground, the original connection between the British Islands and the land of the continent may have more resembled that shown as the boundary of the 600 feet depth, (figs. 65 and 102,) than that which we now find. In such a state of this part of Europe there would be an ample area of continuous dry land for the range of the elephants, rhinoceroses, and their contemporary, but the now extinct, species of hippopotamus, oxen, deer, tiger, leopard, hyæna, bear, and other mammals. Accumulations of bones could readily, as the land became depressed, be washed out of any lacustrine or fluviatile accumulations amid which they might have been embedded, and be mingled with marine remains of the gradually-encroaching seas, sometimes being worn and re-embedded in gravel, at others less mutilated, or even uninjured, amid more tranquilly-formed deposits. Occasionally some, or portions, of the original lacustrine or fluviatile deposits, containing remains of these animals, may never have been disturbed to any great extent, so that the deposits and the included bones became covered by the marine accumulations of the time.*

---

claiming his share of the ivory. One portion was preserved in the collection of Mr. Fitch, of Norwich. A large humerus was, in 1837, trawled up in mid-channel between Dover and Calais, in 120 feet water. A large femur was also found while trawling, about half-way between Yarmouth and Holland in 150 feet water, and the lower jaw of a young animal was dredged up off the Dogger Bank. Other instances of elephant remains, brought up from the sea-bottom off the English coasts are also known. A tusk of the *Hippopotamus major* was dredged up from the oyster-bed at Happisburgh.

* Professor Owen ("Hist. Brit. Fossil Mammals," p. 347), quotes a notice in a Cambridge paper of 26th February, 1845, in which mention is made of high tides having much uncovered the lignite beds at the base of the cliffs near Cromer, Norfolk, and that among the fossil remains of that bed, the lower jaw of a rhinoceros, with seven molar teeth in good preservation, together with the molars of the elephant, hippopotamus, and beaver were discovered. The jaw was examined by Professor Owen, and ascertained to have belonged to a young *Rhinoceros tichorhinus*.

Mr. Strickland pointed out, in 1834 ("Account of Land and Fresh-water Shells found associated with the Bones of Land Quadrupeds beneath diluvial gravel, at Cropthorn, Worcestershire," Proceedings Geol. Soc., vol. ii., p. 111), that "a layer of fine sand, containing 23 species of land and fresh-water shells, with fragments, more or less rolled, of bones of the hippopotamus, bos, cervus, ursus, and canis," reposes on the lias clay of that district. Professor Owen adds the mammoth and urus to this catalogue ("Hist. Brit. Fossil Mammals," p. 258). "The sand passes upwards gradually into gravel, which extends to the surface, and differs in no respect from the other gravel of the neighbourhood, being composed principally of pebbles of brown quartz, but occasionally containing chalk flints, and fragments of lias ammonites and gryphites. The bones, though most abundant in the sand, are interspersed also in the

Upon the hypothesis, that these animals could have spread under such conditions, and prior to the submergence previously noticed, a time would come when the depression of the old land would be such, that, as regards the British Islands, no sufficient or fitting dry land would be found for them, supposing that the diminished temperature did not destroy them. While assuming that such may have been the conditions in this particular case, it by no means follows, with submerging dry land over a large portion of Europe, that abundant space was not left, even in Northern Asia, for the existence and increase of the *Elephas primigenius* and the *Rhinoceros tichorhinus*. The land may not have experienced a contemporaneous depression, or, if so, not one cutting off all the needful feeding-grounds for the support of these mammals. Thus, in several parts of Europe, when the sea-bottom emerged,—the former land, variously modified during its submersion, coated more or less with the detritus drifted over and thrown down upon it, and embedding the remains of such animals as perished during the submergence, there might be many sources whence the elephants, rhinoceroses, and other contemporary animals, could spread over the new land as the fitting conditions obtained. It is not difficult to conceive that these mammals may thus have revisited the area of the British Islands, again connected with the main land, so that their remains may be found in lacustrine and fluviatile deposits above the marine accumulations formed during the interval of depression.* As there is evidence in Western Europe of oscillations, as regards the relative level of sea and land, in the more recent geological time, requiring much attention on the part of the observer, he will have carefully to consider their

gravel; but the shells are confined to the sand." Two of the species of shells were considered to be extinct. From the fluviatile habits of some of these mollusca, Mr. Strickland inferred, that the deposit occupies the site of an ancient river bed He at the same time pointed out "the greater change which has taken place in the mammifera of this island than in the mollusca, since the era when the gravel was accumulated; and the little variation which the climate appears to have undergone since the same epoch." He also adverted to similar deposits, previously known at North Cliff, Yorkshire, Market Weighton, and at Copford, near Colchester.

The section given by Sir Roderick Murchison, M. de Verneuil, and Count Keyserling ("Geology of Russia in Europe and of the Urals," vol., i. p. 502), would appear to show, that as respects a part of Russia, and beneath a covering of "clay drift, containing numerous bones and teeth of the mammoth, 50 feet thick," there was a "band of finely-laminated sand, full of shells, specifically identical with those which inhabit the adjacent river Don." The sand reposes upon a tertiary limestone.

* Localities are mentioned where, in the British Islands, bones of these and of contemporary mammals have been found entombed in fluviatile or lacustrine deposits, supposed to be above the accumulations referred to the period when erratic blocks and other ice-transported detritus were strewed over the sea-bottom in this part of Europe.

influence on the spread of mammals, such as those under consideration. Assuming, however, only one submersion sufficient to disconnect the British Islands, followed by an elevation restoring the connection, it would be inferred that lacustrine and fluviatile accumulations would be the highest amid which we should expect to discover the remains of the *Elephas primigenius* and his contemporary mammals, partly extinct, partly now existing.

Amid any changes arising from the depression and elevation of land and adjacent sea-bottoms, should animals have lived in caves, carrying in their prey, should they had been carnivorous, or have fallen into fissures in the manner previously mentioned (p. 118), their remains, so preserved, would appear the most safe from rearrangement by waves, tidal streams, or ocean currents. Though the bones of extinct bears and other animals found in caves had previously attracted much attention, it was from the discovery of the remains of mammals in a cavern at Kirkdale, in Yorkshire, in 1821, and from the descriptions of all the circumstances attending the mode of occurrence of these remains, and of the condition of the cavern itself, subsequently given by Dr. Buckland, who visited the spot a few months only after the discovery, that ossiferous caves attained a new interest. This cave was found by cutting back a quarry, as many others have also been. Its greatest length was found to be 245 feet, and its height generally so inconsiderable, that in two or three situations only could a man stand upright. The following (fig. 105) is the section of it, as given by Dr. Buckland;*—*a, a, a, a,* being horizontal beds of limestone, in which the cave occurs; *b*, stalagmite incrusting some of the bones, and formed before the mud was introduced; *c*, bed of mud containing the bones; *d*, stalagmite formed since the introduction of the mud, and spreading over its surface; *e*, insulated stalagmite on the mud; *f, f*, stalactites depending from the roof. "The surface of the sediment when the cave was first opened was nearly smooth and level, except in those parts where its regularity had been broken by the accumulation of stalagmite, or ruffled by the dripping of water; its substance was an argillaceous and slightly-micaceous loam, composed of such minute particles as could easily be suspended in muddy water, and mixed with much calcareous matter, that seems to have been derived in part from the dripping of the roof, and in part from comminuted bones."† The

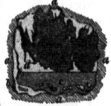

Fig. 105.

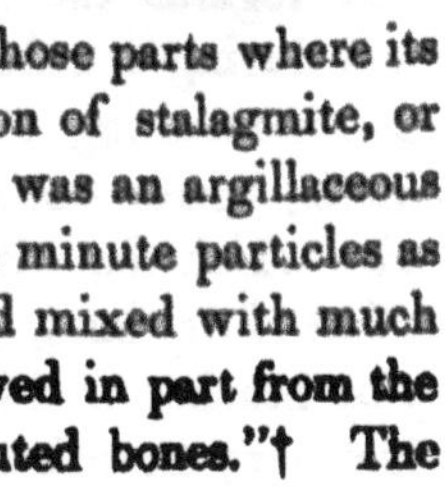

* "Reliquiæ Diluvianæ," 1823. † Ibid.

remains of hyæna, tiger, bear, wolf, fox, weasel, elephant, rhinoceros, hippopotamus, horse, ox, three species of deer, and some other animals, were found to be so strewed over the bottom of the cave when the mud was removed, the proportion of hyæna teeth over those of other animals so great, and the bones of other animals so broken and gnawed, that Dr. Buckland considered the Kirkdale cave to have been the den of the extinct hyænas, the remains of which were found in it, during a succession of years. He further considered that they brought in, as prey, the animals, the bones and teeth of which were mingled with their own, and that these conditions were suddenly changed by the irruption of muddy water into the cave, burying all the remains of the animals, in an envelope of mud, including the fæces of the hyænas, which occurred in the Kirkdale cave, precisely as such now do in the dens of existing hyænas. Many bones were found to be rubbed smooth and polished on one side; a fact showing, Dr. Buckland infers, that one side had been exposed to the walking and rubbing of the hyænas.

There would thus appear to have been a hole or cavern at first raised above common detrital accumulations, and freely communicating with the atmosphere, when the stalagmite *b* was formed; then a change by which water containing fine mineral detritus was introduced, the latter subsiding from the water, which may have completely filled the whole of the cavern; and, thirdly, a time when the cave was out of the reach of water, again freely communicating with the atmosphere—so that stalagmites were thrown down upon the even floor of mud. The stalactites depending from the top may have been partly formed during both periods when the cavern communicated with the atmosphere. Stalactites would not be formed if the cave were full of water, since the solution of the bicarbonate of lime, even supposing such to have passed through into the cavern, would then mingle, in the usual way, with the general volume of the water.

As regards the introduction of fine sedimentary matter into caves during a submersion of previously dry land beneath the sea, the resulting mud not containing the hard parts of marine animals, much would necessarily depend upon the circumstances under which the entrances, or fissures communicating with the old surface of dry land, were placed. Should the entrances be blocked up by beaches or shingles drifted over them (independently of any which may have been closed by the accumulation of fallen fragments before submersion) as the land descended and the coast conditions changed, the shores ranging gradually to higher levels, the matter of fine mud could be water-borne through the shingles or fragments. Such muddy water once in the cavern, either from this source, or entering amid other cracks and chinks, the resulting mud would settle over the floor, enveloping all within its reach in a mass of fine sediment. In either case, any germs of marine animals secreting hard parts, and entering with the water, would scarcely be properly developed in such a situation.

Ossiferous caverns being merely those amid caves in general which, from fitting circumstances, mammals have made their dens, or into which they have fallen or been drifted, all the sinuosities and irregularities of such cavities, both as regards horizontal and vertical range, have to be expected in them. They are found to be variously filled in different localities, so that it becomes difficult to point out any particular arrangement of parts common to the whole. At the same time, the following longitudinal section (fig. 106) may afford somewhat of a general view of many which have been discovered. In it *l*, *l*, *l*, represent the section of a limestone hill (these caverns being like caves in general most

Fig. 106.

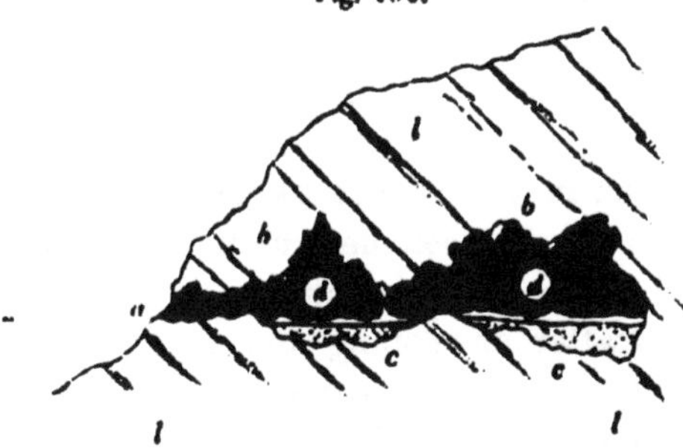

common in limestone rocks), in which there is a cavern, *b*, *b*, communicating with a valley, *v*, by an entrance, *a*. A floor of stalagmite, *d*, *d*, covers bones and fine sediment accumulated in the cavities, *c*, *c*. A column of stalactite and stalagmite is represented between the two chief chambers of the cave, and which may or may not have blocked up the passage from one to the other. Any circumstances having removed a covering of the entrance, *a*, or the latter being even constantly open and well known, an observer, if not informed respecting ossiferous caverns, might easily enter such a cave and remark nothing more than the chambers, the stalactites depending from the roof or covering the walls, and a floor partly rock, partly formed of stalagmite; and even, if the passage between the chambers be closed by stalactite and stalagmite, return from the outer cave without being aware of the chamber beyond it.

It will be, at once, apparent, seeing that the bones in ossiferous caves may either have been chiefly collected by predaceous animals, have fallen into them from openings in the ground above—have been drifted into them, or be the remains of mammals which have entered and died in the caves—that great attention should be paid to the mode in which the bones may be accumulated, and to their whole, fractured, gnawed, or other state. Very careful and complete sections require to be made of the ossiferous accumulations, and these should not be confined to one portion of a cavern; for, during a long lapse of time, an open cave may have been variously tenanted or strewed with bones. If an observer be in search of evidence of ossiferous caves having been the dens of predaceous animals, not only the marks of their teeth upon the remains of such bones as may not have been consumed are valuable, but also the mode of occurrence of fæcal remains, and the rubbing and polishing of portions of the walls, especially in the narrower passages, are important.

With respect to stalactitic and stalagmitic incrustations, they may have happened at all times when a cavern was above the sea or water-drainage of the time, so that the atmosphere entered it, and bicarbonate of lime percolated in solution through the containing rock into the cave. Thus bones, as in the Kirkdale cave, may have been embedded in this calcareous substance, as well prior to the introduction of any fine sediment by means of water, as afterwards. It is the repose of stalagmite upon an even flooring of the sedimentary matter enveloping the bones, which shows an alteration of conditions, one from a state of things when stalagmite could not be accumulated on the bottom of the cave, to that which permitted it.

As the remains of mammals of existing kinds, such as the red deer,* of the roebuck,† badger,‡ polecat,§ stoat,‖ wolf,¶ fox,**

* In Kirkdale Cavern, Yorkshire, and Kent's Hole, Torquay; Buckland, "Reliquiæ Diluvianæ," and Owen, "Hist. Brit. Foss. Mammals."

† Fissure in limestone, with the remains of *Rhinoceros tichorhinus*, Caldy Island, Pembrokeshire; Owen, "Hist. British Foss. Mammals," p. 488. Dr. Buckland mentions an antler, "approaching that of the roe," in the Paviland Cave.

‡ Kent's Hole, Torquay; Owen, "Hist. Brit. Foss. Mammals," p. 110.

§ Belgian Cave, Dr. Schmerling. Berry Head, Devon; Owen, "Hist. Brit. Foss. Mammals," p. 113.

‖ Kirkdale Cave; Buckland, "Reliquiæ Diluvianæ." Kent's Hole, Torquay; Owen, "Hist. Brit. Foss. Mammals."

¶ Kirkdale Cave; Paviland Cave; Oreston, Plymouth; Kent's Hole, Torquay. —Buckland, "Reliquiæ Diluvianæ;" Owen, "Hist. Brit. Foss. Mammals."

** Kent's Hole, Torquay; Oreston, Plymouth.—Owen, "Hist. Brit. Foss. Mammals."

water-vole, field-vole, bank-vole, hare and rabbit,* have been discovered in caves mingled with those which are extinct, and as the remains of man have been detected in similar caverns, it becomes needful most carefully to study the circumstances under which all these remains may occur; so that while, on the one hand, we do not neglect the kind of evidence which might thus show the contemporaneous existence of mammals now partly extinct, and partly living,† and also of man with the same kinds of animals, on the other, the accidents which may have brought such apparently contemporaneous mixtures together may be duly regarded. Thus, had not Dr. Buckland employed the needful caution, human remains (those of a woman) in Paviland Cave, Glamorganshire, might have been regarded as proving the contemporaneous existence of man and of the *Elephas primigenius*, *Rhinoceros tichorhinus*, and *Hyæna spelæa*. In this case, the cave had evidently been employed as a place of sepulture by some of the early inhabitants of that part of Wales, and the ground containing the remains of the extinct animals moved.‡

---

* Buckland, "Reliquiæ Diluvianæ;" Owen, "Hist. Brit. Foss. Mammals."

† The following list of animals, the remains of which have been found in the caves of the British Islands, is given by Professor Owen, in his "History of British Fossil Mammals:"—*Vespertilio noctula*, *Rhinolophus ferrum-equinum*, *Ursus priscus* and *spelæus*; *Meles taxus*, *Putorius vulgaris* and *ermineus*; *Lutra vulgaris* (from Durdham Down, Bristol, on the authority of Mr. E. T. Higgins); *Canis lupus* and *vulpes*; *Hyæna spelæa*, *Felis spelæa* and *catus*; *Machairodus latidens*, *Mus musculus*, *Arvicola amphibia*, *agrestis*, and *pratensis*; *Lepus timidus*, and *cuniculus*; *Lagomys spelæus*, *Elephas primigenius*, *Rhinoceros tichorhinus*, *Equus fossilis* (*caballus?*) and *plicidens*; *Asinus fossilis*, *Hippopotamus major*, *Sus scrofa*, *Megaceros Hibernicus*, *Strongyloceros spelæus*, *Cervus elaphus*, *Tarandus*, *Capreolus*, and *Bucklandi*; *Bison priscus*, and *minor*, and *Bos primigenius*.

‡ The cave in which these remains were discovered is one of two on the coast between Oxwich Bay and the Worm's Head, part of the district known as Gower, on the west of Swansea, and formed, in great part, by carboniferous or mountain limestone. It is known as the Goat's Hole, and is accessible only at low water, except across the face of a nearly-precipitous cliff, rising to the height of about 100 feet above the sea. The floor, at the mouth of the cave, is about 30 to 40 feet above high-water mark, so that during heavy on-shore gales, the spray of the breakers dashes into it. Beneath a shallow covering, Dr. Buckland discovered the "nearly entire left side of a female skeleton." He adds ("Reliquiæ Diluvianæ," p. 88), "Close to that part of the thigh-bone where the pocket is usually worn, I found laid together, and surrounded also by ruddle, about two handfuls of small shells of the *nerita littoralis*, in a state of complete decay, and falling to dust on the slightest pressure. At another part of the skeleton, viz., in contact with the ribs, I found 40 or 50 fragments of small ivory rods, nearly cylindrical, and varying in diameter from a quarter to three-quarters of an inch, and from one to four inches in length. Their external surface was smooth in a few which were least decayed, but the greater number had undergone the same degree of decomposition with the large fragments of tusk before mentioned." Fragments of ivory rings were also discovered, supposed, when complete, to have been four or five inches in diameter. Portions of elephants' tusks were obtained, one nearly two feet long; and Dr. Buckland inferred that the rods and the rings had been made of the fossil ivory, the search for which had caused the marked

In many instances of the mixed remains of extinct and existing species of mammals, independently of the condition and mode of occurrence of the remains themselves, the probable habits* of the animals may offer the observer much assistance. In this manner, certain caves have been inferred to have been the dens of extinct bears and hyænas, in the latter case fæcal remains, considered to be of a very characteristic kind, marking the continued residence of these bone-consuming animals, and a quiet entombment of such bodies with ordinary osseous remains. One kind of animal, such as the cavern bear, may have occupied a cave at one time, while hyænas may have tenanted it at another, and both may have been preceded or replaced by the cave tiger, and its contemporary great feline, the *machairodus*.† During the occupation of the more roomy portions of a cave by such great mammals, smaller animals could have lived in the minor holes and fissures, occasionally feeding upon remnants of the prey brought in by the larger carnivora, and sometimes falling victims themselves to the latter. In certain caves, bats may often have clustered in places in the higher parts of the chambers, secure from the bears, hyænas, or felines, their remains, from time to time, being mingled with the bones of the other animals beneath. With regard to several mammals, the

---

disturbance of the ossiferous ground observed, the ivory being then in a sufficiently hard and tough state to be so worked. Charcoal and pieces of more recent bones of oxen, sheep, and pigs, "apparently the remains of food," showed the cave had been used by man. The toe-bone of a wolf was shaped, and it was inferred that it had been probably employed as a skewer. As regards the date when this cave may have been thus worked for its ivory, and the woman buried, Dr. Buckland calls attention to the remains of a Roman camp on the hill immediately above the cave. Amid the disturbed ossiferous ground there were not only recent bones, but also the remains of edible mollusca, *Buccinum undatum*, *Littorina littorea*, *L. neritoides*, *Patella vulgata*, and *Trochus crassus*.

* No doubt much caution will be required as to any inferences drawn from the habits of existing animals of a particular genus; as, for instance, if the hare were an extinct mammal, and the rabbit only found living, it would be a serious error to infer, from the habits of the latter, that the former always lived in burrows which it dug for itself. At the same time it may not be unreasonable to suppose that animals, such as elephants, rhinoceroses, deer, and oxen, did not make caves their habitations, even when entrances into them were sufficiently large and easy, though they may have occasionally found their way into them, as we have often seen oxen do in England, for shelter from very heavy rains or great heats.

† Mr. Austen, when noticing Kent's Hole and other ossiferous caves of Devonshire ("Geology of the South-east of Devonshire," Geological Transactions, 2nd series, vol. vi., p. 445), calls attention to the habits of the lion and panther, which, after killing their prey, "secure it in their jaws, and bear its weight on their powerful shoulders, retreating with it to these caves." After mentioning the great size of the animals which the African lions carry off, he adds, that "with respect to their usual abodes, we have the authority of all African travellers and hunters, that chasms, caves, overhanging ledges of rocks, and similarly-protected places, are their haunts, and the spots to which they carry their prey."

remains of which are discovered in ossiferous caves, we feel certain that not only would their bulk have prevented them from passing through the only communications, which can either be seen or suspected, between the open air and chambers of many caves, but also that their habits would not direct them to such retreats. As prey to carnivorous mammals inhabiting caves, dragged in piecemeal through comparatively small apertures, when their bodies were dismembered, there appear no difficulties; indeed, there is good evidence on this head. It has been remarked, that the teeth of the extinct elephant found in caves, show that young animals of this kind had chiefly been brought into them. This, however, does not seem to have been the case with respect to the *Rhinoceros tichorhinus*, since the remains of full-formed individuals of this species are, in some cases, sufficiently abundant;* neither does it show that many a large elephant may not have fallen a prey to the great carnivora, especially the feline, its bones and teeth being left elsewhere, and perhaps in great measure consumed on the spot by hyænas.

That men have at various times inhabited caves, and used them as tombs, is well known; and the case of the skeleton of the woman at Paviland, above noticed, is sufficient to show that ossiferous caverns may have been thus employed.† If man had been a contemporary inhabitant of the regions where these extinct carnivora roamed in search of their prey, he might, as well as other creatures, have occasionally formed a portion of such prey. Where pieces of pottery are discovered, which appear to mark the residence of man in the caves, we merely seem to have evidence that he frequented them at some period, perhaps not well defined; unless, indeed, the mode of occurrence of the pottery be such that no doubt of the relative date of its introduction can exist.‡ With flint or other

* Having examined the ossiferous cave of Spritsail Tor, in Gower, Glamorganshire, shortly after its discovery, by the cutting back of a carboniferous limestone quarry, we were much struck by the narrowness of a part of the entrance, where predaceous animals, apparently hyænas (*H. spelæa*), seem to have been stopped, with large portions of the carcasses of the *Rhinoceros tichorhinus*, numbers of the teeth of which, among the other remains, were accumulated close outside it.

† Sir Philip Egerton, "On the Ossiferous Caves of the Hartz and Franconia," (Proceedings of the Geological Society, vol. ii., p. 94), when enumerating the osseous remains which rewarded the researches of himself and the Earl of Enniskillen in the caves of Gailenruth, Kühloch, Scharzfeld, and Baumanns Höhle, mentions that fragments of rude pottery were discovered in these four caves; "old coins and iron household implements of most ancient and uncouth forms in that of Rabenstein," and recent bones of pigs, birds, dogs, foxes, and ruminants, in every cave examined.

‡ The description of the cavern of Mialet, near Anduze, department of the Gard, by M. Tessier ("Bulletin de la Société Géologique de France," tom. ii.), affords a useful illustration of the manner in which human bones may occur with those of extinct

stone arrow-heads and knives, such as have been discovered in Kent's Hole, and elsewhere, there would be more difficulty, if other evidence was not opposed to the inference. When bones of men, as they are stated to have been, are discovered really mixed amid those of the extinct carnivora and other animals found in ossiferous caves,* the subject is one of no slight interest, and requires at least very careful investigation, without prejudgment of any kind.†

Ossiferous caverns may offer greater complication than those previously noticed, in which the apertures or mouths opening to the air are considered to have been more or less lateral, presenting ready ingress and egress to mammals. A cavern of the kind represented, in longitudinal section, fig. 107, may have been of a mixed kind partly composed of a portion, *c*, rising upwards, as is also seen in many which are not ossiferous, and partly having a more horizontal range, *a, a*. If the upright cleft did not reach the surface at the time, in a manner to permit

---

mammals. The cavern is situated 30 yards above a valley, on a steep slope, and in a dolomitic rock. The lowest bed, reposing on the bottom of the cavern, is composed of a dolomitic sand, irregularly covered with thin stalagmite, and here and there by an argillo-ferruginous clay, more than a yard thick. This bed contains the abundant remains of bears. Beneath stalagmite and a bed of clayey sand, from 8 to 16 inches thick, human remains were discovered in different parts of the cavern. At the inmost end they were decidedly mixed with those of bears, which predominated; but at the entrance the human bones prevailed. On the ossiferous clay, and beneath a very rocky projection, a nearly entire human skeleton was discovered, and close to it a lamp and a baked clay figurine; copper bracelets being found at a short distance. In other places were the remains of coarse pottery, worked bones, and small flint tools, exhibiting a ruder state of the arts than the preceding. M. Tessier infers —1. An epoch when the cavern was inhabited by bears. 2. A time when man, little advanced in civilization, inhabited, and probably was buried, in the cave; and 3, the Roman epoch, shown by the remains of more advanced art. As regards the mixed bones of man and the bears, it is inferred that this is accidental, as men and bears could not have lived together in this cavern.

* Dr. Schmerling (Ossemens Fossiles des Caverns de Liége) mentions human bones as decidedly mixed with those of the extinct elephant, rhinoceros, bear, and other mammals in the same clay and breccia in caves near Liége. From the mode of occurrence of the whole, he infers that the human as well as the other bones were all washed into the cave together, men and these extinct mammals being then coexistent. Instances of the mixed bones of extinct mammals and of man, in the south of France, are mentioned by M. Marcel de Serres ("Geognosie des Terrains Tertiaires"), M. de Cristol, M. Tournal, and other geologists, who supported the view that men and these extinct animals had been contemporaneous, a view opposed by M. Desnoyers ("Bulletin de la Société Géologique de France," tom. ii.), who points out that the pottery and weapons discovered in the ossiferous caves correspond with those of the early inhabitants of England, Germany, and Gaul; and that while in the monuments of the latter similar artificial objects occur, no remains of the extinct mammals are discovered, though those of species now inhabiting Europe are detected.

† Professor Owen has pointed out ("Hist. Brit. Fossil Mammals," p. 97) that "of no other quadruped than the bear is the femur more likely to be mistaken by the unpractised anatomist for that of the human subject, especially the femur of the gigantic extinct species commonly found in caves." Figures and descriptions are added in confirmation of this statement.

animals falling through it to the cave beneath, fragments only of the rock in which the whole is situated so doing (and it should be remembered, that in numerous caverns the fall of rocks from

Fig. 107.

various parts of the roofs and sides may have happened at all times), the osseous remains of animals entombed would belong to those which may have entered, lived in, or been dragged into the chambers. If the cleft were sufficiently wide for animals to fall through, as mammals now do similar fissures, there might be two modes of accumulating the remains of the same, or nearly the same creatures; one resulting from the occupation of the cave by predaceous animals, and any others able to live in the same place with them; the other, from the fall of animals through the fissure, sometimes bringing down with them fragments of rocks, and so wholly or partly burying their carcases beneath such fragments.

If we assume the submersion of such a cavern, much, as to the results, would depend on the rapidity or slowness of the submergence. Supposing the latter, and that the mouth of the cave was closed, either prior to it or during its progress, fragments of rock, such as we often see thickly strewed over limestone hill and mountain sides, descending readily over it, the common earth (usually the cementing matter of such fragments on hill sides) would be removed by the wash of the sea, and muddy water, in part, perhaps, thus derived, enter the cave, enveloping with fine sediment the bones in the interior, *a*, *g*. The sediment rising only according to the amount of matter introduced, it might so happen, that an even floor did not surmount the level, *g*, mud alone completely intermingling with fragments of rocks or bones in the lower part of the mass, *h*. Submergence slowly continuing, and the fissure, *c*, still open, animals could, as before, fall through, until finally the whole hill was beneath the water. Much complication might arise in such a case, and more especially if the upper part of a fissure had never been closed over by detritus even to its emergence, or that it had not been covered by water at all, so that it was always open to catch unwary animals, or those hunted by predaceous mammals, during a time when the quadrupeds of the country may have been changed or much modified. Perpendicular fissures in caves are sometimes so filled with fragments of rock, sand, clay, and earth, as to show the necessity of great caution, when inferring that the osseous remains of many caverns had been derived through the lateral mouths alone.

In examining ossiferous caverns, attention should be directed to the kind of foreign detrital matter introduced into them, either occurring amid the bones and fragments of the rock in which the cave is formed, and constituting layers or beds, or which may be strewed about. Let us suppose that in a valley, *v*, of which the following is a section (fig. 108), two ossiferous caverns, *a* and *b*, occur on the side of the hill, *c*, a river, *r*, being of sufficient size to bring down mud, sand, and gravel, especially during floods. The

Fig. 108.

lower cavern, *b*, would from this cause be exposed to deposits, enveloping the bones of mammals which there occurred, floods from time to time surprising and killing animals suddenly caught by them in it. This would not be the case with the higher cavern out of the reach of such fluviatile action and deposits. If the surface of the land had been disposed much as we now find it, anterior to a submergence beneath water, (a supposition by no means necessary,) both these caverns may have had detritus introduced into them, as previously noticed, whatever additions may have been made to the lower cave, *b*, by bones or detritus from the action of the river, *r*. As pebbles of fair size afford evidence which finer sediment may not, it is always important to collect and very carefully examine any found in ossiferous caves, as from them some conclusion may be formed as to the direction whence moving water may have carried them, either from their parent rocks, or from any gravel or shingle accumulation where, for a time, they may have been stationary.*

Much and very proper stress has been laid upon the accumulation of bones with mud, sand, gravel, and fragments of rock in those subterranean and cavernous channels through which streams and rivers so often pass in limestone districts. Into these, animals surprised by floods are often carried, and from them are seldom known to emerge, the passages being commonly so complicated, that even inferring sufficient space for the bodies to pass through, the intricacies and vertical arrangements of the channel are such that the osseous portions of the animals, whatever may become of the flesh, whether eaten or decomposed, remain and accumulate in these cavernous passages. In some tropical and limestone countries, as, for instance, in Jamaica, it is very instructive to watch the effects of a sudden flood hurrying forward a mass of turbid water, and, occasionally, various creatures into great sink-holes.† In more temperate climates, a sudden flood often surprises animals, occasionally large, in low grounds near the entrances into cavernous channels, and, according to the capacity of the channel and the size of the entrance into it, will depend the ready disappearance of the animals thus swept onwards. Sometimes the volume of water is so great, that they are not readily engulfed, whirling about at the entrance, then beneath the level of the water ponded back, until the flood somewhat subsiding, the bodies of the animals enter and become lost in the caverns.

* As regards the contents of the ossiferous cave of Kent's Hole, often mentioned above, Dr. Buckland informed me that Mr. M'Enery found rounded portions of granite and greenstone beneath the stalagmitic crust, as also fragments of sandstone and slate, some of them rolled. The cave itself is in limestone, the sandstone, slate, and greenstone rocks associated with it in the district, but granite does not occur nearer than Dartmoor, 13 miles distant. According also to Colonel Mudge (Proceedings of the Geological Society, vol. ii., p. 400), the pebbles discovered in the ossiferous bed at Yealm Bridge cave, six miles from Plymouth (five distinct deposits being noticed, the highest only containing bones), "are apparently derived from the confines of Dartmoor, and differ from those contained in the bed of the Yealm." The remains found in this bed, 34 feet thick, were those of the elephant, rhinoceros, horse, ox, hyæna, sheep, dog, wolf, fox, bear, hare, water-rat, and a bird of considerable size. Many of the bones were "splintered, chipped, and gnawed," and coprolites were found in the ossiferous bed. Pebbles are found in several ossiferous caverns.

† With the various land mollusca caught by heavy, and often sudden tropical rains, and swept into these sink-holes, land hermit or soldier crabs, are in certain localities also carried in, sometimes bearing marine univalve shells which they have brought with them, occasionally many miles, during their migrations to and from the sea-coast. We have seen these crustaceans at 12 to 14 miles from the sea in the limestone districts of Jamaica, and can confirm the statement of Mr. R. C. Taylor ("Notes on the Geology of Cuba," Philosophical Magazine, 1837), respecting their habits as noticed by him in similar districts of Cuba. Marine shells may thus readily be included in the stalagmites of the caverns, often of large size, common in the white limestone portions of Jamaica.

Even under the somewhat simple conditions of such cavernous channels, as shown in the section beneath (fig. 109), it will be obvious that not only detrital matter and fluviatile molluscs, but also terrestrial mammals may be introduced into a cavern, *b*, *c*, *d*, and the finer sediment, held in mechanical suspension, alone

Fig 109.

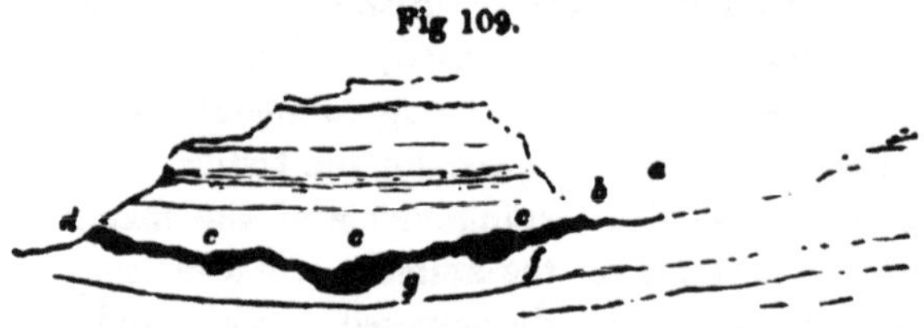

emerge at *d*, supposing the channel to be sufficiently short, and the water be kept in the proper agitation throughout. Under ordinary conditions, a large amount of the elongated cavern would be beneath the level at which the water emerged at *d*, so that the heavier sediment would settle at the bottom of the inequalities, such as *f* and *g*. The bodies of animals could scarcely be forced through even such a comparatively ample passage as that above represented, the general form of the channel, and especially the depending portions of the roof, *c*, *c*, *c*, opposing obstacles to their transport outwards to *d*. Should the impediments to the passage of the water gradually accumulate (and among these large falls of fragments from the roof would be important), an outlet of this kind may be even completely stopped. If we suppose a submergence of the land, such a channel might also be completely filled with detritus, so that, upon a subsequent emergence, the drainage formerly effected through the passage *b*, *d*, being blocked up, it passed elsewhere, and the former outlet, *d*, might form the entrance into an ossiferous cavern on the lower side of a hill.

Many caverns convey out waters which have accumulated amid the rocks of which they form a part, especially in limestone districts. These streams sometimes choke up parts of the cave, so that they cannot be passed during a rise of water, though communicating between chambers still above the level of that water. Such subterranean streams occasionally transport sedimentary matter, and leave it in situations whence it is not easily detached, and where it may cover up the osseous remains of animals, or the works and even the bones of men. This would appear to have been the case, as respects the mode of occurrence of the human remains observed by Dr. Buckland, in one of the branch chambers at Wokey Hole, in the Mendip Hills, near Wells. Human teeth and fragments of bone were "dispersed through reddish mud and clay, and some of them united with it by stalagmite into a firm osseous breccia."* Among the loose bones he found "a small piece of a coarse sepulchral urn." The mud and clay seemed clearly to have been derived from the adjacent subterranean river, which, in its overflowing, reached this chamber.

Ossiferous caverns are sometimes entirely destitute of stalagmite, forming a level crust over a floor, or even any deposits or incrustations of the kind. The ossiferous mass found in Banwell Cave, Mendip Hills, was composed of little else than fragments of the limestone in which the cave occurs, mingled with the bones of the cavern bear and other extinct mammals. Similar caves have been found elsewhere, the bones and fragments of rock only requiring, as M. Therria long since remarked, a cementing substance, such as carbonate of lime, indurated clay, or other mineral matter, to form those accumulations known as *osseous breccias.*† As caverns which may have been the dens of predaceous mammals, occasionally present clefts and fissures filled in this manner, it is important to ascertain, when such are exposed to view by the cutting back of quarries, whether they are merely clefts and fissures, such as represented beneath (fig. 110), and which have probably never formed a portion of a cavern, properly so called, or are parts of an ossiferous cave, which further researches may expose.

In this section (fig. 110) *a*, *b*, *c*, represent fissures filled with ossiferous breccia, an offset at *f*, giving a horizontal character to

Fig. 110.

part of one of them. In limestone districts, and in such countries clefts containing osseous breccias are the most common; a reddish

* "Reliquiæ Diluvianæ," p. 165.

† "Mém. de la Société d'Histoire Naturelle de Strasbourg," vol. i., wherein M. Therria describes the Grotte de Fouvent, in which, according to Cuvier, the remains of elephant, rhinoceros, hyæna, cave bear, horse, ox, and a large feline animal were found. These remains were considered to have entered through a cleft in the rock, laid open by quarrying back a limestone. The cave was completely filled by bones, a yellow marl, and angular fragments of the limestone and rocks of the vicinity.

and calcareous cement is not unfrequent, though not constant, the hardness and consolidation of the general accumulation being very variable. The red colouring substance usually arises from the decomposition of limestones, in or near which the fissures occur, as has long since (1834) been remarked by M. de Cristol.* The carbonate of lime being wholly, or in great part, removed in solution (the needful carbonic acid being present), the remaining portions of the limestone, comprising any carbonate of lime which may have been left, alumina, silica, or other substances, including iron, become coloured by the peroxidation of the latter, as may be frequently observed in the soil of limestone districts, particularly among the carboniferous limestone countries of the British Islands and Belgium. These fissures, when clearly unconnected with caves, are inferred to have been partially filled by the falling in of animals.†

Osseous breccias are found as might be expected, in different countries; their contents variable, and pointing to differences in the time, though always at comparatively recent geological periods, when they were accumulated: indeed, such could scarcely but have happened with these ossiferous accumulations, whether found in caverns or in fissures, since we have reason to infer that the bones of animals are now being gathered together in similar situations in different parts of the world.‡ It is chiefly as regards their possible or probable connection with the inferred interval of increased cold, at a particular time, in the northern hemisphere, that ossiferous caves and breccias are here noticed. Under the hypothesis of this increase of cold being accompanied by the submergence of a large portion of Europe, affecting the area above mentioned, such submergence being gradual, perhaps with oscillations, unequal in different portions of the general area, and

* "Observations Générales sur les Brèches Osseuses," Montpellier, 1834.

† With respect to the falling in of animals into fissures, Dr. Buckland, directing attention to this subject in 1823 ("Reliquiæ Diluvianæ," pp. 56 and 78), mentions that animals now fall into a fissure in Duncombe Park, Yorkshire, as it "lies like a pitfall across the path of animals which pass that way." This fissure was found to contain the skeletons of dogs, sheep, deer, goats, and hogs, "each on the spot on which it actually perished." It is remarked, that if a body of water entered this fissure, the bones and the fragments of the limestone in which it occurs would be all washed to the bottom. Dr. Buckland also referred to the loss of cattle down fissures, and into caves, experienced by the farmers in the limestone districts of Derbyshire, Monmouthshire, and Glamorganshire.

‡ Ossiferous caverns and fissures are found in various parts of Europe. They have been discovered in England, Spain, France, Italy, Sardinia, Dalmatia, Croatia, Carniola, Styria, Austria, Hungary, Poland, and Germany. In the latter, bone caves have long been well known, and Cuvier pointed out, in 1812, that they extended over 200 leagues ("Ossemens Fossiles," 1re Ed.). In 1823, Dr. Buckland took

followed by a rise of the same area, also, perhaps, with oscillations,* and with very considerable modifications of its surface, there are apparently conditions for much movement amid the terrestrial animals of this portion of the northern hemisphere. They would be sometimes isolated and destroyed, as by continued depression, the sea passed over their feeding grounds; at others, they would retreat to regions where they could, for a time, establish themselves and increase, some species being better able to preserve

a general view of the subject ("Reliquiæ Diluvianæ") as far as it was then known. In his "History of British Fossil Mammals and Birds," Professor Owen brought it up, with much new information, to 1846, more especially as regards the osseous remains of this kind discovered in the British Islands; and, in 1848, the Vicomte d'Archiac ("Histoire des Progrès de la Géologie," vol ii., 1re Partie) published abstract statements of the knowledge obtained from 1834 to that date respecting ossiferous caves and fissures, and of their connection with the superficial deposits of the more recent geological accumulations in various parts of the world. Australia has furnished its ossiferous caves and breccias, the remains of the animals detected in them being chiefly of the marsupial character, one so strongly marking the mammalia of that land in the present day. Part of the species of mammals, the remains of which are thus obtained are extinct, while others still live in Australia.

* As regards oscillations, when the caves were situated at a small elevation above a tide-way in an estuary, or at such a distance up a river, tidal at the lower end, that a change in the height of the tides would alter the previous level of the river, there could be oscillations at one time permitting a cave to be inhabited by predaceous mammals, at others so filling it with water that they retreated from it. Where there are alternations of stalagmitic floors, covering *even* surfaces with bone accumulations, as is stated to have been the case at Chockier, on the banks of the Meuse, about two leagues from Liége, it is always desirable to consider the extent to which the presence or absence of sufficient water in the lower parts of caverns may have produced such alternations, the roof and sides always furnishing the needful carbonate of lime, at one time forming the stalagmitic crusts, at another being too much dispersed in the water to afford a deposit.

Though the osseous breccia beneath the Castle Hill at Nice, of which the following (fig. 111) is a section (made in 1827), may not be immediately connected with the northern movement of land noticed in the text, it may yet assist the observer, as showing the kind of evidence which may occasionally present itself. The face of the quarry, *q*, had been cut back beyond the fissure, the sides of which were bored at *l*, *l*, *l*, by the common *Lithodomus*, now inhabiting that part of the Mediterranean, so that it was once an open fissure beneath the level of the sea. This fissure, up to the lip of the cavity on that side, *c*, then became filled with rolled pebbles, chiefly transported from a distance, and afterwards cemented by calcareous matter. Above this was the osseous breccia, *o*, rising up to the top of the fissure on the side, *a*, but whether this was accumulated, like most osseous breccias, on dry land, is not so clear, marine as well as terrestrial shells being mingled with it. Osseous breccias are found in the same vicinity up to the height of, at least, 500 feet above the level of the Mediterranean, and some breccias, not ossiferous, but otherwise similar, solely contain marine remains, so that, perhaps, these fissures may have been partly filled on dry land and partly in the sea. At Cagliari, Sardinia, the remains of a *Mytilus* are found mixed with osseous breccia at 150 feet above the sea.

Fig. 111.

themselves than others. Upon a rise of the sea-bottom, and the consequent formation of new lands, migrations would be effected, according to the relative levels of these lands, as regards the sea, and as passages for the movement of certain animals would sometimes present themselves more favourably in one direction than in another.* Accumulation of the bones and teeth of the same species of elephants, rhinoceroses, and other animals, the remains of which occur in accumulations beneath those formed at the cold or "glacial" time, are considered to have been detected also above them, together with the remains of some mammals not previously inhabiting the area of the British Islands, and adjacent portions of the continent of Europe. This subject offers a fertile field for the labours of an observer. Though much may have been accomplished, much remains to be done, and it will require his especial care to see, that amid the new lakes and river channels formed when the ground took that general configuration which we now find, a rearrangement of bones, washed out of the older deposits containing remains of the *Elephas primigenius*, *Rhinoceros tichorhinus*, and their contemporary mammals, and carried into the newer lacustrine and fluviatile beds, may not occasionally have been such as to mingle the osseous remains of the species of one time with those of another.

As to the ossiferous caves, several closed at their mouths during a time of submergence, may have been reopened by the subsequent removal of the detrital matter then accumulated, so that animals of the later time and of suitable habits again entered or dragged their prey into them; other caverns being laid open for the first time for the entrance or fall of mammals, by the ordinary marine causes of denudation, during a depression and subsequent upheaval of the land. Many a surface is covered over by gravels, concealing former inequalities, amid which there may be caves and

* Mr. John Morris, when noticing the occurrence of mammalian remains at Brentford (Athenæum, Pro. Geol. Society, 5 Dec. 1849), points out as important "that it is generally along those valleys where the present drainage of the country is effected that we find the most extensive deposits of mammalian remains and recent shells, and consequently that very little alteration can have taken place in the physical configuration of the country since their deposition." The remains discovered at Brentford, and giving rise to these observations, were those of the elephant, rhinoceros, hippopotamus, auroch, short-horned ox, red deer, rein deer, and great cave tiger or lion. A few shells of recent freshwater species were found at the same time. As regards the existence of land and fresh-water molluscs, of the kinds still inhabiting Britain, Mr. Searles Wood, in his remarks "On the Age of the Upper Tertiaries in England" (Athenæum, Geol. Society, 5 Dec. 1849), infers, from a list of the mammals at different geological periods, "that a race of animals has arisen and departed whilst the land and fresh-water mollusca have lived on unaltered," and also that "fresh-water mollusca have a greater specific longevity than marine."

fissures, ossiferous or not, according to circumstances. The accompanying section (fig. 112) is one of a quarry wherein limestone, *b*, *b*, presenting a very irregular outline, is covered over by gravels, *a*, *a*, giving a general rounded outline to the surface.

Fig. 112.

The quarry is situated at Waddon Barton, near Chudleigh, Devon, and the limestone is of the kind (Devonian), wherein several caverns and fissures of the district are found (Kent's Hole, Yealm Bridge, Plymouth, and elsewhere), partly ossiferous and partly without bones. Amid such varied modifications of surface as would follow submergence beneath, or emergence from seas, at one time perhaps bounding the area of the British Islands and adjacent portions of the Continent, as represented (figs. 65 and 99) by a line of depth, now no more than 600 feet beneath the surface of the Atlantic; at another cutting existing highlands at about 1,000 or 1,300 feet above that level, and finally producing the present distribution of land and water in Western Europe, it could scarcely happen but that caves and fissures were placed under many modifications of condition. Not only may they have been closed at one time, and open at another, never completely blocked up, or previously laid open, as above noticed,* but they may also be cut

* Mining operations in limestone districts, such, for instance, as that of Derbyshire, afford numerous instances of the irregular distribution of caverns, so that new surfaces of land being produced, changes of this kind would follow. The Speedwell Mine in that county is a good example of a lofty cave, cut into while driving a mining tunnel, this cavern evidently communicating with the great cave of the Peak at Castleton, since the rubbish from the one gets drifted into the other by the water passing through a series of subterranean channels. It was in 1822, while working the Dream lead-mine, near Wirksworth, in the same district, that a cavernous termination to a fissure, communicating with the surface, was discovered; one which was found by Dr. Buckland to contain, among other osseous remains, those of a *Rhinoceros tichorhinus*, so placed as to leave little doubt that they constituted the skeleton of an animal which had fallen from above though a fissure.—*See* "Reliquiæ Diluvianæ," pp. 61–67, and Plate X.

back in such a manner with the adjacent rocks, that though they contain the osseous remains of earlier times, they are now apparently unfit, or at least most inconveniently situated, for the retreat or dens of predaceous mammals, or for trap-falls to them and the animals on which they lived.*

With regard to the migration of the great mammals of the northern hemisphere immediately before, during, and after the time when the cold is inferred to have been such as above noticed, and when, by means of ice, huge masses of rock and other detritus were transported, and thrown down on sea-bottoms, parts of which, upraised, now constitute a large portion of the dry land of Northern Europe, Asia, and America, it becomes interesting to consider the mode of occurrence of the remains of the mastodon, a genus of great proboscidian mammals, approximating to, and of about the bulk and general form of the elephant. Those discovered in England have not been numerous, and have hitherto only been obtained from accumulations in Norfolk,† formed anterior to those, in the British Islands, in which are found the remains of the *Elephas primigenius*, the *Rhinoceros tichorhinus*, and other mammals of that time. As far as can be inferred from negative evidence, the *Mastodon angustidens*, the species which inhabited the British area and

* Dr. Buckland, in 1823 ("Reliquiæ Diluvianæ," p. 95), describing the Paviland cave, occurring in the face of a limestone cliff near Swansea, remarks on this kind of denudation, observing that these caves are analogous to those "in the equally vertical and not less lofty cliffs that flank the inland valleys of the Avon at Clifton (Bristol), of the Weissent river at Muggendorf, of the Bode river at Rubeland in the Hartz, and of the Mur at Peckaw, near Gratz, in Styria;" all these being the truncated portions of ossiferous caverns cut back by denuding influences.

† The "Crag," as these deposits are usually termed, is an accumulation of gravel, sand, and clay, with often an abundance of shells, extending over parts of Norfolk, Suffolk, and Essex. It, and deposits above it, have been described by Mr. Taylor (Geology of East Norfolk, 1827), Mr. Woodward (Geology of Norfolk, 1833) Mr. Charlesworth (London and Edin. Phil. Magazine, 1835; British Association, 1836, &c.), Sir Charles Lyell (Mag. Nat. Hist., *new series*, vol iii., 1839; London and Edin. Phil. Mag., 1840), Mr. Searles Wood ("Catalogue of Crag Shells," Mag. Nat. Hist. 1840–42), Mr. Trimmer ("Geology of Norfolk," Journal of the Agricultural Society, vol. vii.), and others. The lower part of these deposits is known as the *Coralline Crag*, from containing numerous fossil corals, and 400 species of shells are stated to be found in it. Upon this reposes the *Red* or *Norfolk Crag*, in which 300 species of shells have been found, about half of the latter occurring also in the *coralline* crag. Above these are beds known as the *Mammiliferous*, or *Fluvio-marine Crag*, containing fresh-water accumulations. In connexion with the latter, and stated to be rooted upon it, there are the remains of a forest of fir-trees. Mr. Trimmer also notices a marine deposit between the fresh-water beds and the succeeding mass of boulder clay, the parts of which are so strangely contorted and twisted, the effects, it has been inferred, of the action of grounded icebergs and coast ice on a sea-bottom or coast. Though doubts have been expressed as to the beds to which some of the mammalian remains should be referred, it seems agreed that those of the *Mastodon angustidens* occur in the fluvio-marine or fresh-water accumulations, which are also remarkable for containing many existing shells. Dr. Mantell mentions ("Wonders of Geology," 6th edit., 1848, vol. i., p. 224) thirteen teeth of the mastodon as having been obtained from the latter.

parts of Europe,* had passed away, at least in the former, before the mammoth appeared. Wherever this elephant may have retreated during the supposed greater cold in the higher latitudes of the northern hemisphere, it passed, in its subsequent migration, into North America, apparently roaming amid the same districts with a gigantic mastodon (*M. giganteus*), if the inference be correct, that the dispersion of the erratic blocks and associated detritus of that region was contemporaneous with that over Northern Europe. At all events, the surface on which both these mammals fed, appears certainly to have been that which resulted from the dispersion of such accumulations in North America, both animals sometimes lost in boggy ground, as many an animal now is at the present day, and there perishing, their bones, after the decay of the flesh, preserved in a certain amount of original arrangement. If it should eventually be found that the remains of the mammoth do not occur in lower deposits of North America, that the North American is certainly the same elephant with the *Elephas primigenius*, and that the erratic blocks and associated drift of both regions are really contemporaneous, there will have been evidence of a remarkable migration of the mammoth from the west to the east, after an interval of increased cold in the northern regions, and a submergence of them beneath the sea. On the east the mammoths would have become associated with a species of a huge proboscidian which had disappeared, as a genus, from Western Europe, prior to their existence there, but which still continued to flourish on the continent of America. The remains of the mastodon are stated to be found amid the superficial deposits of that continent as far as latitude 66° N., thus bringing them within the climates apparently not unfavourable to the mammoth, though, as Professor Owen has remarked, "the metropolis of the *Mastodon giganteus* in the United States, like that of the *Mastodon angustidens* in Europe, lies in a more temperate zone, and we have no evidence that any species was specially adapted, like the mammoth, for braving the rigours of an arctic winter."†

---

* The remains of the *Mastodon angustidens* have been discovered in France, Germany, and Italy.

† "Hist. of British Fossil Mammals," p. 297.

Respecting the remains of the *Mastodon giganteus*, Bigbone Lick, in northern Kentucky, and about seven miles up a tributary of the Ohio, has long been celebrated for them, and they have also been discovered in several other localities. The "Lick," is so called from the saline springs which various animals frequent. Even the contents of the stomach of the *Mastodon giganteus* have been discovered, containing crushed branches and leaves, grass, and a reed now well known in Virginia. A summary of the knowledge respecting the mode of occurrence of the American mastodons will be found in Sir Charles Lyell's Travels in North America. In his

The observer will readily perceive that much, requiring great care, is needed in investigations of this kind, and that, when endeavouring to trace the paths by which such animals may have migrated, and to ascertain the localities from whence, after retreating, they may again have, in part, been dispersed, districts over which no seas have passed, during the lapse of the supposed geological time, are of no slight value. Hence, among other objects of geological interest, the region of extinct volcanos in Central France is important, inasmuch as it seems to have constituted dry land, during a range of time when several animals which once lived on its surface became extinct, among them the mammoth and *Rhinoceros tichorhinus*. Amid the various notices of the remains of mammals found in situations giving them geological date, may be mentioned that of M. Pomel, wherein he describes an ossiferous fissure in a lava current (near Orbiéres, on the south of Clermont), which had issued from Gravenoire. It was filled with volcanic sand, pulverulent carbonate of lime, and bones which are stated to be the same as those of Coudes and other contemporaneous accumulations, containing the remains of the elephant, *Rhinoceros tichorhinus*, horse, ox, &c.* Land shells, of species now existing in the district, are mentioned by M. Pomel, as associated with these ossiferous deposits, so that in this region also, as in others of Europe and North America, great mammals have become extinct, while land and fresh-water molluscs, living with them, have continued to exist up to the present time.

Paper (Proceedings of the Geol. Society, vol. iv., p. 36, 1848), on the Geological Position of the *Mastodon giganteum*, and associated Fossil Remains of Bigbone Lick, Kentucky, and other localities in the United States and Canada, he pointed out that "on both sides of the Appalachian chain the fossil shells, whether land or fresh-water, accompanying the bones of the mastodons, agree with species of mollusca now inhabiting the same regions." He also concluded that "the extinct quadrupeds, before alluded to in the United States (mastodon, elephant, mylodon, megatherium, and megalonix), lived after the deposition of the northern drift; and consequently the coldness of climate, which probably coincided in date with the transportal of the drift, was not, as some pretend, the cause of their extinction."

* "Bull. de la Soc. de France," tom. xiv., 1842–3. A very instructive lecture was given by Sir Charles Lyell, at the Royal Institution of Great Britain, on this region in 1847, an account of which appeared in the Athenæum of the time. He especially called attention to changes which its mammals had undergone, as shown by the osseous remains preserved in the alluvium associated with volcanic accumulations, "no flood or return of the ocean having disturbed the surface."

CHAPTER XVII.

VOLCANOS AND THEIR PRODUCTS.—HEIGHT ABOVE THE SEA.—CRATERS OF ELEVATION AND ERUPTION.—FOSSILIFEROUS VOLCANIC TUFF BEDS.—SEVERAL CRATERS ON ONE FISSURE.—VOLCANIC VAPOURS AND GASES.—VOLCANIC SUBLIMATIONS.—MOLTEN VOLCANIC PRODUCTS.—FLOW OF LAVA STREAMS.—VESICULAR LAVA.—VOLCANIC CONES.—COTOPAXI.—VOLCANOS OF HAWAII.—EFFECTS OF LAVA ON TREES.

DISTRIBUTED over various portions of the earth's surface, as well in high southern and northern latitudes, as in temperate and tropical regions; at points in the ocean far distant from main masses of dry land, as well as upon the latter themselves, free communications are effected between the interior of our planet and its atmospheric covering, through which molten rock, cinders, and ashes are ejected. That great heat, if not the primary, is at least a chief secondary cause by which these mineral substances are thus upheaved, is rendered evident by the high temperature of the bodies thrown out. The molten rock flows as a viscous fluid, and retains its high temperature for a long succession of years; and mineral substances are volatilized, which, we learn from our laboratories and furnaces, are only raised to that state by great heat. At the same time that these mineral bodies are ejected, vapours and gases, of a certain marked character, are expelled, so that by carefully combining the mode of occurrence of the various products, with the composition of the substances themselves, an observer, by the aid of sound chemistry and physics, may hope so to direct his inquiries as to obtain a fair insight into the causes and effects of volcanic action.

As regards altitude above the level of the sea, volcanic products are accumulated at various heights above that level, doubtless also forming the bases of many volcanos beneath it on the floor of the ocean. The most elevated of known volcanos constitute such an insignificant fraction of the earth's radius, that variations in height do not appear to offer any great aid in ascertaining the causes of volcanic action, though certain of its effects may thereby be somewhat modified, especially when volcanos rise into the regions of perpetual snow. Cotopaxi, the cone of which rises, in the Andes, 12 leagues S.S.E. from Quito, to the height of somewhat more than 19,000 feet above the sea, forms but an insignificant part of the radius of the earth, not constituting so much as 3⅔ miles of that radius, or about $\frac{1}{1100}$th of it.*

With respect to the kind of openings through which the gaseous and mineral substances are vomited forth, there has existed much difference of opinion. While some geologists infer that the rocks through which the volcanic forces found vent had been so acted upon that they were upraised in a dome-like manner, the gaseous products bursting through the higher part, driving the lighter substances into the atmosphere, if the dome were elevated into it, and raising the viscous molten rock, so that it flowed out of the orifice; others consider that there has been a simple fissure or aperture in the prior-formed rocks through which the volcanic products were propelled, the solid substances accumulating round the vent, so that a deceptive dome-like appearance is presented.

The following sections (figs. 113 and 114) may assist in showing the differences between the "craters of elevation," first brought under notice by M. Von Buch, and so ably illustrated by M. Élie de Beaumont and other geologists, and the "craters of eruption," as they have been termed. Fig. 113 represents a portion of

Fig. 113.

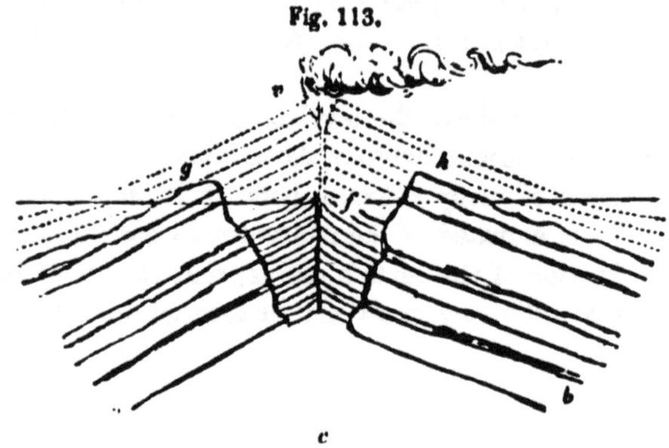

* Humboldt (Kosmos) refers to the relative height of volcanos as probably of consequence if we should assume their seat of action at an equal depth beneath the general surface of the earth. He refers to eruptions being commonly more rare from lofty than from low volcanos, enumerating the following:—Stromboli, 2318 feet (English); Guacamayo (Province of Quiros), where there are almost daily detonations; Vesuvius, 3876 feet; Etna, 10,870 feet; Peak of Teneriffe, 12,175 feet; and Cotopaxi, 19,070 feet.

deposits more or less horizontally arranged, fractured and upraised in a conical or dome-shaped mass, a portion of them, *g a c b h*, being divided and rent at *c*, so that volcanic forces, pressing through, find vent. For the sake of illustration, the rocks broken are assumed to be accumulated in beds. This is by no means essential, the mass disrupted may have been composed of certain crystalline rocks, such as granite, to be hereafter noticed, bearing no marks of stratiform arrangement. If now ashes and cinders be thrown out of this vent, and accumulate in more or less conical layers, one outside the other, until at *g* and *h*, the original and upheaved beds are concealed, and a crater presents itself at *v*, through which similar substances continue to be thrown, it may be very difficult to distinguish such an arrangement of parts from those effected by the propulsion of similar substances through a simple longitudinal crack, as represented in fig. 114. In this section, a

Fig. 114.

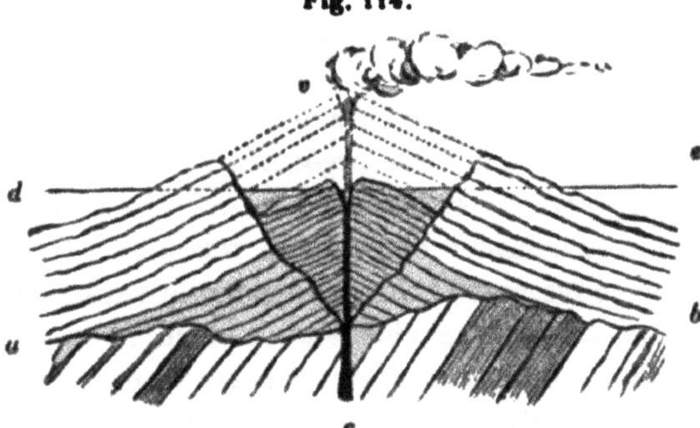

series of beds, *a b* (for more contrast represented as previously upraised in a mass, and as all sloping or dipping in one direction), is traversed by a crack, which, though it divides the beds, has not been accompanied by upheaval or depression of one side or the other. Through this vent, *c*, cinders and ashes are supposed to have accumulated in conical layers, as before, the apex crowned by the crater, *v*.

It will be obvious that in both cases, if the volcanic accumulations had been subaërial, even with the addition of the flow of lava currents, and of cracks amid the ashes and cinders filled with molten rock, (which have been excluded from the sections to render them more simple,) much difficulty would arise from the general similarity in the arrangement of the volcanic products exposed to sight, unless denudation from atmospheric influences, or the sinking or blowing off of a large part of the volcano, afforded a better insight into the general structure of the mass, so that, as shown by fig. 113, portions of subjacent and tilted beds of dissimilar rocks could be seen, as at *g* and *h*, or of similar volcanic accumulations, as in fig. 114.

Evidence of a better kind would be expected, should the ashes, cinders, and molten matter have been accumulated both beneath and above a sea level, the action of the breakers denuding the general mass, so that more illustrative sections would be afforded. Thus, if upraised above the sea level, the original dome or cone-shaped rocks, *a b* (fig. 113), though covered, for a time, by a mass of matter, *g v h*, the result of a high state of activity in the volcano, may finally become visible, and afford the information sought. In the same manner, evidence of another kind may be obtained, as regards the accumulation from simple volcanic eruption, by marine denudation, as shown in fig. 114. In both sections it is supposed, that volcanic action not ceasing, conical accumulations may continue to be formed inside a crateriform cavity, more or less occupied by water, cliffs all round facing an active volcanic vent, as at *f* (fig. 113).

Under even these favourable circumstances, the observer should employ great caution. The facts presented to him may require no little comparison and classification; for in such localities, more especially, he has to consider how far the relative levels of the sea and land may have remained the same since the various accumulations before him have been effected. Let it be supposed, for illustration, that he detects organic remains in beds surrounding the interior basin of water, in which the volcanic island still vomits forth various gases and products. Should the deposits *g* and *h* (fig. 113) be of the more recent geological times, commonly marked by the presence of the remains of mollusca, not much, if at all, differing from those still existing in the vicinity, and should the mineral composition of the including beds not be decisive on the point, the subject may not be so clear. By reference to the section (fig. 114), it will be seen that if the line, *d e*, representing the present level of the sea, be raised, and, consequently, the whole mass of rocks, including the supporting deposits, *a c b*, relatively depressed, the layers now above the sea, being then below it, molluscs may have lived upon and amid these layers while they successively constituted the sea-bottom, as upon any other sea-bottoms, and as many molluscs must now do around volcanic islands. There is no difficulty in considering that, during a long lapse of time, breaker action aided in the re-arrangement of many substances, including animal remains, on the subaqeous slopes of volcanos, the angle of the beds varying according to obvious conditions. Any change in the relative levels of the sea and land, which the observer, as he pursues his researches, will find to have been so frequent, and often so considerable, that should raise the general mass (fig. 114), so that *d e* be the line of sea level, would expose the edges of these fossiliferous beds facing the interior. And it should be borne in mind, that in many localities calcareous beds, and even limestone, may become mingled with such deposits during their submarine accumulation.

When studying the fractures and contortions of rocks, as well on the small as the large scale, there will be frequent occasion to remark, as will be more particularly shown hereafter, the mixture of flexures and fissures, and the extension of the one into the other. The subjoined example (fig. 115) of the termination of a

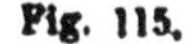

Fig. 115.

fracture and flexure, occurring amid the slightly-inclined beds of lias near Lyme Regis, Dorset, may aid in illustrating a point of much interest connected with the present subject, namely, that in the more marked instances adduced of "craters of elevation," a considerable break or outlet is often found on one side. The plan (fig. 115) shows an alternation of the thin-bedded limestone of the lias of Dorsetshire with shale, the whole broken through by a crack, *a*, *b*, the continuation of one where there is dislocation producing movement on the sides, and which terminates in a boss at *b*, with somewhat diverging small cracks. The interior is composed of limestone, round which shale, covering it, is exposed by the pear-shaped protrusion, outside which is another limestone bed, *c c c*, dipping outwards from the central portion, *b*, the whole taking a more horizontal character towards *a*, where, for a certain length, the plane surface is merely broken by a fissure. With proper forces and resistances employed, a like disposition of parts could be obtained on the large scale.

If, as on the subjoined plan (fig. 116), such a state of things had been brought about on the large scale, and volcanic forces had been enabled to find vent at different points, there may be good evidence of a crater of elevation, and of other volcanos presenting no such evidence, all situated on a great line of fracture ending in a dome-like flexure, with, perhaps, a common communication

Fig. 116.

between them. At *d a c*, the beds broken through would dip, with radiating cracks, around a gorge opening in the direction of the main fissure towards *b*, while surrounding the volcanic vents, *e* and *f*, the strata may be horizontal. Under such circumstances, the volcanic accumulations being continued, so that they may be intermingled, if the whole be regarded with reference to depression beneath the sea, or elevation above its level, the observer will perceive that numerous complications would arise, requiring no slight care properly to appreciate.

As there is every reason to infer that volcanic substances have been, and are ejected at various depths beneath the sea level, as well as above it, the modifications of the products arising under the former conditions have to be properly estimated, more particularly when we have to associate such modifications with changes in the relative levels of sea and land, so that accumulations formed at various depths beneath water may be mingled with those gathered together in the atmosphere. That subaqueous would gradually approximate to subaërial deposits, as the accumulations round volcanic vents rose from different depths in the sea above its level, will readily be understood. When eruptions pierce through the sea level, ashes, cinders, and stones are gathered round a crater, and vapours and gases are evolved, as happened off St. Michael's, Azores, in 1811 (p. 100), and in the Mediterranean, between Pantellaria and the coast of Sicily in 1831 (p. 70). At such times the volcanic forces so accumulate mineral substances around the vent, and so, for the time, overpower the action of the sea, that it is not until these forces have been expended, or greatly abated, that the breakers can abrade the land, and, supposing no subsidence, or falling in of the mass of volcanic matter thus raised, and the latter sufficiently incoherent, level off the accumulations to the depths to which waves can mechanically act.

Fully to appreciate the modifications which may arise in volcanic action at various depths in water, productive of effects which can only be inferred, very careful study of the gases and vapours evolved, of the chemical composition and mineralogical character, and of the mode of occurrence of the solid substances thrown out from subaërial volcanos, is needed. With regard to the vapours and gases evolved, the chief appear to be aqueous vapour or steam, sulphurous acid, sulphuretted hydrogen, hydrochloric acid, and carbonic acid. Steam is a very common product, and, as Dr. Daubeny has remarked, is sometimes omitted for ages from volcanic fissures.* Hydrochloric acid is also common in various parts of the world. Sulphurous acid has been inferred to predominate "chiefly in volcanos having a certain degree of activity; whilst sulphuretted hydrogen has been most frequently perceived amongst those in a dormant condition." † Carbonic acid is observed at the close of eruptions, or in extinct volcanos, and is stated to be emitted more from the bases and neighbourhood of volcanos, than from their craters.‡ Besides these gaseous products, which can be collected when volcanic vents can be approached sufficiently near for the purpose, it is now considered that there is an inflammable gas occasionally evolved from some craters and volcanic fissures, which gives the flame often mentioned, but at one time much doubted. Of what kind this gas may really be, appears as yet uncertain, and may long remain so, inasmuch, as it seems chiefly evolved under conditions, such as violent eruptions, unfavourable to examination. Flame observed by Professor Pilla to be emitted from the crater of Vesuvius, in June, 1833, was of a violet-red colour, and the gas producing it inflamed only when in contact with the air.§ As Dr. Daubeny observes, hydrogen

* "Description of Active and Extinct Volcanos," 2nd edit., p. 607; 1848. There would appear to be a constant emission of steam from Tongariro, New Zealand, a volcanic mountain rising to about 6,200 feet above the sea. From time to time hot water and mud are ejected, and pour down the mountain side, "coupled with ejections of steam and black smoke, with a noise like that of a steam-engine, but no lava or scoriæ."—*Ib.* p. 429.

† Daubeny, "Volcanos," 2nd edit., p. 608. As regards the discharge of sulphurous acid and sulphuretted hydrogen, the one more in some places than in others, Dr. Daubeny remarks, that the presence of the one does not prove the entire absence of the other, since these two gases when they meet decompose each other, forming water and depositing sulphur; and "that merely the portion of either which exceeds the quantity necessary for their mutual decomposition will escape from the orifice; so that the gas which actually appears indicates only the predominance of the one, and not the entire absence of the other."

‡ Daubeny, "Volcanos," 2nd edit., p. 612.

§ Edinburgh New Philosophical Journal, 1843. From observations made by him in the crater of Vesuvius in 1833 and 1834, Professor Pilla concluded that flames never appear at Vesuvius but when the volcanic action is energetic, and is accompanied and its compounds not inflaming when steam or hydrochloric acid are mingled with them in certain proportions, and both these being abundantly evolved in most eruptions, an inflammable gas might escape into the air thus mixed, without being inflamed. Hence, though this gas may be often present during violent eruptions, it may not always be so under conditions for supporting flame.

As regards the sublimations from volcanos, we should anticipate that they would be varied, seeing that the conditions under which volcanic forces and products may find vent could scarcely but be variable also. Among the most common is chloride of sodium, or common salt, one which is important from being found connected with volcanic action in such different parts of the earth's surface. Specular iron ore is often found sublimed in chinks and cavities, as is also muriate of ammonia in certain volcanos. Respecting sulphur, it has been inferred to be derived "either from the mutual decomposition of sulphurous and sulphuretted hydrogen gases, or from the catalytic action exerted upon the latter gas by porous bodies, assisted by a certain temperature."* The sublimations of the sulphurets of iron and copper, chloride of iron, oxide of copper, muriate and sulphate of potass, selenium, and others, though apparently accidental, have been shown by M. Élie de Beaumont† to have an important bearing upon the filling of mineral veins, as will be hereafter stated.

The ashes, cinders, and molten rock ejected, may often be considered as little else than modifications of the same substance, at one time kept in a state of fusion, vapours and gases piercing through it, at another driven off by these vapours and gases in portions of different volume, more or less impregnated with them,

---

with a development of gaseous substances in a state of great tension; that they do not appear when the action is feeble; that their appearance always accompanies explosions from the principal mouth, where, however, they cannot be observed except under favourable circumstances; that they likewise show themselves in the small cones in action, which are formed in the interior of the crater, or at the foot of the volcano; and that, finally, they are not visible except in the openings which are directly in communication with the volcanic fire, and never on the moving lavas, which are at a distance from their sources.

* Daubeny, "Active and Extinct Volcanos," 2nd edit., p. 615. After quoting M. Dumas (Annales de Chimie, Dec. 1846), as having shown that "where sulphuretted hydrogen, at a temperature above 100° Fahrenheit, and still better when near 190°, comes into contact with certain porous bodies, a catalytic action, as it is called, is set up, by which water, sulphuric acid, and sulphur are produced," Dr. Daubeny points out that the vast deposits of sulphur, associated with the sulphates of lime and strontia of Western Sicily, may have been thus produced.

† "Sur les Emanations Volcaniques et Métallifères."—Bull. de la Soc. Géol. de France, 2nd série, t. iv., p. 1249.

so as to be rendered cellular; these portions finally so triturated and worn into fine grains and powder, that while part may fall with the cinders in a conical form around the volcanic vent, another portion may be so light as to be borne great distances by the winds, as from St. Vincent's, in the West Indies, above the trade-winds, far eastward over the Atlantic.

The rock in fusion, while occasionally, but somewhat rarely, uplifted in a volcanic vent to, or so near the lip of the crater as to flow over the outside in a viscous stream, more frequently breaks through different portions of the side; a result which would be anticipated from the pressure of a substance of the kind, and from rents formed in the sides of a volcano during violent eruptions. After ejection its solidification will necessarily depend upon the conditions to which it is exposed, the volume of the molten mass thrown out being duly regarded. Like all other mineral bodies of the like kind, if rapidly cooled, lavas form glasses, commonly known as obsidians, when associated with volcanic products; if slowly cooled, and in sufficient volume, crystallizing, as is easily illustrated by experiment.* The heat of lava currents would appear to vary, a circumstance to be expected, as whatever may have been the temperature of the molten mass when in the volcano, that of its exclusion would depend upon the cooling influences to which it may have been exposed before it flowed outside the volcano, and could be examined. It has been inferred that

* So far back as 1804, the experiments of Mr. Gregory Watt ("Observations on Basalt, and on the Transition from the Vitreous to the Stony Texture which occurs in the Refrigeration of Melted Basalt," Phil. Trans., 1804), proved, as respects basalt (that of Rowley Hill near Dudley), when a mass of it weighing seven hundred-weight was melted and slowly cooled in an irregular figure, that according to the rate of cooling of the various parts, was the structure, one passing from the vitreous to the stony. The silicates forming common glass may, as is well known, by slow cooling, be made to pass into a stony state. We have made many hundreds of experiments upon the melting and recrystallization of igneous rocks, even succeeding in the reduction of certain granites into a glass, and again rendering this glass stony. The varied chemical composition of the substances which may be reduced to the vitreous state is quite sufficient to show that obsidian is a mere rock-glass which can be formed under the requisite condition of comparatively rapid cooling from very different compounds. We have reduced portions of some stratified rocks to this state. This is by no means difficult to accomplish when a moderate amount of lime is present, so that silicate of lime may be produced and act as a flux, as in the ordinary smelting of the argillaceous iron ores of the coal measures. By a little management, slates and shales, with the requisite dissemination of carbonate of lime, may be converted into excellent pumice, intumescence being produced in the melted viscous substance by the carbonic acid. The experiment requires, however, to be carefully watched; for if the crucible be not removed in time, the carbonic acid escapes, and the vitreous substance alone remains, which may readily, if thought desirable, be, by slow cooling, rendered stony. To produce crystallization by very slow cooling requires great care, and, for the most part, a somewhat large portion of rock.

the temperature at which lava will continue to flow is sufficient to melt silver, lead being rendered fluid in about four minutes. Whatever the requisite heat may be,* lava is found to retain it for a long series of years.

Being a bad conductor of heat, as rocks in general are, lava, when subjected to the comparatively low temperature, to which it is exposed after ejection, soon covers itself with a coating of solidified matter. This is necessarily broken as the flow of the viscous mass continues beneath it, and it will be more or less scoriaceous, according, as in cooling, it retains any cellular texture from the passage or dissemination of vapours and gases through it. Hence the surface of lava currents is often broken and rugged, as is represented in the accompanying view of one at Vesuvius (fig. 117).† Under the conditions usually obtaining during the

Fig. 117.

flow of lava, the viscous current, at a moderate distance from the place of its actual discharge, may be considered as moving in a kind of pipe, this breaking from time to time, as the molten rock in the interior tends to drag the parts becoming solid with it. In this manner the pipe will even convey the lava current up rising ground, should the resistance of its sides be equal to the

* In our experiments, ordinary greenstone, when *pounded* fine, and placed in a crucible, usually melted at about the heat required for melting copper: experiments, however, on so small a scale may be very deceptive.

† Taken from Abich's Views of Vesuvius and Etna.

pressure exerted upon them. As a high angle of descent would be unfavourable to the proper slow cooling and quiet adjustment of particles needed for crystallization, MM. Dufrénoy and Élie de Beaumont consider that beyond a moderate angle lava does not take a crystalline texture. That the external character of a lava current should conform to the velocity of its flow, this depending, other conditions being equal, upon the amount of slope, would be anticipated. The observer should, however, be aware that when crystalline minerals may be found in lava, it does not always follow that their particles have separated out from the other component parts of the mass, after the whole has been in a molten state. They seem to have been sometimes formed prior to the outflow of the lava. Of this a good example is stated to have occurred at Vesuvius in April, 1822, when fine crystals of leucite were included in a lava stream which issued from the base of a small cone occupying the crater, the comparative infusibility of the leucite crystals preserving them entire amid the melted rock.* In like manner should the lava be in part composed of a remelted rock containing disseminated minerals, which resisted the heat to which the whole was exposed, such minerals might upon an outflow accompany the lava stream, and be again dispersed amid the new mass, otherwise, perhaps, not crystalline.†

It would scarcely be expected that a molten mass, known to be driven about in a crater by vapours and gases, could either overflow the lip of that crater, or burst out from the sides of a volcano without having some portions of these vapours and gases intermingled with it, ready to escape into the air. This it would accomplish the easier as the lava was the more fluid; and its temperature high, the vapours and gases then striving most to increase their volume. In proportion as the molten rock cooled, and the expansive power of the vapours and gases decreased, cavities would remain, corresponding in size to the equalization of the resistance of the cooling rock on the one hand, and the expansive power of the vapours and gases on the other. As these conditions varied so would the results; and thus according to circumstances the hollows formed

---

* Daubeny, quoting Professor Scacchi, of Naples, "Volcanos," 2nd edit., p. 230.

† In regions where volcanos traverse igneous rocks of an older date, remelting portions of them, it is easy to conceive occurrences of this kind. Should a felspathic porphyry, containing crystals of quartz, or mica, be thus remelted, and the heat be only capable of fusing the felspathic matter, these minerals may be left untouched. In experiments made for this purpose, we have often found this view borne out, and the quartz disseminated through many slags, as, for example, in many of the first copper slags in the furnaces at Swansea, affords another example of the like kind.

would differ from good-sized caves, lined with picturesque stalactites of lava, to small vesicles. Vapours and gases sometimes continue to escape for a long time through the chinks and cracks of cooling lava.

The cavities thus produced in lavas will necessarily take different shapes, according to varying conditions. Lava poured out so as to form a broad and comparatively deep mass, with little movement of importance after its outflow from a volcano, the fluid state long preserved, would have its cavities, large and small, placed under different circumstances, from a stream cooling more rapidly, yet still, from moving on greater slopes, continuing steadily to advance for a long distance. In the latter case the hollows and vesicles would be elongated in the direction of the flow, spherical cavities pulled out into almond-shaped forms, and irregular hollows still exhibiting a stretching in the line of movement. This elongation of vesicles may be so continued that, as in the subjoined section (fig. 118), they may become completely

Fig. 118.

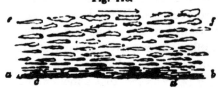

flattened, the tenacity of the lava being of a proper kind. If *c d* be a surface on which a lava stream moves, and *e f* a portion where its viscosity is such that by moving in the direction, *e f*, spherical hollows take almond-shaped forms, the lava becoming more tenacious towards the surface, *c d*, these almond-shaped vesicles would become flatter at *a b*, so as finally to present, in section, mere streaks or lines, giving a laminated appearance to that portion of a lava current when cooled. Upon the solidification of the portion *a b*, the movement continuing, and the upper part gradually taking the tenacity previously possessed by *a b*, the like appearance of lamination might happen there. Thus, as the upper part of a sheet of lava may, as regards loss of fluidity, and the friction of the viscous upon the solid outside portion, be also placed in a somewhat similar condition as the lower part, a laminated character may more or less be given to a considerable portion of a stream of lava. The conditions needed, no doubt, require nice adjustment, but they are such as would appear occasionally to prevail.

Mr. Darwin, describing the laminated obsidian beds of the Island of Ascension, and comparing them with the zoned and laminated character of obsidians and different volcanic rocks of other localities, mentions, with another cause of lamination, the stretching and flattening of vesicles by the flow of those rocks in a pasty state.* It has also been noticed by Humboldt, and other geologists, and is often to be seen in cabinet specimens.

Sometimes vapours and gases escape through molten lava, either for the time occupying portions of craters, or flowing as streams, producing the most fantastic forms. The annexed sketch (fig. 119) represents a somewhat regular accumulation of lava from this cause. It was seen by Mr. Dana, in the crater of Kilauea, in Hawaii, and rose as a whole, to the height of about 40 feet. "It had been formed over a small vent, through which the liquid rock was tossed out in dribblets and small jets. The ejected lava falling around, gradually raised the base; the column above was then built up from successive drops, which were tossed out, and fell back on one another; being still soft, they adhered to each other, lengthening a little at the same time while cooling.†

Fig. 119.

The following is also (fig. 120) an example of the like kind observed by Dr. Abich,‡ at Vesuvius, in 1834, scoriaceous lava being gradually built up into a hollow column by the additions of portions of pasty matter adhering to each other when thrust out of the general molten mass by a current of vapour or gas. In a similar manner, great blisters are sometimes raised, which bursting on one side, parts, sufficiently hard, remain, and any molten lava inside flowing out, singular cavities are left. Indeed, the varieties of hollows left by the consolidation of lava, and arising either from the intermixture of vapours and gases, or from the flow of the fluid rock, partially or wholly, out of inequalities in lava streams, or their tubular cases, would appear to be endless.

Much instruction may be derived from studying the eruptions from small vents, either in the craters of volcanos, when such can

* "Geological Observations on the Volcanic Islands visited during the Voyage of H.M.S. Beagle," p. 62, &c.

† "Geology of the United States' Exploring Expedition," 1838—42, p. 177. Mr. Dana mentions other similar examples, some on a miniature scale, about Mauna Loa. The figure of a man has been added to the original sketch by Mr. Dana, in order to give a general idea of the height of the volcanic projection.

‡ "Geologischer Erscheinungen beobachtet am Vesuv und Aetna," Berlin, 1837. The height of the excrescence represented is only eight feet.

be approached, or on their sides, where they are also sometimes found, not only as respects vapours and gases, but also the discharge and mode of accumulation of fluid and viscous lava, cinders, and

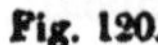

Fig. 120.

ashes. In some vents the molten rock is not much intermingled with the vapours and gases, at others it becomes frothy by intimate admixture with them; the mineral matter occupying much the less portion of the compound. Occasionally the uplifting of the mass merely raises the lava, so that it falls over the accumulations around the vent, not uncommonly more or less conical; at other times portions of the molten mass are suddenly caught and whirled high up into the air, acquiring a spheroidal form by their motion.*

* Respecting these *volcanic bombs*, as they have been termed, Mr. Darwin, remarking on those found in the Island of Ascension (Volcanic Islands, p. 36), which exhibit a cellular interior, inside a shell of compact lava, observes, that "if we suppose a mass of viscid, scoriaceous matter, to be projected with a rapid, rotatory motion through the air, whilst the external crust, from cooling, became solidified, the centrifugal force, by relieving the pressure in the interior parts of the bomb, would allow the heated vessels to expand their cells; but these being driven by the same force

When only ejected short distances, they fall around, squashing into irregular and rough discs, and by their multiplication forming a coating, which may, or may not, be intermingled with scoriaceous cinders, now and then discharged in showers. Small lava streams sometimes burst from these conical accumulations, and the resistances of the sides being overcome, and cracks being formed, the molten matter may be seen to rise in them. Many of the effects of volcanic action on the large scale may thus, in miniature, be conveniently observed.

Although conical accumulations round an aperture mark the effects of volcanic action from it, driving out ashes and cinders, large and small, with patches of frothy molten rock, and streams of fluid and viscous lava, more or less radiating from a vent, the whole braced together by more or less vertical bands of lava, which have entered cracks, effected from time to time in the general mass; this is not necessarily the case with all, nor with all parts of a mountain of which one or more of these conical accumulations may form a part. As respects cones, the accompanying* view of

Fig. 121.

against the already-hardened crust, would become, the nearer they were to this part, smaller and smaller, and less expanded, until they became packed into a solid concentric shell."

* Taken from the Voyage de Humboldt et Bonpland, Atlas Pittoresque, Pl. X, Paris, 1810. Explosions from Cotopaxi are heard at great distances. In 1744, the bellowings from the mountain were heard at Honda, 200 common leagues distant. Humboldt and Bonpland heard them day and night at Guayaquil, 52 leagues distant in a straight line. They were like repeated discharges of a battery. During the

Fig. 122.

Mauna Kea.

Mauna Loa.

Cotopaxi will illustrate the production of one of great size, its beautifully regular shape* showing how well adjusted the volcanic forces, and the substances acted on, must have been for its formation. As to its volume, that of course affords no measure of the time which the cone may have taken for its production, but it shows the great mass of volcanic matter which seems thus heaped by successive coatings into this shape.† The manner in which such a mountain may be braced together by lava currents and dykes we know not. Certainly the general form would lead us to infer a great amount of ashes and cinders, including among the latter large ejected masses of viscous, scoriaceous, and frothy (pumicy) lava, forced through a vent keeping in one place during the accumulations.

Mauna Loa, and Mauna Kea, in Hawaii, also lofty volcanic mountains, the former considered to be 13,760 feet, and the latter 13,950 feet above the sea,‡ appear to afford much modification in structure from that found at Cotopaxi, one also marked by their outlines, as shown by the accompanying sketch of Hawaii (fig. 122), taken from the eastward.§ Maps and descriptions show that Hawaii (which, as Mr. Dana remarks, is one of a group about 400 miles in length, ranging from N.W. to S.E., the islands composing it being merely the higher points of mountains rising above the sea) is a mass of volcanic matter, with three principal elevations, Loa, Kea, and Hualalai.‖ The

eruption of April, 1768, the quantity of cinders vomited from the crater was so great, that in the towns of Hambato and Tacunda night was prolonged to three o'clock on the 5th, and the inhabitants were obliged to go about with lanthorns. The eruption of 1803 was preceded by the sudden melting of the snow on the volcano. For 20 years previously neither vapour nor smoke had issued from it. In a single night, the cone became so much heated, that, the snow being melted, it appeared black from the scoriæ alone.

* Humboldt (Kosmos) points to the form of Cotopaxi as at once the most regular and most picturesque of any volcanic cone which he had ever seen.

† According to Humboldt (Kosmos), Cotopaxi rises to the height of 19,070 (English) feet above the sea.

‡ According to Dana, "Geology of the United States' Exploring Expedition," 1838—42.

§ Ibid., p. 159.

‖ Mr. Dana remarks ("Geology U.S Exploring Expedition," p. 158), "Besides the three lofty summits there are great numbers of craters in all conditions scattered over the slopes, some overgrown with forests, while about others streams of lava, now hard and black, may be traced along their route for miles. Areas, hundreds of square remarkable crater in activity is that of Kilauea, on the flank of Loa, and distant about 20 miles * from its summit.

Ellis† described the crater as situated on a lofty elevated plain bounded by precipices, apparently sunk from 200 to 400 feet below its original level. "The surface of this plain was uneven, and strewed over with loose stones and volcanic rock, and, in the centre was the great crater." * * * "Immediately before us yawned an immense gulf, in the form of a crescent, about two miles in length, from N.E. to S.W., nearly a mile in width and apparently 800 feet deep. The bottom was covered with lava, and the S.W. and northern parts of it were one vast flood of burning matter in a state of terrific ebullition, rolling to and fro its 'fiery surge' and flaming billows. Fifty-one conical islands, of varied form and size, containing so many craters, rose either round the edge, or from the surface of the burning lake; 22 constantly emitted columns of grey smoke, or pyramids of brilliant flame; and several of these at the same time vomited from their ignited mouths streams of lava, which rolled in blazing torrents down their black indented sides into the boiling mass below." * * * "The sides of the gulf before us, though composed of different strata of ancient lava, were perpendicular for about 400 feet, and rose from a wide horizontal ledge of solid black lava of irregular breadth; but extending completely round, beneath this ledge, the sides sloped gradually towards the burning lake, which was, as nearly as we could judge, 300 or 400 feet lower. It was evident that the large crater had been recently filled with liquid lava up to this black ledge."

The descriptions of this crater given by other observers,‡ corresponds generally with that of Mr. Ellis, due allowance being made for modifications, such as might be expected in a volcanic vent of any kind. The following is an eye-sketch plan (fig. 123) of Kilauea, made during the visit of the United States' Exploring

---

miles in extent, are covered with the refrigerated lava flood, over which the twistings and contortions of the sluggish stream as it flowed onward are everywhere apparent; other parts are desolate areas of ragged scoriæ. But a few months before our visit (1840) a surface of 15 square miles had been deluged with lava, which came by an underground route from Lua Pele (Kilauea)."

* Mr. Dana gives the distance as 19·8 miles, and the height of Kilauea as 3,970 feet above the sea, quoting Mr. Douglas ("Journal of the Geographical Society," vol. iv.), as estimating it from barometrical measurements at 3,845·9—3,873·7 feet, and Strzelecki at 4,101 feet.

† "Tour in Hawaii." London, 1826.

‡ Mr. Douglas, "Journal of the Geographical Society," vol. iv.; Captain Kelly, "American Journal of Science," vol. xl.; Count Strzelecki, "New South Wales and Van Diemen's Land," and others.

Expedition, under Captain Wilkes, to Hawaii. It well exhibits the cliffs surrounding the cavity, seven miles and a half in circuit, as also the great ledge above mentioned. Combined with the following section given by Mr. Dana,* it strongly suggests the idea of

Fig. 123.

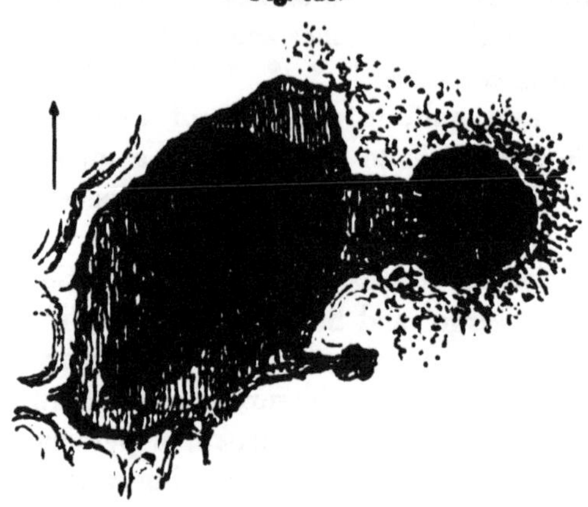

an extensive area of molten rock, rising and falling according to the uplifting force of the time, this somewhat suddenly changing, as is not unfrequent in volcanic action, so that, in some states, vertical walls would be formed from the lowering of the fused mass, while at others this might again fill the cavity, and even overflow. In the following section (fig. 124), which is taken across

Fig. 124.

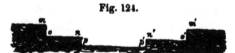

the shortest diameter (scale equal for height and distance), *m m'* is the whole breadth of the crater in that line, *o n*, *o' n'* the black ledge, *p p'* the bottom of the lower pit, *n p*, *n' p'* the walls of the lower pit, 342 feet in height, and *m o*, *m' o'* the walls above the black ledge, 650 feet in height. Mr. Dana describes the beds exposed by the cliffs, as nearly horizontal, and the crater as being, at the time of his visit (November, 1840), somewhat in a tranquil state;†

* "Geology of the United States' Exploring Expedition," p. 174.

† The descriptions given of the arrangement of the beds, and of other facts connected with this crater, have an important theoretical bearing. "These bluff sides of the pit," he observes, "consist of naked rock in successive layers; and in the distance they look like cliffs of stratified limestone. The layers vary from a few inches to 30 feet in thickness, and are very nearly horizontal. They are much fissured and broken, and some have a decidedly columnar structure. Open spaces or caverns and rugged cavities often separate the adjacent layers, adding thus to the broken character of the surface, and at the same time giving greater distinctness to the stratification. The black ledge varies in width from 1,000 to 3,000 feet. With such

one, however, still variable. During subseqent examinations by the United States' Exploring Expedition, in December, 1840, and January, 1841, the lava was observed both to rise and fall in a marked manner, independently of minor oscillations. On one occasion, Dr. Judd had but just time to escape from a sudden uprise and overflow of one of the molten pools, which discharged a mass of liquid lava, not only over the spot where he was standing immediately prior to this upburst, but also over a mile in width and a mile and a half in length. The volume of lava then ejected was afterwards estimated by Captain Wilkes at 200,000,000 cubic feet. The next morning (January 17, 1841) the molten lava of the chief lake was ascertained to have subsided about 100 feet.

Proceeding from Kilauea to Mokua-weo-weo, the crater on the summit of Loa, over a slope described as one, on the average, of only 6°, a volcanic vent is found of much the same general character as that of the former. The annexed sketch (fig. 125) is a reduction of that given by Captain Wilkes.* The deepest part of the crater is nearly circular, and about 8,000 feet in diameter. The walls are described as nearly vertical, stratified like those of Kilauea, 784 feet high on the west side, 470 on the east.† An eruption of this crater occurred in 1843, so that Mauna Loa may be considered as still active. It has been remarked by Mr. Dana, as an interesting

---

dimensions, it is no unimportant feature in the crater. The lower pit is surrounded by vertical walls, which have the same distinctly-stratified character as those above, and are similar in other features. More numerous fissures intersect them, indicative of the unstable basis on which they rest." * * * The south-west extremity formed a partly-isolated basin, of an oval form, and contained a large boiling lake. "The rest of the bottom of the pit, at the time visited by the author, was a field of hardened lava, excepting two small boiling pools, one on the western side, the other near the eastern," p. 174.

Describing the day scene, Mr. Dana states, that the "incessant motion in the blood-red pools was like that of a cauldron in constant ebullition. The lava in each boiled with such activity, as to cause a rapid play of jets over its surface. One pool, the largest of the three then in action, was afterwards ascertained by survey to measure 1,500 feet in one diameter, and 1,000 in the other, and this whole area was boiling, as seemed from above, with nearly the mobility of water." * * * "On descending afterwards to the black ledge, at the verge of the lower pit, a half-smothered gurgling sound was all that could be heard from the pools of lava. Occasionally there was a report like that of musketry, which died away, and left the same murmuring sound, the stifled mutterings of a boiling fluid," p. 171.

"The dense white vapours rose gracefully from many parts of the black lava plain, and the pools boiled and boiled on without any unnecessary agitation. The jets playing over the boiling surface darted to a height of 10 or 12 yards, and fell again into the pools, or upon its sides. At times, the ebullition was more active, the cauldrons boiled over, and glowing streams flowed away to distant parts of the crater: and then they settled down again, and boiled as before, with the usual grum murmur. Thus simple and quiet was the action of Loa Pele," p. 176.

* "Narrative of the United States' Exploring Expedition," vol. vi.

† Dana, "Geology of Exploring Expedition," p. 206.

fact, that this outbreak did not affect Kilauea, though a great lateral crater on the same volcanic dome, 10,000 feet lower down its side. The mass of lava seems to have burst out of cracks around the summit of the mountain, and in one instance a subterranean channel for a portion of the molten rock was observed.*

Fig. 125.

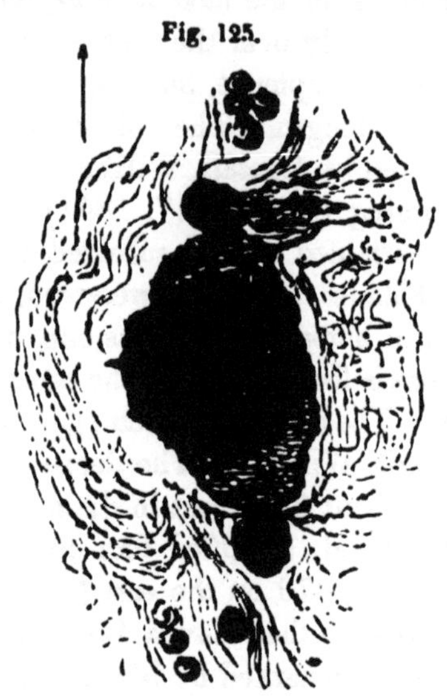

* Mr. Dana quotes from the "Missionary Herald," vol. xxxix, p. 381, and vol xl., p. 44, an account of this eruption, in order to render it more publicly known to scientific persons. For the like reason we insert a portion of it here, not only as it is in itself geologically important, but also as it is stated that only a somewhat limited number of Mr. Dana's valuable work has been printed. The Rev. T. Coan states that, "On the morning of January 10, 1843, before day, we discovered a small beacon-fire near the summit of Mauna Loa. This was soon found to be a new eruption on the north-eastern slope of the mountain, at an elevation of near 13,000 feet. Subsequently, the lava appeared to burst out at different points lower down the mountain, from whence it flowed off in the direction of Mauna Kea, filling the valley between the mountains with a sea of fire. Here the stream divided, one part flowing towards Waimea, northward, and the other eastward towards Hilo. Still another great stream flowed along the base of Mauna Loa to Hualalai in Kona. For about four weeks this scene continued without much abatement. At the present time, after six weeks, the action is much diminished, though it is still somewhat vehement at one or two points along the line of eruption." Mr. Coan ascended the mountain, passing fields of scoriaceous and smoother lavas, and regions at times still steaming and hot. "Soon," he continues, "we came to an opening in the superincumbent stratum, of 20 yards long and 10 wide, through which we looked, and at the depth of 50 feet we saw a vast tunnel, or subterranean canal, lined with smooth vitrified matter, and forming the channel of a river of fire, which swept down the steep side of the mountain with amazing velocity. As we passed up the mountain we found several similar openings into this canal, through which we cast stones; these, instead of sinking into the viscid mass, were borne along instantly out of our sight. Mounds, ridges, and cones were also thrown up along the line of the lava stream, from the latter of which, steam, gases, and hot stones were ejected. At three o'clock we reached the verge of the great crater, where the eruption first took place, near the

From all accounts respecting the mineral volcanic products in Hawaii, the ejection of cinders and ashes would appear to be comparatively rare. They are, however, occasionally thrown out, though in quantities relatively too small to produce much influence in the arrangements of the other and common volcanic products which have accumulated in a molten state. It is stated that, during an eruption of Kilauea in 1789, ashes and cinders were abundantly ejected, darkening the air, and destroying some men forming part of an army then on its march; and Mr. Dana mentions that near Kilauea, "a few miles south and south-east, great quantities of a pumice-like scoria, with stones and sand, are believed to have been thrown out at this time."*

An uplifting of liquid lava in the craters, and a rending of the solid rocks around them, or further down on the flanks of the great volcanic mounds, through which the molten rock is discharged, would appear the characteristic action of the Hawaian volcanos. Mr. Dana has well stated this mode of eruption, which he terms *the quiet mode*. Alluding to Kilauea, he remarks, "The boiling pools of the lower pit have gradually filled this (the lower) part of the crater with their overflowings, each stream cooling, and then, in a few hours or days, followed by another and another overflow in different parts of the vast area, till the rising bottom became as

---

highest point of the mountain. Here we found two immense craters close to each other, of vast depth, and in terrific action."

To queries transmitted by Mr. Dana, Mr. Coan replied as follows:—"The angle of descent down which the lava flowed from the summit to the northern base of Mauna Loa is 6°; but there are many places on the side of the mountain where the inclination is 10°, 15°, or 25°, and even down these local declivities of half a mile to two miles in extent, the lava flowed in a continuous stream. This was the fact, not only during the flow of several weeks on the surface, but also in that wonderful flow in the subterranean duct, described in the "Missionary Herald." There was no insurmountable barrier in the way of the flow from the summit of Mauna Loa to the base of Mauna Kea, a distance of 25 or 30 miles. The stream sometimes struck mounds or hillocks, which changed its course for a little space, or around which it flowed in two channels, reuniting on the lower side of the obstacle, and thus surrounding and leaving it an island in the fiery stream. Ravines, caves, valleys, and depressions were filled up by the lava as it passed down the slope of the mountain, and between the two mountains. In conclusion, I may remark that the stream was continuous for more than 25 miles, with an average breadth of 1½ mile, and flowed down a declivity, varying from 1° to 25°."—Dana, "Geology of the United States' Exploring Expedition," p. 210.

* Mr. Dana ("Geology of the United States' Exploring Expedition," p. 181) quotes from a "History of the Sandwich Islands," published by the Rev. J. Dibble, at Lahainaluna (Island of Maui), in 1843, an account from the lips of those who were in the body of men thus partly destroyed by the eruption. A large volume of cinders and sand is noticed as thrown to a great height, and as falling in a destructive shower for many miles around. Some of the men appeared to have been killed by this shower of cinders and ashes, and others to have perished from an emanation of heated vapour or gas.

high as the black ledge." * * * "The black ledge is finally flooded, and the accumulation reaches the maximum which the sides of the mountain can bear." The pressure increases, and passages are broken out for the molten rock. "In some cases, on the side of the island where the escape takes place, the first indication of the eruption is the approach of the flowing lava. We should not imply that the land is proof against earthquakes, for slight shocks not unfrequently happen, and they have been of considerable force during an eruption. But earthquakes are no necessary attendants on an outbreak at Kilauea. It is a simple bursting or rupture of the mountain from pressure, and the disruptive force of vapours, in consequence of which the mountain, thus tapped, discharges itself." *

The mode of fissuring seems to have been well observed in the eruption from Kilauea in June, 1840. The fissures are noticed as at first small. Those through which the molten lava poured formed series at intervals. Through the last twelve miles there were several rents, two or three in some places running nearly parallel. The mass of lava derived from these several fissures reached the sea, on coming into contact with which it became shivered like melted glass cast into water. Into the sea it continued to flow for three weeks, and the waters were so much heated that the shores were strewed with dead fish for the distance of twenty miles. The depth of the lava is considered to have averaged 10 or 12 feet, though in some places only 6 feet thick. The area covered by it was estimated at about fifteen square miles. The lower pit of Kilauea, calculated to have held 15,400,000,000 cubic feet of molten matter, was emptied by the outflow of the lava through the fissures.† The settling down of the lava in Kilauea would appear always to accompany these eruptions.‡

The lavas of Hawaii seem to have been usually very fluid, judging from the mode in which they occur. That they are so now in Kilauea seems generally admitted. The production of the capillary volcanic glass, known at Hawaii as *Pele's Hair,*§ is an interesting example of this fluidity. Mr. Dana, who witnessed its formation

* "Geology of the United States' Exploring Expedition," p. 195.

† Mr. Dana estimates that this gives the best measure of the amount of lava poured out during this eruption. As measured by the amount of matter observed on the surface, a much less quantity was erupted, estimated in this way at 5,000,000,000 cubic feet.

‡ A very considerable eruption from the mountain, of which Kilauea forms a part, is recorded to have recently taken place, during which a large volume of lava was ejected.

§ Pele is the reputed principal goddess of the volcano.

at one of the pools of melted lava, states that "it covered thickly the surface to leeward, and lay like mown grass, its threads being parallel, and pointing away from the pool. On watching the operation a moment it was apparent that it proceeded from the jets of liquid lava thrown up by the process of boiling. The currents of air, blowing across these jets, bore off small points, and drew out a glassy fibre, such as is produced in the common mode of working glass. The delicate fibre floated on till the heavier end brought it down, and then the wind carried over the lighter capillary extremity. Each fibre was usually ballasted with the small knob which was borne off from the lava-jet by the winds."*

In the flow of the lava outburst from Kilauea in June, 1840, the molten rock, as it passed amid forests, not only enclosed the stems of individual trees, leaving cylindrical holes from the total or partial destruction of the wood, but sometimes also adhered to the branches, descending from them in the form of stalactites. In these latter cases the heat is described as having been barely sufficient to scorch the bark, though the branches were clasped by the molten rock. Should the branch have contained much fluid matter, we can suppose that, the heat and fluidity of the lava being great, aqueous vapour from the bark may have prevented actual contact with it for a time sufficient for the passage of the lava stream at a height at which the branches could be entangled in it.† The lava descending suddenly from this height, by the lowering of the general level of the fluid stream as it passed onwards, the sudden exposure to the atmosphere would preserve, by quickly

---

* "Geology of the United States' Exploring Expedition," p. 179.

† The summary given by Mr. Dana of the effects of this flow of lava amid the forest ground, is highly interesting in many respects. "The islets of forest trees," he states, "in the midst of the stream were from one to fifty acres in extent, and the trees still stood, and were sometimes living. Captain Wilkes describes a copse of bamboo which the lava had divided and surrounded; yet many of the stems were alive, and a part of the foliage remained uninjured. ("Narrative of Exploring Expedition," vol. iv., p. 184). Near the lower part of the flood the forests were destroyed for a breadth of half a mile on either side, and were loaded with the volcanic sand; but in the upper parts Dr. Pickering found the line of the dead trees only 20 feet wide. The lava sometimes flowed around the stumps of trees, and as the tree was gradually consumed, it left a deep cylindrical hole, sometimes 2 feet in diameter, either empty or filled with charcoal. (Mr. Dana refers here to similar facts observed by M. Bory de St. Vincent at the Isle Bourbon, "Voyage aux Isles d'Afrique," 1804). Towards the margin of the stream these stump holes were innumerable, and in many instances the fallen top lay near by, dead, but not burnt. Dr. Pickering also states that some epiphytic plants upon these fallen trees had begun again to sprout." The fact is then mentioned of the lava depending in stalactitic forms from the branches of the trees, "and although so fluid when thrown off from the stream as to clasp the branch, the heat had barely scorched the bark."—"Geology of the United States' Exploring Expedition," p. 191.

cooling, the adhering and depending portions, so that little heat acted on the branches.* The case would be different where the heated lava continued to surround the lower parts of the trees. Whatever may have been the moisture preventing immediate contact, as the cooling proceeded, a time would come for the scorching, if not actual burning of the included stems.

---

* The results attendant on plunging a highly-heated body into a liquid have been long known at the manufactories of crown-glass. These have of late attracted much attention, especially from the experiments and reasoning of M. Boutigny, the vapour or steam preventing actual contact in the first instance, so that the plunged body does not acquire the temperature that might at first be expected. In crown-glass works it has been, from time immemorial, the practice to plunge the melted and very highly-heated glass in some of the first processes, after removal from the melting-pot, into cold water, to reduce the temperature. This does not fracture the glass, the steam produced preventing contact between the highly-heated glass and the water. In after processes with the same piece of glass, a mere drop of water is employed to sever a large attached glass stem, the heat being now so reduced that contact, with its consequences, is immediate. It is not a little interesting, at great crown-glass works, to see both effects, frequently produced at the same time, and within the distance of a few feet.

CHAPTER XVIII.

VOLCANOS AND THEIR PRODUCTS CONTINUED.—VESUVIUS.—ETNA.—VOLCANOS OF ICELAND.—STROMBOLI IN CONSTANT ACTIVITY.—INTERVALS OF LONG REPOSE.—SUDDEN FORMATION OF JORULLO—OF MONTE NUOVO.—FALLING IN OF PAPANDAYANG, JAVA.—SUBTERRANEAN LAKES WITH FISH.—DISCHARGE OF ACID WATERS.—INUNDATIONS FROM VOLCANOS.—CHEMICAL CHARACTER OF VOLCANIC PRODUCTS.—TRACHYTE AND DOLERITE.—COMPOSITION OF FELSPATHIC MINERALS—OF LAVAS—OF OBSIDIANS—OF OLIVINE AND LEUCITE.—DIFFUSION OF MINERALS IN VOLCANIC ROCKS.—FUSIBILITY OF MINERALS IN VOLCANIC ROOKS.—SINKING OF MINERALS IN FUSED LAVA.—FUSION OF ROCKS BROKEN OFF IN VOLCANIC VENTS.

COTOPAXI and the volcanos of Hawaii, though useful in showing how modified the results of volcanic action may be, and pointing to differences in that action of no slight importance to the observer, seeking for facts to guide him to the knowledge of its probable cause, have yet been so recently known to us, that, when studying the changes which may have taken place in volcanic vents, he must look to volcanic lands of which there may be records extending back a few centuries, at least, for the requisite data. Fortunately for this inquiry the volcanos of Italy have engaged attention for many centuries. Vesuvius offers an excellent instance of a volcanic vent which, after remaining long dormant, somewhat suddenly became active, nearly 1800 years ago, and has more or less continued, with intervals of various length, in that state ever since. After a repose, not known to have been interrupted during a long period,* suddenly, on the 24th August, 79, after earthquakes of several days' duration, cinders and ashes were furiously driven out, partly, no doubt, a portion of the old volcanic accumulations. Their

* A very excellent condensation of our information respecting the ancient, intermediate, and modern states of Vesuvius, will be found in Daubeny's "Description of Active and Extinct Volcanos," 2nd edition, 1848.

abundance was so great that three cities, Stabiæ, Pompeii, and Herculaneum, were overwhelmed by them (p. 124). We may assume that lava currents were also vomited forth out of the volcano at this eruption, one, apparently, of the greatest of Vesuvius on record. There then seems to have been a state of repose, or at least of only minor movements insufficient to create attention, for 134 years, when another eruption occurred, succeeded by a similar interval of quiet for 269 years, when there was an outburst so considerable as to cover a portion of Europe with ashes.* There were then intervals between the eruptions, the accounts of which have reached us, of 40, 173, 308, 43,† 13, 89, 1, 167, 194, 131, 29, 22, 12, 3, and 1 years, bringing them down to 1698. "From that time to the present," observes Dr. Daubeny, respecting Vesuvius, "its intervals of repose have been less lasting, though its throes perhaps have diminished in violence; for the longest pause since that time was from 1737 to 1751, and no less than eighteen distinct eruptions are noticed in the course of little more than a century, several of which continued with intermission for the space of four or five years." ‡

Even supposing the earlier recorded eruptions of Vesuvius to be only approximations to the real number, some being omitted which would now not fail to be noticed, the irregularity of the intervals of considerable activity would still be so far marked as to point to inconstancy in the final conditions upon which a marked eruption depends. At the same time, also, the different intensity of the eruptions themselves leads to the same inference. Not only was the crater of Vesuvius so tranquil, prior to the great outburst of 79, as to be clothed with vegetation, that crater occupying the depression now known as the Atrio del Cavallo, (the present Monte Somma forming a portion of its ancient walls,) but also between the eruptions of 1500 and 1631, the crater of the period was covered with herbage,§ as those of earlier times may have been

* When reference is made to the depth of cinders and ashes now found covering Stabiæ, Pompeii, and Herculaneum, it is needful to recollect that a portion of them may have been accumulated during eruptions such as this, and at other subsequent times.

† A great eruption, in 1036, during which much lava was poured out, as is stated, from the crater as well as from the sides.

‡ "Description of Volcanos," p. 226.

§ "In the interval between the eruption of 1500 and 1631, the mountain put on the appearance of an extinct volcano, the interior of the crater, according to Braccini, being, in 1611, covered with shrubs and rich herbage, the plain called the Atrio del Cavallo overgrown with timber and sheltering wild animals, whilst in another part there were three pools, two of hot and one of cold water, and two of these impregnated with bitter salts."—Daubeny, "Description of Volcanos," p. 235.

between other long intervals of repose following the great eruption of 79.

Etna also becomes valuable for the length of time during which its outbursts have been noticed. According to researches respecting the earlier eruptions of this volcano,* the year 480 B.C., or thereabouts, would appear that to which any marked outburst can be traced during historic times. This would give us about 2,332 years, for, if not of all, of at least a considerable number of the chief eruptions of this volcano, the geological records of the activity of which would appear to extend far beyond this comparatively limited time. Taking the early historic notices for the value they may possess, including that of 480 B.C., there have been marked eruptions recorded between that date and the commencement of the Christian era. With the exception of a lapse of time between 396 B.C., and 140 B.C. (256 years), the outbursts noticed occurred at intervals of 53, 31,† 5, 10, 3, 66, 12, and 8 years. They thus correspond in frequency with those recorded between A.D. 1284 and the present time.‡ From A.D. 40, or thereabouts, to 1169, the eruptions from this volcano did not, apparently, receive much attention. If we assume that this lapse of time had not been one of repose more than those which preceded and followed it, Etna seems to have been a somewhat active volcano for the time above mentioned (2,332 years).

The volcanos of Iceland have also been known as more or less in activity during a long lapse of historic time. Of the known marked outbursts of Hecla there have been 23, including that of 1845, since 1004 or 1005. These have varied in intensity and in the length of the intervals of repose between them. The eruption of 1845 appears to have driven out a vast abundance of cinders and ashes, the latter carried, by the movements of the atmosphere to

* A table of the dates of the eruptions of Etna and Vesuvius, taken from Von Hoff's "Geschichte der Veranderungen der Erdoberflache," with some few additions, is given by Dr. Daubeny, in his "Description of Volcanos," 2nd edition, p. 289.

† Respecting this eruption of 396 B.C., Dr. Daubeny mentions ("Description of Volcanos," p. 283), that the stream of lava which then stopped the march of the Carthaginian army against Syracuse, is to "be seen on the eastern slope of the mountain, near Giarre, extending over a breadth of more than two miles, and having a length of 24 from the summit of the mountain to its final termination in the sea. The spot in question is called the Bosco di Aci; it contains many large trees, and has a partial coating of vegetable mould, and it is seen that this torrent covered lavas of an older date which existed on the spot."

‡ From 1284, the intervals of repose have been, in years, 45, 4, 75, 33, 1, 1, and 82. Then a continuance of small eruptions for 58 years (1566 to 1624), after which the intervals were 9, 11, 9, 15, 13, 6, 1, 5, 8, 21, 12, 12, 8, 4, 4, 3, 14, 1, 6, 5, 6, 1, 1, 2, 7, 2, 8, 12, 1, and 10 years (in December, 1842), calculated from the table in Daubeny's "Description of Volcanos," p. 289—291.

great distances.* From Kattlagiau-jökull there have been seven eruptions since the year 900. While thus these volcanos have vomited forth molten rock, cinders, and ashes at intervals for 845 and 950 years, eruptions from other vents of the same great volcanic mass of Iceland, such as Krabla, of which there were four outbursts during the last 125 years, have also taken place. Another great volcano, Skaptar-jökull, previously dormant, as far as the historic records of that land extend, suddenly became active in 1783. During this eruption, which was preceded by earthquakes over the whole of Iceland, and the ejection of volcanic matter in the adjacent sea, considerable masses of lava were thrown out, according to Sir George Mackenzie,† from three different points, about eight or nine miles from each other, on the lower flanks of the mountain, spreading in some places to the breadth of many miles. Of the 20 volcanic vents, as Dr. Daubeny has pointed out, which have ejected lava, cinders, or ashes during the 950 years since Iceland was colonized, "eleven have had but one eruption, and amongst these four only occurred within the last century; whilst of the remaining nine, Myrdalls-jökull, Skaptar-jökull, Sandfells-jökull, Skeidarar-jökull, Reykianes, Hecla, and Krabla alone would appear to be active at present; Trölladyngia having had no eruption since 1510; Orœfa-jökull none since 1362; and others having been for a nearly equal time in a state of quiescence."‡

While Vesuvius, Etna, and volcanos in Iceland thus afford information as to alternate, but irregular, intervals of repose and

* "On the 2nd of September, 1845, the day of the eruption of Hecla, a Danish vessel, near the Orkney Islands, at a distance of 115 Danish miles (about 500 English) from the volcano, was covered with ashes."—Daubeny, "Volcanos," 2nd edition, p. 307. According to Professor Forchammer (Poggendorff's "Annalen," vol. lxvi, 1845), the cinders and ashes, so abundantly discharged, were ejected from three vents on the south-west slope of Hecla, and the lava from a fourth, situated a little distance beneath them. The eruptions continued in force on the 12th of the following month (October), the lava still flowing. The eruption did not finally cease, though there were intervals of repose, until the commencement of March, 1846.

† "Travels in Iceland," 2nd edition. Noticing this eruption from Skaptar-jökull, in 1783, Sir George Mackenzie states, that in January of that year, volcanic eruptions, represented as accompanied by flame, rose through the sea, about 30 miles from Cape Reykianes, and that several islands were observed, as if upraised, a reef of rocks now existing where these appearances occurred. "The flames lasted several months, during which vast quantities of pumice and light slags were washed on shore. In the beginning of June earthquakes shook the whole of Iceland; the flames in the sea disappeared; and the dreadful eruption commenced from the Skaptar-jökull, which is nearly 200 miles distant from the spot where the marine eruption took place."

The eruption of 1783, is stated to have thrown out such an abundance of cinders and ashes that the whole island was covered by them. The ashes were wind-borne as far as Holland.

‡ Daubeny, "Description of Active and Extinct Volcanos," 2nd edition, p. 306.

activity,* the eruptions themselves, differing in intensity, the two former during the lapse of more than 2,000 years, and the latter approaching towards a period of 1,000,† Stromboli, a volcanic vent rising through the sea between Naples and Sicily, has been equally marked, for more than 2,000 years, as exhibiting the same amount of activity. "No cessation," as Dr. Daubeny remarks, "has ever been noticed in the operations of this volcano, which is described by writers antecedent to the Christian era in terms which would be well adapted to its present appearance."‡ There seems a constant boiling of molten matter in the crater, a louder explosion occurring at regular intervals with an escape of steam, and the throwing out of blocks of lava to a considerable height.§ From the smaller and lower of three apertures within the crater, "a small stream of lava, like a perennial spring, is constantly flowing."‖

Not only do ancient and modern records thus afford the needful information respecting both intermittent and continued volcanic action for 2,000 years and more, but also as regards the cessation of the same action for so long a period, that the volcanic vents so circumstanced form a kind of transition from active volcanos to those commonly termed "extinct." The last stream of lava which issued from Monte Botaro, in Ischia, is that of 1302, known as Arso. The only traces of volcanic action now existing in this island are its hot springs. Thus no eruptions of molten rock,

* Selecting Hecla from the table given by Dr. Daubeny ("Volcanos," p. 314) and taken from Garlieb ("Island rucksichlich seiner Vulcane," &c., Freiberg, 1812), with additions, it would appear that its marked eruptions, commencing with that of 1004, have occurred at intervals of 25, 75, 9, 44, 47, 18, 72, 46, 34, 16, 46, 74, 44, 29, 36, 6, 11, 57, 35, 26, 12, 6, and 73 years, the last terminating with the eruption of 1845. The intervals between the outbursts of Trölladyngia, commencing with the eruption of 1150, were 38, 171, 116, and 35 years. For 340 years (since 1510) this vent has been quiet. While Hecla has shown the most constancy in position amid the volcanic vents of Iceland, active at various intervals for the last 846 years, and while single eruptions have only been known at other points, certain vents have shown themselves active during the lapse of the same time for a few years only. Thus eruptions are recorded at Reykianes as occurring in 1222, 1223, 1226, 1237, and 1240, altogether only for 18 years, since which time they have ceased. At Krabla, also, they commenced in 1724, were repeated in 1725, 1727, 1729, and in 1730, after which none have occurred. At Skeidarar-jökull eruptions began in 1725, were repeated in 1727 and 1728, and terminated with one in 1753. The outburst of Sandfells-jökull in 1748 is recorded as continued, probably, with intervals of repose, to 1752, the eruptions being mentioned as annual for that time.

† The volcano of Eldborgarhraun, in Iceland, is inferred to have had an eruption in the year 850.

‡ "Description of Volcanos," 2nd edition, p. 247.

§ Hoffmann, Poggendorff's "Annalen," 1832.

‖ Daubeny, "Description of Volcanos," p. 247. "It flows down the mountain," Dr. Daubeny states, "in the direction of the sea, which, however, it never appears to reach, becoming solid before it arrives at that point. Some portions, however, of the congealed mass are continually detached, and roll down into the sea."

cinders, or ashes have taken place at that old volcanic vent for about five centuries and a half. It may not be improbable, from ancient writings and modern appearances, that at the promontory of Methana (formerly Methone), on the coast of Greece, volcanic forces were in activity, and had not finally ceased in the time of Strabo, though since then that volcanic vent has remained quiescent.* Dr. Daubeny infers, from Livy and Julius Obsequens, that the volcanic action observable in the Alban Hills (Central Italy) may have continued down to historic times.†

The eruption at the promontory of Methana may, it would appear, have been sudden, a considerable mound having been thrust up, or accumulated in a short time, in the manner of Jorullo,‡ which rose above the Mexican plain, in about four months, to the height of 1,600 feet, or the still more rapid production of the Monte Nuovo, near Naples, which, in about two days, attained an altitude of 440 feet, with a circumference of about a mile and a half. These sudden outbursts are important as regards the causes of volcanic action, more especially when no appearance of a previous volcanic vent, seems to have presented itself. It would appear that prior to June, 1759, the area upon which Jorullo now stands was covered by plantations of indigo and sugar, bounded by two brooks, the Cuitimba and San Pedro. In June, subterranean noises, accompanied by earthquakes, commenced, and lasted fifty to sixty days. In September, all appeared again tranquil, but on the 28th and 29th of that month the subterranean noises were repeated, and according to Humboldt, an area of three or four square miles rose up like a bladder. This uprise is considered to be marked by an elevation of 39 feet around the edges of the ground thus moved; one continued to the height of 524 feet towards the centre of the present volcanic district. The subsequent eruption was very violent, fragments of rock being ejected to great heights, cinders and ashes

* Daubeny, "Volcanos," p. 328. It would appear that at that time the volcano was sometimes so hot as to be inaccessible, and to be visible afar off at night, the sea also being heated near it. The hills of the peninsula, according to Virlet ("Expédition Scientifique de Morée," 1839), are 741 mètres (2,431 English feet) above the sea, and he infers that among the igneous rocks of different dates there found, the last volcanic action, here noticed, occurred on the western part of the peninsula, where the trachyte presents a black and scoriaceous aspect.

† He observes ("Volcanos," p. 170) that, "there are indeed some passages in ancient writers which might lead us to suppose a volcano to have existed among these mountains even at a period within the limits of authentic history, for Livy notices a shower of stones, which continued for two entire days, from Mount Albano, during the Second Punic War; and Julius Obsequens, in his work 'De Prodigiis,' remarks that in the year 640, A.U.C., the hill appeared to be on fire during the night."

‡ Daubeny, "Description of Volcanos," p. 327.

thrown out in abundance, and the light emitted being visible at considerable distances. The Cuitimba and San Pedro poured themselves into the new volcanic vent. "Thousands of small cones, from 6 to 10 feet in height, called by the natives Hornitos (ovens), issued forth from the Malpays. Each small cone is a fumerole, from which a thick vapour ascends to the height of from 22 to 32 feet. In many of them a subterranean noise is heard, which appears to announce the proximity of a fluid in ebullition." Six volcanic masses, varying from 300 to 1,600 feet in height, were thrown up from amid these cones, out of a chasm having a N.N.E. and S.S.W. direction. From the north side of the highest (Jorullo) a considerable quantity of lava was ejected, containing fragments of other rocks. The great eruptions terminated in February, 1760.

Respecting Monte Nuovo, the first indications of its production were noticed on the 28th of September, 1538, when, according to an eyewitness,* the sea-bottom near Puzzuoli became dry for 1300 yards, and the fish left upon it were carried away in waggons. At eight o'clock next morning the ground is reported to have sunk, where the volcanic orifice afterwards appeared, about 13 feet. At noon the earth began to swell up, and became as high as the Monte Rossi, and from the vent formed, fire, stones, and ashes were ejected, so that finally the hill took the form now seen. For 70 miles around the volcano the country was covered with ashes, killing birds, hares, and smaller animals, and breaking down trees. Monte Nuovo is 439 English feet high, and has a crater in its centre 420 feet deep, according to M. Dufrénoy. At the bottom there is a cavern, at the extremity of which Professor James Forbes found a spring issuing with a temperature of 182°·5.

These instances of the sudden production of volcanic vents on dry land (and when we consider the chances for observing and recording them, they were probably far more numerous within the last 1,000 years) are sufficient to show that the uprise of volcanos through the sea would be expected amid and around volcanic islands and regions. In the atmosphere they retain their forms, such as are presented at Jorullo and Monte Nuovo: raised through the level of the sea, the stability of such portions depends, as above mentioned (p. 70), upon the power of the volcanic mass to resist the action, first, of the breakers, and, secondly, of the wind-waves, where the former may have cut it down to the proper depths.

---

* Francesco del Nero. A letter of his to Nicolo del Benino of Naples, and sent to Rome in 1538, was first published in Leonhard's "Jahrbuch für Geologie," 1846, and Daubeny gives a translation of it, "Description of Volcanos," 2nd edition, p. 208.

The volcanic outbursts of this kind between Pantellaria and Sicily, off the coast of Iceland, and among the Azores, have been already noticed (pp. 70 and 100). To these may be added (as showing how much depends on the opportunity and ability to have such submarine terminating in subaërial eruptions recorded, and the range of time during which they have been known), the mud, smoke, and flame noticed by Strabo as rising through the sea amid the Lipari Islands, and the flame also rising there above its level during the Social War, as mentioned by Pliny.*

Volcanic accumulations would appear sometimes to rest upon considerable hollows, and also to have large cavities distributed among them, the portions covering or surrounding which being either unable to resist the pressure of the superincumbent weight, even in the tranquil periods of a volcano, or broken through during eruptions, the volcanic matter falls in, or water retained amid the cavities is ejected. Of the falling in of volcanic accumulations, depressions sometimes taking the place of protrusions, many instances are given; but of those which happen to have become known, the disappearance of Papandayang, a volcano of Java, in 1772, would seem to be most remarkable. Papandayang, formerly one of the largest volcanos in Java, was situated on the south-western part of that island. After a short but violent paroxysm, and about midnight, between the 11th and 12th of August, a luminous cloud enveloped the mountain. The inhabitants of the sides and foot of the volcano betook themselves to flight, "but before they could all save themselves, the whole mass began to give way, and the greatest part of it actually *fell in* and disappeared in the earth." This was accompanied by sounds like the discharge of heavy cannon, and an abundance of volcanic substances were thrown out and spread around the adjoining country. The area thus swallowed up was estimated as measuring fifteen by six miles. Forty villages are stated to have been partly swallowed up, and partly destroyed by the volcanic substances thrown out, and 2,957 inhabitants perished. Persons sent to examine the locality found the heat of the substances surrounding it, and piled up to the

* Detailing the evidence on his head, Dr. Daubeny ("Description of Volcanos," p. 253) asks if the comparatively recent origin of the Island of Lipari itself may not be inferred from its present fertility as compared with the sterility ascribed to it by Cicero. He also points to the fresh condition of the craters of this island, as observed by Hoffman, the hot springs and stufes at San Calogero, near the town of Lipari, and the statement of Strabo that this island emitted a fiercer fire than Stromboli, as perhaps showing that an active volcano may have existed in it even within the historical period.

height of three feet, so great, that they were unable to approach the spot six weeks afterwards.*

Cavities amid volcanic accumulations may not only be partially or wholly filled by water, the condensation of aqueous vapours, finding their way into them, or rain or melted snows upon the exterior of a volcano percolating to them, but the waters also may be sometimes of a temperature and kind, permitting the existence and increase of animal life. Humboldt records, that "when, in the night of the 19th June, 1698, the summit of Carguairazo (18,000 French feet in height†), fell in, leaving two immense peaks of rock as the sole remains of the wall of the crater, masses of liquid tufa, and of argillaceous mud (*lodazales*), containing dead fish, spread themselves over, and rendered sterile a space of nearly two square German miles. The putrid fevers, which seven years before prevailed in the mountain town of Ibarra, north of Quito, were attributed to the quantity of dead fish ejected in like manner from the volcano of Imbaburu."‡ The fish here noticed (*Pimelodus cyclopum*), Humboldt further informs us, "multiply by preference in the obscurity of the caverns;" possibly, also, there may be something in the temperature of the waters. He observes, that it was in consequence of these discharges of waters, pent up in volcanic cavities, that the inhabitants of the plains of Quito became acquainted with these little fish, called by them Preñadilla.

That the waters of such hollows and cavities are not always thus fitted for the existence and increase of animal life would be expected, when the observer reflects upon the varied conditions under which they are likely to occur. As an example of the effects produced by the admixture of gaseous volcanic emanations with the waters in such reservoirs, we may adduce the great flow of acid water which accompanied an eruption of the Javanese volcano of Guntur, or Gounung Guntur, in 1800, when not only streams of lava were poured out (a rare circumstance, it would appear, among the Javanese volcanos, commonly ejecting little else than cinders and ashes),§ but also an acid torrent. A river

* Dr. Horsfield, as quoted by Dr. Daubeny, "Description of Volcanos," 2nd edition, p. 406.

† 19,200 English feet.

‡ Kosmos, 7th edition (Sabine's Translation), p. 222. This fact has long since been mentioned by Humboldt in his earlier works.

§ In a letter to Dr. Daubeny ("Description of Volcanos," p. 409), Mr. Beete Jukes, alluding to the almost entire absence of hard rock on the surface of the ground in the volcanic districts of Java, infers, that the Javanese volcanos "had long ceased to erupt lava, and have for ages been burying the previous streams under piles of ashes and powder."

descending from this volcano is described as suddenly swelling, "charged with a large quantity of white, acid, sulphurous mud." On the 8th of October of that year, "the waters came pouring down into the valley, carrying everything before them, sweeping away the carcases of men and sundry animals, and covering the face of the country with a thick coat of mud." On the 12th, a still greater "deluge of mud" came down the valley. Such sudden increases of the volume of water would seem to point to its discharge from extensive cavities where it was, for the time, pent up, and where it became impregnated with sulphuric acid, derived, as Dr. Daubeny points out, from the decomposition of sulphuretted hydrogen gas.*

Without uselessly multiplying examples of the discharge of considerable volumes of water, apparently pent up in the hollows of volcanos, it may be mentioned that, in 1755, a volume of water was suddenly discharged from a cavern below the great crater of Etna, and that, dashing over the snows and side of the mountain, it destroyed and carried before it a large amount of matter. Torrents of water are stated to have issued from Vesuvius during the great eruption of 1631, but whether from caverns amid the accumulations, or as the result of the somewhat sudden condensation of large volumes of aqueous vapour discharged from the crater, is not clear. Be this as it may, the collection of waters amid volcanic accumulations, would appear the needful consequence of the existence of such cavities, and of the condensation of aqueous vapour in, or the infiltration of rain or melted snow into them. These outbursts require to be carefully distinguished from the torrents descending the sides of volcanos more or less covered by snow, either in the higher northern and southern latitudes, or rising above the line of perpetual snow in the temperate or tropical regions. The suddenly-melted snows of Cotopaxi (fig. 121) pour down the furrows on its sides, as in the eruption of 1803, when, in a single night, the snows disappeared from the cone, and the resulting torrents of water transported cinders and ashes into the Rio Napo and the Rio de los Alaques.† Humboldt refers generally to the high volcanos of the Andes as thus, by the sudden melting of their snows transporting smoking scoriæ among the lower lands, and producing great inundations.‡ Similar effects necessarily

* Daubeny ("Description of Volcanos," p. 408), quoting Boon Mesch, "Dissertatio de Incendiis Montium Javæ," 1826, who obtained his information from Reinwardt, the Dutch traveller in Java.

† "Voyage de Humboldt et Bonpland," Atlas, Art. Cotopaxi, Paris, 1810.

‡ Kosmos, 7th English edition (Sabine), p. 221.

follow similar causes in the temperate regions. Probably, however, the consequences of the sudden melting of snow and ice from volcanic action are nowhere so great as in the higher latitudes, where large glaciers, holding, or supporting mineral matter, are broken up, and partly melted and partly in fragments, are hurried onwards to the lower levels. The accounts given of the effects thus produced in Iceland show, that the torrents, so caused and intermingled with ice, are of no slight geological importance. Although, under ordinary circumstances, so little mineral matter appears capable of being moved on Victoria Land (p. 249), it is easy to conceive that, during considerable eruptions of such volcanos as those of Mount Erebus and Mount Terror, great heats may suddenly melt the snows clothing these mountains, producing large volumes of water, which may continue liquid for a time sufficient to furrow into, and carry off scoriæ and ashes, usually bound together by, and, to a certain extent, not unfrequently interstratified with, the great snow covering of those regions.

It is needful well to consider the mineralogical structure and chemical composition of the various volcanic products, whether these may be in the form of lava streams, of molten rock which has risen in, and more or less filled fissures, of scoriaceous substances of considerable bulk, or of those lighter bodies commonly known as pumice, cinders, and ash. Though much has been accomplished, more especially of late years, respecting this knowledge, the discoveries in chemistry greatly advancing such inquiries, and though some apparently sound general conclusions have, from time to time, been formed, it will be evident, before certain of these can be fully admitted, however they may be applicable to the particular localities noticed, that, looking at the distribution of volcanic vents over the surface of the globe, an observer possesses ample opportunities, by careful research in various parts of the world, of still further advancing our knowledge in this respect.

Whether the solid volcanic rocks are crystalline, stony, or vitreous, will, as we have seen (p. 325), often in a great measure depend upon the conditions as to cooling, to which they have been exposed, all other circumstances being the same.* Hence the

* It is very essential, in such investigations, to bear the other equality of conditions in mind, for there may be circumstances much modifying the external parts of lava currents. Thus M. Dufrénoy ("Mémoires pour servir à une Description Géologique de la France," t. iv.) mentions having found that two-thirds of the interior of a lava current near Naples were formed of a mineral which could be acted upon by acids, while the surface was principally composed of one not so attackable. In like manner, also, as has been remarked by Mr. Dana ("Geology of the United States'

chemical composition of volcanic rocks, which have been ejected and flowed in a molten state, may often be the same, notwithstanding the different states of mineral aspect. If the adjustment of the particles composing certain crystalline minerals has been prevented by the absence of the needful conditions, such as by a sudden refrigeration of the mass, these particles would remain diffused.

The solid volcanic products most studied and known to us have been divided into rocks named *trachyte* and *dolerite*. Minerals of the felspar family constitute essential portions of these rocks, entering more extensively into the composition of trachyte than into that of dolerite. Both rocks may be also viewed as silicates, chiefly of alumina, lime, magnesia, potash, and soda. Trachytes are indeed considered "chemically trisilicates, with or without an excess of silica."* Trachyte may, however, according to the definitions given, also contain free silica or quartz, and the separate minerals mica, hornblende, or augite. Dolerite is composed of the felspar known as labradorite and of augite, and the term *augite rock* is sometimes given to this compound. In this latter rock the proportion of silica is diminished, and that of lime and magnesia increased.† This classification of the more solid volcanic products into two main divisions, however convenient as affording facilities for investigation, is found to need such modification, that an intermediate class of rocks, termed trachyte-dolerites, has been proposed by Dr. Abich, in which the composition partakes of the mineral characteristics of both trachyte and dolerite. With respect to changes in chemical composition, Dr. Daubeny remarks, that "the gradual increase of soda is likewise a remarkable circumstance, modern lavas appearing to contain a much larger quantity of it than the volcanic products of ancient periods, and various minerals being hence produced in which this alkali is predominant (natrolite, nepheline, thomsonite, &c.)‡

Exploring Expedition," p. 203), a body of molten and very liquid lava kept long boiling or simmering in a volcanic vent, like Kilauea, in Hawaii, may have certain of its parts separable, the more especially as the temperature may increase in any column of lava in proportion to the pressure upon its parts.

* Daubeny, "Description of Volcanos," 2nd edition, p. 15.

† Respecting the diminution of silica, Dr. Daubeny observes ("Description of Volcanos," 2nd edition, p. 17), that it is "indicated by the substitution of labradorite for orthoclase, or, in other words, of one atom of silica instead of three, coupled with the presence of hornblende or augite, in both which minerals the silica bears a still smaller proportion to the base with which it is combined." Rammelsberg ("Dictionary of Mineralogy," Berlin, 1841) is quoted as pointing to augite as $\dot{R}^3$, $\dddot{S}^2$, where $\dot{R}$ is either lime, magnesia, protoxide of iron, or protoxide of manganese, the silica being sometimes also replaced by alumina, as is also the case in hornblende.

‡ "Description of Volcanos," p. 19.

It is highly needful that the observer should most carefully study the mode of occurrence of these rocks in volcanic districts, as he will readily perceive that if their somewhat general sequence be from the trachytic to the doleritic compounds, as has been supposed, an important change had been effected as to the conditions under which the earlier and later substances have been ejected from volcanos. The subject requires to be regarded on the large scale, and due weight given to those modifications arising, as will be further noticed hereafter, from the admixture of matter derived from various rocks, amid which mineral volcanic products may have had to pass before they were finally ejected.

In such examinations the chemical composition of the rock, more especially when the minerals noticed may be either ill developed, or their component parts have been unable to collect together in definite arrangements, is evidently of importance. The rock-glasses, or obsidians, may as well belong to one class as the other, and so also certain stony varieties, wherein any real development of distinct minerals has not been effected. Dr. Abich has proposed the relative specific gravity of volcanic rocks as affording great aid in ascertaining the amount of silica in them, a view in which Dr. Daubeny would appear to concur, remarking that in these rocks "the specific gravity of the mineral is inversely as the amount of silica, and directly as that of the other bases, so that a near approximation may often be obtained to their chemical composition by merely ascertaining their weight."*

When assuming chemical composition from mineral structure, and that the substances constituting the base of certain definite forms are constant, it is necessary not only to distinguish the minerals themselves, but also to give due weight to the replacement of some substances by others, without altering the form of the

* Description of Volcanos," p. 13. The following table is given in illustration:—

	Specific Gravity.	Silica per Cent.
Trachytic porphyry	2·5783	69·46
Trachyte	2·6821	65·85
Domite	2·6334	65·50
Andesite	2·7032	64·45
Trachyte-dolerite	2·7812	57·66
Dolerite	2·8613	53·09

Clinkstone, with a specific grayity of 2·5770, and containing 57·66 of silica, and glassy andesite, specific gravity 2·5851, with silica 66·55, not harmonizing with this view, it is remarked, that though clinkstone chemically resembles trachyte-dolerite, it "has a different mineral composition, for it appears to be a mixture of a zeolitic mineral with glassy felspar," and that "probably the same may apply to glassy andesite."

mineral.* The amount of matter of different kinds which may, as it were, be entangled with that which gives the form, has likewise to be regarded, the entangled matter being sometimes more considerable than might at first be supposed, compelled, as it were, by that considered essential to the mineral, to take the arrangement of parts belonging to it.†

Carefully searching for facts illustrative of the conditions under which the mineral matter ejected from volcanos may have been derived in the first place, or modified afterwards, it is essential to apply for aid to chemistry as well as mineralogy, important as the latter may be. The passage of vapours and gases through, or their entanglement in, lavas, whether solid, somewhat vesicular, or highly cellular, as pumice, is often sufficient to produce modifications requiring great attention. Again, after cooling, with cavities in them of various sizes, containing matter partly gathered together out of the mass of the containing rock, and partly from extraneous sources, lavas may not only be modified in their composition, but mineral substances may be formed in them of a different character from those which would have separated out from the original fused rock. Again, also, lavas, from exposure to atmospheric influences, may have lost some of the soluble substances originally entering into their composition. Thus no little care is required in the selection of portions of a volcanic rock which shall properly represent its original condition, as regards its chemical character.

As the felspathic minerals enter so largely into volcanic rocks, and indeed constitute a considerable part of igneous rocks, viewed gene-

---

* Before engaging in investigations of this kind, the observer should make himself acquainted with the bodies termed *isomorphous*, or those which replace each other without causing any alterations in the structure of minerals. In inquiries into the chemical composition of rocks a knowledge of these substances is highly important. Thus, for example, magnesia, lime, protoxide of iron, and protoxide of manganese, replace each other in any proportion. As M. Dufrénoy has well remarked ("Traité de Minéralogie," tom. i., p. 19), "it is not necessary, in order to present the same composition, that minerals should exactly contain the same weight of their simple constituent substances; it is sufficient that there is an exact relation between the bases and the acids they contain, or between their isomorphous substances."

† This power of one compound to compel others to take its crystalline form is of no little importance, in estimating the chemical composition of rocks. These admixtures are clearly mechanical in some instances, as, for example, in the well-known crystallized sandstone, as it is sometimes termed, of Fontainebleau, where grains of siliceous sand, in large quantity, are entangled in carbonate of lime, so crystallized as to include them without destroying its form. Artificial compounds may be made, in which large proportions of some substances may be mingled with others, the fundamental crystalline form of the former remaining uninjured; thus, for instance, M. Beudant succeeded in producing crystals of the form of sulphate of iron, which contained 85 per cent. of sulphate of zinc, the remaining 15 per cent. only being the proportion of the substance giving the form to the crystals ("Annales des Mines," 1817, t. ii., p. 10).

TABULAR VIEW of the FELSPAR FAMILY.

NAME.	LOCALITY.	Specific Gravity.	Silica.	Alumina.	Oxide of Iron.	Oxide of Manganese.	Lime.	Magnesia.	Potash.	Soda.	Summary of Ingredients.	Formulæ.
1. Anorthite . .	Monte Somma . .	2·7630	43·79	35·49	0·57	0	18·93	0·34	0·54	0·68	100·34	$\dot{R}\,\dddot{Si} + 3\,\underline{\dddot{R}}\,\dddot{Si}$.
2. Labradorite .	Etna	2·7140	53·48	26·46	1·60	0·89	9·49	1·74	0·22	4·10	98·40	$\dot{R}\,\dddot{Si} + \underline{\dddot{R}}\,\dddot{Si}$.
3. Andesin . .	Popayan (Andes) .	2·7328	59·60	24·28	1·58	0	5·77	1·00	1·08	6·53	99·92	$\dot{R}^3\,\dddot{Si}^2 + \underline{\dddot{R}}\,\dddot{Si}^2$.
4. Oligoclase . .	. .	2·6680	62·61	24·11	0·30	0	2·74	0·55	0·75	8·89	99·95	$\dot{R}\,\dddot{Si} + \underline{\dddot{R}}\,\dddot{Si}^2$.
5. } Pericline . .	Pantellaria . . .	2·6410	67·94	18·93	0·48	0	0·15	0	2·41	9·98	99·90	$\dot{R}\,\dddot{Si} + \underline{\dddot{R}}\,\dddot{Si}^3$. (Nos. 5–8)
6. }			68·23	18·30	1·01	0	1·26	0·51	2·53	7·99	99·83	
7. Potass-Albite .	Drachenfels . .	2·6223	70·22	17·29	0·82	0	2·09	0·41	3·71	5·62	100·16	
8. Albite . . .	. .	2·6140	69·36	19·26	0·43	0	0·46	0	?	10·50	100	
9. Ryacolite . .	Monte Somma . .	2·6180	50·31	29·44	0·28	0	1·07	0·23	5·92	10·56	97·81	$\dot{R}\,\dddot{Si} + \underline{\dddot{R}}\,\dddot{Si}$.
10. Glassy Felspar	Ischia	2·5970	66·73	17·56	0·81	0	1·23	1·20	8·27	4·10	90·70	$\dot{R}\,\dddot{Si} + \underline{\dddot{R}}\,\dddot{Si}^3$. (Nos. 10–13)
11. Adularia . .	St. Gothard? . .	2·5756	65·59	17·97	0	0	1·34	0	13·99	1·01	100	
12. Orthoclase . .	Baveno	2·5552	65·72	18·57	trace.	trace.	0·34	0·10	14·02	1·25	100	
13. Artificial Felspar	Sangerhausen . .	2·5600	65·08	16·84	0·88	0·36*	0·34	0·34	15·26	0·65	100	

* Also oxide of copper 0·3.

rally, the annexed table (page 355) of their chemical composition and specific gravities, by Dr. Abich,* may be found useful.

From the same author has also been compiled the following table of the chemical constituents of several trachytes and other volcanic rocks:—

—	1	2	3	4	5	6	7	8	9	10
Silica	73·46	68·35	67·09	61·19	61·03	67·07	57·76	62·20	53·88	49·21
Alumina . . .	13·65	13·92	15·63	17·18	17·21	13·19	17·56	20·80	12·04	15·76
Oxide of iron .	1·49	2·28	4·59	} 5·46	{ 4·84	4·74	6·73	4·39	9·25	11·84
——— manganese	trace.	..	..		{ 0·17	0·32	0·82	..	..	..
Lime . . .	0·45	0·85	2·25	1·52	1·43	3·69	5·46	2·70	8·83	6·97
Magnesia . .	0·39	2·20	0·97	0·23	2·07	3·46	2·76	1·40	7·96	6·01
Potash . . .	4·39	8·24	3·56	4·37	7·16	2·18	1·42	3·10	} 4·76	{ 4·37
Soda	6·38	4·29	5·07	7·98	4·64	4·90	6·82	5·20		{ 6·06

1. Porphyritic trachyte, with mica, from Ponza. 2. Porphyritic trachyte from Monte Guadia, Lipari. 3. Trachyte from the Drachenfels. 4. Lava from Monte Nuovo. 5. Lava, named Arso, Ischia. 7. Trachyte-dolerite from the Peak of Teneriffe. 8. Rocca di Giannicolo, Val del Bove, Etna. 9. Dolerite of Strombolino. 10. Lava of Vesuvius.

The annexed table has been constructed from the analyses of the lavas from Vesuvius and Monte Somma, as given by M. Dufrénoy:—

—	1	2	3	4	5
Silica	53·10	50·55	49·10	50·98	48·02
Alumina . . .	16·58	20·30	22·28	22·04	17·50
Protoxide of iron .	9·96	8·60	7·32	8·39	7·70
Lime.	3·34	5·20	3·88	5·94	0·24
Magnesia . . .	1·16	1·21	2·92	1·23	9·84
Soda	9·46	8·42	9·04	8·12	2·40
Potash	2·23	2·52	3·06	3·54	12·74
Loss	4·17	3·20	2·40	..	1·56
Increase	..	..	..	0·24	..

1. Lava of Palo. 2. Lava of 1834, taken immediately below the Piano. 3. Lava of Ganatello. 4. Lava from La Scala. 5. Monte Somma, mean of two analyses.

Comparing the composition of the lavas of Vesuvius with those of Monte Somma, M. Dufrénoy points out, that while the latter are almost unattackable by acids, those of the former are in a great measure soluble in them, in about the proportion of 4 : 1 ; and that while the lava of Monte Somma contains a large proportion of potash, in that of Vesuvius soda predominates.†

* "Ueber die Natur und den Zusammenhang der vulkanischen Bildungen," Brunswick, 1841.

† "Parallèle entre les différents products volcaniques des environs de Naples, et rapport entre leur composition et les phénomènes que les ont produit;" Mémoires pour servir à une Description Géologique de la France, t. iv., p. 381, (1838). M. Dufrénoy adds, that this difference of composition is also apparent in the minerals common to the two lavas, the augite of Monte Somma having a base of iron, while that of Vesuvius enters among the calcareous varieties, such as sahlite.

Respecting lava replete with vesicles or cells (pumice), the following analyses are taken from Dr. Abich:*

—	1	2	3	4	5	6	7	8	9
Silica.	60·79	61·08	62·42	62·29	62·04	68·11	69·79	73·77	73·70
Silica and Titanic acid.	1·46	1·45	0·74	..	..	1·23	..	..	..
Alumina.	16·43	17·34	14·72	16·89	16·55	8·21	12·31	10·83	12·27
Oxide of iron	4·26	7·77	6·84	4·15	4·43	8·23	4·66	1·80	2·31
——— manganese	0·23	0·62	0·13	trace	..	trace	..	..	..
Lime	0·62	1·46	5·25	1·24	1·31	0·14	1·68	1·21	0·65
Magnesia	0·79	4·02	3·28	0·50	0·72	0·37	0·68	1·30	0·29
Potash	11·25	2·85	4·74	6·21	6·29	8·32	6·69	4·29	4·53
Soda	2·97	1·82	1·55	3·98	3·06	1·60	2·62	3·90	4·73

1. Pumice from Teneriffe. 2. From the Island of Ferdinandea. 3. From the volcano of Arequipa, Bolivia. 4. From the Island of Ischia. 5. From the Phlegrean Fields. 6. From the Island of Pantellaria. 7. From the Island of Santorino. 8. From Llactacunga.

According to Professor B. Silliman, jun., the modern lava and volcanic glass of Kilauea, Hawaii, not only contain a considerable amount of oxide of iron, but also soda, to the exclusion of potash, all the constituent substances varying much in their relative proportions.†

* "Ueber die Natur und den Zusammenhang der vulkanischen Bildungen," Brunswick, 1841.

† Dana, "Geology of the United States' Exploring Expedition," p. 200, whence the following analyses are extracted:—

—	1	2	3	4
Silica.	39·74	51·93	50·67	59·80
Alumina	10·55	14·07	..	..
Protoxide of iron	22·29	10·91	33·62	31·33
Lime.	2·74	6·20	3·66	..
Magnesia	2·40	1·73	1·13	1·71
Soda	21·62	6·31	10·52	4·83

1. Dark-coloured Pele's Hair. 2. Scoria. 3. Compact vitreous lava. 4. Compact stony lava. 3 and 4 are from the same specimen, the former constituting the exterior portion of the latter.

Mr. Dana also gives the following analysis of Pele's Hair by Mr. Peabody, which agrees with the above as to the large proportion of protoxide of iron, but differs from it by giving potash:—

Silica	50·00
Protoxide of iron	28·72
Lime	7·40
Alumina	6·16
Potash	6·00
Soda	2·00

The following are analyses of rock-glasses, or obsidians, from different parts of the world, showing their variable composition :—

—	1	2	3	4	5	6	7
Silica	60·52	62·70	74·05	74·80	84·00	70·34	69·46
Alumina . . .	19·05	16·98	12·97	12·40	4·64	8·63	2·60
Oxide of iron . .	4·22	4·98	2·73	2·03	5·01	10·52	2·60
——— manganese	0·33	0·39	0·12	..	..	0·32	..
Lime	0·59	1·77	0·28	1·96	2·39	4·56	7·54
Magnesia	0·19	0·82	..	0·90	..	1·67	2·60
Potash	10·63	6·09	4·15	6·40	..	..	7·12
Soda	3·50	4·35	5·11	..	3·55	3·34	5·08

1. From Teneriffe (Abich). 2. Island of Procida (Abich). 3. Lipari (Abich). 4. Telkebanya (Erdmann). 5. Iceland (Thomson). 6. India (Damour). 7. Pasco (Berthier).

As olivine and leucite are minerals often entering largely into volcanic rocks, it is useful that the observer, while estimating the chemical composition of those in which they may occur, should bear in mind that the former is a silicate of magnesia and protoxide of iron $[(\dot{M}g, \dot{F}e)^3 \ddot{S}i]$, and the latter a silicate of potash and alumina $(\dot{K}^3 \ddot{S}i^2 + 3 \bar{A}l \ddot{S}i^2)$.* He should also recollect that

* The following analyses may aid in showing the similar composition of olivine from various localities. Several others might be added of the same kind :—

—	1	2	3	4	5
Silica.	40·09	40·45	41·54	41·44	40·12
Magnesia . . .	50·49	50·67	50·04	49·19	44·55
Protoxide of iron .	8·17	8·07	8·66	9·72	15·32
——— manganese	0·20*	0·18*	0·25	0·13	0·29
Alumina. . . .	0·19	0·19	0·06	0·16	0·14

* Peroxide of manganese.

1. From the Vogelsberg, Giessen (Stromeyer), contains also 0·37 protoxide of nickel. 2. Kasalthoff, Bohemia (Stromeyer), contains also 0·33 protoxide of nickel. 3. Iseweise (Walmstedt). 4. Le Puy, Vivarais (Walmstedt), contains also 0·21 of lime. 5. Monte Somma (Walmstedt).

As respects this mineral, it is highly interesting to find that the olivine found in the meteoric iron of Siberia and Otumba, South America, should possess a similar composition.

—	1	2
Silica.	40·86	38·25
Magnesia . . .	47·35	49·68
Protoxide of iron .	11·72	11·75
——— manganese	0·43	0·11

1. From Siberia (Berzelius). 2. From Otumba (Stromeyer).

With respect to leucite, the two following analyses, the first from Vesuvius, by augite is a silicate of lime and magnesia* ($\dot{C}a^3 \dddot{Si}^2 + \dot{M}g^3 \dddot{Si}^2$), labradorite, a silicate of alumina, lime, and soda† ($\dot{R} \dddot{Si} + \ddot{Al} \dddot{Si}$), orthoclase (potash-felspar), a silicate of alumina and potash‡ ($\dot{K} \dddot{Si} + \ddot{Al} \dddot{Si}^3$), albite (soda-felspar), a silicate of soda and alumina § ($\dot{N}a \dddot{Si} + \ddot{Al} \dddot{Si}^3$).

The chief substances entering into the composition of volcanic rocks are the silicates of alumina, oxide of iron, lime, magnesia, potash, and soda, the fusibility of the different compounds of which, constituting distinct minerals, varies, the rocks into which augite, or that into which silicate of lime enters most largely, being the easiest of reduction to the fluid state by heat. As labradorite likewise contains a considerable proportion of silicate of lime, and is more fusible than orthoclase (the potash-felspar), both the minerals entering into the composition of dolerite render it much more fusible than trachyte, chiefly formed of the potash-felspars. Silicate of lime may indeed be considered as a characteristic substance in the dolerites, while it is comparatively rare in the trachytes, that is, of those in which true orthoclase predominates.|| In localities, therefore, where trachytic have clearly preceded dole-

Arfvedson, and the second from Monte Somma, by Awdejew, will serve to show the proportion of the constituent parts :—

—	1	2
Silica	56·10	56·05
Alumina.	23·10	23·08
Potash	21·15	20·40
Soda	..	1·02
Peroxide of iron .	0·95	..

* In the very numerous analyses which have been made of augite, the silica varies from 47·05 (Arendal, Gillenfelder Maar Eifel) to 57·40 (Tjötten, Norway), the lime from 17·76 (Tyrol) to 25·60 (Achmatowsk), and the magnesia from 6·83 (Finland) to 18·22 (Vallée de Fassa). There is usually protoxide of iron varying from 4·31 (Tyrol) to 26·08 (Tunaberg, Sweden), as also alumina from 0·14 (Dalecarlia,) to 6·67 (Gillenfelder Maar Eifel).

† Ṙ being taken as ¾ lime and ¼ soda, the chemical composition of labradorite is considered to be = 53·7 silica, 29·7 alumina, 12·1 lime, and 4·5 soda. There are usually also small portions of potash varying from 1·79 to 0·3.

‡ The chemical composition of orthoclase is inferred to be 65·4 silica, 18 alumina, and 16·6 potash, a little soda and lime being included in the latter. Nicol, "Manual of Mineralogy," p. 119.

§ Albite is considered to be essentially composed of 69·3 silica, 19·1 alumina, 11·6 soda, part of the last often replaced by lime or potash.—Nicol, "Manual of Mineralogy," p. 124.

|| In those compounds referred to orthoclase, in which soda is more abundant than potash, it may be much doubted how far they really deserve the name, unless it be inferred, with Dr. Abich, that soda and potash are both isomorphous and dimorphous.

ritic rocks, the more fusible have succeeded the least fusible products—a fact of no little theoretical value.

With respect to the diffusion of certain minerals, such as olivine and leucite, through the mass of a volcanic rock, having once been formed, that is, the component particles of the silicates of magnesia and protoxide of iron of the one, and the silicates of potash and alumina of the other, having been placed under the conditions permitting them freely to move and become aggregated in the definite and needful manner, these minerals may become so many comparatively infusible bodies amid a more fusible mass. Hence, by the application of a certain amount of heat, the containing substance, should it, for example, be any of the doleritic mixtures, may be fused, while these bodies may remain unmelted, retaining their forms and general characters, until finally acted upon by the surrounding molten mass, with its large proportion of silicate of lime and alumina, forming a flux, and perhaps by a more elevated temperature. It is easy, therefore, to conceive that, as has been above mentioned, a lava stream may be ejected containing leucites, and olivines derived from the remelting of a previously-formed volcanic rock. Judging from the specific gravity of dolerites, when cold and solid (2·94—2·96), leucite crystals (spec. grav. 2·4—2·5) would easily be upborne, rising towards the top of the rock in its fluid state, ready to be ejected in a lava stream. This would not be the case with olivine, the specific gravity of which (3·3—3·5) is greater than that of the dolerites, so that if the latter, containing disseminated olivine, were remelted, this mineral, from its little fusibility and greater weight, would have a tendency to descend, like any substance mechanically suspended in a fluid lighter than itself. As to augite, disseminated crystals of it would, from their ready fusibility be soon melted, though their specific gravity would be 3·2—3·5.

It may not be out of place to remark, as it has been thought that trachytic may have been formed from felspar-porphyritic and granitic rocks of much older date, that upon the heat to which the one or the other would be exposed, might depend the melting of these rocks either partially or wholly. However silica, if mingled with some other substances, may be readily fusible, when once separated, as quartz, it is highly refractory, even if surrounded by fusible silicates, as may be readily tried in the laboratory, and seen daily in the slags in many great metallurgical works. The felspathic portions, which, in some granitic and felspar-porphyric rocks, contain soda as well as potash, are not difficult of fusion, as may easily also be found by experiment. With the mica, much depends upon whether it is a potash, lithia, or magnesia mica. The second of these fuses more easily than the first, and both more readily than the third, which we have found, in experiments, still crystallized, after the fusion of a felspar-porphyry, through the base or paste of which crystals of it were disseminated. There appears no difficulty in conceiving that, upon the melting of a granite, composed of orthoclase, quartz, and magnesia-mica, the first, after fusion, may again crystallize, and envelope the two latter, held mechanically suspended in the molten fluid during the fusion of the felspathic portion. Even supposing some of the quartz to have been fused (being surrounded by a substance acting as a flux), upon the recrystallization of the orthoclase, we should expect that the extra amount of silica, not required for the formation of that mineral, would be excluded as quartz. As to the position of any unmelted quartz and mica of a granite, the felspathic portion of which was alone fused, if the latter were wholly composed of orthoclase (sp. grav. 2·53—2·58), the quartz (sp. grav. 2·6) might have no great tendency to descend in the fluid body. The mica would more readily fall down, its specific gravity, for the potash kind, being 2·8—3·1, and for the magnesia species, that which is somewhat common in granites, 2·85—2·9. In some felspar-porphyrites mica or quartz, and sometimes both, are, with felspar, and occasionally other minerals, well crystallized, so that supposing the descent of the mica through the molten mass, and the quartz more mechanically suspended in it, an ejected upper portion may contain the quartz crystals, and a subsequent lava, the mica, supposing that it remained still unfused.

The attention of the observer is called to this mode of viewing the subject, so that, even on the minor scale, while some trachytic rocks are before him, he may duly estimate the sinking or rising of certain minerals in a fluid mass of molten rock, the higher or lower parts of which may be poured out of a volcano, as its sides may either hold firm, so that lava overflows the crater, or be fissured, letting off the fluid matter at a lower level. Viewing the whole height of a volcano known to us as a minor fractional part of the depth to which the molten matter, partially from time to time thrown out, may descend, certain minerals which have remained unfused upon the partial melting of felspar-porphyritic or granite rocks, (at first taking their relative positions according to their specific gravities.) may be subsequently melted, their elements mingling with the general mass, to be afterwards elevated and ejected. Thus supposing much magnesia-mica to have descended in the molten fluid, upon the partial melting of the granite which contained it, the magnesia, upon the final fusion of the mica, might wholly, or partially aid in the production of olivine in subsequently ejected lava.* The sinking of minerals in fluid lava long since engaged the attention of Von Buch, who found felspar crystals more abundant in the lower that in the higher part of a current of obsidian at Teneriffe, and Mr. Darwin has more recently directed attention to it.†

Volcanos having apparently pierced through rocks of most varied chemical composition, as shown by the fragments of them so often ejected,‡ it may be assumed that portions of such rocks, when not thrown out as fragments, may be often fused in the interior of the volcano, their elementary substances mingling with the general molten mass. While such fragments are sometimes little altered, as if, after being broken off suddenly from their parent rocks, they had not been exposed to a heat sufficient to effect much change in them, others appear to have been acted upon in various degrees, so that modifications in the arrangement of their component particles are produced, and even additions to, or subtractions of some of the latter themselves are effected.§

* In some of the micas the magnesia amounts to more than 25 per cent. One from Lake Baikal, analysed by Rose, gave 25·97 of this substance, and another from Sala afforded Svanberg 25·39 per cent.

† Von Buch, "Description des Isles Canaries;" and Darwin, "Geological Observations on the Volcanic Islands visited during the voyage of the 'Beagle.'" After quoting the labours of Von Buch, and the experiments of M. de Drée (mentioned by him), in which crystals of felspar in melted lava were found to have a tendency to descend to the bottom of the crucible, Mr. Darwin discusses at length the subject of the relative specific gravities of minerals in fluid lavas. "In a body of liquified rock," he remarks, "left for some time without any violent disturbance, we might expect, in accordance with the above facts, that if one of the constituent minerals became aggregated into crystals or granules, or had been enveloped in this state from some previously existing mass, such crystals or granules would rise or sink according to their specific gravity. Now we have plain evidence of crystals being embodied in many lavas, while the paste or basis has continued fluid. I need only refer, as instances, to the several great pseudo-porphyritic streams at the Galapagos Islands, and to the trachytic streams in many parts of the world, in which we find crystals of felspar bent and broken by the movement of the surrounding semi-fluid matter."

‡ They have long been known on Vesuvius, where the fragments of limestone, on the Monte Somma portion of that volcano, have attracted much attention. A fragment of fossiliferous limestone has there also been found. Without entering generally upon the various instances of the ejected fragments of rocks from volcanos, it may be useful to recall attention to those of limestone, dolomite, and sandstone, thrown out when the volcanic island rose through the sea between Pentallaria and Sicily, in 1831 (p. 70), as showing that they may be sought for in such cases.

§ Dr. Daubeny notices the probable conversion of the ordinary Alpine limestone of the vicinity of Naples into granular limestone by heat, as seen in the fragments of the latter limestone found at the Monte Somma, and he quotes the researches of Dr. Faraday, as showing that carbonic acid cannot be expelled from limestone unless steam be present.

By breaking through, and entangling portions of limestones and dolomitic rocks, much lime and magnesia may be obtained by fusion, useful in affording materials for the production of the silicates of lime and magnesia of augite, and the silicate of magnesia of olivine. So also with other accumulations disrupted and partially melted. Indeed, upon estimating whence the mineral matter of a volcano may have been derived, it becomes not only desirable to consider the probable composition of any igneous rocks which may have been remelted, and the circumstances attending this refusion, but also the aqueous deposits of various kinds which may have become more or less exposed to fusion during the time that a volcano has been ejecting mineral matter, either as molten rock, cinders, or ashes.*

* As respects the volcanic region of Naples, and the fragments of rocks which have been ejected by Vesuvius, it is interesting to consider the modifications of igneous matter which might arise from the addition of lime and magnesia to any fundamental igneous product derived from great depths. Dr. Abich (Ueber die Natur und den Zusammenhang des vulkanischen Bildungen, Explanation of Plates, p. iv.), gives the following analyses of the dolomites and limestones of that vicinity:—

—	1	2	3	4	5	6	7	8	9	10
Carbonate of lime . . .	52·30	56·57	54·10	65·20	96·00	98·04	98·08	98·17	96·72	98·40
Carbonate of magnesia . .	46·97	43·43	39·00	34·79	2·30	1·93	1·78	1·48	1·69	1·51
Oxide of iron and alumina .	0	0	0·94	0	0	0	0	0	0·32	0
Silica and bitumen . . .	0	0	5·25	0	0	0	0	0	1·00	0

1. From Capri. 2. Valle di Sambruo, between Minuri and Majuri. 3. Minuri. 4. Between Vico and Sorento. 5. Valle di Sambrue. 6. Monte St. Angelo, Castellamare. 7. Punta di Lettere, Castellamare. 8. Capri. 9. Vico. 10. Capri.

CHAPTER XIX.

VOLCANOS AND THEIR PRODUCTS CONTINUED.—LAMINATION OF STREAMS OF LAVA.—LAMINÆ OF SPHERULES IN OBSIDIAN.—COMPOSITION OF VOLCANIC ASHES.—VOLCANIC TUFF.—PALAGONITE TUFFS OF ICELAND.—MODIFICATION OF VOLCANIC TUFF BY GASES AND VAPOUR.—SOLUTION OF PALAGONITE TUFF IN ACIDS.—SOLFATARAS.—THE GEYSERS AND THEIR MODE OF ACTION, ICELAND.—SULPHUROUS WATER AND GYPSUM DEPOSITS OF ICELAND.—FUSIBILITY OF VOLCANIC PRODUCTS.—FISSURES IN VOLCANOS FILLED WITH MOLTEN LAVA.—LAVA EJECTED THROUGH FISSURES.—DIRECTION OF FISSURES IN VOLCANOS.

ONE kind of lamination observed in igneous rocks has been above noticed (p. 328) as due to the elongation and compression of vesicles, so that by their extreme flattening this structure is produced. In the cases of minerals ejected in an unfused state, the lava current in which they are included moving onwards, so that they would adjust themselves according to their forms and the different velocities of movement produced by friction against the supporting rocks, or any casing of more consolidated portions of the molten stream, we might expect a certain amount of arrangement in planes, or of lamination to be produced. Mixtures of substances of different kinds may sometimes also be so juxtaposed before ejection, that when flowing as a lava current they formed separate layers, the thinner, other circumstances being the same, when the more elongated.* Looking also to the spherical bodies, commonly

* Mr. Darwin (Volcanic Islands, p. 70), when describing the Island of Ascension, enters largely into the causes of lamination in volcanic rocks, seen there and in many other parts of the world. Among other remarks, he concludes "that if, in a mass of cooling volcanic rock, any cause produced in parallel planes a number of minute fissures or zones of less tension (which, from the pent-up vapours, would often be expanded into crenulated air-cavities), the crystallisation of the constituent parts, and probably the formation of the concretions, would be superinduced or much favoured in such places; and thus a laminated structure of the kind we are considering would be generated."

The lamination of molten matter is often well exhibited in the slags which have flowed from furnaces, especially in some iron-works.

formed of radiating crystals of part of the compound, observed when glasses are passing into the stony form (examples of which are not unfrequently produced artificially), we should anticipate, under the circumstances of lava passing into the stony from its fluid condition, and movement still prevailing in the mass, that the cooling portions, more especially adjacent to the ground over which the whole was passing, might sometimes have their parts so acted upon that planes, composed of little spherules, might be formed; even alternations of them produced as successive portions of the fluid lava became exposed to similar conditions. Obsidian is but the vitreous state of melted rock, and all the conditions obtaining when artificial glasses are passing into the stony state, such as those producing separate crystals of certain silicates, and the arrangement into spherules, has to be looked for as well in the one as in the other, the modifications depending on the kind and abundance of the different silicates, with due regard to the conditions under which the general mass may have moved or remained quiet. The obsidians, in certain volcanic countries, are especially advantageous for studies of this kind, and will well repay the attention of an observer.* He will also find examples of lamination in volcanic rocks which have passed the vitreous state, or intermixture with that state in cooling, and it will be desirable that such, as well as

* Dr. Daubeny points out (Description of Volcanos, p. 256), with respect to the obsidian of Lipari, that "some of its varieties possess a remarkable resemblance to certain products obtained by Mr. Gregory Watt (Philosophical Transactions, 1804) during the cooling of large quantities of basalt, an incipient crystallization beginning to manifest itself in the midst of the vitreous mass in the appearance of white or lighter-coloured spots, which appear to be made up of points radiating from a common centre. In many of the Lipari obsidians, however, these round spots are composed of concentric laminæ, and are disposed in general in lines, so as to give a resemblance of stratification to the mass. In other cases, the whole mass is made up of globules of this kind, which are hollow internally, and are sometimes cemented by black obsidian."

Mr. Darwin gives (Volcanic Islands, p. 54—65) an interesting account of laminated volcanic beds alternating with and passing into obsidian at the Island of Ascension. After describing these beds, he remarks, that "as the compact varieties are quite subordinate to the others, the whole may be considered as laminated or striped. The laminæ, to sum up their characteristics, are either quite straight, or slightly tortuous, or convoluted; they are all parallel to each other, and to the intercalating strata of obsidian; they are generally of extreme thinness: they consist either of an apparently homogeneous, compact rock, striped with different shades of gray and brown colours, or of crystalline felspathic layers in a more or less perfect state of purity, and of different thicknesses, with distinct crystals of glassy felspar placed lengthways, or of very thin layers chiefly composed of minute crystals of quartz and augite, or composed of black and red specks of an augitic mineral and of an oxide of iron, either not crystallized, or imperfectly so." Mr. Darwin also mentions the occurrence of layers of globules or spherulites in the transition of one class of beds into the other, one kind of spherulites white, or translucent, the other dark-brown or opaque, the former distinctly radiated from a centre, the latter more obscurely so.

the obsidians, should be well examined for evidence either of movement while consolidation was being effected, or for the simple and very gradual crystallization of parts during any long period which the whole body of rock may have taken to cool.* In such researches the observer will have to recollect, that the top of a lava stream is so far differently circumstanced from the lower portion, that, while the former is exposed to the atmosphere and all its changes, the latter rests upon a bad conductor of heat, so that somewhat modified effects may often be produced, as regards the arrangement of the component substances, in the one part and the other.

With regard to the cinder and ash accumulations on the sides of volcanos, the adjacent country, and the far-distant regions to which the latter may be borne, it would be expected that their chemical composition would be similar to the lavas, for the time, of their respective volcanos, should any be thrown out, subject to such modifications as their more complete exposure to the vapours and gases rushing out might occasion. We should anticipate that during the eruptions of trachytic lavas the cinders and ashes would be likewise trachytic, and so with the other kinds of volcanic rocks. Thus, should trachytic have preceded doleritic eruptions, in any localities, the ashes and cinders of the one would have preceded the other.† Ashes and cinders being so exposed, particularly the former, to be intermingled with, and surrounded by, these volcanic vapours and gases, much would depend, as to any modification or change in the original mineral substance, upon the time during which this action might last, as also upon the kinds of the vapours

---

While remarking on the spherulites in obsidians and in artificial glasses, Mr. Darwin calls our attention to the observations of M. Dartigues (Journal de Physique, t. lix, pp. 10, 12, 1804), on the difficulty of remelting spherulitic and devitrified glasses without first pounding them and mixing the whole well together, the separation of certain parts from the general compound in the spherules or crystals rendering this necessary.

* In all such researches the slow cooling of a lava stream has to be well considered. Dr. Daubeny mentions, that he found the temperature of the lava stream, ejected from Vesuvius in August, 1834, to be 390° Fahr., four months after its outflow, the thermometer placed upon the lava, after the scoriæ on the surface had been removed. Daniell's pyrometer gave similar results when introduced into a cavity of the lava (Description of Volcanos, p. 229).

† M. Dufrénoy (Examen chimique et microscopique de quelques cendres volcaniques; Mémoires pour servir a une Description Géologique de la France, t. iv.) considers that volcanic ashes are most frequently composed of distinct minerals, therein differing from the powder produced by the trituration of rocks, usually formed of the union of several minerals. He therefore infers that volcanic ash "is rather the result of a confused crystallization, produced under the influence of brisk agitation, such as in the saltpetre prepared for the manufacture of gunpowder, than the product of the trituration of lavas in volcanic vents, though the ashes, collectively, do not the less represent the composition of the lava."

and gases to which the ashes or cinders may be exposed. In all cases it would be expected that where the cinders and ashes were the most abundantly and speedily accumulated, as upon the cone or sides of a volcano, the effects arising from an intermixture of the acids and vapours with the ashes and cinders would be the most considerable. For instance, where hydrochloric acid is much mingled with the ashes and cinders, the whole piled around a crater in a hot moist state, such portions as were soluble in that acid might be much acted upon. The like also with sulphurous and carbonic acids.

In considering the original composition and subsequent modification which any mass or layers of volcanic cinders may have sustained, it is also needful for the observer to search for evidence as to the probability of these cinders and ashes having been arranged, as now found, either in the air or beneath water, such, for instance, as is afforded by the occurrence of shells or other organic remains among them,* or by layers of detritus or chemically-deposited matter, showing a subaqueous accumulation. Ashes and cinders descending into water, and afterwards arranged by it, would probably be well washed, so that little change would be effected afterwards by any acids adhering to, or mingled with them.

The term tuff, or tufa, is not uncommonly given to the ash and cinder accumulations of volcanic regions. Dr. Abich has given the following analyses of the tuff of the Phlegrean Fields, Posilippo, and the Island of Vivara, the two former being termed trachytic tuff, the last basaltic tuff:—

—	1	2	3	4	5	6	7
Silica	51·65	52·80	54·41	54·57	56·63	45·50	51·08
Alumina	15·08	15·83	15·40	17·93	15·33	16·05	13·71
Oxide of iron . .	6·21	7·57	7·74	5·49	7·11	11·69	13·16
Lime	5·43	3·13	3·17	0·77	1·74	5·03	7·09
Magnesia	1·18	0·84	1·50	0·77	1·36	3·20	4·72
Potash	6·19	7·86	7·54	5·23	6·54	4·12	2·94
Soda	1·01	2·90	2·87	6·40	4·84	2·28	2·94

1. Yellow tuff, from Nola. 2. Yellow tuff, from Posilippo. 3. White tuff, from Posilippo. 4. Tuff, from Epomoeo. 5. From the crater of Monte Nuovo. 6. Yellow tuff, from the Island of Vivara. 7. Grey tuff, from Vivara.†

* So long since as the time of Sir William Hamilton, shells were detected in the tuff of the vicinity of Naples. They have also been noticed in other localities in that vicinity, and are described as those of species still living.

† Mr. Dufrénoy (Mémoires pour servir a une Description Géologique de la France, t. iv., p. 384) observes, that the tuffs of Posilippo, Pompeii, and Ischia (the two former analysed by M. Berthier, the last by himself), present nearly the same general characters, with the exception of that of Pompeii, which contains nine per cent. of

Looking at the varied manner in which ashes and cinders may be accumulated, either wholly in the atmosphere or beneath water, to the substances with which they have been mingled in the crater of a volcano, and which may more or less coat or impregnate them afterwards, and to the infiltrations through beds and masses of them subsequently to their deposit, either adding to, abstracting from, or modifying the arrangement of their component substances, we should expect that at times even very solid rocks may be produced, at first sight presenting little of the aspect of an accumulation of fine powder and cinders. Mr. Darwin describes a tuff, apparently of this kind, at Chatham Island (Galapagos Archipelago), one evidently formed at first of cinders and ashes, but now having a somewhat resinous appearance resembling some pitchstones. He attributes this alteration to "a chemical change on small particles of pale and dark-coloured scoriaceous rocks; and this change could be distinctly traced in different stages, round the edge of even the same particle."*

In Iceland, a tuff apparently also in a changed or modified condition from that of its original accumulation, and named palagonite-tuff,† would seem to be of much importance. According to Professor Bunsen (of Marbourg), the palagonite-tuff of Iceland has a density of 2·43, and contains nearly 17 per cent. of combined

---

carbonate of lime, a substance which he infers was infiltrated, adding weight to the opinion, that the entombment of Herculaneum and Pompeii was produced by an alluvion of the tuff forming the flanks of Monte Somma, water having greatly aided the filling up of the edifices in the two towns. Remarking on the trachytic tuff of the Phlegrean Fields, Dr. Daubeny observes (Description of Volcanos, p. 16), that the analysis of it proves that, "like pumice, it is only a metamorphosed condition of trachyte." He considers tuff, pumice, and obsidian, as all modifications of the same basis, the two former containing "water chemically combined, namely,—yellow tuff, three atoms; white tuff, two atoms; pumice, one." "Now lava," he continues, "although commonly accompanied at the time of its eruption by abundance of steam, and containing, even for several months afterwards, entangled with it a large quantity of this and other volatile matters, holds no water in chemical combination, so that the fact with respect to tuff and pumice shows, that these formations have been placed under circumstances of another kind than those of molten lavas."

* Volcanic Islands, p. 99. Mr. Darwin describes this tuff, where best characterized, as "of a yellowish-brown colour, translucent, and with a lustre somewhat resembling resin; it is brittle, with an angular, rough, and very irregular fracture, sometimes, however, being slightly granular, and even obscurely crystalline; it can easily be scratched with a knife, yet some points are hard enough just to mark common glass; it fuses with ease into a blackish-green glass. The mass contains numerous broken crystals of olivine and augite, and small particles of black and brown scoriæ: it is often traversed by thin seams of calcareous matter. It generally effects a nodular or concretionary structure. In a hard specimen, this substance would certainly be mistaken for a pale and peculiar variety of pitchstone; but when seen in mass, its stratification, and the numerous layers of fragments of basalt, both angular and rounded, at once render its subaqueous origin evident."

† From Palagonia, in Sicily, where a similar tuff is found.

water. The following is the composition assigned to this rock by him:—

Silica	37·947
Sesqui-oxide of iron	14·751
Alumina	11·619
Lime	8·442
Magnesia	5·813
Potash	0·659
Soda	0·628
Water	16·621
Residue	4·106*

It will be obvious, that in volcanic-tuff accumulations much will depend, as respects subsequent modification and change, upon any foreign matter with which they may be mixed, so that when, as beneath water, calcareous matter (often, perhaps, derived through animal life,) as well as clay or other fine sediment, not directly derived from volcanic eruptions, is mingled with them, and the whole is heated or raised above the water, effects would be produced not precisely corresponding with those where the modifying action has been alone exercised upon the direct products of volcanos. Tuffs of this kind can scarcely but be often formed, and their examination in connexion with volcanos now in action, or which, geologically speaking, have recently been in that state, will be found important as explaining the origin of certain mixtures of igneous and sedimentary rocks, even amid very ancient deposits.

In regions, such as Iceland, where volcanic action is widely spread amid its mineral products rising above the level of the sea, and where modifications due to the action of vapours and gases passing through lava streams, cinders, and ashes, may be so great, there would appear good evidence of the changes to which such mineral products may be exposed. Professor Bunsen has pointed out several which he considers to be now in progress in Iceland. "The Icelandic mineral springs," he remarks, "to which belong all the systems of geysers and suffiones, are distinguished from all others in Europe by the proportionally large quantity of silica which they contain; and, if we except the acidulous springs which are confined to the western part of the island, the so-called beer-springs (*ölkilder*) of the natives, we may divide the springs of Iceland into

* On the intimate connexion existing between the pseudo-volcanic phenomena of Iceland:—A Memoir translated by Dr. G. E. Day, Chemical Reports and Memoirs, Works of the Cavendish Society, 1848. From the chemical composition noticed, Professor Bunsen derives the formula—

$$\left.\begin{matrix}\dot{Mg}\,3\\ \dot{Ca}\,3\\ \dot{K}\,3\\ \dot{Na}\,3\end{matrix}\right\}\dddot{Si}_2 + 2\left\{\begin{matrix}\dddot{Fe}\\ \dddot{Al}\end{matrix}\right\}\dddot{Si} + H.$$

two main groups, according to their chemical properties, one of which would comprise the acid and the other the alkaline silica springs."

Whether the water of these springs has been derived directly from the atmosphere by means of rain, or melted snow and ice, or from sea-water finding its way to the interior of volcanos, the aqueous vapours thence thrown off being condensed in their rise upwards, to it and to the substances with which it can mingle, we have to refer many modifications which evidently take place in the mineral matter through which it passes. The experiments instituted on this subject, and the conclusions deduced from them, and from a personal examination of the springs of Iceland, by Professor Bunsen, are highly valuable. With respect to the action of pure heated water alone for some hours upon the palagonite-tuff above noticed, he found that at the temperature of 212° Fahr. (100° centigrade) or 226°·4 (108° cent.,) silicic acid, potash, and soda, were dissolved.* When the water was saturated with carbonic acid, and allowed to act upon pulverized palagonite, all the constituents, with the exception of alumina and oxide of iron, were dissolved in the form of bi-carbonates.† When the palagonite was heated for ten hours, in water saturated with sulphuretted hydrogen, sulphide of iron was formed, and the solution contained silica and the sulphides of calcium, magnesium, sodium, and potassium.‡ Pala-

* Cavendish Society's Works; Chemical Reports and Memoirs, 1848, p. 364. "1,000 grammes of water after 12 hours' digestion yield, in this manner, a solution containing the following proportions:—

	Grammes.
Silica	0·03716
Soda.	0·00824
Potash	0·00162
Total . .	0·01702"

† "1,000 grammes of this water, after four hours' digestion, yielded the following constituents:—

	Grammes.
Silica	0·09544
Bi-carbonate of lime . .	0·16893
,, magnesia . .	0·05333
,, soda . .	0·06299
,, potash .	0·00189
Total . . .	0·38368"

‡ The solution contained, for 1,000 grammes:—

	Grammes.
Silica	0·1175
Sulphide of calcium . .	0·2748
,, magnesium .	0·0727
,, sodium . .	0·0438
,, potassium . .	0·0410
Total . .	0·5498

gonite was found to be "entirely dissolved in hydrochloric and sulphurous acids, except a small quantity of silica left as a residue."* Thus many and great modifications and changes may be effected in this variety of volcanic tuff; pointing to those which may take place in other volcanic regions, the results in each depending on local conditions.

In many districts, as well those in some portions of which volcanic action is now well exhibited, as in those where it is becoming extinct, as far as respects the ejection of molten rock, cinders, and ashes, discharges of aqueous vapours are effected; sometimes alone, at others accompanied by some of the usual volcanic gases.

Some mention has already been made (pp. 15 and 18,) of thermal or warm springs found to rise as well in regions not marked by volcanic action on the surface as in those where that action is now apparent, or may be inferred to have existed at no very distant geological period.† In some volcanic countries the various modifications under which aqueous vapour, and the gases connected with volcanic action are emitted, can be well studied. The observer can readily suppose that while in the great eruptions these are so driven off as to have effected little combination while in the crater, minor action would leave sufficient time for the condensation of the aqueous vapour into water, and the combination of the latter with

* "We see," observes Professor Bunsen, "from the relations existing among these salts themselves (alluding to those mentioned in the text and previous notes), and with the silica, that the constituents of palagonite take very different parts in the decomposition which is induced by hot water, carbonic acid, and sulphuretted hydrogen respectively; whilst, as we have already seen, this mineral is entirely dissolved in hydrochloric and sulphurous acids, except a small quantity of silica left as a residue. The alkaline siliceous springs, in which there is a smaller quantity of this volcanic gas, assume, consequently, a very different character from the waters of the suffiones; since it is evident, that the composition of the water and the nature of the argillaceous deposits produced from these actions, must stand in a definite relation to the greater or smaller resistance opposed by the separate constituents of palagonite to the action of the weaker volcanic acids, that is to say, to the water, carbonic acid, and sulphuretted hydrogen gas." . . . . "When the alkaline silicates, removed by the heated water from the palagonite, are brought into contact with carbonic, hydrochloric, and sulphuric acids (the latter of which is formed by the oxidation of the sulphurous acid through the oxide of iron in the palagonite), these alkalies must be converted into carbonates, sulphates, and chlorides, whilst the silicic acid remains dissolved in the alkaline carbonates and in the water, and is partially separated from them by evaporation, as siliceous tuff,—a fact already observed by Black in 1792."

† While noticing the dispersion of hot springs, and their issue from all kinds of rock, Humboldt (Kosmos) mentions that the hottest permanent springs yet known are those discovered by himself, "at a distance from any volcano—the 'Aquas calientes de las Trincheras,' in South America, between Porto Cabello and New Valencia; and the 'Aquas de Comansillas,' in the Mexican territory, near Guanaxuato." The first of these has a temperature of 97° centigrade (206° 6′ Fahr.), according to M. Boussingault, who visited this spring in 1823.

volcanic gases, the whole acting upon the rocks through which it has to pass, abstracting matter from them as above noticed.

The Solfatara, near Puzzuoli, has long been known in the volcanic region of Naples, from the emission of aqueous vapour and certain gases, manifesting a kind of subdued volcanic action unaccompanied by the ejection of lava, cinders, or ashes.* Dr. Daubeny found the gas evolved to be sulphuretted hydrogen, with a minute portion of muriatic acid.† Solfataras, or modifications of them, are noticed as existing in many volcanic regions in different parts of the world. Professor Bunsen has shown the connexion of the solfataras of Iceland (the *Námar* of the Icelanders), with the acid springs of that country. He remarks that they "owe their slight acid reaction more commonly to the presence of a small quantity of ammonia-alum, or soda, and potash-alum, than to their inconsiderable traces of free sulphuric or muriatic acids."‡

While such springs in Iceland thus illustrate the condensation of some of the aqueous vapours, mixed with gases discharged in that volcanic region, the Geysers also well illustrate that of the aqueous vapours under other conditions. Allusion has been previously made (p. 15) to those long celebrated discharges of steam and water, the Geysers, and to siliceous deposits from them. According to Professor Bunsen, the thermal group to which the Geysers belong, occurs southward from the highest point of Hecla, and about 20 geographical miles from it. Their main direction is about N. 17° E., "almost parallel with the chain of Hecla, and with the general direction of the fissures." The rock beneath the incrustations of the springs is palagonite-tuff, a vein of clinkstone running lengthwise from the western margin of the springs. The following are analyses by Dr. Sandberger and M. Damour, of the water of the Great Geyser:—§

* An ancient lava current, of a trachytic kind, is supposed to be traceable from the mountain to the sea.

† "Description of Volcanos," p. 211. After pointing out the probable effects of the two gases upon the trachyte of the mountain, the sulphuretted hydrogen uniting with the bases of the several earths and alkalies, and its consequent decomposition, Dr. Daubeny accounts for the absence of muriatic compounds with these bases, by noticing that, "if they existed they would be immediately decomposed by the sulphuric acid generated; and that muriatic acid itself is incapable *per se* of decomposing trachyte, except it be concentrated, and the rock pounded, as shown from the fact of its continuance during so many ages in the domite of Auvergne in a free condition."

‡ Cavendish Society Works; Chemical Memoirs and Reports, 1848, p. 327.

§ The cause assigned by Professor Bunsen for the alternate states of repose and activity of this great natural fountain, is very different from that usually inferred. By very careful experiments by M. Descloizeaux and himself, it was ascertained, 1. "That the temperature of the column of the Geyser decreases from below upwards,

	Sandberger.	Damour.
Silica	0·5097	0·5190
Carbonate of soda	0·1939	0·2567
Carbonate of ammonia	0·0083	..
Sulphate of soda	0·1070	0·1342
Sulphate of potash	0·0475	0·0180
Sulphate of magnesia	0·0042	0·0091
Chloride of sodium	0·2521	0·2379
Sulphide of sodium	0·0088	0·0088
Carbonic acid	0·0557	0·0468
Water	..	998·7695

Not only do the vapours and gases escaping from volcanic vents decompose, under variable conditions, the rocks through which they

as had already been shown by Lottin and Robert. 2. That, setting aside small disturbances, the temperature goes on increasing regularly at all points of the column from the time of the last eruption. 3. That the temperature in the unmoved column of water did not, at any period of time up to a few minutes before the great eruption, reach the boiling-point that corresponds to the atmospheric and aqueous pressure at the point of observation; and 4. That it is at mid-height in the funnel of the Geyser, where the temperature approaches nearest to the boiling-point, corresponding to the pressure of the column of water, and that it approaches nearest to this point in proportion to the approximation of the period of a great eruption." Diagrams are given in illustration, and indeed are almost necessary to the view taken. It may, however, be stated, that there is a constant addition of heated water below in the tube or funnel, and an evaporation of the water above, and that the whole is in such a condition that every cause that tends to raise this column of water only a few metres would bring a large portion of it into a state of ebullition. Vapour is generated, and it is calculated that an excess of 1° (centigrade) over the corresponding boiling-point of the water, "is immediately expended in the formation of vapour, generating in the present case a stratum of vapour nearly equally high with the stratum of water 1 metre in height. By this diminution in the superincumbent water a new and deeper portion of the column of water is raised above the boiling-point; a new formation of vapour then takes place, which again occasions a shortening in the pressing liquid strata, and so on, until the boiling has descended from the middle to near the bottom of the funnel of the Geyser, provided always that no other circumstances have more speedily put an end to this process."

"It appears from these considerations, that the column of water in the funnel of the Geyser extending to a certain distance below the middle, is suddenly brought into a state of ebullition, and further, as may be shown by an easy method of computation, that the mechanical force developed by this suddenly-established process of vaporization is more than sufficient to raise the huge mass of the waters of the Geysers to that astounding elevation which imparts so grand and imposing a character to these beautiful phenomena of eruption. The amount of this force may easily be ascertained by calculating from the temperature of the preceding experiments (those above alluded to), the known latent and specific heat of the aqueous vapour, and the height of the column of vapour, which would be developed by the ascent, to the mouth of the Geyser, of a section of the column of water. If we designate the height of such a column of water in the funnel of the Geyser by h; its mean temperature expressed in centesimal degrees by t; the latent heat of the aqueous vapour by w; the density of the latter compared with that of the water by s; and the co-efficient of the expansion of the vapour by d; we shall find that the excess of heat of the water above the boiling-point under the pressure of one atmosphere is $t - 100$. But the height h, of the section of the column of water, which at the mouth of the Geyser, that is to say, under the pressure of one atmosphere, would be converted into vapour by the quantity of heat, $t - 100$, would be to the whole height of the water column h, as $(t - 100) : w$. A column of water of the height $\frac{h(t-100)}{w}$ would therefore be evaporated at the mean temperature t, if the water were under the pressure of one rise or against which they may be driven in the atmosphere, and often the latter extends to some distance from the place of escape (well shown in the case of winds prevailing in particular directions), but deposits of different kinds are effected, thrown down from the waters containing them. Professor Bunsen has carefully investigated the well-known siliceous deposits from the Geysers of Iceland. Referring to the analysis of the water of the Great Geyser, above mentioned, and remarking that the silica is dissolved in the water by alkaline carbonates, and in the form of a hydrate, he observes that "no trace of silica is precipitated on the cooling of the water, and it is only after the evaporation of the latter that silica is deposited in the form of a thin film on the moistened sides of the vessel where evaporation to dryness takes place, whilst the fluid itself is not rendered turbid by hydrated silica until the process of concentration is far advanced." Professor Bunsen then points out, that, in consequence of these circumstances, the incrustations increase in proportion as the surface of evaporation expands with the spread of the water.*

The same land presents us with other deposits from waters and

---

atmosphere. Hence it directly follows, that the height H, of the column of vapour sought at 100° (centigrade) and 0·76 metre (29·921 English inches) will be

$$H = \frac{h\,(t - 100)\,(1 + 100\,d)}{w\,s}.$$

On applying this formula to the value of the numbers found by observation, we obtain the remarkable result that, in the period of time immediately preceding an eruption, a column of water of only 12 metres (39 feet 4·442 in. English) in length, which projects 5 metres (16 feet 4·851 inches English) to 17 metres (55 feet 9·294 inches English) above the base of the tube, generates for the diagonal section of the Geyser, a column of vapour 638·8 metres (2093 feet 2·245 inches English) in height (assumed to be at 100° centigrade, and under the pressure of one atmosphere), this column being developed continuously from the upheaved mass of water, as the lower strata reach the mouth of the Geyser. The whole column of the Geyser, reckoned from the point where the temperature amounts to 100° centigrade down to the base, is capable, according to a calculation of this kind, of generating a similar column of vapour, 1041 metres (3,415 English feet) in height."—Bunsen, On the intimate connection existing between the Pseudo-Volcanic Phenomena of Iceland; Works of the Cavendish Society, Chemical Reports and Memoirs, pp. 346—349.

* Chemical Reports and Memoirs; Works of the Cavendish Society, 1848, p. 344. Professor Bunsen remarks, that the peculiar forms of the Geysers result from certain conditions. "As," he observes, "the basin of the spring has no part in this incrustation, it becomes converted into a deep tube as it is gradually inclosed by a hillock of siliceous tuff, combining, when it has reached a certain height, all the requirements necessary to convert it into a Geyser. If such a tube be narrow, and be filled with tolerable rapidity by a column of water strongly heated from below by the volcanic soil, a continuous Geyser must necessarily be produced, as we find them in so many parts of Iceland. For it will easily be understood that a spring, which originally did not possess a higher temperature at its mouth than that which would correspond to the pressure of the atmosphere, may easily, when it has been surmounted by a tube, formed by gradual incrustation, attain at its base a temperature of upwards of 1000° centigrade (212° Fahrenheit), owing to the pressure of the fluid resting in the tube."

gaseous emanations, of importance geologically, showing some of the modes in which mineral matter may be accumulated or disseminated. Here again the researches of Professor Bunsen supply us with valuable information. He points out that the acid silica springs, besides inconsiderable traces of hydrochloric and sulphurous acids, and small quantities of ammonia-alum, or potash and soda-alum, contain "sulphates and chlorides of calcium, magnesium, sodium, potassium, and iron, also silica and sulphurous acid, or in the place of the latter, sulphuretted hydrogen gas. They are especially characterised by deposits of gypsum and sulphur. Professor Bunsen found the composition of the water, in 10,000 parts, taken from the Reykjahlider solfatara, in August, 1846, to be:—

Sulphate of lime	1·2712
Sulphate of magnesia	1·0662
Sulphide of oxide of ammonium	0·7333
Sulphate of alumina	0·3261
Sulphate of soda	0·2674
Sulphate of potash	0·1363
Silica	0·4171
Alumina*	0·0537
Sulphuretted hydrogen	0·0820
Water	9995·6467

The clay and gypsiferous accumulations resulting from these waters, or rather from the general action of their constituent parts upon the rocks traversed by them, and upon each other, possess much interest. The palagonite-tuff is decomposed, and clay, often variegated in colour, is deposited, and sulphate of lime is also formed. The gypsum occurs as isolated crystals, and "in connected strata and floor-like depositions, which not unfrequently project as small rocks, where the loose soil has been carried away by the action of the water. These depositions are sometimes sparry, corresponding in their exterior very perfectly with the strata of gypsum so frequently met with in the marl and clay formations of the trias."†

* It is remarked respecting the alumina, "the small quantity of which brings it within the limits of the errors incidental to the experiment," that it may have been dissolved in excess by the alum of the water.—Chemical Reports and Memoirs, 1848, p. 332.

† Bunsen, Works of the Cavendish Society, Chemical Reports and Memoirs, 1848, p. 336. "Their deposition," the Professor adds, "is owing to the fact that has not hitherto been sufficiently regarded in the explanation of geological phenomena, viz. :— that substances crystallizing from solutions are more readily deposited on a surface identical with their own (although at a considerable distance from the limits of their solubility), than on substances different from themselves. These depositions of gypsum increase, therefore, in these formations, in the same manner as we observe small crystals to enlarge in a solution, without any deposit being formed on the sides of the vessel; much salt being removed from the solution (not by a change of temperature, but owing to the cohesive force emanating from the crystal), so that no further deposit can be made on the particles of bodies of a different nature. The

In the clay deposits from the Icelandic fumeroles iron pyrites is found in small crystals, mentioned by Professor Bunsen, as "often very beautifully developed." The sulphur accumulations of Iceland, the Professor attributes to the same cause, as Dr. Daubeny does generally (p. 324), namely, the reciprocal reaction of sulphurous acid and sulphuretted hydrogen gas.

With regard to the presence of nitrogen, ammonia, and their compounds, so frequently observed in connexion with volcanic action, opinions seem somewhat divided: while some consider that the nitrogen is actually evolved from the craters and other volcanic vents, part, perhaps, of the air disseminated in water finding its way to volcanic foci, others infer that ammoniacal products, found in connexion with volcanos, have had a different origin. Dr. Daubeny, treating of volcanic action, remarks:—"Nor is the access of atmospheric air to volcanos more questionable than that of water; so that the appearance of hydrogen united with sulphur, and of nitrogen, either alone or combined with hydrogen, at the mouth of the volcano, seems a direct proof that oxygen has been abstracted by some process or other from both."* On the other hand, Professor Bunsen seems disposed to consider the sublimations of muriate of ammonia as due to the overflow of vegetation by lava currents, at the same time referring to nitrogen and its compounds, and to their being scarcely ever absent from volcanic exhalations, adding that "they undoubtedly belong originally to the atmosphere, or to organic nature, their occurrence being due to the water which holds them in solution and conveys them from the air to the subterranean depths."† If we assume that water from the atmosphere, or sea,

---

process of crystallization here comes within the domain of mechanical forces, since it causes, by the expansive growth of the layers of gypsum, the upheaval of the moistened clay deposit, or compresses it towards the exterior, as the first-named masses increase in quantity."

* Daubeny, Transactions of the British Association for the Advancement of Science, vol. v., for 1837, and Description of Volcanos, 2nd edition, 1848, p. 655.

† Psuedo-Volcanic Phenomena of Iceland, p. 330. "In July, 1846," observes Professor Bunsen, "only a few months subsequent to the eruption of the volcano (Hecla), when I was sojourning in that district, the lower portion of the lava stream appeared studded over with smoking fumeroles, in which so large a quantity of beautifully-crystallized muriate of ammonia was undergoing a process of sublimation, that, notwithstanding the incessant torrents of rain, hundreds of pounds of this valuable salt might have been collected. On surveying the stream from the summit of Hecla, it was easy to perceive that the formation of muriate of ammonia was limited to the zone in which meadow lands were overflowed by the lava. Higher up, where even the last traces of a stunted cryptogamic vegetation disappear, the formation of this salt likewise ceased. The large fumeroles of the back of the crater, and even of the four new craters, yielded only sulphur, muriatic and sulphuric acids, without exhibiting the slightest trace of ammoniacal products. When we consider that, according to Boussingault, an acre of meadow land contains as much as 32 pounds of

more or less containing air (p. 145), does find a passage into volcanos or their foci, it may not be improbable that both causes can contribute to the effects recorded.

It will have been seen that volcanic products are, as a mass, easily melted, notwithstanding that, during times when the particles of matter could freely adjust themselves in a definite manner, certain mineral bodies were formed of a less fusible character, and that from the decomposition of ejected lava, cinders, and ashes, certain other substances are produced, such as siliceous and purer argillaceous deposits of a more refractory kind. Looking, however, at the mass of matter, it continues readily fusible, whether partaking of the trachytic or doleritic character, or of a mixture of both. Sheets of trachytic or doleritic tuff could be easily acted on by heat, so that even altered as they may have been previously, a considerable area, if a sufficiently-elevated temperature could be applied to it, would begin to yield, rising upwards if any elevatory force were acting from beneath, a fracture finally being effected, so far as any sufficient coherence of parts remained, where the resistance became unequal to the force employed.

That volcanic action does work through points, however these may be complicated, as from time to time changed, volcanic craters show, and that it has at least sometimes continued to find vent through the same points, or thereabouts, for long periods, is not only attested by volcanos which have been observed during the historic period, but also by those to the products of which, intermingled with other accumulations, geological dates may be assigned.

That the temperature of volcanos may even change externally, so that snows reposing upon them at one time are suddenly melted from them at another, we have evidence both in high northern latitudes (Iceland) and in the warm regions (Cotopaxi). Volcanic products being, like rocks generally, bad conductors of heat, these changes are sufficient to show the variable amount of temperature to which portions, at least, of volcanic mountains may be exposed from changes in the conditions to which it may be due. That the

---

nitrogen, corresponding to about 122 pounds of muriate of ammonia, we shall hardly be disposed to ascribe these nitrogenous products of sublimation in the lava currents to any other circumstance than the vegetation which has been destroyed by the action of the fire. The frequent occurrence in Southern Italy of tuff decomposed by acid vapours containing muriate of ammonia, likewise confirms the hypothesis regarding the atmospheric origin of this salt. For the same body of air which can annually convey to a piece of meadow land a quantity of ammonia corresponding to these large nitrogenous contents, must at least be capable of depositing an equal quantity of alkali on tuff-beds saturated by acid water; which may be actually observed in some rare instances both in Southern Italy and Sicily."

layers and variably-shaped masses of substances composing volcanos, no matter how accumulated, have been exposed to tension and subsequent fracture, is proved by the rents in them which have been filled by molten matter. In the view beneath (fig. 126), in the

Fig. 126.

Val del Bove, Etna,* exhibiting the now hard lava protruding from that weathering which the whole has experienced from atmospheric influences, we see that the original volume of the previously-deposited volcanic products has been increased by the amount of the molten matter so introduced. The following section (fig. 127).

Fig. 127.

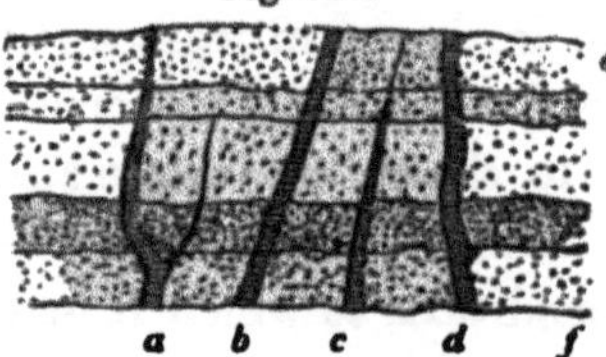

showing the same thing, is not unfrequent amid volcanic craters and ravines. In it, *e f* represent a thickness of volcanic tuff or lava layers, traversed by dykes (as they are termed), *a b c d*, of lava which has entered fissures made by the rupture of the beds *e f*, the force acting from beneath, so that while, for the length seen, some fissures have their walls equidistant from the top to the bottom

* Taken from Dr. Abich's "Views of Vesuvius and Etna."

of the section, at *a* and *c* they diminish upwards. In all these cases, the layers or beds are not disturbed further than by fracture, those on either side of the dykes preserving their continuous lines of original accumulation. This need not always be the case, as will be readily inferred, since after a fracture is made, should the liquid or viscous lava rise with much force from considerable pressure of a column of molten rock with which it may be connected, with a comparative cooling of the upper part of the lava as it rose, increasing the solidification of the particles, the upper layers of tuff or lava broken through may be heaved upwards by the friction of the uprising lava, this even overflowing, as appears to be frequently the case. Of this kind the section beneath (fig. 128), taken from a view by Dr. Abich, of a dyke exposed by the fall of part of the crater of Vesuvius, is probably an instance.

Fig. 128.

The observer has next to consider the magnitude and direction of these fissures. Perhaps volcanic islands, such as those in the Atlantic and Pacific, are as favourable situations for studying volcanic fissures as are to be found, not, however, that great facilities do not present themselves elsewhere.* The rent or series of rents which traversed the side of Kilauea, during the ejection of lava in 1840, is not only remarkable for its length, but also for the com-

* Respecting the Hawaiian group, Mr. Dana, (Geology of the United States' Exploring Expedition, p. 282), infers:—1. That there were as many separate rents or fissures in the origin of the Hawaiian Islands as there are islands. 2. That each rent was widest in the south-east portion. 3. That the south-easternmost rent was the largest, the fires continuing there longest to burn. 4. That the correct order of extinction of the great volcanos is nearly as follows (leaving out Molokai and Lanai, which were not visited by Mr. Dana)—*a*, Kauai; *b*, Western Oahu; *c*, Western Maui, Mount Eeka; *d*, Eastern Oahu; *e*, North-western Hawaii, Mauna Kea; *f*, South-east Maui, Mount Hale-a-Kala; and *g*, South-east Hawaii, Mauna Loa. "This order," he observes, "is shown by the extent of the degradation on the surface. Each successive year, since the finishing of the mountain, has carried on this work of degradation, and the amount of it is, therefore, a mark of time, and affords evidence of the most decisive character." (p. 283.)

paratively tranquil manner in which it was effected, the inhabitants not being aware of its formation by any earthquake motion, but from finding a torrent of liquid lava poured out.* Fissures so produced would seem to point to much softening of the subjacent rocks, so that when fractures were formed, though many miles in length, comparatively little resistance, from cohesion, remained in the rocks. From an examination of Maui (Hawaiian group), Mr. Dana infers that at its last eruption, a huge segment of that volcano must have been broken off, by which two great valleys were formed (one two miles wide), through which great opening the lavas were poured out.† Here would appear to have been far greater resistance and a more sudden overpowering of it by the force exerted. According to the accounts given, Jorullo was the result of an uprise of ground, finally traversed by a fissure (p. 346). With respect to fissures traversing active volcanos from which lava has issued, there is abundant evidence. The great outflow of lava from Skaptar-jökull, in 1783 (p. 344), was from fissures at the base of the mountain. Great fissures have been made in Etna, and the numerous subordinate craters seem little else than points in those formed at various times through which volcanic matter has been ejected.‡ Respecting the fissures on Etna, M. Élie de Beaumont remarks, that they occur, for the most part, in nearly vertical planes, often so cutting the crater, as it were, to star it, the lower part of the fissures usually filled with lava, the higher with scoriæ, and with pieces of tuff and lava fallen from the upper part. He mentions that the fissure formed in 1832, was so far a shift, or fault, that one of the sides of the dislocation rose about a yard higher than the other.§ The great fissure, which in 1669 traversed the slope of the great gibbosity of Etna and the Piano del Largo, is described as having ranged from near Nicolosi to beyond the Torre del Filosofo, and to have been about two yards wide at the surface, a livid light being emitted from the incandescent lava rising in it.‖ An observer should carefully ascertain the directions of such rents or fissures whether large or small, and

* Dana, "Geology of the United States' Exploring Expedition," p. 217.

† Ibid, p. 259.

‡ It is stated that there are 52 of these subordinate volcanic hills on the west and north of the summit of Etna, and 27 on the east side; "some covered with vegetation, others bare and arid, their relative antiquity being probably denoted by the progress vegetation has made upon their surface."—Daubeny, Volcanos, 2nd edition, p. 272.

§ Recherches sur la Structure et sur l'Origine du Mont Etna, par M. L. Élie de Beaumont; "Mémoires pour servir a une Description Géologique de la France," 1838, t. iv., p. 111.

‖ Ibid. p. 108.

always with reference to the complication which may arise from variable resistances, even in the prolongation of the same fissures, to the force employed, seeing especially if the greater fractures have continued to preserve any definite directions at different intervals of time.

M. Élie de Beaumont has called attention to the fact that these fractures so often starring Etna, and into which the molten lava is introduced, there hardening and remaining, must produce a tumefaction or elevation of the whole, each eruption of the mountain, so characterized, having a tendency to elevate the mass of the volcano.* The same reasoning is applicable to all volcanos rent and fissured in a similar manner, and the abundance of the resulting dykes of lava is often a prevailing feature in many. They are sometimes interlaced so as to show differences in date, hose of one time cutting those of another, exhibiting proofs of repeated fractures through the same general mass of volcanic matter.

In examining some volcanic regions, the observer will have to consider, as above noticed (p. 323), the probable differences which would arise in the structure and arrangement of the accumulations from a part or the whole of them having been produced beneath water. At considerable depths in the ocean, beyond those at which an equal temperature of 39°·5 (p. 96) would appear to prevail, not only the pressure but also the constancy of that temperature and the mass of water possessing it have to be borne in mind. Assuming a communication made, whether by an elevation of the sea-bottom and the bursting of a tumefaction formed by forces acting from beneath, or by one of those adjustments of the earth's surface by which more or less considerable fractures are produced, he will have to recollect that a great volume of water, with a low temperature, would be at once brought to bear upon it, and that not only are the usual volcanic gases absorbed by water, but that the very pressure itself might tend to drive them into the liquid state.† It would be out of place here to enter into the probable effects produced under such conditions, further than to notice that, supposing the communication made, and the elevatory force sufficient to lift a body of molten lava, so that it could pass out of the volcanic orifice or crack, the observer has to consider the effects which would follow. However any intense heat might permit the existence of

* "Mémoires pour servir a une Description Géologique de la France," t. iv., p. 118.

† Dr. Faraday has shown (Philosophical Transactions, 1823), that sulphurous acid gas becomes liquid under the pressure of two atmospheres, at a temperature of 45° Fahr.; sulphuretted hydrogen under that of 17 atmospheres at 50°; carbonic acid under 36 atmospheres, at 32°; and hydrochloric acid gas under 40 atmospheres at 50°.

the vapours and gases observed at subaërial volcanos, before the rupture was effected, as soon as the water came into contact with them, a ready supply of that at 39°·5 pouring in, the more heated water ascending, as so heated, they would disappear as they rose. If disseminated amid the lava thrown out, the great pressure upon the latter would produce its effects upon them, while the low temperature would soon act on the external liquidity of the lava itself.

From such a state of things to the minor depths and surface of the water, great modifications would be expected, solid lava, (supposing the struggle between the forces brought into action to be such, as, on the whole, to permit a gain on the side of the volcanic products), probably prevailing beneath, while there was an admixture of more scoriaceous matter above, as the accumulations rose into the atmosphere. The whole mass would be liable at all times to be cracked and fissured, molten lava rising into the rents according to the pressure of the time upon it, and the tumefaction mentioned by M. Élie de Beaumont progressing.

The extent to which sheets of matter could be spread in various directions and at different times around submarine volcanic vents, would necessarily depend upon circumstances; among them, the absence of piles of cinders and ashes into cones, such as are formed by the discharge of vapours and gases through lava into the atmosphere, being important, so that when fissures were produced, molten matter flowed more freely out, in the manner, so far as liquidity and the absence of cinders and ashes are regarded, of the streams which poured out on the flanks of Mauna Loa, in 1843.

It has been seen that volcanic vents may remain for a long time dormant or closed, and then lava, cinders, and ashes be driven out of them. The probable differences which would arise in such cases with volcanic accumulations beneath the sea and those above its level, require attention. It may be inferred that, beneath given depths of water, where the rents have been more frequent, and the lavas have more frequently been thrown out, there may be such a mechanical resistance to the new application of an elevatory force, that an increase of heat may soften and even melt some of the prior-formed accumulations, for the most part readily fusible. Thus a dome-shaped mass may be raised, not finally splitting, in given localities, at its surface, until even above water, and the quiet cracking of Mauna Loa for the length of many miles, would appear to show us that conditions may arise, even in subaërial volcanos, permitting the heating and softening of a volcanic crust to within comparatively moderate distances from that crust.

CHAPTER XX.

VOLCANOS AND THEIR PRODUCTS CONTINUED.—THE CALDERA, ISLAND OF PALMA.—SECTIONS OF ETNA AND VESUVIUS.—FORM AND STRUCTURE OF ETNA.—ORIGIN OF THE VAL DEL BOVE, ETNA.—FOSSILIFEROUS VOLCANIC TUFF OF MONTE SOMMA.—MIXED MOLTEN VOLCANIC ROCKS AND CONGLOMERATES.—MODIFICATION OF SUBMARINE VOLCANIC PRODUCTS.—PEAK OF TENERIFFE.—SANTORIN GROUP.—ISLAND OF ST. PAUL, INDIAN OCEAN.—DISTRIBUTION OF VOLCANOS IN THE OCEAN—ON CONTINENTS AND AMID INLAND SEAS.—VARIABLE PROXIMITY OF VOLCANOS TO WATER.—EXTINCT VOLCANOS.—MINERAL AND CHEMICAL COMPOSITION AND STRUCTURE OF BASALT.

THAT dome-like elevations of volcanic products have been formed, MM. Von Buch, Élie de Beaumont, Dufrénoy, and others consider they have sufficient proof. Some notice has been above given (p. 318), of the equivocal appearances which may be presented by the "craters of elevation" and the "craters of eruption." The Caldera, in the Island of Palma, Canaries, is adduced by Von Buch as a good example of the "craters of elevation." A large precipitous cavity or crater is there surrounded by beds of basalt and conglomerate, composed of basaltic fragments, dipping regularly outwards, and is broken only by a deep gorge on one side, through which access can be obtained to this central cavity. White trachyte, and a compound of hornblende and white felspar, are also noticed among these rocks. There being no mixture of scoriæ or ashes, and the beds of molten rock as well as the conglomerate presenting a uniform stratification, it is inferred that the whole was formed under different circumstances, such as beneath water, from the ordinary eruptive accumulations of a volcano, and had been upraised in a dome-like manner, until finally the rupture was effected, and the least resistance being in one direction, the lateral gorge was produced, the whole presenting the appearance of the pear-shaped termination of the fissures in figs. 115 and 116.

M. Élie de Beaumont has given a valuable description of Etna,

Fig. 129.

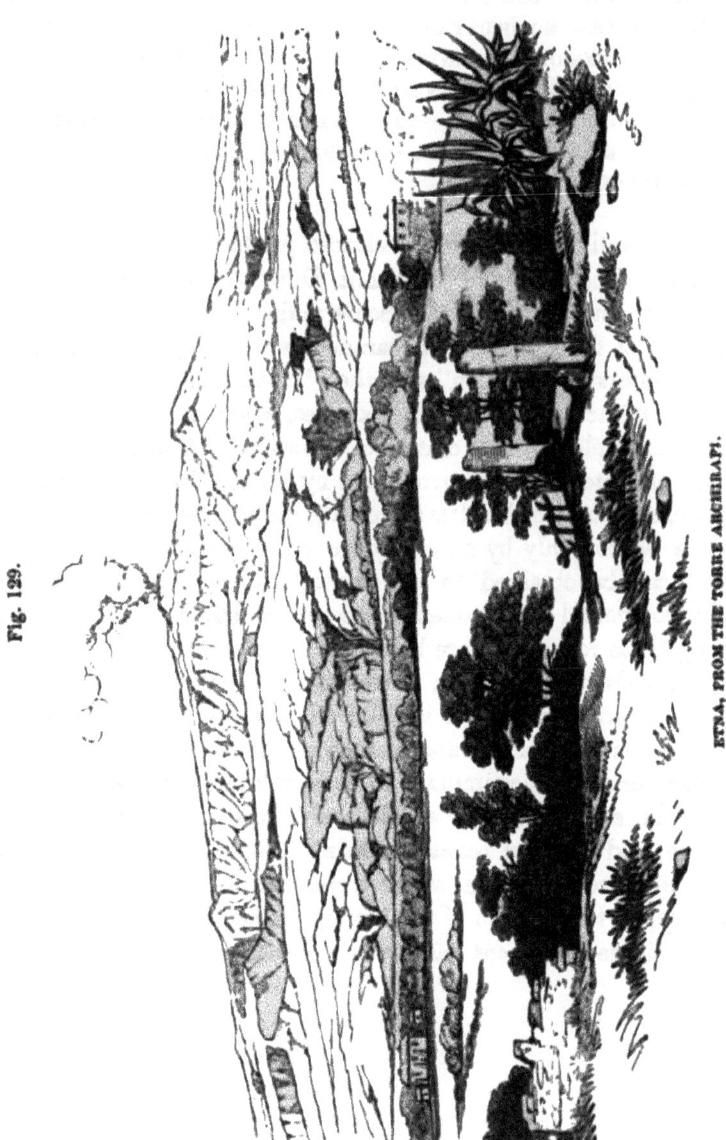

ETNA, FROM THE TORRE ARCHIRAPI.

Fig. 130.

SECTION OF ETNA FROM WEST TO EAST.

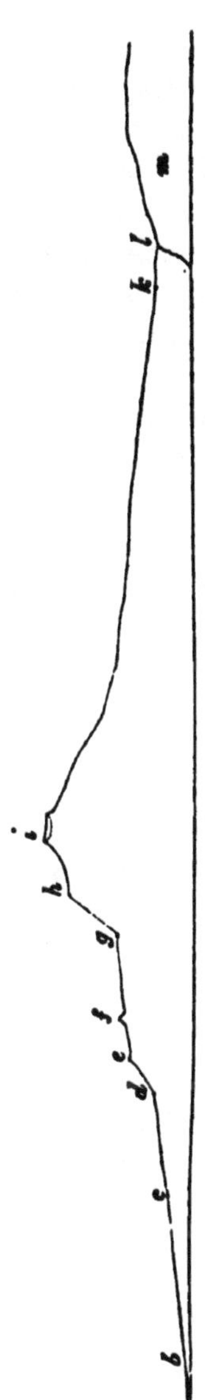

a, Punta Secca; *b*, Giarre (203 French feet); *c*, Zaffarana (1859); *d*, Portella (2972); *e*, Salto di Giamento (4417); *f*, Rocca Musarra (4871); *g*, Val del Bove (5292); *h*, Cima della Valle (8808); *i*, Crater (10,210); *k*, Bronte (2549); *l*, River Simeto (1915); *m*, Non-Volcanic rocks.

Fig. 131.

SECTION OF VESUVIUS FROM SOUTH TO NORTH.

b, Somma (413 French feet); *c*, Fontana dell' Olivella (994); *d*, House of the Cancaroni (1588); *e*, Punta di Nasone (3430), Monte Somma; *f*, Atrio del Cavallo; *g*, Punta del Palo (3640); *h*, Crater; *i*, Punta St. Angelo (3468); *k*, Bocche Nuove (1515); *l*, Camaldoli (534); *m*, Torre d'Annunziata.

illustrating the same views.* The Val del Bove shows an accumulation of many hundred layers of fused rocks, somewhat resembling the modern lavas of this mountain, interstratified with others composed of fragmentary and pulverulent substances, the beds varying in thickness from half a yard to several yards, those of fused rock commonly thinner than the fragmentary deposits. The surfaces of the former are rough, and their outer part is penetrated by cells to the depth of 8 or 12 inches; whence beds only 20 inches or 2 feet thick are cellular throughout. The thicker beds are solid in the middle and resemble certain trachytes, labradorite replacing orthoclase. The fragmentary beds are true tuffs, their structure similar to those found in the Cantal and Mont Dore, the component fragments composed of the same substances as the fused beds, sometimes scoriaceous, at others compact. The beds of the Val del Bove undulate in different directions, varying from horizontality to inclinations of 20° and 30°, without their structure or thickness being altered in a constant manner. They are traversed by a multitude of lava dykes, the rock of which is of the same kind as that of the fused beds. Though these take various directions, it is considered that there is a tendency on the whole to one ranging E.N.E.

The general form of Etna will be seen from the accompanying view (fig. 129) taken from one of those in the atlas of M. von Walterhausen.† By it the very slight rise of the general mass to the central crater will be observed, as also the great break in front, known as the Val del Bove, where the more ancient volcanic accumulations are considered to be exposed. While this view may thus be useful in showing the general outline of the mountain, so far as views embracing considerable areas may do so, the annexed section from west to east (fig. 130),‡ will more correctly give the real shape of Etna, taken through the Val del Bove, therefore somewhat across the outline represented in the view, and exhibit the steep descent through the cliffs of that great break or depression on the flank of the mountain.

* "Mémoires pour servir à une Description Géologique de la France," tom. iv. According to M. Élie de Beaumont the order of the Etna formations is as follows:—1. Granitoid rocks, known only by ejected fragments. 2. Calcareous and arenaceous rocks. 3. Basaltoid rocks. 4. Rolled pebbles, forming a line of hills at the junction of the Plain of Catania and the first slopes of Etna. 5. Ancient lavas, forming the escarpments round the Val del Bove; and 6. Modern eruptions (p. 53).

† Maps and Views of Etna.

‡ Taken from Dr. Abich's "Erläuternde Abbildungen Geologischer Erscheinungen beobachtet am Vesuv und Aetna," pl. 9, Berlin, 1837. In this section, the scale for height and distance is the same.

M. Élie de Beaumont quotes M. Mario Gemellaro as pointing out that the central mass of Etna is composed of two cones passing into each other; one, interior, formed of ancient volcanic products, the other, exterior, formed of modern accumulations. These two cones are upon the same axis, the ancient nearly east from the modern cone, whence they do not completely embrace each other, the modern not altogether covering the ancient, and the products of the latter being exposed on the eastern side of the mountain, especially in the Val del Bove*. This considerable and sudden gap in Etna, M. Élie de Beaumont agrees with Sir Charles Lyell† in referring to a great subsidence of that part of the mountain, in the same manner as the much larger volcanic mass of the Papandayang fell in at Java in 1772, and that of Carguairazo subsided on the 19th of July, 1698; a volcano considered previously to have rivalled its neighbour Chimborazo in height. M. Élie de Beaumont refers to the possibility of the lava which lifted the ancient mass of Etna having been abstracted, so that the needful previous support of the portion now occupied by the Val del Bove being removed, depression was the result. He considers Etna to have been an irregular crater of elevation, the uplifting force not having there acted in the same simple manner as at the Isle of Palma, Teneriffe, and Monte Somma (Vesuvius).‡ M. Élie de Beaumont infers that the first-formed deposits were nearly horizontal, successive fissures presenting channels for the outpouring of very fluid lava which spread round in various directions, in the manner that sheets of basalt are seen to have done in many countries, and especially in Iceland, cinders being also ejected, so as to alternate with the fluid rock. Upon the accumulations thus produced, the elevatory force is considered to have acted in the manner described.

In like manner, the part of Vesuvius known as Monte Somma has been inferred to be the remains of a crater of elevation, which has been broken through, so that it became, in a great measure,

* "Mémoires pour servir a une Description Géologique de France," t. iv., p. 124.

† "Principles of Geology," 4th edition.

‡ "Mémoires pour servir a une Description Géologique de France," t. iv., p. 188. M. Élie de Beaumont further observes, that "the force which raised the gibbosity of Etna appears to have acted not on a single and central point, but in a straight line, represented by the axis of the ellipse, of which the southern, northern, and eastern flanks of the Val del Bove form a part; and it seems to have acted unequally on the different parts of this straight line, so that its western extremity, answering to the actual volcanic vent, has been more raised than the rest. An elevation of this kind could not be produced without the upraised masses being broken, and the rents ought chiefly to coincide with the line of elevation, or diverge, radiating, from its extremities."

covered by the eruptions of more modern times, those chiefly which followed the great outbreak of 79. The section* (fig. 131) will serve to show the general outline of this volcano, as well as the portion of the cone (Monte Somma) which existed prior to that eruption. After describing the volcanic tuff of the environs of Naples, showing that it is composed almost entirely of the débris of trachyte, the greater portion of the fragments contained in it pumice, that it was, in part at least, fossiliferous,† and inferring that it was arranged in beds under water,‡ M. Dufrénoy observes that the tuff of Monte Somma is the continuation of the same accumulations, and quotes M. Pilla as having discovered fossil shells in it.§ He points out that among the limestone fragments in the tuff of Monte Somma some are covered by small *serpulæ* of the same species as those which adhere to the rocks on the coast of Sicily, and that these fossils not being in the least altered, they prove, even more than the general disposition of the tuff, that it has been formed beneath water, and subsequently raised to its present height on the Monte Somma. This tuff, with its fragments or pebbles of limestone (commonly saccharoid) and of micaceous rocks, M. Dufrénoy considers, with the lava associated with it on Monte Somma, to have been upraised, in a vaulted manner, by volcanic forces acting from beneath, the eruptions finally finding vent through this elevated mass, and forming the present Vesuvius. The rocks composing Monte Somma and Vesuvius are pointed out as different. While the lavas of Monte Somma resemble crystalline rock, such as granite and trachyte, the Vesuvian products are scoriaceous. The former are composed of leucite, augite, labradorite, and some

* From Abich's "Erläuternde Abbildungen Geologischer Erscheinungen beobachtet am Vesuv und Aetna," pl. 9. Berlin, 1837. The scale for this section is the same for height and distance, but it differs from that of the section of Etna on the same page.

† M. Dufrénoy (Mémoires pour servir a une Description Géologique de la France, t. iv., p. 240) adds to the evidence of Mr. Poulett Scrope, who pointed out (Geol. Transactions, 2nd series, vol. ii., p. 351) that this tuff contained the remains of *ostrea*, *cardium*, *buccinum*, and *patella*, of species now living in the Mediterranean; and of Sir Charles Lyell (Principles of Geology) respecting the fossiliferous tuff of Ischia, that *ostrea*, *cardium*, and *pecten* have been obtained from the quarries in the hill of Posilippo, an *ostrea* and a *pectunculus* at St. Elmo, above Naples, and fossils in other places.

‡ M. Dufrénoy mentions (Mémoires pour servir, &c., t. iv., p. 238), that this tuff often presents cavities from 6 inches to 2 feet in height, almost always taking a vertical direction. They are numerous in the tuff of Naples itself, and in the escarpments on the high road to Nesita. Their parallelism leads him to infer that they have been caused by the abundant escape of gas which traversed the beds before their solidification. M. Dufrénoy (p. 241) also notices concretions in the tuff, chiefly in the argillaceous beds.

§ The fossils found by M. Pilla are *Turritella terebra*, *Cardium ciliare*, *Corbula gibba*, and a portion of an Echinite.

rare nodules of olivine; while the latter, when compact and crystalline, are formed of crystals of the order of felspar, but differing from ordinary felspar, albite, and labradorite. They, moreover, contain crystals of green augite, some nodules of olivine, and some rare plates of mica. As with the lava and tuff beds of the Val del Bove, Etna, the lava and tuff of Monte Somma are traversed by numerous lava dykes.

Without multiplying examples of mixed beds of conglomerate, tuff, and lava, so occurring as to render it probable that these volcanic accumulations had been effected beneath water, mention may be made of the evidence on that head obtained by Mr. Dana among the islands of the Pacific, as the descriptions given of such aggregations at Oahu, and Maui (Hawaiian group), and Tahiti, possess much interest in their geological bearings. Some of the rounded masses in the conglomerate of Kauai are stated to contain 30 cubic feet, lying against each other, the interstices between filled in with pebbles and finer matter. At Oahu there are some finely-laminated tuffs, which as much point to their formation beneath water as the conglomerates.* Molten rocks and conglomerates are mentioned as alternating at Tahiti, some of the stones in the latter being six inches in diameter.† In such islands we have merely the upper portions of volcanos above the level of the sea. As respects the evidence of their uprise from situations where large rounded blocks and pebbles could be formed, the simple tumefaction of the volcanic mound, from the causes above noticed (p. 381), would alone aid the uplifting of beaches and various deposits, in minor depths, to heights proportionate to the introduction of the matter filling dykes traversing the mass, and to the extension of the various deposits of ashes and cinders, and of lava, by heat, as covering after covering was accumulated, independently of any great force applied from beneath, and tending to dome out and perhaps throw off the flanks.

The observer will have to consider the probable figures which beds would take round volcanic islands. If we are to suppose some volcanos, now inland in various parts of the world, to have been once islands, the depth of water around them at different times would much influence the arrangement of their mineral products. We should expect the deposits which now take place

* Dana, "Geology of the United States' Exploring Expedition," pp. 239, 261, 267, 268.

† Ibid., p. 295. Mr. Dana describes the general dip of these beds to be from the central part of the island outwards, the central rocks being more compact than those on the exterior and less vesicular, more trachytic and syenitic (p. 296).

around and amid the Hawaiian Islands to be much modified, as regards general arrangement, from those of any volcanos in shallow seas. We must refer to previous notices of volcanic ash and cinder accumulations in tideless (p. 69) and tidal seas (p. 99), and the working out of soft from hard volcanic matter by the breakers (fig. 80, p. 192, as pointing to these modifications. To these may be added the condition of volcanic islands, such as those in various parts of the ocean exposed to almost ceaseless breaker action, separating the harder from the softer volcanic products, the volcanic mass gradually rising, from time to time; great subaërial eruptions, and perhaps submarine also, being effected. Very complicated arrangement of parts could scarcely but arise, and much admixture of molten matter with conglomerates and finer volcanic sediment, be the result, and this independently of any accumulations at considerable depths, which, as they rose, would become acted upon by the breakers in a similar manner; to be afterwards covered with subaërial accumulations, should volcanic action continue above water in the elevated mass. The accompanying view of the Peak of Teneriffe (fig. 132), by M. Deville,* taken from

Fig 132.

near Santa Ursula, may serve to illustrate the slope of that mountain, considered fundamentally due to the elevation of beds of tuff and molten rock around the central portion, upon which the sub-

* "Études Géologiques sur les Iles de Ténériffe et de Fogo," 1847.

aërial eruptions have formed the present Peak, as also the cutting back of the mass at the level of the sea by the breakers.

As to the elevation of volcanic accumulations, the island of Santorin has attracted considerable attention, not only from the discovery of organic remains in the tuffs, now raised above the sea, but also from its general form and its history. Dr. Daubeny infers—with respect to this island, or rather islands, the larger known to the ancients as Thera, and the second in size as Therasia, though there may be some uncertainty as to the whole of the little group having been thrown up in historical times—that some considerable convulsions may have occurred, to which early ancient writers refer.* Whatever obscurity may hang over the exact times at which the Santorin group, or parts of it, were upraised above the level of the sea, there appears none as to changes having been effected in its isles and islets by volcanic action from remote historical times.

Nearer our own times, it seems certain that a portion of this volcanic mass was raised above water in 1573, forming a rock known as the little Kaimeni; that there was an eruption of pumice near Santorin in 1638; and that, in 1707, a new rock rose between the Little and Great Kaimeni, "which increased in size so rapidly, that in less than a month it became half a mile in circumference, and had risen 20 or 30 feet above the level of the water, constituting a third island, which was called New Cammeni, a name which it still bears."† Some persons landing on the upraised rock to collect oysters adhering to it, were compelled to leave it from the violent shaking of the ground. The commencement of the island was first observed on the 23rd of May, 1707. In July, black smoke accompanied the upheaval of the rocks, and much sulphuretted hydrogen appears to have been discharged. Stones, cinders, and ashes were shortly afterwards ejected; showers

---

* Daubeny, "Description of Volcanos," 2nd edition, p. 320. Dr. Daubeny refers to the statement of Pliny, that 130 years after the separation of Therasia, the island of Thera was thrown up; "a statement," he observes, "confirmed by Justin and Plutarch, as to the fact, though not as to the date." * * * "It is to this event that Seneca seems to refer, where he speaks of an island thrown up in the Ægean Sea, by an accumulation of stones, of various sizes, piled one upon another." * * * "Pliny also speaks of another phenomenon of the same kind, as happening in his own time, for he tells us that in the reign of Claudius, A.D. 46, a new island called Thia appeared near Thera. But as he mentions it as only two stadia distant from Hiera, it is possible that the island may have been joined to the latter by a subsequent revolution, as by that recorded to have taken place in the year 726, by which Hiera is said to have been greatly augmented in point of size."

† Daubeny (Description of Volcanos, p. 321), who quotes from Father Goree, an eyewitness of the fact, seen from Scaro and all that side of Santorin.

of the two latter spreading to considerable distances. The volcanic action continued for nearly a year, more or less; indeed, during ten subsequent years.* As Dr. Daubeny remarks, this elevatory action has not yet ceased, inasmuch as a reef found by the fishermen to have been raised, during a short time, to within 30 or 40 feet of the surface, was in 1829 ascertained, by M. Lalande, to have no more than 9 feet water over an area of 2,400 by 1,500 feet, the ground gradually sinking around from the centre, and less water, by two feet, was obtained about two months afterwards by M. Bory de St. Vincent.

The accompanying (fig. 133) is a map of these islands, with a

Fig. 133.

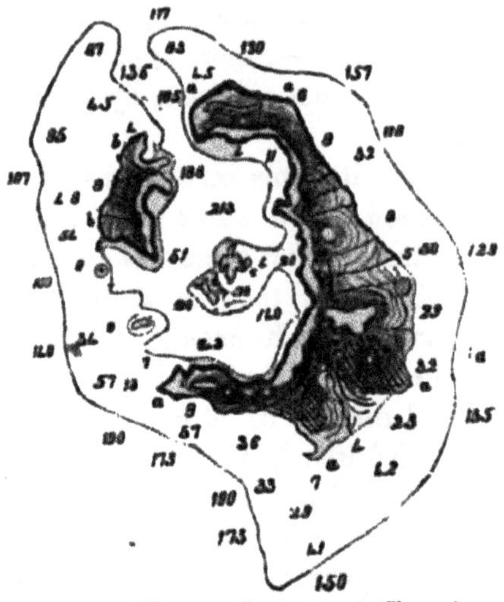

a, a, a, a, Thera or Santorin; b, Therasia.

* "In July, the appearances were more awful, as all at once there arose, at a distance of about 60 paces from the island already thrown up, a chain of black and calcined rocks, soon followed by a torrent of black smoke, which, from the odour that it spread around, from its effect on the natives in producing headache and vomiting, and from its blackening silver and copper vessels, seems to have consisted of sulphuretted hydrogen. Some days afterwards the neighbouring waters grew hot, and many dead fish were thrown upon the shore. A frightful subterranean noise was at the same time heard, long streams of fire rose from the ground, and stones continued to be thrown out, until the rocks became joined to the White island originally existing. Showers of ashes and pumice extended over the sea, even to the coasts of Asia Minor and the Dardanelles, and destroyed all the products of the earth in Santorino. These, and similar appearances, continued round the island for nearly a year, after which nothing remained of them but a dense smoke. On the 15th July, 1708, the same observer (Father Goree) had the courage to attempt visiting the island, but when his boat approached within 500 paces of it, the boiling of the water deterred him from proceeding. He made another trial, but was driven back by a cloud of smoke and cinders that proceeded from the principal crater. This was followed by ejections of red-hot stones, from which he very narrowly escaped. The mariners remarked, that the heat of the water had carried away all the pitch from their vessel."—Daubeny, "Description of Volcanos," p. 322.

sketch of the banks and form of the ground beneath the sea, as shown by the late survey of Captain Graves, R.N. The crateriform cavity in the centre, with its depth of 213 fathoms (1,278 feet), will be at once apparent, with the shallow depth on the west, dividing it from the Mediterranean in that direction. Equally interesting is the deep channel running to the northward, with as much as 990 feet in it, reminding us of those figures and descriptions of the craters particularly brought under notice by Von Buch, where a great rent appears to have been effected on one side, forming a ravine entrance to their central, and often otherwise inaccessible interiors. Such a form also will again remind the observer of the pear-shaped termination of fissures noticed above (figs. 115 and 116), one which would so readily accord with the power of a force acting from beneath upwards, so that if this was exerted in the centre of the group of Santorin, and a fissure extended northerly through the deep channel there presenting itself, there appears no mechanical difficulty in supposing a somewhat yielding covering, rendered so, in a great measure, by heat, opening outwards in such a manner that the chief fissure would suffice for any required separation of parts, a certain amount of cohesion still remaining. It is needful, in estimating the effects which would be produced in this, or in a somewhat similar manner, to regard the whole mass with reference to proportion and real sections,* so that no undue value should be attached to heights or distances; and also to the masses of limestone of Mount Elias and the hill on the north of it, a range of that limestone on the eastern side of the island (marked by straight lines on the map, fig. 133) having to be taken into account. It is also easy to conceive—indeed, the variable intensities of different eruptions from the same volcanic vent point to the fact—that the action which at one time may elevate a considerable mass, may at another, and after time, be unable to cause more than a central movement in a volcanic vent.†

Professor Edward Forbes and Captain Spratt, R.N., who visited Santorin in 1841, state that "the aspect of the bay is that of a great crater filled with water, Thera and Therasia forming its

---

* Those constructed with the same scale for heights and distances.

† When noticing the uprise of ground and eruptions witnessed at Santorin by the Father Goree, in 1707 and 1708, Dr. Daubeny calls attention, as important to the natural history of volcanos, "that in this case, as in many others, the mountain appears to have been elevated before the crater existed, or gaseous matters were thrown out. According to Bourguignon, smoke was not observed till 26 days after the appearance of the raised rocks."—"Description of Volcanos," p. 322.

walls, and the other islands being after-productions in its centre."* In the Little Kaimeni they found the elevated sea-bottom, formed of fine pumiceous ash, to be fossiliferous.† They were informed that similar beds of shells were found on the cliffs of Santorin itself. "In the main island, the volcanic strata abut against the limestone mass of Mount St. Elias, in such a way as to lead to the inference, that they were deposited on a sea-bottom, on which the present mountain rose as a submarine mass of rock."‡ The following view (fig. 134), kindly communicated by Captain Spratt, R.N., is highly illustrative of the general appearance of the interior of the Santorin group, of the position of the central islets, and of the kind of stratification which occurs around the central opening.

In a group of this kind, independently of any eruptions through the central cavity or crater, which it would appear have taken place even in later historic times, breaker-action upon the interior cliffs, upon the softer substances especially, would tend to degrade them, and deposit the detritus, so derived, in the central depression, a deep cavity in a tideless sea, as the Mediterranean may be considered, as far as regards geological effects. In like manner, also, heavy seas rolling over the gap facing the south-west, between Therasia and Cape Akroteri (Santorin), where Aspro Island rises above the shallow bank connecting the chief islands (the brim of the volcanic basin, slightly covered with water),§ tend to force in detrital matter. Deposits at the bottom of the central cavity, varying in depth from 960 to 1,278 feet in its curved passage round the Kaimeni, would seem well at rest, except as regards upheaval or depression from volcanic action there prevailing. In cases where animal life might become extensively destroyed by the boiling of the waters, a ready supply of the germs, even of

* In a letter from Professor E. Forbes to Dr. Daubeny, quoted by the latter in his "Description of Volcanos," 2nd edition, p. 324, 1848.

† Professor E. Forbes informs me that the following shells were there obtained:— *Pectunculus pilosus, Arca tetragona, Cardita trapezia, Trochus ziziphinus, T. fanulum, T. exiguus, T. Coutourii, Turbo rugosus, T. sanguineus, Phasianella pulla, Turritella tricostata, Neæra cuspidata, Cerithium Lima, Pleurotoma gracile.*

‡ "My own impression is," adds Professor E. Forbes, "that this group of islands constitutes a crater of elevation, of which the outer ones are the remains of the walls, whilst the central group is of later origin, and consists partly of upheaved sea-bottoms and partly of erupted matter, erupted, however, beneath the surface of the water."

§ Commencing with the southern point of Therasia, the soundings show this submerged brim of the crater to be successively 42, 36, 54, 66, 54, and 42 feet (depths at which wind-wave action would cause much disturbance on the bottom, especially during heavy gales), a point, marked as a sunken rock, rising higher between the south end of Therasia and Aspre Island.

Fig. 134.

VIEW OF SANTORIN FROM THE ENTRANCE.

a, Apanomeria; b, North-west Cape of Santorin; c, Mount Elias; d, Mikro Kaimeni; e, Neo Kaimeni; f, North-east Cape of Therasia.

those inhabiting deep water, could be furnished by the deep channel opening to the northward; one also through which the heated waters could flow outwards on the surface, while cooler water supplied their place by inflow beneath. As regards the exposure of this channel to wind-wave action, a glance at the charts of the Ægean Sea will show that it is comparatively well sheltered by Nio, Sikino, and Polykandro, with the bank connecting those islands, on the north and north-west, while Siphano, Paros, and Navos prevent any great range of sea exposure still further beyond those islands.

Examining the charts of coasts or islands where volcanos occur, various instances will be found by the observer in which the sea more or less enters the central cavities or craters, or where, by a little more cutting away of the remaining walls of the crater by breaker action, or a slight change of level, bringing some lower part of its lip further down, a similar entrance of the sea would be effected. The island of St. Paul may be taken in illustration of a crater just so far laid open by breaker action, that the portion of the brim of the basin through which the sea enters is nearly dry at low tide. It will be seen by the accompanying plan (fig. 135),

Fig. 135.

that this little island, scarcely 2¼ geographical miles from N.W. to S.E., and about 1¼ mile broad from N.E. to S.W., rising in the Indian Ocean, between the south of Africa and the west of Australia, Madagascar being the nearest mass of dry land, and more than 2,000 miles distant, is the mere summit of a volcano. With its companion, Amsterdam Island, it forms a remarkable protrusion through the ocean amid a mass of waters, an excellent example of the uplifting of volcanic matter through them. That such mere points should be easily removed by breaker action, unless some hard rocks should be accumulated, would be expected, and the foregoing plan (fig 135) and accompanying view (fig 136),* would seem to show that its abrasion by such means is now

Fig. 136.

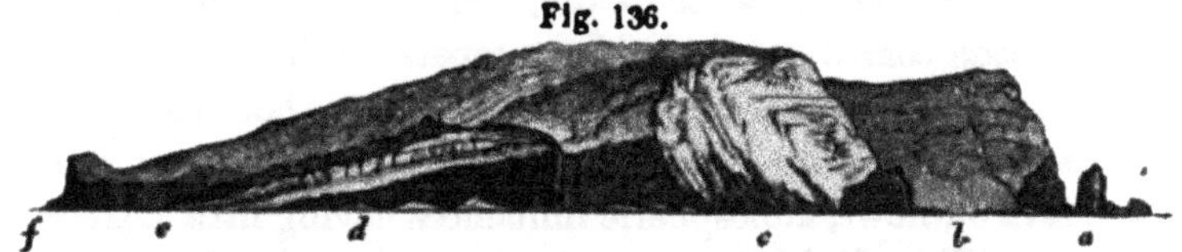

a, Nine-Pin Rock; *b*, Entrance to crater-lake; *c*, Cliff, well exhibiting the united action of breakers and atmospheric influences; *d*, Dark-coloured rock, dipping seaward; *e*, Section of a dark-coloured bed; *f*, The southern point of the Island.

being accomplished. The high cliffs, several hundred feet in elevation, appear the remains of the accumulations cut back by the breakers, so ceaselessly at work in such a situation, the amount of removal on the N.E. of the island being, to a certain extent, shown where the anchorage (*a*) of the surveying ship, the "Fly," is marked in that direction on the chart, as upon sand and stones, and beyond which the depth suddenly increases seaward. The greatest height of the island is stated to be 820 feet. The Nine-Pin Rock (*a*, fig. 136) is merely a harder portion left by the breakers, and it may be inferred, the present breaker action continuing to remove the matter of St. Paul's Island, and no new eruptions adding to its mass, that in time such points might alone remain above water to attest the former presence of a volcanic crater, the walls of which once rose several hundred feet above the level of the sea.†

The distribution of volcanos over the face of the globe will strike the observer as necessarily important when searching for their cause and studying their effects. It will also at once be apparent to him that independently of the modifications which may arise in local conditions, so that the same vent may be active at one time, with variable degrees of intensity, and dormant at another, there may be great general conditions becoming so permanently changed, that a region once marked by volcanic action may so far be considered as ceasing to be so, that some very considerable modification in such portion of the earth's crust must be effected to produce a return of that action. In consequence of such complicated conditions it be-

* From that accompanying the Admiralty chart of the island of St. Paul, by Captain Blackwood, R.N.

† In the view above given there is an apparent interstratification of differently-shaded rocks, perhaps of lavas and tuff. During the abrading process by the breakers, these would necessarily be acted upon according to their relative hardness and positions.

comes almost impossible to separate active volcanos from those termed extinct, inasmuch as we can scarcely be certain that many of the latter may really be such, and not in a comparatively dormant state. As a matter of convenience, therefore, rather than of principle, it has usually been considered desirable to separate volcanos known to have been active from those never recorded to have been so, though often preserving the forms which they took under subaërial eruptions and lava outflows, atmospheric influences having little changed such forms. This division, however convenient in the present state of the subject, should not mislead the observer, nor will it do so when he regards volcanic action on the larger scale, and igneous products generally, during the long lapse of geological time of which we can obtain any relative records.

It has long been remarked that active volcanos are chiefly, though not altogether, situated amid, or at moderate distances from, oceans and seas,* a circumstance also considered important as regards their products, especially as respects aqueous vapours and certain of the gases evolved. It would be out of place to enter upon the hypotheses framed in consequence, further than to call attention to points which appear important for the effective study of the subject. First, however, as respects the facts recorded. Upon glancing at any of the maps of the world whereon the existing knowledge of volcanos, considered sufficiently active, is laid down, we find them in the ocean separating America from Europe and Africa, ranging in points, or groups of points, from Jan Mayen's Island on the north, by Iceland, the Azores, Canary, and Cape de Verde Islands, Ascension, and Trinidad (South Atlantic), to Tristan da Cunha on the south. On one side of the same waters the line of the West Indian volcanos presents itself. In the Indian Ocean appear the somewhat scattered groups of the Islands of Bourbon, Mauritius, and Rodriguez, and the small isolated points of St. Paul and Amsterdam Islands. In the central portion of the Pacific are the Hawaiian, Marquesas, and Society Islands groups, with Easter Island. On the north of the same ocean is the range of the Aleutian vents, having a somewhat W.S.W. and E.N.E. direction, (as if upon a great fissure,) into the north-west of North America. To the westward of the Aleutian volcanos a range of vents, commencing with the somewhat lofty volcanos in Kamschatka, proceeds in a south-west direction through the Kurile Islands to and beyond Japan. Southward of the latter, and having

* M. Arago pointed out in 1824 (Annuaire), that about five or six only of the 1[illegible] reputed active volcanos of the world were not so circumstanced.

a north and south direction, are the volcanos of the Bonin and Mariana Islands. On the S.E. of Japan a series of vents commences, which, ranging down by Formosa and the Philippines, passes round in the form of a huge fish-hook, by the N.E. point of Celebes, Gilolo, the volcanic isles between New Guinea and Timor, Floris, Sumbawa, Java, and Sumatra, to Barren Island, where a central active cone rises amid water, entering from the sea on one side, this interior water bounded, except on that side, by a series of cliffs.* Returning to the Pacific we find the volcanic group of the Galapagos in that ocean, off the coast of Quito.

As respects volcanos on continents, or in seas more or less intermingled with them, the vents of South America constitute by far the most important range. After quitting the volcanos of Tierra del Fuego, a space intervenes northward for several degrees of latitude in which they have not been noticed. Then succeeds the long line of the volcanos of Chili, several rising to considerable heights above the sea. Another break then occurs, after which volcanos appear in the Andes of Quito, Cotopaxi, being one of them. Passing the Isthmus of Darien, the vents of Guatemala come in, succeeded more northerly by those of Mexico. Continuing in the same direction volcanos become scarce in North America, a few points only being noticed in California, upon the Columbia, on the island of Sitka, and in Russian America,† where the Aleutian range joins in. Active volcanos in Europe are confined to the Neapolitan States and Santorin, the latter being inferred to be merely for the time, dormant. On the continent of Africa no active vents are known, Jebel Tarr, in the Red Sea, being as much Asiatic as African. With respect to Asia, the great mass of that continent appears (omitting the Kamschatkan peninsula), to be at least, in a great measure, without active vents. Though doubts are expressed respecting such vents, the statements regarding volcanos in Central Asia are deserving of every attention, and numerous warm springs would appear there to be found, under circumstances which may connect them with volcanic action.‡

* Von Buch views this island as highly illustrative of a cone of eruption in the midst of a crater of elevation; and Dr. Daubeny observes, that "if we compare this locality with Santorin, there will be found nothing but the absence in the latter case of an active vent in the centre of the bay whereby to distinguish it; and from Monte Somma it chiefly differs in its lower level, which causes the bottom of the crater to be sunk beneath the waters of the ocean."—"Description of Volcanos," p. 413, where a view of Barren Island is given.

† Wrangell's Volcano, on the Atna.

‡ With respect to the volcanos of Central Asia, Humboldt, after observing that Abel Rémusat first called the attention of geologists to them (Annales des Mines, t. v., p. 137), remarks (Kosmos, Sabine's, 7th edition, p. 232), that there is "a great

It will be perceived that the statement as to these communications with the surface of the earth being chiefly found amid oceans and seas, or not far from them, seems borne out. Volcanos in Mexico and those of Central Asia, assuming that there are still active volcanos there, would appear the principal exceptions.* As respects the former, it has been held by the advocates of the necessity of water as one of the causes of volcanic action, that there may be a connexion along a great fissure extending in an east and west direction across Mexico, vents being established upon it at Colima, Jorullo, Pococatepetl, and Orizaba. With regard to Central Asia, it can be inferred that it is a region which may have once been in a great measure occupied by an inland sea, waters being then supplied to the volcanic foci, as is now supposed by some to happen in the volcanic regions of the Mediterranean.†

Should the presence of water be considered only a secondary cause of volcanic action, it would follow that during the change of levels which have taken place over large areas on the earth's surface, circumstances may arise that should at one time permit the easy access of water to great fissures or apertures of any form, and at another prevent it. Thus aiding in changing active volcanic regions into those termed extinct, independently of the termination of other and perhaps more general conditions from which volcanic action may [illegible]. If we regard the variable amount of dry land which would be exposed above water by changes of the relative level of sea and land, such as has been above noticed in the British Islands (fig. 99), it would appear that parts of France might constitute islands at one time to which sea-

---

volcanic chain, the Thian-schan (celestial mountains), to which belongs the Pe-schan, from whence lava issues, the Solfatara of Urum-tsi, and the still active 'fire mountain' (Ho-tscheu) of Turfan, almost equidistant from the shores of the Polar Sea and of the Indian Ocean (1,400 and 1,528 miles). Pe-schan is also fully 1,360 miles from the Caspian Sea, and 172 and 208 miles respectively from the great lakes of Issikoul and Balkasch." * * * "It is impossible not to recognize currents of lava in the descriptions given by the Chinese writers of smoke and flame bursting from the Pe-schan, accompanied by burning masses of stone flowing as freely as 'melted fat,' and devastating the surrounding district, in the first and seventh centuries of our era."

* Respecting the range of the Mexican volcanos, Humboldt remarks (Kosmos, Sabine's, 7th edit., p. 232), that Jorullo, Pococatepetl, and the Volcano de la Fragua, are respectively 80, 132, and 156 geographical miles from the ocean.

† Reference to the remarkable fact that in the island of Cephalonia a stream of sea-water, in sufficient quantity and volume to turn a mill, is constantly flowing from the sea inland, where it becomes swallowed up, has been made, as illustrating the employment of water in volcanic action, that supply of water being converted into gases and vapours, producing earthquakes and volcanic eruptions.—See Mr. Strickland, "Geol. Trans.," 2nd Series, vol. v., p. 408, and "Geological Proceedings," vol. ii., pp. 220, 393; also Daubeny's "Description of Volcanos."

water could have more ready access (from proximity) to any volcanic foci beneath, than at another. Thus, supposing such supply needed, if it were stopped by changes removing the sea to greater distances, a region containing active volcanos at the one period, might present only extinct craters at another. This, even under the hypothesis of the water being essential, might not necessarily be always the real cause of change, inasmuch as the general conditions, productive of volcanic action in such districts, may have been exhausted, so that whether the supply of water was or was not altered, that action became extinct.

Whatever may have caused volcanic fires to have ceased, there are whole regions, independently of portions of districts in parts of which active volcanos still exist, or have been known in historic times, which offer clear evidence of volcanic action having prevailed, sometimes extensively, in them at no very remote geological period, loose piles of cinders and ashes, and lava streams being found nearly as fresh as when ejected. The district of Auvergne, in central France, has for about a century engaged attention, as one of extinct volcanic action.* This seems to have continued for a long period, various points of communication having been established between the interior of the earth and the atmosphere at different times, and changed and modified results the consequence. In addition to Auvergne, similar accumulations are found in France, in the Cantal, the Velay, the Vivarais, the Cevennes, and in the vicinity of Marseilles and Montpellier. In Germany they are seen in the districts of the Eifel, the Siebengebirge, and other places. Hungary, Transylvania, and Styria, present their trachytic and other igneous products. Extinct volcanic action is also traced in Spain and Sardinia; Italy offers its extinct as well as active volcanic accumulations, as does also

* It is now about a century since (1752) that Guettard published his "Mémoire sur quelques Montagnes de la France, qui ont été des Volcans (Mémoires de l'Académie des Sciences)." As regards general views of the extinct volcanos of France, the observer will find them in Mr. Scrope's "Geology of Central France," 1827; in the "Mémoires pour servir a une Description Géologique de la France," (1830–38), and the "Explication de la Carte Géologique de la France," 1841, by MM. Dufrénoy and Élie de Beaumont, and in Dr. Daubeny's "Description of Volcanos," 2nd edit., 1848. More than 70 years since M. Desmarest published a map of Auvergne, always adverted to with satisfaction by those who have visited that region. As Sir Charles Lyell remarks ("Principles of Geology," 7th edit., p. 51), "Desmarest, after a careful examination of Auvergne, pointed out, first, the most recent volcanos which had their craters still entire, and their streams of lava conforming to the level of the present river-courses. He then showed that there were others of an intermediate period, whose craters were nearly effaced, and whose lavas were less intimately connected with the present valleys; and, lastly, that there were volcanic rocks, still more ancient, without any discernible craters or scoriæ, and bearing the closest analogy to rocks in other parts of Europe."

Greece, when including its islands. In Asia Minor, extinct volcanos are found, and more especially in the wide district of the Katakekaumene.* With respect to the Holy Land, the destruction of Sodom and Gomorrah has been attributed to volcanic eruptions, and volcanic accumulations are elsewhere noticed in the same land, and in Persia, and its adjoining countries. Doubtless, also, many other regions, not yet explored by the geologist, will be found to present similar accumulations, and indeed they have been noticed in the great continent of America.

In various parts of the world, as well in regions where lava streams intermingled with ash and cinders, either piled up conically, or more evenly distributed, are not apparent, as in those where active or extinct volcanos exist, certain rocks are found to which the name *basalt* has been given. In the application of this name, care has not always been taken to distinguish the same compound considered chemically and mineralogically, so that in the matter of fusibility alone, substances so termed differ somewhat materially.† Fine varieties of greenstone (diabase), consisting of orthoclase and hornblende have as often been termed basalt, as those of labradorite and augite (dolerite). If, with M. Rose, hornblende and augite be considered only modifications of the same mineral, this would leave the difference of these two varieties of basalt to consist in that of the two felspars. The basalt of the Mont Dor has been stated to contain both the augite and hornblende forms of this mineral. Basalt has again been supposed essentially to consist of augite, magnetic iron, and a mineral of the zeolitic family.‡ The un-

* Messrs. Hamilton and Strickland (Geol. Trans., 2nd series, vol. vi., 1841, and "Travels in Asia Minor," 1842, by the former geologist) consider the volcanic products of the Katakekaumene, as referable to three periods. The volcanic accumulations of the last period are as fresh as amid active vents, the ashes and scoriæ still loose and piled up as after immediate ejection, the lava streams rugged, a few straggling plants alone finding fitting conditions for their growth.

† During the experiments on the fusibility of rocks, to which allusion has been above made, we found marked differences in that of the so-termed basalts. Allowing for changes by the different conditions under which substances, originally similar, may have been placed, so that while one may have been deprived of certain substances, another may have mineral matter added to it, there were still evidently original differences. It has been stated by De Saussure (Journal de Physique), that basalt melts at 76° of Wedgwood. The experiments of Sir James Hall (Trans. Royal Soc. of Edinburgh, vol. v.) went to show that whinstone, or basalt, as it has been called, from the vicinity of Edinburgh, became soft at a temperature from 28° to 55° Wedgwood, a heat, as Dr. Daubeny remarks (Description of Volcanoe, p. 616), inferior to that of a common glass-house.

‡ Referring to the composition of this zeolitic mineral, Dr. Daubeny observes (Description of Volcanos, p. 18), that it "is such as to imply that it may have been formed out of labradorite by the addition of water, the presence of which in all zeolites is the cause of the bubbling up under the blowpipe, which has occasioned them to be distinguished by that general appellation." Following out this view, it seems highly

certainty in the employment of the term basalt, would appear to require attention. Thus the rocks which encircle the Peak of Teneriffe, and usually noticed under that head, are referred by Dr. Abich to his class of trachyte-dolerites. While endeavouring to trace the sources whence certain igneous rocks may have been obtained, even sometimes with reference to the melting of masses which may have been accumulated by means of water, or have been intermingled with such deposits, mineralogical and chemical distinctions, as far as they can fairly be carried out, would appear very desirable.* While at times sheets of basalt cover extensive areas, at others they are mingled with ordinary volcanic products, apparently, therefore, ejected under dissimilar conditions. Basalt is sometimes highly vesicular, at others very compact; these modes of occurrence are observable over areas of different extent, both considerable and limited.

As regards the relative antiquity of basalt, we find it noticed as well among the ancient as the more modern volcanic products of central France, and among the more modern of the Vivarais in the south of France, as also in those of the Eifel.† It is noticed as intermingled among the ancient volcanic rocks of the Siebenge-

desirable to consider how far a change may be brought about in a compound of augite, magnetic iron, and labradorite, so that the latter became modified by water after ejection. The vesicles of basalts, as, for example, those of the north of Ireland, are often filled with zeolitic minerals, the results of infiltrations into them, quite as much as agates, &c., also found amid the same rocks. In fact, incer tain districts, the vesicles are filled with a variety of substances, the zeolites forming only a part of them.

* As respects the chemical composition of basalt, including that of Teneriffe (trachyte-dolerite of Dr. Abich), the following table of basalt, from Saxony, (1), by Mr. Phillips, from Banlieu (2), by M. Beaudant, and from Teneriffe (3), by Dr. Abich, may be useful:—

—	1	2	3
Silica	44·50	59·5	57·76
Alumina	16·75	11·5	17·56
Protoxide of iron . .	20·00	19·7	4·64
Peroxide of iron. . .	..	0·5	2·09
Oxide of manganese .	0·12	..	0·82
Lime	9·50	1·3	5·46
Magnesia	2·25	..	2·76
Potash	..	1·6	1·42
Soda	2·60	5·9	6·82
Chlorine	..	..	0·30
Water	2·00	..	trace

† On this point, Dr. Daubeny remarks (Description of Volcanos, p. 42), when mentioning the occurrence of basalt with the fresh-water limestones, near Clermont, and the proof by M. Élie de Beaumont of this basalt forming dykes amid the fresh-water formation of the Limagne (Mémoires pour servir, &c., tom. i.), that while it occasionally underlies the trachyte and subjacent tuffs of the districts, "its general relation to both these rocks indicates that it is of more modern eruption."

birge, as also of later date in the same district. Basalt is described as among the ancient igneous rocks of Iceland. It occurs in many parts of the world where its relative date is not so apparent, sometimes forming the isolated caps of hills, and resting upon other rocks, in a manner pointing to the considerable or partial destruction of some great sheet of this rock. This has been supposed the case with the basaltic hills in parts of Germany. The largest area occupied by basalt seems to be in India, where rocks of this class appear to occupy one of 200,000 square miles.* With respect to this rock, a fine exhibition of it is found in the north of Ireland, where the Giant's Causeway and the adjacent country have long attracted attention. Though on a much smaller scale, the island of Staffa, Hebrides, has also long been equally celebrated for its basalt. In the north of Ireland, its eruption was posterior to the formation of the chalk of the same district, but the portion of the tertiary period to which this should be referred is not clear.

Though by no means confined, among igneous rocks, to basalt, the spherical and columnar structures often developed in that rock have also long attracted much attention. The minor spherical structure seen on the small scale in some volcanic rocks, and also in artificial glass, and which has been previously noticed, would appear to have been produced on the larger scale, under certain conditions, in basalts. Sometimes this globular structure, as shown during the decomposition of the rock, is irregular, so that the whole has the appearance of balls of various dimensions piled up without much order (fig. 137); at others, a great order prevails, and the concretions are either roughly arranged above one another in wide spheroidal shapes, or so pressed against each other as to produce prisms, sometimes of very symmetrical forms. In 1804, Mr. Gregory Watt showed by his experiments on basalt, that when, in the cooling of a molten mass of that rock, this structure was developed, and "two spheroids came into contact, no penetration ensued, but the two bodies became mutually compressed and separated by a plane, well defined and invested with a rusty colour," and he observed, when several spheroids met, that they formed prisms.†

Fig. 13

* Lieut.-Colonel Sykes (Geological Transactions, 2nd series, vol. iv., p. 409) observes, that in the Dukhun there are proofs of a continuous trap formation, covering an area of from 200,000 to 250,000 square miles.

† Observations on Basalt, and on the transition from the vitreous to the stony texture which occurs in the gradual refrigeration of melted basalt, Phil. Trans. 1804.

From the arrangement observed by Mr. Gregory Watt, he inferred that "in a stratum composed of an indefinite number, in superficial extent, but only one in height, of impenetrable spheroids, with nearly equidistant centres, if their peripheries could come in contact on the same plane, it seems obvious that their mutual action would form them into hexagons; and if these were resisted below, and there was no opposing cause above them, it seems equally clear that they would extend their dimensions upwards, and thus form hexagonal prisms, whose length might be indefinitely greater than their diameters. The further the extremities of the radii were removed from the centre, the nearer would their approach be to parallelism; and the structure would be finally propagated by nearly parallel fibres, still keeping within the limits of the hexagonal prism with which their incipient formation commenced; and the prisms might thus shoot to an indefinite length into the undisturbed central mass of the fluid, till their structure was deranged by the superior influence of a counteracting cause."

It will require the careful study of this class of rocks, more particularly in a decomposed state, for the observer to ascertain the extent to which the view of Mr. Gregory Watt may be applicable. Where one plane of a sheet of basalt may have been exposed to cooling influences, so that the spheroidal structure could be first developed in it, and in the manner suggested, and also so that no other spheroidal bodies could be developed in the general body of the rock, and thus interfere with the extension of the original spheroid, there would not appear much difficulty in following this view. In those basaltic dykes that are sufficiently common in some districts, where we may suppose that the walls of the fissure, which had been filled by the molten rock, presented equal cooling conditions, we sometimes see, as in the subjoined section (fig. 138), that the prisms shoot out at right angles to the walls of the containing rock (b c), as if each set commenced at the sides (d and e), confusion arising at the central portion (a f).* In cases, also, when

Fig. 138.

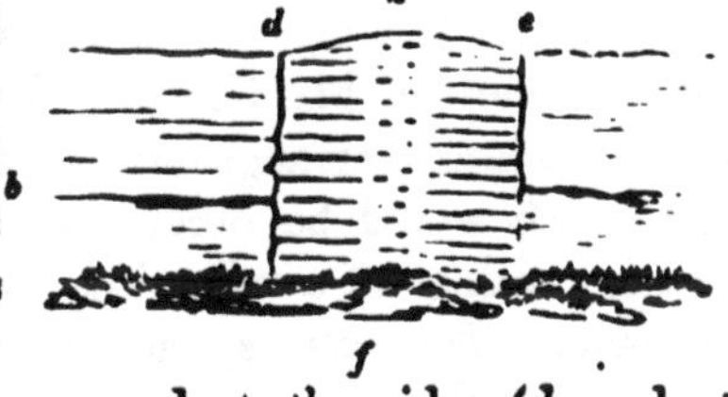

* It sometimes happens that the central portions of a basaltic dyke are more prismatic than the sides, as if the cooling had been too rapid at the sides for the production of this arrangement of parts. Again, the prisms are sometimes found ranging from wall to wall of the fissure, as if artificially-cut prismatic blocks of rock had been piled in it on their sides.

not a trace of joints can be observed, as in the annexed section (fig. 139), where the columns (*c*), are seen to rise at right angles

Fig. 139.

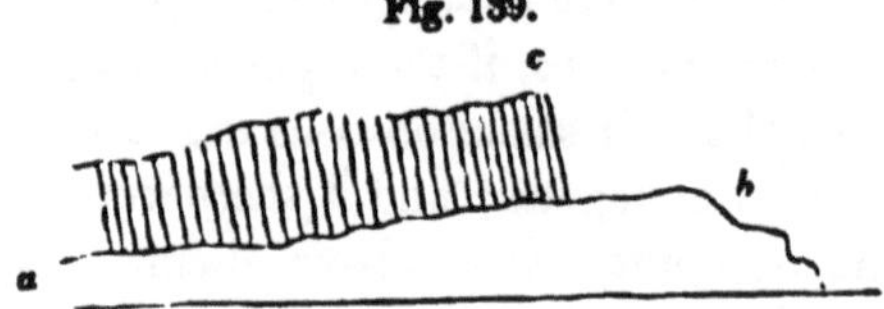

to the supporting rock (*a b*), which may be of any kind, igneous or accumulated in water, the prisms reaching to the height of 100 feet or more, an original cooling lower plane may have produced the prisms throughout. Also in those curved columns of basalt, where, as in the following sketch (fig. 140), no joints are apparent,

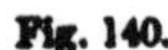
Fig. 140.

even upon the weathering of the rock, we may suppose that some tendency of an original set of spheroids to develop themselves more in one direction than another, from some local cause, has been so continued as to produce the general curve observed.

When the jointing of the prisms is marked, though, no doubt, upon the view of Mr. Gregory Watt, the prolongation of additions to the radiating arrangement of parts would render the pauses of that which would be otherwise concentric coatings of a spheroidal mass, somewhat flat plains, across the prisms, so that the annexed structures (fig. 141), might be thereby accounted for, facts are occasionally seen, where the decomposition of the joints would rather point to the production of separate centres of radiation. Certain joints of the great bed of prismatic basalt which, dipping into the sea, forms the well-known Giant's Cause-

Fig. 141.

way, in the north of Ireland, would seem to countenance such a view. These joints are observed to have minor pieces, *a*, *a*, *a*, supplemental, as it were, to the main joints, filling up corners; giving an idea of each joint having been a separate sphere, the minor pieces completing the arrangement of particles in the corners, where sphere pressing against sphere, these remained to be filled up. At times the minor pieces constitute more of the whole sharp corners of the prisms, as represented in the annexed (fig. 143).

Fig. 142.

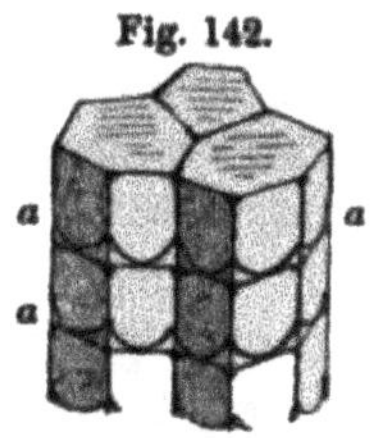

Fig. 143.

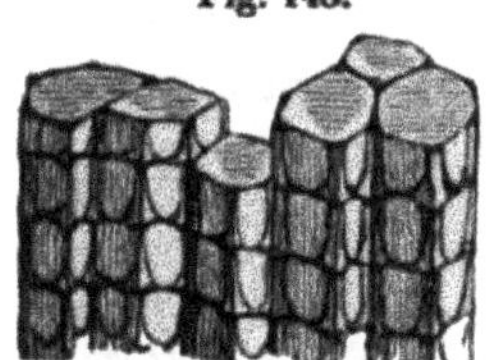

Of the intermixture of conditions producing flows of melted rock at one time from the same general vent, or system of vents, which should take the prismatic form, and at another exhibit no tendency to that structure, the Giant's Causeway and adjacent district in the north of Ireland will afford the observer a good example. The same mixture of prismatic and more solid basalt is also to be found in the Island of Staffa, where, as shown

Fig. 144.

in the annexed sketch (fig. 144),* the action of the Atlantic breakers has worn out the celebrated Fingal's Cave.

* Reduced from Macculloch's "Western Islands of Scotland."

CHAPTER XXI.

SALSES OR MUD VOLCANOS.—GASEOUS EMANATIONS UNCONNECTED WITH VOLCANOS.—RESULTS OF DECOMPOSING IRON PYRITES AMID BITUMINOUS SHALE.—MUD VOLCANOS IN THE BAKU DISTRICT—OF THE NEIGHBOURHOOD OF TAMAN AND KERTCH—OF MACULABA, GIRGENTI.—BORACIC ACID LAGUNES OF TUSCANY.

MINERAL matter is raised from beneath and thrown out upon the surface of the ground, and vapours and gases are evolved, the latter sometimes inflammable, in a manner which so differs from, or forms a modification of, the volcanic action previously noticed, as to merit separate attention. Amid the changes effected during the modification of ordinary volcanic action, it may readily happen, as has been seen, that aqueous vapours and certain gases alone escape from old volcanic vents, and masses of mud may be ejected, as from Tongariro, New Zealand (p. 323). In these cases, the gases evolved would tend to show the observer the connexion between volcanic action, such as it is manifested, with a very general resemblance, in so many situations scattered over the face of the globe, and any localities he may be examining; more especially if volcanic rocks prevail in them. As the subject at present rests, it requires more attention than has always been assigned it, inasmuch as somewhat similar appearances may be brought about by different means. While a modification of volcanic action may connect certain of these salses or mud volcanos, as they have been termed, with the general cause of that action, others may depend upon causes which, though producing effects of local importance, could scarcely, as regards the crust of the globe, be considered as exerting any great geological influence; at the same time, as manifesting alterations in the condition of the matter composing even limited portions of the accumulations on the earth's surface, they require consideration.

With respect to gaseous emanations, they are not only found so connected with volcanic regions, that their origin can scarcely be doubted; but also in localities where that action is either not apparent, or where other sources may be reasonably assigned them. Of the latter kind are those discharges of carburetted hydrogen, which rise in several coal districts; this gas occasionally evolved in such volume as to be economically employed.* In these cases, our experience in working collieries shows us that such gases are abundantly produced from certain coal beds and associated carbonaceous shales, the result of a decomposition of those bodies by which, among other changes, a portion of the constituent carbon and hydrogen is evolved in a gaseous state. That fissures, or other natural rock channels, should permit the escape of this gas to the surface, and that, the causes for its production continuing, it should have been known during a long lapse of time, would be expected. Emanations of carburetted hydrogen are well known in the coal districts of Europe and America.

When beds of lignite, coal, or shales highly impregnated with bituminous matter, can be acted on by heat, so that these substances may be placed under somewhat of the conditions of the coals in a gas-work, we should expect results corresponding with the resistance to the escape of the gas which any associated or superincumbent rock-deposits may offer, with the additional force exerted by any steam which may be derived from disseminated water, the latter sometimes forming no inconsiderable power for overcoming superincumbent resistance. In such instances, the heat produced by the decomposition of iron pyrites, so often disseminated amid carbonaceous and bituminous deposits, should scarcely be neglected, a sufficient supply of air and water being

* At Fredonia, State of New York, this gas has long been collected in a gasometer for the lighting of the place; and, according to Humboldt (Kosmos), it has been used in the Chinese province of Tse-tchuan, for more than 1,000 years. M. Imbert states, that an inflammable gas is employed in evaporating saline water at Thsee-lieou-tsing. "Bamboo pipes carry gas from the source to the place where it is to be consumed. These tubes are terminated by one of pipe-clay, to prevent their being burnt. A single source (of gas) heats more than 300 kettles. The fire thus produced is exceedingly brisk, and the caldrons are rendered useless in a few months. Other bamboos conduct the gas intended for lighting the streets and great rooms or kitchens."—"Bibliothèque Universelle," and "Edinburgh Philosophical Journal," 1830. The wells whence this inflammable gas rises were sunk for the purpose of obtaining the saline water. This they first afforded; but the water failing, they were sunk much deeper, when, instead of water, the gas rushed out suddenly with considerable noise (Humboldt, "Fragmens Asiatiques"). This seems a good instance of tapping, as it were, a supply of inflammable gas, pent up in a compressed state.

effected. Indeed, the "burning," as it is usually termed, of bituminous shales exposed in cliffs,* and through which easily-decomposed iron pyrites are disseminated, is sufficient to show that this circumstance should receive attention, however exaggerated the views taken respecting the effects of such causes may once have been.

The country around Baku, a port on the Caspian, would appear instructive, not only as respects the emanation of inflammable gas, but also with regard to the production of one class of salses, or mud volcanos. That district is described as impregnated with petroleum and naphtha, to such an extent that the inhabitants of Baku employ no other fuel. About ten miles N.E. from the town there are many old temples of the Guebres, in each of which inflammable gas, burning with a pale flame, and smelling strongly of sulphur, rises in jets from the ground.† A large jet is stated to issue from an adjoining hill-side, and the whole country around, for a circumference of two miles, is so impregnated with this gas that a hole being made in the ground it immediately issues, the inhabitants thrusting canes into the earth, through which the gas rises and is used in cooking.‡ It was near Jokmali, to the east of Baku, that, on the 27th November, 1827, flame burst out, where flame had not previously been known, rising to a considerable height, for three hours, after which it became lowered to three feet, burnt for 20 hours, and was then succeeded by an outburst of mud, covering an area of more than 1,000,000 square feet to the depth of two or three feet.§ Large fragments are mentioned as having

* The Kimmeridge clay of the Weymouth coast, in whicht here is much shale, in places so bituminous as to have been distilled for the bitumen in it, offers from time to time a good example of the "burning" of a cliff from the decomposition of iron pyrites amid bituminous shale by the action of the weather. The heat generated has been occasionally so considerable as to fuse some of the clay or shale.

† It would be expected that these natural jets of inflammable gas would be utilised, wherever ascertained to be emitted, by those to whom a perpetual fire could be of importance in their religious rites. Captain Beaufort (Karaman'a) describes a jet of inflammable gas, named the Yanar, near Deliktash, on the coast of Karamania, probably once thus used. "In the inner corner of a ruined building, the wall is undermined, so as to leave an aperture of three feet in diameter, and shaped like the mouth of an oven; from thence the flame issues, giving out an intense heat, yet producing no smoke on the wall." Though the wall was scarcely discoloured, small lumps of caked soot were found in the neck of the opening. The Yanar is considered to be very ancient, and possibly the jet described by Pliny. The hill whence it issues is formed of crumbling serpentine and loose blocks of limestone. A short distance down it there is another aperture, whence, from its appearance, another jet of a similar kind is inferred once to have risen.

‡ "Edinburgh Philosophical Journal," vol. vi.

§ Humboldt, "Fragmens Asiatiques."

been thrown out and hurled around.* A column of flame rose so high at an eruption near Baklichli, west of Baku, that it could be seen at the distance of 24 miles. The country is considered to afford other traces of similar eruptions.

While these eruptions have taken place near Baku, on the east of the Caucasus, similar outbursts of flame and mud have occurred, under similar circumstances, in the neighbourhood of Taman and Kertch, at the western extremity of the same range. These have been long known, and taking place in an area which comprises the Cimmerian Bosphorus, where the Sea of Azof communicates through a shallow channel with the Black Sea, they become important in effecting surface changes, tending still further to close this channel upon the outflow of the river waters poured into the Sea of Azof, chiefly by the Don and its tributaries, and not evaporated in it. These salses, or mud volcanos, are found on both sides of the strait, and are situated, like those of Baku, in a district replete with bituminous matter. M. Dubois de Montpéreux gives sections showing the area to be principally composed of a highly-bituminous (tertiary) shale, sometimes with lignite, alternating with sands. From these bituminous beds asphalte is prepared, and there is evidently much bituminous matter, including naphtha, disseminated in its various forms; indeed, naphtha springs are mentioned as rising near the crater-cavity of Khouter. In some situations the salses seem to have vomited forth flame and mud from the same spots at different times, at others these suddenly rise from places not previously known.† The gases evolved from the salses at Baku, Taman, and Kertch, and from the vicinity of Tiflis, where similar facts are noticed, chiefly consist, (Dr. Abich, who has personally examined them, informs me), of carburetted hydrogen, an important circumstance connected with

* Humboldt, "Kosmos,"—Mud Volcanos.

† Dubois de Montpéreux (Voyage autour du Caucase, t. v., p. 51), mentions, respecting these mud volcanos, that Koukou-oba was in eruption in February, 1794; and Koussou-oba on the 26th April, 1818; that the chief eruption of Gnila-gora, near Temrouk, was in February, 1815; that an island appeared in front of the Isle of Tyrambe, on the 10th May, 1814; and that the mud Volcano of Taman was never in a greater state of activity than in April, 1835. He comments on these eruptions having occurred at one time of the year, remarking, with Pallas, that the only known autumnal eruption was on the 5th September, 1799, when the first island was thrown up. In the Geological Atlas accompanying the "Voyage autour du Caucase," there is a plan (pl. xxvi.) in which the various salses or mud volcanos of Taman and Kertch are laid down; and in the Carte Générale Géologique of the same work, the districts of Baku and Tiflis are included. Section, pl. xxv., shows the alternations of bituminous shales and sands whence a mud volcano broke out near Koutchougourei, bordering the Sea of Azof.

their origin. M. Dubois de Montpéreux mentions the emission of sulphuretted hydrogen when the mud of Khouter was disturbed, and that there was also a sulphurous spring not far distant. M. de Verneuil gives an elevation of 250 feet to some of the conical mud accumulations of Taman and the Eastern Crimea.* Iron pyrites seems to be found amid the ejected mud. As might be expected, these jets of flame, smoke, and mud occur as well in the shallow water adjoining the dry land as upon the latter, even adding to and modifying its form. In 1814, flames rose through the Sea of Azof, mud was thrown out, and an island gradually produced. Among the stones ejected at these eruptions are limestones and shales not known among the surrounding strata.

The mud volcanos of Maculaba, near Girgenti, whence mud and bituminous matter are thrown out, Dr. Daubeny attributes to the combustion of the beds of sulphur there associated with the blue clays, amid which these mud eruptions take place.† He ascertained that the gases given off consisted of carbonic acid and carburetted hydrogen. At the time of his visit the cavities were small and filled with water, somewhat above the usual temperature of that in the country, mixed with mud and bitumen, through which the gases bubbled up.‡ Dr. Daubeny refers similar phenomena at Terrapilata, near Caltanisetta, and at Misterbianco, near Catania, to the same causes.

To ascertain how far such salses or mud volcanos may arise from other than strictly volcanic causes, or be merely some secondary effects produced by them, it becomes very desirable not only that the geological structure of the country should be well examined, but also that the gases evolved should be very carefully ascertained. According to Humboldt and M. Parrot, almost pure nitrogen is found among the gases evolved from the mud volcanos of the peninsula of Taman, and the former mentions hydrogen mixed with naphtha as emitted from salses of this kind. The stones

---

* "Bulletin de la Soc. Géol. de France."

† "Description of Volcanos." Alluding to the combustion of the sulphur, Dr. Daubeny remarks, that "the sulphurous acid being retained by the moisture of the rock, and gradually converted into sulphuric acid, would act upon the calcareous particles, and give rise to the extrication of carbonic acid gas, whilst if any bituminous matters were present, the heat generated might cause a slow decomposition, and resolve them into petroleum and carburetted hydrogen," p. 267.

‡ It is stated that at times "the mud has been known to be thrown up to the height of 200 feet, accompanied by a strong odour of sulphur."—Daubeny, "Volcanos," p. 266.

mentioned by Sir Roderick Murchison* as ejected in the Taman and Kertch district differing from those forming the adjacent rocks, he infers an action more deep-seated than the combustion of the bituminous beds amid which the salses are found, one of a more true volcanic kind.

Naphtha and the thicker bitumens are at present so scattered over various parts of the world, that though certain localities may abound with them more than others, they appear to show little beyond the conversion of some organic matter, accumulated under variable conditions, into that form.† Inflammable gases have also been found evolved from the earth, not only in connexion with bituminous and coal deposits, but under other circumstances, where no volcanic action is required for their production, as, for example, at the salt-mines at Gottesgabe, at Reine, in the county of Tecklenberg,‡ and from borings for salt in America,§ and China.‖

With regard to the rise of boracic acid with the steam at the lagunes near Volterra, in central Italy, accompanied by carbonic acid and sulphuretted hydrogen, it has been referred to volcanic action beneath the rocks in which the lagunes are situated.¶ That great heat exists beneath is certain, but how far this heat may now be considered volcanic and distinct from a more general dispersion

* "Geology of Russia in Europe and the Ural," vol. i., p. 576.

† Naphtha springs seem to continue, in some cases at least, in the same state during a long lapse of time, pointing to the long duration of the needful conditions. Thus, according to Dr. Holland ("Travels in the Ionian Isles, Albania," &c.,) the petroleum springs of Zante are in the same state as when described by Herodotus. The pitch lake of Trinidad is a good example of a considerable collection of the more solid bitumens. It is estimated at about three miles in circumference, though its exact boundaries are difficult to trace, in consequence of the soil which covers parts of it, from which crops of tropical productions are obtained. (Nugent, Geol. Trans., vol. i.) According to Captain Alexander (Edinburgh Phil. Journal, January, 1833), masses of this pitch advance into the sea at Pointe la Braye. The same author notices an assemblage of salses or mud volcanos at Pointe du Cac, 40 miles southward from the pitch lake, the largest about 150 feet in diameter.

‡ The gas is obtained from the abandoned pits, and is considered to consist of carburetted hydrogen and olefiant gas. It was employed by M. Röders, the inspector of the mines, for lighting and cooking, being conveyed to the houses by pipes.

§ While boring for salt at Rocky Hill, in Ohio, near Lake Erie, the borer suddenly fell after they had driven to the depth of 197 feet. Salt water immediately rushed forth, and was succeeded by a considerable outburst of an inflammable gas, which, being ignited by a fire in the vicinity, consumed all within its reach.

‖ At Thsee-lieou-tsing (previous note, p. 409), according to M. Klaproth, a jet of inflammable gas from a locality also producing salt water, was burning from the second to the thirteenth century of our era, at about 80 li S.W. from Kioung-tcheou.

¶ A Memoir on the mode of occurrence of the boracic acid lagoons of Tuscany, by Sir Roderick Murchison, entitled "On the Vents of Hot Vapour in Tuscany, and their relations to ancient lines of Fracture and Eruption," will be found in the Journal of the Geological Society of London, vol vi., p. 367

of an elevated temperature beneath the surface of the ground, more intense at some points than at others, seems not so certain. The boracic acid is found in combination with ammonia, as well as free, and Dr. Daubeny remarks that its presence in the steam may arise from the aqueous vapour passing over this substance, and carrying it upwards in mechanical suspension, as steam, by experiment, has been found capable of effecting.*

---

* By employing the heat of the superabundant vapour, the water collected in artificial ponds is sufficiently evaporated to dispense with fuel, and the boracic acid obtained at small cost. These lagunes furnish about 1,650,000 lbs. of boracic acid annually, sufficient, when purified and mixed with soda, forming borax, nearly for the supply of Europe.—Daubeny, "Volcanos," p. 156.

CHAPTER XXII.

EARTHQUAKES.—CONNEXION OF VOLCANOS AND EARTHQUAKES.—EXTENT OF EARTHQUAKES.—MOVEMENT OF THE EARTH-WAVE DURING EARTHQUAKES.—SEA-WAVES PRODUCED DURING EARTHQUAKES.—COMPLICATED TRANSMISSION OF EARTHQUAKE WAVES.—UNEQUAL TRANSMISSION OF EARTHQUAKES.—LOCALLY EXTENDED RANGE OF EARTHQUAKES.—EARTHQUAKES TRAVERSING MOUNTAIN CHAINS.—FISSURES PRODUCED DURING EARTHQUAKES.—SETTLEMENT OF UNCONSOLIDATED BEDS ADJOINING HARD ROCKS DURING EARTHQUAKES.—BREAKING OF GREAT SEA-WAVE, OF EARTHQUAKES, ON COASTS.—EFFECTS OF EARTHQUAKES ON LAKES AND RIVERS.—FLAME AND VAPOUR FROM EARTHQUAKE FISSURES.—SOUNDS ACCOMPANYING EARTHQUAKES.—ELEVATION AND DEPRESSION OF LAND DURING EARTUQAKES.—COAST OF CHILI RAISED DURING EARTHQUAKES.—EFFECTS OF EARTHQUAKES IN THE RUNN OF CUTCH.

It has been seen that prior to, and sometimes during volcanic eruptions, the country in the vicinity has been disturbed by vibrations, as if from time to time certain resistances to volcanic forces were suddenly overcome. The rending of rocks by fissures, such as have been previously noticed, could scarcely but produce vibrations, supposing the needful tension and cohesion of parts. It is by no means required that these fissures should always rise to the surface of the ground; indeed, in many volcanic accumulations, the rents formed, and subsequently filled with molten rock, are observed to terminate before they reach it. From the absence of the proper cohesion of parts amid great masses of ashes and cinders, these may so yield, that though a fissure might be suddenly produced in more solid matter beneath them, they could adjust themselves above in a very general manner over its upward termination.

It would be anticipated that, all other things being equal, vibrations of the ground around volcanos would be more intense after a vent had long been closed and dormant, so that time for the consolidation of tuff beds had elapsed, the whole well braced together by lava streams of various dimensions, than when the vent was still open, the volcano active, and the ashes and cinders incoherent. It may also be inferred that a certain thickness of trachyte, dolerite, or basalt, if not too much divided by columnar, or other joints, would offer greater resistance to any given volcanic force employed than tuff beds, unless these were so changed and consolidated as to assume the character of palagonite, or others of that class. Again, different effects would be expected from the resistance of intermingled sheets of tuff and rocks which had been in fusion, such as those described as occurring in the Val del Bove, Etna, and where similar substances are mixed, as narrow lava streams and irregular piles of matter, in both cases prior fissures, more or less filled by dykes of lava, considerably modifying the effects produced.

A connexion has often been inferred to exist between volcanic eruptions and vibrations of the ground at distances far beyond the immediate vicinity of the former, as if the volcanos were great safety-valves, through which, under ordinary circumstances, a certain amount of force escaped, mere local disturbances being thereby produced; while at others, from the overloading of the valves, or a greater exertion of power, larger portions of the earth's crust were shaken. Without including dormant or extinct volcanos, active vents are so widely dispersed over different parts of the world, that considerable areas may readily be disturbed by vibrations more or less depending upon general conditions, of which the discharge of molten rock, vapours, and gases, at certain points, is only one of the effects thereby produced. Hence, as respects this mode of viewing the subject, volcanic eruptions and earthquakes may be intimately connected, volcanic eruptions being equally regarded in the same general manner, and other adjustments of the earth's surface included, by which great fissures have been formed, and huge masses of rocks squeezed, broken, and thrust up into great ridges and mounds of varied forms and magnitude.

Many instances are given of the inferred connexion between earthquakes and volcanic eruptions, as, for example, the sudden disappearance of smoke in the volcano of Pasto, when the province of Quito, 192 miles distant, was so violently shaken by the great earthquake of Riobamba, on the 4th of February, 1797, and the sudden tranquillity of Stromboli from its otherwise constant activity, during the great earthquake in Calabria, in 1783. As we are quite assured that in minor areas there is often much vibration of the ground prior to such eruptions, and that subsequently to them tranquillity is restored, at least for a time, an observer would be led to inquire how far such apparent causes and effects may be extended. Herein caution is much needed, so that, from a preconceived opinion, accidental circumstances may not have an undue value assigned them, some of the inferences drawn respecting the immediate connexion between given earthquakes and the eruptions from certain volcanos being scarcely borne out by the facts adduced.

It would be anticipated that in regions of volcanos, such as those of South America, great vibrations of the ground should be experienced, these vibrations extending to variable distances, not only according to their intensity, but also to the kinds of rocks through which they are transmitted. In certain regions earthquakes are sometimes of such frequent occurrence, that except when of particular intensity they are so little regarded, that these, and similarly circumstanced portions of the earth's surface, may be considered in a more unstable state than others. The great earthquake of Chili, in 1835, was merely one of a more intense kind in a district often shaken by such vibrations. It is described as having been felt from Copiapo to Chiloe in one direction, and from Mendoza to Juan Fernandez in another; and the volcanos of that part of the Andes are noticed as having been in an unusual state of activity immediately prior to, during, and subsequent to it. In a previous earthquake (1822) the same region of South America was shaken through a distance, from north to south, of about 1,200 miles.

With respect to the areas actually disturbed by earthquakes, as waves are necessarily raised by them in the sea adjoining the lands shaken, or by the vibration of the rocks beneath it, attention has to be directed as to the amount of dry land moved, and the extent to which any adjoining portion of the sea-bottom may have been simultaneously shaken. For instance, this has to be done with the great earthquake of Lisbon, the area disturbed being represented as spread over a large portion of the Northern Atlantic, and comprising a part of North America, with some of the West India Islands (Antigua, Barbadoes, and Martinique) on the one side, and a part of Northern Africa and a large portion of Western Europe on the other. In such a case the extent to which the sea-wave produced by earthquakes may have been propagated, has to be well considered.* The known amount of dry land shaken

* Sir Charles Lyell (Principles of Geology, 7th edit., p. 344), calls attention to the great Lisbon shock, as having come in from the ocean, remarking that "a line drawn

in Europe was alone very large, comprising Portugal, Spain, France, the British Islands, the southern portions of Norway and Sweden, Denmark, Germany, Switzerland, and the north of Italy.

As respects earthquakes, the transmission of the vibrations has to be regarded with especial reference to the kind of substances through which an earthquake-wave may have to pass, so that, even, for illustration, assuming the impulses given to be equal, the extent of the vibrations and their amount might be very materially modified.* Mr. Mallet infers that an earthquake "is the transit of a wave of elastic impression in any direction, from vertically upwards to horizontally in an azimuth, through the crust of the earth, from any centre of impulse, or from more than one, and which may be attended with tidal and sound waves, dependent upon the impulse, and upon the circumstances of position as to sea and land." At the same time, he admits that the truth of this view has not yet been fully and experimentally demonstrated.

The movement of the great earth-wave† is commonly classed as undulatory or vertical, as the ground may be observed to roll onward in a given direction, or simply rise and fall in a nearly perpendicular manner. We have descriptions, in the one case, of the surface of the ground moving in a wave-like manner, and in the other, of a mere sudden rise and fall, as far as regards a particular locality. Of the latter the great earthquake experienced at Riobamba, in 1797, would appear an excellent example, many bodies of the inhabitants having, according to Humboldt, been hurled to a height of several hundred feet on the hill of La Cullca, beyond the small river of Lican.‡ We may readily infer that these two classes of earthquake movements are only modifications of the

---

through the Grecian Archipelago, the volcanic region of Southern Italy, Sicily, Southern Spain, and Portugal, will, if prolonged westward through the ocean, strike the volcanic group of the Azores;" hence inferring, as probable, their submarine connection with the European line.

* Mr. Mallet (Naval Manual of Scientific Enquiry, *Art.* Earthquakes, p. 197), in order to illustrate the transmission of waves through different materials, supposes a person to stand upon a line of railway, near the rail, and that a heavy blow be struck upon the latter a few hundred feet distant. "He will," Mr. Mallet remarks, "almost instantly hear the wave through the iron rail; directly after he will feel another wave through the ground on which he stands; and, lastly, he will hear another wave through the air; and if there were a deep side-drain to the railway, a person immersed in the water would hear a wave of sound through it, the rate of transit of which would be different from any of the others—all these starting from the same point at the same time."

† (Admiralty Manual of Scientific Enquiry, *Art.* Earthquakes).—Mr. Mallet defines the "great earth-wave" as the "true shock, a real roll or undulation of the surface travelling with immense velocity outwards in every direction from the centre of impulse."

‡ "Kosmos," *Art.* Earthquakes.

same thing, and that while a spot, such as the town of Riobamba, situated immediately over that where the impulse was given, should be lifted suddenly upwards, the same shock would appear to travel outwards to various distances around, in the manner, as often noticed, of waves on the surface of water into which a stone has been cast.

With respect to the vorticose movement, which has been often regarded as another class of earthquake motion, we may also, with Mr. Mallet, consider it as only a modification of the same kind of shock. With regard to the two obelisks at the Convent of St. Bruno, at Stefano del Bosco, the stones of which were twisted on a vertical axis in a similar manner, without falling, during the great Calabrian earthquake of 1783,* and inferred well to illustrate this movement, Mr. Mallet has shown, that this, and other cases of a similar kind, may be explained by the transmission of the ordinary shock, under a modification of circumstances by which the rectilinear is converted into a curvilinear motion.† In the same manner, when the complicated structure of some part of the earth's surface is considered, particularly where igneous rocks have been extended among, or otherwise much intermingled with, other accumulations, the observer may have reason to infer that, during the transmission of an earthquake-wave, the various parts of the whole may sometimes be so circumstanced, that a kind of twist may be locally given to considerable masses.

Taking the great earth-wave as the base of all the movements, however modified this may be according to conditions, the waters of seas, lakes, or rivers, resting or flowing upon the solid crust of the globe, will have the shock communicated to them. When we look at the present distribution of land and sea, and consider earthquakes in their generality, these are quite as likely, if not more so, to have been produced by impulses received beneath parts of the great ocean as beneath dry land. As the rate at which the earth-wave would travel, under such circumstances, would be greater than that at which the vibration transmitted to the water would proceed, two waves, as Mr. Mallet has pointed out, will result. One will arise from the vibration along the surface of the

* Figures and descriptions of these obelisks are given by Sir Charles Lyell in his "Principles of Geology," and in Dr. Daubeny's "Description of Volcanos," taken from the Transactions of the Royal Academy of Naples.

† Mr. Mallet remarks (Admiralty Manual of Scientific Enquiry, *Art.* Earthquakes), that "this motion arises from the centre of gravity of the body lying to one side of a vertical plane in the line of shock, passing through that point in the base on which the body rests, in which the whole adherence, by friction or cement of the body to its support, may be supposed to unite, and which may be called the *centre of adhesion*."

ground, in contact with the bottom of the superincumbent water, and becoming apparent in shallow water; the other from the heaping up of the water above a vertical uprise of the sea-bottom, such as we may suppose given if Riobamba, in 1797, had been beneath the sea. The first is named by Mr. Mallet, the "forced sea-wave," seen when the shock or earth-wave passes beneath or into shallow water, whether the earthquake travels from seaward inland or the reverse: the second he terms the "great sea-wave."

The geological importance of the "forced sea-wave," would seem much to depend upon the distance at which any shore or shallow water may be from the spot where a chief vertical movement, either inland or beneath the sea, has been given. If this were in the ocean far distant from the land, or shallow water, the movement communicated to the sea would be small, as also if the shock came from the dry land with little intensity, either from the original impulse having been unimportant, or of its force being nearly expended. Should, however, the vertical movement of the earth-wave be close to a coast, whether on the sea or land side, or beneath shallow water, then the "forced sea-wave" may merge in the "great sea-wave," sufficient distance not existing to permit much distinction. The one wave would precede the other under ordinary conditions, the "great sea-wave" throwing huge masses of water upon the land, mechanically disturbing sea-bottoms to a great extent, and often producing effects of considerable geological importance. As Mr. Mallet has remarked, while a "great sea-wave" may be so broad and low in deep water as not to be observed in the open ocean, it could break with great force on a coast or in shallow water.

It will be convenient, as has been pointed out by Mr. Mallet, so to classify observations on earthquakes, that things accessory may be separated from those which are material. Unfortunately, as has been remarked by Sir Charles Lyell,* it is only in comparatively recent times that earthquake phenomena have been studied with reference to their real geological bearing, accounts of the lives and properties destroyed, with now and then a notice of a new lake or island produced at the time, having chiefly occupied attention. Whatever may cause the shock, whether from a portion of the earth being suddenly thrown into motion, without violent rupture, viewing the subject on the large scale, or from sudden and violent fracture, we have to consider not only the depth beneath the surface, where the impulse may be given, but also the

* "Principles of Geology," 7th edit., p. 431.

mineral masses through which the waves have to be transmitted, both as regards the kind and relative position of those masses.

During violent volcanic eruptions, when, as for instance, in that of Tomboro, in Sumbawa, on the 5th April, 1815, the detonations were heard as far as 970 miles, and with such distinctness, and so loud at Macassar, 217 miles distant, that a vessel of war was sent out with troops in search of supposed pirates engaged in the neighbourhood; it may be assumed that vibrations of the earth would radiate around, as from any point, *a*, in the annexed plan (fig. 145), which may represent any district having such a centre

Fig. 145.

of disturbance. Assuming the roar of any great volcanic eruption, such as that at Tomboro, to arise from the violent discharge of the vapours, gases, cinders, and ashes through the crater, the vibrations thereby produced in the adjacent mineral accumulations would be felt more or less horizontally, according to the variable composition and solidity of the substances shaken. Should the cause of the earthquake-waves be deep seated, the vibrations on the surface would correspond with the radiation of the waves from their centre of origin, so that there would be a point where the shock would be felt vertically.* If *b c* (fig. 146) be supposed a

Fig. 146.

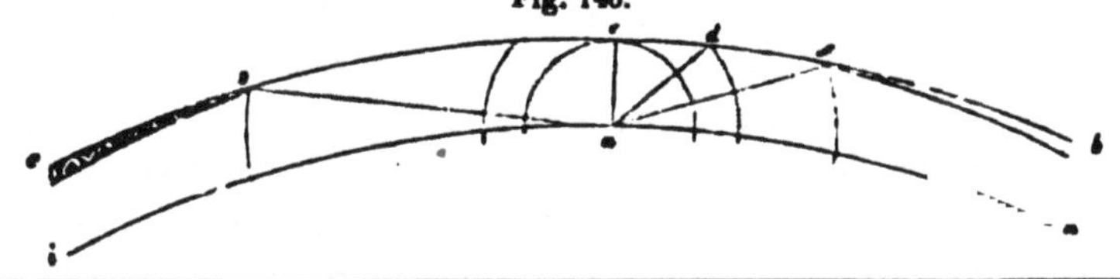

* The great Lisbon earthquake of 1755, felt so severely around a space near that city, has been considered a good example of a radiating earthquake with a deep-seated source. The earthquake of 1828, experienced in the Netherlands and Rhenish Provinces, is inferred to have been radiating, though less deeply seated. The area most shaken formed an ellipse, comprising Brussels, Liége, and Maestricht, and the shocks radiated to Westphalia, and to Middelburgh and Vliessingen. Referring to the great Calabrian earthquake of 1783, also considered somewhat central, Dr. Daubeny remarks (Description of Volcanos, 2nd edit., p. 515), after mentioning certain movements noticed, that such earthquakes may have "the impelling force situated along a particular line of country, although at the points at which it is exerted in its greatest intensity, the vibrations are propagated with greater or less violence in all directions around."

section of part of the earth's crust, and *a* a point in a curve, 250 miles beneath the surface, where an impulse is given producing earthquake-waves, these would strive to radiate around, so far as resistances or facilities would permit, in spherical shells. If the substance through which the wave passed was homogeneous, as, for example, a piece of iron, the wave would first traverse the distance *a e*, then *a d*, *a f*, and *a g*, in succession, the shock being felt most vertically as regards the surface of the iron at *e*, and more laterally at *f*, and most so as regards the section, at *g*. Geological investigations show us that the composition, state of solidification, and mode of accumulation of the mineral substances, forming so much of the earth's surface as we have the power of examining, are very variable. Hence, if in the foregoing section, instead of a homogeneous body, we suppose a great mass of mineral matter, granite for example, supporting two accumulations, one at *b*, arranged in beds of a hard coherent substance, such as compact limestone, and another at *c*, formed of strata slightly cemented, or loose sand and pebbles, it will be seen that the shock striking at *f* might be transmitted readily along the planes of the limestone beds, while, though the shock would strike the loose accumulations at *e* more laterally, the wave might be there more complicated from the want of sufficient coherence of parts.

Numerous modifications of the arrangements above noticed will readily suggest themselves, more particularly as regards the interruptions to the course of earthquake-waves by contorted and variably-intermingled masses of solid and loosely-aggregated rocks in mountainous districts, by the long wide-spread sheets of interstratified and dissimilar substances in some regions, by the fractures and alterations of mineral masses in others, and by the mixture of active volcanic districts with those of very different origin. It would be inferred that, on the minor scale, a shock may be modified in apparent direction and intensity when felt amid horizontal, or nearly horizontal beds, composed of different rocks, such as in the following plan (fig. 147), where *f* may represent a limestone,

Fig. 147.

e g a clay, *d h* a sandstone, and *c i* a conglomerate, resting in a trough-shaped cavity, as shown in the annexed section (fig. 148) of the same piece of country, formed of hard slates and limestones, which had, previously to the deposit of the first-named beds, been

Fig. 148.

thrown into a vertical position. Taking the shock to pass in the direction *a b*, it could easily traverse the line of vertical rocks beneath in that direction, while both the duration and intensity may be found modified in any town situated upon the central limestone, perhaps a stripe many miles in length, joining finally with a considerable sheet of the same substance. It sometimes happens that earthquakes do not affect certain upper beds, while the shock is continued beneath and transmitted onwards. Humboldt states, that such upper strata, rarely if ever shaken, are by the Peruvians termed *bridges*.*

Careful observation shows that shocks are more readily transmitted in certain lines in particular localities than others, much necessarily depending on the direction, either vertically or laterally, from whence these vibrations come, the minor adjustment of parts so lost occasionally amid the whole mass shaken, as not to be very readily appreciated. This could scarcely otherwise than happen when the source of the shocks remains for any length of time sufficiently fixed, and the relative position and structure of the rocks composing a region, continue unaltered.† Changes in this arrangement have been noticed even within the last 60 years, sufficient to show that, either from local modifications in the causes of earthquakes, or in their effects, adjustments of this kind may become permanently altered. Humboldt mentions that since the destruction of Cumana, on the 14th of December, 1797, the range

* "Kosmos," (Earthquakes). Remarking on this circumstance, Humboldt observes (Notes), that "these local interruptions to the transmission of the shock through the upper strata, seem analogous to the remarkable phenomenon which took place in the deep silver mines of Marienberg, in Saxony, at the beginning of the present century, when earthquake shocks drove the miners in alarm to the surface, where, meanwhile, nothing of the kind had been experienced. The converse phenomenon was observed in November, 1823, when the workmen in the mines of Fahlun and Persberg felt no movement whatever, whilst above their heads a violent earthquake shock spread terror among the inhabitants of the surface."

† The Cordilleras, extending from north to south, and a transverse line ranging from the Island of Trinidad to New Granada, are considered to be shaken in a marked manner. "In a line with both these ranges," observes Dr. Daubeny, "frightful earthquakes have occurred, as at Lima, Callao, Riobamba, Quito, Pasto, Cumana, Caraccas, &c., by which 40,000 persons have been known to be at one time destroyed. In all these cases the greater effects have not only been confined to the range of the mountains, but have pursued the direction of the coast."—"Description of Volcanos," p. 516.

of earthquake vibration in that district has so changed, that every shock has since that time extended to the peninsula of Maniquarez, which did not previously happen. He also points to the gradual advance of the almost uninterrupted earthquake shocks from south to north, up the valleys of the Mississippi, the Arkansas, and the Ohio, between 1811 and 1813, as showing that the subterranean obstacles to the propagation of the earthquake waves had been as gradually removed.*

When earthquake-waves traverse mountain chains, as they have been known to do, across the lines of their general range, the composition of such mountains requires much attention. If merely long ridges of a homogeneous rock, such as granite or the like, that may, as in the subjoined section (fig. 149), descend beneath various subaqueous accumulations, *c* and *d*, an earthquake-wave could readily

Fig. 149.

be transmitted across the ridges *a* and *b* from *e* to *f*, in preference to lines corresponding with them, should this be the general direction of the wave in accordance to the impulse given. In estimating the transmission of an earthquake-wave through any portion of the earth's crust, the observer will thus have, as it were, to dissect the portion shaken, endeavouring to separate the minor from the major effects, duly weighing the probability of the undulation passing through, or along such mountain chains as the Alps, Andes, and Himalaya, according to the depth of its cause. He has also to see if the shocks experienced along great lines, corresponding with those of accumulations, however contorted and broken these are, may be merely regarded as subordinate to a major motion, modified according to conditions, or be conformable to the general range of the earthquake-wave, regarded with reference to the total mass shaken.† The rocks of the same region may be differently

* "Kosmos," *Art.* Earthquakes.

† As regards the range of earthquake-waves along or across mountain chains, Humboldt remarks (Kosmos, *Art.* Earthquakes), after adverting to mountains transmitting shocks in lines corresponding with the walls of the fissures along which they may be raised, that earthquake-waves sometimes "intersect several chains almost at right angles; an example of which occurs in South America, where they cross both the littoral chain of Venezuela and the Sierra Parime. In Asia shocks of earthquakes have been propagated from Lahore and the foot of the Himalaya (22nd of January, 1832), across the chain of the Hindoo Coosh, as far as Badakschan, or the Upper Oxus, and even to Bokhara." As regards earthquake-waves traversing mountain ranges, Dr. Daubeny (Description of Volcanos, 2nd edit., p. 516) quotes also that of 1828, which crossed the Apennines from Voghera, by Bochetta, to Genoa.

affected if the wave be propagated from a great depth, than when the undulation has been produced by a less deep-seated cause. The transmission of the wave amid them might, in the first case, be a mere modification of some great movement, common, as in the Lisbon earthquake, to a large portion of the earth's crust, while in the second the same rocks may be directly acted upon in the first instance. Hence the importance of observations as to how far, during any given earthquake, particular districts, even great mountain ranges, may be considered to transmit a primary wave, or some modification of it.

As the earthquake-wave would pass with different velocities through different rocks,* it would follow that while the particles may so yield in some that fracture may not be produced, cracks and dislocations could be effected in others. Even in the simple arrangement of sheets of the one class above the other, the whole acted on laterally by an earthquake-wave, one set of rocks may be dislocated, the other returning to its original state, in the same manner as if the observer were to cover a sheet of copper with plaster of Paris, and throw both into vibration, when the latter would be broken, while the copper remained sound. It is easy to conceive, independently of the different conditions of the upper to the lower beds of rock, composing a series of horizontal or nearly horizontal deposits, as regards difference of pressure upon them, that the lower may be, from heat beneath, not in so fragile a state as those above, and be capable of more ready vibration without fracture. Thus many cracks and fissures may be made, not penetrating to great depths, and yet extending sufficiently beneath the surface to permit the ejection of water, mud, sand, or other easily-expelled body, out of them, and some of these may again so close as to envelope any substances which may have fallen into them, while others continue permanently open, the new adjustment of parts produced by the earthquake-wave not permitting a perfect return to the old conditions. Of such fissures formed during earthquakes there is abundant evidence, their forms very variable, as would be anticipated from the complicated rock accumulations frequently shaken, their complexity of structure often concealed by coverings of deposits, perhaps only a few hundred feet thick,

* Mr. Mallet points out (Admiralty Manual of Scientific Enquiry, *Art.* Earthquakes) that "an erroneous notion of the dimensions of the great earth-wave must not be formed from its being called an undulation—its *velocity* of translation appears to be frequently as much as 30 miles per minute, and the wave or shock moving at this rate often takes 10 or 12 seconds to pass a given point; hence its length or amplitude is often several miles."

while the mass thrown into vibration may extend downwards many thousands of feet, if not many miles. Inverted conical cavities have been so frequently noticed after earthquakes in plains and loosely-aggregated deposits, that they deserve attention. Water is usually mentioned, as having risen through them, as if, during the earthquake, it had been violently driven through points in the loose ground.*

That at the junction of loose or slightly-consolidated deposits, such as sands and gravels, with hard rocks, the latter rising through the former, so that when the whole became violently shaken there should be settlements of incoherent substances, with fissures and mounds of adjustment, would be anticipated, and is on record. During the great earthquake of Calabria, in 1783, this seems to have occurred to considerable extent. In the great Jamaica earthquake of 1692, this shaking off, as it were, of loose materials, appears to have produced the "swallowing up," as it has been termed, of Port Royal. Documents which have been preserved fortunately show that the part of that town which then disappeared was built upon sands accumulated against and around a rock, which, though shaken by the earthquake, retained its place as respects the level of the adjoining sea. The darkly-shaded parts, P and C, in the annexed plan (fig. 150), represent those which

Fig. 150.

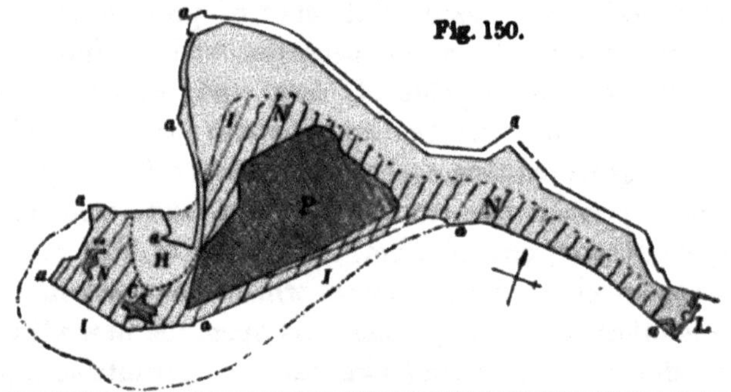

* Circular cavities were formed in the plains of Calabria during the earthquake of 1783. They are described as commonly of the size of carriage-wheels, sometimes filled with water, more frequently by sand. Water appears to have spouted through them. (Lyell's Principles of Geology, where a view and section of these cavities are given). During the earthquake of 1829, in Murcia, numerous small circular apertures were formed in a plain near the sea, whence black mud, salt water, and marine shells were ejected. (Lyell's Principles, and Ferussac's Bulletin, 1829.) After the earthquake of 1809 at the Cape of Good Hope, the sandy surface of Blauweberg's Valley was studded with circular cavities, varying from six inches to three feet in diameter, and from four inches to a foot and a half in depth. Jets of coloured water are stated to have been thrown out of these holes during the earthquake to the height of six feet. (Phil. Magazine, 1830.) During the Chili earthquake of 1822, sands were raised up in cones, many of which were truncated, with hollows in their centres.—Journal of Science.

remained standing after this earthquake, and are considered to be based on a white compact limestone, common in that part of Jamaica. *a*, *a*, *a*, *a*, and L form the boundary of Port Royal prior to the earthquake; N, N, N, the restoration by sand, drifted by prevalent breaker and wind action, at the close of the last century, and I, I, L, and H, subsequent additions effected by a continuance of the same causes to about the first quarter of the present century.* The settlement of the loose sand, combined with the sea-wave caused by the earthquake, appears to have produced all the effects observed during this earthquake at Port Royal, no mention being made, amid the details extant, of any permanent change in the relative level of the sea and the part of the town preserved.†

In like manner landslips take place on the sides of mountains and from sea-cliffs during earthquakes, some often of considerable magnitude. When the numerous slips of this kind which occur in mountainous and even hilly districts and along coasts are considered, as well as the frequent fall of rocks from the effects of ordinary atmospheric influences, it could scarcely otherwise than happen that when such districts are violently shaken, settlements of varied kinds are effected. Looking at the sources of springs, and especially of those which rise through joints and fissures, that these should be disturbed, and that matter should be subsequently thrown out mechanically suspended in the water, would also be anticipated.

The "great sea-wave" produced by earthquakes, sometimes

* There are documents to show the rate at which the long stripe of sands, known as the Palisades, was prolonged, so as to join the mainland of Jamaica with the ground on which Port Royal is built. From the evidence of Captain Hals, who accompanied Penn and Venables to Jamaica in 1655, it appears that the sands of the Palisades (the drift of the prevalent winds and breakers, as noticed in the text) were separated from the town by a narrow ridge of sand just appearing above water, an accumulation within about 17 years, for at that time Port Royal formed an island. Prior to the earthquake the junction was complete, as represented on the plan.

† Heavy brick houses were built on the sand; and it is stated (Philosophical Transactions, 1694), that "the ground gave way as far as the houses stood, and no further, part of the fort and the Palisades on the other end of the houses standing." Sir Hans Sloane says, "The whole neck of land being sandy (excepting the fort, which was built on a rock, and stood) on which the town was built, and the sand kept up by the Palisades and wharfs, under which was deep water, when the sand tumbled, on the shaking of the earth, into the sea, it covered the anchors of ships riding by the wharfs, and the foundations yielding, the greatest part of the town fell, great numbers of the people were lost, and a good part of the neck of land, where the town stood, was three fathoms covered with water." Long (History of Jamaica) says, "The weight of so many large brick houses was justly imagined to contribute, in a great measure, to their downfall, for the land gave way as far as the houses erected on this foundation stood, and no further." Dr. Miller, of Jamaica, was informed that it was a tradition at Port Royal, prior to 1815, among the descendants of the early settlers, that the great damage was produced by the slipping of the sand during the earthquake.

aids materially in the modification of the coasts shaken, seizing and transporting before it masses of matter which could not be moved under ordinary circumstances, and tearing up deposits thrown down in, or raised to, shallow situations. The magnitude of these waves is occasionally very considerable, though no doubt this may often have been much exaggerated from the terror of those endeavouring to escape from them. In the Jamaica earthquake of 1692, "a heavy rolling sea" followed the shock at Port Royal, and the 'Swan' frigate, which was by the wharf, careening, was borne by it over the tops of houses, and some hundreds of persons escaped by clinging to her. The sea-wave of the Lisbon earthquake of 1755 rose to the height of 40 feet in the Tagus, leaving the bar dry as it rolled inwards, followed by others, each less in importance, until the water again returned to its ordinary repose. The sea-wave of the same shock was 60 feet high at Cadiz, 18 feet at Madeira, and, under modified conditions, was felt on the coasts of Great Britain and Ireland, rising 8 to 10 feet on the coast of Cornwall. The shock was experienced at sea so severely, that vessels were thought to have struck the ground, and it is worthy of remark, as regards the locality over which the "great sea-wave" may have had its origin, that on board a ship 120 miles west of St. Vincent, the men on deck were violently thrown perpendicularly upwards to the height of a foot and a half.* The coasts of Chili† and Peru have suffered from similar waves; and in the great Calabrian earthquake of 1783 the shore of Scilla was inundated by one rushing 20 feet high over the low grounds. Such waves are, indeed, sufficiently common, though seldom prominently noticed unless productive of considerable effects. The sudden rise and fall of the sea observed in so many harbours of the world, as well in tidal as tideless seas, evidently independent of the tides in the former and not due to wind-wave undulations prolonged to the shores, often seem little else than the continuation of these waves reaching coasts where the earthquake itself has not been noticed.

While the earthquake movement is thus communicated to the waters of the ocean, minor volumes of water, even small lakes and rivers of all kinds, cannot be otherwise than more or less affected by it. According to the form of the bottom, situation as regards

---

* Lyell, "Principles of Geology," 7th edit., p. 475.

† Sir Charles Lyell remarks (Principles of Geology, 7th edit., p. 478), respecting the destruction of the ancient town of Conception (called Penco), in 1751, an earthquake sea-wave rolling over it, that "a series of similar catastrophes has also been traced back as far as the year 1590," including one in 1730. In 1835, the town also suffered from a "great sea-wave."

the range of the shock, and size of the lakes and enclosed seas, the intensity of the earthquake-wave being the same, will necessarily be the effects produced. The inland seas and lakes would be like so many basins or troughs of varied forms filled with water. We can conceive important geological modifications on the shores of districts adjoining Lake Superior, for example, when situated immediately above such an impulse as threw up men vertically to considerable heights at Riobamba, in 1797, or jerked sailors upwards off the decks of a vessel, 120 miles from shore, during the earthquake of Lisbon in 1755. In connexion with the earth-wave around the centre of the great Lisbon shock, the waters of Loch Lomond, even though this earth-wave was then transmitted so far, are represented to have been thrown two feet four inches high on the shores. As respects rivers, should the shocks pass up their courses, and the undulations be considerable, their waters would be precipitated onwards, or rolled back into the troughs or hollows formed, as the vibration passed onwards, gushes of water rolling afterwards down their channels in accordance with the temporary interruptions to their usual flow. Should fissures be formed during the undulation, and not remain more permanently open, the river waters rushing into them might be suddenly discharged out of them upon their again closing.*

Accounts of earthquakes contain such frequent mention of gases and flames evolved from fissures during shocks, that although there may be many exaggerations and mistakes on this point, there would appear little doubt of their occurrence. The emission of flame is interesting, whether it be produced by the escape of gases simply inflaming by rising into the atmosphere, or from causes more resembling those observed in volcanos (p. 323). In the latter case we should have to infer the fracture of rocks down to the needful supply of volcanic gases. The emission of steam as well as flame would seem still more to show that the fissure was opened down to depths where considerable heat existed. In the instance of the earthquake of Cumana in 1828, where the water hissed and

* The effects produced by the earthquakes in the Valley of the Mississippi in 1811-1812 are highly instructive. Sir Charles Lyell has not only collected valuable information respecting them, but has also personally examined the region then shaken. The ground near New Madrid is mentioned as having been so disturbed that the Mississippi was arrested in its course, and a temporary reflux produced Large lakes were formed in the course of an hour, twenty miles in extent, and others were drained. Hundreds of deep chasms were produced, which remained open many years afterwards, and during the shock large volumes of water and sand were thrown out of them. Sir Charles Lyell found, in 1846, the remains of many of these fissures extending for half a mile and upwards.—Principles of Geology, 7th edit., 1847.

bubbled up round a vessel in the harbour, as if a hot iron had been thrust in it, and when, on weighing the anchor, it was found that the links on part of her chain cable had been elongated from two inches in diameter to the length of three and four inches, there would appear proof of some sudden communication by a fissure with great heat.* In regions composed either wholly or in part of such accumulations as those of the great coal deposits of Europe and North America, and where fissures descended to depths whence great heat could rise upwards through them, not only might such gases as carburetted hydrogen, disseminated amid such deposits, and to a certain extent liberated, be inflamed, but even from the access of atmospheric air for a time, the broken parts of the coal beds themselves might be burnt, producing certain secondary effects in such districts.†

The shocks are often, but not always, accompanied by noises, transmitted through the ground. These are necessarily of very different kinds, from the varied conditions under which they may be transmitted. According to Humboldt, the great shock of Riobamba (4th February, 1797), was unaccompanied by any noise, while at the cities of Quito and Ibarra the great detonation of the same shock occurred eighteen or twenty minutes afterwards. As an example of the great distance to which subterranean noises may be transmitted, without earthquake shocks, he adverts to the noise like thunder heard over an area of several thousand square miles in the Caraccas, the plains of Calaboso, and on the banks of the Rio Apure during the eruption of St. Vincent, in 1812, this being, as Humboldt remarks, in point of distance, as if an eruption of Vesuvius should be heard in the north of France. He also points out, that in the great earthquake of October, 1746, at Lima and Callao, a noise like a subterranean thunder-clap was heard a quarter of an hour later at Truxillo, unaccompanied by any shaking of

* "During the earthquake of 1828 at Cumana, an English vessel in the harbour was suddenly enveloped in mist, and noise like distant thunder was heard. At the same time a shock was felt, and the surrounding water hissed as if a hot iron had been introduced into it, sending up a number of bubbles, accompanied by a smell of sulphur. Multitudes of dead fish floated on the surface. On weighing anchor, it was found that one of the chains which connected it with the vessel, lying on soft mud, had been melted, and the rings, which were two inches in diameter, had been stretched to the length of three or four inches, and become much thinner than before." Daubeny (Description of Volcanos, p. 528.)

† Any accumulation of gas, or of substances rendered liquid by pressure ready to assume the gaseous form when this is removed, would be expected to escape upwards should earthquake fissures traverse or extend to them. Humboldt notices (Kosmos) that during the earthquake of New Granada (16th November, 1827), carbonic acid issued from fissures in the Magdalena River, suffocating snakes, rats, and other animals living in holes.

the ground. These are merely, as will be apparent, the transmission of the earth-wave through fair conductors, such as most solid rocks, beyond distances where any tremulous motion of the ground is apparent. When noises precede earthquake shocks of importance, and these are sometimes noticed, they would chiefly appear to arise from vibrations insufficient to be termed earthquakes, succeeded by those which arrest attention, the greater earth-waves being alone regarded. The continued subterranean sounds heard during a month at Guanaxuato, in 1784, afford a good example of such noises, unaccompanied by vibrations sufficiently sensible to be termed earthquakes.*

The permanent elevations and depressions of land accompanying earthquakes require to be well considered, apart from the great earth-waves and their consequences, since such waves may be merely movements resulting from the cause producing these permanent relative changes of level, sometimes extending over considerable areas. It will be readily seen that a force acting from the interior of the earth outward, rending and otherwise disturbing portions of its solid crust, could throw such portions into motion, causing earth-waves, which, though often so terribly disastrous to man and his works, are nevertheless insignificant when measured by a very minor part of the earth's radius. We have seen (p. 381), that molten matter raised upwards into cracks formed in the relatively small mass of a volcano will increase its volume, raising the ground around in a manner which may produce changes of importance when near shores. An observer would therefore be prepared to expect that where there may be no very ready outlet, such as a crater or the sides of a volcano may present, for a greater mass of molten matter pressing to overcome superincumbent obstacles to its escape, greater fractures, extending over wider areas, may be formed, throwing the fractured and adjoining rock masses into movement, molten rock remaining in its new position as far as circumstances will permit. In such a case, the earthquake would be merely a secondary effect consequent on the exertion of force

* "Kosmos," *Art.* Earthquakes. Humboldt obtained good evidence on this subject. The noise began on the 9th January, 1784, at midnight. From the 13th to the 16th of the same month "it was as if there were heavy storm-clouds under the feet of the inhabitants, in which slow rolling thunder alternated with short thunder-claps. The noise ceased gradually as it commenced: it was confined to a small space, for it was not heard in a basaltic district at the distance of only a few miles." "Neither at the surface nor in the mines, 1,598 English feet in depth, could the slightest trembling of the ground be perceived." "Thus," he adds, "as chasms in the interior of the earth close or open, the propagation of the waves of sound is either arrested in its progress, or continued until it meets the ear."

raising the ground upwards. Although allusion has been made to molten matter raised upwards over a large instead of a minor area, the surface of the earth might be rent, earthquakes produced, and land permanently elevated, as will be noticed hereafter, by the mere expansion of a considerable portion of the earth's crust, the resistances upwards being in the end somewhat suddenly overcome.

From the adjustment of the minor volume of a volcanic mountain, to that of great masses of the earth's crust, by which parts may be either raised or depressed, and this by such sudden movements that earth-waves of various magnitudes are communicated to the adjacent rocks, no slight modifications would be expected. The geological importance of the rise or depression of land, especially on sea-coasts, at the time of earthquakes, being fully recognized, it is very desirable that, whenever opportunities present themselves, exact researches as to the amount of rise or depression above or beneath a somewhat permanent datum level should be undertaken. The mean tide level on oceanic coasts is very desirable for this purpose, when available, and may often readily be obtained with sufficient accuracy. In certain estuaries an alteration in the bottom of the seaward portion might influence the tides, so that a greater or less amount of water could flow upwards to situations where no real change of the relative level of land and the main sea had been effected. An observer will see, by reference to charts of estuaries, especially those with extensive bars at their mouths, how materially tides might be influenced in their action by moderate elevations or depressions at their mouths. In some, where the amount of water entering with the flood tide is so important in keeping a channel clear upon the ebb, especially where a shallow coast is exposed to heavy breaker action, the volume of water passing up or down might be most materially modified.

Modern observations on the western coasts of America, which, fortunately for these researches, is so truly oceanic, uncut by great rivers, have successfully established the rise of extensive lines of coast during earthquakes. At the time of that of November, 1822, felt from north to south for a distance of about 1,200 miles, the coast was raised four feet at Quintero, and three feet at Valparaiso above its former level; and Mrs. Graham records that oysters and other molluscs were elevated out of the sea, becoming offensive as they decomposed.* Dr. Meyen found, nine years afterwards,

* Geological Transactions, 2nd series, vol. i.

sea-weed and shells adhering to the coast thus raised, and infers it was so to the height of about four feet along central Chili. Sir Charles Lyell, detailing the evidence as to the rise of land at the time of this earthquake, considers that if the estimate of the mass moved be correct, namely, that superficially it extended over 100,000 square miles, the area elevated would be equal to half that of France, and five-sixths that of Great Britain and Ireland, so that only giving two miles for the depth of the mass raised, 200,000 cubic miles of mineral matter were elevated above their previous position at that time.*

At the time of the earthquake on the coast of Chili in 1835, when the towns of Conception, Talcahuano and Chillan suffered so seriously from the shocks,† much land was also raised. Captain Fitzroy, who was then engaged in a survey of the coast, states that the sea did not for some days fall, by four or five feet, to the usual marks; and that "even at high-water, beds of dead mussels, numerous chitons, and limpets, and withered sea-weed, still adhering, though lifeless, to the rocks on which they had lived, everywhere met the eye."‡ The amount of rise gradually diminished, so that about two months afterwards, the coast was within two feet of its former level, a kind of settlement after the first upheaval having been effected.

During the earthquake of Cutch, in June 1819, the surface of a wide area was so acted upon, that part became depressed beneath and part elevated above, its former general level. The Runn of Cutch, as it is termed, is the lowest part of a considerable district situated between the eastern branch of the delta of the Indus and the Loonee river. The area is estimated at about 7,000 square miles, and is so slightly above the level of the sea, that during the monsoons sea waters are driven up from the Gulf of Cutch and the creeks at Luckput, overflowing a large part of the Runn, the subsequent evaporation of the waters sometimes leaving a deposit of

* "Principles of Geology," 7th edit., p. 436.

† Though there was one chief shock, there are considered to have been more than 300 minor shocks subsequently, between the 20th of February and the 4th March.—Lyell, "Principles," 7th edit., p. 433.

‡ Captain Fitzroy adds (Voyages of Adventure and Beagle, vol. ii.), with respect to the Island of Santa Maria, south-east from Conception, that its southern extremity "had been raised eight feet, the middle nine, and the northern end upwards of ten feet." * * * "An extensive rocky flat lies around the northern parts of Santa Maria. Before the earthquake this flat was covered by the sea, some projecting rocks only showing themselves; now the whole flat is exposed, and square acres of it are covered with dead shell-fish, the stench arising from which is abominable. By this elevation of the land, the southern port of Santa Maria has been almost destroyed, little shelter remaining there, and very bad landing."

salt about an inch thick. It is also described as liable to be occasionally overflowed in parts by river water. As a whole, it seems to be a district peculiarly favourable for having any modifications of its surface marked by changes in the position of water flowing over, resting upon, or bounding it. From the facts accumulated respecting the earthquake of 1819, by Sir Charles Lyell,* it would appear that immediately after the chief shock† a mound was found to be raised across the eastern branch of the Indus more than 50 miles in length from east to west, and in some places 16 miles in breadth, with a height of 10 feet. This was named by the inhabitants the Ullah Bund, or Mound of God. At the same time a submersion of land was effected on the south of the Ullah Bund, into which the sea flowed up the eastern channel of the Indus, converting an area of 2,000 square miles of land into a great sea lagoon. The village of Sindree, situated on the land bordering the river prior to the earthquake, was submerged, the tops of the fort and houses being alone visible above the waters.‡ At Luckput, further down the Indus, the river which was there fordable at low water, being then only about a foot deep, became afterwards 18 feet deep at the same time of tide. Other portions of the channel were also found to be deepened. The course of the Indus is described as much unsettled after the earthquake, and the river finally cut through the Ullah Bund in 1826, throwing such a body of water into the salt lagoon, formed during the earthquake, as to render the water fresh for many months, though it became again salt in 1828.§ Being in the course of such a river, it would be expected that this submersion would be obliterated by the usual transport of detritus into it, a change now in progress, the lagoon having been found diminished both in size and depth in 1838.

* "Principles of Geology," 7th edit., pp. 437–441.

† Shocks are mentioned as having been felt from the 16th of June, the day of the great earthquake, to the 20th, when it is said an eruption broke out at the volcano of Denodur, 30 miles N.W. from Bhooj, the vibrations then ceasing. The chief shock was felt destructively at Ahmedabad, and feebly at Poonah, 400 miles more distant.—Lyell, "Principles," p. 437.

‡ Remarking upon the houses not having been thrown down (Bhooj, the principal town of Cutch was converted into a heap of ruins by this earthquake), Sir Charles Lyell observes that, "had they been situated, therefore, in the interior, where so many forts were levelled to the ground, their site would, perhaps, be regarded as having remained comparatively unmoved. Hence we may suspect that great permanent upheavings and depressions of soil may be the result of earthquakes, without the inhabitants being in the least degree conscious of any change of level."

§ It is represented as having been more salt than the sea, and the natives, according to Sir A. Burnes, supposed that it was so from a solution of the salt with which the "Runn of Cutch" is impregnated.—Lyell, "Principles," p. 439.

CHAPTER XXIII.

QUIET RISE AND SUBSIDENCE OF LAND.—RISE AND DEPRESSION OF COAST IN THE BAY OF NAPLES.—ELEVATION AND DEPRESSION FROM INCREASE AND DECREASE OF HEAT.—GRADUAL RISE OF LAND IN SWEDEN AND NORWAY.—RAISED COASTS IN SCANDINAVIA.—GRADUAL DEPRESSION OF LAND IN GREENLAND.—MOVEMENTS OF COASTS IN THE MEDITERRANEAN.—UNSTABLE STATE OF THE EARTH'S SURFACE.

In volcanic regions where there is sufficient activity to show that the vents are merely in a half-dormant state, or where, from time to time, though the eruptions may occur occasionally after even considerable intervals of comparative repose, volcanic action produces very marked effects on the surface, we should expect that there would sometimes be quiet elevations or depressions of the ground. Differences in the relative level of the sea and land could be caused by the variations of heat to which the hard rocks or other mineral accumulations may be exposed, such differences producing effects likely to be appreciated by the inhabitants of coasts only in proportion as the areas acted upon may, or may not, be more or less covered by water, or be left dry. Changes of temperature which could in so short a time deprive a volcanic mountain, such as Cotopaxi in the hot, or such as those in Iceland in the cold regions, of their snows, could scarcely but be attended with the expansion of the accumulations acted upon. How far minor volcanic areas may permanently, so far as regards a certain amount of time, remain elevated or depressed, would depend upon the conditions under which such areas may be generally placed. A minor volcanic area exhibiting considerable activity at one time may present a mass of mineral matter more heated, and be, consequently, more expanded than at another when this activity may cease, even only for several centuries.

In tracing back the elevation or depression of a coast by means of the human works which appear to have risen or to have been submerged, relatively to the level of an adjoining sea, assuming that there are no difficulties respecting the permanency of the latter, as might happen, especially as regards tidal seas, there may be much uncertainty as to how far the one or the other has been slow and tranquil The sudden uprise or depression of land during earthquakes does not necessarily suppose such undulations and vibrations of the ground as always to overthrow the works of man, though on coasts the resistance offered by them to a great sea-wave, rolling furiously over the shore, in consequence of the earth-wave, may often be very limited. Great caution is evidently needed on this head, so that a slow continuous elevation or depression of the land, relatively to the level of the sea, be well separated from its sudden rise or fall at the time of an earthquake.

As regards a minor area in a volcanic district exhibiting relative changes of level within the historic period, the coasts of part of the Bay of Baiæ, Naples, have been regarded as affording sufficient proof. Whether these changes may have been more or less sudden, or were gradual and continued through a somewhat long time, has not been altogether settled. Looking at the kind of country acted upon, a change of level, sometimes slow, at others sudden, would not appear inconsistent with the facts noticed. With respect to the probable dates at which the changes of level were effected, the Temple of Jupiter Serapis, at Puzzuoli, has been considered as affording good approximations. The main fact is, that three marble columns, somewhat more than 40 feet high, slightly out of the perpendicular, are smooth and uninjured to the height of 12 feet, above which, for 9 feet, they are perforated by the *Lithodomus*, a common and existing boring mollusc of the Mediterranean. The remainder of the columns, all of which exhibit the same fact, at the same heights, only show the usual effects of atmospheric exposure. On the pavement of the temple are other broken columns of marble perforated in certain parts, some of them bored not only on the exterior, but also in the cross fracture. The inference from these facts has been, that the lower parts of the column were protected by some deposit during submersion beneath the sea, the columns standing erect, or nearly so, while the part above was perforated, and consequently in water sufficiently clean for the animals to live in, bore, and obtain their food, the remainder rising above the sea, or only submerged to a depth beneath which the *Lithodomus* usually lives. This supposes the building of the temple on dry land, its submersion beneath the sea to between 20 and 30 feet, and its subsequent emergence, as now seen; so that the platform of the temple is about one foot, or thereabouts beneath the high water mark of the small tides of the Bay of Naples. From the various circumstances connected with this locality, Sir Charles Lyell infers, respecting the ground forming the foundation of the temple, that "first, about 80 years before the Christian era, when the ancient mosaic pavement was constructed, it was about 12 feet *above* its actual level, or that at which it stood in 1838; secondly, towards the close of the first century after Christ it was only six feet above its actual level; thirdly, by the end of the fourth century it had nearly subsided to its present level; fourthly, in the middle ages, and before the eruption of Monte Nuovo, it was about 19 feet *below* its present level; lastly, at the beginning of the present century it was about two feet two inches above the level at which it now stands" (in 1838.) *

The evidences of recent changes of the relative level of sea and land, even as respects the works of man in the vicinity, are not confined to the temple of Serapis. Mr. Babbage mentions that at the sixth pier of the Bridge of Caligula, at Puzzuoli, a line of perforations by the *Lithodomus*, and other indications of a water level, are found four feet above the sea, as also at ten feet above the present sea level on the twelfth pier, and points to the broken columns of the Temples of the Nymphs and of Neptune, as remaining now standing in the sea.† With respect to the columns of the latter temple, Sir Charles Lyell observes, that as they now stand erect in five feet water, just rising to the surface of the sea, their pedestals buried in the mud, if the sea bottom be raised, and the covering accumulations removed, they might exhibit similar appearances to those observed at the Temple of Serapis.‡ Roman roads are mentioned as under water, one between Puzzuoli and the Leucrine Lake, and another near the Castle of Baiæ. A road with some fragments of Roman buildings

* "Principles of Geology," 7th edition, in which Sir Charles Lyell gives the results of his personal examination of the district, as published in the early editions of the same work, and the chief facts mentioned by other authors.

† Proceedings of the Geological Society of London (March 1834), vol. ii., p. 74.

‡ "Principles of Geology," 7th edition, p. 491.

is beneath the level of the sea on the Sorrento side of the Bay of Naples; and in the island of Capri, one of the palaces of Tiberius is covered by water.*

Independently of these evidences connected with the works of man, of changes of the relative level of sea and land, there is also geological evidence of the same movements within a comparatively recent period. Mr. Babbage mentions a line of perforations by the *Lithodomus*, like those on the columns of the Temple of Serapis, 32 feet above the present level of the sea, in an inland cliff opposite the island of Nisida.† Sir Charles Lyell points to this cliff and other facts as capable of proving these changes, even if human works in the Bay of Naples had not afforded the evidence above noticed; and which, taken in connexion with that furnished by the geological facts observed, would appear to show an unequal elevation and depression of the land in different parts of an area comprising this bay.

In accounting for the gradual sinking and rising of the ground on which the Temple of Serapis is based, and of which he concludes there is sufficient evidence, Mr. Babbage adverts to the changes of volume which might be produced in the subjacent accumulations by the difference of heat in them at different times; an important consideration, not only as respects a minor area of this kind, but also the elevation and depression of great masses of land, constituting even considerable portions of continents. He observes, that "in consequence of the changes actually going on at the earth's surface, the *surfaces* of equal temperature within its crust must be continually changing their form, and exposing thick beds, near the exterior, to alternations of temperature;" and, that "the expansion and contraction of these strata will probably form rents, raise mountain chains, and elevate even continents."‡ With respect to these greater results, Mr. Babbage refers (1) to the increase of temperature found as we descend beneath the surface of the earth; (2), to the expansion of solid rocks by heat, while clay and some other substances contract under the same circumstances; (3), to different mineral accumulations conducting heat unequally;

* Professor James Forbes, "Physical Notices of the Bay of Naples;" Brewster's Edinburgh Journal of Science, vol. i., new series.

† "Observations on the temple of Serapis, at Puzzuoli, near Naples; with remarks on certain causes which may produce geological cycles of great extent."—Proceedings of the Geological Society of London (March, 1834), vol. ii., p. 47.

‡ Proceedings of the Geological Society of London (March 1834), vol. ii., p. 75.

(4), to the different radiation of heat from the earth, or at different parts of its surface, according as it is covered with forests, with mountains, with deserts, or with water; and (5), to existing atmospheric agents and other causes constantly changing the condition of the surface of the globe. Applying these views to the ground on which the temple of Serapis is placed, Mr. Babbage supposes it to have had an elevated temperature when this temple was first erected, and that it "subsequently contracted by slowly cooling down; and that when this contraction had reached a certain point, a fresh accession of heat from some neighbouring volcano, by raising the temperature of the beds, again produced a renewed expansion, which restored the temple to its present level."*

Quitting the minor area of Naples, where complications may arise from the volcanic character of the district, it fortunately occurs, that in Northern Europe observations have been sufficiently long and carefully continued to prove that a mass of land in Norway and Sweden has been slowly and tranquilly rising above the level of the sea during historic times. About a century and a half since, facts were known which induced Celsius to infer that the level of the Baltic and Northern Ocean was sinking, as was likely to be concluded at that time with respeet to any relative change of the levels of sea and land. Although Playfair may have pointed out that, in accordance with the views of Hutton, it was more probable that the land had risen, it was not until Von Buch had personally visited the district in 1807 that the latter inference became established as a fact. He concluded, "that the whole country from Frederickshall, in Norway, and perhaps as far as St. Petersburgh, was slowly and sensibly rising;"† inferring that the northern portion was rising faster than the southern. Referring to the marks cut upon rocks during calm weather, considered to represent the standard level of the Baltic, it was concluded by officers charged with the examination in 1820-21, that there had been a relative change of that level, though the rise had not been generally to the same extent. In 1834, Sir Charles Lyell examined the marks then cut by these officers, and concluded that the land had risen four or five inches in certain localities on the north of Stockholm. He convinced himself at the time, "after conversing with many civil

* Proceedings of the Geological Society of London, vol. ii., p. 75.
† "Travels in Norway."

engineers, pilots, and fishermen, and after examining some of the ancient marks, that the evidence formerly adduced in favour of the change of level, both on the coasts of Sweden and Finland, was full and satisfactory. The alteration of level evidently diminishes as we proceed from the northern parts of the Gulf of Bothnia towards the south, being very slight around Stockholm "*

The elevation of the area noticed is considered to extend to the North Cape, so that further traces of it become lost beneath the Northern Ocean. Taking a general view of the evidence, Sir Roderic Murchison has concluded,† that, assuming an east and west line traversing Sweden in the parallel of Solvitsborg, there has been in recent times on the north, and continues to be, an elevation, and on the south a depression. As regards the slow depression of Scania, Professor Nilsson infers, that this has been in progress for several centuries;‡ and Professor Forchhammer considers that the isle of Saltholm has not sensibly changed its level, with respect to the sea, for 600 years; while the isle of Bornholm appears to have risen one foot in a century, this elevation having been continued for 1600 years.§

With respect to very exact measurements, as regards small changes in the relative level of sea and land in inland seas, such as those of the Baltic, from the configuration of which and their mode of communication with the main ocean, disturbing influences may arise, no doubt, without reference of the general area to some more constant level, such as that of mean tide in some adjoining ocean, there may be difficulties; but looking at the evidence as a whole, it would appear decisive of a slow change in the

---

* "Principles of Geology," 7th edition, p. 300, and "On the Proofs of a Gradual Elevation of certain parts of Sweden," Philosophical Transactions, 1835.

† Address to the Royal Geographical Society of London, 1845.

‡ Communication to Sir Charles Lyell (Address of the latter to the Geological Society of London, 1837). Professor Nilsson mentioned, among other circumstances, that a large stone, the distance from which on the shore of Scania was measured by Linnæus in 1749, was, in 1836, one hundred feet nearer the water's edge, and that in the sea-port towns, "all along the coast of Scania, there are streets below the high-water level of the Baltic, and, in some places, below the level of the lowest tide. Thus when the wind is high at Malmö, the water overflows one of the present streets; and some years ago some excavations showed an ancient street in the same place, eight feet below, and it was then seen that there had evidently been an artificial raising of the ground, doubtless in consequence of that subsidence. There is also a street at Trelleborg, and another at Skanör, a few inches below high-water mark; and a street at Ystad is just on a level with the sea, at which it could not have been originally built."

§ "On Changes of Level which have taken place in Denmark in the present times," Transactions of the Geological Society of London, vol. vi., 1841.

relative level of the sea and land in the manner inferred.* Geological evidence supports the views derived from the circumstances mentioned; for, while the oceanic coast shows deposits raised above the present level of the sea, containing the remains of shells still existing in it, even barnacles and small zoophytes adhering to the rocks on which they fastened while beneath the water, on the Baltic side there are also raised accumulations containing shells characteristic of that sea. Although these facts might not show that the land had been raised in historic times, they are important, as proving a relative change of level at a recent geological period.†

Changes in the relative level of sea and land, and which can be measured by that of the ordinary tidal wave of an oceanic coast, are not confined to the north of Europe. Facts appear to show, that there has been a gradual sinking of the west coast of Greenland during at least a century. Dr. Pingel has shown that in a firth, called Igalliko (lat. 60° 43′ N.), a house built on a small rocky island is now submerged; that the foundations of a storehouse of the colony of Julianahaab, founded in 1776, are only now dry at low water; that near the village of Fiskenäss (lat. 63° 4′ N.) they, have been obliged to shift the poles for the women's boats, the old poles still standing in the sea; and that to the north-east of Godthaab (lat. 64° 10 N.), the remains of a winter house are now beneath high water. Dr. Pingel mentions, that no original Greenlander builds his house so near the water's edge. This author adds, that from information highly deserving of credit, ruins of ancient Greenland winter-houses at Napparsok, 45 (English) miles north of Nÿ-Sukkertop (lat. 65° 20 N.), are to be seen under water.‡ Thus

* The average rate of rise in Sweden is estimated at about three feet four inches in a century. With regard to the various authorities on the subject of this change, we would refer, for his usual impartial statements, to the Vicomte D'Archiac's "Histoire des Progrès de la Géologie," chap. v.; Soulèvements et Abaissements Contemporains, t. i., p. 645.

† M. Alex. Brongniart found *balani* still on the rocks, beneath a mass of shells, of the same species as now live in the adjoining sea, and 216 (English) feet above its level, near Uddevalla (Tableau des Terrains qui composent l'Écorce du Globe, p. 89). Sir Charles Lyell had, in 1834, an opportunity of verifying this observation, not only by discovering *balani* adhering to the rocks, but also small zoophytes (*Cellepora?*) beneath a mass of similar shells at Kured, two miles north of Uddevalla, at more than 100 feet above the adjoining sea. With respect to the raised accumulations on the Baltic side, the same geologist found them more than 100 feet above the adjoining sea at Söderteljе, 16 miles south-west from Stockholm. The shells in these deposits are well characterized as Baltic, and Sir Charles Lyell points out that the marine mollusca found in the Baltic, though "very numerous in individuals, are dwarfish in size, scarcely even attaining a third of the average dimensions which they acquire in the salter waters of the ocean."—Principles, 7th edition, p. 503.

‡ Pingel, Proceedings of the Geological Society of London, vol. ii., p. 208.

for about 368 English miles there would appear evidence of this subsidence, and it is considered to extend to Disco Bay, about 256 miles further north.

It has been supposed that the movement noticed in the Bay of Naples has not been confined to it, and that, however local some of the oscillations of the ground may be, in consequence of the volcanic action connected with it, there is a slow elevation in progress affecting Italy from the neighbourhood of Naples to Venice. It has been inferred that there is a change of the relative level of sea and land near the latter city of about six inches in a century, and that, extending to Naples, this elevation (varying in the proportion of 155 to 660 southwards), is felt for at least the distance of 520 miles.* With respect to elevations in the Mediterranean connected with the works of man, Captain Spratt and Professor E. Forbes mention an antique sarcophagus in the water of the Bay of Macri (the ancient Telmissus), perforated by boring molluscs up to a third of its height, showing a depression and subsequent elevation of the coast.† Not only are there traces of terraces on the limestones of Greece, with lines perforated by boring molluscs, such as now inhabit the adjoining sea, but M. Boblaye also points out a cavern near Napoli di Romania, raised five or six yards above the level of the Mediterranean, containing a breccia, the formation of which he refers to historic times, inasmuch as fragments of antique pottery are included in it.‡ Continuing researches of this kind in the Mediterranean, we find, on the authority of M. de la Marmora, that on the coast of Sardinia there is a deposit now raised above the sea, in which, mingled with terrestrial, fluviatile, and marine shells, are the remains of ancient pottery. The bed is described as sloping gently seawards, so as to represent part of an ancient coast with a portion of its adjoining sea-bottom. The remains of pottery are found where an ancient coast, inhabited by man, may be supposed to have ranged, the marine shells of the same species as now found in the adjoining sea becoming abundant outwards where the old sea-bottom occurred. At about 150 feet on the north-west of Cagliari, oysters (*Ostrea edulis*) are found adhering to the rock on which they grew; and M. de la Marmora discovered, also on the north-west of Cagliari, among the pottery, a round ball of baked

* MM. Ant. Niccolini and Em. Campo-Lonzé, as quoted by M. D'Archiac, Histoire des Progrès de la Géologie," t. i., p. 659.

† "Travels in Lycia, Mylias, and the Cibyratis," vol. ii., p. 189, 1846.

‡ "Journal de Géologie," tom. iii.

earth, about the size of an apple, with a hole in the centre, as if to pass a cord through. M. de la Marmora considers that this ball may have belonged to fishermen following their calling on this coast, and who used such balls instead of lead before the change of level elevating the deposit to its present situation.*

The circumstances above noticed will be sufficient to show that movements of the ground as well gradual as somewhat more sudden, have taken place since the localities mentioned have been inhabited by man, and that there may have been oscillations of the land in certain situations. These movements cannot be termed permanent in a strictly geological sense, since the history of the surface of our planet is one of change and modifications, with respect to the distribution of land and water; but they may, for the most part, be so regarded with reference to the lapse of many centuries, during which man may modify or change his mode of existence on the areas so acted upon. Whatever the cause of these movements may be on the great scale, and however the action which is commonly termed volcanic, may merely constitute a modification of the effects due to some general influence by which whole continental masses are upraised or depressed beneath the sea-level, we have, in earthquakes and the slow elevation and depression of land now taking place, manifestations of the unstable support on which the present mineral surface of the earth reposes. That earthquakes on the large scale may be due to the rending of portions of the earth's crust so acted upon that some previous resistance to an upraising or depressing force is suddenly overcome, while, in the gradual movements of elevation or depression, this resistance is quietly overpowered, may not be improbable. To the cause of this unstable state of the earth's surface, the observer will, no doubt, be induced to inquire more particularly when, searching amid the various accumulations which he will find recording the past history of our planet, he sees proofs of elevations and depressions of old surfaces to which those above mentioned are almost as nothing. It would be out of place here to enter upon the hypotheses which have been framed respecting it; at the same time, it may not be undesirable to recall attention to the results produced by changes on the earth's surface, by which dry land is lowered and sea-bottoms raised higher, and which Mr. Babbage has pointed out when accounting (p. 438) for the oscillations of the ground on which the Temple of Serapis is based, inasmuch as,

* "Journal de Géologie," tom. iii.

whether the explanation be sufficient or insufficient for all the phenomena observed, it can scarcely be disregarded, if we look for any source of heat beneath the surface of the earth either partial or central.*

---

* Mr. Babbage (Proceedings of the Geological Society of London, February 1834, vol. ii., p. 75) observes, that "whenever a sea or lake is filled up by the continued wearing down of the adjacent lands, new beds of matter, conducting heat much less quickly than water carries it, are formed; and that the radiation, also, from the surface of the new land, will be different from that from the water. Hence any source of heat, whether partial or central, which previously existed below that sea, must heat the strata underneath its bottom, because they are now protected by a bad conductor. The consequence must be, that they will raise, by their expansion, the newly formed beds above their former level, and thus the bottom of an ocean may become a continent. The whole expansion, however, resulting from the altered circumstances, may not take place until *long* after the filling up of the sea, in which case its conversion into dry land will result partly from the filling up by detritus, and partly from the rise of the bottom. As the heat now penetrates the newly-formed strata, a different action may take place; the beds of clay or sand may become consolidated, and may contract instead of expanding. In this case either large depressions will occur within the limits of the new continent, or, after another interval, the new land may again subside and form a shallow sea. This sea may be again filled up by a repetition of the same processes as before, and thus alternations of marine and fresh-water deposits may occur, having interposed between them the productions of dry land."

CHAPTER XXIV.

SUNK OR SUBMARINE FORESTS AND RAISED BEACHES.—SUNK FORESTS OF WESTERN EUROPE.—SUNK FORESTS BENEATH THE SEA IN ROADSTEADS.—MODE OF OCCURRENCE OF SUNK FORESTS.—LOCALITIES OF SUNK FORESTS.—REMAINS OF MAMMALS AND INSECTS IN SUNK FORESTS OF WESTERN ENGLAND.—FOOT-PRINTS OF DEER AND OXEN AMID THE ROOTED TREES OF SUNK FORESTS IN SOUTH WALES.—SUNK FORESTS AS REGARDS VARIABLE HEIGHTS OF TIDE ON TIDAL COASTS.—RAISED BEACHES CONCEALED BY DETRITUS.—RAISED BEACHES OF PLYMOUTH, FALMOUTH AND NEW QUAY.—RAISED DUNES OF BLOWN SAND, PERRAN BAY, CORNWALL.—DISTRIBUTION OF DETRITUS IN ENGLISH CHANNEL, AND IN SEA SOUTH OF IRELAND.—CARE REQUIRED IN TRACING RAISED COAST LINES.—RAISED COAST LINES, SCANDINAVIA.

THE names of sunk or submarine forests and raised beaches for comparatively recent changes in the relative levels of land and sea, since the vegetation of the former and the animal life in the latter have been much the same as now found adjacent on the one or in the other, though not perhaps too well chosen, since there have been many depressions and elevations of land marked by the submergence of terrestrial vegetable life and the emergence of marine remains in beaches at various geological times, are here retained as convenient for the present. The evidence is merely negative as to the absence of man from some of the coasts where these changes have been effected, certain conditions being needed for the preservation either of his remains or those of his works. Certainly some of them may readily have occurred since man was created on our planet, though no traces of human existence either in the contemporary accumulations themselves or in their mode of occurrence may have been detected. While alterations in the relative levels of land and sea have occurred in countries long inhabited by civilized races, and are now known to be effected where sufficient interest is taken to record them, it is scarcely to be supposed that the like have not taken place in regions inhabited by man in a less advanced state, the more especially as the study of geology teaches us that such, as will be hereafter seen, have occurred during a long lapse of geological time.

It can rarely happen that, without some historic record of the event, the submergence of a coast to any marked depth can be well ascertained. The water, except under very rare and favourable circumstances, would remove the traces of the old coast lines from our view, and new accumulations, mechanical or chemical, would tend still further to conceal them. As regards the evidence of a submergence of the shores of Europe for a considerable extent on its western and oceanic front, we fortunately possess good evidence in those trees and accumulations of other plants around them, which have been termed *Sunk or Submarine Forests.* These are to be found under the same general conditions, from the shores of Scandinavia to those of Spain and Portugal, and around the British Islands. So common to the whole are their general characters, that without supposing an absolute contemporaneous submergence, or one of equal amount throughout, there still remains a change of the relative level of the sea and land of a marked kind over the whole of this area. These 'forests' sometimes occur on the seaward front of a minor valley, and of others of far larger dimensions, even beneath the accumulations of a considerable estuary, and are found stretching inland for considerable distances under deposits of gravels, sands, and clays, the latter sometimes slightly elevated above the sea, and occupying somewhat large tracts of country.

The slopes on which the 'forests' rest are variable, though usually dipping seaward at a very slight angle. If the observer will imagine, that during low water, on any tidal coast, a change of relative level of land and sea were effected, so that the low water line became that of high water, he may form an idea of the varied slopes and different areas on which the trees and other plants may have grown, and which, now partially or wholly submerged, constitute 'sunk or submarine forests.' While they are often wholly beneath the level of high water, at others they are partly beneath it, and partly rise to, or above it. The following section (fig. 151) will illustrate the mode of occurrence of several, where, after a submergence, other accumulations have been effected over a portion, if not the whole; and so that while a part may be laid bare by the action of the breakers, others may be concealed seaward beneath the water, or be covered by gravels, sands, or clays inland.

Let *a*, *b*, represent the level of high water, and *c*, *d*, that of low tide, *e*, *f*, a line marking the general plane of the 'forest,' *g*, a

Fig. 151.

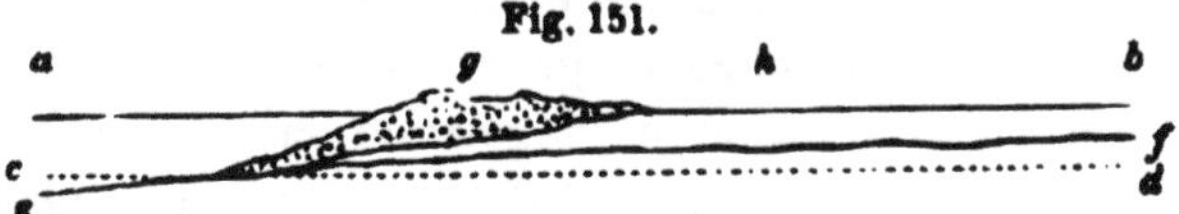

beach thrown up in the usual manner, and *h*, sand, clay, or any other accumulation covering up the 'forest;' then it usually happens, especially after such a state of the tides and weather as shall remove a part of the beach, that the trees and other vegetation are alone visible on the shore at levels corresponding with those at which the tide may cut the general plane of the 'forest.' The extension of the trees and other vegetation seaward may never be known except in the case of a roadstead for shipping such as at the Mumbles, near Swansea, or on fishing-grounds, where the anchors or nets may bring up portions of them. In like manner inland their spread in such directions may only be made apparent by canals, docks, or other works cutting through the superincumbent accumulations, as has been done in many localities.

Although this movement over so considerable an area may not always have been tranquil, the very common state of the vegetation preserved would lead to the inference that it had very frequently been so; for, as in the following section (fig. 152), the trees *a*, *a*,

Fig. 152.

a, *a*, *a*, are in their actual places of growth, though prostrate trees, *b*, may be often found among them, and the matted remains of branches, leaves, and various plants, as well as certain animal remains, such as the horns of deer and oxen, *c*, *c*, intermingled with the roots or accumulated round them, and constituting part of the old ground, *d*, *d*, are undisturbed. For further illustration the supporting rocks, *e*, *e*, which may be, and are of all kinds, as also some covering beds, *f*, *f*, supposed inland, are also represented.

When an observer is studying any of the numerous situations where these 'forests' are to be seen, it will be desirable that he should do so with reference to the locality, and its connexion with any larger area; to the mode of growth of the trees, and distribution of the other remains of vegetation mingled with them, and to their agreement with, or difference from, any plants of a similar kind now found in the vicinity, whether as regards kind or mode of accumulation; to the remains of animals found intermingled with the vegetation, and to the probable form of the area occupied by the 'forest,' as well inland as seaward. Various nooks and corners of oceanic bays, where we may suppose vegetation could have flourished under differences of level, so that more dry land was exposed, should be examined as well as very sheltered situations in places less open to the ravages of the sea. Thus a part of the coast of Tiree,* Hebrides, and of another in the Bay of Skaill,† Mainland of Orkney, though both exposed to the ocean, furnish the remains of these 'forests' as well as the ramifications of old estuaries amid the shores of the British Channel and Severn,‡ and the low grounds of Lincolnshire and Cambridgeshire, now bounded on the eastern coast of England by 'the Wash.'§ As regards the tideless Baltic, trunks of oaks and pines (*Pinus sylvestris*) and other trees, the roots in their natural positions, often several times above each other, and the whole five feet beneath the level of that sea, are found on different points of the coast near Greifswald, near Gnageland, on the south-east side of the Haff in the island of Usedom, and in the vicinity of Colberg. They are separated from the sea for variable breadths of coast by sandy

* The Rev. C. Smith, "Edinburgh New Philosophical Journal," 1829.

† Watt, "Edinburgh Philosophical Journal," vol. iii. Stems of small fir trees, ten feet long and five and six inches in diameter, were here found partly imbedded in and partly resting on the vegetable matter, chiefly composed of leaves.

‡ The "forest" passes beneath a considerable portion of the flat low land commonly known as the Bridgewater Levels, and it is to be found in numerous other portions of the old area of the estuary. The part exposed on the coast of Stolford has been described by Mr. L. Horner (Geological Transactions, vol. iii., p. 380) who pointed out that many of the remains of trees were rooted as they grew, while others were prostrate, some 20 feet in length. Remains of the *Zostera oceanica* were dispersed amid the vegetable matter in which the trees occur. Dr. Buckland and Mr. W. D. Conybeare (Geological Transactions, 2nd series, vol. i., p. 310) mention oak, fir, and willow trees, sometimes of large dimensions, partly rooted as they grew and partly prostrate, 15 to 20 feet beneath the surface of the Bridgewater Levels. Furze bushes and hazel trees with their nuts are intermingled with them.

§ The vegetable accumulations of this kind have long been known in Lincolnshire and Cambridgeshire. In 1799, M. Correa de Serra described (Philosophical Transactions) the "submarine forests" of Lincolnshire as composed of roots, trunks, branches, and leaves of trees and shrubs, intermixed with aquatic plants, many of the roots still standing in the position in which they grew, while the trunks were laid prostrate. Birch, fir, and oak were distinguishable.

dunes, under which they do not extend, there gradually disappearing. In the vegetable mass accompanying the trees, terrestrial, marsh, and fresh-water plants, with their seeds, are alone discovered, remains of marine vegetation not being found.*

Occasionally the bones of quadrupeds, and the traces of their foot-prints, are discovered in these 'forests,' as also the remains of insects, which are important as enabling the observer to consider the distribution of the terrestrial animal life as well as that of the plants of the time. Thus among the vegetable accumulations apparently of this date on the banks of the Humber remains of the red deer (*Cervus Elephas*) and the fallow deer (*C. Dama*) have been detected; and in the 'submarine forest' of Minehead, Somersetshire, the bones and antlers of the red deer are discovered amid the upright stumps of trees (chiefly oaks) now below the level of the sea and covered by it at high water, the trees rooted as they grew. The latter is especially an interesting circumstance, as the red deer are still found wild in the adjoining forest of Exmoor, so that the change of level has been effected since the red deer inhabited the district. Extending our researches into Cornwall, we find that a change of level may have happened, submerging vegetation in its place of growth, even after the introduction of man into Western England; for, at the Carnon tin streamworks, north of Falmouth, whence pebbles of tin ore have been extracted from beneath the bottom of an estuary, human skulls are stated to have been discovered with the bones of deer, among the trees and other vegetable remains covering the stanniferous gravel. Trees, partly in their places of growth, their roots descending among the tin pebbles, have been found 48 feet below high-water mark at the Pentuan tin stream works, Cornwall, covered by estuary and fluviatile accumulations, and which may be the equivalent of the Carnon bed,† not far different in depth beneath the

* German Translations of De la Beche's Geological Manual. "In some places the *Arundo phragmites* so abounds that the peaty mass seems entirely composed of it. The lower layers contain *Ceratophyllum demersum*, *Potamogeton pusillum*, *Najas major*, and *Nymphæa lutea*. *Scirpus palustris* and *Hippuris vulgaris* are also found with the *Arundo*. Seeds, especially of the *Menyanthes trifoliata*, are also frequent in the lower layers. The ground beneath the peat contains *fresh-water* shells; *Paludina impura*, Lam., *Planorbis imbricatus*, *Cyclostoma acutum*, and *Limneus vulgaris*.

† The section showed a bed, about 18 inches thick, of wood, leaves, nuts, &c., beneath about 50 feet of silts and sands, with shells, the vegetable accumulation, with its human skulls and remains of deer, resting on the pebbles of tin ore, and of quartz, slate, granite, &c., commonly termed the *tin ground*.—Henwood, "Trans. Geol. Soc. of Cornwall," vol. iv., p. 58.

At the Pentuan tin stream-works, where mining operations were continued under the sea-level for the extraction of the tin-ore pebbles, the vegetable accumulation, the same level. Whatever may have been the relative date when the skulls were entombed, supposing the Carnon accumulation in which they were discovered not to be precisely equivalent to the 'forest' disclosed by the mining operations at Pentuan, it would still appear, as we have elsewhere remarked,* "that after the causes which produced the tin ground or stanniferous gravel (of Cornwall) had ceased, the relative levels of sea and land were such in this district that a growth of plants and trees, not dissimilar from that of the present day, took place upon the gravel, and that subsequently these levels became altered, so that the sea covered the lower parts of the valleys previously above water. In the creeks thus formed, silt, mud, and sand were deposited, entombing the remains of marine and estuary shells of the same species as those which now exist on the coast, and finally, from the continued drift of alluvial matter down the valleys, river detritus covered up these marine or estuary deposits when they had accumulated to the necessary height."

As regards the British 'sunk or submarine forests,' they not only show that red and fallow deer, species now living, roamed among them when they were above water and in full growth, and possibly that man may have been an inhabitant of Western England at the time, but also that they were tenanted by species of at least one

---

roots of trees passing down to the '*tin ground*,' was (at the Happy Union Works) about 30 feet below the level of low water, and 48 feet beneath that of high-water spring-tides. The trees had been submerged, so that oyster-shells were found attached to their stumps. "The roots of the oak are in their natural position," observes Mr. Colenso, "and may be traced to their smallest fibres (in the tin ground) even so deep as two feet; from the manner in which they spread, there can be no doubt that the trees have grown and fallen on the spot where their roots are found." Resting upon this accumulation is a bed of silt, about two feet thick, in which there are also wood and hazel nuts, and with these vegetable remains the bones and horns of deer, oxen, &c. Mr. Colenso further states, that the shells dispersed through this bed, commonly in layers, present the appearance of their animals having lived and died in the places where their remains are now discovered. Above this accumulation follow in ascending order;—*a*, a bed of sand, four inches thick, containing marine shells; *b*, silt or clay, two feet thick; *c*, sand, 20 feet thick. ("In all parts of this sand there are timber trees, chiefly oaks, lying in all directions, and also the remains of animals, such as parts of red deer, &c. Human skulls have also been found in it, as also those of whales.") *d*, a bed of rough river sand and gravel, here and there mixed with sea sand and silt, about 20 feet thick, extending to the surface. Mr. Colenso states, that a short time before he described the section (1829), the remains of a row of wooden piles had been found in this sand, sharpened for the purpose of driving, and that they appeared to have been used in the construction of a wooden bridge for foot passengers. They crossed the valley, and were about six feet long, their tops being about 24 feet from the present surface, just on a level with the present low water at spring tides. He remarks, that if the relative sea level had been then as now, such a bridge would have been useless.

* "Report on the Geology of Cornwall, Devon, and West Somerset," p. 406, (1839).

large quadruped which is now extinct. Of this evidence has been obtained in the 'submarine forests' on the coast of South Wales. Among other places where these are found on the shores in that district, there is a considerable tract of low ground extending from the mouth of the Neath river eastward beyond Port Talbot, fringed by a covering of blown sand-hills (fig. 64, p. 88). Beneath these and the low ground, natural and artificial operations have occasionally exposed the vegetable accumulation, the stumps of trees with their roots standing as they grew, with prostrate trunks, and the usual characteristics of the 'forest.' On the surface of the clay in which the trees are rooted, foot-prints have been here and there detected, as if in passages amid the trees by which animals found their way through them, these foot-prints of various forms and sizes, some clearly those of deer, while now and then a large impression would be observed resembling that of some gigantic ox, having feet spreading far more widely than any domestic ox, even of the largest size, now known. This is not an isolated fact, for more westward, (about 28 miles,) while docks were being constructed at the port of Pembre, Caermarthenshire, and some covering sands removed, the 'submarine forest' which there occurs beneath much of the estuary of the Burry and Llwchwr was exposed, and similar foot-prints were found, some of a great ox mingled with those of the deer. Having attracted attention, drawings of these impressions were made at the time. As the horns and skull of the *Bos primigenius* were discovered near the same place, apparently derived from the same beds, it may be that the foot-prints mentioned might have been those of this large animal.

We would thus seem to arrive at a period for the growth of these 'forests' in England, when not only species of existing British animals then wandered among them, but also one, if not more, of the now extinct mammals,* leading into the times when elephants, hyænas, and other extinct quadrupeds also tenanted this country, Indeed, when contemplating from any of the adjacent heights the range of country which includes the estuary of the Burry and Llwchwr, with its 'submarine forest,' and also one of the limestone caves of that part of the country, wherein the remains of hyænas, rhinoceroses, and other animals are found, the cave's mouth fronting, and not far above the range and level of

* It becomes interesting, as connected with the subject, to ascertain how far any of the localities where the antlers and bones of the *Megaceros Hibernicus* are found may be connected with the tracts of 'submarine forests.' The general evidence respecting this gigantic and extinct deer would appear to be, that its remains are discovered in fresh-water shell marls or gravels beneath existing bogs.

the 'forest,' an observer has some difficulty in very clearly defining the time when the forest grew and the red deer of the present time, the great extinct ox, and the rhinoceros may have ceased to be contemporaneous, anterior to the submersion of the land beneath the level of the adjoining ocean, in such a manner that not only the stumps of trees remained rooted in the ground in which they grew, but the foot-prints of mammals which roamed amid the forest of this period also remained uninjured during the time when they were covered over by silt and sand.

While thus there is evidence of a change in the relative lovel of sea and land, by which the latter has been lowered several feet beneath the former along the oceanic shores of Europe for about 20° of latitude, there is also evidence of changes of levels on the same coasts of the reverse kind, beaches and worn cliffs affording proofs of them, and the remains of molluscs showing that such changes occurred after these were of the same speciés as those which now inhabit the adjoining seas. Reference has been previously made to the mollusc remains of existing species found entombed in deposits, of the inferred comparatively recent and very cold condition of Northern Europe; a time when molluscs of an arctic character reached more southwards than at present. Still referring to the same period, and to the evidence pointing to a submergence and emergence of the lands of the British Islands to the amount of 1,000 to 1,500 feet, and probably also of much of Western Europe to variable depths and heights, many tracts of old coasts and beaches would be expected, their greater or less state of preservation depending upon local circumstances as well as on the more general influences of different climates. Amid the varied cliffs and beaches left by so considerable an emergence, if we are to suppose it slow, intervals of comparative stability intervening, the observer would anticipate much difference of level in the cliffs and beaches he may discover, expecting nevertheless, all other circumstances being the same, that the cliffs and beaches would be the less injured in proportion as they were the more recent.

The coasts of Europe present many examples of cliffs and beaches elevated above the present level of the adjoining seas, the beaches containing fragments of the shells of molluscs still inhabiting the latter. The coasts of the British Islands, from their position, and the variable conditions under which they occur relatively to exposure to or comparative shelter from the Atlantic, and the variable rise and fall of tides, afford excellent opportunities for the study of these cliffs and beaches. And with respect to the consideration of such changes of level, the alterations that may be effected by the conversion of an estuary, facing the tidal wave coming in from the Atlantic or any other ocean, into a more spread area of water by submergence of the land, should be borne in mind, as also the wearing away of cliffs and the accumulation of beaches up to high-water mark, should an estuary be converted into land by an emergence. For example, if the land of New Brunswick and Nova Scotia were depressed beneath the ocean (and no very considerable submergence would be required), so that the tidal wave flowed freely over from the present Bay of Fundy to the Gulf of St. Lawrence, there would be an end of the causes (p. 78) producing the very high rise of tide in that bay, and, consequently, its plane of lines of cliffs and beaches. The same would also happen, though on a minor scale, if the land bounding the British Channel, and its continuation, the Severn, was so depressed beneath its present relative level, that the great rise of tide (46 to 50 feet) at King's Road (Bristol) and Chepstow was no longer produced, the tidal wave sweeping onwards without much obstruction, and passing round on the north and south of Wales, then becoming an island. In such cases the inclined plane corresponding to the high-water mark would be depressed at different depths beneath the general level. In like manner the observer should well weigh the changes and modifications by which similar estuaries or bays during emergence from the sea may have such tides produced in them as are now found, so that after having cliffs worn out, or beaches thrown up at some more equal level, these more inclined planes of the one or the other may be formed. The modifications of the relative heights at which cliffs and beaches may be contemporaneously formed on all tidal coasts, according to the general level of land and sea for the time, require very great care, as also the probable conversion of tidal into tideless seas, and the reverse, tideless seas (employing that term with reference to tides capable of producing very appreciable geological effects, and not strictly,) affording as a whole (due reference being made to the disturbing influences of winds) a better general level than the high-water line on coasts variably affected by the action of tides upon them. *

From the effects, chiefly of atmospheric influences, by which

* It is much to be desired, that the governments of different countries having sea-coasts would, at convenient points, ascertain the level of mean tides (not a difficult operation), connecting the spots where this may be accomplished (as marks on the

the sides of hills and mountains are decomposed and the disintegrated portions descend downwards into the valleys and low grounds, as in the following section (fig. 153), where certain rocks,

Fig. 153.

b,b, slates, for example, decomposed on the surface of the hills, *a,a*, are more or less covered by this detritus, accumulating in depressions, such as the valley *c*, many a cliff and beach is covered up, so that inland the opportunities are less frequent usually for observing them than near the sea, where a coast may be so cut back by breakers as to exhibit the beach and cliff beneath this kind of covering. Let, for illustration, the following section (fig. 154),

Fig 154

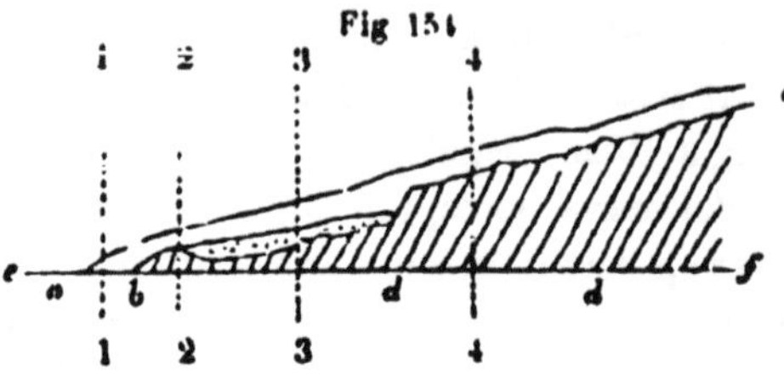

one which is not uncommon in Western England, represent a raised beach, concealed by a covering, *a, a*, composed of decomposed rock and other detritus, descending from an adjoining hill; *e, f*, being the level of high tide. Should there be a large modern beach at *e*, so that the breakers have little access to the lower part of the modern detritus, *a, a*, even the subjacent rock may be covered at that point, *b*; but should the breakers act freely, so as to cut back a cliff, then neither the first distance, 1, 1, nor the second, 2, 2, would expose the concealed beach, the latter only showing the subjacent rock at *b*. When, however, the cutting

coast itself at the actual level found may be in time obliterated from the action of the sea or atmospheric influences), with copper bolts, or other *bench marks* in, or on some inland cliff, religious edifice, or other building likely to be preserved. By connecting such original bench marks, and others inland, by a carefully-considered system of levels, not only might any variations in the relative levels of sea and land be hereafter detected, but also movements of the like kind on the great, though tranquil, scale be ascertained inland, the means of obtaining the needful evidence even extending considerable distances into the great continents. With this view the British Association for the Advancement of Science had lines of level run, in 1837–8, uniting bench marks connected with the tides in the English Channel at Axmouth, Devon, and in the Bristol Channel, at Porteshead, near Bristol, and at Minehead. The careful levels worked out during the progress of the Ordnance Survey in the British Islands permit excellent connections with the level of mean tides around.

back had reached the distance 3, 3, the beach may be well exposed; but should the breaker action still further wear away the cliff to 4, 4, then no trace of the beach would be left. The subjoined section (fig. 155) of the Hoe, Plymouth, may serve to show how

Fig. 155.

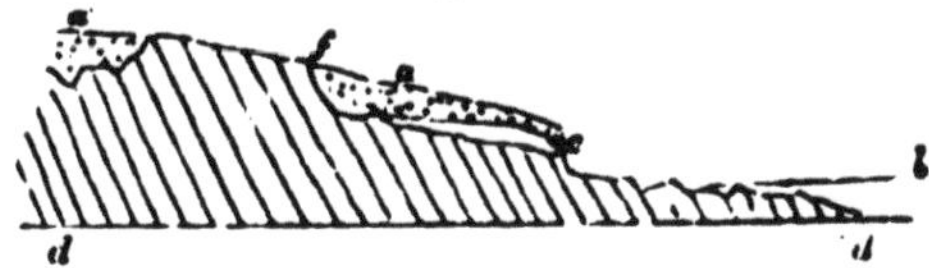

this can really happen. In it, *d*, *d*, represent the Devonian limestones of the locality, on a part of which the beach, *c*, reposes, about 30 feet above the present high-water mark, containing the remains of shells of the same species as are now found in the adjoining sea. At *a*, this is covered by angular fragments of the limestone of the hill, derived from the decomposition of its upper part, of the same kind which fills up a cavity above at *a'*. At *f*, the old cliff is seen behind the beach, *c*. This section was exposed by blasting away the limestone rock, taken away for use in large quantities.*

The following section of part of the Cornish coast near Falmouth

Fig. 156.

affords a useful illustration of the manner in which a raised beach may be covered by the detritus falling over from the hill above; in this case over the face of an ancient cliff, which would be concealed except from the wearing away of the coast by the breakers. The section is exposed between Rosemullion Head and Mainporth, and the angular detritus, *c*, of slate and more arenaceous beds, clearly derived from the hill, *h*, is well seen to cover over the cliff, *b*, and the beach, *a*; in all respects corresponding with those in the adjacent coves and bays. In this section, the observer also finds a low level of rocks, *e a*, formed at the time

* The section is given as seen in 1830. The raised beach was composed of pebbles of limestone, slate, reddish porphyry (occurring in places in another part of Plymouth Sound), and red sandstones, all rocks of the vicinity. Beneath the Plymouth Citadel, where a sandy prolongation of this raised beach occurs, it is chiefly formed of fragments of molluscs, of the same kinds apparently as those in the Sound adjoining. Other raised beaches are seen on the coasts of Plymouth Sound, as under Mount Edgecumbe, at Staddon Point, and nearly opposite the Shag Rock, on the eastern side, angular detritus of the adjacent hills covering them all.

when the breakers, at another relative level, were cutting back the ancient cliff, *b*, as similar planed portions of rocks are being now cut back on the same coasts at a lower level. Not far distant also, on the same coast, at a place named Nelly's Cove, the sub-

Fig. 157.

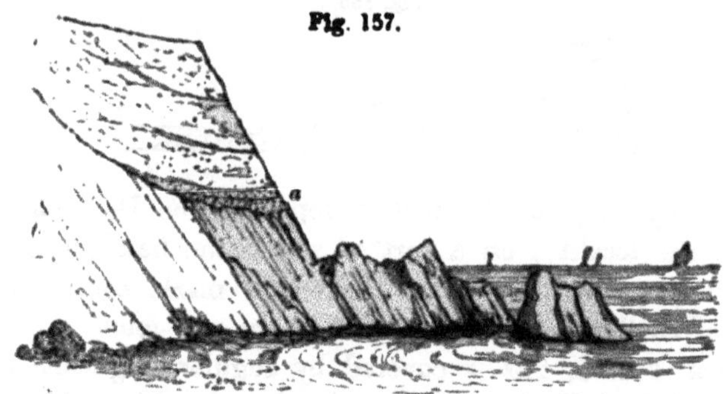

joined section is exposed, wherein *a* is the raised beach, *b* the supporting rock, and *c* the angular deposit derived from the rocks above, and which, as it accumulated, slid into a form corresponding with that of the beach beneath and the old cliff behind, the covering detritus, the beach, and the supporting rocks being all now in the process of being cut back by the heavy breakers of the adjacent sea. These in time will obliterate all traces of the beach, its covering, and the old cliff, leaving nothing but a bare wall of the rocks now behind the whole.

When formed of calcareous substances, either limestone pebbles of various sizes, or of comminuted sea-shells, raised beaches are sometimes as highly consolidated as the rocks which may support them, carbonate of lime thrown down under fitting conditions from a solution in water of the bicarbonate by means of carbonic acid (p. 13) cementing the whole together. Of the consolidation of a raised beach formed chiefly of comminuted sea-shells, that at New Quay, on the north coast of Cornwall, has long been celebrated. The following (fig. 158) is a section seen on the Look-out Hill, *a*, *a*, *a*, being slaty and arenaceous beds (dipping at a considerable angle) upon which the beach, *b*, composed of rounded pebbles of the adjacent rocks, cemented by consolidated sea-shell sand, reposes. At *c* are layers of the same comminuted sea-shell and sand, not uncommon on the shores and blown sandy dunes of the neighbouring parts of Cornwall, the lowest layers being much consolidated.* These are covered at *d* by an accumulation of angular fragments of rocks derived from the hill above. The present level

* The consolidation of these sands has been previously mentioned, *note*, p. 62.

of high tide is shown by the line *e, e*. In this case there would appear to have been some modification in the condition of this part

Fig. 158.

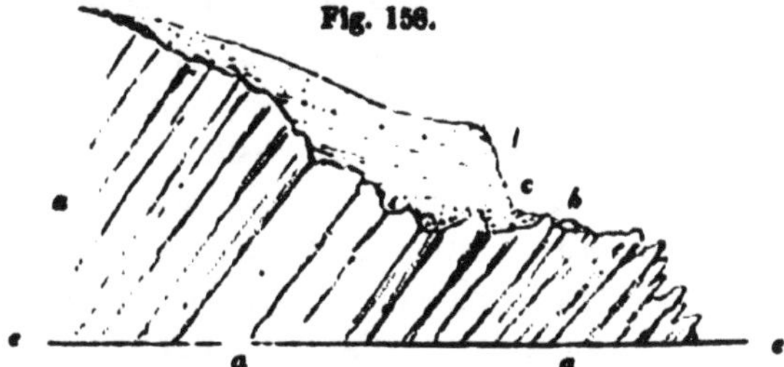

of the coast, permitting the deposit of the layers of comminuted sea-shells after the time during which a shingle beach was formed, and prior to the accumulation of the covering of angular fragments; perhaps, a time when blown sands were drifted over it, as in parts of the adjacent coasts at the present day, where such sands are driven over the shingle of ancient beaches now removed from the action of the sea. This view is supported by a section (fig. 159) in Fistral

Fig. 159.

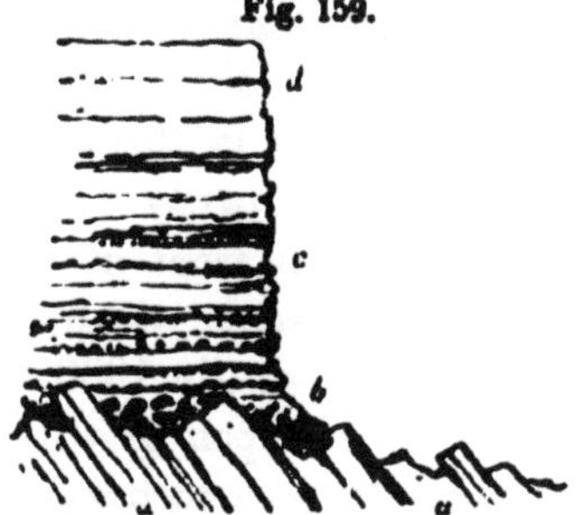

Bay, part (on the western side) of the projecting land on which the other section (fig. 158) is exposed, and where slates and more arenaceous beds, *a, a*, forming a portion of the same mass with those exhibited beneath the Look-out Hill (fig. 158, *a, a, a*), support rolled pebbles, often of large size, mingled with smaller gravel and sand, the whole constituting a kind of beach, *b*. This is surmounted at *c* by frequent alternations of fine gravel and sand, some of the layers of the latter being more consolidated than others. At *d*, the sand is less indurated, and at the extremities of the dunes on the north and south become mingled with angular fragments of rocks derived from the adjacent hills. In this instance there would appear evidence of a portion of a sea bottom, adjacent to the coast, having been elevated when the beach at the Look-out Hill was uplifted.

Still keeping to the north coast of Cornwall, as it appears useful to illustrate changes of level of this kind, where various modified effects, arising from them, are well exhibited in very accessible localities within moderate distances, the observer will find good examples even of raised sandy dunes; thus obtaining an insight into the condition of a range of oceanic coast, with its modifications of shingle beaches at the foot of cliffs, shallow shores with their prolongation of blown sands, and accumulations in shallow coast waters of the time, all upraised and variously acted upon at the present level of the breakers. At St. Ives' Bay and Perran Bay, sandy dunes, accumulated when the relative level of the Atlantic ranged along this land 30 or 40 feet higher than it now does, the latter having been since upraised, are seen perched where existing conditions could not place them, their old supporting rocks, previously removed from breaker action, now cut into cliffs by it. This is especially well shown in the former bay, near Gwythian, where a cliff of hard rocks, rising 35 or 40 feet above the present high-water mark, is surmounted by part of an ancient beach, with old sandy dunes above it. After this uprise, the slope of the coast was such that on the south-west, in the direction of Hayle and Lelant, conditions for the production of sandy dunes still continued, so that in this mass of blown sands, three miles in length, modern are partly driven over the older accumulations on the sides and in front of the valley between Gwythian and Godrevy Head, towards which, near Godrevy, an excellent section of a raised beach was, in 1838, to be found.

Masses of sand on coasts, acted upon by winds, and apparently not produced by existing conditions on such coasts, have not always been accumulated as blown sands and then elevated; as, for example, at Porth-dinlleyn, on the coast of Caernarvonshire, where a mass of sand covers a clay and gravel, of the deposits termed *glacial*, (p. 280,) and might, at first sight, be referred to raised sandy dunes. Careful investigation shows that this sand is an elevated sea-bottom, layers of a harder and more argillaceous kind being interstratified with the more loose sand, and retaining all the perforations made by marine animals when these layers were at the bottom of the sea. Breaker action is now removing these sands, brought within its influence by the elevation of the land, and does not assist, with the wind, in forming sandy dunes. These sands only constitute a portion of raised sea-bottoms, formed of either clay, sand, and gravels, with larger blocks of rock, dispersed over the adjoining land.

A raised beach of a very instructive kind was long since (1822) described by Dr. Mantell,* as occurring near Brighton, where one, elevated several feet above the sea, rests upon chalk, the rock of the coast, in the same manner as the beach at Nelly's Cove, Falmouth (fig 157), reposes on the old slates and accompanying beds. The beach near Brighton is backed by an ancient cliff of chalk, and, above the beach, chalk rubble, loam, &c., obscurely bedded, contain many teeth and bones of the fossil elephant, whence the name "Elephant Bed" has been given it. Rolled pieces of chalk and limestone are discovered among the pebbles, "full of perforations made by boring shells."† In this case the beach would appear to have been formed prior to, or during the existence of, the mammoth in Britain.

With regard to the fossil contents of these beaches, they afford much information as to the exposure of the coasts of the time to differences in the range of sea to which they may have been open; tidal streams and ocean currents being modified by alterations in the distribution of land and water. Professor E. Forbes informs me that the fossil shells of the raised beaches on the shores of the Clyde are, in many cases, those of species which, though still living in the British seas, present a more southern character than the molluscs now existing in them, and that they are confined to districts more southern and western than the Frith of Clyde. He thence infers a change in the direction of the currents from the south, (especially in that known as Rennell's current,) this change being probably due to the conformation of the coast lines of the time.

It is desirable, as has been done by Mr. R. C. Austen,‡ to connect these raised beaches and elevated sea-bottoms of the same geological dates, and the submarine or sunk forests, with the present state of the seas adjoining or covering them. After carefully considering the subject, Mr. Austen shows that although the distribution of the detritus derived from the present coasts of France and England, in the English Channel, and from England and Ireland on the sea-bottom to the south of the latter, with the sediment brought down by the rivers to those coasts, is in accordance with the arrangement which would be expected from breaker and wind-wave action and tidal streams; there are, especially in the central parts of the English Channel and on the outer range of the 100 and 200 fathom soundings towards the

---

* Fossils of the South Downs, 1822.

† Mantell, "Wonders of Geology," 6th edit. (1848), p. 113, where a section and detailed description are given of the raised beach at Brighton, east of Kemp Town.

‡ "On the Valley of the English Channel;" Journal of the Geological Society of London, vol. vi., p. 69.

Atlantic, bare rocks, shingles and coarse ground, and the shells of littoral molluscs so occurring as to point to the submergence of former coasts and shallow water adjoining them. With regard to the "submarine or sunk forests," Mr. Austen calls attention to the necessity of not limiting their extension beneath the sea outwards to the shores where they are now discovered, but to take a more general view of them as parts of submerged dry land; one which would better accord with the coarse detritus at depths, or in situations, where existing wind-wave action and tidal streams would not transport it, and also with the remains of littoral molluscs, *Patella vulgata, Littorina littorea, &c.*, found in similar situations. The evidence adduced shows very uneven ground outwards, especially towards the Atlantic; strewed over inwards by varied detrital deposits, partly the adjustment of existing circumstances, partly the mixed result of these and former conditions when the present sea-bottom was more elevated, even forming dry land connecting the British Islands with the continent. Without a proper chart * showing the condition of the sea-bottom around the British Islands, it would be difficult to convey a correct idea of the inferences to be derived from it; but, as illustrating a portion of this bottom, Mr. Austen remarks, that "within a distance from the summits of the Little Sole Bank, (parts of which rise to within 50 and 60 fathoms on the south of Ireland and off the mouth of the English Channel,) not so great as from the top of Snowdon to the sea, soundings have been obtained of 529 fathoms (3,174 feet); in other words, the Sole Bank rises from that level to nearly as high, and more rapidly, than does the mass of Snowdon from the sea level of the Caernarvon coast by the Menai Straits." †

* The observer should consult the chart appended to Mr. Austen's Memoir, in which a large amount of valuable information relating to the sea-bottom of the area noticed is gathered together.

† "The character of the greater part of the channel area," continues Mr. Austen "if laid bare, would be that of extensive plains of sand, surrounded by great zones of gravel and shingle, and presenting much such an admixture and arrangement of materials as we may observe at present over the Bagshot district of deposits; whilst along the opening of the channel there is an obvious configuration of hill and valley, and an amount of inequality equal to that of the most mountainous part of Wales."—Journal, &c., vol. vi., p. 85.

Referring to the examination of the range of the 200 fathom line from Cape Finisterre to the parallel of the Lisard, undertaken by Captain Vanhello in 1828 and 1829, Mr. Austen points out that the irregularity of soundings at this line, which runs at a comparatively short distance, as we have elsewhere remarked (Researches in Theoretical Geology, p. 190), outside that of 100 fathoms, (represented in figs. 65 and 99, pp. 91 and 261,) is far from being confined to one spot, but ranges not only southward, as shown by Captain Vanhello, but also to the northward. A reference to fig. 99, p. 261, will show that the Rockall Bank, westward of Ireland, much resembles an island under water; an uprise of only 600 feet would make it one.

Amid the complications which may arise in coasts where there has been gradual elevation of the land above the mean tidal level of the ocean, from the tidal differences above mentioned (p. 453), from the variable exposure to breaker action, as the shores become sheltered at one time and more exposed at another, from the amount of concealment of sea action on the surface of land caused by atmospheric influences, combined with running waters, and from unequal elevation of the land itself, the observer will, no doubt, require much caution while endeavouring to trace the line of coast of any one particular time. This will especially be the case when there have been oscillations, as there is frequently reason to conclude there have often been, during a time when the molluscs of adjoining seas continued to be much the same as now found in them.

In the Scandinavian region, where a slow rise of land (p. 439) is now taking place more on the north than on the south, and where surface changes, of no great geological magnitude, by which the land now separating the Baltic from the Atlantic could so easily convert a tideless sea into a branch of the ocean (a tide rushing up the Gulf of Bothnia and producing its effects in the same manner as is now found in the Bay of Fundy), traces of elevated ranges of coast are seen, which are the more interesting, as they serve to connect former movements of this kind with that now taking place. Respecting the evidence on this head, a very valuable summary and general view will be found in the observations of M. Élie de Beaumont on the researches of M. Bravais (connected with this subject) in Scandinavia.* Shells of molluscs now found living, as littoral species, on the shores of Norway, are discovered raised 518 (English) feet above the sea in the province of Drontheim, 482 feet at Skiöldal and Hellesaön, 360 feet around Lake Odemark, and 206 feet at Uddevalla. Lines of erosion are also inferred to mark the former relative levels of sea and land on the Norwegian coasts. In Finmark traces of an ancient line of sea-coast were followed from Alten Bay to Hammerfest. These consisted of beaches and worn lines of rock, forming the section of a plane so inclined that while on the south of Altenfiord it rose 221 feet above the sea, it descended to 94 feet near Hammerfest. Beneath this first line was a second, 88 feet above the sea in the former locality, 46 feet at the latter, both these lines falling from south to north, the reverse of the movement now taking place in northern Scandinavia. M. Bravais

* "Comptes Rendus," vol. xv., p. 817 (1842). Report on the Memoir of M. Bravais, Voyage de la Commission Scientifique du Nord en Scandinavie, en Laponie, &c.

considers that an intermediate line of ancient coast occurs between these more marked lines, which are not exactly parallel with each other, though they may appear so for short distances, showing the observer the necessity of exact measurements in researches of this kind.

With reference to the erosion of rocks in connexion with raised beaches in an oceanic situation, and where sea levels are not likely to have been much disturbed by changes, altering tidal action during the amount of elevation of land inferred, attention may be called to one of the earliest observations of this kind by Captain Vetch, at the Island of Jura, Hebrides. He there found six or seven lines of raised beaches, the highest about 40 feet above the present high-water mark. The beaches are composed of shingles of quartz rock (that of the island), of about the size of cocoa-nuts, and they are precisely similar to those which constitute the present beaches on the Loch Tarbert side of Jura, where these raised beaches are well seen. Their aggregate breadth varies "according to the disposition of the ground: where the slope is precipitious, it may be a hundred yards; where gentle, as on the north side of the loch, three-quarters of a mile from the shore." * The beaches repose partly on bare rock, and partly on a compound of clay, sand, and angular pieces of quartz rock. Captain Vetch observed that caves are found at the same level on the north side of Loch Tarbert, at a considerable height above the sea, and as he had never seen caverns formed in the quartz rock of Isla, Jura, or Fair Island (Hebrides), except on the shore, he considers these to have been formed at the time when the relative levels of sea and land were such as to cut the line of the caves.

* Geological Transactions, 2nd series, vol. I.

CHAPTER XXV.

TEMPERATURE OF THE EARTH.—TEMPERATURE OF DIFFERENT DEPTHS IN SIBERIA.—TEMPERATURE FOUND IN ARTESIAN WELLS.—HEAT OF WATERS RISING THROUGH FAULTS AND OTHER FISSURES.—VARIABLE TEMPERATURE FROM UNEQUAL PERCOLATION OF WATER THROUGH ROCKS.—TEMPERATURE OF WATERS IN LIMESTONE DISTRICTS.

As the temperature of the earth may have an important bearing upon conclusions which an observer might feel disposed to form respecting the causes of certain phenomena which he may be investigating, it is desirable that he should carefully direct his attention to it, so that its full value, as a geological agent, may be duly appreciated. Mention has been above made (p. 206), of the heights above the level of the sea at which, with certain modifications, water remains in a solid state. Independently of the well-known action of the sun on the surface of the earth, it is found that, after due allowance has been made for the temperature thus produced, there is another temperature, commencing at certain distances beneath that surface, the cause of which appears to require another explanation. Diurnal variations of temperature are considered not to extend, viewing the subject generally, to a greater depth than about three feet, and annual variations are inferred to cease at from 65 to 70 or 80 feet. Beneath depths not much differing from the latter, the temperature of rocks has been found to increase in mines, as also in the perforations into the ground commonly termed *artesian wells*. The rate of this increase of temperature has been found to vary, as might be expected, from certain local causes, such as the relative exposure of the mass of ground examined with respect to the form in which it may project into the atmosphere, should it be a mountain, its proximity to any particular source of heat, such as a volcanic region in activity, and the different circulation of water amid its parts, either among fissures or through beds of rocks of variable porosity.

We have seen (p. 291) that in the colder regions of the

northern hemisphere frozen ground may descend to very different depths. While in Siberia ice is still found at a depth of from 300 to 400 feet, in the same latitude (62° N.) in America the frozen ground does not extend beneath 26 feet; so that very modified conditions must exist for the temperature of the two regions. While, however, this may be the case, it is interesting to find (note, p. 291) that the mineral accumulations passed through in Siberia show an increase of temperature downwards, so that, while at 75 feet beneath the surface it was 21°·2 (Fahrenheit), at 378 feet it was 31°·1. The circulation of water in mines can scarcely otherwise than produce modifications of an important kind as to the exact rate of increase of heat downwards, though the fact itself, in its generality, may be evident, since the excavation of the mine, and the necessity of keeping it clear of water while its works are in progress, will cause water to descend from the surface downwards, more or less bringing its first temperature with it, to the depths whence it is again raised to the surface.* In districts, such as those of mines often are, broken by numerous fissures, this introduction of surface water, continued for a long time, may produce very modifying influences. Still much may be accomplished, with care, in mines, by selecting portions of rocks of the more solid kinds, and in situations where the temperature produced by the miners, their lights and their works, especially those carried on by blasting with gunpowder or gun-cotton, may cause the least amount of error in the needful experiments.†

---

* In some mining districts, the descent of water from the surface downwards, and the stoppage of its progress upwards from depths beneath the mining operations, have been found so to intercept the outflow of the previous natural springs, as to be productive of much inconvenience to the inhabitants, with respect to their supply of water for household purposes.

† M. Cordier, who has paid great attention to this subject, adopted the following method of obtaining the temperature of the rock itself, in certain coal mines in France. The thermometer was loosely rolled in seven turns of tissue-paper, closed at bottom, and tied by a string a little beneath the other extremity of the instrument, so that so much of the tube might be withdrawn as might be necessary for an observation of the scale, without fearing the contact of the air; the whole contained in a tin case. This was introduced into a hole from 24 to 26 inches in depth, and 1½ inch in diameter, inclined at an angle of 10° or 15°, so that the air, once entered into the hole, could not be renewed, because cooler, and consequently heavier, than that of the levels or galleries. The thermometer was kept as nearly as possible at the temperature of the rock, by first plunging it amid pieces of rock or coal freshly broken off, and by holding it a few seconds at the mouth of the hole, into which it was afterwards shut, a strong stopper of paper closing the aperture. The thermometer usually remained in the hole about an hour.—"Essai sur la Temperature de la Terre;" Mémoires de l'Academie, tom. vi. Other observations have been made on the temperature of the rocks in mines in various ways, and among them, holes, a yard or more in depth, have been drilled in convenient situations, and the temperature observed for a given period, such as a year or more.

Artesian wells have been found very valuable in affording information as to the temperature of the earth at different depths in certain localities, the bore-holes and the waters rising in them showing an increase of temperature with the depth. Though numerous experiments have been made upon the heat found at different depths in these wells, few have attracted more attention, from the care taken in conducting them, the kind of ground perforated, and the depth of the well itself, than those at Grenelle, near Paris. The rocks traversed consist of successive beds of various kinds, a thick one of chalk being among the most prominent, and are situated far distant from any volcanic vent or any known disturbing cause of that kind. After exhibiting an increase of temperature downwards, as the work proceeded, it was found by MM. Arago and Walferdin that when the well had reached the cretaceous clay known as the *Gault*, at the depth of 1,657 feet, the temperature was 79°·5 (Fahrenheit), and it became 81°·7 lower down, at 1798 feet.*

With respect to the temperature obtained in artesian wells, it is desirable that the mode of occurrence of the rocks of the district in which they may be situated should be carefully considered, so that the rise of thermal springs beneath the mineral accumulations traversed may not complicate the heat found. As, for example, should it happen that in a country affording a section similar to that beneath (fig. 160), a series of nearly horizontal deposits *b*, *c*,

Fig. 160.

rests upon a previously-disturbed assemblage of beds, *d*, traversed by faults, *e* and *f*, formed prior to the accumulation of the upper beds, *b*, *c*, and that thermal waters rise through these faults, as they often do, and as previously noticed (p. 22), the ordinary supply of rain water entering at *g*, and passing to a lower porous bed *c*, the temperature of the earth at any artesian boring, situate at *h*, might, to a certain extent, be obtained, while another well sunk at *i*, being immediately near a supply of thermal water through the fault *e*, would give a more elevated temperature, the higher in

* It is calculated that, taking the constant temperature (53°) of the caves of the Paris Observatory, 91⅓ feet beneath the surface, these temperatures would give an increase in heat at the rate of 1° centigrade (1°·8 Fahrenheit), for 32·3 metres, nearly 106 English feet (105 feet 11·659 inches).

proportion to the volume and velocity (p. 22) with which the previously-suppressed ready out flow can nowmore freely find vent. This is by no means an unnecessary circumstance to be regarded, since in districts such as that of the neighbourhood of Bath, there is evidence of faults, some of them of considerable size, having been formed anterior to the accumulation of the new red sandstone and oolitic series of that country, the surface of the fractured and contorted rocks being covered by the nearly horizontal beds of these deposits, and as the thermal waters of Bath (116° Fahrenheit) appear to rise through one of the old fissures, a ready vent for them occurring through the superincumbent beds, as at *l* (fig. 160). The observer would do well to search in such suspected districts for the temperature of waters pouring abundantly through any faults as at *k*. And it is worthy of remark, that in the district above noticed, a thermal spring (temp. 74°) appears among the older broken and disturbed rocks at the Hotwells, Bristol.

Regarding the temperature found at different depths in artesian wells, and the variations sometimes observed therein, it will have to be borne in mind that the different seams of rock whence water may be obtained, though not in sufficient abundance for the supply sought, will, from any different porosity in them, only permit waters to permeate or flow through them in such a manner that a given quantity can pass through each in a given time, thus influencing the circulation of any heat which they may carry with them from one part of a series of beds to another. If the following section (fig. 161) represent that of certain beds of rock traversed in sinking an artesian well, *a*, *a* being a clay, such as the London clay; *b* a porous bed of sand and gravel, gathering surface waters at *c*; *d* chalk; *e* sands receiving surface waters from *f*; *g* clay or marl, and *i* other sands or gravel gathering surface waters at *h*, we have very different porosities of the beds

Fig. 161.

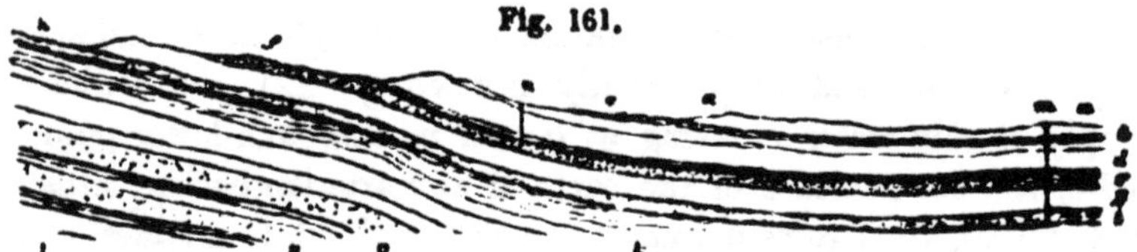

which can permit water to pass somewhat freely through them. Upon perforating through these beds, as at *m*, their relative permeability to water would influence the temperature in such an artesian well, a highly-porous bed, *b*, carrying its surface waters more readily to the well, to rise through it, than the chalk, *d*, beneath; as would probably also happen with the sands lower down at *e*. In all cases of this kind, an observer has to allow for the friction of the water through, and capillary action in the rock, which can only permit the circulation of this water according to needful conditions, so that it can only be delivered into the artesian well at a certain rate in each case.* In the section (fig. 161), a smaller well is represented as sunk at *n*, to the sands, *e f*, and on the bend of the beds, where, in consequence, the heated waters are more able freely to ascend, and be replaced by heavier and colder water, always supposing the whole to have a temperature above 40° Fahrenheit. In this case it might be inferred that, from the greater facility of percolation from the surface, *f*, to the bottom of the well, *n*, on the one side, the water would produce a lower temperature in the rock through which it passed, than in the same bed of rock at *m*. At *o* and *p*, part of the curves of two porous, interstratified with less permeable beds, *k l*, are represented, (forming thus, as it were, flat pipes,) for the purpose of showing that, if the curve were continued downwards on the left, water percolating in them, heated beneath, and not easily escaping upwards, might possess a somewhat higher temperature than at the same depths from the surface in the adjoining beds, (supposed, for illustration, to be equally porous,) not having the same facilities offered for obtaining, by circulation according to temperature and densities, colder waters from above.

To whatever extent water, permeating amid rocks, may modify their temperature, the greatest density of water will have its influence. In all regions where the annual temperature is such as to exceed that of the greatest density, however the surface, and corresponding depths beneath, may be acted upon, after 60 or 80 feet, the hotter waters would tend to rise and the colder to descend. In those, however, where the temperature is such that the water takes a contrary course, instead of cool waters descending to modify any heat which the containing rock might otherwise possess, they would ascend. For example, in the Siberian shaft (p. 291), descending beneath the 378 feet at which a temperature of 31°·1 was obtained, and allowing the same rate of increase as was found

* It is often practically found, in borings for common wells, that the relative porosity of the rock or rocks traversed, and the consequent possible delivery of water into them, have not been sufficiently regarded. Though certain loose sands and gravels may afford a volume of water considered most abundant, so far as the supply sought is regarded, rocks generally are but filters of various degrees of porosity, and only capable of permitting water to pass through them in a given quantity and time.

from the depth of 75 feet, namely, about 1° (Fahrenheit) for each 30 feet,* it would only be at a depth of about 630 feet that water at 39°·5 (Fahrenheit) would be found.

As to the springs which rise from fissures, such as those above mentioned (p. 465), a lower temperature, without due precaution, will often be obtained than should be assigned them, even supposing that the waters, as they flow upwards from various depths, lose much of their original temperature, and acquire that of the rocks amid which they rise.† The heat of ordinary springs has also to be carefully considered with reference to the kind and mode of occurrence of the hard rocks or less coherent accumulations of matter whence they issue. If we suppose *a* in the following section (fig. 162) to be a porous sandstone, resting upon a bed of

Fig. 162.

clay, *b b'*, the rain-waters, absorbed by the former, are prevented from permeating downwards by the latter, so that the water not retained amid the sandstone, issues as springs, on the side of the hill, at the top of the subjacent clay. Should another sandstone, or any other rock, through which water may readily percolate, *c c'*, occur beneath the clay, this porous stratum also based upon an impervous bed, *d d'*, the atmospheric waters, with any water derived from the springs above and absorbed by the lower porous rock, could alone find a natural outflow, as springs, on the side of the hill *d t*; while in the opposite direction, *d' t'*, they would saturate that portion of the bed, laterally aiding, by their superabundance, if we infer the needful facility of passage, the springs between *c d*. The dotted line *t t'*, representing any depth beneath which an uniform temperature is preserved throughout the year, should water percolate slowly to the surface, there would be —all other things being equal—a tendency between *a* and *b*, and *c* and *d*, on the one side, and *a* and *b'* on the other, to have springs issue with nearly equal temperatures. At *w*, also, if a well be sunk, the temperature of the water being within the depth of

* This rate of increase of temperature very nearly coincides with that obtained at Grenelle, namely, 1°·8 (Fahrenheit) for 53 feet.

† It is commonly needful to clear away the ground, so that a thermometer may be plunged in the water where it rises amid the rocks themselves. And this is especially necessary when the volume of water is far from considerable, and flows away slowly. Those thermometers in which the bulb and a portion of the glass project beyond the graduated scales, when handled carefully, will be found the most useful instruments.

variable temperature, we should expect it to be much of the same kind, the supply being derived laterally from the same reservoir which supplied the springs between *c* and *d*, and the impervous bed *d* (impervious so far as regards the ready passage of water through it) preventing appreciable communication with, and circulation of, waters of a higher temperature beneath. Should waters find their way, as springs, by means of joints or fissures, from the reservoirs in both porous beds, *a* and *c c'*, beneath the line of variable temperature, more rapidly in some places than in others—or the beds themselves differ materially in the facility with which water can pass through—variations may be expected, important or not, according to circumstances, in the temperature of springs issuing from them. All other things being equal, the lower reservoir—assuming that the temperature increases from the surface downwards—would be expected to supply the water with the more elevated temperature. It becomes needful, therefore, that after other conditions have been ascertained, the quantity of water delivered by a spring in a given time, and the rapidity with which it flows, should be duly regarded.

With respect to the temperatures of those waters which, in limestone districts especially, rush out, often in considerable volume and with much force, from subterranean channels, and which result from the loss of many minor streams and of rain-water amid fissures and cavernous rocks, they may be often very deceptive. Should the waters have been absorbed partly as streams, previously exposed to the temperature of the climate of the region, and partly derived from slow percolation through chinks, joints, and the minor cavernous structure of the rock, a mixed heat would follow, affording no correct data as to the temperature of the subterranean channels through which the waters have passed. When, also, the whole is derived from the absorption of atmospheric waters by channels of various kinds, the rapidity of passage of the waters downwards to the great drainage stream, and the differences in this respect have to be considered, as also the chances, not uncommon in some districts, that great fissure waters, derived from considerable depths, may not be mingled with the general volume of those discharged. Hence much care is required in investigating the temperatures of waters thus discharged, however desirable it may be that they should be properly ascertained.

While on the one hand the observer has to regard the adjustment of water, permeating amid the fissures and joints, or the mass of rocks, to its greatest density, and the variable mechanical manner in which this may be effected, he has also to consider the depths at which water itself may cease to exist; assuming the increase of temperature from the surface downwards, whatever its rate, locally or generally, to be certain, as the general evidence would lead us to believe. Should it be inferred that the rate of increase of heat usually supposed probable, namely 1° Fahrenheit for each 50 to 60 feet, is too great, and that sufficient information as to this rate has not yet been obtained, if we take only 1° for every 100 feet, we still seem to obtain a comparatively minor depth, allowing for increase of pressure from the superincumbent water, with the friction on the sides of any fissures, for that portion of the earth's crust in which water may be considered to circulate under the most favourable conditions. Taking the ordinary mode of calculation, allowing for pressure at increased depths, and assuming every facility of movement of the waters in a fissure, it may be estimated that at a comparatively moderate depth steam would be found instead of water.

Waters in fissures, rushing upwards with a rapid rate of outflow and in considerable volume may (as noticed p. 19) bring with them a greater temperature than those finding their way upwards with less velocity and in smaller quantity, the one heating the waters communicating with them laterally in their course upwards, beyond the temperature due to the containing rocks themselves, and the ordinary percolation of water through them; the others being cooled by these lateral waters. In certain districts, such as those where volcanic fires have once found vent, and which may be now concealed by various overspreading aqueous accumulations, there may be influences of this kind much modifying the exact depths at which certain temperatures would otherwise be found. No doubt, supposing a general source of heat to exist in the earth governing the outer temperature of its crust on the great scale, these would be merely local variations, yet, when endeavouring to ascertain the distribution of heat in the globe, all such variations require attention, so that the disturbing circumstances may be duly separated from the essential causes of the increase of heat downwards from its surface.

CHAPTER XXVI.

MODE OF ACCUMULATION OF DETRITAL AND FOSSILIFEROUS ROCKS.—DETRITAL ROCKS CHIEFLY OLD SEA-BOTTOMS.—MIXTURE OF BEDS WITH AND WITHOUT FOSSILS.—VARIABLE MODE OF OCCURRENCE OF ORGANIC REMAINS.— SEA-BEACHES OF THE SILURIAN AND DEVONIAN PERIODS—OF THE NEW RED SANDSTONE TIME, MENDIP HILLS, ETC.—OF THE LIAS PERIOD.— VARIED MODES OF OCCURRENCE OF THE LIAS.—BORING MOLLUSCS OF THE INFERIOR OOLITE TIME.—OVERLAP OF INFERIOR OOLITE, MENDIP HILLS.—LIAS CONGLOMERATE PIERCED BY BORING MOLLUSCS.— LAND OF THE TIME OF THE LIAS.—EVIDENCE OF LAND FROM FRESH-WATER DEPOSITS.—EFFECTS PRODUCED ON COASTS, RIVERS, AND LAKES, BY CONTINUED ELEVATION OF LAND ABOVE SEA.—ELEVATION OF LAND OVER A WIDE AREA.—EFFECTS OF CLOSING THE STRAITS OF GIBRALTAR.—UNEQUAL ELEVATION OF LAND.—LAKES ON THE OUTSKIRTS OF MOUNTAINS.—MIXTURE OF ORGANIC REMAINS OF DIFFERENT PERIODS FROM SUBMERGENCE OF LAND.—VARIABLE EFFECTS OF SURMERGENCE OF PRESENT DRY LAND.

THE observer, having well considered the manner in which the accumulations of mineral matter are at present effected, chemically and mechanically, through the agency of water, as also the mode in which the remains of animal and vegetable life may be entombed amid such accumulations, has to study the various layers, beds, or other forms of mineral substances formed by aqueous means, and in which organic remains are more or less distributed in various parts of the world. In one respect he has an advantage over his previous investigations, inasmuch as while he could then often only infer that which takes place beneath seas and lakes, he has in these rocks frequent opportunities of obtaining direct evidence of that which actually occurred beneath them, the large proportion of these beds being the bottoms of various seas or bodies of fresh water, deposited over each other, and subjected to variation from local causes.

Inasmuch as the dry land of the world is thus little else than the bottoms of seas and lakes, intermixed with igneous matter vomited upwards at different times from beneath the surface of the earth, some of the latter spread at once on this surface, at other times only laid bare by the removal of superincumbent deposits, the observer will have to dismiss from his mind the existing dry lands and waters of the world and substitute such other distributions of them as may best accord with the evidence which, from time to time, he will obtain. No matter how highly raised into mountains, or slightly elevated in plains, these ancient bottoms of oceans, seas, and bodies of fresh water may now be, they did not constitute dry land when formed, and consequently waters once occupied the areas where they now occur. We have seen that, to produce detrital accumulations, certain conditions of dry land are needed, whence their component parts have to be derived; and, therefore, to form the ancient sea-bottoms of any given time, dry land appears required out of an area so circumstanced, and yet so near to it as to afford the materials found. Considerations of this kind demand an enlarged view of the physical geography of different geological times, and such a disregard of the existing distribution of land and water that while all due weight is allowed for the employment of a given amount of mineral matter, over certain large areas, in the production of detrital accumulations of different dates—the wearing away of one portion raised above the ocean presenting materials for an equal and subsequent deposit beneath it in an adjacent situation; and consequently, that oscillations in the relative levels of the existing areas of our present continents may keep such matter much in one large area, the mind of the observer must not be too much occupied by the present arrangements of land and water on the surface of the earth.

While evidence is sought amid detrital or fossiliferous accumulations, of the mode in which the mineral matter of rocks has been chemically or mechanically gathered together, and the observer endeavours to trace among them former beaches, estuaries, bays, promontories, shallow and deep seas, fresh-water lakes, and the other modifications of water around and amid dry land, he has at the same time most carefully to study the mode of occurrence of any organic remains found in these accumulations. He will have to see if there be evidence that the animals or plants lived and died in or upon the beds where their remains are now found; or whether, after death, such remains were drifted into these situations. He will also have most carefully to refer to the distribution of the animals and plants existing at any given geological time, according to conditions, regarding that distribution as well on the large scale as with respect to any minor area.

With respect to the class of rocks usually named *fossiliferous*, this term has to be regarded in an extended sense. It is by no means required that the various beds composing any given series of sea-bottoms, should all contain organic remains in certain localities. Frequently, as in the subjoined sketch (fig. 163), representing a series of beds of rock, *a b c d* and *e*, exposed on a cliff, one of them only, such as *d*, may contain them, the others

Fig. 163.

not affording any animal or vegetable exuviæ. These beds are not, however, the less interesting on that account, inasmuch as some cause for this difference may present itself by diligent investigation, of importance as bearing upon the conditions, or their modifications, under which the whole series may have been formed. Should the beds be of different substances—as, for example, should *a b* and *e* be formed of sands of different kinds consolidated, as hereafter to be noticed, into sandstones; *c*, of gravels now hardened into a conglomerate; and *d* be composed of mud, now constituting a shale; the mode of accumulation of the non-fossiliferous beds have to be studied, as well on the small as large scale; and this study may tend to show how it probably occurred that the mud contained the remains of life, which has existed on or in this sea-bottom of the time, while no such remains are found in the sands and gravel.

Some rocks are only seen to be fossiliferous at rare intervals, a depth of perhaps only two or three inches affording organic remains, these occurring amid a great mass of mud, silt, and sand, as, for example, among the lower of the oldest fossiliferous deposits, the Silurian,—a class of rocks for the due appreciation and knowledge of which geologists stand so much indebted to Sir Roderick Murchison. In certain parts of this series, as developed in the British Islands, there are hundreds of feet in depth, in some localities, where no trace of an organic remain is found, and then a thin seam, replete with the remains of animal life, may be seen, showing that the portion of the sea-bottom which it represents was a mere thin sheet of little else than the crustaceans, molluscs, and corals of the time—partly, perhaps, living, partly dead, or all in one state or the other. Nevertheless, the whole series of deposits of which such seams constitute a portion (forming, as it were, rare streaks in the general mass), is the section of a certain minor area in the sea-bottoms of the time and locality, wherein the fitting conditions for the development of the germs and subsequent existence of the perfect animals occasionally presented themselves. The probable cause for this distribution and mode of occurrence has to be sought and well considered; and herein the non-fossiliferous portion of the general mass becomes important for the solution of the problem.

At other times very variable kinds of beds are all full of organic remains, as, for example, in such a section as that beneath (fig. 164),

Fig. 164.

where a cliff may afford a view of the various sea-bottoms which have succeeded each other in that locality, when the whole was beneath water. If, for illustration, *a* be a calcareo-siliceous sandstone; *b*, a coarser-grained siliceous sandstone; *c*, a marl or clay; and *d*, a fine argillo-siliceous sandstone; then, so far as the section extends, a silty bottom was first formed, to which succeeded mud, which in its turn was covered by siliceous sand; over which, as the general accumulation proceeded, a finer sand, with the addition of calcareous matter, was deposited. There is here evidence that the physical conditions affecting the area wherein these deposits were effected must have been modified or changed, and an observer would in consequence search for that showing how far any modification or change in the animal life, the remains of which are detected in the various beds, may have been contemporaneously produced.

In seeking the boundaries of any ancient land which may have furnished the mud, silt, sand, and gravels of accumulations around it, of whatever geological date the one or the other may be, it becomes evidently important to look for traces of ancient beaches, inasmuch as these show the actual margins of the seas of the time. From the desire at present manifested of following out investigations of this order, such beaches have been more frequently detected than might at one time have been expected. From the researches of Professor Ramsay it has been ascertained that during the deposit of the Silurian rocks of Wales and Shropshire, there was a time when the older accumulations, now forming the district of the Longmynds, rose above the sea, and were bounded by beaches; while a part of the Silurian series, named the Caradoc sandstones, was being deposited adjacent to them.* Again, in the Malvern district, the labours of Professor John Phillips have shown that at about the same geological date, a portion of the syenites of the Malvern hills must have been above the sea; a beach deposit, in which there are angular fragments of the pre-existing rocks, occurring at the Sugar-loaf Hill, on their western flank.† In both cases organic remains are detected mingled with the shore accumulations, and Professor Edward Forbes considers that those which he has examined in the Longmynd district are of a coast character.‡ These may not be among the oldest beach and littoral deposits in the British Islands, inasmuch as where conglomerates are found among the beds of the Cambrian rocks near Bangor, North Wales, such may also have constituted the shores of still more ancient lands, furnishing the materials for these conglomerates.

In the ascending series of fossiliferous rocks, the materials of which were furnished at succeeding geological times, the like kind of evidence, if carefully sought for, is to be obtained in many localities. To take the old red sandstone series, as shown in the British Islands, while part of it may merely constitute a portion of deposits formed one after the other beneath the waters of a sea, as in Devonshire, other parts point to a littoral origin. This may be well inferred from the mode of occurrence of the old red sandstone series in parts of Ireland, North Wales, and Scotland, where shingles of various sizes, sometimes large, are arranged around masses of older rocks, and follow many sinuosities of the more ancient land against which they were piled. Portions of the older land of the time have sometimes an insular character, as, for exam-

* "Journal of the Geological Society of London," vol. iv., p. 294.
† "Memoirs of the Geological Survey of Great Britain," vol. ii., p. 67.
‡ "Journal of the Geological Society of London," vol. iv., p. 297.

ple, in the county Kildare in Ireland, where the range of heights, chiefly known, from one of them, as that of the Chair of Kildare, seems to have risen above the sea of the time, its rocks furnishing materials for the shingle on its shores, the whole having been subsequently covered, or nearly so, by the calcareous deposit known as the carboniferous limestone. The removal or denudation of this limestone (in its turn furnishing an abundance of the shingles or gravel at other and later geological times) has in a great measure disclosed this arrangement of parts, and left the range of the Chair of Kildare, even now rising like an island above a level district.

Quitting these more ancient accumulations, and still not passing beyond the area of the British Islands, in order to show how much of this kind of observation may be carried out in such a minor portion of the earth's surface, we again find marked evidences of beaches at the time commonly known as that of the new red sandstone series, one in these islands following much new adjustment in the relative distribution of land and water, by which the former bottoms of seas and of fresh-water deposits were irregularly upraised. In South-western England and in South Wales, the beaches of this time, though they are by no means absent or indistinct in many other districts, are particularly well exhibited.

Among the Mendip Hills (Somerset), in various parts of Gloucestershire, Monmouthshire, and in Glamorganshire, we have, from the removal of subsequent accumulations by denuding causes, evidence of ancient shores, as is the case near the Chair of Kildare, though the latter are of earlier geological date. As at the latter also, from a repetition of similar causes, we seem to have islands with their beaches before us, much as they existed at this subsequent time. In investigations of this kind it sometimes happens that sections are presented, or information obtained, justifying the construction of sections, by which it is shown that, during the submergence of the dry land, while the mud, silt, sand, and shingles were accumulating along the shores and the islands of the time, beach after beach, became buried up, a long wide-spread patch of shingles covering some subjacent ground, as it gradually sank beneath the sea level of the period. The subjoined sections (figs. 165, 166) of the ancient island of old red sandstone and carboniferous limestone of the Mendip Hills, and also of another island of a somewhat similar character, one of several in the vicinity of Bristol, may serve to show this circumstance; as also how shingles of the same general character, and derived from subjacent or adjoining rocks, under similar general circumstances, may be accumulated as a beach, on sloping ground, during the lapse of much geological time, while the dry land of a particular locality became gradually submerged beneath the sea. Both

Fig. 165.

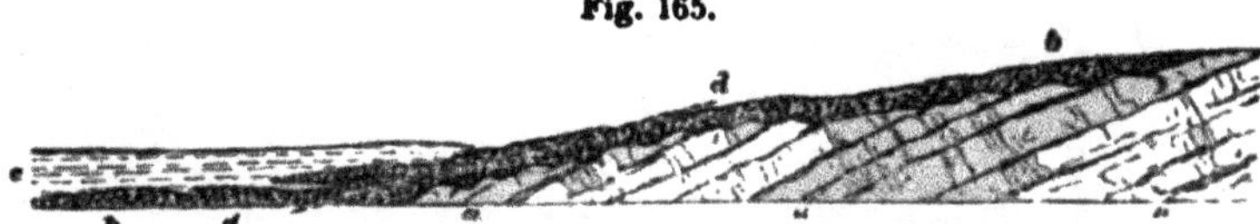

a, a, a, disturbed beds of carboniferous limestone; *b, b*, conglomerate of pebbles derived from the subjacent or adjoining rocks, cemented by magnesio-calcareous matter; *c*, red marls; *d, d*, line showing how denudation might cause successive accumulations to appear as of one time.

sections exhibit the beaches, usually composed of shingles of carboniferous limestone, now cemented by magnesio-calcareous matter, jutting, as it were, into the red mud of the time, (now red marls,) having extended over one portion of it, during the submergence, and having been covered by another as this proceeded. One section (fig. 165) shows only the accumulation of the red mud (marl), while the other (fig. 166) exhibits a subsequently-formed deposit of dark mud, sometimes calcareous, alternating with an argillaceous limestone, together known as the *lias*. In the red

Fig. 166.

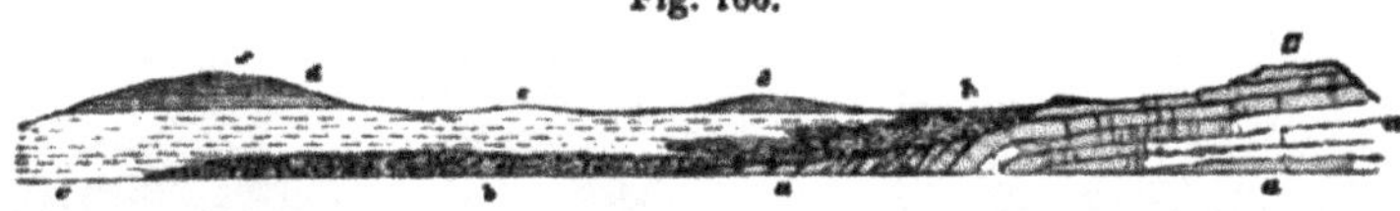

a, a, beds of disturbed carboniferous limestone; *b, b*, conglomerate of pebbles derived from the subjacent or adjoining rocks, cemented by magnesio-calcareous matter; *c*, red marl; *d*, lias; B, Blaize Castle Hill, near Bristol; *s*, Mount Skitham.

mud no traces of a marine organic remain have been detected in that district, though more northerly streaks of them are found; but even supposing some rare organic remains may hereafter be discovered, there is still evidence of a beach resting on a coast, this beach thrown up by seas during a period when the dry land was becoming gradually submerged, and a change was effecting in the existing conditions, so that the adjacent sea was no longer without animal life, or at least only contained a small portion of that affording harder parts for preservation amid the deposits of the time, but swarmed with molluscs, fish, and reptiles. The manner in which the filling up was effected is well shown in the section near Blaize Castle (fig. 166), though evidence of this kind is to be found as well elsewhere; one patch of lias (*d*, on the right of the figure) nearly reaching over the old beach, as it actually does at no great distance on the westward, where it covers up the carboniferous limestone on the margin of the coal-field from

Fig. 167.

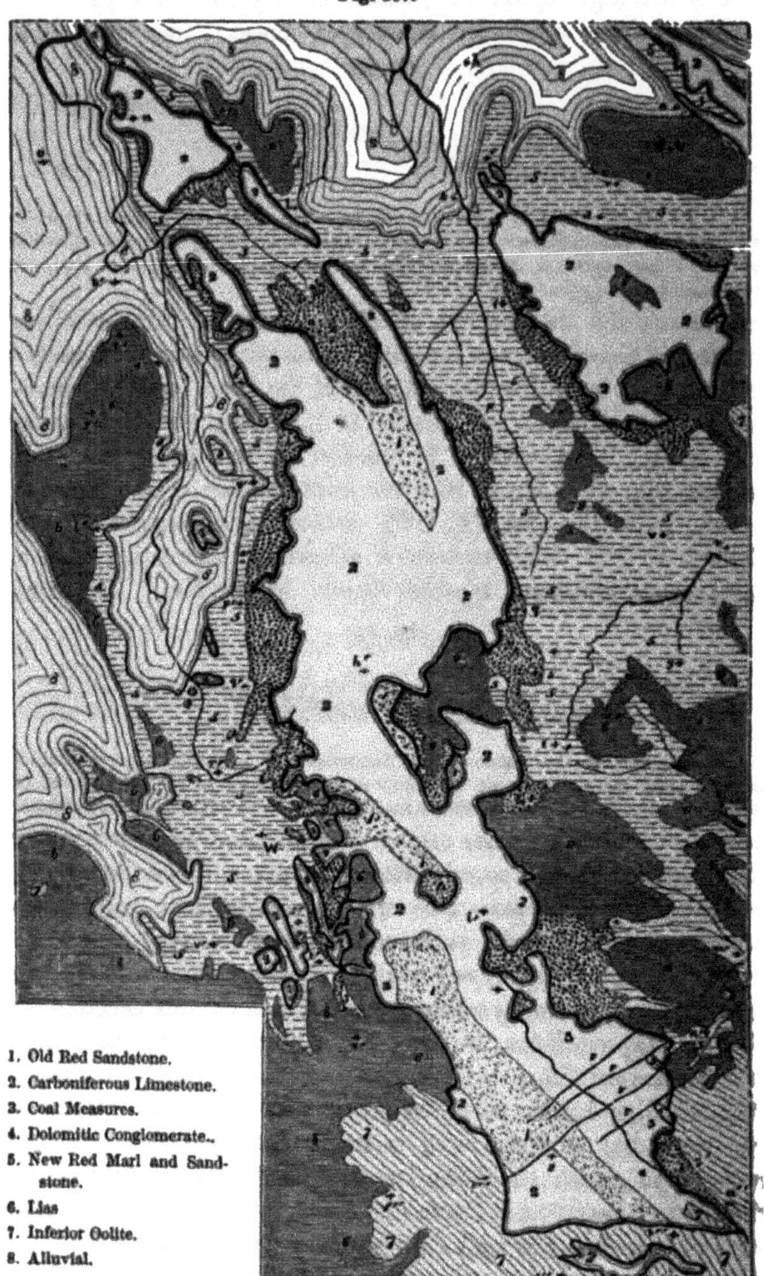
1. Old Red Sandstone.
2. Carboniferous Limestone.
3. Coal Measures.
4. Dolomitic Conglomerate.
5. New Red Marl and Sandstone.
6. Lias
7. Inferior Oolite.
8. Alluvial.

Redland, near Bristol, to Alverston, on the north. It will be perceived that if the denuding causes, which have removed so much of the deposits of a later geological date than these old shingle beaches, had carried off all traces of them in the section near Compton Martin, Mendip Hills (fig. 165), so that a surface corresponding with the line *d d*, had only been exposed, there would have been great difficulty in assigning the different parts of this shingle (now conglomerate) covering to their relative geological dates; though, with the old mud and sands outside of them, deposited at successive times, the relative date of the parts is sufficiently obvious.

While on the subject of this district, it may not be uninstructive, as it is one fertile in information, within so very small an area, to call attention to the successive coatings of fossiliferous accumulations as they followed one another, each spreading over a part of a preceding deposit, as the dry land of the Mendip Hills and adjacent country sank, and as it would appear, gradually, beneath the sea. For this purpose the accompanying map (fig. 167) may be useful. In it the different deposits represented consist, in the ascending series, of (1) old red sandstone; (2) carboniferous or mountain limestone; (3) coal measures; (4) dolomitic or calcareo-magnesian conglomerate and limestone; (5) the new red sandstone and marl; (6) lias; (7) inferior oolite, and others of the lower part of the series, known as the oolitic or Jurassic; and (8) alluvial accumulations, deposits from branches of the adjacent British Channel, where these found their way amid the sinuosities of the land, often covering a plain whereon forests once grew, at a higher relative level of sea and land than now exists, the outcrops of these sheets of concealed vegetable matter and trees forming the "submarine forests" of Stolford and other places on the present coast (p. 448).*

The darkly-dotted patches in the map (conglomerates, 4) will serve to show the mode of occurrence of the beaches surrounding

* The names of the various places marked by crosses and letters in the map (fig. 167) are as follow:—*a*, Tickenham; *b*, Nailsea; *c*, Chelvey; *d*, Brockley; *e*, Kingston Seymour; *f*, Wrington; *g*, Nempnet; *h*, Congresbury; *l*, Banwell; *m*, Locking; *n*, Bleadon; *o*, Lympsham; *p*, Burington; *q*, Compton Martin; *r*, Hinton Blewet; *s*, East Harptree; *t*, Litton; *v*, Chew Stoke; *x*, Chew Magna; *y*, Stowey; *a'*, Shipham; *b'*, Biddesham; *c'*, Badgworth; *d'*, Weare; *e'*, Axbridge; *f'*, Chapel Allerton; *g'*, Chedder; *h'*, Priddy; *i'*, Binegar; *k'*, Chewton Mendip; *l'*, Wedmore; *m'*, Radstock; *n'*, Kilmersdon; *o'*, Draycot; *p'*, Stoke Rodney; *q'*, Westbury; *r'*, Wookey; *s'*, Dinder; *t'*, Crosscombe; *v'*, North Wooton; *w'*, Wells; *x'*, Shepton Mallet; *y'*, Downhead; *z'*, Mells; *a''*, Elm; *b''*, Whatley; *c''*, Nunney; *d''*, Cloford; *e''*, East Cranmore; and *f''*, Chesterblade.

the older rocks of the Mendip Hills, and an adjoining portion of country near Wrington, *f*. Although, from the travelling upwards of continuous portions of these beaches during the gradual submergence of the dry land, and the subsequent wearing of the rocks, including all in the district, up to the time of its alluvial plains inclusive, they may not give the exact representation of the beaches of one time, they will still serve to show the manner in which they were accumulated round this old portion of dry land. Taken in connexion with similar facts observable even so near as Gloucestershire and Glamorganshire,* and looking at the size of the rounded fragments sometimes found in them, the effects of considerable breaker action is observable on the shingles, and they seem to have been well piled up at the bottom of old bays and other localities where favourable conditions for their production existed. The following section (fig. 168) will show one of these

Fig. 168.

a, a, limestone, intermingled with sandstones and marls, of the upper part of the carboniferous limestone series of the district, brought in by a large fault, on the N.W. of the Windmill Hill, Clifton; *b*, boulders and pebbles, in part subangular, of the subjacent rocks, cemented by matter in part calcareo-magnesian, variably consolidated; *c*, conglomerate or breccia, in which the magnesio-calcareous matter is more abundant, becoming more so at *d*, where it further assumes the character of the more pure dolomitic limestone in which pebbles and fragments do not occur.

ancient beaches facing the gorge of the Avon, near Clifton, Bristol, in a depression between Durdham Down and Clifton Hill, in which some of the rounded portions of the subjacent rock cannot be much less than two tons in weight, requiring no slight force of breaker action to move them and heap them up as now seen.

The submergence of this dry land continuing while geological changes were being effected over a wide area, (in which this district occurred as a mere point,) and so that, without reference to the modifications of deposits produced elsewhere, the red sediment of the seas near the shores of the land, then above water in the area of the British Islands, was succeeded by others in and above which animal life swarmed, the beaches moved upwards on

* See the Geological Map, "Memoirs of the Geological Survey of Great Britain," vol. i., pl. 2, in which a large area occupied by accumulations of this class and time will be found represented.

the slopes of adjacent rocks. Thus the rolled pebbles of the latter, and of the cliffs of the time, were occasionally intermingled with the remains of the animal life then existing. Near Shepton Mallet (*x'*, in the map, fig. 167), where the lias (6) rests both on the old red sandstone (1), and the carboniferous limestone (2), there is much of this old shingle (now conglomerate).* The following (fig. 169) is a section, exhibited close to Shepton Mallet, on the Bath road, wherein a line of pebbles (*b*) is strewed over the previously upturned edges of supporting carboniferous limestone (*a*, *a*), and constitutes a continuation of some more arenaceous and pebble beds, presenting much the appearance of a shore, not far distant.†

Fig. 169.

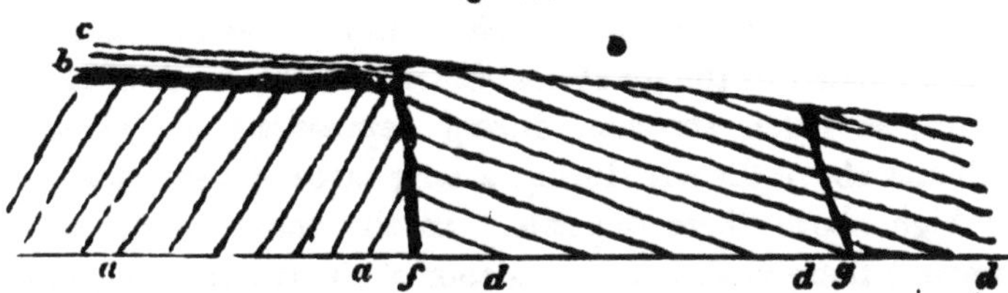

The lias at *c*, covering this pebble or shingle bed, has been thrown down (as it is termed) by a dislocation, or fault *f*, so that beds above that at *c*, are seen at *d*, *d*, *d*, the latter again broken through by a dislocation at *g*, and the whole surface of the hill being so smoothed off by denuding causes, that a gently-sloping plane is alone seen. Before we quit this section, it may be mentioned, that an observer in search of the different conditions under which fossiliferous deposits may have accumulated will here see that much less mud must have been mixed with the calcareous matter of the lias than is usual in the district, and which is to be found not far distant from this locality. The lias limestone beds (*d*, *d*, *d*) are here thick, for the most part, and in purity more resemble the carboniferous limestone (*a*, *a*) on which they rest, showing a cleaner state of the sea where they were formed than in those areas over which the usual mud, and muddy and silty limestones of the lias were accumulated. Coupled with the evidence of beaches, this greater freedom from mud would seem to point to the greater proximity of a shore with minor depths of sea, near and at which the waters were generally more disturbed, so that the lighter sub-

* These conglomerates, which are abundant, and wherein the pebbles are chiefly derived from the adjacent carboniferous limestone, have been long since pointed out by Dr. Buckland and the Rev. W. Conybeare (1824), "Observations on the South-Western Coal District of England;" Geological Transactions, 2nd series, vol. i., p. 294.

† This was well seen further up the road, in 1845, at which time some new cuttings were in progress.

stances being readily held in mechanical suspension, they were easily moved away by tides and currents to more fitting situations for deposit.

This character of a less muddy condition of the lias is by no means confined to the vicinity of Shepton Mallet; it is to be seen in several places in that part of England and South Wales. It is well shown in parts of Glamorganshire, where, indeed, as in the vicinity of Merthyr Mawr care is required not to confound some of the lias with the carboniferous limestone to which it there bears no inconsiderable mineralogical resemblance. Here, again, the observer finds this character in connexion with old conglomerates, resembling beach accumulations of the time of the lias, pointing to the probable proximity of dry land, such as may be readily inferred to have then existed in the great coal district on the north of it, even now, after so much abrasion, during depressions and elevations beneath and above the sea during a long lapse of geological time, rising high above these deposits. In the same neighbourhood (Dunraven) there is also good evidence of the lias reposing upon a clean surface of carboniferous limestone, as will be seen in the annexed sketch (fig. 171) and in the subjoined section (fig. 170),

Fig. 170.

wherein *a* represents disturbed strata of the latter, and *b* beds of the former, resting on their edges. In the section (fig. 170) the lower beds (*b*) of the lias are light-coloured, and contain fragments from the subjacent carboniferous limestone, these succeeded by argillaceous grey limestones at *c*. *f*, *f* are dislocations or faults, traversing the beds. In this case, though an observer might suspect the vicinity of a coast from the fragments in the lower lias, he would desire further evidence, and by search he would find,

Fig. 172.

round the point *d*, in the sketch (fig. 171), a conglomerate (*b*, *b*, fig. 172), reminding him of a beach interposed, to a certain extent

Fig. 171.

Lias resting upon carboniferous limestone, Dunraven Castle, Glamorganshire.

and level, between the beds of lias (*d, d*) and a worn slope of supporting carboniferous limestone beds (*a, a*), which here, from a local curvature are brought into a horizontal position. At *c*, in this section, the whitish variety of the lias of the district is found in a great measure free from muddy admixture. It is even occasionally dolomitic, and somewhat crystalline in this vicinity.*

Returning to the minor area of the Mendip Hills for evidence respecting the dry land and shores of the locality and period, we find, as the land became more and more depressed beneath the sea, that the lias, as it were, crept up the sides of the steeper shores, accumulating more muddy matter outwards, depressions being filled up, sometimes even on the shores when sufficient tranquillity permitted such a deposit, fine sediment accumulating above fine sediment, so that there was a kind of passage of the lias into the fine red marls beneath (fig. 173).†

Fig. 173.

a, gray lias limestone and marls; *b*, earthy whitish limestone and marls; *c*, earthy white lias limestone; *d*, arenaceous limestone; *e*, gray marls; *g*, red marls; *h*, sandstone, with calcareous cement; *i*, blue marl; *k*, red marl; *l*, blue marl; and *m*, red marls.

The observer next finds limestone beds (7), known as the *inferior oolite*, resting (map, fig. 167) from Cranmore (*e''*), on the south, to Mells (*z'*) on the north, upon various older accumulations; old red sandstone (1), carboniferous limestone (2), coal measures (3), and lias (6), passing over the nearly-horizontal beds of the latter as well as the variously-curved beds of the three former. The remains of animals of marine character show that this accumulation was effected in a sea, and therefore, that the depression of the land above mentioned had continued; but, as no distinct beaches have yet been discovered in connexion with this

* Part of these lower beds of the lias of the district is known at Sutton Stone, and has been employed for architectural purposes during many centuries, being well fitted for them.

† The section selected is from the vicinity of Shepton Mallet, reference to which has been previously made in the text, p. 481.

calcareous deposit, the probable boundaries between the sea and the land are not so apparent. The whole of the Mendip Hills may have been beneath the waters, though the relative levels of the different parts of the general masses of rock, notwithstanding the changes in these levels produced by various dislocations, effected during a long lapse of geological time, would lead us to infer that portions of dry land may still here and there have risen above the sea in that minor area. Be this as it may, when this overspread of calcareous matter (inferior oolite) took place, passing over the old margin of the lias, there were bare patches of carboniferous limestone (2) in the sea, and into these the boring animals of the time burrowed. Their remains are now found in the holes worked by them; and when good surfaces are exposed, an observer might imagine himself walking on limestone rocks, dry at low tides, in which the lithodomous molluscs of the present day were in the cells hollowed out by them. Not only are these old surfaces thus bored by the rock-burrowing molluscs of that period (the time of the inferior oolite deposit), but here and there —as, for example, near Nunney (c''),—the oysters of the same date are still found adhering to the bare limestone submarine surfaces of the time. There may be doubts as to the depths at which the boring molluscs worked, and the oysters adhered to those bare carboniferous limestone rocks; the whole of the dry land may, as before mentioned, have been then under water; but while the observer thus loses his traces of dry land, he has evidence that it still continued to descend, so that, at least, the movement in that direction had not ceased.

The following section (fig. 174) will serve to illustrate the

Fig. 174.

manner in which the older beaches (*d*, *m*) were overlapped, as it is termed, by the inferior oolite (*g*, *n*). The sands, commonly known as the inferior oolite sands (*f*), separating the calcareous beds of the inferior oolite from the lias (*e*), including in the latter the upper beds of that rock termed the marlstone (an accumulation replete with organic remains), is also overlapped. *a a* is carboniferous limestone, *b* its lower shales, and *c* old red sandstone, all moved prior to the deposit of the other beds. *h l* is the clay known as Fuller's earth, *i* its limestone.

With regard to the mode of occurrence of the inferior oolite (7, map, fig. 167), the annexed sketch, taken near Frome, on the road to Mells, will show not only the manner in which it (*a*)

Fig. 175.

frequently rests on subjacent carboniferous limestone (*c*), but also the very even surface which the latter often presents over comparatively large areas in that vicinity. These surfaces are usually drilled not only by the large boring mollusc of the time, occasionally found in their holes, as shown beneath (*a*, fig. 176), but also, in a

Fig. 176.

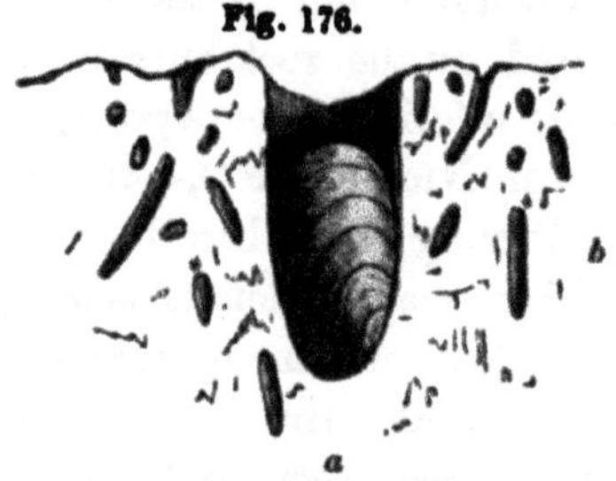

more tortuous manner, by another animal, sections of the holes made by which are seen at *b*, *b*. At *b* (fig. 175) there is a somewhat arenaceous parting, covering over the bored surface of the limestone, an introduction of matter which may have served to render that surface no longer fitted for the habitation of the lithodomous animals.

To mark the date of these borings still more perfectly, the same vicinity fortunately presents us with evidence (fig. 177) of a portion of the shingle (*a*), accumulated at the time of the lias (organic remains characteristic of that deposit as it occurs in the vicinity being found in it), having been consolidated and planed down, by denuding causes, to the same level with the carboniferous limestone, *b*, *b*, in a cleft of which it occurs, and having been bored by the same marine animals anterior to the deposit of the inferior oolite, *c*, *c*. Still further affording the observer relative dates for these perforations, he will find that the beds of the inferior oolite itself are thus bored, and by the same kinds of animals, as can be seen at the quarries of Doulting, on the south of the Mendips, and near Ammerdown, on the north, where, as the beds became successively consolidated, they afforded the needful conditions, on

Fig. 177.

their upper surfaces, for the existence of these marine animals, requiring shelter in hard rocks.

Assuming, for illustration, that the older rocks of this small district became totally submerged at this time, the geologist will, as it were, have traced the state of a minor area from one where there may have been dry land intermixed with sea, to another where the latter overspread all traces of the former. What may thus be done on the small scale can sometimes be readily effected over areas of far wider extent. It can be so, with care, over much of the British Islands, as respects the probable intermixture of land and sea, at the time of the accumulation of the new red sandstone series. To a certain extent the study of any general geological map of these islands will show this, though not altogether, since denuding causes have sometimes so acted as to remove these rocks and subsequent deposits from localities which would, judging from the mode of occurrence of the rocks in them, have been beneath the waters at the time that other portions of these beds were formed.*

* It becomes extremely interesting to consider the wide-spread distribution, over a portion of Western Europe, of the conditions prevalent at the period when the marls (termed new red sandstone marls, upper trias marls, marnes irisées, and keuper) ceased to be thrown down on the sea-bottom of the time with such a common character, and the mud, and calcareous mud of a succeeding accumulation, the lias, also occurred over much of the same area with its common character, carefully looking for the probable lands whence the needful mineral supply was derived, and around or on which the terrestrial plants, insects, flying, coast, and oviparous reptiles, including the marine animals whose preserved hard portions correspond with littoral

Extending his views, it behoves the observer still further to consider the probabilities of dry land over wider areas, such as would embrace large portions of our continents, if he expects to arrive at comprehensive conclusions as to the conditions which may have governed the production of any given detrital or fossiliferous accumulation which he may have under examination. As we have seen, it is highly important for him to obtain good evidence of beaches wherever they can be observed, inasmuch as these give him the boundaries of land and sea at certain geological times. These, however, he cannot always detect, since their preservation will depend upon favourable circumstances, such, for example, as their consolidation at the time of their production, as is now effected in certain localities, or their occurrence in such sheltered situations, under altered conditions of sea and land, that they could be covered up and not be removed (p. 454).

Though fresh-water deposits, as they are termed (from the remains detected in them being limited to those of terrestrial and lacustrine or fluviatile life), may not give any definite boundaries of the land and sea for any given geological time, they nevertheless prove the existence of dry land surrounding them at that time. Supposing the great lakes of North America to be filled up, mechanically or chemically, by mineral matter, enveloping the remains of the life inhabiting them, or drifted into them by rivers, and these accumulations to be elevated into the atmosphere, forming hills and dales, or even crumpled and broken into parts of high mountains and deep valleys, though they would not afford defined land boundaries, collectively taken they might prove the existence of no inconsiderable tract of dry land at a given period.

To estimate the probable preservation of such deposits amid movements of the earth's surface, depressing dry land beneath the

---

creatures of the present time, existed. The careful sketching, however approximate it might at first be, of the probable area occupied by land and sea in the European area at this time, with the distribution of organic remains found in the lias, distinguishing its upper and lower parts, would alone furnish matter for much useful progress in inquiries of this kind. The scarcity of animal remains in the upper red or variegated marls, and the distribution of the little areas where these are detected, with the subsequent abundance of life, as shown by its remains, constitute in themselves interesting inquiries for those tracing out effects to their probable causes, and who reason from the known to the unknown. The unequal distribution of the saurians of the time, so numerous and so varied in certain localities, and the patches where drifted terrestrial plants are discovered, afford much information that is valuable, it appearing sometimes needful to suppose that the shores formed of the mud and sands of immediately preceding accumulations had been upraised above the sea level, to account for these distributions. Hence the probable oscillation of the land at the time, above and beneath the sea, becomes also a part of such researches.

level of the sea, attention should be directed to the effects which would follow such a change. If any large area of dry land, wholly or partly bounded by seas, were now elevated higher above the latter, there would probably be a fringe of submarine deposits upraised at the same time, while the various outflows of river waters would have to adjust themselves to the new sea level. Many estuaries would cease to be such (if the boundary seas were tidal), and the courses of their feeding rivers would require adjustments in accordance with the change of level; much additional velocity and consequently scouring power, the volume of these waters remaining the same as a whole, being given to those parts of their channels, the adjustments of which had been more or less regulated by the former sea level. Under such conditions, it would happen as beneath (fig. 178), that many a river course was so lowered that

Fig. 178.

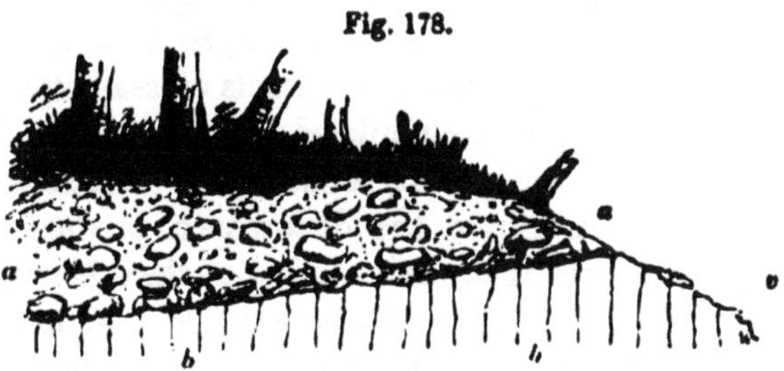

old river beds, *a*, *a*, would be left perched on the sides of valleys, above those which the river had formed for itself, either by removing loose materials accumulated in the valley, *v*, during its former adjustments, or by cutting through some rocky barrier, *b*, *b*, which the new velocity of its course, and power to transport the means of effective friction, have permitted. The new coast line would be variably acted upon by the breakers. In cases of cliffs of hard rocks, previously descending into the sea, so that upon the uprise of the land the breakers still acted upon the cliffs, no material difference would take place, except that the now shallower depths adjoining may be more disturbed than previously by waves, so that fine sediment, formerly at rest, might be removed, and its further accumulation in that locality prevented. With the fringes of any littoral sea bottoms upraised, the effects would be very different. The adjustment to the previous breaker action on shallow shores, especially of those where the surplus sands were driven on land, and accumulated in sandhills (p. 59), would be destroyed, and the sea would remove the loose materials before it, until a new balance of force and resistance had been again accomplished, in the manner represented beneath (fig. 179). If in this section *a* be the surface of the sea, after an elevation of the adjoining land, so that a

Fig. 179.

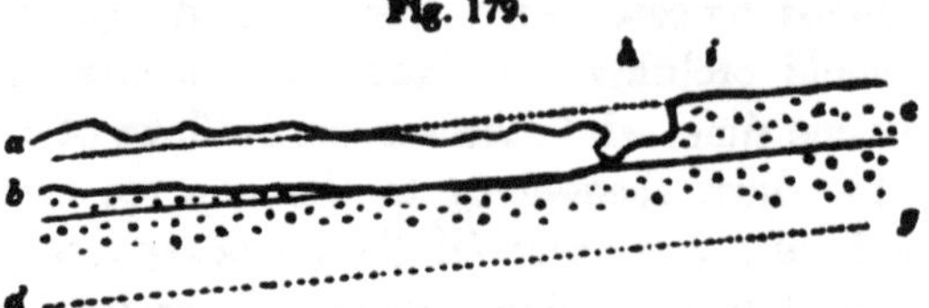

sea bottom, *b*, *e*, be brought within the action of the breakers, *h*, under which its slope had been previously adjusted, then that action would readily remove the unconsolidated mud, silts, or sand exposed to it, as well as the remains of any animals, entombed in them during their accumulation, or destroyed when upraised. Should the elevation of the land be continued, the subjacent detrital bed, *d*, *g*, with any organic remains it also might contain, would, in its turn, be subjected to the same action, so that upon exposed ocean coasts, with heavy breakers, vast tracts of sea-bottoms might be removed, their materials accumulated elsewhere in the localities where they could find rest. As, under such conditions, there would be depths, where, from the disturbance produced by the waves above, a mere sifting of the materials of these bottoms would be effected, it is desirable that an observer should bear in mind the mixture which might thence arise of the hard remains of marine animal life, entombed in the older sea bottoms, with those of the animals inhabiting the seas of the time. Should any change have been effected in the life of the area by the changes of relative levels, during the interval of the production of the respective sea-bottoms, the organic remains of both might be much mingled, especially when accomplished in a quiet manner, so as not to injure the entire shells or other remains of the older accumulations.

Upon such an elevation of land, if it were horizontal, as regards a particular area, upon which any fresh-water lake might be situated, the conditions affecting the deposits in it might remain much the same, except in such cases as where a discharging river into some adjoining sea was so much changed, as regards its outflow, that instead of a moderate velocity, it should acquire one so considerable that the surface of the lake itself became lowered from the cutting down of the new channel, by which an adjustment of fall had been effected to the relatively new sea level. Under such circumstances, the area and volume of the lacustrine deposits thence exposed would depend upon the shallowness or depth of the lake. When we regard great regions of lakes, such as those in

North America, and refer to the probability of the unequal lifting of the land, so that one portion may be tilted more than the other, it is easy to conceive that the waters of great lakes can be so removed that much of their old deposits may be exposed as dry land, while accumulations may continue to proceed in the portions of their basins still submerged. In great lakes, where breaker action produces appreciable effects, there would be the same tendency to remove loose upraised bottoms, thus brought within its influence, as on the sea-coasts, calcareous deposits necessarily offering resistances in proportion to their hardness, amounting even to that of compact limestone.

Turning now to the elevation of land over a wide area, so that the communications between seas intermingled with it and the main ocean may be cut off, great geological changes may be effected, not only in the life of the waters enclosed by, or running amid the land, but in the conditions of the dry land itself. As we have seen (p. 71), the evaporation in the Mediterranean is so far beyond the supply of water it receives through the Dardanelles, from the Black Sea, and from the rivers flowing into it, that if any elevation of the land took place, so that the Straits of Gibraltar were closed, a change which the geologist will learn from his researches to consider as one of no great comparative amount, this sea would have to adjust itself to its evaporation and supply of water, so that the one should balance the other. The result would be a great exposure of the present shallow sea bottom of the Mediterranean, and a change of level by which the outfalls of the rivers would acquire additional velocity, one which would also be communicated to the current through the Dardanelles, with a new adjustment of levels in the Black Sea and the rivers flowing into it, extending to the Sea of Azof.

The great rivers pouring their waters into the Mediterranean would continue to produce their present effects long after a stoppage to a free communication with the ocean was occasioned at the Straits of Gibraltar, a vast mass of detritus being borne into it as now, entombing the remains of animal and vegetable life. It is interesting to consider, that should the adjustment required for evaporation and supply of water lower the level of the Mediterranean sufficiently far (450 feet), two basins would be formed by a barrier passing between Sicily and the coast of Africa,* and the mouth of the Dardanelles would present dry land after the depres-

* See Captain Smyth's "Charts of the Mediterranean."

sion had continued to 222 feet, so that either the descent of the waters supplied by the Black Sea would be over a rocky channel, or the removal of the matter in the Dardanelles and Bosphorus would effect a free communication with the Black Sea, lowered to such an extent as to produce the most marked changes on its shallower coasts, and most materially to reduce its area.

Thus by the mere uprise of land over a moderate area, one comprising Spain and the opposite land of Africa, very modifying effects would be produced over a wide range of land and intermixed water. A glance at maps of the world will show how readily other important modifications of great areas might be effected by comparatively local elevations of land, such as closing the outlets of the Baltic or the Red Sea, the one, as at present, continuing to obtain an outlet for any waters which may find their way into it, the supply being greater than the evaporation, while the reverse might be expected in the latter, bounded by coasts to which little river water flows, so that a larger area than the Dead Sea (its level 1312 feet beneath that of the Mediterranean), would be presented, all the shallow portions of the Red Sea exposed, and such extension of the sand-drifts permitted as the new conditions might offer.

From such conditions to a complete removal of water, so that wide tracts of desert sands are produced, the results obtained are but the needful consequences of an inadequate supply of water to compensate for the evaporation. Thus, while in Central Asia there are still the remains of the waters which once covered so wide an area in that part of the world, as above noticed (pp. 73, 103), whole regions are strewed over with unconsolidated sea-bottoms driven about by the winds.*

All such surface changes, with the various modifications resulting from the unequal tilting of considerable tracts of country from one course of general drainage to another, as can often be so easily effected by comparatively moderate and unequal elevations of portions of dry land, should be well considered when weighing the probabilities of the existence of such land in any particular part of

* The observer will do well to study the evidence adduced by Sir Roderick Murchison and his colleagues ("Geology of Russia in Europe and the Urals," vol. i., p. 297), respecting the character of the wide-spread deposits of the great region in which the Caspian is included, showing the change of life during a time when the area passed through a condition of brackish water to the mixture of dry land and water which now presents itself. An important addition to our knowledge of the changes which the earth's surface has undergone, over a wide-spread region, in comparatively recent geological times.

the earth's surface at some given geological time. It will be evident that many complicated and intermingled deposits, containing the remains of marine, fresh-water, and terrestrial life, may be formed without the submersion of a great area of dry land beneath the waters of the ocean, a point of no small importance when the contemporaneous spread of animal and vegetable life, intermingled with the mineral accumulations of the time, is under consideration.

Though upon an elevation of land much of the shallow sea-bottoms adjoining it may be exposed to the destructive action of the breakers, as above mentioned (p. 490), the same movement would cause many portions of a littoral sea bottom to be brought up to the height most favourable for the preservation of its slope seaward, and the accumulation of sand-hills beyond the new line of shore inland. Another effect would be the conversion of many arms of the sea into lakes, the shallower depths, found in many localities at the outer or seaward part of such inlets of the sea amid the land, forming a low barrier between the sea and the more inward and deeper parts then separated from it. In such cases the newly-included portion of the sea would gradually become less saline from the continued supply from the rivers which are generally to be found in such localities, so that, finally, a fresh-water lake, such, for example, as Loch Ness, in Scotland, might still preserve a considerable depth, its surplus waters finding their way to the sea by some river. To illustrate this, let, in the subjoined plan (fig. 180), *a*, *a*, represent some arm of a tidal sea,

Fig. 180.

such as is to be found in many parts of the world, and *b* a submarine bank, in part formed by the check given to on-shore waves, stirring up detritus seaward, in part by drifts produced by prevalent winds, as in the manner previously pointed out (p. 55, fig. 50), and in part also from some check given to the outflow of the ebb tide carrying detritus brought down by feeding rivers, *t*, *t*, in seasons of flood, where it reached the general coast line. Submarine bars of this kind are far from uncommon. Upon the elevation of such a coast, so that its previous line, *m*, *m*, becomes shifted to *n*, *n*, a low piece of ground, *f*, *f*, of a breadth depending upon its shallowness beneath the former relative sea level, and its general slope might extend to a considerable distance outwards, the surplus river waters finding their way as a river, *r*, amid the new low ground, to the sea at *o*, *o*. Lakes at the outskirts of mountainous countries, bounded by low ground at their outlets, this low ground continuing to some neighbouring sea, would often appear to have been thus produced. In like manner numerous rivers of considerable size would have their lower portions converted into lakes, their present bars (p. 77) with much of an adjoining shallow sea-bottom being upraised, so that the river formed a new channel between the old bar and the new coast line, where, assuming conditions to be still similar, a new bar would be formed, and the spread of water behind the old bar might continue as a kind of lake until filled by detritus, its waters, after the change, becoming gradually fresh from the absence of the daily inflow of the sea upon the flood tide. These and other modifications of coasts by elevations of land, some on the small and others on the large scale, will readily be seen by an inspection of the charts of various parts of the world, whereby it will be found that lakes would be the frequent consequence of such changes, especially upon mountainous shores, where the continuations of many valleys are beneath the sea level, and where, at their termination seaward, the bottoms become somewhat raised, forming more of a portion of a general slope in connexion with the adjoining sea-bottom outwards than with the arm of the sea continued inwards.

The production of the lakes on the outskirts of mountainous districts, in the manner last mentioned, would often seem to involve the necessity of a previous submersion of a portion of the same region beneath the sea level, the arms of the sea being mere continuations of that level amid depressed land. When such a submersion happens—and there can be little doubt that it has often done so during the lapse of geological time—the filling up of the submerged portions of such valleys will be modified according as the land may be situated in a tidal or tideless sea. In the one case there would generally be estuaries and their results (p. 77), in the other the mere discharge of detritus outwards, much as at the head of lakes, due allowance being made for the mode in which river waters, bearing detritus in mechanical suspension, may flow over the sea water in such situations (p. 63).

We have, while noticing the accumulation of beaches (p. 56), the condition of estuaries (p. 58), the preservation of foot-prints (p. 129), coral reefs and islands (pp. 195, 200), distribution of erratic blocks and superficial gravels (pp. 258, 260, 263, 287), ossiferous caves (p. 305), and the uplifting of the subaqueous parts of volcanos (p. 389), been compelled to mention some of the effects which would be produced by the elevation and depression of land above or beneath the sea, and even to advert to the material changes which would be produced by the depression of the Isthmus of Panama, and the elevation of the sea-bottom between the north coast of Australia and the Malayan peninsula (p. 136). The effects thus caused have to be more or less regarded with respect to the mineral and fossiliferous accumulations of all geological periods to which the modifications and changes now in progress on the earth's surface are applicable.

As with an elevation of the dry land above, so with its depression beneath the sea, the steepness or gently-sloping character of the mineral mass moved has to be duly regarded. While amid a mountainous region the depression of the land for about 200 or 300 feet may merely somewhat more intermingle the sea with the land, arms of the sea extending further into the country, the same depression in lower lands may cover whole districts, the tops of some higher grounds only rising as scattered islands amid a wide-spread sea. The effects produced in one region would form no measure of those following such a depression (in a geological sense of a very minor kind) in another. The change might, indeed, so far affect a mixed region of mountains and low plains, that some old state of things may often, to a certain extent, be reproduced, the mountains forming islands, or groups of islands, in a part of the ocean, and so that ravages on their flanks from heavy breaker action are recommenced. While, in the one case, the area occupied by terrestrial life, animal and vegetable, was comparatively little circumscribed, in the other, large tracts would be laid waste, and many a plant and animal, peculiar to the low districts, might, under certain conditions, be entirely swept away.

Whether we contemplate the submersion of a large area of dry land, much intermingled with lakes, such as Northern America, partly overspread by deserts, such as portions of Africa, and Asia, or under an ordinary condition of the growth of trees and other terrestrial plants, and an adjusted distribution of animal life, there would appear few geological changes so effective in bringing deposits marked by dissimilar organic remains into contact, than such submersions. Even horizontal or nearly horizontal accumulations may be thus superimposed, after the lapse of considerable intervals of geological time, should one sea-bottom have been long horizontally raised above water until a change in the relative levels of sea and land, in that area, produced the conditions for its submersion and subsequent coverings by new deposits. Under such circumstances it may require some caution on the part of the observer not to conclude that the two kinds of sea-bottom have not succeeded each other quietly beneath the sea. If any low region be regarded with reference to the effects which might be produced during its tranquil submersion, such, for example, as where wide tracts of sand-hills border such a district, it will be obvious that these latter would soon disappear before the action of the sea, and be readily spread over the low grounds inland as these became depressed. As the land descended, its surface would be acted upon by the sea, the looser and lighter parts readily taken up and removed, to be deposited elsewhere in fitting situations according to circumstances, the larger and harder parts often, as it were, sifted from the finer and lighter, and occasionally enveloped by new detritus not far distant from their places of first accumulation. When a geologist considers the decomposition of the surfaces of ancient and upraised sea and lake bottoms, forming soils, and the frequent dispersion of their organic contents either in these soils or subjacent decomposed rocks, he will perceive that under favourable conditions, these remains may sometimes be preserved with little injury, and be mingled with those of the animal life introduced over the old land-surface by the new submersion beneath the sea. The lower part of the new accumulations and the upper surface of the old deposits may thus become mixed, and lead, without due care, to the supposition that the two were marked by a passage of the organic remains found in the one into the other. This is no useless caution, as the observer, when studying the effects produced during the submersion of some ancient land, will have occasion to remark. Some good examples of weathered fossils of the carboniferous limestone, even so much so as to be completely detached, may be seen in the dolomitic or magnesian limestone deposits (of the new red sandstone series) in Somersetshire, Gloucestershire, and Glamorganshire, and these might readily, without due care, be considered as organic remains of the later geological time. It is as if, after abstracting the turf or soil covering the carboniferous limestone of a district, and the quiet removal of the intermingled earth in place, a deposit of magnesio-calcareous matter was thrown down from solution amid the fossils and fragments of the older rock. The fitting of deposits of this kind into the inequalities beneath is well shown in the district to which allusion has been made, and may be illustrated by the following section (fig. 181), seen at Pen Park, north of Bristol, where the dolomitic

Fig. 181.

or magnesian limestone, *b*, rests upon the edges of upturned beds of carboniferous limestone, *a*, *a*, into the superficial inequalities and interstices of which it enters, covering up blocks of the same rock, *c*, reposing on the surface before the deposit of the newer rock. In such localities, the weathered portions, including fossils, of the subjacent rock may sometimes be found penetrating or intermingled with the subsequent accumulations, occasionally marking a state of much tranquillity, and as if an overspreading deposit had been effected on an old surface of land which, in favourable localities, had not been much subjected to breaker action, rounding the fragments and destroying the old weathered surface of the rock.

That a mixture of the organic remains of former geological times with those of molluscs and other marine animals, the species of which at present exist in the seas adjoining, is now taking place, a walk along many a coast will show, shells especially being seen washed out of sands and clays forming the cliffs, and being mixed with those now cast on shore. We are, therefore, well prepared to expect that when, during a submersion of dry land, the loose surfaces of ground, with distributed organic remains, are exposed to similar action, the results will be the same, with this difference, that while, on an exposed coast, the ancient and modern organic remains may often be all ground down together into one common mass, in a submerging land, more sheltered localities may frequently present themselves where the ancient organic remains could be more quietly sifted out of the loose earthy matter surrounding them, and be intermingled with the exuviæ of animals, the habits of which lead them to prefer equally tranquil situations.

Fully to appreciate the varied geological effects which may be produced by the submersion of differently circumstanced dry land, it may not be uninstructive to consider those which would follow the re-establishment of the sea over the many thousand square miles now occupied by the great desert or deserts of Northern Africa, Sahara and others. Judging from such observations respecting the heights of parts of these deserts as appear deserving of credit, a submersion of the kind mentioned as probable for the British Islands during the inferred cold period preceding the present state of that area, namely, from 1,200 to 1,500 feet, would place at least a large portion of them beneath a continuation of the Atlantic. As the sea moved inwards, according to its level, however this might present itself with respect to the variation from horizontality, wholly or partially, of the submerging land, the sifting of hard and coarser parts from the lighter and softer would be effected. Thus the remains of men, camels, and the ordinary desert animals, here and there mingled with the additions to the former which the oases produce, might be mixed with those of the marine animals introduced with the sea as it advanced over the land. Should there still be organic remains amid the sands of the deserts, entombed when the whole had previously been beneath water, these also might be mingled with the animal exuviæ of the new sea-bottom.

When we consider the depression of land occupied by many and perhaps great lakes, such as those in North America, the amount of submersion more in one part of the general area than in another has to be duly regarded; as also the consequent different conditions under which these bodies of fresh water may be placed. While the progress of depression may in some cases be such that the outflowing waters were gradually shortened in their courses, until the time arrived when the sea entered into the lakes, a mere overtopping of the fresh-water basin being accomplished; in others the unequal tilting of the ground may have occurred so that the sea was introduced and covered a deeper portion of the lake basin in one direction than in another. Inferring the usual mode of distribution of matter by the combination of wind-wave action beneath the sea at the proper depths, and breaker action on the shores, with the effects of tidal streams in tidal seas, the accumulations might so far differ under these conditions that while, in both instances, the animal life gradually became adjusted to the sea, the greater part, if not the whole, of the previous deposits in the simply overtopped lake might be preserved and be covered by the brackish water, and finally by the marine accumulations. The unequally-tilted lake banks might permit a part of the older deposits to be so exposed to breaker action that they were partially removed, the component mineral matter and its organic contents partially also rearranged with the new accumulations. If, during a re-establishment of part of North America beneath the sea, it so occurred that Lakes Erie and Ontario were depressed more rapidly than Lakes Huron, Michigan, and Superior, the sea finally over-spreading the whole, the relative positions of the lakes to the direction of the greatest depression would much influence the results. Lakes Erie and Ontario would present their breadths to the movement, while Lakes Michigan and Huron would be acted upon in their lines of length, Lake Superior presenting a more complicated form. Under such a movement, the entrance of the sea would necessarily depend upon the varied surface and levels for the time opposed to it; but it may readily happen that while Lakes Ontario and Erie were beneath the sea, and Lake Huron brackish water, Lake Superior might continue as fresh water, the contemporaneous deposits in each containing the remains of animals capable of living in the various kinds of water respectively, such of the original lacustrine creatures remaining in the brackish water as could adjust themselves to it, mingled with those marine animals which could support life under the same conditions, the terrestrial vegetation drifted into all the deposits being of the same general kind.

CHAPTER XXVII.

MODE OF ACCUMULATION OF DETRITAL AND FOSSILIFEROUS ROCKS CONTINUED. —EVIDENCE AFFORDED BY THE COAL-MEASURES.—STEMS OF PLANTS IN THEIR POSITION OF GROWTH.—FILLING UP OF HOLLOW VERTICAL STEMS, AND MIXTURE OF PROSTRATE PLANTS WITH THEM.—GROWTH OF PLANTS IN SUCCESSIVE PLANES.—THICKNESS OF SOUTH WALES COAL MEASURES.—FALSE BEDDING IN COAL MEASURE SANDSTONES.—SURFACE OF COAL-MEASURE SANDSTONES.—DRIFTS OF MATTED PLANTS IN COAL MEASURES. —EXTENT OF COAL BEDS.—PARTIAL REMOVAL OF COAL BEDS.—CHANNELS ERODED IN COAL BEDS, FOREST OF DEAN.—LAPSE OF TIME DURING DEPOSIT OF COAL MEASURES.—PEBBLES OF COAL IN COAL ACCUMULATIONS.—MARINE REMAINS IN PART OF THE COAL MEASURES.—GRADUAL SUBSIDENCE OF DELTA LANDS.—FOSSIL TREES AND ANCIENT SOILS, ISLE OF PORTLAND.—WEALDEN DEPOSITS, SOUTH-EASTERN ENGLAND.—RAISED SEA-BOTTOM ROUND BRITISH ISLANDS.—OVERLAP OF CRETACEOUS ROCKS IN ENGLAND.

WHILE the remains of drifted terrestrial plants, large or small, may not give very exact information as to the area occupied by dry land, whence they have been derived, since they could have floated from considerable distances (p. 125), according to the currents of particular geological times,* where these remains occur either in their places of growth, or so that we may rightly conclude that they have not been removed far from them, they become important. Those deposits of vegetable matter interstratified with shales, sandstones, and conglomerates, which occur in a particular portion of the geological series of accumulation in Europe and America, and to which the term *coal measures* has been assigned, from abundantly furnishing the fuel which has become so important to the progress of civilization, afford the observer the means of in-

* The Gulf Stream, as before pointed out, is an excellent example of a body of water capable of transporting the vegetable products of the tropics to the temperate regions of the north across an ocean.

ferring the existence of land in particular portions of the northern hemisphere at that time. When carefully examined, a large proportion of the coal beds have been found, in the British Islands (and the evidence would also appear to justify similar conclusions in many other countries), resting upon others immediately beneath, in which the roots of particular plants are found to extend in a manner showing that these are actually in their places of growth, as respects the beds of mineral matter containing them. These roots were at one time considered as separate plants (*Stigmaria*), but now, from the researches of Mr. Binney and other geologists, it seems established that they belong to other plants (*Sigillaria*, if not also to other genera). With this advance of knowledge, we find that great sheets of vegetable matter were based upon a mud or silt, in which the amount of sand varied considerably in different situations, even in the prolongation of the same bed, and that into this mud or silt the roots of at least some of the plants of the time and locality spread as in ground for which they were suited.

Upon further investigation, it has been found that roots of this character are to be seen attached to stems of plants still vertical, or nearly so, to the beds of shale or sandstone (formerly mud, silt, or sand), in which they are enclosed. Though the attachment of such roots may be rarely seen, the examples of vertical stems of plants, apparently in their places of growth, are sufficiently common, so much so that if certain parts of the coal measures of the British Islands could have the detrital matter removed, various and extensive areas would be found covered by the stumps of plants in such positions. These stumps are so numerous in the ordinary detrital deposits reposing on some coal beds, that they become dangerous in the collieries, (unless great care be taken in the works,) from being merely sustained aloft by the coaly matter representing the former outer portion of the plants, so that when this is insufficient to retain them, they fall on the heads of the miners. The following sketch (fig. 182), at Cwm Llech, towards the head of the Swansea valley, Glamorganshire,* may serve to illustrate the manner in which these plants may sometimes be exhibited in quarries or natural cliffs, rising amid the beds which have enveloped them in their places of growth. The largest of the two stems was 5½ feet in circumference. They merely formed a part of a surface more or less covered by stems of

* Made by Mr. Logan, by whom and the author the locality was carefully examined. The stems were subsequently removed to the Royal Institution of South Wales, at Swansea, where they now are.

this kind, as others were to be seen in similar positions in the same bed of rock higher up in the same valley. Upon uncovering a

Fig. 182.

shale beneath the sandstone, in which these plants (*Sigillariæ*) stood, an abundance of fern leaves, and fragments of other plants, commonly seen in these deposits, were found distributed around in the same manner as leaves and other parts of plants may be dispersed around stems of trees in muddy places at the present day.

It sometimes happens that the vertical stems of the plants rise through different kinds of beds, the component parts of which accumulated around them, while the vegetable matter still held together. The following (fig. 183) is an example of this kind, as it was exhibited at the Killingworth Colliery, Newcastle district. In this section *a* represents the high main coal of the district, *b* argillo-bituminous shale (formerly carbonaceous mud), *c* blue shale (mud or clay), *d* compact sandstone (sand), *e* alternating shales and standstones (beds of mud and sand), *h* white sandstone (clean sand), *i* micaceous sandstone (sand with mica), and *k* shale (mud or clay). In such cases various changes were effected in the kind of mineral matter transported to, and deposited amid, the vegetation there standing. Though we do not know the extent to which such plants may have been covered up before they died, an attentive study of the mode in which the mud, silt, or sand has been accumulated round the stems often shows the observer that the water bearing or moving the detritus was very shallow.

Around the stems at Cwm Llech (fig .182), the laminæ of the sandstone were so arranged as forcibly to suggest that they represented

Fig. 183

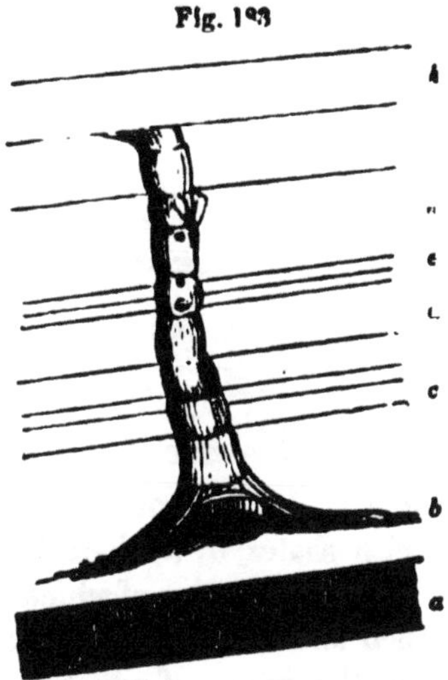

the washing up of sands around the plants in shallow water agitated by slight waves. Such an arrangement may frequently be seen, as also occasionally, when the stems are carefully uncovered, an adjustment of the laminæ of the original sand or silt, in a manner pointing to the passage of a slight current of water by them. When this can be found, the direction whence the current came may be inferred by the position of the laminæ marking the place of the eddy, behind the stems.

From the manner in which these vertical stems are so frequently terminated upwards, it would appear that while, for a time, their lower portions continued to resist the pressure both of the water in which they were immersed, and the gradually-accumulating detritus borne or drifted by it, their tops became decayed, and were removed, so that finally sheets of detritus uninterruptedly spread over the localities where such plants may have grown. We seem, indeed, to have evidence, in the manner in which so many of these stems have been filled with mud, silt, sand, and the remains of other plants, that before such sheets of continuous detritus were spread over their tops, they were hollow, like so many open and vertical tubes, in which, when overtopped by waters bearing detrital matter, and the leaves and fragments of plants, these were deposited in the same way that sediment and the remains of vegetation are accumulated in the hollows of upright and decayed or broken stems of bamboos, and other plants on the side of rivers, or amid low grounds, during and upon the subsidence of floods. That the interior and exterior deposits, in and around the vertical stems are not the same, different minor layers being found in the stems not corresponding with those outside, may often be seen, as shown in the annexed section (fig. 184), where a stem, *a*, *a*,

Fig. 184.

covered by a sandstone bed, *b*, is surrounded by other sandstones, *c*, *c*, *c*, interstratified with shales, *d*, *d*, their lines of deposit abutting against the stem, the only remains of which are usually formed of coal from half an inch to two inches in thickness, according to the size of the plant, in the inside of which other layers, *e*, *e*, of shale and sandstone, with or without leaves of fern or other plants, occur arranged in a manner showing that they were accumulated independently of those outside.

Strewed amid the same accumulations (those of the coal measures), prostrate stems, sometimes measuring thirty feet in length, and of proportionate breadth, considered by botanists to be of the same and similar genera, and frequently even species as those found vertical, would often appear to show that they have not undergone violent transport in waters, being so little, if at all, injured. Indeed, occurring, as they sometimes do, among the stumps of stems, these apparently in the positions in which they grow, they far more resemble those prostrate trees found amid the stumps of the rooted trees in the "submarine or sunk forests" (p. 447). In some collieries an observer may, as it were, see beneath such an accumulation of plants in muddy ground, the ends of the upright stumps, like so many irregular rings, scattered over head, the long prostrate stems strewed among them, and a multitude of ferns of various kinds, *Lepidodendra*, and other plants, matted together, the whole presenting the appearance of a growth of plants in soft or wet ground, if not shallow water, mud mingled with various portions of them. Often the plants appear to have partly grown in the same locality, and partly to have been drifted into it, sometimes from an adjoining situation, at others from more distant places.

While areas of fair size are known by colliery workings to have had numbers of vertical stems tranquilly covered over by detrital matter on a particular geological plane, so that a forest of this kind of vegetation has been contemporaneously entombed, it sometimes occurs that there is good evidence of similar conditions having produced similar results more than once over the same area. Of the facts brought to light on this head, though it may be well known in many coal districts that vertical stems of plants are found at more than one geological level, the occurrence of one series of vertical stems above others seems to have been hitherto, in no artificial or natural sections, better exhibited than in the coal districts of Nova Scotia and Cape Breton, where several of these planes of vegetation, the stems of plants still standing in their places of growth, are seen above each other. Sir Charles Lyell describes ten forests of this kind, as occurring above each other, in the cliffs between Minudie and the South Joggins, at the head of the Bay of Fundy. The thickness of the mass of beds containing the upright stems is estimated at about 2,500 feet, and the usual height of the trees is from six to eight feet, but one was seen apparently 25 feet high and four feet in diameter, with a considerable bulge at the base. All these stems appeared to be of the same species.* We are indebted also to Mr. Logan for a very detailed account of these coal measures. In his description of the Sydney coal-field, Island of Cape Breton,† Mr. Richard Brown notices many upright stems of plants in different beds. Among the sections given, the annexed (fig. 185) will be useful, as

Fig. 185.

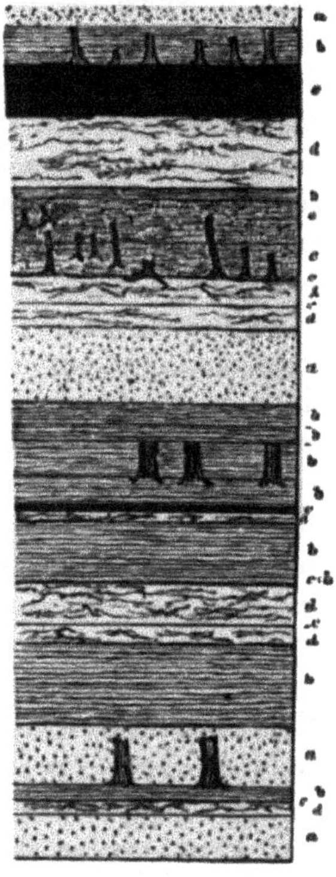

* Lyell, "Travels in North America," vol. ii., pp. 179–188.

† Brown, "Section of the Lower Coal Measures of the Sydney Coal Field, in the Island of Cape Breton," (Quarterly Journal of the Geological Society of London, vol. vi., p. 115). After adverting to the descriptions of the coal measures of Nova Scotia by Sir Charles Lyell (Travels, &c.) and by Mr. Logan (Section of the Nova Scotia Coal Measures at the Joggins), Mr. Brown estimates the productive coal measures of Cape Breton at more than 10,000 feet in thickness. The Sydney portion, described in this communication, was, by measurement, 1,860 feet thick. The dip is mentioned as at an angle of 7°.

showing this occurrence of many vertical stems above each other.* In it, *a* represents sandstones, *b* shales, *c* coal, and *d* the beds, usually argillo-arenaceous, in which the roots (*Stigmaria*) are in their positions of growth. The total thickness of the deposits amounts to 92 feet, and in it occur four planes of upright stems, (the second showing different levels of growth in it,) and six ancient soils, surmounted by as many seams or beds of coal of very different depths, the most considerable being six feet, and the least seam, one of mere carbonaceous matter, one half-inch thick.

It will no doubt at once suggest itself that such accumulations of mud, silt, sand, and sometimes gravel, intermingled with layers of fossil vegetation, these layers based upon a soil, probably moist or wet, in which the roots of certain plants freely grew, while vertical stems occurred, as much sometimes as 15 or 20 feet high, and two to four feet in diameter, even planes of these old forests being found above each other in limited sections, must have been gradually submerged, so that, at intervals, the soil was sufficiently exposed to, or near the atmosphere, that the plants entombed amid them could come under their proper conditions of growth. A trough or other cavity, or slightly-inclined plane of shore, gradually filled up to the level of the atmosphere, would only give one layer of vegetation, whereas, in some coal districts, where the seams of coal are reckoned with the soils on and in which their constituent plants grew, 50 or more intervals for growth may have to be accounted for. A submersion of the ground on which the plants flourished, so that at times the mud, silt, or sand of the time accumulated at a greater rate than this submersion could keep them beneath the level of water, or during which, though the descent of the land may have been, as a whole, constant, there were minor amounts of movement (by which after a subaqueous area had been filled up to the atmosphere, there were pauses when the plants could grow), would alike appear to explain the facts observed. The section of the 1,860 feet in which the upright stems of the Sydney beds (Cape Breton) occur, shows that there were more than 40 periods in the general descent of the mass when there were soils in which the roots (*Stigmaria*) of the plants of the time and locality found their needful conditions for growth, those for the accumulation of

---

* In this section the beds are reduced to horizontally, and are on a proportional scale, the relative thickness of the beds being taken from the detailed description of them by Mr. Richard Brown (Journal of the Geological Society, vol. vi., p. 120.)

the vegetable matter above them having varied materially.* When we turn to the sections of the European coal-fields of this kind, similar evidence presents itself.† In the section of the Bristol coal measures between the Avon and Cromhall Heath, there were no less than 50 periods during which the conditions for soils obtained, and roots (*Stigmaria*) were freely developed in them, these soils topped by a growth and accumulation of plants, apparently requiring contact with the atmosphere for their existence. The general thickness of that series is about 5,000 feet and it is based upon an accumulation, chiefly sandy, about 1,200 feet thick. The Glamorganshire coal-field gives a still greater deposit of mud, silt, sand, and gravel, intermingled with soils in which roots of some, at least, of the plants of the time spread out freely, most frequently, though not always, covered by beds or seams of coal, the thickness of which necessarily depended upon the duration of the conditions needful for the growth and accumulation of their component plants. The mass of these various beds in the neighbourhood of Swansea may be estimated at about 11,000 feet; so that if accumulated by subsidence, horizontal beds piled on each other, it would have to be inferred that in this part of the earth's surface, and at that geological time, there had been a somewhat tranquil descent of mineral deposits, sometimes capable of supporting the growth of plants requiring contact with the atmosphere, but most commonly beneath water, for a depth by which the first-formed deposits became lowered more than two miles from

---

* The detail of the general mass is thus summed up by Mr. Brown:—

	Feet	in.
Arenaceous and argillaceous shales	1,127	3
Bituminous shales	26	5
Carbonaceous shales	3	3
Sandstones	562	0
Conglomerate	0	8
Limestone	3	11
Coal	37	0
Underclays	99	6
Total	1,860	0

From this it would appear, that while the calcareous matter (limestone), gravel (conglomerate), and mud-mingled organic matter (bituminous and carbonaceous shale), were of little importance, the mass was composed of silt and mud (arenaceous and argillaceous shales), and of sand (sandstone), the former double the thickness of the latter. The more pure vegetable matter (coal) amounts to about $\frac{1}{50}$th part, and the soils (underclays) to somewhat less than $\frac{1}{18}$th part.

† See the detail of the coal-fields of South Wales, Monmouthshire, and Gloucestershire (Vertical Sections of the Geological Survey of Great Britain, Sheets 1-11), and descriptions of portions of the same districts (Memoirs of the Geological Survey of Great Britain, vol. i. pp. 161-212).

their original position. It may be inferred that this thickness is not really that of the general mass, as the component beds might have been accumulated one against each other, as happens in single sandstone and conglomerate beds (figs. 38, 57), and as no doubt has more often to be taken into account than it has been, in the calculations of thickness. It may, however, be remarked, that in these coal deposits, where planes of vegetation of a peculiar kind seem so frequently to have been based on very soft soils, and the whole has been so intermingled with continuous accumulations of mud, that the general sections appear often to point to great thickness, more particularly when the component beds are, after dipping downwards, found rising with similar characters at a considerable distance. Nevertheless, the unevenness in many of the deposits should be well considered, and the probable value of the general decrease of the whole thickness from such causes be duly estimated.

Though the fine mud of the time (now argillaceous shales) gives little information as to deep or shallow water in which it may have been deposited from mechanical suspension, the sandstones of the coal measures very frequently show that they have been far more the result of sands drifted along the bottom of moving water, than of having been mechanically suspended in it. Indeed, the accumulation of the sands is much that which would be expected from a pushing forward of the bottom detritus into a shallow depression, where the conditions may have been so changed by alteration of levels that the sand of a higher situation, and nearer its source of supply, was readily transported into it. Sections of the subjoined kind (fig. 186) are of the commonest occurrence

Fig. 186.

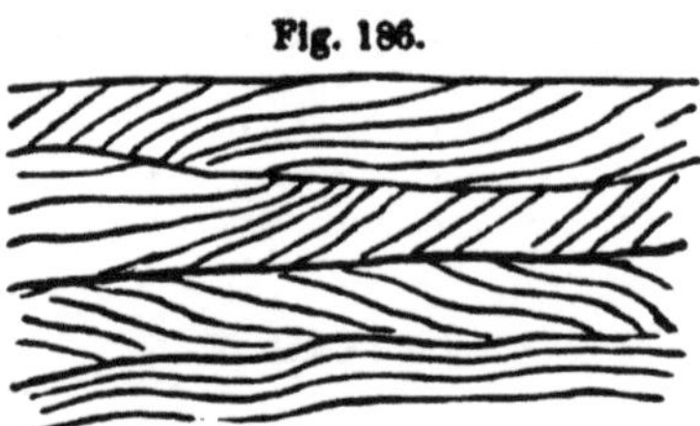

in many parts of the British coal measures, and they would appear not less common in the great coal deposits of North America and parts of Europe, the geological age of which has been considered somewhat equivalent. By careful removal of the upper surfaces of these beds, the overlaps of the differently-drifted laminæ may be seen, and occasionally still better in coast exposures.

The following (fig. 187) is a sketch* of the upper surface of a bed of sandstone exposed on the coast near Nolton Haven, Pembrokeshire, showing the different margins of the sand, as its various drifts proceeded.

Fig. 187.

An observer having thus obtained evidence of the apparent growth and accumulation of terrestrial plants in place, and the rooting of at least some of them in soils beneath of such a character that fine rootlets could spread freely amid their parts, has to look carefully into the species of this and other plants entombed in the general mass, endeavouring to see if there may not be some drifted amid the mud, silt, and sands, and even included among the coal itself, which may differ from those inferred to have grown on the spot. There would appear much to accomplish on this head, at the same time, however, it seems probable that while some plants have thriven in the planes of vegetable matter now converted into coal, others, even trees, have been borne into the general mass of vegetation, by water transporting them, as many a river now does. Matted masses of plants are often discovered among the sandstones, as if drifted by some stream, transporting such plants on its surface, while it pushed onwards the sands beneath it, streaks of such intermingled vegetation sometimes extending many yards in length, and occurring amid sandstones, the component sands of which have been thus accumulated. The following is a sketch (fig. 188) of the upper surface of part of one of these vegetable drifts at Pembrey, Caermarthenshire, in which multitudes of the stems of *Sigillariæ* and *Lepidodendra*, chiefly the former, and now converted into coal, are crossed and matted together in all directions.

These drifts of plants, now forming streaks of coaly matter in

* By Professor John Phillips, when examining that part of South Wales with the author.

the sandstones or shales including them, are sufficient to show that though numerous coal beds may be the result of the growth

Fig. 188.

of a peculiar vegetation in place, the roots of which required and penetrated a suitable soil beneath, it might so happen that extensive and deep accumulations of drifted plants may wholly form coal beds under favourable circumstances, so that an observer, while investigating coal deposits, should carefully weigh any evidence of this kind, as well as that pointing to the growth of plants in the situations where their remains now constitute coal. The two modes of accumulation are by no means incompatible with each other. On the contrary, they may be often intermingled, sometimes conditions prevailing more, or even entirely, in favour of one instead of the other. At the same time it may be remarked that, as careful investigations have proceeded, the evidence of the growth in place of the mass of plants now constituting extensive coal beds, during the time when the chief coal accumulations of Europe and America were effected, has been gaining ground, inasmuch as the soils beneath most of the coal beds and containing roots (*Stigmaria*) have been very commonly found.*

---

* These soils, though far from having been acknowledged as such, have long been known, and employed as guides by the working colliers, whose experience taught them their frequent occurrence beneath beds of coal, the more especially where they constitute, as they frequently do, excellent materials for the fire-bricks so often required in our coal districts, for the different metallurgical and other uses for which that fuel is employed. The name given to these ancient soils varies in different districts—*underclay*, *bottomstone*, and *undercliff* are not uncommon names in South Wales and the west of England. The *ganister* of Yorkshire and Derbyshire is a bed or beds of this kind. Though so long known to the coal miner, they have been rarely noticed until lately in colliery sections.

An observer will not long have been engaged in the examination of extensive coal districts without usually finding that, while certain beds of coal can be traced outcropping for long distances, and found also beneath the surface at various depths, according to circumstances, others are more local, mere patches, as it were, amid sheets of vegetable matter far more persistent over wider areas. In like manner some of the former mud, silt, or sands, accumulated at the same time, present a more common character, scattered over extensive districts, than others, the muds usually, as might be expected from their component parts having been diffused in a fine state of mechanical suspension in water, being the most persistent. Taking the chief sheets of coal as guides, duly weighing the kind and amount of distribntion of the accompanying ancient mud, silts, sands, and gravels, and reducing the section and plan, so that all embarrassments of contorted or simply tilted beds, with any fractures or dislocations which the whole accumulation may have sustained, be removed, it will be seen how far these sheets of interstratified matter may extend in a manner requiring an even, or nearly even surface, over a wide space. To accomplish such an object, it will be obvious that an observer should free himself from mere local variations, and attend to the evidence presented on the large scale. Thus it may be required that all the coal districts of Great Britain and Ireland, whether remaining as patches, reposing on older rocks, or simply exposed by the action of denuding causes which have removed some covering of subsequent deposits, should be regarded as a whole, and with reference to any portion of dry land of which they may have constituted an addition, and from which the needful supply of mud, silt, sands, or gravel, now forming its accompanying beds of shale, argillaceous and arenaceous, sandstone and conglomerate, was derived.

With regard to the sheets of vegetable matter, now constituting coal beds, they sometimes present traces of water action on their surfaces, much reminding us of the erosion to be seen upon extensive areas of bog, channels being cut out by drainage and running waters. Sands have been sometimes drifted above such sheets of vegetable matter, before they became consolidated, mud, or even sands, first covering them, being removed, as in the following section (fig. 189)—

Fig. 189.

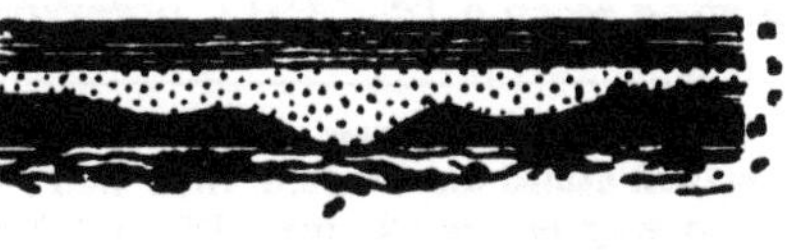

where *d* is a coal bed reposing on an ancient soil, *e*, full of roots

(*Stigmaria*), and *c* is mud (shale) first covering the vegetable matter (coal), but which was subsequently cut into by the water drifting the sand (sandstone) *b*, a deposit covered subsequently by mud (shale) *a*. In this manner many a portion of the bed once resting on coal may be found swept away in parts, even to the removal of portions of the coal beds themselves. The Forest of Dean presents an excellent example of channels cut in the vegetable matter (now forming coal) of a particular portion of the coal measures there seen. The chief channel represented in the annexed plan (*a*, *b*, fig. 190), has long been known to

Fig. 190.

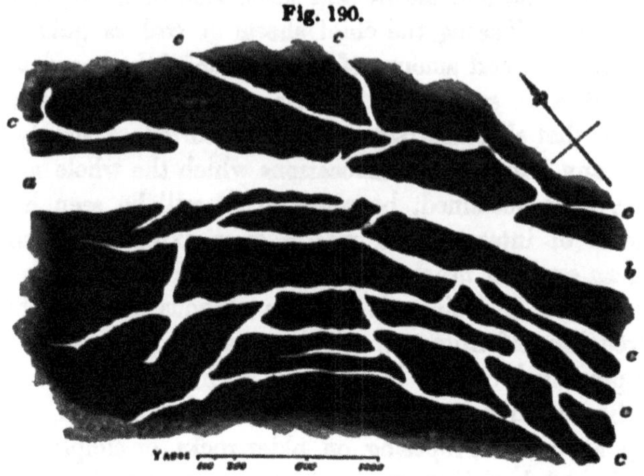

the colliers of the district as the "Horse." Mr. Buddle very carefully examined the circumstances connected with the "Horse" and its tributaries (*c*, *c*, *c*), known as the "Lows," whence it would appear that when the vegetation was in an easily-removable state, like that of some bogs, drainage water had cut out a main and subsidiary channels, into which a subsequent deposit of sand was thrown down, covering over the whole surface, as any sand deposit might now do a great area of bog if submerged.*

As proving that the unequal action of water was not confined to that on the surfaces of the sheets of vegetable matter, it is needful to remark that careful observation will frequently show this to have happened with other portions of the coal measures. The following section (fig. 191), observed on a cliff, composed of these

* The "Horse" has been followed in the working of the coal-bed in which it occurs (that named the Coleford High Delf) for about two miles, and it has been found to vary in breadth from 170 to 340 yards. Quartz pebbles are observed in some portions of the sandstone covering up the "Horse" and the "Lows," as also fragments of coal and ironstone.—Buddle, "Geological Transactions," vol. vi.

rocks, between Little Haven and Gouldtrop Road, Pembrokeshire, may serve to illustrate this circumstance. Herein a deposit of

Fig. 191.

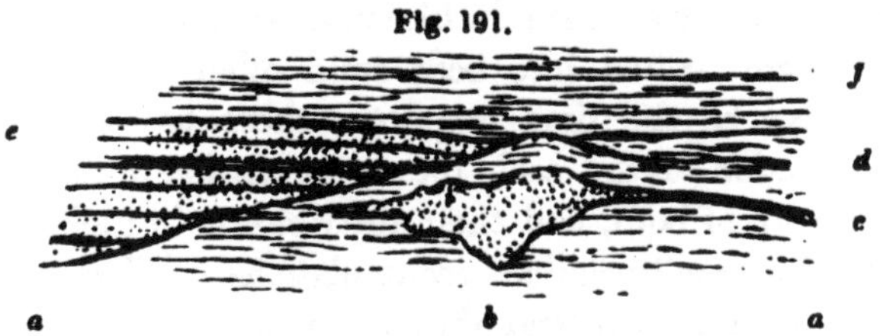

mud (shale), *a a*, seems to have been cut into by a furrow at *b*, extending to *c*, the water which made it bearing in sand, and mud being again accumulated over the sand at *d*. A sweep of the surface appears now to have occurred, and on the side *e* sands were thrown down from mechanical suspension, (the component layers being quite flat, and unmarked by diagonal drifting,) into a cavity formed in that direction, by which the previous mud deposit, *a a*, was worn away. Circumstances connected with the local mode of deposit then changed, and mud, *f*, was again spread over the surface of the first accumulation, its modifications, and the deposits which followed those modifications.

While adverting to various changes produced by the removal and deposit of the mineral matter of coal-bearing deposits, it may be desirable to notice the evidence often afforded by the coal-measures as to the lapse of time during which their accumulation was effected. The various growths of plants upon different soils, and the general thickness of the mass may, no doubt, be taken as evidence of a long lapse of time, though the rapidity of the growth of such plants as are found entombed in these beds may have been considerable; the sand and mud deposits may also have been somewhat readily effected, and, from a rapid mode of accumulation, the soils (underclays) may also have been soon formed. When, however, pebbles and small grains of coal itself, are discovered amid the sand-drifts and deposits of the period, we seem to advance somewhat further in the evidence of a considerable lapse of time. We certainly do not know that required, under fitting conditions, for converting the vegetation of the kind and period into the coal, so that beds of it, partially broken up, could be used as a portion of the higher deposits of the general mass. Herein there may be somewhat of a difficulty. Still, viewing the subject generally, and with due reference to the action of running water on land, or breaker action on the shores of waters, also required, no little lapse of time would appear needed for the changes in the vegetable matter, its removal in part, and its redeposit. It sometimes happens in certain coal-measure districts, that the ironstone also of previously-formed strata have in like manner been broken up, and pebbles of them drifted into beds amid other detrital deposits. Whatever may be the time required, there has been sufficient for the production of the coal, the consolidation of the ironstone, the breaking up of both, and their distribution in higher portions of a series of generally similar accumulations. When sufficiently large, the pebbles of coal (and they are sometimes discovered two or three inches in diameter) often exhibit the jointed or cleavage structure of the beds whence they were derived, their planes of cleavage taking various directions in the coal-pebble beds of which they now form parts, while the cleavage of the outside portions of the stems of *Sigillaria*, occasionally drifted with them, and converted into coal, have a constant direction in the same beds. Moreover, rounded portions of coal of a distinct character, and known in lower portions of the general deposits, have been found higher in the series, and little doubt can exist that at the time they were detached, they had undergone the same order of change as their parent beds, and that, even if these have been still further modified, the same modification from similarity of structure had extended, under the same general influence to which the whole mass of these deposits has been exposed, to these pebbles also. Certain beds, well exhibited amid the quarries of the Town Hill, Swansea, are highly illustrative of the pell-mell drift of such coal-pebbles with stems of Sigillaria, the latter showing the forms of many a coal-pebble beneath, the plants having conformed in a soft state to the hard pebbles of the coal, itself a substance probably derived from plants of the same genus, and often also of the same species as the stems, intermingled and entangled in the common drift.

The necessity of land for the sufficient supply of the detrital matter of the "coal measures" would appear a somewhat needful condition carefully to be borne in mind, since the mass of the coal measures of the British Islands would require its contents to be measured by no small amount of cubic miles of mineral matter, worn away from some other position which its parent rocks, even themselves, perhaps, detrital, may have occupied at a distance whence they could have been moved.

An observer has next to inquire how far the removal of this large amount of detritus has been accomplished by breaker action, or by other means, for distribution at the bottom of water. Here the great sheets of vegetation, based upon old soils in many situations, and often so frequently repeated, afford him important aid, inasmuch as they are not composed of marine plants, neither are the numerous upright stems, in their places of growth, marine. Over some wide spaces, and through considerable thicknesses of deposits, no trace of a sea-bottom is found, though the remains of mollusca, inferred to be forms similar to those now detected in rivers or fresh-water lakes, have been discovered. While this may be true in many districts, and through considerable thickness, it is not so as a whole even for the comparatively limited area of the British Islands, for here and there the forms of marine molluscs are discovered amid the other deposits. Proceeding from south to north over this area, it is found that the remains of other marine animals, as well as molluscs, are entombed in beds interstratified with the coal deposits, even somewhat thick limestones affording evidence of the presence of the sea for a time sufficient, at intervals, for the growth and continued increase of different marine creatures.* Duly flattening out all the present inequalities of the British coal districts, and reducing the whole towards horizontality, several thousand square miles of tolerably even ground would appear to present themselves, much reminding us of some great delta, such as those of the Ganges, the Quorra, or the Mississippi, in a state of descent as regards the level of the ocean, in such a manner that, as the land was depressed, and the fall and velocity of some great river or rivers for the time increased, detritus was borne readily onwards over sinking sheets of vegetation.

That some sheets of vegetation should be more extensive than others could scarcely otherwise than happen under such conditions; or that occasionally also the sea waters became introduced, should there be any partial subsidence so great that these waters entered areas of different dimensions, while lakes of fresh water were tenanted by suitable inhabitants, and even limestones were formed,

* Except in some rare and higher part of the carboniferous limestone series, even small coal-seams cannot be traced in that rock in South-Western England and South Wales. The mass of the coal measures of the same district, notwithstanding its great thickness, exhibits no admixture of marine remains with those of terrestrial vegetation and of the molluscs possessing forms resembling those now inhabiting fresh waters. The same general conditions appear to have reached [illegible] north, in the British Islands, as Northern Wales and Derbyshire. Still further north, however, coal-beds become more intermingled with the mass of supporting calcareous deposits (mountain or carboniferous limestones), so that the latter include among them shales, sandstones, and coals; thus showing that, in the northern portion of this area, the conditions for the growth and entombment of this kind of vegetation commenced at an earlier geological period than in the southern.

embedding their remains. That the general conditions should be introduced earlier at one portion of a given area than another, might be anticipated, if some general sea-bottom, preceding any extension of a delta or accumulation of that order, had been sufficiently raised either by the amount of deposits thrown down upon it, or by general movements in the mass of such sea-bottom and adjoining dry land, so that the vegetation of the low flat grounds of the time could flourish. To whatever extent this or any other view of a similar kind may assist observation with respect to the general circumstances connected with these coal deposits, the geologist, in search of evidence of dry lands in certain portions of the earth's surface at given geological times, should carefully attend to any which may present itself in favour of terrestrial plants having grown at, or near, the place where their remains are now discovered. It will readily be inferred that circumstances may have occurred at different geological dates, in fitting situations, under which vegetation may have been entombed, producing layers of carbonaceous matter in different conditions of change, so that anthracite, bituminous coals, or lignite may now occur among the mud, silt, sand, and gravel, accumulated at those different dates. This is now well understood; and the deposits to which the term "coal measures" has been especially assigned in Europe and North America, have only been selected for notice, because of easy access in several parts of those continents. Coal deposits of importance are now well known in Asia, Australia, and some other regions. How far there may be proof of the growth, in place, of the plants which have furnished the materials for the carbonaceous portions of these accumulations, becomes a matter of no slight geological interest, as supplying information not only of the dry land of the relative time which the general evidence may lead us to infer most probable, but also as to the kind of vegetation which, under certain conditions, flourished at such times in given regions.

To return to the comparatively limited area of the British Islands for the purpose of again illustrating how much may sometimes, under favourable circumstances, be observed in minor portions of the earth's surface, we find two other instances at different geological dates; one, during the accumulation of the group of beds known as the *oolitic series*, and the other, at the close of its deposit, when vertical stems so occur that we have further evidence of plants entombed in their places of growth. The coal-beds of the oolitic series in Yorkshire have been long known as occurring on a "geological horizon," to adopt the term of Humboldt, with limestones, and clays, replete with marine organic remains, on the south of England; and Sir Roderick Murchison pointed out in 1832, that the vertical stems of the *Equisetum columnare*, apparently in the positions in which they grew, were not only found in the shale and sandstone of these deposits on the coast, but also at a distance of 40 miles on the north-western escarpment of the Yorkshire moor-land, pointing to the submersion of many square miles of ground in such a manner that the plants were quietly entombed in the mud or sand accumulating round them.*

The Island of Portland, on the coast of Dorsetshire, also affords evidence of trees in place, some standing as they grew, with the soil preserved in which they spread their roots. These soils have long been known by the quarrymen of the island as the "dirt beds."† While some trees are rooted in their ancient soils, others are prostrate, in the manner represented in the following section (fig. 192); one much reminding us of the "submarine or sunk forests" (fig. 152, p. 447) so frequent on the shores of Western Europe. In this section‡ the erect and prostrate remains of trees,

Fig. 192.

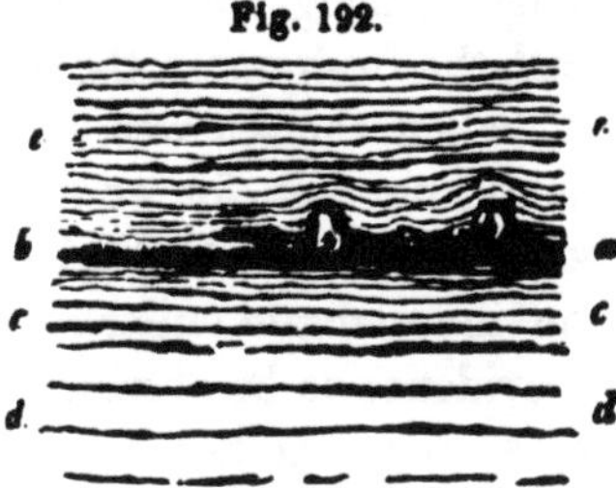

among which occur those of cycadeous plants, with the soil of the period (*a*, *b*), repose on a calcareous rock (*c*, *c*) containing the remains of fresh-water animals, and resting upon the marine oolitic limestones (*d*, *d*), commonly known as the Portland oolite. Above the remains of the trees and cycadeous plants there are other calcareous deposits (*e*, *e*), also containing animal remains pointing to their accumulation in fresh waters, and known as the Purbeck beds, from being well exhibited at that locality, on the coast eastward from Portland.

Thus the vegetation and the soil upon which it flourished are

* Murchison, "Proceedings of the Geological Society," vol. i., p. 391.

† These beds were first described by Mr. Webster, "Geol. Trans.," vol. ii., p. 41.

‡ As many as three of these "dirt-beds" have been noticed in some parts of this series of deposits in Portland—different remains of successive soils, perhaps not always of exactly the same equal data, though representing general conditions of the time. Only one of such "dirt-beds" is represented in the section, for more clear illustration of the general circumstances under consideration.

included in an accumulation effected in fresh water, implying that dry land existed somewhere in the vicinity anterior to the growth of the trees. From an attentive examination of the district,* Professor E. Forbes found that the fresh-water animals, the remains of which occur in the lower part of the covering beds, were not changed by the conditions permitting the production of the "dirt-beds," and the growth of the plants, being the same as in the calcareous beds immediately beneath. He found, moreover, that there had been three successions of species in the Purbeck deposits. As to the general character of those beds, the Professor ascertained that while the higher and lower accumulations bear evidence (from their organic contents) of having been deposited in fresh waters, the central portion points to alternations of fresh water, brackish water, and sea. Altogether a highly-interesting series of facts, showing a disappearance of the sea, and the formation of dry land, by which animals inhabiting fresh water could obtain the conditions for their existence, the actual evidence of this dry land in particular portions of the area, and the continuance of the fresh-water accumulations by some change, during which, while the soil or soils became submerged beneath the fresh water, the sea was not admitted. A time came, however, when the sea was let in, brackish water also occurring; but this did not last, for we again find fresh-water deposits above these beds. Professor E. Forbes mentions, that so far as the remains of the invertebrate animals extend, it would be impossible, without the evidence to be obtained from superposition of other accumulations, to say whether the fresh-water deposits belonged to the oolitic, cretaceous, or tertiary series of rocks.† Referring back to the time (p. 480), when a depression of the lands then above water in the area of Southern England was in

* The ancient soil, with its trees, some prostrate, and others in their place of growth, is not confined to the Isle of Portland. It may be also well seen amid beds of the Purbeck series, in the east cliff of Lulworth Cove, a few miles to the eastward. With regard to the further extension of these conditions at that geological time, it should be observed, that Dr. Fitton mentions an earthy bed in the same geological position in Buckinghamshire and the Vale of Wardour, as also in the cliffs of the Boulonnais. Silicified wood is found in a bituminous bed from Boulogne to Cap Gris-nez ("Geological Sketch of the Vicinity of Hastings," 1833, p. 76.) A "dirt-bed" is noticed by Dr. Buckland as occurring, in its geological place, near Thame, in Oxfordshire; and Dr. Mantell mentions one as found at Swindon, Wiltshire, on the top of the Portland beds, fossil coniferous wood being seen in abundance, with a few examples of Mantellia. "Wonders of Geology," 6th edit., vol. i., p. 390.

† Among other important observations, Professor E. Forbes found that although a bed of oysters (*Ostrea distorta*), occurs as the most conspicuous feature of the middle division of the Purbeck beds, the fresh-water fauna of the time was not interrupted.

progress, so that the lower part of the oolitic series of deposits (various limestones, sometimes oolitic,* sands, and clays), spread over the submerged rocks, the animals of the period even boring into them under favourable conditions (p. 486), the depression apparently ceased not long after that geological date. Whether the sea-bottom and adjacent lands then took a contrary movement, rising gradually, so that the area occupied by sea was diminished, and the shores extended, or that, remaining stationary, the detrital and animal accumulations so filled up the seas around that the shores were thrown back, or that both these causes were in operation, it would appear that the remainder of the limestones, sands, and clays of the oolitic series, with their animal remains, was formed within a gradually-diminishing area, as far as that of the British Islands was concerned, so that finally, in a particular portion of it, the conditions prevailed which produced the results observed in Dorsetshire, and by which the existence of dry land in particular spots is proved, the remains of trees being found rooted in the soil in which they grew.

The change from sea to dry land conditions would appear to have further continued, for upon these lower (Purbeck) accumulations marked by the remains of fresh-water animals, a very considerable depth of deposits is found, pointing to the presence of some large river or body of fresh water in the area of South-eastern England. These accumulations, with the Purbeck beds, are now commonly known as the Wealden series, a name derived from the beds of that geological time found in the Weald of Sussex, for our first knowledge and numerous subsequent illustrations of which we are indebted to Dr. Mantell.† These beds, consisting of ancient mud, sands, and calcareous accumulations, are not only marked by the remains of fresh-water molluscs, but also contain those of remarkable reptiles (*Iguanodon*, &c.), of gigantic size,‡ and of terrestrial plants growing in the banks of, or swept down by a river, the matter borne

---

* The calcareous grains so united together as to resemble the roe of some fishes, whence also the name roe-stone for this description of rock.

† The Tilgate beds were described by Dr. Mantell in 1822, in his "Fossils of the South Downs," and the same year he communicated the joint observations of Sir Charles Lyell and himself as to the extension of these beds over the Weald. The observer will find an excellent summary of the Wealden series, as known in England, and on the continent of Europe, in Dr. Mantell's "Wonders of Geology," 6th edition, vol. i., pp. 360-449. He should also consult the works of Dr. Fitton on the lower part of the cretaceous series (greensand, &c.), contained in the "Geological Transactions and Proceedings," and he will find much instruction in his "Guide to the Geology of Hastings."

‡ For the knowledge of these, also, geologists are indebted to the labours of Dr Mantell.

in mechanical suspension in it covering the whole up, as fitting circumstances for the deposits occurred. That an elevation of a mass of land, and its adjoining sea-bottom might first produce variable mixtures of lakes and minor estuaries, and, finally, some larger rivers, will readily be seen, by considering the effects which would be produced by an elevation which should extend the coast line of the British Islands and the continent of Europe from Norway to the Pyrenees (figs. 65 and 99), so that the present drainage of Western Europe from Ushant to Norway, and from the Land's End, by the east coast of Great Britain, to the north of Scotland, should be thrown into two chief drainage depressions, divided at the Straits of Dover, or thereabouts. At first, as the sea-bottom gradually rose, there would be many minor admixtures of estuaries and of bodies of water subsequently rendered fresh, until, finally, all the rivers draining into the Baltic, with those now finding their way into the North Sea (Elbe, Weser, Ems, Rhine), would have to flow outwards, more or less uniting at different distances, together with the drainage of the new area of dry land, into the Atlantic, between the Shetland Isles and Norway, perhaps somewhat about the submarine gulf stretching down southerly between them (fig. 65). While this happened on the north, all the rivers in the English Channel would be more or less united, and flow out into the Atlantic by the greatest depression between the Land's End and Ushant, the drainage waters of the new dry land being also added to them. In both cases marine deposits would be succeeded at first by many intermingled estuary and fresh-water accumulations of various extent, and, finally, by those marking at the mouth of the English Channel, and between the Shetland Isles and Norway, the presence of far greater rivers than those which now discharge their waters into any of the seas bounding Western Europe from Norway to the Pyrenees. While the Loire and the Garonne might readily extend their courses without union over the new dry land, (a portion of the Bay of Biscay,) more complication would arise amid the rivers of the west part of Great Britain and around Ireland. Looking, however, to the charts, there would be a tendency to gather waters together into great rivers outwards between Northern Ireland and Scotland, and between Southern Ireland and the Land's End.

While thus so far advanced upon the changes which have occurred with regard to the presence and disappearance of dry land in so limited an area as that which has been noticed, it may not be undesirable to advert to the great change which subsequently converted a very extended portion of the same part of the earth's surface again into a sea-bottom, upon which a considerable thickness of mud and sands (greensands and gault), with a thick covering of calcareous matter (chalk) was accumulated. This was apparently accomplished by a somewhat gradual depression of a sea-bottom making way for the detritus borne to, and over it, in addition to so much of the volume of deposit as was due solely to the accumulation of the hard parts of marine animals, for the evidence is in favour of a greater general area being gradually covered, as this portion of geological time advanced, so that the higher beds overlapped or overspread the lower, the upper members of this series of deposits (the cretaceous), thus reaching beyond the lower in Northern and in South-western England. Again, conditions changed over the same area, and in the supra-cretaceous or tertiary time we find deposits according with such altered circumstances, and showing that dry land was then intermingled with sea; that there were estuaries and fresh-water lakes; and moreover, that there were oscillations of the land and sea-bottom, producing submersion beneath and emergence above the level of the adjacent ocean. These oscillations and their consequences have been, as we have seen (p. 445), continued up to the present adjustments of land and water, when we have atmospheric influences and the sea wearing away the former, the matter thus removed variably dispersed along the shores and over the adjacent sea-bottom, no doubt entombing a mass of the remains of the vegetable and animal life of the time and area—the whole, with the dry land and its lakes, rivers, and estuaries, ready to be elevated above, or depressed beneath, the ocean level, as has happened over the same area at previous geological times.

CHAPTER XXVIII.

MODE OF ACCUMULATION OF DETRITAL AND FOSSILIFEROUS ROCKS CONTINUED. —CRACKED SURFACES OF DEPOSITS.—FOOT-PRINTS OF AIR-BREATHING ANIMALS ON SURFACES OF ROCKS.—RAIN-MARKS ON ROCK SURFACES.—EFFECTS OF ELEVATION OR DEPRESSION OF THE OCEAN BOTTOM.—CHARACTER OF ROCK SURFACES.—FRICTION MARKS ON ROCK SURFACES.—EFFECTS OF EARTHQUAKES ON SEA-BOTTOMS.—DIAGONAL ARRANGEMENT OF MINOR PARTS OF BEDS OF DETRITAL ROCKS.—MODE OF OCCURRENCE OF ORGANIC REMAINS.—DISTRIBUTION OF ORGANIC REMAINS.—EFFECT OF THE RISE AND DESCENT OF LAND ON THE DISTRIBUTION OF ORGANIC REMAINS.—DISTRIBUTION OF ORGANIC REMAINS WITH RESPECT TO KINDS OF SEA-BOTTOMS.—INFUSORIAL REMAINS.—CHEMICAL COMPOSITION OF ORGANIC REMAINS.—CAUTION AS TO FORMS OF LIFE SUPPOSED CHARACTERISTIC OF DIFFERENT GEOLOGICAL TIMES.

THE footprints of air-breathing animals and cracked surfaces of beds afford the observer the means of judging of the presence of dry land at particular times. These have of late received their well-deserved share of attention. Although, as in the plan beneath

Fig. 193.

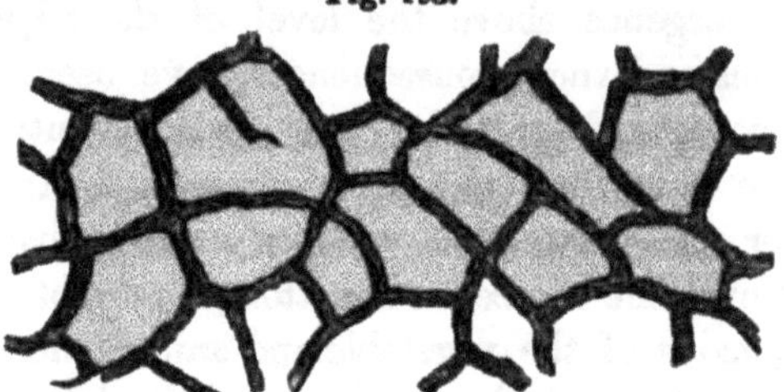

(fig. 193), when uncovering a clay or shale bed, he detects a splitting of parts corresponding with that seen upon the drying of any mud or clay surface exposed to the sun and air, he would be led to infer the contact of the atmosphere with such a surface, and the consequent presence of land, so as at least to permit a space to be exposed for a time sufficient to produce this amount of desiccation; such, for example, as on somewhat flat shores upon which there were great differences in the spring and neap tides (p. 78), the evidence becomes more perfect, by the addition of the well-marked footprints of animals. Such footprints have now been found in various parts of the world—Europe, Asia, and America—with and without the evidences of the cracks pointing to exposure of the atmosphere, and are highly important, as showing the tread of animals on shores or in waters so shallow and tranquil, that creatures breathing in the air and walking on soft ground left the prints of their footsteps uninjured behind them. The following sketch (fig. 194) is taken from the figure by Dr. Sickler, of foot-

Fig. 194.

prints in the red sandstone quarry at Hessberg, near Hildburghausen, Saxony,* and well illustrates both such impressions and cracks from desiccation. While these footprints have been considered as those of reptiles, some of gigantic Batrachians, others have been discovered of forms from which they have been attributed to birds of different species and sizes. To these Professor Hitchcock long since called attention as occurring in a red sandstone series in the valley of the Connecticut, United States.† The following sketch

* These footprints appear to have attracted attention at Hessberg, about 1833 or 1834, when they were described by Dr. Hohnbaum and Professor Kaup, the latter of whom gave the animals considered to have formed them the name of *Chirotherium*. Dr. Sickler published a further account of them in a letter to Blumenbach, in 1834. Prior to this discovery (1828), Dr. Duncan gave an account (Transactions of the Royal Society of Edinburgh, vol. xi.) of similar footsteps found in the new red sandstones of Corn Cockle Muir, Dumfriesshire, and in 1834, Dr. Duncan informed Dr. Buckland (Bridgwater Treatise, vol. i., p. 259) that like impressions had been found in the same series of deposits, 10 miles from the former locality, and 2 miles from the town of Dumfries.

† Professor Hitchock described these footprints under the name of *Ornithichnites*,

(fig. 195) is taken from among the illustrations given by the Professor:—

Fig. 195.

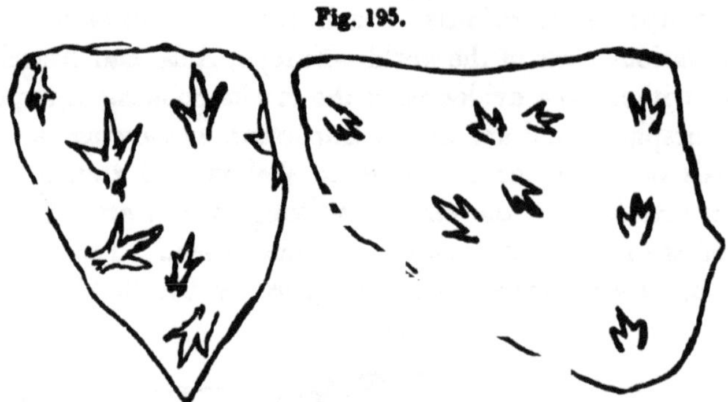

The footsteps attributed to reptiles have, in part, been assigned as probable to the *Labyrinthodon*, one whose bones have been discovered in the same series of deposits. As still further showing contact of the air with mud or sand where these or other animals have left the imprints of their feet, marks in the same as well as other surfaces of associated beds have been discovered, strongly resembling those left on clay or sand by a heavy fall of rain, such as may often be observed on coasts when the tide is out.* These various impressions have usually been made upon layers of clay or mud, sand having been tranquilly accumulated over the hardened surface retaining the footprints and other marks. As the resulting marl, clay, or shale is frequently broken by the removal of the sandstone bed covering it, the lower surface of the latter usually reveals the condition of the upper surface of the former, before it was overspread by the sand. At the same time we have seen impressions on the upper surfaces of sandstones themselves, which, though not so well defined, resemble footprints on sand subsequently and quietly covered over by mud.†

in the American Journal of Science, vol. xxix., 1836, and also in his Report on the Geology of Massachusetts. Sir Charles Lyell also gives an account of them in his "Travels in North America," chap 12. The footprints are of various sizes, some not longer than those of our common sanderlings, while others exceed that of the ostrich, measuring 15 inches in length, exclusive of the largest claw, two inches long. Dr. Buckland, remarking on the dimensions of this supposed bird, observes (Bridgewater Treatise, vol. ii., p. 40), that "in the African ostrich, which weighs 100 lbs., and is nine feet high, the length of the leg is about four feet, and that of the foot ten inches."

* An illustrative figure of the impression of rain drops upon the same slab with that of a biped, from the red sandstone series of Massachusetts, is given by Dr. Mantell, in his "Wonders of Geology," vol. ii., p. 556.

† The footprints, noticed in the text as discovered in Asia, were found impressed upon red sandstone in India, by Lieut. Pratt.

Of whatever animals the footprints may have been, with the cracks from the exposure of surfaces of mud and clay to desiccation in the air, and the marks resembling rain drops—for these, however singular they may appear, are not to be neglected—they show us that, during the deposits of the layers or beds of sand, silt, or mud in which they occur, dry land was there at hand also, and that the beds themselves may have formed part of its shores, as those in the Bay of Fundy,* in the Bristol Channel,† and in numerous other localities, where similar and fitting conditions present themselves, now do. The mere piling of layer upon layer on shores of this kind has been found sufficient to preserve such marks (p. 128); and when this is combined with a quiet submersion of the locality, so that the layers of deposit are little, if at all, broken up, a considerable thickness of beds marked in this manner may be, as they apparently have been, accumulated in succession, until finally the fitting conditions cease, and the preservation of such impressions can no longer be effected.

However desirable it is for an observer thus to trace, by means of beaches, fresh or brackish water deposits, the footprints of animals on shores and the remains of plants rooted in their places of growth, the presence of dry land on different parts of the earth's surface (for the circumstances which have been noticed by way of illustration are applicable, with certain modifications, to many other regions), in some districts he finds himself so completely surrounded by ancient sea-bottoms, piled up in various modes in succession, that he cannot avail himself of the aid which this knowledge of the probable position of the dry lands of given geological times may afford him. Although aware that the wearing away of the mineral masses forming dry land, furnished, with the stirring up of sediment from shallow depths by wave action, the materials for the detrital accumulations he may have before him, and which may alone contain the remains of marine life, should the arrangement of their inorganic component parts merely point to a deposit from mechanical suspension in water, he might still be at a loss as to the direction or character of the dry land of

---

* Sir Charles Lyell has figured the recent footprints of the sandpiper on the shores of the Bay of Fundy, in his "Travels in North America," vol. ii., pl. vii., and has presented specimens illustrative of the preservation of these footprints in different layers, deposited in succession, to the British Museum, and to the collections of the Museum of Practical Geology.

† We have frequently collected good examples of footprints of different kinds preserved in the muddy banks of this channel, left dry and hardened in hot summer weather, on the wide spaces between the lines of neap and spring tides.

the time. A study of the charts of many different regions will show that mud is found as well near coasts as remote from them, according as the required tranquillity for deposit and subsequent rest may prevail, though as a whole wind-wave action upon sea-bottoms, at depths where it can have influence, tends so to sift those bottoms as to remove muddy sediment further away from land than sand.

When all the modes of distributing detrital matter, above-mentioned as now in progress in tidal or tideless seas (pp. 63, 77), are combined with movements of dry land and sea-bottoms, sometimes upwards, at others in the contrary direction, it is evident that, in addition to the consequences of such movements on coasts and sea-bottoms adjoining them, it might often happen that considerable areas may be elevated or depressed in the sea itself without rising above its surface into the atmosphere. Mere points constituting their higher portions may now and then be protruded and be acted upon by breakers and atmospheric influences, the detritus thence arising being scattered around, and arranged by tidal streams, or transported, especially the finer matter, in a broader and more distant manner by ocean currents, still able to force their way amidst these minor obstacles to their courses. The floor of the ocean is not yet so well known as probably it will be at some future time, when systematic researches in this direction may be deemed important by maritime nations. The depths, nevertheless, of certain points have been ascertained, more especially of late years, sufficient to render it probable that very important aid to geological inferences would be obtained by more extended information on this head.

While, on the one hand, the distribution of detritus outwards by the great rivers of the world, draining large portions of continents and pushing forward their deltas, has to be well borne in mind, on the other, such changes as shall raise a mass of sea-bottom, scattered higher portions of which may or may not now rise into the atmosphere, should receive their due attention. If the extent of sea-bottom, above which various points rise and form the multitude of isles and islets of the Polynesian groups in the Pacific, were to be gradually elevated so as to constitute some great continent of dry land, no great deltas would be raised—at least none now in progress—whatever former conditions of that part of the earth's surface may have produced; and it would only be by degrees that the drainage of the new land formed rivers, these uniting into larger streams as the dry land became extended, some, perhaps, finally of the magnitude and importance of the present great rivers of the world.*

Although detrital matter deposited from mechanical suspension in water, and arranged in layers and beds, may not, from the structure of the interior portions of the layers or beds themselves, present much information as to the depths of the water beneath which they have been accumulated, while they may, as they often do, exhibit the proof of a multitude of very thin layers having been thrown down above each other (as many, perhaps, as twenty or thirty of these in one inch of depth), their surfaces often aid most materially in affording valuable information on this head. It frequently happens that the under surfaces may be useful as well as the upper, inasmuch as they often give the imprint of the former condition of the surfaces of layers or beds which they cover, the materials of which were of a perishable kind when raised into situations where the percolation of water either softened or even removed them. Thus the upper surfaces of shales, or hard clays, may be converted into mud or soft clay, in which all traces of their original state are lost, while some sandstones above them, the consolidated sand which covered over the impressions left on these surfaces of clay or hard mud, preserve reversed impressions of the state of the old sea-bottom before it was covered up. Under the conditions which so frequently present themselves, while alternating or intermingled beds of shales, clays, and sandstones are under examination, and occasionally, also, limestones, and it is considered desirable, if possible, to trace the state of the upper surfaces of the mud or clay before they were covered up, the under surfaces of the present hard beds above them should be carefully studied. The search will frequently reward the geologist with an excellent picture of such old surfaces of sea-bottoms, with their various markings, even to the impressions left by the crawlings or way-tracks of the molluscs of the time. There is a class of surface-conditions on consolidated layers of sand and silt (sandstone and arenaceous shale), to which the term *ripple-mark* has been applied, from a supposed

* If the observer will follow out this supposed uprise of the area in question, he will find numerous subjects of interest connected with it, which, though many may be sufficiently obvious, such as the mode of occurrence of the coral accumulations now in progress, their modifications as the dry land became extended, the effects of tides and altered courses of ocean currents during the change, the modified distribution of animal and vegetable life on the land and in the seas adjoining, the chances of salt or fresh water lakes, mediterranean seas, or the like, are yet, collectively, of importance to be well borne in mind while he may be occupied upon the geological effects which would thence arise.

resemblance to the ripple produced by light winds on water, or the condition of many tracts of sand on the retreat of the tide, that would afford most valuable information as to the depths at which the layers or beds were situated beneath water when any such markings were produced, were it not that such kinds of surface might frequently arise from similar conditions at different depths. We have previously mentioned (p. 90), the friction of streams or currents of water on sandy surfaces beneath them, ridging and furrowing the yielding matter. Such may be often seen on the surfaces of sandstone beds, the ridges and furrows well preserved, as beneath (fig. 196), so that by carefully studying the steep sides of

Fig. 196.

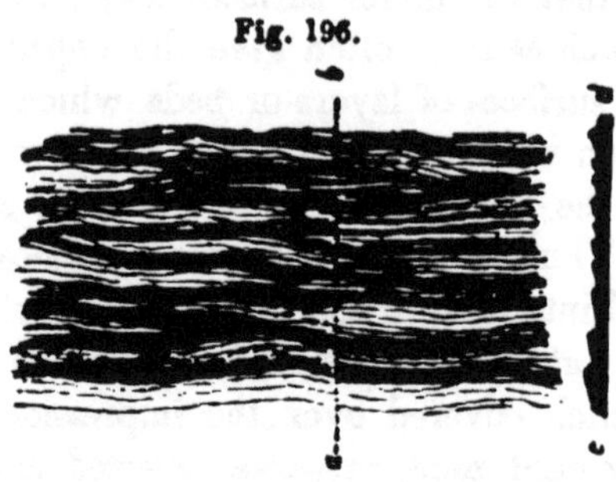

the ridges, the direction taken by the moving water at the time may be determined. In this case it is assumed that a section taken across at *a b* would give that shown by *c d*, one pointing to the course of the moving water from *a* to *b*. If we were sure of the depths at which existing ocean currents swept sands at the sea-bottoms beneath them, producing surfaces of this kind, some guide would be obtained to the range of depths, from a few feet downwards, at which, only measuring hy the amount of force now in action, these effects could follow. Herein, however, there is much uncertainty. From the experiment of Sir Edward Belcher, off the west coast of Africa (lat. 15° 27′ 9″ N., and long, 17° 31′50″ W.), it would appear that a current there found moved with nearly the same velocity (0·75 nautical miles per hour) at the depth of 240 feet (40 fathoms) as at the surface. When we regard the great ocean currents of the world, with the probable masses of water put into movement in given directions at the same time, it may not be improbable that comparatively considerable depths are exposed to conditions where the ridging and furrowing of sand and silt sea-bottoms may be produced. It has also to be recollected that as large surfaces of sea-bottoms may be raised or depressed, from some of the more general movements of the solid parts of the earth's surface, very considerable areas could be brought up to the action of ocean currents, or removed beneath their influence.

Upon carefully studying the surfaces of great banks and flat tracts of sand which are somewhat suddenly drained by a retiring tide, so that they are not much altered by the action of the small waves or heavy breakers of the time, as the case may be, the geologist will frequently find, as already noticed (p. 90), a mixed adjustment of inequalities, partly due to the movement of the waves before the superincumbent water passed away, and in part to the friction of this water draining off the banks and sandy flats. These ridged and furrowed surfaces are occasionally somewhat extensive when the sea deserts a considerable area in a short time, so that friction is produced rather suddenly in some general direction. This will often happen when there may be a heavy sea on shore, as the great waves break at a proportionate distance outwards upon the shallows during the progress of the ebb-tide, minor action only taking place nearer the coast, where, the great body moving outwards, the ridging and furrowing by friction on the sands may point to the chief movement, with the sharp escarpment of the furrows often seaward, though the wash of the breakers would tend to drive sand before them while rushing on shore.

Where, as on many great banks dry at low tides at the mouths of estuaries, there may be a complication of surface arising from the wave movements anterior to the removal of the sea from above them and from the friction of waters left to drain off them, it will be remarked, as might be anticipated, that much will depend upon the state of the weather and tides of the time. Calms would leave friction-markings, such as might arise from the movement of a stream of water over a sand-bank before it was left by the tide, more than gales of wind, since the wash of the breakers, as its action was felt, would pass over and tend to obliterate the ridges and furrows due simply to the stream of tide. The more sudden retreat of the sea during the chief spring-tides, from the same depths, would tend also to leave the surface of a sand-bank more marked by any furrowing from the previous flow of a stream of tide over it, other circumstances being equal, than a neap-tide, during the descent of which, wave-action might be continued for a longer time after the stream of tide ceased to be felt on the surface of the bank.

The surfaces of some layers and beds of rock so resemble those which are seen in the last-mentioned situations, particularly when sufficiently large portions of them are exposed, either on coasts or amid highly-inclined strata in mountainous regions, even to the apparent minor drainage of waters off sand-banks, that the inference of these surfaces having been produced on or near tidal coasts (p. 90) somewhat forces itself upon an observer. At the same time he will have properly to weigh the probable effects due to wind-waves on sea-bottoms at different depths beneath (p. 89), and the power thus brought into action of disturbing such bottoms, occasionally sifting their constituent parts, so that a tranquilly-formed deposit of mud may cover any unequally-disposed surface of sand, produced while the agitation of the sea continued. Many surfaces of rocks strongly remind us of loose matter thus moved about by the to-and-fro action of an agitated sea above, in the same manner as sand may be readily acted upon by agitating water above it in conveniently-formed vessels of sufficient dimensions. Such approximations to the ridges and furrows of friction upon sands and silts in one given direction should be well distinguished from the latter. These sections, instead of being as above represented (fig. 196), are usually more undulating or evensided, the surfaces varying from obscure ranges of depressions, *a*, *b* (fig. 197), and

Fig. 197.

those somewhat resembling the sharp ridges and furrows of current or stream action, *c*, to unequally-distributed and variably-formed elevations and depressions (fig. 198), which require also to be well separated from concretions, to be noticed hereafter, and which, sufficiently juxtaposed, may present a somewhat similar appearance.

Fig. 198.

With regard to the surfaces of sea-bottoms, now consolidated into hard layers and beds of rock, attention should be paid to the probable modifications of them, even at great depths, by the passage of earthquake movements, shaking these surfaces in contact with the superincumbent water. In some regions, such movements can scarcely be otherwise than frequent, the force employed being sometimes so considerable, and its application repeated in such quick succession, that the finer sediment may be shaken up into a mechanical suspension whence it would require some lapse of time again to settle and cover over the heavier bodies, taking superficial arrangements according to the vibrations produced hy the earthquake, the kind of substances acted upon, and their mode of previous distribution. In cases of fissures produced beneath the sea, as on land, during earthquakes, the consequent disturbance of adjoining sea-bottoms has also to be regarded. Thus the effects of the transmission of earthquake vibrations both on the large and minor scales, those of the great sea-wave, and of the smaller movements produced by the contact of the sea-bottom and water above it, the earthquake vibration travelling faster through the former than the latter, have also to be borne in mind when the surfaces of sea-bottoms of even the oldest geological times are under consideration, and the geologist is endeavouring to deduce from them the probable depths of water beneath which they took the forms presented to his attention. Submarine areas thus disturbed, and the surfaces of the sea-bottoms moved, could scarcely often be otherwise than considerable, the effects, no doubt, modified by relative depths of the water, the facility with which the vibrations could be transmitted through the various supporting bodies, and the like. Ridges and furrows may be raised in certain localities by the onward courses of chief sea-waves in the shallower waters, and not be again wholly obliterated, though often modified in form before they were finally covered up and secured in shape until constituting a portion of hard rock.

While there may often be much uncertainty as to the depths at which the component parts of layers and beds of rock, even with ridges and furrows on their surfaces, have been thrown down from the waters in which they have been previously held in mechanical suspension, when unaided by other evidence, the arrangement of parts resulting from the pushing of detrital matter forward on the bottom often seems to point to somewhat shallow waters. In this case, again, as the depth is uncertain to which currents may act on sea-bottoms, these unequal, like those at the edge of the soundings of 1,200 feet (200 fathoms), from Spain round the British Islands to Norway (p. 460), so that sedimentary matter derived from adjacent lands is transported and pushed along the bottom into the hollows, the like effects may be produced at far greater depths than is usually inferred. Supposing an ocean current or tidal stream so to act as to push forward detrital substances from some land affording the required amount of increase to the general mass of previous accumulations, much in the same manner as that to which Mr. Austen has called attention,* so that after spreading over a somewhat level sea-bottom, the general increase had to be provided for still further outwards over uneven ground, the matter thus shoved on would have to fall over into deeper waters, and arrange itself much as on the outskirts of rivers delivering their detritus into deep and tideless seas or other still waters (p. 44). In this manner even sandy beds, affording sections of the component layers arranged diagonally to their upper and under surfaces, might, as before mentioned (p. 67), extend over large and flat accumulations of mud, thrown down from mechanical suspension.

Diagonal arrangements of the minor parts, resulting from this pushing action along the bottom, are very common in many sandstones, as well as those which, from their occasional organic contents, leave little room to doubt were formed beneath the sea, as in those so frequent in many parts of the coal-measure accumulations (p. 508). These arrangements are sometimes diversified in a way to show, that while some of the sandy matter has thus been forced or brushed onwards on the bottom, the same kinds of sand were, at other times, thrown down in horizontal layers, more pointing to deposit from mechanical suspension. Instances of this kind are not rare. The following section (fig. 199) of the arenaceous beds, forming a kind of passage from the old red sandstone in parts of Ireland to the lower and usually shaly beds of the carboniferous limestones (the *yellow sandstone series* of Mr. Griffith), may be found useful.†

Fig. 199.

* Austen, on the Valley of the English Channel; Journal of the Geological Society of London, vol. vi.

† The section was obtained at Clonea Bay, County Waterford, an interesting locality for the study of the upper part of the old red sandstone series and the lower part of the carboniferous limestone, more especially when taken in connexion with the development of equivalent accumulations at the Hook Point, County Wexford, on the eastward, and the country near Cork, and extending thence by Cape Clear to Bantry Bay. The whole is highly illustrative of contemporaneous accumulations of this geological date, modified by the conditions under which they have been formed, such as varieties of the sedimentary matter carried, pushed forward, or thrown down, according to distance from its supply, and different depths of water.

In this section, forming only a portion of a far more considerable thickness and extent of beds, exhibiting general evidence of the like kind, a horizontal deposit of sand (*e*), probably from mechanical suspension, is covered by a silt (*d*), apparently also accumulated in the same manner. To this layer succeed two beds (*c b*), pointing to an accumulation from sands pushed or brushed along the bottom, there having been a sufficient pause to make a surface between them. This condition changed, and horizontal layers (*a*) were again formed.

The following section (fig. 200) will serve to show that the

Fig. 200.

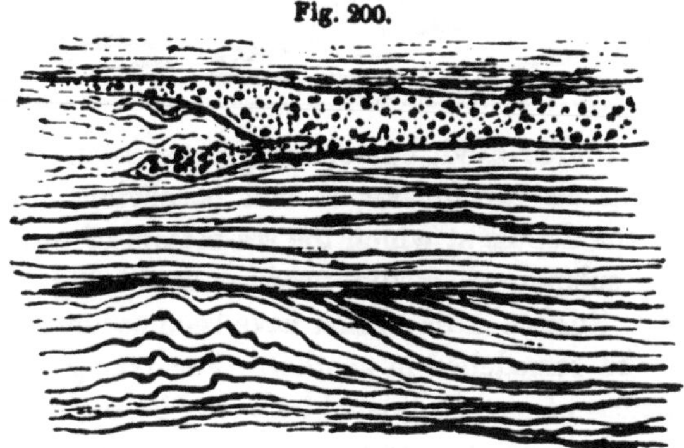

like unequal distribution of component parts, even extending to gravel drifts amid sandy and muddy sediment, is to be found among still older fossiliferous deposits, being one among many others to be seen on the ascent of the Glydyr Vawr, on the north-east of Snowdon, where the lower Silurian rocks are much mingled with volcanic accumulations of that geological time. Among some rocks the exposed surfaces as well as sections point so much to the shifting of minor streams or currents, sufficient to carry forward pebbles of fair size, the general accumulations pointing to repeated action of this kind, that, looking at the forces of existing currents, these deposits would generally seem referable to shallow waters. The accompanying section (fig. 201) of old red sandstone at Ross, Herefordshire,* of a kind common to much of the same series of deposits in that and adjoining districts, will illustrate this mode of occurrence.

When the disturbing power of wind-wave action upon sea-bottoms, to whatever depth that power may sometimes extend, and the modification of surfaces which may be produced during

* Part of a more extended section by Captain James, R.E.; Memoirs of the Geological Survey, vol. i., pl. 3.

earthquakes, are regarded, as also any greater vibrations of the sea-bottom, should larger masses of water be thrown into movement

Fig. 201.

by forces of a similar kind, but of far greater intensity than anything known as an earthquake, it will be obvious that tranquil alterations of the depths at which the sea-bottom of any geological time may be submerged, would produce modifying effects of a marked kind. Surface beds which have been accumulated in one manner, may be remodelled in another. For example, diagonally-arranged portions of unconsolidated beds may be worked backwards and forwards when exposed to the to-and-fro motion of water disturbed by the winds above, or by the tides brought into action, so that their streams are rendered more or less sweeping by intermixture with shallow-depths and the unequal configuration of adjoining lands. Though these causes may only modify the surfaces for the time being, a repetition of them, with oscillations in the movement of the sea-bottoms, would often produce complicated effects, so far as the original mode of deposit of any beds may, in part, be subsequently altered; even the organic remains contained amid the detritus being sifted and re-arranged without much injury.

It is when the structure of the beds of detrital rocks, and the forms of their surfaces are viewed in connexion with any organic remains they may contain, that the observer has increased opportunities of inferring the depth of water beneath which the layers or beds, have acquired the general character they now present. With respect to the mode in which organic remains generally may be entombed beneath fresh waters or the sea, whether the latter be tidal or tideless, we would refer to the previous remarks on this subject, (pp. 112—205). Amid the detrital matter, of whatever kind this may be, piled up in successive layers or beds, every variety of manner in which organic remains have been enveloped by it occasionally presents itself. While, on the one hand, we see the shells of molluscs precisely in the positions in which these animals buried themselves in the mud, silt, or sand, of the time, according to their habits; at others, we find the fragments of shells or corals in multitudes, dealt with and arranged like ordinary mineral substances, precisely as may often be found at the present day, and especially amid coral accumulations on the large scale, such as among the Great Barrier Reefs of Eastern Australia.

The occurrence of organic remains in the situations where the animals lived and died, affords direct proof that the fluviatile, lacustrine, estuary, or sea-bottoms, thus containing these remains, have not been broken up and re-arranged, but that, independently of consolidation or certain other modifications of structure, they exhibit plans and sections of the fresh or brackish water, or sea-bottoms of a particular geological time. Careful search shows that this manner of entombment is by no means so rare as might once have been considered. The occurrence of the remains of boring molluscs in the holes formed by them in rocks, has been already noticed, as resembling those of any *Pholas* in limestones of the present day on the British coasts (fig. 176); and it has been also stated, that during calcareous deposits of the same date (inferior oolite) several beds in succession were drilled at their surfaces by the same species, the shells still in the holes made by their animals (p. 486). With respect to molluscs piercing mud, silt, or sand, we may point to the observations of Mr. Prestwich, as to the shells of *Panopæa*, found abundantly in the vertical position common to the habits of the existing species, in the beds of the London clay at Clarendon Hill, and of *Panopæa*, *Pholadomya*, and *Pinna*, at Cuffnel; as also to those of Dr. Fitton, on a similar mode of preservation of the shells of *Panopæa* and *Pinna*, in the lower greensand of Southern England.* In cases of this kind, it certainly is not always clear that the animals, thus in the positions which their habits required, were suddenly destroyed by any physical change in the water or the kind of sediment deposited above them, though this may be surmised, since in all beds containing burrowing molluscs, their shells may be found in the positions where they died under ordinary circumstances. Be this, however, as it may, it proves that these animals of the time lived and died in the mud, silt, or sand, now perhaps beds of hard rock, in which their remains are found.

* Journal of the Geological Society of London, vol. i.

Many shells of molluscs so occur that even the direction of the stream or current which drifted them according to their weights, volumes, and forms, may be inferred, having been, in all probability, empty shells at the time, and the same with the exuviæ of other fresh-water and marine animals. Sometimes, as beneath (fig. 202), whole and broken shells are found drifted along the bottom with intervals of repose, during which mud was alone thrown down.

Fig. 202.

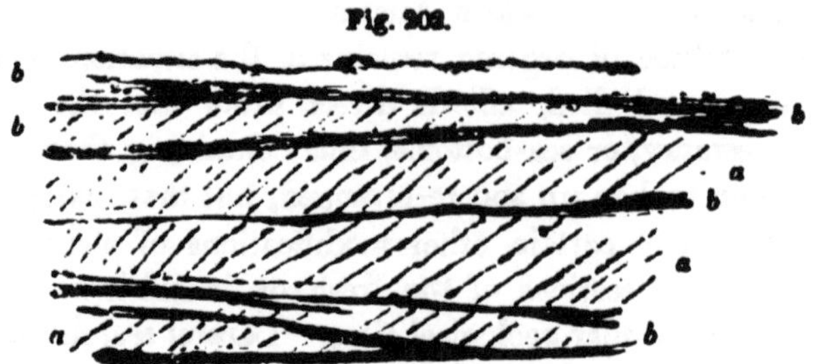

a, a, a, beds formed of diagonal layers, composed of broken shells, fish-teeth, pieces of wood, and oolitic grains, sometimes mere rounded pieces of shells, the various substances lying in the planes of the diagonal layers, and presenting every appearance of having been shoved or pushed over the more horizontal surfaces formed during the intervals between the mud deposits, b, b, b, b.*

With regard to the destruction of marine animals in place, that in volcanic regions, certain gases, such as carbonic acid, sulphuretted hydrogen and others (p. 323), when suddenly discharged into waters through subaqueous fissures or volcanic vents, should destroy the animals to which they find access, would be expected at all geological times as well as at present. In like manner, subaqueous fissures formed at all periods during earthquakes, and from which gases have been evolved, destructive of animal life, would be inferred to have been always followed by the same results. So also with the heat communicated to waters during submarine volcanic eruptions, or when fissures, formed during earthquakes, reached the depth of very considerable temperatures. With regard to waters impregnated with deleterious gases, so long as these

* This section was taken at a quarry of Forest marble, part of the oolitic series, at the Butts, Frome, Somersetshire. The Bath oolite, into which this part of the oolitic series graduates, as may be well seen in Somersetshire, is often, in some of its beds, nothing else than a modification of the same thing, the rounded grains of shells and corals, mixed with those of the true oolite (having concentric coatings of calcareous matter), being drifted in a similar manner. Broken shells, fish-teeth, and other organic remains are seen in the sections of the same neighbourhood, occurring as streaks in clay, conditions from time to time having occurred, during which the deposit of the mass of mud of the beds termed Fullers' earth, was locally interrupted by these shell drifts.

remained sufficiently disseminated in them, predaceous animals, not included in the areas so affected, would be prevented from entering in order to feed upon the multitudes of fresh-water or marine animals which may have been killed, until their remains were covered over by fine sedimentary matter; or, being burrowing creatures amid mud, silt, or sand, until the sedimentary matter was so adjusted around them, with probably also a certain decomposition of the softer parts, as to be no longer desirable food, if even they were attainable.

Multitudes of fossil-fish are sometimes so found in rocks, that their sudden destruction, with the preservation of their bodies from predaceous and scavenger creatures, seems needful in order to account for their mode of occurrence, it appearing also necessary that their entombment in the containing substances was sufficiently rapid, subsequently to their death, to prevent the distribution of the various harder parts of their bodies after decomposition. In all cases where volcanic action can be inferred at various geological times, at or near the localities where the observer may have aqueous deposits under examination, and which present the remains of animals in a condition showing that whole creatures have been preserved without injury to the arrangement of their harder parts, he will do well to recollect the modes of entombment which may now be in progress in similar regions of the present day. He will thus see organic remains among the volcanic ashes of different geological times, even amid the old accumulations of the Silurian deposits, (Ireland, Wales,) and in such positions as very forcibly to remind him of the causes of destruction and preservation which he finds, or can fairly infer, are now in action.

Independently of any sudden destruction and entombment of animal life in connexion with volcanic eruptions or earthquake movements, the study of the old fresh-water and sea-bottoms presents us with the occurrence of animal remains so preserved, and amid such substances that the sudden influx of waters, charged with much fine matter in mechanical suspension, may have destroyed multitudes of aqueous animals in some given area. At least, their remains are so entangled amid this matter as to lead to this inference. That fixed creatures or others of slow movements could thus readily be overwhelmed, would be expected under such conditions at all geological periods. When, for example, in the vicinity of Bradford, the *Apiocrinites* of that locality is found rooted upon a subjacent calcareous bed (one of the oolitic series) and entangled in a seam of clay, its parts sometimes beautifully preserved, it may be inferred that it was destroyed by an influx of mud, from which it could not escape. In like manner, also, the preservation of long uninjured stems of various encrinites found amid the Silurian and other older deposits, on the surfaces of limestone and other rocks, and having had a covering of fine sediment, would appear to be explained. Sometimes, as in the lias of Golden Cape, near Lyme Regis, multitudes of belemnites, some with even the ink-bag of these molluscs preserved, so form a seam of organic remains, that the geologist is led to infer a sudden destruction of thousands of them over a moderate area. Ammonites are also sometimes found in great numbers, distributed in a depth of only a few inches, over areas of a square mile or more, as if suddenly destroyed. The beautiful bed of myriads of ammonites occurring amid the lias of Marston Magna, Somerset, was a good case of this kind. It sometimes happens that the shells of molluscs show that when their animals were entombed, the space occupied by their bodies prevented the entrance of the sediment which enveloped them. The following section (fig. 203) of an ammonite (lias,

Fig. 203.

Lyme Regis) may be taken as an example of this mode of occurrence. All the chambers of the ammonite are filled by carbonate of lime, infiltrated into their hollows, beyond which there is a space apparently occupied by the animal when overwhelmed by the surrounding calcareous mud, now argillaceous limestone; this space terminated outwards by sedimentary matter (*a*) which entered so much of the shell as the retreat of the animal permitted. In this case the intruding sediment has become highly impregnated with dark-coloured matter, as if effected by the decomposition of the animal within. Such deposits as clays and argillaceous limestones, the latter especially, from the usual consolidation, without much pressure, of the matter around the organic remains, are very favourable for observations of this kind, numerous shells of molluscs appearing to show that their animals may have been in them at the time of their entombment. In such researches attention should be paid to the positions of the shells in the beds, and the forms of their interior cavities, so that the entrance of sediment might be prevented by such positions and forms. Multitudes of examples are found in certain areas and deposits where the presence of the animals in their shells would seem required. When we consider the probable voracity of numerous creatures in fresh and sea waters, and the multitudes of scavenger animals consuming decayed animal matters at all geological times, the discovery of certain aqueous reptiles preserved entire amid rocks, even with the contents of their intestines preserved, leads us to infer that their entombment, if not also their death, was sudden. And this appears the more probable when we find, as often happens, that in the same deposits the same kinds of aqueous reptiles are dismembered, as if by predaceous animals feeding upon them. While at times, in the lias of Western England, the skeletons of Icthyosauri and Plesiosauri are so well preserved that all, or nearly all, the bones are in their proper relative situations; even their skins preserved, and the contents of their intestines, at the time of death, in their right places, at others the bones of these reptiles are dispersed, though not always far removed from the place where the animals died. In fact, the appearances presented are precisely those of decomposition having been so far advanced that the scavenger animals could feed upon some of the carcases, and drag the bones short distances, so as somewhat to scatter them.

Every mode of the occurrence of organic remains should be carefully considered, and viewed with reference not only to the district, as regards the depths of water, and the probable form of any neighbouring land, but also as to the general distribution of marine life at equivalent geological times over much more extended areas. The endeavour to obtain a general view of the distribution of life over great surfaces at given geological times, as well as of the deposits effected during those lapses of them to which given names have been assigned, would appear especially needful. Experience has taught geologists that many a genus of marine animals, the remains of which were at first found only in particular beds of various districts, have been discovered in the deposits, both of more ancient and more modern periods; as also that, as regards species, these will be observed in certain districts to have a wider range than in others, through a section of consecutive sea-bottoms.

It would seem essential that an observer should well weigh the evidence of the distribution of the animal and vegetable life of different geological times, as exhibited by organic remains, with reference to the probable distribution of land and water of those times, and the consequent variation of depths of seas, kinds of bottom, forms of coasts, discharge of rivers into the sea, fresh-water accumulations, and the like. He should refer to the depths at which animal and vegetable life is now known to exist in the sea (p. 145), with the forms and kinds of both found under the different conditions of heat, light, and shelter from violent movements. He can scarcely neglect the views of naturalists, as to the distribution of existing animal and vegetable life, over the surface of the globe; the spread of the different kinds under the circumstances fitted for each respectively; the overpowering, as it were, of some by others, the centres or localities whence species are inferred, under favourable circumstances, to have been diffused, and the representatives of different species in different regions.* The various supposed equivalent accumulations, chiefly sea-bottoms, have to be carefully dissected to ascertain the probable conditions under which the remains of life entombed in them have been gathered into the situations where they are now discovered, and the life itself was then adjusted. At all geological times when waters existed, they would arrange themselves according to the laws now governing their position as to temperature, and they would possess the same properties with respect to light and pressure.† All masses of water would also tend to be moved, as now, by the great causes of ocean currents and tidal streams, however modified these may have been by the manner in which dry land presented itself amid the ocean, at any particular geological period.

At the same time that all due attention is paid to these circumstances, it becomes also necessary to bear well in mind the modifications and changes which would arise from the movements of the crust of the earth elevating or depressing mineral masses, so that sometimes they were above the sea level, sometimes beneath it. To

---

* As regards works on the distribution of animal and vegetable life, the observer may conveniently consult the text and maps of Johnston's "Physical Atlas." For the geographical distribution of plants, reference can be made to the works of Humboldt and Shouw, and the Reports by Griesbach (translated by the Ray Society). The distribution of fishes, a subject of considerable geological interest, has received much attention from Sir John Richardson in his "Fauna Boreali-Americana," and British Association Reports.

† Assuming that the saline contents and their proportion to the waters have not been materially different during the lapse of time, since animals existed in the seas and fresh waters of the world. As respects the adjustment of marine animals to light, the eyes of *trilobites*, crustaceans found among the oldest fossiliferous deposits, have often been pointed out as satisfactory proof. Valuable remarks on this head will be found in Dr. Buckland's Bridgewater Treatise, vol. i., p. 396.

advert again to the change produced by the submersion of the Isthmus of Panama, and the junction of the Atlantic and Pacific Oceans, the land descending to the moderate (geological) depth of 1,000 feet, relatively to these oceans. By recent observations it would appear, that the summit level (named Baldwin's) is 299 feet above the sea, so that when the depression had continued to 600 feet, there would be a channel above this height deeper than between Dover and Calais; and when the submersion to 1,000 feet had been completed, one deeper than any part of the North Sea, or the channels between Great Britain and Ireland, or these islands and France, one nearly corresponding with the line of 100 fathoms, extending from Spain, outside the British Islands, to Norway (fig. 65).

By comparing a map of the Americas, with the land which would be under the ocean, if this movement of depression were carried out, gradually diminishing no further than 20° of latitude on each side of the isthmus, the great modification likely to arise from the free passage of the waters of the Atlantic into those of the Pacific, and the difference of the surface of dry land, will be obvious.* It is by carefully considering a few areas of the present dry land in this manner, with regard to the effects of depression or elevation, as the case may require, that the observer will readily perceive how needful it is for him, when endeavouring to trace the distribution of the life of any particular geological time, well to weigh the consequences of such changes; whether, on the one hand, they permit a mingling of species previously separated, or separate some given area, distinguished by the presence of some marked species into two parts, one or both of which were subsequently subjected to different conditions.

Inasmuch as we find marine animal life adjusted to certain conditions, among which, from the labours of Professor Edward Forbes, and other naturalists, depths of water may be considered, all other circumstances being the same, to have an important influence; reasoning from the known to the unknown, we should expect an adjustment of a similar kind to have extended back to the earliest state of the earth's surface, when water, fitted for life, washed the shores of continents and islands. Even under the hypothesis of a

* With regard to the differences in the levels of the Pacific and Atlantic on the shores of the Isthmus of Panama, the researches of Col. Lloyd would give a higher relative level to the former, to the amount of 3·52 feet. High-water mark at Panama is stated to be 13·55 feet above that of the Gulf of Mexico, at Chagres; while it is only 6·51 lower at low water on the Pacific side.—Philosophical Transactions, 1830.

heat of the earth's solid crust at former times, sufficient to keep the waters dispersed as oceans and seas above it, at equal temperatures at certain depths, independently of solar heat, littoral, shallow and deep water conditions would be expected to have had their influence, more particularly when combined with differences in sea-bottoms, and position as to shelter from wind-wave action, tidal streams, or ocean currents. At all events, it would appear most desirable that the observer, having before him the advantage of the sea-bottoms of different geological times, with organic remains variously distributed among them, should endeavour to trace out differences and resemblances of this kind, carefully considering the evidence afforded by the physical structure of the fossiliferous rocks in connection with that presented by the contained organic remains. It may be that certain forms of the shells of mollusca, for example, are deceptive, so that the palæontologist may not always reason safely when referring some to animals similar to those now living near shores or in shallow or deep waters; and that these last may be found to vary at the present day more than is now known; still the investigation can scarcely but be productive of an approximation to the knowledge sought, the general evidence, be it what it might, pointing out those modes of occurrence which may be ultimately seen to be somewhat constant; while others, though they present themselves in a more uncertain manner, may yet be important as regards the general subject.

As the researches of naturalists show that whether we rise high into the atmosphere, or descend deep into the sea, the conditions for the existence of life, under various adjustments and modifications, terminate; it follows that the great mutability of the earth's surface, as respects both conditions, could scarcely fail to produce great changes in that life, independently of those made inherent to it as created. The separation of great areas of dry land into minor portions has been above mentioned, as producing even the extinction of certain kinds of terrestrial life, while at others it may have preserved parts of it and mingled some together. Upon the descent of a continent beneath the sea (and the researches of the geologist teach him the necessity of such submersions, as well as that the dry lands for the time have been chiefly raised from beneath seas into the atmosphere), any terrestrial life peculiar to it would be destroyed, though evidence of its existence might be preserved amid mineral matter where circumstances permitted. In like manner when, upon submersion, shores ceased to present themselves, the littoral marine animals, previously inhabiting them, and moving to the coasts as these retreated upon the descent of the main mass of land, would be expected also to have disappeared, unless able wholly or in part to have adjusted themselves to the new conditions. When, however, zone after zone of the marine vegetation disappeared as the circumstances fitted for its growth ceased, the animals which fed upon the plants would perish, and with them those which lived upon the vegetable eaters, unless they could escape to other localities where food of the same kind, or of others which they could substitute for it, was to be found, and was sufficient for them. If in the annexed section (fig. 204),

Fig. 204.

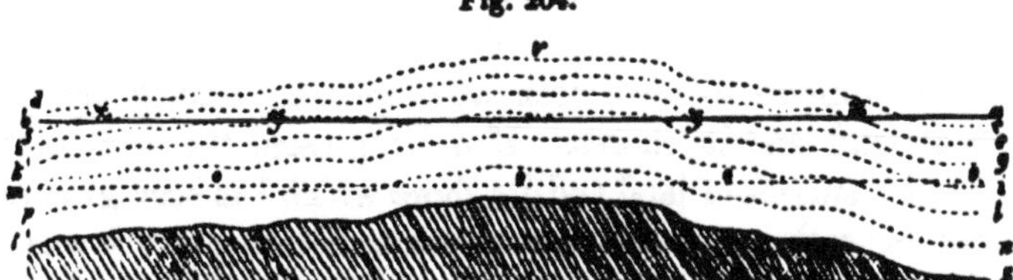

a b represent the level of the sea, remaining constant, or nearly so, and *c v d* the outline of any mass of land, partly in the atmosphere and partly beneath the sea, and *o, o, o, o*, the depth at which marine plants supporting the life of a certain portion of the marine fauna grew, the littoral portion of that life would be shifted to *x x′*, upon submersion of the land to *e f*, and at the same time a portion of the sea-bottom inhabited by animals at greater depths would be brought lower down, so that these would probably also move over the ground of others previously adjusted to minor depths. Submersion continuing, when it reached the line *g h*, the shores would still further be shifted to *y y′*, with the same general consequences as before, and so also with the submersion to *i k*. When it reached *l m*, the whole of the land, previously above water, would be beneath it, and littoral life may be considered to have disappeared when it reached *n p*. At the amount of submersion represented by the line *s t*, the whole of the former dry land, with its shores and any shallow seas adjoining, would be beneath the depths of marine vegetation. In this section the probable consequences of breaker action on the descending land, tending to plane it off, as the great Banks of Newfoundland may have been land to a certain extent levelled out during a gradual submersion, have not been included, in order to render the illustration simple. During, however, such depression of the land, this action has to be well borne in mind, so that the detritus thence arising, distributed over the sunk land, and entombing the remains of the animal life of the time, with its variations, according to circumstances, those of deep-water creatures ranging over those of shallow water, and littoral species, be not neglected.

With respect to this covering of detrital deposits containing the remains of littoral species by others entombing those species which contemporaneously inhabited deeper waters adjoining, the following section (fig. 205) may serve to show the manner in which this

Fig. 205.

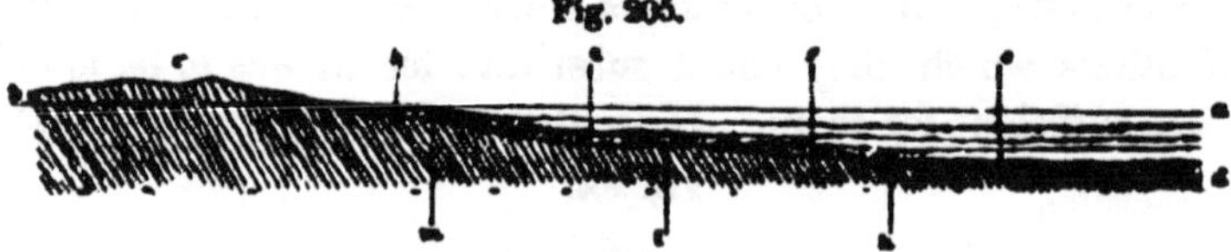

may be, and appears to have been accomplished, during the submersion of land and its shores. If *a b* represent the level of the sea, and *c d* a surface of land and its shore which has been gradually depressed beneath it, *e*, the littoral accumulations when *k* was the coast, *f* and *g*, those when the sea boundaries were at *l* and *m*, as *h* is supposed to be at the time of the section, there would always be deeper waters outwards in the direction *a d*. Hence, though a certain thickness of deposits, *e f g h*, variable according to circumstances, might cover the surface of the descending land, entombing the remains of the littoral marine animals, these would be covered, in the direction outwards, and as the land descended, by detrital deposits of kinds which could be transported to and formed there, a corresponding series of marine animal remains intermingled with them, differing as far as the deeper water differed from the littoral marine life of the time. Thus numerous species which had been really contemporaneous with those entombed beneath may appear, in certain sections, to have succeeded them as creations in the progress of geological time, this appearance extending even to the remains of those living in the deepest waters of the period and locality, as any large mass of dry land became submerged.

With respect to the emergence of land, should this be gradual, large areas might be laid dry, presenting sheets of sedimentary matter not contemporaneously produced, yet containing littoral species of molluscs in great abundance, these species being of the same kinds, should no change have been effected in that portion of the animal life of the locality and time during the rise of the land and sea-bottom. If the sea-bottom around the British islands were gradually raised, so that the boundary line extended to not more than 100 fathoms in depth (fig. 65, p. 91), the remains of littoral molluscs would be scattered amid the accumulations of the time, as the shores became extended, covering over those of other and contemporaneous molluscs. If *a b*, in the following section (fig 206), represent the level of the sea, *c d* a surface of rock,

Fig. 206.

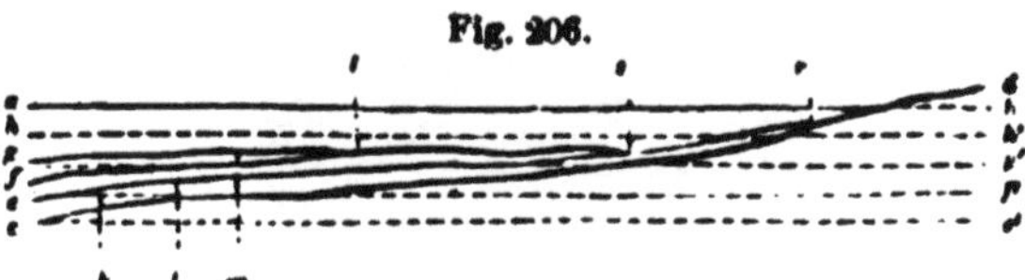

partly above the sea and partly beneath it, and *e* a deposit extending to *r*, it would contain the remains of molluscs inhabiting the different depths, including those at which wind-wave action could drive them onwards towards the coast. If the land be now raised, so that the relative sea level be represented by *h h'*, a deposit *f*, extending to *s*, would be under similar conditions as that previously formed and extending to *r*, and so with an accumulation *g* extending to *t*. Successive beds *k*, *l*, *m*, are thus produced, probably containing the remains of molluscs, allowing the mingling of many by the action of the waves in shallow situations, and corresponding with the depths *h h'*, *g g'*, *f f'*, *e e'*, so that, other things being equal, these exuviæ are similar in sections of the detrital accumulations which do not correspond with the general planes of those deposits, but with others representing their littoral, shallow, or deep-water conditions, as the case may be, of succeeding times.

These modifications, from the causes noticed, have to be well considered when certain organic remains are viewed as characteristic, as it has been termed, of the accumulations of a particular geological time, those to which some name may have been assigned. When any such are found more in abundance in, or seem confined to, the deposits of some particular area, and appear to be the exuviæ of animals which have lived at or near the localities where they are obtained, the kind of bottom, probable depth of water, and proximity to or distance from the dry land of the time have to be sought, so that the conditions under which the creatures themselves flourished may be duly appreciated. In such researches it will be often found that the kind of bottom appears to have materially influenced the abundance and distribution of these particular animals, so that, when a change was effected in the sedimentary matter deposited, they moved elsewhere, even returning in the same abundance as before to the same area, should

2 N

the conditions fitted for them have been re-established. If, in the following plan (fig. 207), the shaded portions represent minor areas

Fig. 207.

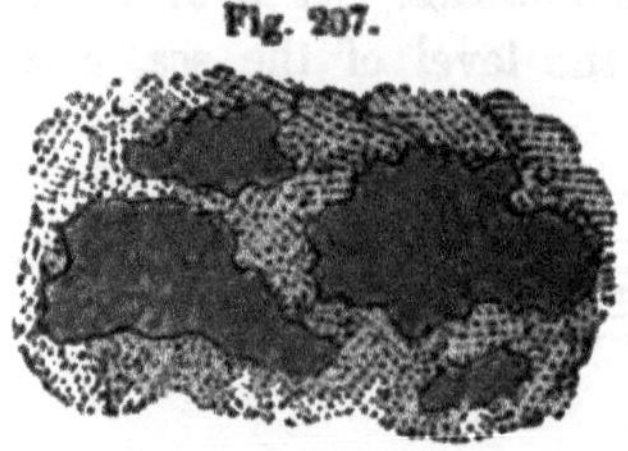

of mud, distributed amid sands, it would be expected that the creatures whose habits induced them to prefer the one to the other would keep within the respective variations of sea-bottom, so that if, in the course of accumulation, this bottom became modified, sands drifting or being thrown over the mud, or the latter over the former, the animals would follow the modifications according to their habits. Thus in any given sections of these sea-bottoms streaks of different kinds of them may be found accompanied with peculiar organic remains, the animals from which they were derived merely shifting their ground as circumstances arose, thus introducing interlacings, as it were, of different kinds of sea-bottoms. Looking at the conditions which at the present time appear to govern the existence of marine life both as regards the relative position of different portions of it and the distribution of similar animals, very great care seems to be required in assuming particular species as characteristic of particular geological periods without reference to their mode of occurrence at the time. It would seem very needful that the probable habits of these species should be well considered, so that proper importance should be assigned to other and contemporaneous species whose remains may be equally of value in continuous or contemporary accumulations formed under modified conditions elsewhere. Unless this be done, it may often happen that littoral species, very characteristic of the shores of a particular region, will be uselessly sought for amid contemporaneous accumulations in the deep seas of other regions, while not a trace can be found of deep-sea species, abundant elsewhere at the same geological time, amid shallow water and littoral deposits.

The calcareous and fossiliferous accumulations of different dates are frequently of so mixed a character as to require much care. They are often mere beds of organic remains; these cemented together by the carbonate of lime, which, after the deposit, has been formed at the expense of the organic remains themselves. At other times, however, they have been clearly produced by deposits from solutions of the bicarbonate of lime, in the manner previously mentioned (p. 106). Some limestones require very careful examination in order to ascertain their mode of formation. Thus it has been observed that beds presenting no appearance of organic remains to the naked eye, may yet be found to be almost wholly composed of them when the microscope is employed and due precautions taken. In this manner many beds of the mountain limestone series of the British Islands have been found replete with the remains of life where none were at first suspected. Even when upon exposure to atmospheric influences fossils of far larger dimensions, readily visible to the naked eye, and extending to half an inch or more in length or breadth, are found in fair abundance, it sometimes occurs that the ordinary fracture of the limestone bed may not readily show them. We do not here include the remains of encrinites, echinites, and some other fossils, which, from their rhomboidal fracture, a little practice will enable an observer readily to distinguish; but others, where they are far from being easily detected. The most beautiful shells will occasionally thus present themselves upon searching a weathered surface, not a trace of which can be obtained by ordinary observation.

It is now known that certain beds, as well siliceous, calcareo-siliceous, as calcareous, are made up almost wholly of minute organic remains, far too small to be seen hy the unassisted eye. For our great progess in this order of investigation geologists are indebted to M. Ehrenberg, who has shown how much infusorial remains are diffused, even producing deposits of considerable importance, and most materially adding to the volume of others. Whether or not some of these microscopic minute bodies may be vegetable instead of animal, their geological importance remains the same, if indeed it be not increased from such deposits, altogether or in a great measure composed of myriads of microscopic organisms, being referable to both animal and vegetable life.*

While on the subject of deposits chiefly formed of organic remains, the probable chemical composition of these remains, when first introduced amid the accumulations in which they are found, should not be neglected. In this manner it may be seen that the magnesia, so much more commonly distributed amid limestones than has often been inferred, may sometimes be due to such remains,

* As to some of these supposed infusoria being vegetable, see note, p. 238.

particularly where many corals are present.* The like also with the phosphates of lime, silica, and other substances. Whole layers may be formed of the harder parts of infusoria, so that when these are siliceous, they, and the spiculæ of many sponges, may serve to diffuse no small amount of silica amid deposits of a different character.

By careful investigation of the conditions under which the remains of various fresh-water or marine animals may be found in rocks, the deposit of which by means of water is evident, and also by well-directed attention to the mode in which the remains of terrestrial life, not forgetting those of insects,† may have become intermingled with them, the observer will frequently find himself most materially aided in a knowledge of the probable physical geography of different areas, often considerable, at given geological times. With this knowledge and a due regard to the varied distribution of the life of the time, and the abundance and kind of mineral matter deposited at the same period, he may be enabled to trace the changes and modifications which have taken place contemporaneously in the rivers or lakes, amid the lands, or in the seas, at different times in such portions of the earth's surface. Regarding that surface as a whole, it is difficult to conceive that the distribution of life, allowing for great changes in that distribution during the lapse of time, could not have been adjusted to conditions as they successively arose, and which modified it more in one locality than another, so that great care seems required properly to separate the local from the general effects produced at assumed equal periods, or during a long succession of them. Modern investigations, while they, on the one hand, lead us to infer many great changes in animal and vegetable life during the accumulation of the various deposits in which its remains have been preserved,‡ teach us, on the other, that forms once supposed

* In some investigations undertaken by Mr. Maule at the Museum of Practical Geology, for the purpose of tracing the various changes which organic remains may have undergone under different conditions of entombment, he found magnesia, even to the extent of 6 and 7 per cent. in some recent corals.

† In countries, and especially in tropical islands, such as the West Indies, where the off-shore or land-winds are at times somewhat strong, multitudes of insects are often borne out to sea, where, though the greater proportion may become the food of marine creatures, some fall in situations to be entombed amid mud, silt, or sand. Those accustomed to pass along such coasts are familiar with this fact. Our own coasts in summer weather present many instances of insects surprised and drawn off coasts seaward by the sudden setting in of the land-wind in the evening.

‡ Referring to various general works containing lists of the remains of animal and vegetable life considered characteristic of the different deposits which it has been thought convenient to separate and class under particular names, it is only required

only confined to the more modern accumulations have existed in far more remote times.* While it is probable that the evidence of great changes having taken place during the lapse of geological time in the vegetation and animals which have existed on the earth's surface will be only confirmed by extended research, it seems equally probable that investigations carried out with proper regard to the varying physical geography of different geological periods will show the necessity of tracing the probable causes productive of new adjustments of lands and waters at those different times, and of studying the distribution of the life† of such times

---

to point to such animals as the trilobites, among the more ancient accumulations, and to the ammonites of the middle portion of the fossiliferous series, to show that certain marine creatures which have now ceased to exist once swarmed in particular areas at given times, and have not lived after those times. No doubt the preservation of the parts of many terrestrial animals requires a combination of favourable circumstances, so that no great surprise is to be experienced when we obtain few traces of such animals amid the contents of the old sea-bottoms usually presented to our examination. The remains of the marsupial mammal (*Phascolotherium Bucklandi*, Owen) and of the insectivorous mammals (*Amphitherium Prevostii*, and *Am. Broderipii*, Owen) in the Stonesfield slate (oolitic series), near Oxford, are sufficient to introduce caution into general reasoning as to the existence or non-existence of terrestrial mammals at different geological times. Speaking of the conditions under which these remains occur, Dr. Buckland remarks (Bridgewater Treatise, vol. i., p. 121) that "at this place (Stonesfield) a single bed of calcareous and sandy slate, not six feet thick, contains an admixture of terrestrial animals and plants with shells which are decidedly marine; the bones of Didelphis (*Amphitherium* and *Phascolotherium*), Megalosaurus (a great saurian 40 or 50 feet long, partaking, according to Cuvier, of the structure of the monitor and crocodile), and Pterodactyle (a flying saurian), are so mixed with ammonites, nautili, belemnites, and many other species of marine shells, that there can be little doubt of this formation having been deposited at the bottom of a sea not far distant from some ancient shore." With respect to the wing-covers of insects found in the same deposit, Dr. Buckland remarks (Bridgewater Treatise, vol. i., p. 411) that they are all coleopterous, "and in the opinion of Mr. Curtis, many of them approach nearly to the *Buprestis*, a genus now most abundant in warm latitudes."

* As regards the forms of mollusca, the genera *Avicula*, *Modiola*, *Terebratula*, *Lingula*, and *Orbicula* are found from the Silurian rocks upwards to the present day. The like with *Turbo*, as a restricted genus, and also with *Nautilus*, with slight variations in form. With respect to those remarkable and beautiful animals the starfishes, Professor Edward Forbes states, that species of the genus *Uraster* are found in the Silurian rocks closely resembling the existing northern forms (Decade I., "Memoirs of the Geological Survey"), and that in the lias, *Uraster Gaveyi* (Decade III.) is only critically to be distinguished from the common *Uraster rubens*, now inhabiting the British seas. According also to the Professor, the *Terebratula striatula*, of the cretaceous series, cannot be distinguished specifically from the *Terebratula caput-serpentis* of the same seas.

† It is very desirable, in the enumeration of organic remains discovered in different beds and localities, properly to represent the abundance of the individuals of each species. This mode of investigation has received careful attention during the progress of the Geological Survey of the United Kingdom. Without due precaution of this kind, the remains of a single individual figures as prominently as those of many hundreds, and a correct view of the correspondence or difference between the various portions of contemporaneous accumulations as to the life entombed in them becomes

in accordance with those laws which appear to govern that distribution at the present time. At the same time we should not neglect those conditions which would follow a gradual decrease in the heat of the earth, should it eventually be found that a temperature more equal over the earth's surface than that afforded by the sun would appear required for the distribution of animal and vegetable life in the earlier periods of its existence on our planet.

---

much impeded. A careful study of the comparative numbers of individuals often shows how much some species of marine molluscs have preferred one kind of sea-bottom to another, while others seem to have flourished equally well through varied changes in the sea-bottoms. It is well to bear in mind that the researches of naturalists teach us that many an area is now little, if at all, tenanted by marine mollusca, such areas being unsuited to their habits, while others adjoining them may be covered by multitudes of various mollusca.

CHAPTER XXIX.

IGNEOUS ROCKS OF EARLIER DATE THAN THOSE OF MODERN VOLCANOS.—SIMPLE SUBSTANCES FORMING IGNEOUS ROCKS.—VOLCANIC PRODUCTS AMID THE OLDER FOSSILIFEROUS ROCKS.—FOSSILS AMONG OLD IGNEOUS PRODUCTS.—VOLCANIC TUFF AND CONGLOMERATES AMONG THE SILURIAN DEPOSITS OF WALES AND IRELAND.—OLD VOLCANIC PRODUCTS INTERMINGLED IN THE DEVONIAN ROCKS OF SOUTH-WESTERN ENGLAND.—IGNEOUS ROCKS ASSOCIATED WITH THE CARBONIFEROUS LIMESTONE OF DERBYSHIRE.—RELATIVE DATE OF THE WICKLOW AND WEXFORD GRANITES—OF THE DEVON AND CORNWALL GRANITES.—UNCERTAIN DATES OF SOME IGNEOUS DYKES.—ELVAN DYKES IN CORNWALL AND DEVON.—IGNEOUS ROCKS OF THE NEW RED SANDSTONE SERIES OF DEVONSHIRE.—DATES OF CORNISH AND DEVONIAN ELVANS.—ELVANS OF WICKLOW AND WEXFORD.—CHEMICAL COMPOSITION OF IGNEOUS ROCKS.—EFFECTS OF SILICATE OF LIME IN IGNEOUS ROCKS.

As has been previously remarked, the distinction between the products of active and extinct volcanos is rather one of convenience than of fact; and the same may, to a certain extent, be also observed as to the differences between those above noticed (pp. 317—407), and the products about to be mentioned. By arranging igneous products according to the different geological dates to which they may be assigned, the observer has the means of studying not only their modes of occurrence, but also the constancy or change of the elementary substances entering into their composition during the lapse of geological time.

The igneous rocks known to us by their appearance on the surface of the globe, have been found sufficiently well distributed to be available for an approximative estimate of their component elementary bodies. Viewed as a whole, they are chiefly oxides of substances commonly considered simple, one of the oxides, that of silicon, acting as an acid, and combining with a large portion of the other oxides. Silicic acid (silica), free or combined, may be seen more to prevail in certain rocks than in others; but there are few igneous products found in any abundance, which do not mainly consist of silica, or the silicates. The simple substances, with silicon, constituting this mass of matter, whence the sedimentary deposits have been, with minor exceptions, more or less directly or indirectly derived during the lapse of time, have not been found numerous. They are chiefly aluminium, potassium, sodium, calcium, magnesium, iron, and manganese, making with silicon eight substances, considered elementary, all combined with another, oxygen, and forming the great volume of the igneous rocks, such as they are known to us. Of other elementary substances entering into their composition on a minor scale, probably sulphur, boron, lithia, and fluorine, may be regarded as the principal bodies, with the addition of hydrogen, so far as it may enter into the composition of any waters that can be regarded as a real component part of these rocks. Numerous other simple substances, no doubt, may be detected amid these products in different localities, even sufficiently abundant in some to be remarkable; but viewed in the mass, the nine elementary substances above mentioned, with the four others in a minor manner, appear to constitute the great mass of the igneous rocks of all ages.

That so much of the great volume of these rocks should consist of the combination of oxygen with a few simple substances, and that the union of oxygen with one of them should constitute such an important compound for further union with the other oxides, are in themselves circumstances of no slight interest to a geologist anxious to trace some connexion between the igneous products of all geological periods and the substances beneath the exterior and consolidated portion of the earth during the same lapse of time. We have elsewhere* estimated silica as constituting 45 per cent. of the mineral crust of the globe, hence the oxygen contained in silica alone would form at least 26 per cent. of that crust.† If the amount of oxygen in the other oxides be included, the percentage becomes largely increased; so that when this substance is regarded as free from its union with the matter forming rocks, and in a gaseous form, its volume becomes enormous.‡

In studying these rocks it may be assumed that the observer would be desirous of ascertaining how far there may be evidence

* "Researches in Theoretical Geology," 1834, p. 8.

† According to Berzelius, silica is a compound of 48·4 parts silicon and 51·6 parts of oxygen.

‡ The volume of oxygen would be obviously still farther augmented by the addition of that contained in the various waters on the surface of the globe, water being a compound of oxygen and hydrogen.

of igneous products having been thrown out in the manner of those ejected from active volcanos at different geological times. As it so happens that certain portions of the earth's surface appear to have remained in a state undisturbed by igneous action from very early periods to the present day, while other portions seem frequently to have been subjected to this action during the same lapse of time, all regions, however interesting they may otherwise be geologically, do not present the needful conditions for this kind of investigation. In the British islands, presenting so many coast and other natural sections, as also so cut and pierced by the operations of the miner and engineer, it fortunately happens that amid the older fossiliferous deposits there is evidence of igneous products having been contemporaneously ejected. Igneous rocks are so entangled with detrital accumulations of the Silurian series, especially well exhibited in Wales, and in the counties of Wicklow, Wexford, and Waterford, on the opposite shores of Ireland, that a geologist has excellent opportunities afforded him for observation. He finds that the igneous products, thus associated with these old fossiliferous deposits, may be divided into those which have occupied their relative positions in a molten form, and those which have been mingled with them through the agency of water, with also certain accumulations which may even have been piled up in a mechanical manner in air.

Having in view the manner in which the products of existing volcanos are thrown out into the air or water, and are commingled, it is desirable that the geologist should endeavour to trace any differences or resemblances he may find when opportunities of the kind noticed present themselves. In the first place he does not possess the advantages of the surfaces usually presented in active volcanic districts, or of those, such as in France (p. 401), which have not been disturbed by the action of seas upon them, but finds masses of mineral matter, of which the igneous products only constitute a part, usually thrown out of the positions in which they were originally accumulated, the igneous often bent and contorted with the aqueous deposits with which they are associated. These districts, moreover, are often the mere wrecks of the mud, silt, and sand of former sea-bottoms, combined with the igneous products, large portions having been removed by denuding causes, so that not only has the general mass been squeezed, bent, contorted, and sometimes broken, but portions of it (occasionally to be measured by cubic miles) entirely removed. Hence, no slight care and exact research are required to collect the needful evidence, so that all the parts may, mentally, be again restored satisfactorily to their places. This may often nevertheless be sufficiently accomplished.

In examining the igneous products associated with the Silurian rocks in Wales and Ireland, two kinds become somewhat prominent, one in which the matter constituting felspar prevails, another in which that forming hornblende is mingled with the first, to an equal and even greater amount. Those accustomed to active volcanic regions might be disposed to see in this circumstance a general resemblance to trachytes and dolerites (p. 352) therein distinguished, as also to those mixed products which have been named trachyte-dolerites. Proceeding still further in the inquiry, it will be found that certain of these old products are mingled mechanically with substances that have once formed ordinary mud, silt, sand, and even conglomerates, reminding us of the mixture of the ashes and lapilli thrown out of existing volcanos, and intermixed with the detrital accumulations forming under the fitting conditions around or near many active igneous vents of the present day. Even lapilli may be detected amid beds which are composed of something more than the igneous substances themselves. These appearances would alone lead an observer to consider the old igneous products before him with reference to certain of the results of modern volcanic action, and he would probably be not the less induced to take this course when he found, as in such districts he often may, organic remains amid them, either alone or mingled with common detritus, preserved precisely as in volcanic tuff associated with the common mud, silt, and sand deposits of the present time. He will sometimes find the organic remains arranged in seams, as amid ordinary detrital accumulations, representing in like manner the bottoms of ancient seas strewed with the harder part of mollusca, and other marine animals of the time.

The geologist may occasionally discover organic remains, thus arranged in seams, the matter of the shells of molluscs still preserved in beds of hard and solid rocks, ringing under the hammer, and at first sight appearing as if they had flowed in a molten state. Such beds of consolidated igneous matter, arranged in water, are frequently very deceptive, requiring no slight care not to confound them with the rocks which have really flowed in a molten state amid those with which they are often associated. Not only in cases where such beds contain organic remains, but also where no trace of them can be detected, much caution is needed. For the most part microscopic observation will show that they are composed of fragments of igneous molten products, in which the component parts of felspathic or hornblendic minerals have variously prevailed, and that these fragments are angular. When lapilli, especially those having the aspect of pumice, are mingled in these accumulations, there is usually little difficulty in determining their true character, and they then assume the appearance of those resulting from the deposits of the ashes and cinders thrown out of volcanic vents at the present day, more especially of such as have been formed in water by the fall of volcanic ashes and cinders in it, and are now elevated above it. They thus resemble the tuffs in the vicinity of Vesuvius, Etna, and some other active volcanos, which have showered ashes and cinders into seas adjoining them, a change of relative level of the land and sea having been effected, and parts of former sea-bottoms having been upraised.

Those who have devoted much close attention to the structure of the valcanic tuffs of the present time can scarcely fail to be struck, particularly when they are regarded on the large scale, with the resemblance of many of the accumulations of igneous products associated with the Silurian rocks of Wales and Ireland to certain of them, especially to those such as palagonite (p. 368) and some others which have become consolidated and modified in appearance, so that the original small grains of ashes and fragments thrown out of volcanos have become one general and, at first sight, almost homogeneous substance. While in many localities the laminated character of the beds, and the presence of marine organic remains occurring in the same manner as in any other detrital and associated deposit, point to their accumulation beneath the sea, ashes, and sometimes lapilli, vomited forth from the volcanos of the time and locality, and arranged in extensive and comparatively thin beds; at others the conglomerates and breccias of igneous rocks, mingled confusedly with ancient volcanic tuff, the whole interlaced with dykes and veins of felspathic and hornblendio molten rocks of different kinds, remind the observer of a confused mixture of substances in the body of a volcano itself, partly subaërial and partly subaqueous, the general mass buried up by other accumulations as the volcanic rocks gradually descended beneath them. Of the natural sections exposed in Wales and Ireland, though there are many excellent opportunities in various places, the most instructive is probably on the coast of the county of Waterford, between Tramore and Ballyvoil Head, a distance of fifteen miles. Huge masses of these igneous products are there found in great variety, and molten rocks of different kinds and shapes will be seen sending out veins, and cutting as well the ordinary detrital deposits, formed prior to these outbursts, as the igneous substances of the time. Conglomerates and breccias are found piled in various forms, cemented by igneous matter, apparently thrown out as ashes, and ancient tuffs formed of smaller fragments are observed in different places, while examples are to be seen of deposits of various kinds changed in aspect and character where molten matter has burst in among them. There is also a variety of minor, but collectively important, objects of interest bearing upon the igneous products and their mode of accumulation at this early geological period. As regards the intrusion of molten matter amid the conglomerates and breccias, the following section (fig. 208) on the west of Kilfarrasy Point may be found illustrative, others of equal interest being, however, sufficiently common.

Fig. 208.

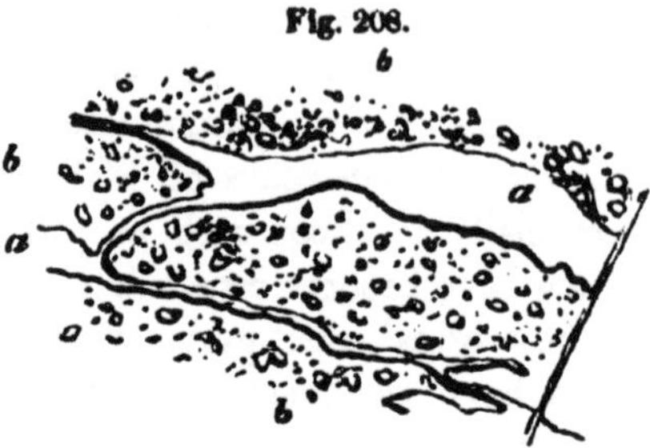

a, a, compact igneous rock, in which the substances composing felspar prevail; b, b, b, conglomerate and breccia formed of various portions of igneous products, chiefly felspathic, cemented by matter resembling that of volcanic tuff.

Such districts require obviously to be studied on the large scale. For example, the section exhibited on the Waterford coast, excellent as it may be, would scarcely afford the needful evidence, taken by itself. It would be necessary to consider it with reference to the mode of occurrence of similar accumulations along their whole range, thence northward through the counties of Waterford, Wexford, and Wicklow. When this is accomplished, the sections afforded on the coast of Waterford are seen to form part of the general evidence pointing to the relative age of the igneous accumulations at this time; one shown, moreover, to have been anterior to the formation of the old red sandstone of the south of Ireland, inasmuch as these igneous rocks, as also the beds of the Silurian series with which they are associated, were disturbed, bent, and contorted before its accumulation, and this old red sandstone not only reposes quietly upon the disturbed rocks, but also contains worn fragments of the latter, the igneous substances included.

The geologist, seeing this considerable resemblance to the products of modern volcanos, and also the general similarity of the elementary substances found in them, as far as researches have yet extended,* (considering both igneous products in their masses,) would be prepared to find evidence of igneous action also bearing a resemblance to that of volcanos at the present time amid any accumulations of intermediate geological date, should the fitting conditions prevail. For evidence of this action he would look necessarily in very different regions; for not only is it required that there should have been igneous products of this kind at all geological times in various parts of the earth's surface, sometimes in one locality, sometimes in another, but also that in the present arrangement of land and seas they should be attainable for observation. In this research, however, the geologist may again find opportunities in the British islands. In Devon and Cornwall he may obtain evidence of a continuance of the like igneous action at a subsequent period, amid deposits, some of which may be referred to the date of the old red sandstone series of other parts of the British islands, while some are more modern, and others perhaps more ancient. Amid the Devonian and Cornish accumulations of this date he will detect beds apparently also formed of the volcanic ashes of the time, and other arrangements of igneous matter, some rocks evidently poured forth in molten masses, and breaking through ordinary and previously-formed deposits. Among numerous localities good sections are afforded at low tide on the Tamar, near Saltash, showing an association of the Devonian rocks, molten products, and other accumulations of igneous origin.† In many situations, the igneous ash of the time graduates into the ordinary

* Investigations are now in progress in the laboratory of the Museum of Practical Geology for the purpose of ascertaining the chemical composition of the igneous rocks of different dates, obtained during the progress of the Geological Survey of the United Kingdom.

† As we have elsewhere mentioned ("Report on the Geology of Cornwall, Devon, and West Somerset," 1839, p. 63), there is an abundant mixture of igneous and ordinary sedimentary rocks in the vicinity of Saltash and St. Stephen, and thence across the Tamar to St. Budeaux on the east, and towards St. Urney on the west, and in the creeks which run up from the Lyhner to Manaton Castle and St. Urney. The schistose varieties are certainly contemporaneous with the associated sedimentary deposits, while dykes of greenstone and other compounds of hornblendic and felspathic matter are seen to cut through the various accumulations, "analogous to those which are produced in the beds of ash, and filled by lava on the flanks of volcanos, in cases where the latter are partly submarine; traversing shales, clays, and other aqueous deposits, as well as the ash, which in such cases may readily have become interstratified among them." In the continuation of the beds near Saltash, many of the schistose accumulations of ash so graduate into the common kinds of deposit of mud and silt that no correct distinctions can be drawn between them.

detrital matter associated with it, as well in the continuation of a contemporaneous deposit as in successive deposits, in the one case pointing to a gradual removal from the source of supply, such as a volcanic vent; in the other, to an unequal supply over the same area, occasionally intermittent, so that common deposits were effected on a sea-bottom at intervals. Good examples of these kinds of igneous accumulations, with an intermixture of solid molten matter, some of the latter showing large-grained compounds of felspar and hornblende, are to be found in the direction of Davidstow and St. Oletha. In certain of the ash-beds much calcareous matter is sometimes found, assisting as a cementing substance. They are so calcareous near Grylls, on the south of Lesnewth, that attempts have been made to burn the compound rock for lime.

Upon attentively examining the composition of those beds of igneous substances which are arranged amid the ordinary deposits of the time, it is found to vary much as to the molten products associated with the general mass of deposits. While some, like the trachytic tuffs of modern times, are chiefly formed of the component parts of felspar, others are more like the dolerite tuffs, and contain substances usually found in augites and hornblendes, while others again partake of the character of both. Seeing that, like the ordinary muds, silts, and sands with which they are associated, they have become consolidated, and like them also have been exposed to the passage of water through them, as well when buried deep (by depressions of the general area) beneath their present levels, as when exposed, as now, to atmospheric influences, many of these rocks may not now contain all the substances originally distributed in them, while they, like the other deposits associated with them, may have received additional mineral matter. It will readily be inferred that soluble substances, such as the silicates of soda and potash, may have been removed during the long lapse of geological time in which they may have been exposed to modification and change. Though there is this difficulty, much may yet be accomplished by accurate analysis of portions carefully selected.

In Derbyshire the observer will again see igneous rocks associated with ordinary deposits; in this case with limestone, known as the carboniferous or mountain limestone, in such a manner that their relative geological antiquity can be ascertained. Careful investigation shows that in that area, at least, and probably much beyond it (beneath a covering of the sands, shales, and coals, known as the millstone grit and coal measures), and after a certain amount of these limestones had been accumulated, there had been an outburst and overflow of molten rock, irregularly covering over portions of them. And further, that after this partial overflow, the limestone deposit still proceeded, probably spreading from other localities where the conditions for its accumulation had continued uninterruptedly. Occasionally water action upon the igneous products may be inferred prior to the deposit of the calcareous beds upon them, if not also a certain amount of decomposition of the former, the limestones immediately covering them containing fragments (some apparently water-worn), and a mingling of the subjacent rock, such as might be expected if calcareous matter had been thrown down upon the exposed and decomposed surfaces of the igneous rock. In some parts of the district another outflow of the same kind of igneous rock again took place, and was again covered by limestone beds, so that in such portions of the area, two irregularly-disposed sheets of once molten rock are included among the mass of the limestone beds.

The following section (fig. 209) of part of this district,* by Professor John Phillips, may serve to illustrate the mode of occurrence of these beds of igneous rock, the areas of which do not coincide, so that one outflow did not exactly cover that overspread by the other. In this section, *a a* are the igneous rocks, locally known as *toadstones*,† and *b b* the limestones, *c* being the covering beds of millstone grit, and *f f* faults.

Fig. 209.

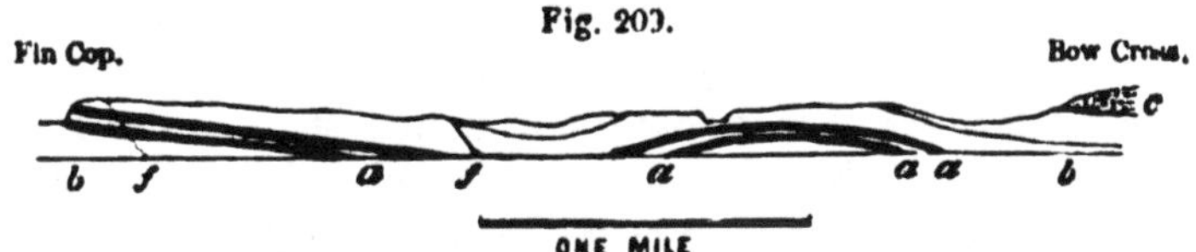

Natural sections (many of which are excellent) and mining operations show that as regards thickness these overflows vary considerably, so much so as to aid the observer in forming some estimate of the localities whence the molten matter, when ejected, may have been distributed around.

* A reduced portion of one (No. 18) of the horizontal sections of the Geological Survey of Great Britain.

† Professor John Phillips suggests that this name is a corruption of the German word *todtestein*; *rothe todte liegende* (red dead, or unproductive bed), being a term applied by German miners to the unproductive rocks subjacent to the copper-bearing slate of Mansfield and other localities. In like manner, the name *Barmaster*, given to those who superintend the distribution of the mines and collect the dues or royalties, has long been considered a corruption of *Bergmeister*.—*See* Pilkington's "Derbyshire," 1789, p. 110.

Although there are clays amid the limestones in the relative positions of the igneous rocks, and some of these seem clearly little else than such rocks in a highly-decomposed state, retaining the arrangement of their component mineral substances, as, for example, at the isolated boss of limestone at Crich, protruding (at a distance of 3½ miles from the main mass) from the squeezing action to which these rocks and the coal measures above them have been subjected, through the lower part of the latter, known as *millstone grit*, it would scarcely be safe to conclude that all lying nearly in the same general geological levels were so, inasmuch as some of them may be clays of another character. Care on this head is rendered necessary by finding a clay—a true underclay of the coal-measure kind,—supporting a thin bed of impure coal in the higher part of the limestone series near Matlock Bath.*

In the case of Derbyshire, though there may have been a removal of a portion of the igneous beds by the action of water upon their exposed surfaces (and an attentive examination of the upper overflow likewise shows a quiet adjustment of the limestone beds formed upon it), no deposits resembling the ash and lapilli beds above mentioned as found in Devon and Cornwall, Wales and Ireland, have yet been detected. There is no evidence showing an accumulation of ash and cinders in the manner of subaërial volcanos. If there had been such, and this had been attacked by breaker action and currents, the geologist would expect to discover some portions included amid the limestone beds, and such have not been found. It may readily have happened, therefore, that the igneous matter was thrown out in a molten state, without any accompaniment of ash and cinders; and this might have taken place as well beneath the level of the sea as above it.

Upon examining the structure of the igneous rock, it is found to be partly solid, and confusedly well crystallized, a compound of felspar and hornblende, with, sometimes, sulphuret of iron. It is partly vesicular, in some localities highly so; the vesicles, as usual, filled with mineral matter of various kinds,† where the rock has remained unaffected by atmospheric influences, but exhibiting the original and vesicular state of the molten rock where these have

---

* This impure bed of coal was cut while driving the tunnel through the High Tor, for the railway running by Matlock Bath, and is to be well seen, dipping rapidly, with the other beds, in the drift cut into the cavernous mine, part of which is shown by the name of the Rutland Cavern, at the Heights of Abraham.

† Carbonate of lime, as might be expected, is a very common substance in these vesicles.

removed the foreign substances in them. In some localities the scoriaceous character of the rock is as striking as amid many volcanic regions of the present day. Like more modern igneous products, also, it will often be found decomposed in a spheroidal form. The following (fig. 210) is an example of this decomposition at Diamond Hill, on the south side of Millersdale, where the concretionary structure has been developed somewhat on the minor scale, and the size of the spheroidal bodies is about that of bombshells and cannon-balls.

Fig. 210.

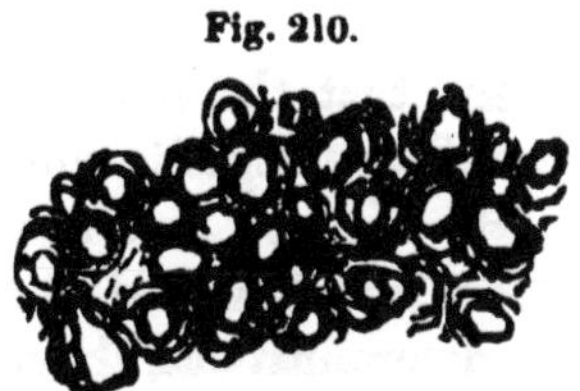

It will be thus seen that amid the older fossiliferous deposits igneous rocks may be so associated as to give the relative dates of their ejection, even in such a manner as to lead to the inference that in some cases there have been subaërial volcanic vents at hand, whence molten matter, cinders, and ashes may have been thrown out, as in the present day, the elementary substances of which this ejected matter is composed, reminding the geologist very strongly of those thrown out in a similar manner in modern volcanos. As has been stated (p. 553), it will require the observer to readjust in his mind the various parts of countries, like those noticed in Cornwall and Devon, Wales and Ireland, replacing the portions now removed by denudation, properly to consider this subject with reference to the relative times when the various igneous products were ejected and accumulated amid the ordinary sedimentary deposits of that early geological time. Let the following section (fig. 211) be one of a volcano, so situated that while lava

currents and dykes of molten matter (*a a a*) were thrown out and became mingled with subaërial tuff and volcanic breccias (*b b b*), subaqueous deposits were formed near and over these products, mingling volcanic and ordinary detritus in the same or associated beds.

If the volcanic action ceased, and the general area were depressed so that new and ordinary chemical or detrital deposits, *d d d*, were effected, and the whole was merely tilted, not complicating the subject with squeezing and contortion, and some new surface, *s*, *s*, be given to the general mass, as shown beneath (fig. 212), the ob-

Fig. 212.

server will at once perceive that the mode of occurrence of the igneous rocks amid the ordinary deposits will require careful consideration and study. He will see that a hasty investigation is not likely to afford the requisite data, and that prolonged research is needed for very exact determinations, though he may often find sufficient in a short time, if the natural or artificial sections be favourable, for a just general view of the subject.

When igneous products are not associated with ordinary fossiliferous deposits in the manner mentioned, and often, unfortunately, they cannot be so favourably studied, a geologist may still obtain certain relative dates by their mode of occurrence on the great scale. Fortunately, we may again take the British isles for illustration, as showing how much may be found connected with the subject even in that minor area. When the granite range of Wicklow and Wexford, and which also includes portions of adjacent counties, is examined with reference to the rocks in contact with it, it is seen that certain Cambrian and Silurian rocks, the range of which it traverses in a slanting manner, are upturned, much modified in their mineral structure, when in contact with the granite, and often much broken at the junction; even huge masses of them included in the latter, granite filling the cavities and fissures thus produced; so that little doubt is left that these rocks were formed prior to the intrusion of such granite. Thus far the observer merely obtains evidence of no very definite kind as to the actual period of this intrusion, though in the district noticed he would see that this kind of igneous action took place after that which in the same area produced an out-throw of felspathic and hornblendic products as above noticed (p. 553). He only discovers that the one set of igneous products has been uplifted by the other. Continuing his researches, he sees certain conglomerates of the old red sandstone reposing quietly upon the granite, and, when this happens, containing rounded portions of that rock, as well as much finer detritus from it. He also finds where the same conglomerate stretches over the disturbed older rocks, with their included igneous products, that rounded and angular fragments of these products are imbedded in it. He has now the approximate relative date of the granite of the district, so far that it rose up after that portion of the Silurian series was formed which is there disturbed, and prior to such portion of the old red sandstone series as is represented by this conglomerate. We will suppose that he has obtained evidence of a portion of the Silurian series disturbed being the lower, and of the conglomerate representing some higher or middle portion of the old red sandstone series, as found developed elsewhere in the British islands. There would then, no doubt, be something of an interval in the geological series, during which the uprise of the granite may have taken place, nevertheless the observer has, by the means employed, arrived at a certain approximation of no slight value as to the real relative date of its protrusion.

To show this value, it is only needful to turn to Devon and Cornwall, where at such a comparatively trifling distance, the geologist finds a granite of much the same general character protruding through the equivalents of those accumulations which have quietly covered the Irish granite mentioned, after its consolidation, the disturbance caused by the uprise of the Devonian and Cornish granite extending to the lower portion of the coal measures, as may be seen around the northern part of Dartmoor, where veins extend from the granite in that direction into these sedimentary rocks, in the same manner as into the Silurian deposits of Wicklow and Wexford. In the case also of the granites of Cornwall and Devon, it becomes necessary to seek for evidence as to any deposits so occurring as to show the geological dates between which their uprise was effected. Throughout the greater part of the district, evidence of the kind required is not to be found, but on the eastward of Dartmoor, and of the continuation of the deposits which have been disturbed at the time these granites were intruded, beds are found, known as the new red sandstone series, reposing quietly on the disturbed rocks, the lower portion of them containing rounded and angular fragments of the latter. It would thus appear that the approximative date for the elevation of the Cornish and Devonian granites amid the accumulations effected up to that time, was somewhere between the lower part of the coal-measure series (including the millstone grit of central England in that series), and the lower portion of the new red sandstone deposits.

Thus in south-eastern Ireland and south-western England there is evidence of two protrusions of granite at different geological periods, different rocks of known relative ages being disturbed on the one hand and unmoved on the other, so that approximative dates are obtained for both protrusions. If in the annexed section (fig. 213) *a*, *a*, be a mass of granite thrust upwards through sedi-

Fig. 213.

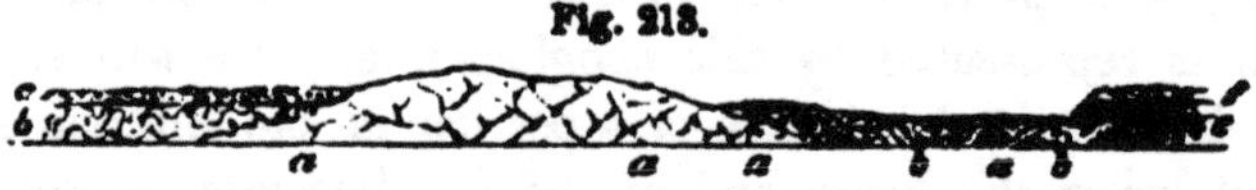

mentary beds *b b*, sending veins into fractures effected in them, as well as modifying their mineral structure at the junction, and *c* be an accumulation containing rounded or angular fragments of *a* and *b*, it follows that the relative geological dates of *b* and *c* being known that of the protrusion *a*, *a*, would be known, also, within greater or less limits as the formation of *b* and *c* may be separated or approximate to each other in the geological series. This would be the case of south-eastern Ireland. In that of Devon the disturbed beds *b b*, altered as before, and with granitic veins *a*, *a*, in them, would be covered by beds *f* reposing quietly on them, and also containing fragments of them, with here and there igneous rocks, *e*, interposed.

Usually the relative dates of the rise of molten mineral substances into fissures of prior-formed rocks, such portions of igneous matter, known as *dykes*, cannot be obtained when these are uncovered by accumulations of which the position in the geological series is known; as, for example, if, in the subjoined section (fig. 214), *a* and *b* be dykes of any igneous rocks cutting through

Fig. 214.

some sedimentary deposit *c d*, and these be uncovered by any accumulation of ascertained geological date, the exact relative time when the cracks were effected and the molten matter rose in them would remain uncertain. It sometimes happens, however, that some evidence as to relative date may be obtained, of a fair approximative kind, even with respect to dykes of this character. It would not be sufficient that they cut one set of rocks, and not another, in some given district, without further general evidence, so as to refer them with certainty to a particular time, anterior to the formation of the beds not cut by them, since it may have happened that contemporaneous causes did not act beyond a given area, though in certain of these cases there may appear much to support an inference to that effect. For example, numerous greenstone dykes are found to traverse the Cambrian rocks in Merionethshire and Caernarvonshire, while these are not observable amid certain upper Silurian deposits in Denbighshire and Flintshire, and contemporaucous igneous rocks are associated with intermediate accumulations in Caernarvonshire, and other adjacent counties. It might hence be inferred that, when the igneous eruptions producing the latter were effected, fissures were formed in the still more ancient deposits (Cambrian) and molten matter injected into them, and that igneous action ceasing, the adjoining higher parts of the Silurian deposits were undisturbed by the intrusion of any igneous matter. It is far from improbable that this inference would, in a great measure, be correct; but that it is not wholly so, the inspection of dykes of the same kind traversing various parts of Anglesea, and seen to cut into the coal measures of the Menai Straits, between Bangor and the great suspension bridge, at once shows. It may readily have happened that igneous matter had been thrown into fissures formed at these different times in even the moderate area of Caernarvonshire and Anglesea, and hence it would be hazardous, without other evidence, to decide upon one dyke being separable in geological time from another, even when not far distant from each other, at the same time that many probabilities might seem to exist as to the relative date of some of them.

The granitic and porphyry dykes in Cornwall and Devon, known locally as *elvans*, may be taken in illustration of the approximation to relative geological dates occasionally attainable. It has been seen that the granites of that district were upraised posterior to the deposit of the lower part of the coal measures, and anterior to that of the new red sandstone series. Subsequently to the protrusion of the granite, and to, at least, its partial consolidation, fissures were formed traversing both the granites and the various disturbed sedimentary rocks adjoining them, and into these fissures molten matter was introduced, as shown previously (fig. 7, p. 9), and as may be further illustrated by the following section (fig. 215), seen on the cliffs at Trevellas Cove, near St. Agnes, where an elvan *a*, *a*, cuts through the slates *b*, and is traversed by dislocations *f*, *f*, one of which materially shifts the rocks, and thereby displaces the elvan dyke, near the sea. With respect to the same fissures having traversed both the previously-consolidated rocks and the granite, the

Fig. 215.

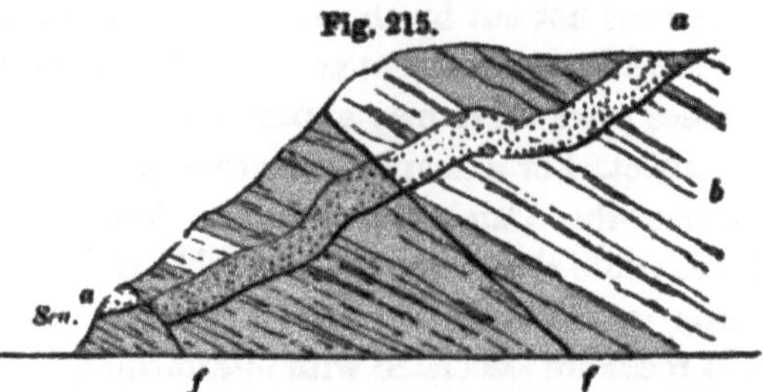

following map (fig. 216) of part of the mining district of Gwennap, Cornwall, may be useful, *a*, *a*, being the granite, *c*, *c*, the schistose

Fig. 216.

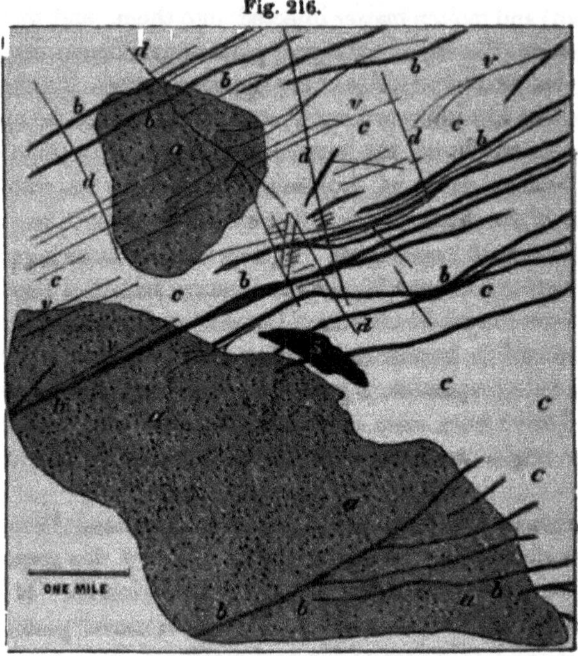

rocks broken through by it, *b*, *b*, the elvan dykes, and *s*, greenstone. The fissures *v*, *v*, *v*, and *d*, *d*, *d*, were produced at different subsequent periods, some of them variously filled by the ores of tin and copper, or other substances, and known (locally) as *lodes* and *cross courses*.

Upon examining the composition of these elvans, they are found to be formed of matter similar to that of the granites of the districts, usually corresponding with any modifications observable in patches of that rock exposed nearest on the surface. Indeed, they seem merely portions of the same general matter which rose in fissures formed by the cracking of the adjacent granite, only consolidated in its higher parts, such cracks also extending through the various rocks above the granite. The relative date would be only so far thus obtained as to show that the filling of the fissures was posterior to the intrusion of the main masses of granite, some of the latter rock, in its molten state, readily rising into such fissures, formed both in its own higher parts, and in any covering rocks.*

Proceeding eastward from the Dartmoor granite to the boundary of the new red sandstone series, where this reposes on the uneven surfaces and indentations of the older and previously-disturbed fossiliferous deposits in that direction, igneous rocks are found associated with its lowest part in some localities, pointing to local igneous action, while these lowest beds were accumulating. Not only are some of these lower accumulations so entangled with the igneous rocks, that there appears difficulty in not considering them of contemporaneous production as a whole;† but there would also appear to be traces of subaërial action. The latter seems to occur near Calverleigh, where, as in the annexed section (fig. 217), *a*, *a*

Fig. 217.

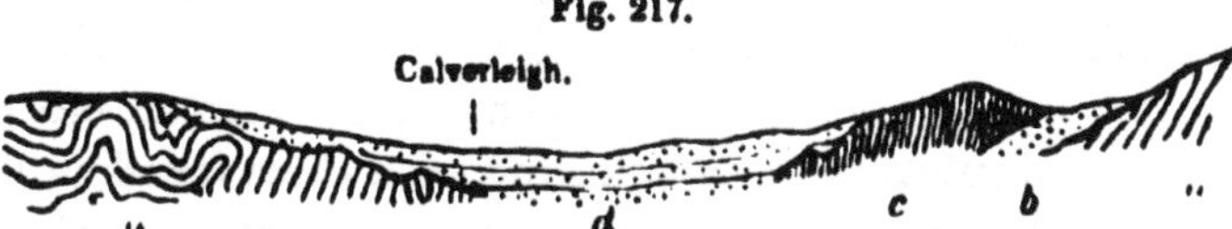

represent the disturbed beds of the lower coal measures, at part of an ancient gulf amid those rocks; *b*, a conglomerate wholly composed of portions of these subjacent deposits, cemented by red sandstone and argillo-arenaceous matter, without any fragments of igneous rocks; *c*, felspathic porphyries, and more compact felspathic rocks, some scoriaceous; and *d*, conglomerates and sandstones, fragments of the igneous rocks, and others of a similar character, being contained in the conglomerates. Along the range

* Occasionally fragments have been detached from the adjacent rocks, and enveloped in the molten matter of the elvans. That at Pentuan is among the best examples of this circumstance. This elvan is a fine-grained compound of felspar and quartz, with crystals of mica. Fragments of the slate rocks traversed are found in it. Occasionally, though rarely, there are portions of quartz which appear to have been broken off some quartz vein in the slates, and thus became, like the other fragments, included in the molten rock. In a branch of the Pentuan elvan, taking a course alongshore to the Black Head, the fragments derived from the adjoining rocks are very numerous, decreasing in abundance from the sides of the dyke towards its central part, in which they are rarely detected.

† The intimate connexion of igneous rocks and the red sandstone series at Thorverton and Silverton was pointed out, in 1821, by the Rev. J. Conybeare, "Annals of Philosophy," new series, vol. ii., p. 161.

of the igneous rocks, particularly on the north of them, there is an arenaceous deposit, here and there mingled with the ordinary sandstone, which bears a great resemblance to a volcanic product, so much so as to lead to the inference that it had been ejected in the manner of volcanic ash, and that, falling into water, it had been mingled with the mud, sand, and gravel, adjoining some volcanic vent of the time.*

The igneous rocks of this date can be well studied at the base of the new red sandstone series from Exeter to Haldon Hill. They are seen at Pocombe Hill, resting directly on the edges of the disturbed and subjacent coal measures, and are chiefly formed of a siliceo-felspathic compound, with occasional though not numerous vesicles. These igneous rocks are also well exhibited between Ide and Dunchidiock, resting on similar accumulations. Near Western Town, the intimate connexion between them and the red sandstone and conglomerates can be seen. By reference to geological maps, it will be observed that the igneous rocks thus associated with the lower portions of the new red sandstone series near Exeter, Crediton, Thorverton, Kellerton, Silverton, and even near Tiverton, have been thrown out in a prolongation of the general direction of the granite bosses and elvans extending from the Scilly Islands to Dartmoor. By examining their component parts, they are observed to be formed of substances corresponding with those found in these granites and elvans. While many of them present a porphyritic character, others are more homogeneous in structure, and sometimes vesicular. Much of the lower new red conglomerates and breccias in the neighbourhood of these igneous rocks is composed of fragments derived from them, so that these fragments, if again gathered together, would constitute no inconsiderable mass. Among them many porphyries are found, as well containing quartz as felspar. Masses of the igneous rocks from which they are derived are not often observable, though in such a district the portions visible on the surface afford no measure of the igneous masses which may be buried beneath a thick covering

* The facts in this locality would appear to show, that along a range of ancient coast, of a date corresponding to the first production of the new red sandstone series of Devon and Somerset (see Maps of the Geological Survey, Sheets 20, 21, 22), there was (1) a subaqueous valley, or depression, among the disturbed coal measures, there occurring, the partial abrasion of which, by breakers on the shores adjoining, produced (2) the shingles, and other detritus, now forming a conglomerate. Subsequently (3), igneous products were accumulated, probably ejected from a neighbouring vent, which, with others in South Devon, were then in action; and finally a partial destruction of these rocks affording (4) some of the materials for a conglomerate, afterwards formed.

of detrital matter. It is, therefore, important to observe porphyries in place, sometimes only containing quartz crystals; at others, these mingled with crystals of felspar, associated with the lower part of the new red sandstone series, at Ideston and Knole.

Weighing all the facts thus observable, the geologist might be led to infer that the date of at least some of the elvans of Cornwall and Devon, though they are uncovered by deposits affording direct means for approximating to the time when they rose in the fissures where they are found, might not very materially differ from the commencement of those accumulations which constitute the lower portion of the new red sandstone series of that part of England, granitic matter constituting the base of the various rocks ejected, and being merely modified in its aspect according to the varied conditions to which it had been subjected. So much denudation has taken place in this region since these ancient igneous rocks were ejected, that no doubt many a mass showing any connexion which once existed between such igneous rocks as those near Exeter and other adjacent parts of Devonshire has been swept away. As illustrating a denudation of deposits of the new red sandstone series in Devonshire, so that a portion of them only now remains, we have already noticed the Thurlestone rack in Bigbury Bay (fig. 47, p. 52), a detached piece of the small patch there occurring. Proceeding still further westward to Plymouth Sound, a porphyritic rock, of the same general kind as those which are found near Exeter, is seen cutting through the Devonian rocks at Cawsand, and on the coast thence towards Redding Point,* forming, as it were, a sort of connecting link between the elvans more westward and the igneous rocks above noticed, and appearing to constitute the denuded remains of the lower part of the new red sandstone series, extending, with an admixture of igneous products, in this direction, a small patch still remaining of the old continuous deposit at Bigbury Bay, and at Slapton, in Start Bay.

With respect to the elvan dykes in the counties of Wicklow and Wexford, which in their mode of occurrence and aspect resemble those of Devon and Cornwall, though an observer does not appear to possess the same opportunities of inferring their relative dates, inasmuch as igneous rocks, composed of similar substances with

* This porphyritic rock is a compound of felspar and quartz, containing crystals of mica, and, more rarely, of felspar. It is of a somewhat earthy character, probably, from the effects of decomposition. The colour is reddish, as a mass, mixed occasionally with spots of bluish green.

these elvans, have not hitherto been detected in the lower part of the old red sandstone covering up the disturbed rocks in which the fissures, filled by them, have been effected; still, as the old red sandstone contains portions of the granite of the district, and is uncut by the elvans, it might be inferred that the date of these elvans was not only posterior to the granite, but also anterior to the old red sandstone. They are to the granites of this part of Ireland what the elvans of Devon and Cornwall are to the granite of that part of England. They seem the result of cracks from the cooling and solidification of a crust, so to speak, of the molten mass beneath, such cracks passing through superincumbent rocks adhering to this cooled and solidified crust. The elvans of Wicklow and Wexford can be well studied, not only inland but on the coasts. Good examples of their mode of occurrence at the latter are to be found at Seapark Point, Wicklow.

Of the two classes of igneous rocks above noticed, the one chiefly differs from the other chemically, in the presence, in part of one class only, of a larger proportion of lime and magnesia, these sometimes replaced by oxide of iron. This difference is principally confined, as a whole, to that portion of one class which contains the mineral named *hornblende* in which the silicates of lime and magnesia, though somewhat variable in quantities, form marked ingredients, the lime alone mounting to from 10 to 15 per cent. of that mineral, and the magnesia varying from 15 to 25 per cent. Usually in these rocks the protoxide of iron more or less replaces some of the lime or magnesia of the *hornblende*. The presence of hornblende, when in proportions extending even to $\frac{1}{4}$th or $\frac{1}{3}$rd of the mass, renders the rock in which it thus occurs far more fusible than the compounds of felspar and silica, or of felspar, quartz, and mica, a difference due probably, in great measure, to the silicate of lime acting as a flux.

In the other igneous rocks, those which have been ejected in a molten state (not referring to those which have been noticed under the head of modern volcanic products), and in the first place confining our attention to the great mass of them composed of two or more of the minerals named quartz, felspar (whether orthoclase or albite), mica and hornblende, as chief and prevailing substances, neither in the compounds of quartz and felspar, nor in that of quartz, mica, and felspar (orthoclase or albite), is there the same amount of lime as when hornblende enters into the mass. The prevailing mica in such rocks seems to be that commonly termed potash-mica, from that substance being a marked ingredient in it.

In this mineral the lime is usually in very small quantity, commonly under 1 per cent. In the lithia and magnesia micas it is rare, and, when found, has been so only in very small proportion. In the felspar, also, when either orthoclase or albite, members of this family apparently much distributed amid the older igneous rocks, lime has only been detected hitherto in small quantities, rarely in proportions equal to 1 per cent.* In compounds wherein labradorite is found, the case would be different, since this is a felspar in which lime usually occurs in comparatively large proportions, from 10 to 15 per cent. Silicate of lime, therefore, would appear to constitute a marked source of difference between the igneous rocks with and without hornblende and labradorite. With respect to the magnesia in many hornblendes, this also would be a substance of importance when compared with compounds of quartz, felspar (either orthoclase, albite, or labradorite), and mica, unless the latter were magnesia mica, into which this substance is found to enter in proportions varying from 10 to 15 per cent., or those varieties of felspar which have been referred to orthoclase, and yet contain from 10 to 20 per cent. of magnesia.

As the trachytes of more modern geological times have been inferred to be some modifications of granites (p. 360), the observer might be induced to inquire how far the old igneous products noticed as occurring like certain of those of the present time may, in like manner, have been modifications of granitic matter beneath them; how far, in fact, certain of the molten felspathic rocks of the British islands associated with the older fossiliferous deposits may have been the trachytes of those times, and have been derived from granitic matter below them, such granitic matter afterwards upheaving these earlier modifications of portions of it, when geological time advanced and with it conditions for such a movement. With the hornblendic compounds there would be the same difficulty as with the modern dolerites and lavas of that class, so far as the silicate of lime was concerned, though in both cases, supposing silicate of lime to form a marked part of a fused mass, that it should be ready, as a substance aiding in fluxing others, to be thrown upwards, might be anticipated from the conduct in our furnaces of the slags into which silicate of lime largely enters.

* Dr. Abich found 1·26 of lime in the orthoclase of the trachyte of Pantellaria, and 2·06 in the basis of the Drachenfels trachyte. The orthoclase of the older igneous rocks has not hitherto afforded any proportion of this kind, though at the same time, it must be confessed, that the igneous rocks of that date have not, as yet, received sufficient extended examination to arrive at any accurate results as to the chemical composition of the greater masses.

Respecting the compound of the matter of felspar and an additional quantity of silica, beyond that required for the silicates in the minerals of that family, good opportunities are often afforded for studying the variable aspect it assumes as the conditions for cooling may have been favourable, or otherwise, to the crystallization of the felspar. When cooled so that the crystallization is not apparent, the compound has a homogeneous aspect, and is commonly known as *compact felspar;* when confusedly crystallized and silica is well separated, as quartz, from the other ingredients, it forms one of those binary granitic mixtures sometimes termed granitello and pegmatite. Occasionally crystals of felspar being developed while the remainder of the rock retains its homogeneous character, a variety of hornstone porphyry is produced.* The variable aspect of the less crystalline varieties may be seen in numerous situations, the complete crystallization not so frequently.† In countries in which granitic matter has upheaved the prior superficial accumulations of this class, the resemblance of some kinds of products is sometimes so considerable as occasionally to lead to much ambiguity respecting their relative dates.

* Of a columnar mass of the latter, the columns in part somewhat bent, a good example may be seen among the igneous products associated with the Silurian series west of Knock Mahon, on the coast of Waterford.

† A good example of a binary compound of quartz and felspar may be found among the igneous rocks amid the Cambrian series, close to the town of Caernarvon, on the northward; part of a portion of molten matter, in which the more common homogeneous mode of occurrence of the silicates of the felspar combined with the quartz, prevails.

CHAPTER XXX.

MODE OF OCCURRENCE OF GRANITES IN SOUTH-WESTERN ENGLAND AND SOUTH-EASTERN IRELAND.—GRANITE VEINS.—CHEMICAL COMPOSITION OF GRANITIC ROCKS.—SCHORLACEOUS GRANITES OF CORNWALL AND DEVON.—SLIGHT COVERING OF OTHER ROCKS OVER THE GRANITES OF WICKLOW, WEXFORD, AND CORNWALL.—SERPENTINE AND DIALLAGE ROCK OF CORNWALL.—SERPENTINE OF CAERNARVONSHIRE—OF ANGLESEA.—CHEMICAL COMPOSITION OF SERPENTINE—OF SERPENTINE AND OLIVINE COMPARED.—COMPOSITION OF GREENSTONE AND SYENITE.—RESEMBLANCES AND DIFFERENCES OF ORDINARY GRANITIC AND HORNBLENDIC ROCKS.—GRANITIC, FELSPATHIC, HORNBLENDIC, AND SERPENTINOUS ROCKS EJECTED AT VARIOUS GEOLOGICAL TIMES.—RELATIVE FUSIBILITY OF IGNEOUS ROCKS.—MODIFICATION OF THE MATTER OF IGNEOUS ROCKS.—ADDITIONAL MINERALS ENTERING INTO THE COMPOSITION OF ORDINARY IGNEOUS ROCKS.—GENERAL CHARACTER OF IGNEOUS ROCKS.

WHETHER the observer studies the granite of south-western England, or that of prior elevation in south-eastern Ireland, he finds the same general mode of occurrence, one very different from that of the igneous products associated with the Devonian rocks in one district, and the Silurian rocks in the other. There is no interstratification and contemporaneous intermingling of parts, but, on the contrary, evident protrusion in mass, and a subsequent filling of fissures traversing the beds of pre-existing deposits. In both districts, the granitic protrusions appear the accompaniments of great contortions, foldings, and even dislocations of prior accumulations of all kinds, as if, amid this squeezing and new adjustment of such accumulations, molten matter beneath rose upwards (there being sufficient pressure upon it), and occupied areas where the resistance of any prior superficial covering was insufficient to resist this intrusion.

Upon examining the boundaries of the granitic masses observable on the surface, the amount of fractures affected around them, and in the various rocks adjoining, is found to be considerable. Indeed, where opportunities are afforded either by natural exposures or artificial sections, they are seen to be common. Thus, independently of any great movements or dislocations of prior-formed rocks of all kinds, the margins of the granitic intrusions are themselves marked by abundant fractures on a minor scale, as if those intrusions had themselves in some measure been connected with their production. As to the extent of the fractures into the adjoining and prior-formed rocks, it may be considered as somewhat insignificant when regarded with reference to their mass and that of the granites. In the range of the Wicklow and Wexford granite, not only are these cracks found abundantly, but evidence is also afforded of huge detached masses of detrital rocks being apparently embedded in the external parts of the same granite. This can be well studied in Glenmalure, where such great masses seem as if partly contained in the granite, having floated on that rock when in a molten state, like great icebergs in the sea, and like them also in part submerged. No doubt this may be only appearance, as the parts connecting these masses may have been removed by denudation. At the same time when sections are made of the whole on a scale equal for height and distance, and all the foldings of the older rocks are considered, a great breaking up of the latter seems needed to account for the mode of occurrence of all the rocks. No doubt that much of both the older rocks and the granite of south-eastern Ireland has been removed by denudation effected during a long lapse of geological time, often by abrasion from heavy breaker action, while rising above or descending beneath the ocean level;* yet there still appears to have been disruption of the prior-formed Cambrian and Silurian rocks. The curve, which agrees with the

* It is not a little interesting, in this part of Ireland, to study the denudation with reference to the exposure of both the granite and altered sedimentary rocks (for they and certain associated igneous products of that date are much modified and altered, as will be hereafter noticed) to the same degrading forces. The granite is often of a decomposing kind, while the altered rocks are, for the most part, tough; hence the exposure of both to the same abrading force has caused the softer substance to be worn away more than the harder. In consequence the tough altered rocks have been the means of preserving much of the granite beneath them from removal. Lugnaquilla, the highest of the range, is capped by these altered rocks, now chiefly mica slates; and many other examples of heights and flanks of mountains thus preserved may be seen. When this denudation is also studied with reference to an Atlantic exposure, the interest is not lessened, inasmuch as the western flanks of the mountains point to more abrasion on that side than on the east, just as would happen from the destructive influence of Atlantic breakers, rolling in, as now, from the westward. It requires very little imagination, when standing upon some parts of this range, to fill up the lower ground with sea, so that the Atlantic may break upon the cliffs beneath, facing the west. In the range of mountains near that named Blackstairs, the cliff character of the western flanks is very marked.

upraised masses of the prior-formed rocks, and fortunately many of these are still preserved, showing the probable extent to which they have been so raised to a height above those crumpled and folded on either side, is of the kind represented beneath (fig. 218). This may not be considerable, yet it seems difficult to obtain the effects produced without much separation as well as disruption of parts of the older rocks. In this section, upon the same scale for

Fig. 218.

heights and distances, *a a*, is the intruded granite, *c c*, the contorted and older rocks on either side, altered near the granite, and *b b b*, portions of them uplifted, a large mass forming the summit of (L) Lugnaquilla.

Upon examining the contents of the cracks in the prior-formed rocks surrounding the granitic masses, they are found filled with the granite in such a manner as to show the comparative liquidity of that substance when the cracks were made and filled, for even fine threads may be occasionally seen, branching out of the main cracks, with granitic matter in them. Though, as might be anticipated, the crystallization of this matter is modified in the finer fissures, from differences of the rate of cooling alone, the contents of the granitic veins generally would point to long-sustained heat among the intruded rocks, the whole having probably required a considerable lapse of time for solidification. The following sketch (fig. 219) of some granitic veins at Wicca Cove, or Pool, near

Fig. 219.

Zennor, Cornwall, may serve to illustrate the mode of occurrence of many of them, and the annexed section (fig. 220) will show their connexion with an adjacent mass of granite behind the rocks exposed in the sketch (fig. 219), *a a*, being the granitic veins, *b b*, altered slate, and *c* the main mass of granite.

The west coast of Cornwall, exposing the junction of the granite of that district with the sedimentary deposits and the igneous rocks

Fig. 220.

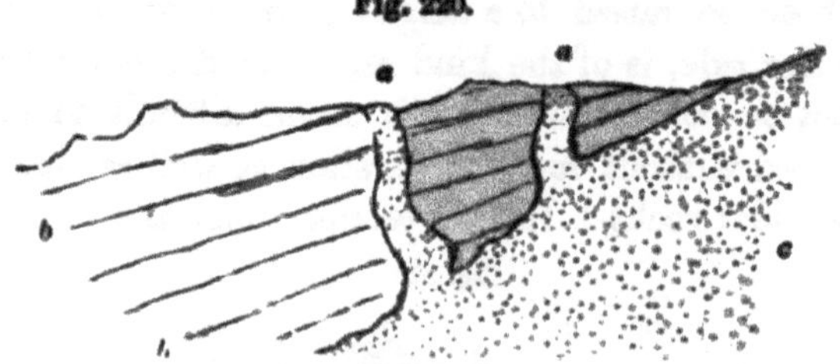

associated with them, offers many other illustrative instances, as at Pendeen Cove, Cape Cornwall, Tetterdu Point, Mousehole, and other places. They are also as well and easily seen at St. Michael's Mount. The following sketch (fig. 221), exhibits the section of a

Fig. 221.

somewhat complicated fracture, the shaded parts (*a a a*) being altered slates, and the dotted portion granite, the mass of the latter occurring on the side *b*, *b*. Looking at these veins as a whole, it would often appear as if the prior-formed rocks had not yielded very slowly to the force applied, but in a comparatively sudden manner, the granitic matter being driven into the cracks, formed by heavy pressure, so as to fill up the fine fissures.* There often also seems evidence of cracks having been formed after only a mere comparative

* Portions of these rocks are sometimes found completely isolated in the matter of the granitic vein.

film of the main mass of granite had been consolidated, granitic veins similar to those amid the prior-formed rocks, and clearly merging into the main mass of granite, being found alike to traverse a certain amount of the external parts of the granite and these other rocks.* In general such veins are easily to be distinguished from the elvan dykes (the result apparently of subsequent action) by their tortuous courses, and by their general resemblance to those first formed amid the older rocks at their junction with the granite.†

The chemical composition of granitic masses will necessarily engage the attention of the observer, more especially when he considers that so much of the detrital deposits of all ages have been derived from granitic matter; indeed, the volume thus distributed as detrital accumulations must be enormous. As has been seen, the elementary substances forming the chief part of the volume of this rock do not appear to be numerous. For certain of the modifications of mineral structure it may be again desirable to refer to the portions of the British islands already noticed, since the relative ages of the igneous rocks in them are so well shown. Fundamentally, the constituents of the granites in south-western England and south-eastern Ireland seem little different. The chief variations may probably consist in the greater admixture of schorl with the other constituent minerals in the former than in the latter; indeed, generally speaking, schorl is rare in the granites of southern Ireland. Such differences can readily be considered as merely local, the same molten matter beneath having supplied the portions upraised at different geological times. Be this as it may, the presence of a mineral in any abundance which contains boracic acid as an essential ingredient,‡ is one of importance, more particularly when we refer to the researches of M. Ebelmen, he having shown that by

* Instances of this kind are not uncommon, both in south-eastern Ireland and south-western England. They are well exhibited at Killiney Hill, near Dublin, and the large masses of granite brought from thence for the harbour at Kingstown often show them. They are also to be well seen in the granite of the Scilly Islands, and the exposed granites of the Land's End coast, as at Tol-Pedn-Penwith and Lamorna Cove.

† In examining granitic countries it is very needful not to confound the filling of joints in granite with quartz, felspar, and mica, in the manner of fissures, including mineral veins, with the granitic veins noticed in the text; such modes of filling being very deceptive, unless due care be employed. They can, however, be usually well distinguished by the manner in which the minerals occur in them, showing a deposit from solutions against the walls of the granitic fissures, the crystals pointing inwards, and arranged in the manner of many common and mineral veins.

‡ The analyses of M. Hermann give about 10 per cent. of boracic acid in schorl, 39 of silica, 31 of alumina, a variable quantity of protoxide of iron (4 to 12 per cent.), 2 to 9 of magnesia, with a few other subordinate, and, probably, accidental substances, such as lithia, soda, and potash.

employing that acid as a solvent, at an elevated temperature, minerals may be produced by the evaporation of this solvent, some of them gems, such as rubies, which are usually termed insoluble, and infusible in our furnaces, a result having a considerable bearing upon the production of many igneous compounds.*

Cornwall and Devon present frequent and good opportunities for the study of schorlaceous granites and rocks composed of schorl and quartz (usually termed *schorl rock*) in connection with them. As might be expected from the comparatively easy removal of boracic acid by considerable heat, the chiefly schorlaceous compounds are found at the extreme parts of the granitic masses. They vary from a simple binary compound of schorl and quartz to mixtures of schorl, felspar, quartz, and mica; the latter is, however, not an usual ingredient in the granitic rock when schorl is present in any abundance. Complete passages may frequently be traced between the ordinary compound of quartz, felspar, and mica, by the gradual loss of the felspar and mica, into the simple mixture of quartz and schorl, the mica being commonly the first to disappear. The schorl sometimes presents itself in radiating bunches of crystals, especially amid the quartz.† Here and there different arrangements of schorl

---

* The researches of M. Ebelmen on this subject are marked by the true spirit of philosophic investigation. He sought for a substance which at a high temperature acts like water, as regards others dissolved in it. As by the evaporation of water certain crystalline bodies might be formed, so, he inferred, that by employing those which could be volatilised at high temperatures, yet at a given heat, while in fusion, be capable of dissolving the greater part of metallic oxides, certain calculated proportions of some oxides would crystallise, when the dissolving body was evaporated in open vessels at a great heat. Acting upon this view, and selecting boracic acid as the solvent, he was completely successful, producing rubies, sapphires, spinels, chrysoberyl, chrysolite, chromate of iron, and others. Crystals of emerald were formed from pounded emeralds, when fused with boracic acid and a little oxide of chromium. The crystals of chrysoberyl were sufficiently large to have their optical properties tried, and these were found to be identical with those of the natural mineral.

† Good examples of nests of schorl in quartz, the crystals radiating, may be seen in the Dartmoor granite, as above Bowdley, near Ashburton. Schorlaceous granite and schorl rock can be also well seen in the same granitic district at Holne Lee, and on the south of the moor, as also near Tavistock. The granite of the Brown Willy mass is not so schorlaceous, though schorl is found, especially towards the south. Near St. Cleer, there are compounds of schorl, felspar, quartz, and mica, similar to some found on Dartmoor. The St. Austell granite is much more schorlaceous, veins of that mineral being common in it. The decomposed granite of that district, furnishing so much clay to the porcelain works of England, is extremely schorlaceous. Singular stripes of schorl rock are found at its outskirts, as between Watch Hill and Longlane; on the north and south of Burthy Row, near St. Enoder, and at the long-celebrated Roche Rock. Near Meladore there is an interesting mixture of schorl and quartz, containing large crystals of felspar, some of these decomposed, and crystallized schorl introduced into the cavities left by them. At Calliquoiter Rock there are variable mixtures of schorl, quartz, felspar, and mica, the outside portions formed of the two former. The granite of St. Dennis Hill is in like manner a compound of these four minerals. The Carn Meneles granite is not so schorlaceous,

and of the other minerals are observed. The following (fig. 222) is a somewhat marked instance of the adjustment of varied com-

Fig. 222.

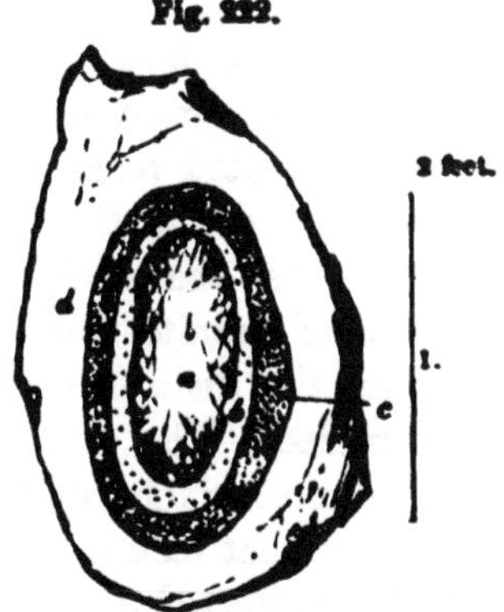

pounds round a kind of central nucleus. It occurs in the Dartmoor granite, towards Carnwood. *a* is a cavity not quite filled by long crystals of schorl, crossing in many oblique directions, but with a general tendency towards the centre; *b* is an envelope of quartz and schorl, the former predominating; *c*, another covering of the same minerals, the schorl being more abundant; and *d*, a light flesh-coloured granite, the felspar predominating.

Large crystals of felspar are not uncommon in the granites of Cornwall and Devon, rendering the rock a porphyritic granite. That of Dartmoor is not unfrequently of this character, as is also the granite of the Brown Willy district, and the same variety may be seen in many other localities.* The granites of south-eastern Ireland are also occasionally porphyritic, from the distribution of felspar crystals amid the ordinary triple compound of quartz, felspar, and mica.

Throughout these districts, though the granite may enter the fractures of the adjacent and prior-formed rocks, there is no trace of an overflow of the igneous matter in a molten state, so that the observer is led to infer that, when the intrusion was effected, the

---

though schorl is found, and more especially at the confines of the mass. The Land's End granite is schorlaceous to a considerable extent. A variety of schorl rock, composed of a base of schorl and quartz, with large crystals of felspar, is found close to Trevalga, near St. Ives. Here also, in some parts, the crystals of felspar have been decomposed and removed, and the cavities more or less filled with crystals of schorl.

* As chiefly differing from the ordinary granite, that of St. Austell is probably the most marked, a steatitic mineral therein replacing mica to a great extent, particularly in the portions which are found in a decomposed state. Much pinite (a silicate of alumina and magnesia, the latter partly replaced by protoxide of iron) is mingled with a part of the granite near the Land's End.

igneous rock was not as now exposed to the atmosphere, or beneath waters in such a manner that it could pass beyond the broken portions of the deposits now forming its superficial boundaries, and flow over them in the manner of lava discharged from a volcanic vent. If any portion of these granites did so pass over prior-formed, consolidated, and disrupted rocks, all traces of such overflows have been removed by denudation. Molten matter in a sufficiently fluid state to enter the smaller ramifications of the cracks around the masses of granite, would readily, if elevated sufficiently high, overflow the disrupted and contorted deposits amid which it was protruded. The covering of the granite of south-eastern Ireland is comparatively slight; the whole district adjoining the main masses of that rock is so pierced and cut by it, as to show upon the surfaces exposed, that the whole of the prior-formed accumulations has been upborne, so that upon the denudation of the various inequalities, the granite was unequally exposed.* The same may be said with reference to south-western England between Dartmoor and the Scilly Islands..

There is yet another igneous product in a part of this limited area to which a relative geological date may be assigned. This product is serpentine, which is chiefly found in considerable abundance in the Lizard district, in Cornwall. It is seen among the Devonian rocks in a manner reminding us of the mode of occurrence of some of the contemporaneous compounds of felspar and hornblende, which have been associated, in a molten state, with the sedimentary deposits of that date. That it was vomited forth anterior to the granite of the district, would appear from its being traversed by veins of that rock, in the same manner that other rocks of the district are traversed by them. Even allowing that these veins may be of no greater antiquity than the elvans of the same county, this would limit the fissures for their introduction to about the age of the lower new red sandstone deposits of that land. At Clicker Tor, south of Liskeard, serpentine is found amid Devonian slates, and near Veryan, *diallage rock* (diallage and felspar) is seen associated with similar serpentine, and in a manner pointing to an ejection of these rocks in the same way as certain greenstones amid accumulations of the igneous products of the

* The granite of the island of Anglesea, probably of about the same date, is also interesting, as showing how readily it might be concealed from superficial exposure, by a somewhat more thick envelope of the Silurian rocks through which it has risen. Indeed some of the portions exposed are merely minor inequalities cut into by denudation.

district. The position of the Lizard serpentine, and the diallage rock found with it, seems much the same with these minor portions of serpentine more eastward. The Lizard serpentine occupies a somewhat large area, reposing upon hornblende slates and rock, which appear little else than the ordinary volcanic ash-beds above mentioned as intermingled contemporaneously with the ordinary detrital deposits of the time and locality (p. 558).* There is often an apparent passage from the diallage rocks into the serpentine,† while also there seems an intrusion of serpentine amid the former, as between Dranna Point and Porthalla.

Though there may be some intermixtures of the serpentine and the diallage rock rendering their relative antiquity a little doubtful in places, as a whole, the latter would appear to have been thrown up after the former. At the junction of the diallage rock of Crousa Downs and St. Keverne, with the serpentine at Coverack Cove, veins of the former cut through the latter.‡ On observing, also, the connexion of these two rocks, in a range extending from Careglooz through Gwinter towards Goonhilly Downs, the diallage rock seems to have cut through and disturbed the serpentine. Near Landewednack, also, the diallage rock appears to rise through the hornblende slates and cut into the serpentine. This diallage rock, as between Coverack Cove and St. Keverne,

* It is not altogether clear whether this alteration may not be due to the influence of some granitic mass beneath, with which the granite veins, traversing the serpentine, may be connected, such granitic mass closer to the latter than might be inferred from the natural sections, inasmuch as beneath the hornblende rocks and slates, there are talco-micaceous slates to a certain extent interstratified with the latter, much reminding the observer of the various alterations effected in the proximity of the granites of the district. A glance at the Geological Survey Maps (Sheets 23, 24, 25, 30, 31, 32), or at the Index Map in the Report on the Geology of Cornwall, will show that there may readily be a line of granite concealed beneath the sea, and ranging in a somewhat general manner with the granite from Dartmoor to the Land's End, which has caused the alteration of the rocks into the mica slate and gneiss of the Start Point, and Bull Head, Devon, and produced the gneiss on which the Eddystone Lighthouse, in front of Plymouth Sound, is erected, and the talco-micaceous slates of the Lizard Point. The connexion of the hornblende slates with the latter may be conveniently seen near Poltreath, on the west of the Lizard Town.

† As we have elsewhere remarked (Report on the Geology of Cornwall, &c., p. 30), "whatever the cause of this apparent passage may have been, it is very readily seen at Mullion Cove, at Pradanack Cove, at the coast west of the Lizard Town, and at several places on the east coast between Landewednack and Kennick Cove, more particularly under the Balk, near Landewednack, and at the remarkable cavern and open cavity named the Frying Pan, near Cadgwith. It will generally be found that, at this apparent passage of one rock into the other, there is calcareous matter, and a tendency to a more red colour in the serpentine near its base than elsewhere.

‡ The veins of diallage rock in the serpentine between the rivulet in Coverack Cove and the pier at the village will repay examination. Some of them are large grained, the crystals of diallage of considerable size, reminding the geologist of the larger-grained *gabbro* of Italy.

passes occasionally into a compound, in which hornblende also enters; so that while in some places it appears a mixture of diallage and felspar, in others it more resembles one of hornblende and felspar. Regarding a mixed mass of matter in which the proportions of the chief substances, silica, magnesia, lime, alumina, and oxide of iron, may be unequally disseminated, such changes may be readily appreciated, the conditions for the adjustment of the substances in crystalline forms being variable.*

With respect to the serpentinous rocks in Anglesea and Caernarvonshire, the relative and approximative dates are not so certain. At Porthdinlleyn, a rock, which has been commonly termed serpentine from its appearance, though not altogether agreeing with the usual varieties of that rock, has apparently traversed the chloritic and micaceous slates of that part of Caernarvonshire; but being only covered by a raised sea-bottom of comparatively recent geological date (p. 458), the time when this may have been effected remains doubtful, though an impression of its intrusion being even referable to the date of some of the older rocks of the district might exist. In its greenish and red colours, it much resembles the ordinary serpentines. The component parts are much gathered together in some situations in irregular nodules;

Fig. 223.

between which much red jasper is frequently found,† as in the

---

* Looking at the principal ingredients in hornblende and diallage, as given by a mean of three analyses of the former by Göschen, Bonsdorff, and Struve, and by a mean also of three analyses of the latter by Köhler, Regnault, and Von Kobell, the differences between these minerals would be as beneath:—

	Hornblende.	Diallage.
Silica	40·86	52·00
Magnesia . . .	13·54	15·91
Lime	12·35	19·59
Oxide of Iron . .	14·54	7·47
Alumina	15·96	3·18

† Judging from the frequency of jasper fragments of precisely the same kind in the superficial drift of the district, fragments of even several hundredweights being found (Aberdaron), there would appear to have been much destruction of rocks similar to that of Porthdinlleyn, perhaps of a softer kind, the jasper, from its hardness, being preserved and included amid the other hard detritus.

annexed sketch (fig. 223) taken towards the north-western point of the roadstead, where the dark portions represent the jasper, or other siliceous matter between the nodules, sometimes of large size. Of the serpentine in Anglesea, the aspect of which presents much the usual characters of that rock, though some of it may have been in a molten state when included among the beds where it is now found, other portions much remind the geologist of some mingling of calcareous and serpentinous matter, altered from the state of the original accumulation of their component parts. This may be the case with part of the serpentine at Cerig-moelion, as also at Rhoscolyn. There are also some appearances near Amlwch, amid the bedded rocks there found, as if certain of the contemporaneous beds had taken a serpentine character from the conditions for the adjustment of their constituent ingredients, to which the whole of the associated beds had been exposed, having been favourable to such a modification of parts. As to an accumulation of serpentinous matter in the manner of the felspathic and hornblendic rocks so common in North Wales, contemporaneously with the Silurian deposits, there would not appear any particular difficulty, since, even without supposing an outburst of serpentinous matter in the manner of volcanic ashes and cinders, (though why this may not also have happened does seem clear), the wearing away of serpentine rocks, formed at an earlier date, may readily have supplied the detrital materials for deposits, which when consolidated presented the character above mentioned. At all events this appears a mode of occurrence which it would be desirable that the observer should bear in mind, and the more so that in some other localities for serpentine in the British Islands, as, for example, in the county of Galway, there are some interlaminations and other modes of occurrence of serpentinous and calcareous matter, suggesting to the geologist that such mixtures may have been arranged in water, the accumulations subsequently acted upon so that the present structure of the rocks was produced.*

This brings us to consider the chemical composition of the serpentines mentioned, viewed geologically. They are of very varied mixtures of a kind of base of silicate of magnesia with silicate of alumina, and occasionally of soda and potash, as also of oxide of

* An examination into the chemical composition of some large pilasters of this serpentinous rock, in the Museum of Practical Geology, London, showed that it was a mixture of silicate of magnesia and carbonate of lime, with minor quantities of oxide of iron and alumina. The interlamination of the chief portions of the mixture is often most marked in parts of this rock.

iron. Water is likewise a marked ingredient. Amid all this variety, among which those serpentines may be included through which diallage may be disseminated (a compound common in parts of the Lizard district), more pure serpentine (as it is inferred) is to be found; that is, the serpentine which has been often considered as a distinct mineral species (how far correctly remains to be determined), and which is a silicate of magnesia combined with water, and a minor portion of oxide of iron.* Looking at the chemical composition of the common igneous product olivine, the observer finds that it also is essentially a silicate of magnesia with oxide of iron, the presence of water as an essential ingredient in serpentine being the marked difference between it and olivine.† This is an interesting circumstance, pointing to the very moderate modifications of constituent parts which may produce mineral aspects of such a varied kind. Taken in the mass, the serpentine of the Lizard seems often a compound into which alumina enters as a marked ingredient, thus more resembling, in that respect, the substance named soapstone, occurring in veins in it, and which is a compound of silicate of magnesia and alumina.‡ As a substance also worthy of notice, since so frequently occurring in small veins in portions of the rock, asbestus should not be neglected, its component parts being apparently derived from the mass of serpentine amid which it is found. Though the minerals so named appear of varied chemical composition, and have been regarded as members of the hornblende family, the asbestus of the Lizard seems chiefly a silicate of magnesia, more like the selected serpentine inferred to be a mineral species, without its water.

Quitting this minor area, mentioned merely because the igneous products noticed may be there referred approximately to certain

---

* The chemical composition of these selected portions of serpentine is inferred to be $\dot{M}g^3 . \dddot{S}i^2 . + 2\dot{H}$.

† Taking the composition of the serpentine and of olivine from the 13 analyses of each by several chemists, such as are given by Professor Nicol, in his Manual of Mineralogy, the similarity or difference would be as follows:—

	Serpentine.	Olivine.
Silica	41·99	41·92
Magnesia . . .	40·24	46·67
Oxide of Iron . .	3·38	10·75
Water	12·68	..

The small quantities of alumina, lime, soda, and carbonic acid, in a few of the selected serpentines, and of alumina, lime, and the oxides of manganese, tin, nickel, and chrome in some of the olivines are not here noticed.

‡ According to Klaproth, a soapstone from the Lizard district, contained, silica, 45; alumina, 9·25; magnesia, 24·7; peroxide of iron, 1; potash, 0·75; and water, 18. Svanberg found in a soapstone from the same locality, silica, 46·8; alumina, 9; magnesia, 33·3; peroxide of iron, 0·4; lime, 0·7; and water, 11.

geological dates, and the localities can be easily visited, and passing to more extended and distant regions, the geologist will scarcely fail to be struck with the similarity of various igneous products in each, these being to a certain extent classified. Those which have been termed volcanic and extinct volcanic, with reference to the present time, have already been noticed as presenting certain marked resemblances in different parts of the earth's surface. The same general resemblance will be found in those products in which the minerals of the felspar and the hornblende families prevail, with or without an excess of silica (occurring as quartz), in various regions. Though their real modes of occurrence may not always have been properly ascertained in the numerous and different localities, whence specimens and notices of them have been obtained, and though certain accounts of their manner of association with other rocks may require more attention to the methods of investigation which the progress of knowledge now requires, there is, nevertheless, frequently sufficient to show the great mineral resemblance of many of these igneous products in widely-distributed parts of the earth's surface. Viewed chemically, there is yet much to be accomplished respecting them, particularly with regard to any modifications as to the prevalence of some simple substances more at one time than at another, as also more in certain regions than in others. Of the class of igneous products to which the name *greenstone* has been given—from that crystalline state wherein the constituent minerals, felspar and hornblende, are distinctly seen associated in variable proportions, to the rock wherein the matter of these minerals has not been exposed to the conditions fitted for its separate adjustment in that crystalline form—there are endless varieties. With an excess of silica, beyond that required for the silicates of the component minerals, *syenite* is produced, quartz being then distinctly added to the other two minerals. Again, it sometimes happens that while there is a granular arrangement of the felspar and hornblende, even occasionally with the addition of quartz, crystals of felspar are disseminated through the mass, forming a *greenstone* or *syenitic porphyry*, as the case may be. Some of the compact varieties, termed *compact felspar*, have already been noticed (p. 572). Altogether the shadows and shades of modification have been found so numerous, depending on variations of chemical composition on the one hand, and on different conditions for cooling on the other, that there has been a disposition to seek some term for the whole, which shall leave the exact composition of the rock open to description, while a kind of generic name is preserved. The name of *trappean rocks** has been somewhat adopted of late, particularly by British geologists, for this class of igneous products. It is one, no doubt, open to objection if regarded as a name to be preserved; but in the present state of knowledge, this or some other general term has its convenience as massing together certain products of a family character.

This class of igneous rocks appears to be found amid accumulations of all geological ages, from the older deposits to the accumulations which approximate to the date of those amid which the basalts and associated products, previously mentioned (p. 402), are seen, having been thrown out from some points on the earth's surface, however these may have varied in position. Seeing that their mode of occurrence is such, even amid the old Silurian deposits, as to remind us of the products of modern volcanos, it may be inferred also as probable that from that geological date to the present time, rocks of a similar kind have formed portions of the products discharged from igneous vents, similar to those now scattered over the surface of the earth.

Looking at the granitic rocks as a class, they also are found to present a great family resemblance in different parts of the world, though sharp distinctions between them and those previously mentioned cannot always be found, the one class passing into the other, especially when the hornblendic minerals are absent, in a manner resembling the modifications only of some general amount of given substances. When these minerals are present as is sometimes the case, the chief chemical differences between such mixtures and more ordinary granites, appear to consist in the abundance or scarcity of the silicates of lime and magnesia, these substances forming comparatively a small portion of the granitic rocks, viewed on a large scale, while they enter conspicuously into the composition of the hornblendic rocks.† Where the two classes are found passing into each other, it often becomes desirable to see how far the hornblendic rocks may have been previously thrown out and

* This term has been derived from the Swedish word *trapp*, a stair, it having been once supposed that an arrangement in stair-like forms, on the large scale, was characteristic of these rocks.

† With reference to the difference or resemblance between granites and greenstones, as we have elsewhere remarked (Researches in Theoretical Geology, p. 397, 1834), "granites, no doubt, vary in their chemical composition, and so do greenstones, yet they always so differ from each other as masses of matter, that the one can never become the other from mere differences in cooling." If we suppose the felspar to be of the ordinary potash kind, and a granite to be formed of two-fifths of such felspar, of two-fifths of quartz, and one-fifth of mica (containing fluoric acid), and a greenstone

consolidated, and have been remelted by the granitic rocks, so as to have thus formed an addition to their original molten mass, the whole, upon cooling, having its constituent parts so adjusted as to present the appearances observed.

As a common character, the granitic rocks seem to be chiefly formed of silica and alumina, after which come, as principal ingredients, potash and soda, the latter sometimes more prevalent probably than has been usually inferred. The silica and alumina often constitute 80 per cent. of the whole mass, thus leaving only 20 per cent. for the other substances. In cases where labradorite is the member of the felspar family present in granitic rocks, either altogether replacing other felspars, or associated with them, lime would form an ingredient of importance,* though silica and alumina would still constitute the most marked substances in such rocks. Sufficient examination has not yet been given to granitic rocks to show us the relative prevalence of soda, potash, or lime (in cases of labradorite), during the progress of geological time. Taking the granite of Wicklow and Wexford, above noticed, it would appear that soda occurred in some fair abundance in the granitic rocks, protruded in that part of the world, anterior to the accumulation of the old red sandstone.

As to the geological times when granitic rocks have risen through prior-formed, and usually disturbed, deposits accumulated

to be composed of the same kind of felspar and an equal proportion of hornblende, the calculated differences may be taken somewhat as follows (Geological Manual, 3rd Edition, p. 448-50):—

	Granite.	Greenstone.	Difference.
Silica	74·84	54·86	19·98
Alumina	12·80	15·56	2·76
Potash	7·48	6·83	0·65
Magnesia	0·99	9·39	8·40
Lime	0·37	7·29	6·92
Oxide of Iron	1·98	4·08	2·10
Oxide of Manganese	0·12	0·11	0·01
Fluoric Acid	0·21	0·75	0·54

* The presence of lime amid igneous products, though it may there occur as a silicate, is interesting as affording the base of a supply for some, at least, of the calcareous matter required by animal life, or distributed as ordinary limestones. However powerful silica may be, acting as an acid where heat, and especially great heat, is employed, at the lower temperatures it is comparatively weak. As for example, at great heats the silicates of potash and soda are readily formed, whether carbonic acid be present or not, but at low temperatures, solutions of the silicates of potash or soda are easily decomposed by the carbonic acid. So also with silicate of lime, if that substance were in contact, in the presence of water at a moderate temperature, with carbonic acid, it would be decomposed, forming carbonate of lime, and if the carbonic acid were in sufficient abundance, bicarbonate of lime, ready to be removed in solution.

by the agency of water, they would appear to include all from the earliest, even to the production of comparatively recent beds of the tertiary series. Of the latter kind, Mr. Pratt has found instances in Catalonia.* Thus there is no conclusion to be drawn as to the relative antiquity of these rocks from the mere fact of their occurrence in any particular locality. This has to be sought in the manner in which they may be found associated with other accumulations, the relative geological dates of which are determinable.

The serpentines, also, and their not unfrequent associate diallage rock, seem to have appeared with somewhat common characters through a long range of geological time. They have been above mentioned as probably of early date in Wales. In Cornwall, though not of equal antiquity, they are apparently still referable to the earlier geological times. In Ireland, also, they seem to have been formed at a remote geological period. Various lands show that they were not confined to those times, but became associated with accumulations of less antiquity; and in Italy, where there are many good opportunities of studying these rocks, they have been found amid deposits up to those of the tertiary times included, it being inferred that the rocks in that land which contain the fossils named *nummulites* were, as pointed out by Sir Roderick Murchison,† accumulated at a time when the lower deposits of the tertiary series were effected in several other parts of Europe. The occurrence of serpentine and diallage rocks amid the Alps, and among the various accumulations of the Jurassic and cretaceous series, usually cutting through them in Italy, and in the continuation of the same accumulations, eastward, in different localities into Asia, is a marked circumstance. These rocks were probably ejected from beneath, at various geological times, over the area of Europe, from the early fossiliferous deposits up to some part of the tertiary series included. So much of various parts of the world remaining to be examined geologically, it would be premature to conclude that these rocks have not been ejected at more recent geological times in some localities.

It has been seen that into the serpentines, magnesia enters largely, the relative amount of that substance being somewhat characteristic, as lime and magnesia combined, are among the hornblendic rocks. It would not, however, be right to infer that silicate of magnesia is alone to be regarded, since the mixtures in which diallage is dis-

---

* Pratt, MSS.

† Journal of the Geological Society of London, vol. v., p. 157.

seminated and even prevails, show that other marked substances have entered into the composition of the mass when in a molten state. In such arrangements of parts of the compound, the ingredients needful for diallage have merely separated out from it under the fitting conditions, the lime, oxide of iron, and alumina having probably been in a more disseminated state previously.* Sometimes the base of the rock, still termed serpentine, from its general aspect, and the diallage crystallized out from the general mass, appear of nearly the same composition.†

With respect to the fusibility of the igneous rocks generally, they no doubt present considerable differences. At the same time, it is needful to bear in mind, that experiments upon them, in the condition in which we find them, do not exactly give us the measure of their fusibility when they were in a molten state. Prior to the adjustment of the parts of many into minerals of a definite kind, they must often have been far more fusible, as can be shown by again placing them under their old condition of a molten mass, producing the vitreous adjustment of parts, so that these definite compounds be not again formed.‡ It hence becomes desirable to view the fusibility of these rocks, with reference to a complete mixture of all their constituent parts, anterior to the separation of any, or the whole of them into crystalline compounds.

* M. Berthier found the diallage from La Spezia, a locality very favourable for the study of serpentine and diallage rock, to be composed of—

Silica	47·2
Magnesia	24·4
Lime	13·1
Protoxide of iron	7·4
Alumina	3·7
Water	3·2

† According to Dr. Köhler (Thomson's Mineralogy, &c., vol. 1., p. 174), the composition of the diallage, and of the rock containing it at Harlzburg, is as follows:—

	Diallage.	Rock.
Silica	43·900	42·364
Magnesia	25·856	28·903
Protoxide of Iron and Chromium	13·021	13·268
Protoxide of Manganese	0·535	0·853
Lime	2·642	0·627
Alumina	1·280	2·176
Water	12·426	12·074

‡ In experimenting upon the fusibility of igneous products, we have often found very considerable difference in that fusibility, after some crystallized and compound rock had been formed into a glass, from that which it had exhibited when first acted upon by the same amount of heat employed. In the same manner, artificial glasses which have been melted and cooled slowly, so as to form a stony mass, or merely exposed to a temperature at which a certain crystallized arrangement of their constituent parts is produced, become more difficult of fusion than when in their first state.

If we are to regard certain of these rocks to have been ejected from volcanic vents in the manner of modern volcanos, it seems also needful to consider that they have been accompanied by outbursts of vapours and gases, sublimations of different kinds having taken place at those different times as now. As a substance very common among the compounds of felspar and hornblende, even of those contemporaneously thrown out amid the older fossiliferous rocks, sulphur combined with iron is very common, indeed, sulphuret of iron is often a marked ingredient among those which are commonly termed greenstones. Not, however, that it is confined to them, for the more felspathic products often also contain it.*

Amid the various modifications and changes of structure to which the deposits associated with certain igneous products have been often subjected, it is to be expected that the latter having been exposed to similar conditions, would, in like manner, have their parts also much modified. Indeed, those igneous products which have been versicular, show, by the various mineral substances found in them, that mineral matter has often been in movement in proper solvents, and passing through its pores, had adjusted itself in the cavities of the versicular rock as definite mineral compounds. Numerous soluble substances, once disseminated amid the general mass of such rocks, may readily have been transported elsewhere, and aid in forming, by new combinations, less soluble substances. Thus many are found disseminated amid modern volcanic products, which, assuming that they were once disseminated amid those of ancient times, would scarcely be now detected in the latter. As regards the conditions to which igneous rocks of ancient geological date may have been exposed during the lapse of time, it would scarcely be expected, when they may have been subjected to the influence of long continued heat, from any depression to considerable depths, especially beneath a thick covering of other deposits, that any obsidians would preserve their vitreous character, such disap-

* It sometimes happens that iron pyrites is found in prior-formed deposits of ordinary detrital matter, adjacent to protrusions and dykes of these igneous rocks, in such a manner, as if either the sulphur, or sulphur and iron had been derived from them. A good example of this mode of occurrence may be seen at Bettws Disserth, on the north of Builth, South Wales, where spheroidal pieces of iron pyrites occur in a Silurian slate adjacent to some hornblendic rocks; these spheroids somewhat abundant in places, and the slate having all the appearance of having been altered by the intrusion of the igneous rock. At the falls of the Wye, near Builth, much iron pyrites is also seen at the contact of some igneous rocks intruded among slates, in like manner altered, certain fossils in them being likewise coated with the same mineral near the contact of the two rocks, though this is not observed at a short distance from it.

pearing from the usual causes productive of devitrification, the component substances taking a stony form.

As to the minerals which appear, as it were, additions in different localities to the general masses of granite, and even to those rocks where hornblende and felspar chiefly constitute the component minerals, they are often very various, and, as M. Élie de Beaumont has remarked with respect to granite, much distributed outside their masses.* While they are often merely some other arrangements in different proportions of the simple substances contained in the general mass, † at others, they appear as if in some manner the result of an addition derived from the rocks, against which the molten mass has been thrown, and thus formed during the long continuance of those conditions (among which great heat is prominent) that have prevailed after the uprise of such igneous rocks in different localities. Among these minerals, garnets of different kinds may be remarked, as occurring as well in the igneous as in the prior-formed, and subsequently modified rock, against which the former has been thrust. When we consider the various substances which analyses seem to show are, as it were, entangled amid those constituting the chief mass of the igneous matter ejected,‡ it would be anticipated that when these were relatively abundant, and could make their own adjustments more freely, less controlled by the influences of those forming the chief minerals, compounds would be effected of a definite kind and be separated from the main mass. Thus, occasionally, mixtures would be formed of more than the usual substances, even constituting masses of importance in parts of the earth's surface, where, though the usual free silica and silicates of ordinary granite and other compounds were still the most prevalent substances, others are present, giving a somewhat modified character to the general rock.

With respect to the occasional component parts of granitic

* Sur les Émanations Volcaniques et Métallifères. Bull. de la Soc. Géol. de France, 2nd série, t. iv. (1847).

† In talc, a mineral sometimes associated with others in granites, we seem to have magnesia in a certain relative abundance, separating itself from a main mass in which it may usually have been a subordinate substance, talc being essentially a silicate of magnesia. Its formula is considered to be $3\,\dot{M}\,\dddot{Si} + \dot{M}g^{2}\,\dddot{Si}^{2}$.

‡ M. Élie de Beaumont, in his table of the distribution of simple substances in nature (Bulletin de la Soc. Géol. de France, 2nd série, t. iv.), considers the following to be found in granite, viz.:—Potassium, sodium, lithium, calcium, magnesium, yttrium, glucinium, aluminium, zirconium, thorium, cerium, lanthanium, didymium, uranium, manganese, iron, cobalt, zinc, tin, lead, bismuth, copper, silver, palladium?, osmium, hydrogen, silicon, carbon, boron, titanium, tantalum, nobium, pelopium, tungsten, molybdenum, chromium, arsenic, phosphorus, sulphur, oxygen, chlorine, and fluorine.

rocks, chlorite should be mentioned as one of some importance, inasmuch as while it shows a modification of the mixture and relative proportions of some of the ordinary constituent ingredients of granitic minerals, silica, alumina, magnesia, oxide of iron, and oxide of manganese, it also points to water as an essential ingredient. When disseminated, therefore, among granitic rocks, as it is in the Alps, Scandinavia, and some parts of the British Islands, chlorite becomes a combined mineral of no slight interest, from the addition of water to the other substances present.*

As respects the various minerals, which are, as it were, additional to those usually constituting the mass of the chief divisions of the igneous rocks, not only has the dispersion, in variable proportions, of other substances than the usual ingredients to be regarded, considering these likewise in their greater or less local proportions, but also the additions which may be derived from the melting of parts of prior-consolidated accumulations, even of those thrown down from solutions in water, and fused by the intrusion of the igneous rocks. Though the chief portions of the ordinary detrital deposits are but abraded parts of previously-consolidated igneous rocks which have been worn away, and then dispersed as above noticed (pp. 63—101), this has been most frequently so accomplished that a remelting of the deposits thence formed, would not reproduce the original rock, the various parts having been separated mechanically into different beds, and decomposition having deprived certain of even the separated substances of portions of their original ingredients. With respect to the latter, for example, should the silicates of soda or potash have been removed in solution, as has often happened, from a felspar of which they once constituted a part, the matter again fused might not contain any of those silicates, so far as the felspar is regarded, silicate of alumina being then the prevailing substance.† Igneous matter, the usual granite compounds, for instance, melting limestone rocks, the lime might be introduced into the molten mass, and the carbonic acid being thrown off, the silicates of lime be formed, ready for combination in other minerals than those resulting from the mass of the granite, as it rose from beneath. So also with dolomite, which

* Taking various analyses, from 10 to 11 per cent. of water enters into the composition of chlorite. The formula for chlorite is considered to be $(\dot{M}g^{3}\,\ddot{S}i + 3\,\ddot{R}\,\ddot{S}i) + 9\,\dot{M}g\,\dot{H}$.

† This substance constitutes the base of the clays employed in the manufacture of porcelain, and which are formed from decomposed felspars in districts where that mineral has been distributed in sufficient abundance.

could thus furnish not only the lime, but also the magnesia for the production of hornblende, should the other ingredients of that mineral be near and not drawn elsewhere. In this manner it will be obvious very material additions may be made to an original and general mass of rocks in a molten state.

There appearing so much of a general character in the various igneous products of different geological times, to call the attention of an observer towards some general cause, which, though much modified under certain circumstances, has yet always exerted an important geological influence, he has carefully to consider the subject, so that, while a proper and close attention may be given to local sources of modification, the great cause of these igneous products, taken as a whole, be not neglected. Whatever may have been the conditions under which substances were probably ejected in the manner of modern volcanos in past geological ages, from time to time molten matter of a very common general character seems as if always ready to have been upheaved, in larger masses, whenever there were great disruptions of prior-formed accumulations on the earth's surface. Thus, while the minor and perhaps modified manifestations of the conditions for throwing out igneous substances generally, were constant in different points of the earth's surface for the time being, these substances mingled with the ordinary accumulations of the day, from time to time a greater amount of molten matter was upheaved, lifting such igneous products as well as their associated sedimentary deposits, as if the former action, however intense, was but superficial as compared with that from which the more wide-spread and important movements were derived. Be this as it may, the igneous products form objects of the greatest interest, whether regarded as the source whence so large a proportion of the detrital accumulations are derived, for the modifications they have so frequently effected in the deposits against or amid which they have risen, or been protruded, for the differences and resemblances they exhibit among themselves, or for the proof they afford that during the long lapse of geological time of which we can obtain traces, and up to the present day, there have been conditions for uplifting mineral matter in a molten state, that matter chiefly composed of the oxides of a few simple substances—two of them especially (sodium and potassium)—being not only remarkable for their comparative lightness, but also for an avidity for oxygen so great that they will decompose water in order to obtain it.

CHAPTER XXXI.

CONSOLIDATION AND ADJUSTMENT OF THE COMPONENT PARTS OF ROCKS.—ADJUSTMENT OF COMPONENT PARTS OF CALCAREO-ARGILLACEOUS DEPOSITS.—ARRANGEMENT OF SIMILAR MATTER IN NODULES.—CENTRAL FRACTURES IN SEPTARIA.—NODULES OF PHOSPHATE OF LIME.—SPHEROIDAL CONCRETIONS IN SILURIAN ROCKS.—CRYSTALS OF IRON PYRITES IN CLAYS AND SHALES.—MODE OF OCCURRENCE OF SULPHATE OF LIME.—MODIFICATION IN THE STRUCTURE OF ROCKS FROM CHANGES OF TEMPERATURE.—CHLORIDE OF SODIUM DISSEMINATED AMONG ROCKS.—IMPORTANCE OF SILICA AND SILICATES IN THE CONSOLIDATION OF DETRITAL ROCKS.—ALTERATION OF ROCKS, ON MINOR SCALE, BY HEAT.—FORMATION OF CRYSTALS IN ALTERED ROCKS.—CRYSTALLINE MODIFICATION OF ROCKS.—ALTERATION OF ROCKS NEAR GRANITIC MASSES.—READJUSTMENT OF PARTS OF IGNEOUS ROCKS.—PRODUCTION OF CERTAIN MINERALS IN ALTERED ROCKS.—MINERAL MATTER INTRODUCED INTO ALTERED ROCKS.—MICA SLATE AND GNEISS.

WHEN the gravels, sands, silts, clays, or mud of various geological times are presented to the attention of the geologist in the form of conglomerates, sandstones, arenaceous and argillaceous slates and shales, their component parts, originally drifted, or otherwise borne into the relative situations where they are now found, have either been joined together by mineral matter, subsequently introduced among them, or by a change in the condition of some part or parts of the original deposit which should permit such portions, in an altered form, to cement the remainder. With carbonate of lime, the oxides of iron and manganese and occasionally with silica, as substances cementing fragments of rocks, either angular or rounded, on hill sides or other subaërial localities, where springs containing and depositing those substances occur, we may consider the observer as familiar. That various breccias, conglomerates, and even sandstones so formed, occasionally constitute parts of a series of geological products, may be considered probable. It is easy also to infer that during geological changes, gravels, sands, and mud constituting the margins and bottoms of lakes and seas, may be so placed beneath isolated portions of water, to which the access of rivers or streams may be insufficient to meet the loss by evaporation, that certain substances held in solution may be slowly deposited amid such subjacent gravels, sands, or mud, so as to produce modification, change, or even consolidation of various kinds in them.

Independently, however, of these effects, the observer will have to direct his attention to modification, change, and consolidation of a far more general kind, and for which some more general cause appears to be required. He will, in the first place, have to dismiss the view that the relative age of rocks is alone a sufficient cause for the effects noticed; though, taken as a whole, the relative geological age of deposits is so far important, that, other things being the same, there may be a greater chance of the older rocks being consolidated or modified in their structure, inasmuch as they may have been more exposed, during the lapse of time, to the causes productive of such consolidation and change.

It may, in the first place, be desirable to consider the modification of parts which might arise in a bed or mass of mud, or clay after its deposit, the component parts of such mud or clay being variable. We may take, by way of illustration, those alternations of argillaceous limestones and shales, often calcareous, which are observable in the lias of some parts of Western Europe, and which appear the result of an unequal supply of mud and calcareous matter, sometimes the one and sometimes the other predominating. Examples of irregular deposits of this kind must not, however, be considered as confined to any particular age, since among the older

Fig. 224.

as well as newer geological accumulations, this kind of deposit may often be found. The above (fig. 224) may be taken as illustrating alternations of this kind, the surfaces of the beds being irregular.

In itself such a section may merely present us with the evidence of alternating conditions, by which carbonate of lime was more thrown down at one time than at another, though, with care, forms of the surfaces are often traced which would seem to point to an abstraction of calcareous matter from the adjacent original clays or mud; a circumstance which becomes more evident where the calcareous matter in the general deposit has decreased, and many irregular patches of the argillaceous limestone, and nodules of it, are arranged in lines or are more dispersed through the deposit, as shown in the subjoined section (fig. 225). In such cases the cal-

Fig. 225.

careous matter of given times of deposit, irregular like those where whole sheets of argillaceous limestone were produced, seems gathered to different points in or about the same plane, that upon which the general deposit was accumulated, the matter arranged round these points, thus variously dispersed on the plane, so that two or more nodules may be joined together while others remain isolated. This gathering together of similar matter distributed through a soft muddy or clay mass, would be anticipated, and the more so, when we remember the manner in which similar matter may be gathered together from solutions, dragged away, as it were, forcibly to points where some of it may have been first deposited, as noticed by Professor Bunsen (p. 375).

Facts of this kind are as well seen among the carbonates of iron, of so much value in the coal measures of the British Islands, as amid the accumulations above noticed; and they, in like manner, point to a separation of the carbonates from the muddy mass, and, for the most part, in planes corresponding with the relative times of their original deposit in the general accumulation, one chiefly detrital, and thrown down from mechanical suspension. It occasionally happens that this gathering together of similar matter from amid a mass through which it was originally dispersed, usually in certain planes and thicknesses, can be seen to have taken place so that a certain original lamination of parts is not destroyed. Instances of this kind are to be found in one or two of the ranges of nodules in the lias of Lyme Regis, Dorset, where, as beneath (fig. 226), these are seen still preserving the lamination of the general deposit;

Fig. 226.

an arrangement of parts easily ascertained by breaking the nodules in this plane. In these nodules some organic remain, such as a fish, nautilus, ammonite, or a piece of wood, not unfrequently seems to have formed a point around which the carbonate of lime was aggregated, though this has by no means been always the case, since some are occasionally found without organic remains, or only contain them in a dispersed state.

Such aggregations and separation of parts are at the same time a modification of the original deposit, and a partial consolidation of it. As a proof that the mass was soft when the nodules were formed, it will be often found that while the same kinds of organic remains, and especially thin shells, are flattened, in the same planes, in the associated and adjoining clays, marls, or shales, they are comparatively well preserved, uncompressed, in the nodules, the consolidation of the latter having protected them from the pressure to which those had been subjected in the remainder of the deposit, then in a yielding condition.

With regard to the relative time and mode of consolidation of the nodules, the observer may be frequently enabled to study it in those commonly known as septaria, where, after the aggregation of the similar matter, such, for example, as the carbonate of lime in many clay or shale deposits, and the carbonate of iron in the coal measures and some other rocks, a splitting of the interior has taken place, and subsequently to a certain amount of consolidation, since the fractures are usually sharp, pointing to a sufficient amount of cohesion of parts. The subjoined section (fig. 227) will show the ordinary

Fig. 227.

manner in which such nodules are broken in the interior, the cracks not extending to their exterior surfaces, as if there had been a shrinking of parts from the centre outwards, so that the resulting largest openings were central. In the nodules of this kind, not uncommon in many clays, marls, and shales, the cracks are usually filled according to the character of the general deposit of which the nodules constitute a part; thus carbonate of lime is frequent in those where that substance is much disseminated, and carbonate of iron where the latter is not uncommon. Occasionally other substances are introduced, such as, in the ironstone nodules of many parts of the British Islands, the sulphurets of lead, zinc, and iron, copper pyrites, and certain other minerals.

Nodules and other formed bodies of phosphate of lime, also sometimes occur in a manner pointing to the aggregation of their component parts from previous dissemination amid surrounding detrital deposits. Many of the nodules and other forms of phosphates of lime in the lower parts of the cretaceous series of south-eastern England and in parts of France, seem thus produced. Mr. Austen has informed us,* that the nodules he examined had a concentric arrangement of parts, like agates, and he points to the probability that the phosphoric acid may have constituted part of the fæcal or coprolitic matter accumulated with other organic bodies, at the period of the original deposit, and had been disseminated among the sand and ooze of the locality and time. Modern researches have shown that phosphate of lime is far more diffused among rocks than was at one time supposed. When free carbonic acid is present in water, the phosphate of lime is, like the carbonate, soluble, though not to the same extent as the latter; so that conditions may readily arise not only for its dissemination, but also for its aggregation into various forms amid rocks through which its particles could move. Not only waters impregnated with free carbonic acid, in the usual manner, would afford the common means of transport for such particles, but also, in the cases referred to by Mr. Austen, for the mixture of coprolitic with vegetable matter, the decomposition of the latter, and often, indeed, of the fæcal matter itself, might produce the carbonic acid needful in the required solution.

The association of similar matter in nodules, is also sometimes well seen amid deposits of siliceous sands, these aggregated so that the nodules protrude as marked objects on weathered banks or cliffs. Sometimes the nodules are dispersed among the arenaceous accumulations, while at others they range in certain general planes, corresponding with those of deposit, and thus, in their mode of

* Journal of the Geological Society of London, vol. iv., p. 257, 1848.

occurrence, resemble the nodules of the carbonates of lime and iron, above mentioned. In certain of the arenaceous deposits the cementing substance of the nodules is occasionally calcareous, apparently aggregated from that matter once more dispersed amid the sands, and deposited amid the grains from solution, as a bicarbonate. The oxides and hydrated oxides of iron are also observed gathered in nodules, either dispersed or in planes, aggregating portions of sands.

Even amid the older detrital accumulations with which geologists have become acquainted, this structure is observable. The separation of calcareous matter into nodules from among the component parts of an original mud deposit, can be as well seen in the old series of rocks, known as Silurian, such as in portions of the Wenlock shales and limestones of that series, as it occurs in parts of Wales and the adjoining English counties, as in far more modern geological accumulations. So also with the aggregations of siliceous matter in the nodular or spheroidal forms, showing that similar conditions for these arrangements and adjustment of parts have continued to prevail through a long range of geological time. The following section (fig. 228) of part of the upper portion of the

Fig. 228.

Silurian series (Ludlow Rocks) of Brecknockshire, to be seen at a considerable development of that portion, in Cwm-ddu, near Llangammarch, will exhibit the arrangement of parts of this arenaceous rock, in certain beds, in a spheroidal form; layer after layer, as the decomposition of the rock shows, having been arranged round somewhat central points of aggregation dispersed in certain lines of beds. Aggregations of this kind occasionally measure many feet in diameter. Such aggregations are sometimes only to be detected on the face of rocks by lines arising from the stains of peroxide of iron, which, when followed out, are found to correspond with spheroidal surfaces.

When, geologically, these adjustments of the parts of deposits may have been effected, it is not easy to infer, since in the instances of those in the older accumulations, they may have been produced, as many of those in certain more modern accumulations are have been, before the solidification of the sandy portions the spheroidal aggregations and nodules, the whole of the bed, or beds, having been submitted to further conditions for consolidation, after the separation of certain portions of them into such aggregations of similar matter.

There are certain other separations of the original portions of a deposit, where the particles have possessed such free movement and powers of adjustment, that they have been enabled to gather themselves into crystals. Of this the crystals of the sulphuret of iron amid the mud deposits of all geological ages is an example, as also the crystals of sulphate of lime in numerous clays. Cubes and other forms of iron-pyrites are as common amid the oldest fine sedimentary accumulations, occurring in a manner to leave little doubt of the aggregation of their component particles from the mud in which they were diffused, as among the clays of tertiary deposits. That iron-pyrites should be gathered round organic remains in rocks of different ages, particularly in those, such as have been mud and clays, where the movement of its component particles may be inferred to have been, as in the case of the crystals above noticed, somewhat easy, would be anticipated, inasmuch as the production of iron-pyrites in connexion with decomposing animal matter is well known.* Thus we frequently find the sulphuret of iron incrusting organic remains, as crystals, and in more irregular lumps and patches, particularly amid clay and shale accumulations.

Regarding sulphate of lime, irrespectively of its distribution in crystals, as selenite, amid clays and shales, it often constitutes considerable nodules, and dispersed irregular masses, as if, independently of original deposit, or change from the carbonate by the introduction of sulphuric acid amid particles of limestone, it had separated out from the body of the rock, and became aggregated amid a soft muddy deposit, thrusting aside the latter. Certain

* Mr. Pepys, in 1811 (Transactions of the Geological Society of London, 1st series, vol. i.), was among the first to publish a very illustrative case of the production of iron-pyrites from the decomposition of the bodies of some mice in a solution of sulphate of iron. Another illustrative instance of the formation of iron-pyrites upon animal matter in a decomposing state, occurred at the bottom of a mine-shaft, near Mousehole, Cornwall, where a dog had fallen into a solution of iron, and its body was found surrounded by iron-pyrites. In these, and other well-known cases, the hydrogen evolved from the decomposition of the animal matter, is considered to take the oxygen both from the sulphuric acid and oxide of iron, so that iron-pyrites, or bi-sulphuret of iron, is formed.

nodular portions so occur in particular lines, that we may suppose them to have been produced much in the same way by segregation as the nodules of the carbonates of lime and iron, above noticed. At the same time beds of gypsum, both on the large and small scale, also so occur amid clays, marls, and shales, especially well seen amid portions of the red and grey marls of the upper new red sandstone series, or trias, that there is much difficulty in deciding as to the probability of their original production from solutions, amid the clays or mud, in a manner similar, as regards general principles, to that noticed by Professor Bunsen, or partly in that manner, and partly by segregation into veins formed subsequently to the general accumulation and its partial induration. The section beneath (fig. 229), seen at Watchet, Somersetshire, amid the marls

Fig. 229.

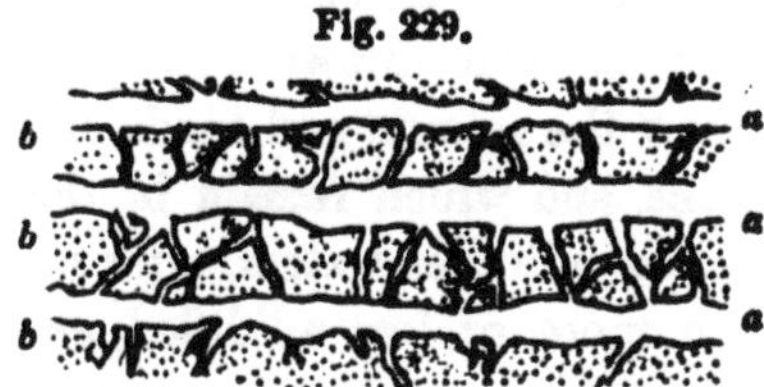

of the trias, will illustrate a mode of occurrence of not an uncommon kind, wherein beds of gypsum *a*, *a*, *a*, are united by strings of the same substance traversing the intermediate marls *b*, *b*, *b*, in various directions, and having somewhat the appearance of cracks filled, inasmuch as the fibrous gypsum in them has the fibres usually at right angles to the walls of the containing marls, as if crystallization had taken place against those walls. No doubt this appearance may be deceptive, but at all events, it becomes an interesting object of inquiry, to ascertain how far, under such modes of occurrence, the evidence may be in favour of an original separation and deposit of the sulphate of lime, contemporaneously with the matter of the marls, or of a segregation of, at least, part of the same substance into veins, from a dispersion of the sulphate of lime amid the body of the accumulation.

When the observer reflects upon the different conditions, to which the various deposits in seas and bodies of fresh water may have been subjected, posterior to their original accumulation, he will not fail to appreciate the modifications which the whole mass of many may have sustained. The mere change from being superficial, on the bottoms of seas and other bodies of water, to being buried beneath many, and sometimes varied additional accumulations, is alone a condition under which new adjustments of parts may arise, and this without a change in the relative distance between the surface of the sea, or other waters, and the deposit itself. Should the accumulation above it be thick, changes (p. 444) arise in its temperature, with their consequences as regards the motion of aqueous solutions distributed through beds of different degrees of porosity.

The geologist should direct his attention to the still greater causes of modification and change which would follow the sinking of such deposits, as regards the crust of the earth, when they descended into comparatively elevated temperatures, so that their component parts, and the various solutions with which they may be moistened, become affected by that temperature. The springs which issue from various rocks, and for which the supply is derived by the simple percolation of atmospheric waters through porous beds of different kinds, until thrown out by less pervious beds (p. 16), suffice to show the amount and kinds of substances soluble under such conditions, and which remain in the various deposits effected beneath the sea or other waters, after many of these accumulations have been more or less solidified, and raised into the atmosphere, where they now constitute portions of land above the level of the sea. In the various borings or sinkings for mine-shafts, the driving of extensive tunnels and levels, and in wells of various kinds, especially of those termed artesian, he has also the opportunity of ascertaining the soluble contents of the waters which may be disseminated among the rocks traversed; and where such waters may be considered as in a somewhat stagnant state, except so far as movement through any fissures, joints, and the pores of the rocks themselves, may be induced by differences of temperature from the surface of the earth downwards towards the interior. There does not exist so much exact information as to the substances in solution among the waters disseminated amid rocks in this manner as is desirable; neither are the soluble contents of the various waters rising through faults on the surface of the ground, or flowing up at the bottoms of mines, with a temperature sufficiently elevated to render it probable that they rose from greater depths, so well known as is required for properly estimating the amount and kinds of substances, which may be thus circumstanced; but there still exists sufficient knowledge on the subject to show the observer the value of investigations in this direction.

The waters rising from the chalk at the artesian well in Trafalgar-square, London, and which are obtained from their dissemination in that rock, show, that in 68·24 grains of solid matter in an imperial gallon, 18 grains are composed of carbonate of soda; while the carbonate of lime contained among the solid matter above mentioned, only amounts to 3·255 grains; and thus the waters resting, to a certain extent, stagnant in the chalk beneath London, with its thick covering of (London) clay, exhibit a very different character, as to the substances in solution, from that of the spring waters which flow out of the chalk on the surface, where that rock arrives at or adjoins it.*

Among the various substances found in solution, either disseminated among the pores of rocks, or which become, as it were, washed out of them in solution, by waters percolating through them and issuing as springs, the observer will do well to recollect the amount of chloride of sodium so often obtained. That it should be a somewhat abundant substance would be expected in deposits of mud, silt, sand, and gravel effected beneath the sea; as also that, when such accumulations were elevated into the atmosphere, and rain-waters found their way to the chloride of sodium, it should be removed by any springs thence resulting. It will be seen that in the waters disseminated amid the chalk beneath London, this substance was found to constitute somewhat more than two-sevenths of the whole solid contents obtained from it.† Looking at chloride of sodium alone, and its dissemination among beds of quartz or other siliceous sands, and the descent of the whole to some very elevated temperature by depression of the earth's surface in any given region,

---

* The following are the substances contained in an imperial gallon of the waters of the Trafalgar-square well, according to Messrs. Abel and Rowney:—

	Grains.
Carbonate of lime	3·255
Phosphate of lime	0·034
Carbonate of magnesia	2·254
Sulphate of potash	13·671
Sulphate of soda	8·749
Chloride of sodium	20·058
Phosphate of soda	0·291
Carbonate of soda	18·049
Silica	0·971
Organic matter	0·908

In the cases of soluble mineral matter disseminated in rocks, such as the chalk beneath London, it should be borne in mind, that when there is a movement of the contained water among their pores or fissures to supply that raised to the surface by pumping, or rising from boring and overflowing, the original condition of somewhat stagnant dissemination becomes changed by the amount of the water thus required, so that when many wells reach into the chalk, as beneath London, a movement of water amid the body of that rock is occasioned towards the various wells, which would not have taken place under ordinary natural circumstances.

† As sea, or rather estuary waters, are inferred partly to percolate into the chalk beneath London, some caution is needed as to the source of all the chloride of sodium in the chalk so situated.

some effect might be anticipated from the production of a silicate of soda, aiding a consolidation of the sands, in the same manner as a salt glaze is produced by the potters.

While studying the variable amount of consolidation of rocks, the geologist cannot fail to have his attention arrested by the different states, in this respect, in which he sometimes finds the beds amid a series of deposits, grouped together, and which have evidently been subjected to the same general conditions. It would strike him, probably, that the original condition of the deposits could not fail to produce marked differences in this respect. He would anticipate that a bed of pure quartz sand, unmingled with other and muddy matter, might, if cemented by somewhat pure silica, form a substance of a harder and more solid kind than when ordinary sand was deposited, mingled with a certain portion of mud, or when the grains were composed of different substances, so that they could be variably acted upon by the matter forming the cement. In the one case, there may be a rock, commonly known, from its composition, as *quartz-rock*, wherein it is sometimes even difficult to trace the original grains of sand, their surfaces having been more or less acted upon by the mode in which the infiltration of the cementing silica has been effected;* while in the other, a sandstone of the ordinary amount of consolidation has been alone produced. The occurrence of certain quartz rocks among the accumulations of all geological ages, and amid other and contemporaneous beds, can be often well studied; and sometimes the passage of an ordinary sandstone bed into a quartz rock can be easily traced. Of this, a quartz rock, amid the new red sandstone series near Bridgend, Glamorganshire, may serve for an example, as the same bed can be readily followed from its ordinary sandstone character on the north of the town, to that of quartz rock on the road to Pyle Inn. Changes of a similar kind are sufficiently common in the course of numerous rocks, as well in single and marked beds, as in numbers of them collectively; and the observer will, no doubt, have to seek for the causes of these differences as well in the unequal or variable supplies of the cementing matter, according to

* The arrangement of parts in certain of these quartz rocks is sometimes such that it requires very careful examination, and even occasionally a thin slicing of a part, so that it can be studied through transmitted light, in order to distinguish the original grains of quartz sand, the cementing and external parts of these grains having become so much blended. For the most part, however, the detrital origin of the quartz grains is sufficiently evident. In examining these rocks, as they are often traversed by veins of quartz, it is needful carefully to distinguish between the latter, which are merely the ordinary infiltrations of silica into cracks and fissures, from the body of the rock itself, a circumstance that has not always received attention.

subordinate local influences, as among the different original compositions of continuous deposits; the latter often, nevertheless, appearing a sufficient cause, in the same way that, in a series of beds, wherein varieties of this kind are very striking, much original differences are apparent. Certain hard quartzose beds beneath others of coal, between Swansea and the Mumbles, may be taken in illustration of a probable change effected by the introduction of silica, or some silicates, after their original deposit. In these beds, the roots of a plant (*Stigmaria*), existing when the coal measures, of which they constitute a portion, were accumulated, once as freely grew, spreading out their finest parts in the evidently yielding ground of the time (p. 501), as in any other of the similarly-circumstanced beds of the same district supporting seams of coal, and known as *under-clays* (p. 510), though now they are bound up in a hard siliceous rock, upon which atmospheric influences have as little action as on ordinary quartz rocks, the original silty and loosely-aggregated substance of the beds being converted into a hard quartzose substance.

Looking at the mass of detrital matter, more or less consolidated by silica or the silicates, the study of the manner in which this may have been effected by them, becomes a matter of no slight interest to the geological observer. He finds silica in a pure or nearly pure state in cavities of various rocks, especially of those of igneous origin, wherein hollows and vesicles have been left, it being seen more or less filling such cavities with agates, onyxes, chalcedony, and rock crystals, and he can have little doubt that this silica was introduced into the hollows and vesicles by infiltration and in solution. Indeed, the stalactitic forms of the silica often sufficiently show this, certain agates, as is well seen upon their decomposition, being merely forms of this kind eventually filling hollows. At other times, the layers of the siliceous deposits occur in planes, apparently horizontal at the time they were effected. These modes of occurrence show him that silica has been, and can be, disseminated amid the pores of rock, often hard and (so called) compact, its particles finding their way for deposit in a pure or nearly pure state into the vesicles and cavities of such rocks.

In investigations of this kind it will be desirable that the observer should bear in mind that certain silicates are not difficult of decomposition, as, for example, those of potash and soda, when free carbonic acid may be present. Upon looking at this subject generally, such conditions may be inferred not to be so rare as might at first be supposed. In certain regions, the decomposition of the felspar alone in granitic and some other igneous rocks, gives rise to solutions of the silicates of potash and soda, and the introduction of waters having free carbonic acid, derived from the atmosphere, in them, would separate the potash or the soda, as the case might be, from the silica, the latter being deposited under favourable conditions for dissemination amid the pores of rocks.* When we regard the manner in which carbonic acid may arise from the decomposition of organic bodies, be mingled with water, and act upon certain silicates, it is also to be inferred that favourable conditions may arise under which silica could be thus thrown down, even when vegetable matter afforded the carbonic acid, amid the pores and cavities of a certain part of the plants themselves, preserving their finest structures.†

Though silicic acid may thus, under favourable conditions, to which it is here sufficient to direct careful attention, be easily separated from certain silicates under the common temperatures which are known on the surface of the earth, or found at moderate depths, circumstances with regard to this substance become changed when the heat to which it is exposed in connexion with others is considerable. For instance, instead of decomposing the silicates of potash or soda, in the manner above mentioned, the carbonic acid would be driven off, and the silicic acid would remain combined with the alkalies. Again, it is now known that while pure silica, so very important geologically, may be very difficult to dissolve in water at the temperature commonly termed ordinary, when the heat of water is much increased beneath the requisite pressure, it may be considered simply, like many other substances, as more soluble in highly-heated waters than in those of more moderate temperatures. Hence, when the observer regards the facility with which pressure, and elevated temperature may be obtained, by descent beneath the surface of the earth, he will see that no slight modification and change may be effected by the mere lowering of beds, moistened with water to situations where such water could act upon the silicates of the rocks among which it may be disseminated, and even upon silicic acid itself, existing as grains of pure quartz, this solution ready to be effected by, and to produce various modifications and changes among the substances forming the

---

* Mr. Henry informs me that when experimenting upon silica he found that a silicate of soda was decomposed even by the carbonic acid of the atmosphere and the silica deposited, its state and appearance being much affected by the degree of concentration of the solution.

† The fine structure of fossil siliceous wood is often beautifully preserved.

original deposit, or other matter subsequently introduced amid its parts.*

The action of considerable heat upon rocks, producing change and modification of their component parts, can often be so studied among a minor mixture or juxtaposition of igneous rocks, and those evidently produced by chemical or mechanical deposit in water, as much to assist inquiries into the manner in which more general changes and modifications may be aided, and even sometimes effected on a great scale. In volcanic regions, substances, such as clays, become hard, in fact baked, as any tile or brick may be, hy the overflow of a lava current among them, the result being the same as might be expected from our knowledge of the action of heat upon different varieties of clays in our potteries and porcelain manufactures, some clays burning or baking well, others ill. In such cases the usual result is the production of certain changes by the action of the heat communicated from the liquid lava. A still further modification of parts is effected when, without loss of the original form of the deposit acted upon, some of the constituent particles have separated from the main mass in which they were disseminated, and, joining together, have produced crystals, there having existed a power of movement in these particles, similar, so far as regards conditions for separation from the main mass, and the movement obtained, to that above mentioned as having taken place in yielding deposits, such as clays.

This modification in the arrangement of the component parts of rocks is common to the igneous action of all geological times. It can be as well seen amid the accumulations of igneous matter deposited with the old Silurian series of the British Islands, as in various regions among the volcanic products of the present time, and is one requiring some attention, since it might otherwise much interfere with the conclusions of an observer as to the conditions under which the component parts of a rock may have been originally gathered together. A porphyritic character, from the dissemination of certain crystals, as, for example, those of some of the

---

* It is to be hoped that investigations in this direction may more occupy the attention of chemists than has hitherto occurred. The subject is full of interest, and appears one likely to reward the labours of those who, taking a certain class of geological facts for their guide, unite with them the conditions of high temperature beneath great pressure, as also exclusion from the atmosphere, such as may be inferred to exist beneath given depths in the earth, upon the hypothesis that heat increases downwards towards the central portions of the earth, for at least the distance at which water, should it continue to exist as such, can be heated up to a very elevated temperature.

felspars, may be too hastily assumed as indicating [illegible] characterised, to have been in a complete molten state.* [illegible]

In those instances where the rocks have evidently been [illegible] prior to the introduction of the igneous matter, which [illegible] a simple *dyke*, as it is usually termed, or some tortuous form of vein, the observer would necessarily infer consolidation sufficient for the production of the fracture, so that any change or modification found in the rock fractured would have taken place after such consolidation. Cases of this kind of alteration are far from uncommon. In studying them it becomes needful to recollect that not only the mere action of heat may be brought to bear under such circumstances, as it might be with regard to the clay of a brick or porcelain vase, but also that moisture and solutions would probably be disseminated in the usual manner amid the pores and cracks of the rock so acted upon, and this often beneath much pressure; exposure of the changes thus produced being often due to some of, or all, the causes of denudation. removing former, and considerable, pre-existing and covering portions of rocks.

Changes and modifications in such cases must necessarily depend much upon the substances acted upon, and the manner in which their component parts may have been arranged. The most simple forms of modification are those where some substance, such as common limestone, may have its parts so modified that a crystalline adjustment of them is effected; the portions of rock in contact with the igneous matter being thus altered, the greatest modification effected nearest the igneous rock and becoming less as the distance from it is increased. Of this kind of modification the often-quoted instance of the chalk in the Isle of Raghlin may be taken as an example. In this case, as shown beneath (fig. 230),

Fig. 230.

dykes, *a a a*, of basaltic rock traverse the chalk of that part of Ireland (so much broken up by eruptions of igneous matter at a period subsequent to the chalk), converting that rock, between and adjoining them, into a more crystalline substance, *c c*, this character gradually disappearing on each side, *b b*. The alteration at the contact of dykes of igneous rocks is not confined to the more

* Modifications of this kind, by which crystals of felspar have been developed in rocks which still preserve their original planes of deposit, are not uncommon.

crystalline arrangement of the traversed and adjacent beds, certain minerals being very often formed by the movement of their component particles, under conditions when they could adjust themselves into crystals, the surrounding matter giving way to their forms. These minerals vary much according to the chemical composition and physical structure of the deposits acted upon, and also according to the volume as well as kind of the igneous rocks introduced.

A far larger amount of modification and change is necessarily effected when the mass of igneous rock, introduced amid prior accumulations, is considerable, and when it may be inferred, as it often can be, that this intrusion has been effected at depths beneath the surface where there was no contact with the atmosphere; but where, on the contrary, any water distributed amid the pores or crevices of the previously-formed rock, consolidated or otherwise, as may have happened, could not escape, with any solutions it contained, having been confined to a certain range, beyond which, a continuation of the same, or other rocks, with their disseminated moisture, remained much in its condition prior to the intrusion. As has been previously remarked, certain of the granitic intrusions appear to have effected much change in adjacent accumulations. In various parts of the world, such modifications of previously-formed rocks of all kinds in contact with the intrusions and upheavals of granitic matter, are most marked, the altered rocks being traceable to their more usual forms of ordinary limestones, argillaceous and arenaceous slates, sandstones, or the like; some even of these rocks being fossiliferous, and so occurring, that their relative geological age can be readily assigned them.

Changes and modifications of this kind can be well seen to have been produced upon the Cambrian and Silurian rocks, prior to the accumulation of the old red sandstone, in parts of Ireland (in the counties of Wicklow, Wexford, &c.); and in south-western England, upon deposits of a later date (effected anterior to the new red sandstone); the rocks acted upon being of varied composition, including different igneous accumulations, as well thrown out in a molten state, as deposited as ashes and lapilli beneath water (p. 554). In such situations, the observer will find, as he might anticipate, the consolidation by silica and the silicates often very considerable, beds of ordinary sandstone sometimes exhibiting their component grains as if passing into the matter cementing them. Judging from the solvent effects of water and steam, at high temperatures, upon the usual silicates employed in glass, when moisture is disseminated in rocks, and raised to a very high temperature, under the conditions above noticed, the silicates, so common among various argillaceous and arenaceous slates and sandstones, would be acted upon, so that considerable consolidation by them became frequent among these accumulations.

Such conditions could be scarcely otherwise than favourable to the aggregation of certain substances into a crystalline state upon a more extended scale than in the case of the smaller bodies of molten rock intruded among, or rising through, fractures in prior and consolidated accumulations. Certain igneous rocks seem sometimes to have had the volume of their component minerals increased, as, for example, the crystals of hornblende and felspar to have become enlarged near the contact with the granite. Sometimes either with, or without increase of volume, the particles of hornblende and felspar of an ordinary greenstone become so adjusted as to present far greater brilliancy of aspect, so that the rock takes the appearance of that commonly known as *hornblende rock*. Good instances of this kind are to be seen in the county of Wicklow. In like manner, the old Silurian volcanic ash-beds of the same district will be seen, under similar conditions, with the brilliant aspect of *hornblende slates*. The hornblende rock and slate of the Lizard district, Cornwall, seem, in like manner, little else than ordinary greenstone, and the volcanic ash of the Devonian series, modified either by the action of a great mass of serpentine which has flowed over, and remained upon them, or by that of granitic matter beneath. Seeing the slight chemical differences usually noticed between hornblende, augite, and hypersthene, it would be expected that changes would be effected in the aspect of the rocks containing them, under the circumstances mentioned. The hypersthene rock of some localities, as for example, that of Cocks Tor, near Tavistock, Devon, appears to come under this head. While on this subject, it may be remarked that other modifications are sometimes observable, as if the substances composing the rocks being originally more varied, or certain others having entered among them from without, after exposure to change from the consequences of juxtaposition to great masses of molten matter, minerals appeared not found beyond the limits which may be assigned to these alterations.

In some regions, as, for example, in the counties of Wicklow and Wexford, in Ireland, the manner in which *andalusite* has been developed, forcing off other portions of the rock, such as mica; the old stratification of the deposit being still retained, is highly instructive. Near the intruded granite, the crystals of this mineral are occasionally found of large size; and while they have, as it were, shouldered off the other substances in the way of their formation, they sometimes exhibit portions of entangled matter, such as mica, as might be expected in such a mode of production. This mineral, not uncommon under similar conditions, is precisely one of those which would be expected to be thus formed, being essentially a silicate of alumina (the base of the clays), a compound forming a prominent part of the original deposits in which these andalusites become developed. *Chiastolite* is also a form in which the silicate of alumina appears amid altered rocks, and is one not uncommon among the old sedimentary deposits modified in contact with granite in Devon and Cornwall. *Staurolite*, another common mineral developed amid mechanically-formed deposits, when acted upon by masses of granitic and of other igneous rocks in a molten state, is again one which might be expected, being essentially composed of silica, alumina, and peroxide of iron, with the addition of a small portion of magnesia. *Cyanite* is also another form in which silicate of alumina occurs developed amid altered rocks. Garnets are often very common in those which have suffered modification in the adjustment of their component parts. When the observer refers to the chemical composition of these minerals, as shown by various analyses, he will see, by duly considering the isomorphism of certain substances that their component parts may be readily gathered together, under conditions for their movement, from amid rocks apparently of different kinds. While peroxide of iron constitutes a prominent portion of some garnets, it is replaced by alumina in others; and while lime forms an important substance in most garnets, it may be considered as replaced by protoxide of iron in others.* Amid the various altered rocks in which garnets have been developed, the pushing aside, as it were, of other parts

* The following, among the numerous analyses of garnets, may show their varied composition, chiefly due to isomorphism :—

No.	Silica.	Alumina.	Iron.		Manganese Protoxide.	Magnesia.	Lime.	Potash.
			Protoxide.	Peroxide.				
1	39·85	20·60	24·85	..	0·46	9·93	3·51	..
2	35·64	..	..	30·00	3·02	..	29·31	2·35
3	42·45	22·47	9·29	..	6·27	13·43	6·53	..
4	36·45	2·06	..	24·48	0·28	0·06	30·76	..

1. Greenland (Karsten); 2 Altenau, Hartz (Trolle-Wachtmeister); 3. Arendal (Trolle-Wachtmeister); 4. Beaujeux (Ebelmen).

of associated mineral matter, when their crystallization was effected, may be well studied. Among their modes of occurrence, those where they have been developed amid sandstones, as, for example, near Killan; in the county of Wexford, are highly interesting, the grains of sand being forced asunder to permit the development of the crystals of garnets.

The transmission of mineral matter from the igneous and heating body into the prior-formed rocks, whether these were or were not consolidated, seems well shown when boracic acid is present among the former. This appears to have been the case around much of the granite in Devon and Cornwall, where schorl, as above mentioned (p. 578), is not only found disseminated around the general mass of granite in numerous localities, but also constitutes the outer portion of that rock in many situations. The matter of the schorl (chiefly silicic acid, boracic acid, and alumina)* has passed into the pre-existing and mechanically-formed rocks in many places, among which Fatwork Hill and Castle an Dinas, near St. Columb, Cornwall, may be noticed as localities where this circumstance can be well seen. The boracic acid might, indeed, have solely escaped out of the granitic mass, and meeting with the other essential parts of schorl, have produced the latter mineral amid the grains of the mechanically-formed rocks thus acted upon.

With respect to mica, also, its component parts often appear as if introduced from the igneous rocks into many of the sedimentary deposits against, or amid which they have been intruded, at the same time that certain mechanically-formed rocks, such as arenaceous deposits containing detrital mica, seem merely to have had

* The analyses of schorl, by M. Hermann, gave for the composition of black schorl from Gornoschit, near Katherenenburg (1), of brown, from Mursinsk (2), and of green, from Totschilnaja, Ural (3), and of black, from Monte Rosa, by Le Play (4):—

—	1	2	3	4
Silica	39·00	37·80	40·54	44·10
Boracic Acid	10·73	9·90	11·79	5·72
Alumina	30·65	30·56	31·77	26·36
Protoxide of Iron	1·58	0·50	..	11·96
Peroxide of Iron	6·10	12·07	3·65	..
Manganese Protoxide	..	2·50	0·90	..
Magnesia	9·44	1·42	6·44	6·96
Lime	..	..	..	0·50
Lithia	..	0·50	2·09	..
Potash	..	..	..	2·32
Soda	..	2·09	..	..
Carbonic Acid	2·50	1·66	1·66	..
Chrome	..	..	1·17	..
Water	..	..	..	0·60

the micaceous matter so acted upon, as to form better developed films of that mineral, its component parts having adjusted themselves in a manner more resembling an original formation of mica. In like manner, also, ordinary felspar seems to have been thus produced, so that an original deposit in which grains of quartz, felspar, and mica have been accumulated, (the detritus of some pre-existing granitic rock,) would become the laminated compound of that character known as *gneiss*, while one formed only of grains of quartz and mica would become *mica slate*.* Paying due regard to the original composition of the rocks acted upon, with proper reference to the conditions under which mineral matter may be passed out, and move among the parts of the igneous rocks, gradually cooling down during a long lapse of time, and also amid the pores and fissures of the prior-formed rocks, the observer would anticipate very numerous combinations, producing great modifications and changes in the original composition of deposits. He will find the study one of great interest, well repaying the time he may devote to it.

* When considering the various kinds of modification of parts which deposits may sustain from the contact of great masses of igneous rocks in a fused state, or from descent towards the interior of the earth, so that somewhat similar conditions may be produced, it is not a little interesting for the geologist to direct his attention to those which certain of them would, in consequence, present. If, for example, the thick beds of the *millstone grit*, forming the lower part of the coal-measure series (as this grit occurs in the midland and northern counties of England, composed of quartz, mica, and felspar, the latter usually decomposed), were exposed to the conditions for alteration above noticed, the compound would have a granitic appearance, the more especially if the silicates of potash or soda, or both, were introduced, and again united with the remains of the previous felspar. This rock, even as it is, has often the appearance of a somewhat decomposed granite, as is the case with many sandstones of the coal measures of the British Islands generally.

CHAPTER XXXII.

CLEAVAGE OF ROCKS.—CLEAVAGE IN SANDSTONES AND SHALES.—CL IN SHALES AND LIMESTONES.—MINOR INTERRUPTIONS TO CL ACTION.—CLEAVAGE ON THE LARGE SCALE.—RANGE OF CL THROUGH CONTORTED ROCKS.—DOUBLE CLEAVAGE.—RELATIVI OF CLEAVAGE.—ELONGATION AND DISTORTION OF ORGANIC REMA CLEAVAGE ACTION.—DIFFERENT DIRECTIONS OF CLEAVAGE IN TI OR JUXTAPOSED DISTRICTS.—JOINTS IN ROCKS.—DIRECTIONS AND OF JOINTS.—JOINTS IN GRANITE.—JOINTS IN SEDIMENTARY R JOINTS AMONG CONGLOMERATES—JOINTS IN LIMESTONE.—MOVEM ROCKS AFTER JOINTING.—COMPLICATION OF BEDDING, JOINTIN CLEAVAGE.

WHETHER modifications and changes in rocks have been pr by the depression of deposits to depths beneath the earth's s where high temperature may cause the effects noticed, be b about by similar action from the juxtaposition of great i masses elevated or intruded in a state of igneous fusion, or c dation be produced chemically in any other way, an observe not be unprepared to consider that the matter thus acted upon sometimes exhibit an arrangement of parts corresponding some adjustment on the great scale. Whether those arrang of the parts of rocks to which the terms *cleavage* and *join* been applied, may have taken place during their first consoli may be due to some action on the large scale after their consol as a whole or in part, or sometimes have been produced und conditions by an influence acting independently of consoli the subject is one which would appear to require more ex observation, and a better-digested body of facts than has as y obtained.

When a geologist has before him, as in the slate qua North Wales, a mass of mineral matter which he has re conclude has once been a bed of clay or mud, now not only consolidated, but also rendered highly fissile in planes which do not correspond with that of the original deposit; such planes constituting a part of certain others traversing the rocks of a district generally, that he should be led to infer, with Professor Sedgwick and some other geologists, that the finely-divided, yet mechanically-deposited matter had been gathered together by some force, reminding us of that which unites the particles of crystals of certain given combinations of substances, might appear probable.

Upon studying the districts where the cleavage of rocks occurs, the observer usually finds a considerable degree of uniformity of direction in the course of the cleavage planes, where they cut the horizon, through a variety of rocks, and over considerable areas. In certain of the masses or beds in which this has taken place, the effect appears to have been more easily produced than in others, and it may be sometimes seen that while, in associated coarse and fine-grained beds, the latter are beautifully cleaved, the former appear unaffected, as if the fine particles of the one could be easily acted upon, while the coarser grains better resisted a new adjustment. This conclusion requires, however, much caution, since though the coarser beds may not, at first sight, appear affected by the arrangement of parts producing cleavage, they may be found when broken to exhibit divisional planes, though these may not be so numerous, in the direction of those in the finer-grained beds associated with them. Thus, in the following section (fig. 232),

Fig. 232.

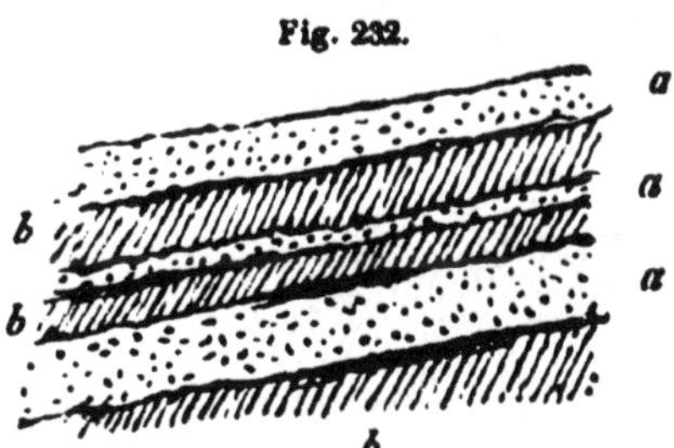

representing sandstone beds *a*, *a*, *a*, associated with argillaceous beds, *b*, *b*, *b*, while the cleavage may be well developed and easily seen in the latter, in the former it may also be found by fracture of the beds, or by their decomposition from atmospheric influences. This, however, by no means constantly occurs, the conditions under which the fine-grained beds were cleaved, not having apparently been favourable to the adjustment of the parts of the other beds, so that they became thus divisible.

In certain of the carboniferous limestone districts in Ireland, where shales are associated with the limestone beds, the cleavage of the former is apparent, while the latter are either unaffected by it, or so obscurely as not to have their component parts adjusted to the same amount by this action. The following section (fig. 233), at

Fig. 233.

Clonea Castle, county Waterford, will serve to illustrate this circumstance, *a*, *a*, *a*, being beds of limestone which do not exhibit cleavage, while the shales, *b*, *b*, usually more or less calcareous, and associated with them, are cleaved. The same locality also shows that where the limestone beds become somewhat argillaceous, partaking partly of the character of the shales, and partly of the more pure limestones, they exhibit marks of cleavage action, as if the particles, being then less coherent, had more readily given way before that influence.

The following sketch (fig. 234) may be found useful in illustration of the modified passage of cleavage through dissimilar substances, or of those, the coherence of the parts of which, at the time of the cleavage action, may have been different. It represents a portion of the Devonian series, on the east of Hillsborough,

Fig. 234.

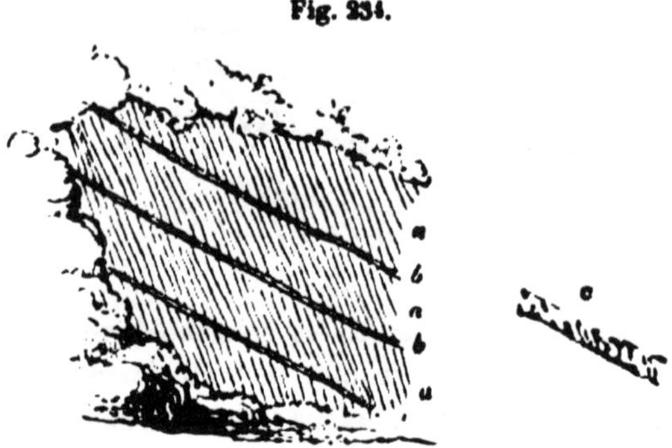

near Ilfracombe, North Devon, *b*, *b*, *b*, being thin seams of limestone, about two or three inches thick, *a*, *a*, *a*, argillaceous slates; the argillaceous slates being made so by cleavage. The true planes of deposit are shown by those of the interstratified seams of limestone. The planes of cleavage lamination traverse the whole with a general southern dip, slightly interrupted at the seams of limestone, where their course is modified; and though the limestone is divided in the same general direction, it is so in a somewhat contorted manner, as shown by the carbonate of lime which has been subsequently infiltrated and deposited in the fissures so formed. The somewhat contorted modification of the cleavage in the limestone is shown at *c*. In such cases as these the geologist may consider the limestone to have been consolidated so as to have been capable of a certain amount of fracture, while the mud or clay, now consolidated as well as cleaved, so as to form hard slates in the direction of the cleavage, had the whole of its component parts more or less rearranged in certain directions by the cleavage action.

At times an interruption may be traced even between argillaceous beds themselves, when these are piled upon each other in a manner marking a pause in the deposit, so that a clear surface has been formed between the production of one bed and the accumulation of another. The following section (fig. 235,) seen near Wiveliscombe, West Somerset, shows that while the planes of cleavage

Fig. 235.

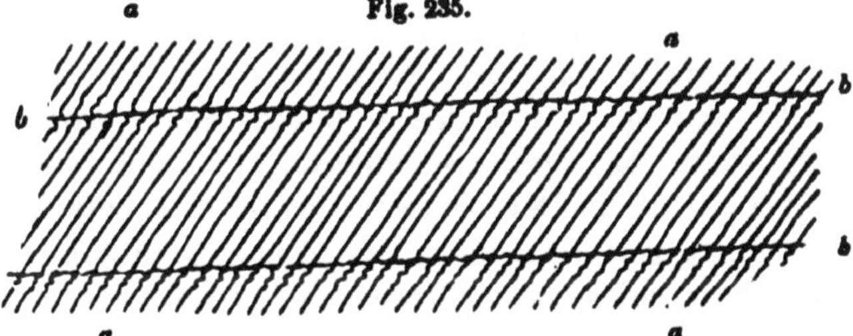

take a general direction, *a a*, they are slightly bent, and undulate where the partings of the beds, *b b*, occur. Whether, at the time of the cleavage the mere break in the continuity of the argillaceous matter was sufficient to produce this effect, or whether any kinds of substances may have been in solution in water, occupying the interstices between the beds, causing the effects observed, becomes matter for investigation.

These interruptions and modifications of cleavage, though important for the study of its cause and mode of action, and which are far from being always observable, become lost when cleavage is traced through considerable masses of [illegible] on the large scale. We find it, as Professor Sedgwick [illegible] pointed out, in a large portion of North Wales,* traversing all kinds of rocks in given directions, over wide areas, notwithstanding the varied position of their beds. Indeed, it may sometimes be readily traced through them when contorted in all directions, as well horizontally as vertically. As the cleavage thus cuts through even contorted rocks, it must clearly often happen that it passes through their planes of direction at various angles. This will be found to occur more frequently than might, without very careful examination, at first sight appear; however it may arise that, in certain districts, the general direction of the true bedding, or planes of deposit, may be found to coincide with that of the cleavage, as regards horizontal range, whatever variation the dip of the cleavage may exhibit as regards itself, or the true bedding of the rocks. With respect to a mass of rocks of varied kinds, detrital and igneous, with some admixture of limestone, traversed by cleavage ranging diagonally to the general bedding, the chain of hills known as the Chair of Kildare, Ireland, may be taken as a good example, easily visited. The range of the cleavage coincides generally with that of the neighbouring districts in the counties of Wicklow and Wexford; while the beds have a direction diagonally across the cleavage. Among these hills highly-crystalline porphyries will be found cut by cleavage, as well as the argillaceous and arenaceous beds of the Silurian series of which they are principally composed. As illustrative of variation in the dip of the cleavage, and of the true beds seen vertically, and especially when the latter are contorted, the following section (fig. 236) of part of the Devonian series, on the coast between Morte and Bull Point, near Ilfracombe, Devon, may be useful, inasmuch as the true beds, chiefly of argillaceous matter, have been so cleaved in a constant direction, that while the cleavage planes, *a*, *a*, some-

* "The whole region," the Professor observes, with reference to the country near Rhaiadr to the gorges of the Eolan and Towy, "is made up of contorted strata; and of the true bedding there is not the shadow of a doubt. Many parts are of a coarse mechanical structure, but subordinate to them are fine crystalline chloritic slates. But the coarser beds and the finer, the twisted and the straight, have all been subjected to one change. Crystalline forces have rearranged whole mountain masses of them, producing a beautiful crystalline cleavage, passing alike through all the strata. And again, through all this region, whatever be the contortions of the rocks, the planes of the cleavage pass on generally without deviation, running in parallel lines from one end to the other, and inclining at a great angle, to a point only a few degrees west of magnetic north."—"On the structure of Large Mineral Masses, Trans. of the Geological Society of London," 2nd series, vol. iii., p. 477.

times cut the bedding at right angles, at others they coincide with it, as at *b*, *b*.

Fig. 236.

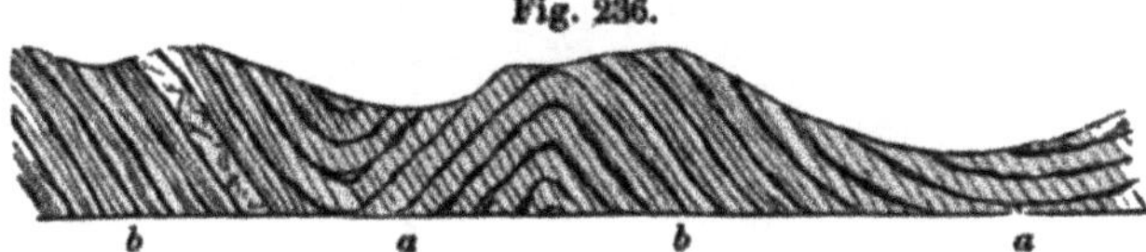

Among the contorted and cleaved rocks, some will be found which may lead an observer to consider how far the cleavage took place after their consolidation into the hard sandstone, and even quartz rocks which he may find, or had occurred while that consolidation was in progress. The following sketch (fig. 237) of the

Fig. 237.

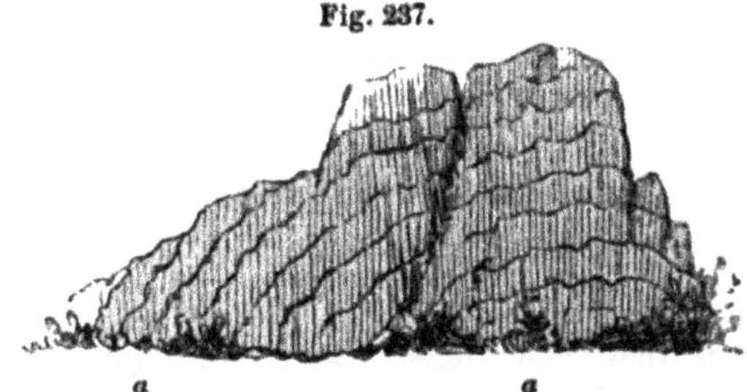

cleavage of hard sandstone beds, with some slate, part of the Cambrian series, at Bwlchhela, nearly opposite the Penrhyn slate quarries, North Wales, may serve to illustrate this subject, *a*, *a*, being the lines of cleavage traversing the disturbed beds. Cleavage of this kind is exhibited on a much larger scale at the Holyhead Mountain, Anglesea, amid its quartz rocks, thus rendered fissile in a great measure across the bedding, as is shown in the following section (fig. 238) seen on the cliff opposite the South Stack Light-

Fig. 238.

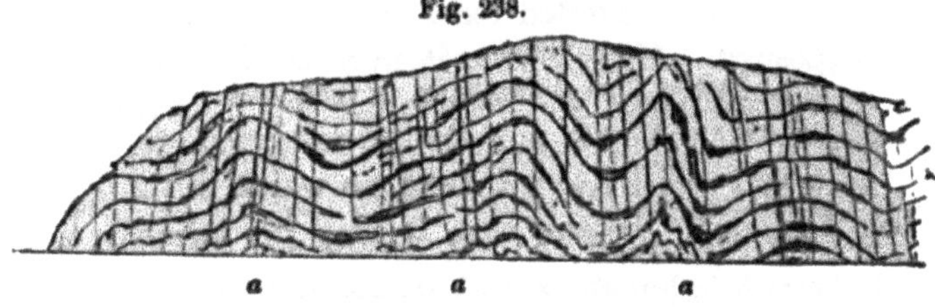

house ; the nearly vertical lines, *a*, *a*, *a*, representing the direction of those of cleavage, varying somewhat in their amount of dips, the contorted beds being composed of sandstone and slate ; the former, for the most part, converted into quartz rock.

Occasionally a double cleavage will be found, one set of planes cutting another, as in the annexed section (fig. 239), where two the mineral matter of the deposit into elongated forms of a pris-

Fig. 239.

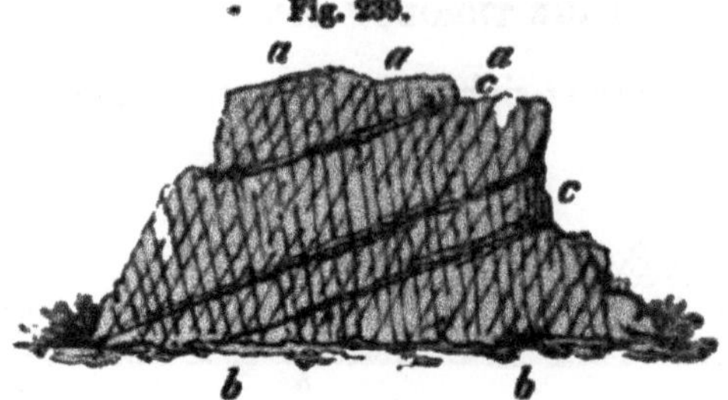

matic character.* To the same kind of action we may, perhaps, assign those somewhat regularly-formed solids into which arenaceous beds are occasionally divided in countries where cleavage has been effected. In such cases the cleavage planes cut those of deposit or true bedding, either at right or considerable angles, so as to produce forms of the following kind (fig. 240). The result-

Fig. 240.

ing portions of rock vary considerably in size. We have seen some not much larger than four times that above represented, though they are usually much larger.

As to the relative date of the cleavage, the observer may sometimes obtain evidence of importance. Thus in Ireland, in the old red sandstone series of the counties Waterford, Kerry,† and Cork, which affords such excellent examples of cleavage, it will probably have been effected after that of the Silurian rocks of Wicklow, Wexford, and Waterford; inasmuch as portions of Silurian slates are to be found in the conglomerates of the old red sandstone of Waterford, clearly worn off, in a cleaved condition, from the subjacent, upturned and contorted rocks, on which this conglomerate has been deposited. Hence, also, it may be inferred that the cleavage of the mountain or carboniferous limestone shales found in some parts of Ireland, occurred after that of the Silurian rocks above mentioned. In the following section (fig. 241), representing

* When the joints, to be hereafter noticed, are numerous and somewhat close to each other, the rock becomes broken up into a multitude of short irregular prismatic portions, of a marked character.

† In the road, now so much visited for the beauty of its scenery, between Killarney and Glengariff, excellent examples of cleavage will be seen.

veins of a porphyritic rock, *a*, *a*, traversing some Devonian slates, *b*, *b*, between Cawsand and Redding Point, Plymouth Sound, the

Fig. 241.

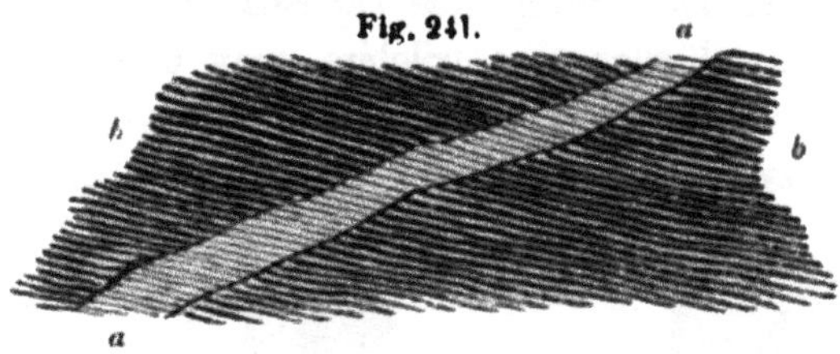

slates and the igneous rock presenting one common lamination from cleavage, the latter would be effected after the intrusion of the porphyry. Upon studying the mode of occurrence of the rocks in the district, it will be found, as above noticed (p. 569), that porphyries of this kind might belong to the period of those known as *elvans* in Cornwall and Devon.

By examining the conglomerates usually forming the lower part of the new red sandstone series of Devonshire, the observer detects portions of the prior-formed rocks laminated in a manner so agreeing with cleavage, that he may infer that the cleavage of the older Devonian rocks was effected before the deposit of the new red sandstone in that district, and subsequently to the intrusion of the granite.

Upon following out the modification and adjustment of parts in rocks traversed by cleavage, it will in many districts be seen that there has been a movement and rearrangement of them in directions corresponding with the planes of cleavage. There have often been elongations in those directions, so that any organic remains

Fig. 242.

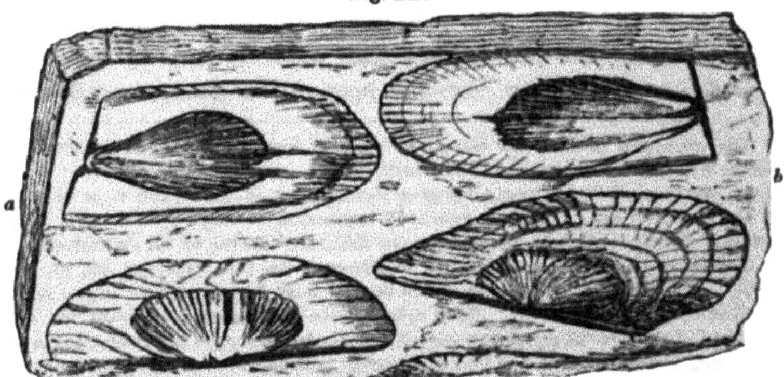

contained in the beds become distorted, and seem as if pulled out, as in the preceding sketch (fig. 242), where several shells of

Strophomena expansa have suffered this elongation, in the direction of the plane of cleavage, *a b*, at Cwm Idwal, Caernarvonshire, the real form of this shell being that represented beneath (fig. 243). Sometimes a fossil, such as a trilobite, may be doubled down on both sides, as over a ridge, in the following manner (fig. 244), the

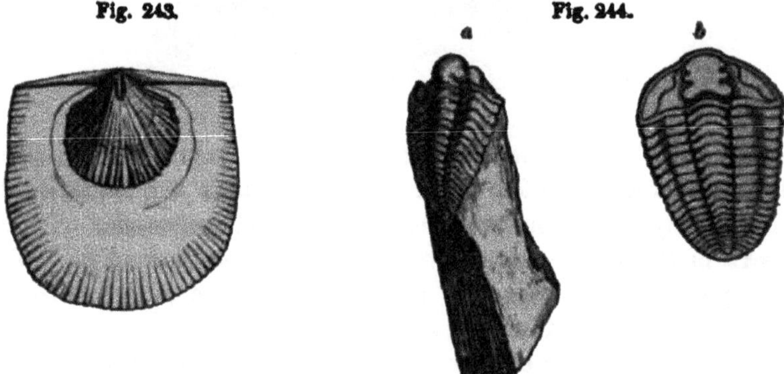

Fig. 243. Fig. 244.

sides of a *Calymene Blumenbachii*, *a*, having been, as it were, pulled down by the cleavage, the real form of this trilobite being that shown at *b*.* This adjustment of a fossil to the planes of cleavage has been regarded by some geologists as effected by a purely mechanical movement, cleavage being referred to a pressure of the component parts of rocks productive of the effects seen. The observer will do well to consider the evidence adduced in support of this view.† At the same time he will have to direct his attention to the lamination of clays effected by electrical action, as shown by Mr. Robert Were Fox,‡ and Mr. Robert Hunt,§ and bear in mind that great masses of rocks are often extended layers of dissimilar or variously aggregated matter, moistened by saline solutions, in which common salt frequently occupies a prominent place. He has also to recollect that these layers, as has been pointed out by Professor Rogers,‖ may, under certain conditions,

* The specimen whence the sketch is taken is from the lower Silurian rock... of Hendre Wen, near Cerrig y Druidion, North Wales.

† See writings of Mr. Sharpe, Geological Journal, vol. ii., p. 74; vol. v., p. 111.

‡ Mr. Robert Were Fox having produced lamination of clays by means of long-continued Voltaic electricity, pointed out (in 1837) the bearing of his experiments in the Report of the Polytechnic Society of Cornwall for that year (pp. 20, 21, and 68, 69). He found that the planes of the laminæ were formed at right angles to the direction of the electric forces. With reference to the cleavage of rocks, Mr. Fox considered "the prevailing directions of the electrical forces, depending often on local causes, to have determined that of cleavage, and the more or less heterogeneous nature of the rock to have modified the extent of their influence."

§ Experiments carried on by Mr. Robert Hunt, at the Museum of Practical Geology, showed similar results; Memoirs of the Geol. Survey, vol. i., p. 433.

‖ Athenæum, Proceedings of British Association, Birmingham, 1849.

be differently heated, one portion descending to the depths beneath the surface of the earth, where a high temperature could act on such parts of the rocks, and the solutions in them, so that thermal electricity may be brought to influence the arrangement of their component parts. Viewing the subject in this light, the particles, all other things being equal, which were the most readily moveable, such as those of clays or unconsolidated accumulations of silt, would appear to be the deposits most readily acted upon. There appears little difficulty in inferring that the particles of matter, such as those which have composed slight organic bodies, or represent such particles at the time when the cleavage action was in force, would yield, like those amid which they were placed, to the same influence, so that, at the final adjustment of all the component parts of a cleaved rock, they would be found so arranged, as regards their original position, as to present an elongated form.

Cleavage will sometimes be found to occupy a somewhat isolated position in a given area, as also a portion unaffected by any action of the kind to occur in a generally-cleaved district. Facts illustrating the causes of these differences are very desirable, as are, indeed, all careful observations on the subject of cleavage, considered as a whole. Whatever view may be taken of its cause, extended research is required as to the direction of cleavage through many and variously-situated regions, the composition and mode of occurrence of the rocks traversed, and the changes in the dips of the planes of cleavage, as well as in their directions. In certain districts, where different directions of cleavage seem much to correspond with that of the ranges of the rocks themselves, and these are of different geological ages, as, for example, the north-east portion of North Wales, where certain Upper Silurian deposits have been accumulated upon rocks of the Lower Silurian series, upturned, with others of the Cambrian series supporting them, it becomes desirable to ascertain how far one direction of cleavage [illegible] limited to the rocks of one general range and not found in the other. In other words, endeavouring to ascertain how far the cause of cleavage, or its mode of action, may have depended upon the general position of the beds in the rocks themselves. Now, in the case mentioned, the cleavage of the older rocks (Lower Silurian and Cambrian) differs considerably from that found in those more recently deposited (Upper Silurian). When the observer connects this with the circumstance that in the same country (North Wales) one set of rocks (the former) had been disturbed and contorted, even in parts probably constituting dry land, (certain of its rocks, at least, so consolidated that they could be broken off and form the materials for beaches, with vein quartz, pointing to the filling of cracks and fissures,) he may be led to inquire not only into the evidence of the cleavage having been effected in the older deposits anterior to that in the more recent, but also into the probability of this action having coincided with the different geological times during which the consolidation of both may have taken place.

In some districts, it requires no slight care on the part of a geologist to ascertain whether the lamination of a rock before him may be due to that of original deposit or to cleavage, far more, without some experience, than might at first be thought probable. This difficulty is occasionally also increased by such arrangements of parts, apparently produced during the action effecting cleavage, that the matter of the rock is gathered under somewhat different forms in the planes of cleavage, causing a diversified kind of lamination of a very deceptive kind. Instances of this fact may be well seen in Wales and in southern Ireland. Where such difficulties present themselves, the observer should very carefully search for lines of organic remains, which usually afford clear evidence of the true planes of bedding; a slight seam of such remains may often suffice to place him right with respect to the true bedding of a mass of cleaved rocks. Sandstone, limestone, or other rocks marking the bedding, should also be carefully sought, so that errors, easily committed in some regions, leading to the confusion of the range and direction of the true bedding, may be avoided. In a section such as the following (fig. 245), a stratum

Fig. 245.

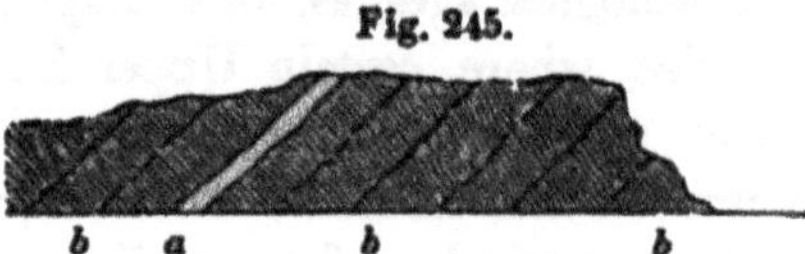

of this kind as at *a*, may show the true bedding, perhaps otherwise very indistinct, the cleavage, *b*, being so prominent as to give a false appearance of deposit lamination in another direction.

Independently of the arrangement of the component parts of rocks into cleavage, under certain conditions, there is another adjustment of them, to which the term *joint* has been applied. It is one to which the observer, among consolidated deposits and igneous matter, will often have to direct his attention. Joints are of far more extended occurrence than cleavage, though they are to be found as commonly in districts affected by the latter, as in others. They are seen as frequently among rocks which have been ejected or protruded in igneous fusion, as among those which are detrital, or which may have been deposited from solution. They traverse the coarsest conglomerates as well as accumulations of the finest sediment, such as once may have been common argillaceous clay or mud. The distinction between coarse cleavage and closely-approximated joints may be sometimes difficult to determine, as, for example, in the section beneath (fig. 246), representing a por-

Fig. 246.

phyry traversed by planes, dividing it into slabs of moderate thickness in a given direction, on the summit of the mountain on the south of Clynog Vawr, Caernarvonshire. As a whole, however, the joints so traverse all kinds of rocks, cutting through such varied bodies, those, for example, often gathered together in a coarse conglomerate, and in such definite and perfect planes, that powerful as cleavage has often been, the action productive of joints sometimes appears to have been more capable of dividing the mineral matter brought within its influence.

It is often by means of the minor solids into which many rocks are divided from the intersection of joints, or by that of the latter and the planes of true bedding, that they can be employed for useful purposes. Though from the last circumstance long known, joints have only attracted attention as among objects of geological interest within the present century. The joints of granitic rocks appear to have been among the earliest of those observed as having definite, or nearly definite directions for considerable distances in given areas.* The planes of joints in these rocks not only present

* With respect to the directions of joints in south-western England, Professor Sedgwick remarked in 1821 (Cambridge Philosophical Transactions, vol. i.) that "whenever any natural section of the country (Devon and Cornwall) exposes an extended surface of the granite, we find portions of it divided by fissures, which often, for a considerable extent, preserve an exact parallelism among themselves." He further adds, that "these masses are not unfrequently subdivided by a second system of fissures, nearly perpendicular to the former, in consequence of which structure the whole aggregate becomes separated into blocks of rhomboidal form." In 1833, Mr. Enys pointed out that the vertical joints of the Penryn granite ranged from

much interest in themselves, as dividing the matter of the rock, so that in some which contain large crystals of felspar, as in the south-west of England, and other districts, the parts of the divided crystals exactly face one another on each side of the joints, but also as coinciding with a kind of cleavage, usually found ranging parallel to the fissures of jointing, so that the quarry-men will work off minor portions of the granite by, as they term it, taking the *grain* of the stone, this grain being parallel with the planes of the joints. Though the joints are sufficiently obvious, this grain may not be perceptible to the eye of an observer, at the same time that the quarry-men, working from experience in certain directions and planes, produce the effect desired by forming holes and driving numerous wedges in such planes.

The columnar appearance of granite, produced by the occurrence of block upon block, as if artificially piled on each other, is often to be seen on exposed mountain peaks, bosses protruding from more rounded and less elevated masses of that rock, and on sea cliffs. The following (fig. 247) may serve to illustrate the appearance of

Fig 247.

jointed granite on sea-coasts, such as those of the Land's End district, Cornwall. The horizontal planes shown, are due to the structural arrangement of the granite of that county, and are to be

---

N.N.W. to S.S.E., varying but a few degrees from those points.—("On the Granite District near Penryn, Cornwall," London and Edinburgh Phil. Mag., May, 1833.)

From very careful research in the granitic districts of Cornwall and Devon we found, as elsewhere stated (Report on the Geology of Cornwall and Devon, 1839), that many hundreds of observations gave about 80 per cent. of cases in which the great joints differed only 14° from N. 25° W., and about 15 per cent. of instances in which they varied between 14° to 20° from that point, leaving 5 per cent. of cases in which the northerly and southerly joints more approximate to the cross joints. The prevailing direction of the joints in the serpentine district of Cornwall ranges within a few degrees of N. 25° W. In this body of rock, as in the various granitic portions of the same district, there are numerous variations in direction, but viewed as a whole the general range of joints is as above stated.

seen in many other parts of the world, in accordance with the external form of the general mass. It may be, that in the joints of granitic rocks we only have the structural arrangement of parts effected at their consolidation, so that the different origin of the horizontal from the vertical planes may be somewhat imaginary. The more or less vertical joints of some granitic areas are certainly often continued, as for example in Cornwall, into the sedimentary rocks amid which they may occur, but how far it can thence be inferred that these were produced subsequently to the horizontal planes in the granite may be questionable, inasmuch as the jointing of the whole area may have been effected at one geological period. Indeed, as above noticed (p. 580), the granite may often be inferred to be at comparatively slight depths beneath such districts, supporting the sedimentary rocks that have been upraised.

Though it may be sometimes doubted, regarding the more or less vertical jointing of granitic rocks, how far this may be considered as originally structural, like the divisions in certain felspathic and hornblendic rocks, giving a columnar character to them, in the manner of basalts (p. 405), there can be no doubt as to the joints in the sedimentary rocks. Here the observer certainly sees the matter of consolidated gravels, sand, silt and clay, or mud, distinctly divided by planes cutting through it in marked directions for considerable distances. Such appearances as the following (fig. 248), are often presented to our attention amid sedimentary

Fig. 248.

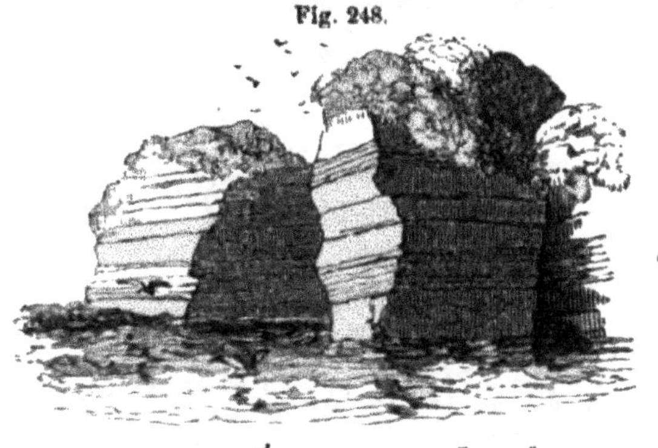

accumulations, particularly when these are well consolidated, two sets of joints, shown by the planes *a* and *b*, intersecting at *c*, and a joint parallel to *a* appearing at *d*.

The most striking illustrations of the action of the power productive of joints are to be seen in conglomerates, where a great variety of pebbles, and of different sizes, is sometimes found divided as smoothly, in given planes, as if these pebbles had been formed of soft yielding substances, and had been cut by some thin sharp instrument, dividing them asunder in one plane. Good illustrations of this circumstance may be seen in the conglomerates amid the older rocks, and perhaps are nowhere better exhibited than among the old red sandstone conglomerates in the county of Waterford. Huge masses of the conglomerate, composed of quartz pebbles, and of portions of older arenaceous and other deposits, as also of igneous rocks, in certain localities, may be found smoothly cut through, and separated by joint planes. In the Commerachs, as for example, in the cliffs rising several hundred feet above the lakes, they seem to divide the mass of conglomerate into huge columns. Upon careful examination, the division presents no trace of dislocation or movement, the faces of the divided parts of the pebbles fitting each other exactly. Joints of this kind are very accessible, and readily seen in the old red sandstone conglomerate resting upon upturned Silurian rocks, opposite the town of Waterford. Of the manner in which these divisional planes pass through conglomerates, without the slightest trace of movement of the beds, or of the pebbles in them, the best opportunities are sometimes afforded on sea-coasts, especially where the beds may be nearly horizontal, and well defined, and where the tide may recede considerably from the shore. Of this kind, the following sketch (fig. 249) of a joint, *a*, *b*, traversing a remarkable conglomerate amid the mountain limestone series on the coast, near Skerries, county

Fig. 249.

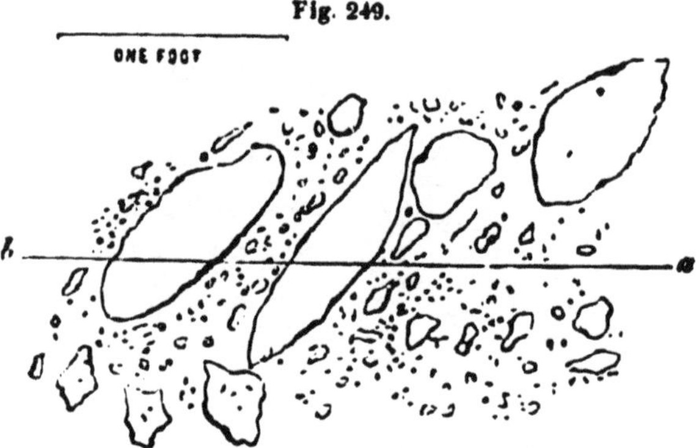

Dublin, the pebbles being of considerable size, may be found illustrative. The surfaces of the divided pebbles, composed of portions of Cambrian rocks, probably derived from masses of them still in part remaining in the vicinity, are as smooth as if no divisional plane of the kind passed through them, yet it is one not only cutting through this conglomerate, but the limestones with which it is associated.

Joints in limestones are often of the most marked kind. In many cases there is no difficulty in distinguishing the bedding from the joints. In others, however, the observer will not find it so easy to determine between the two surfaces, without much care. It sometimes happens, that the joints have a much more marked appearance than the divisions of true bedding. As, for example, in the annexed sketch (fig. 250), wherein the joints are prominently shown, one in particular being somewhat opened at *a*, while the

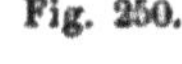

Fig. 250.

true bedding, *b b*, is more obscure. In such cases, the observer has carefully to search for lines of organic remains, dissimilar beds, or partings of shale or other substances, in order to be sure of the true bedding.

The courses of joints, though often of a marked kind through various rocks in the same district, and in the same general directions for long distances, as if the power producing them had been brought into action under some great leading influence, affecting a great mass of mineral matter in that district, however modified in character its parts may be, appear not a little adjusted, as in cleavage, to the main position of the component beds, there being frequently a tendency in joint divisions to take courses at right angles, as a whole, to it. As in cleavage, also, divisions resembling jointing, so far as their distance from each other is concerned, appear to run through certain beds of a general accumulation more abundantly than in others. Of this kind, the divisions through parts of the shales of the lias near Lyme Regis may be taken as an example. Though joints are not there observed in the mass of the argillaceous limestones composing that deposit, in certain beds of shales, on the west of the town, divisions, perpendicular to the beds, may be seen to run like so many planks on a floor, stretching as far as the beds are exposed at low water.

As there appears little reason to doubt that joints, like cleavage, have been formed, under suitable conditions, at different geological times, and as these cleaved or jointed rocks may readily have been moved after they were divided in this manner, it would be expected that, sometimes, the position of the one and the other, as regards their direction with the horizon, is not that in which either the cleavage or jointing was effected. Cleaved and jointed rocks are sometimes found in positions to render such subsequent movements probable. For example, the old red sandstone series of southern Ireland reposes upon Silurian rocks probably cleaved, if they were not also jointed, prior to the accumulation of the former, and the same series is also traversed by similar divisions. Upon studying that portion of Ireland, the observer finds that the old red sandstone, with also the carboniferous or mountain limestone series resting upon it, has been also disturbed since its deposit; hence, the lower rocks having been again moved to permit the rolling and bending of the great mass of matter resting upon them, their original planes of cleavage, if not of joints also, can scarcely be in their original position. The probability of such movements may, therefore, somewhat interfere with first views as to the original position of cleavage and joints, and the geologist should bear in mind, that the movement of a body of rock, divided in this manner, into flexures, might be accompanied by the friction of some of the surfaces of the divisional planes upon each other, thus embarrassing his researches into the original condition of such surfaces. Movements of this kind may give an uncertainty to the slightly-inclined planes of joints which are sometimes found, though there is, as yet, no evidence to show that joints have originated in a manner to render divisions in the mineral matter improbable at these angles with the horizon. Such planes of joints require to be well distin-

Fig. 251.

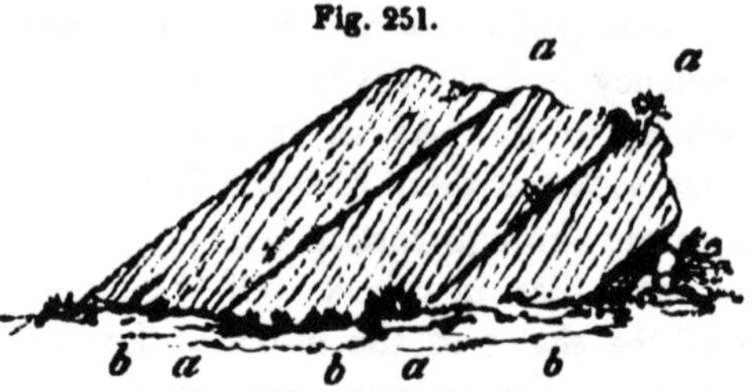

guished from those of true beds, which they often much resemble, as, for example, in the preceding section (fig. 251), where a mass of argillaceous matter, originally a thick accumulation of clay or mud, though now consolidated into hard rock, shows joint lines, *a a a*, and sections of the planes of cleavage, *b b*, but does not exhibit a surface sufficiently large to show the planes of true bedding.

Occasionally, the division of an original deposit of clay or silt, by cleavage and joints, becomes most complicated, requiring no slight care on the part of an observer to arrive at the surfaces of the true beds, more especially when organic remains are absent, and the mineral matter is of a common character throughout. Of this kind of complication, the following sketch (fig. 252) of a quarry at

Fig. 252.

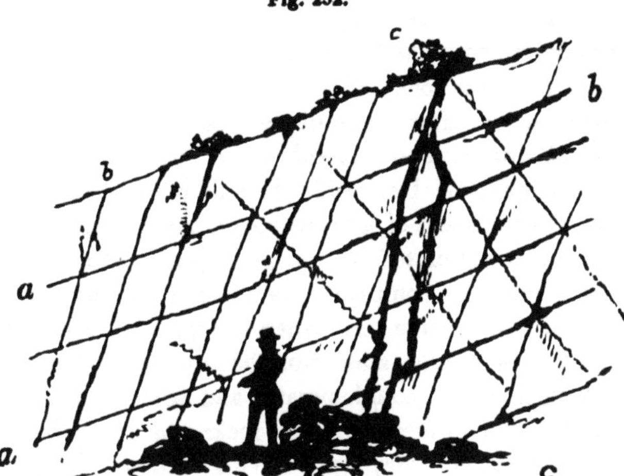

Brewer's Hill, county Wicklow, may be useful as an illustration. The true bedding is a plane, facing the reader, while there are divisional planes ranging in the direction *a a*, in that of *b b*, and in that of *c c*.

CHAPTER XXXIII.

BENDING, CONTORTION, AND FRACTURE OF ROCKS.—EARTH'S RADIUS COMPARED WITH MOUNTAIN HEIGHTS.—VOLUME OF THE EARTH WITH MOUNTAIN RANGES.—EFFECTS OF A GRADUALLY COOLING GLOBE.—MOUNTAIN RANGES VIEWED ON THE LARGE SCALE.—DIRECTIONS OF MOUNTAIN CHAINS.—CONDITIONS EFFECTING THE OBLITERATION OR PRESERVATION OF MOUNTAIN CHAINS.—LATERAL PRESSURE ON BEDS OF ROCK AMID MOUNTAINS.—BENDING AND FOLDING OF DEPOSITS IN THE APPALACHIAN ZONE, NORTH AMERICA.—FLEXURES AND PLICATIONS OF ROCKS IN THE ALPS, AND OF THE RHINE DISTRICT.—IGNEOUS ROCKS AMID CONTORTED BEDS.—CONTORTED COAL MEASURES OF SOUTH WALES.—CONTORTION OF THE COMPONENT PARTS OF BEDS.

THOUGH it has been necessary to allude to the disturbance of various accumulations, as well igneous as those formed by means of water, while noticing rocks of different kinds which have been more or less moved after their deposit or intrusion, it may be desirable to call attention to this subject as one which may also be conveniently considered by itself. It will have been seen, when pointing out the intrusion of igneous rocks, that the disturbance of mineral matter accumulated at one geological period, while the deposits of another were comparatively unmoved, assisted in affording evidence of the relative time when the igneous rock may have been elevated in a molten state from beneath (p. 564); and also that the arrangement of conglomerates and sandstones against or around beds of prior-formed disturbed rocks was useful in showing the probability of ancient dry lands having occurred in particular situations, edged by beaches and coast cliffs, (p. 475).

Though mountains hy no means present us with the only means of studying the bending, contortion, and fracture of rocks on the large scale, they become important from the masses of matter raised in them comparatively high into the atmosphere and sometimes continuous for considerable distances, from the frequent adjustments

Fig. 253.

of lower grounds to them, and the opportunities afforded for obtaining illustrative sections in various planes. A glance at any artificial globe of fair dimensions will be sufficient to show the *ranges or chains*, as they have been termed, of those mountains which constitute marked ridges upon the surface of the earth. With such a globe before him, and bearing in mind the heights of the various ranges or chains of mountains as compared with the diameter of our planet, an observer may, probably, be led to infer that, however elevated and important these may be considered by those wandering amid their depressions, or striving to ascend their heights, viewed as ridges on the surface of the earth, they constitute very minor protrusions, interfering little with the general form of the world. It is somewhat important, in searching for facts illustrative of the production of mountains, that their relative proportion to the volume and diameter of the earth should not be neglected. If, in the annexed diagram (fig. 253), *a b e*, represent a section of a portion of our planet. from its surface *a*, *b*, to its centre *e*; the thick line, *a*, *b*, would be the elevation of even the highest mountains as compared with the radius of the earth. Hence it is not difficult to conceive that the rending of any portion of consolidated or partly consolidated mineral matter, distributed in various ways over the surface, *a*, *b*, and the squeezing of the sides of these rents or fissures against each other, (with or without the propulsion upwards of any molten substances amid interstices in the squeezed masses of consolidated or partly-consolidated mineral matter,) would present ridges of varied forms more or less corresponding with the lines of the fissures.

It has been seen that igneous rocks have been ejected in various ways, that mineral matter worn from them by the action of the sea and atmospheric influences, or obtained in solutions, has been spread over differently-sized areas, that these have sometimes moved up and down as regards the surface of the ocean, and that, considering rocks to have met with more elevated temperatures when depressed, (particularly when covered by superadded mineral matter above,) than when raised into the atmosphere, modifications have been effected in the arrangement of their component parts. Bearing all this well in mind, and giving considerable latitude to views of the thickness of the earth's surface which may thus have been moved, if we assume this thickness to extend in depth even to 100 miles (*a c*, *b d*, fig. 253), we merely arrive at the relative proportion of volume and thickness of the exterior of our planet shown in the accompanying diagram (fig. 254), wherein the depth of the 100 miles is represented by the thick line forming the circle.

Fig. 254.

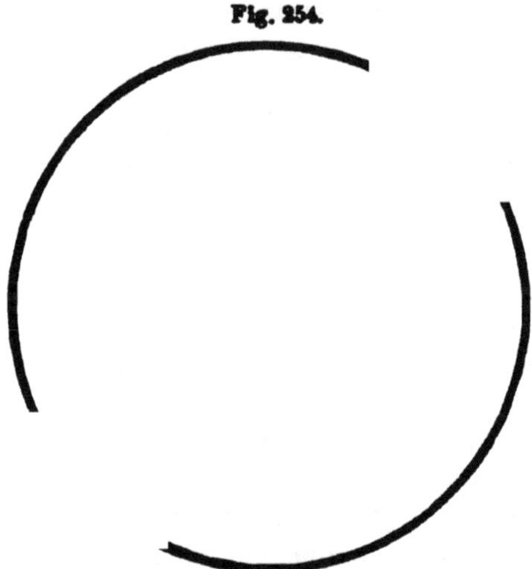

Having prepared himself by this general view of the relative importance of the volume and diameter of the earth, and of the mountain ridges on its surface, the geologist will probably feel also disposed to regard the contortions and fractures of various rocks which he may discover in such ridges with reference to some cause acting generally over the surface of our planet, since he finds marked mountain ranges in all extensive areas of dry land. If, upon further investigation, he obtains evidence, as he will not fail to do, that all mountain ranges have not been elevated to their present positions contemporaneously, the deposits of particular geological periods resting upon prior-formed and disturbed beds in some, while in others, equivalents, in geological time, to these unmoved deposits are themselves disturbed and broken, even, perhaps, covered tranquilly by subsequently-formed beds, he may be induced to conclude that whatever the cause of mountain ranges, it may have continued in action during a long lapse of geological time, and may still exist.*

Upon connecting the form, volume, and diameter of the earth

---

* The following section (fig. 255) may probably be useful in showing the relative age of disturbed beds of rock in mountain ranges. If the rocks *a a*, are found resting quietly on the upturned strata, *b b*, it is inferred that *b b* have been disturbed prior

Fig. 255.

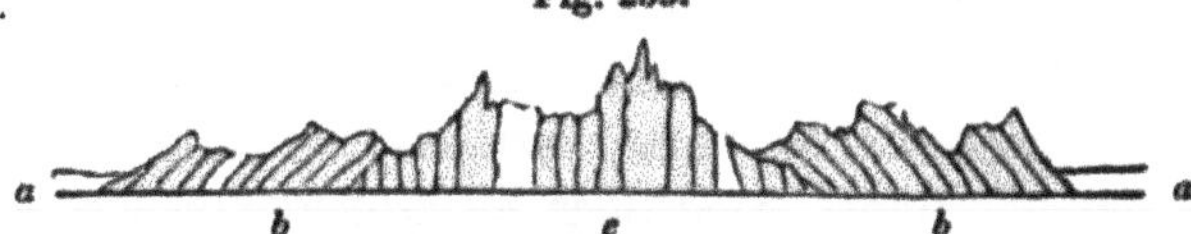

to the accumulation of *a a*; and, consequently, if *a a* be a known rock in the geological series, a relative date is obtained for the movement of the beds *b b*, so far as relates to *a a*. If it should so happen that there are no commonly known deposits absent between them, the approximate relative date of the uplifting of *b b* is obtained. Should it also occur, in any range of mountains or disturbed country, that other accumulations, *c*, are, in like manner, so placed relatively to the deposits *b b*, that another and anterior movement of rocks can be inferred, then, in such range of mountains or disturbed district, there would have been two distinct movements, one prior to the production of *b b*, the other anterior to the accumulation of *a a*. In the case of beds covering contortions, it becomes very needful carefully to observe them sufficiently on the large scale. For example, let beds *a a*, in the annexed section (fig. 256),

Fig. 256.

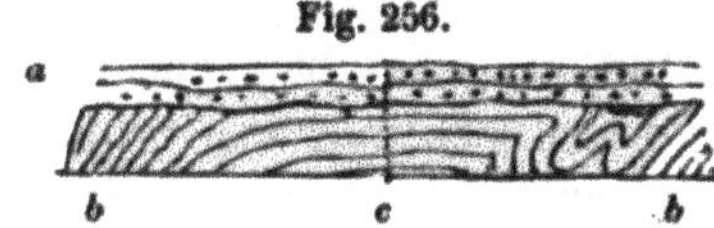

repose quietly on the contorted strata *b b*, and let the only portion exposed to view be where they are cut by the line *c*; then all the beds would appear undisturbed, and it would be only by moving to the right or left, and where the disturbed strata beneath might chance to be fairly exposed, that the real mode of occurrence may be found. This is by no means so needless a caution as might, at first sight, be supposed, particularly when the bends and contortions are upon the large scale. While on this subject it may be useful to notice the imperfect knowledge of the dip, or inclination

Fig. 257.

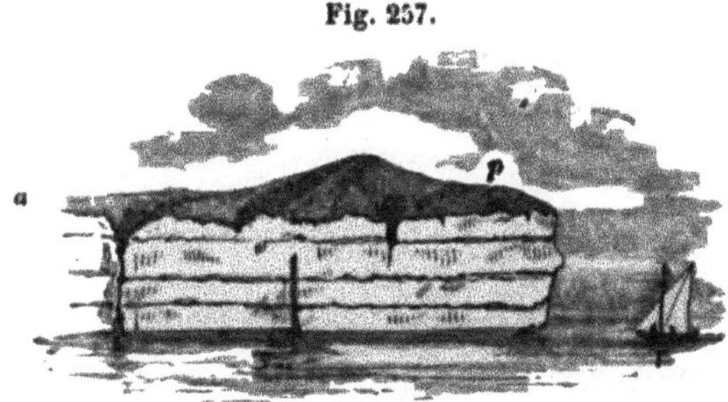

of beds, from one view of them only, since they may even appear horizontal, as in the annexed sketch (fig. 257), while in reality they have been much disturbed, forming a with the relative proportion of the volume and height of mountain ranges, such as those of the Alps, Andes, and Himalaya, it may suggest itself to the observer to consider how far some general cause for these comparatively trifling ridges and rugosities, little interfering with the even character of the surface of the world, may not have followed some change in the volume of the earth itself. Should he try the hypothesis of a spheroid, such as that of the earth, losing heat by radiation into surrounding space, by which a given volume of matter parted gradually with its temperature, one sufficient at first to keep the whole in a liquid state, perhaps he might be led to infer that an oxidised and comparatively cooled superficial covering of solidified mineral matter, having a prevailing crystalline arrangement of parts, especially in its lower portion, might be brought under conditions by which it would have to crack and ridge up, with various adjustments as to foldings and fractures, in order to adjust itself to a mass below, gradually ceasing to occupy some originally-supporting space beneath it. Upon this hypothesis, the oxidation of the various elementary substances constituting the mass of the mineral matter known to us on the surface of the globe, has to be regarded, inasmuch as such oxidation would add to the volume of the elementary substances on that surface, and thus alone aid in altering the exact fitting of a crust of mineral matter upon the remaining portion of the earth beneath, the elementary substances in which had remained unchanged.

Be this as it may, and whatever the hypothesis employed to arrive at the cause of mountain chains, it appears desirable so to examine into the facts connected with the arrangement of the masses of mineral matter of which mountain ranges may be composed, that, while all due regaid be paid to individual chains, observation should

---

portion of some bent or contorted rocks, as is shown in the following view (fig. 258),

Fig. 258.

supposed that of the same cape (*p*, in both figures) on a coast projecting from the mainland, *a a*.

also be directed to the subject on the larger scale. The earth so little differs from a sphere in form, that, in investigations of this kind, it may be regarded as one composed of matter upon which some general action, tending to ridge its surface, might also produce results on that surface of a definite general kind, supposing forces and resistances, and all other circumstances, equal. It is in this supposition of exactly equal conditions that there may be much difficulty with such relatively minor volumes of matter as mountain chains, so that even inferring some constant action, it may be so modified by circumstances as to be materially concealed from observation. To the direction of lines of disturbances on the earth's surface, productive of mountain chains, or otherwise, as may have occurred, much attention has been given of late years, in consequence of the labours of M. Élie de Beaumont on this subject.* He has inferred that there is evidence to show that, during the lapse of geological time, the disturbances of the earth's crust have been effected in given directions, at certain times, and that these disturbances have taken place along considerable fractions of the great circles of our planet.† He has further considered that there

* The first account of the views of M. Élie de Beaumont on this subject was communicated to the Academy of Sciences of Paris, in June, 1829.

† M. Élie de Beaumont remarks, in a communication to the author, in 1831 (Geological Manual, 1831), "Pursuing the subject, as far as my means of observation and induction will permit, it has appeared to me that the different systems (of mountains and disturbed rocks), at least those which are at the same time the most striking and recent, are composed of a certain number of small chains, ranged parallel to the semicircumference of the earth's surface, and occupying a zone of much greater length than breadth; and of which the length embraces a considerable fraction of one of the great circles of the terrestrial sphere. * * * The secular refrigeration, that is to say, the slow diffusion of the primitive heat to which the planets owe their spheroidal forms, and the generally-regular disposition of their beds from the centre to the circumference, in the order of specific gravity,—the secular refrigeration, on the march of which M. Fourier has thrown so much light, does offer an element to which these extraordinary effects (the elevation of mountain chains) may be referred. This element is the relation which a refrigeration so advanced as that of the planetary bodies establishes between the capacity of their solid crusts, and the volume of their internal masses. For a given time, the temperature of the interior of the planets is lowered by a much greater quantity than that on their surfaces, of which the refrigeration is now nearly insensible. We are, undoubtedly, ignorant of the physical properties of the matter composing the interior of these bodies; but analogy leads us to consider, that the inequality of cooling above mentioned would place their crusts under the necessity of continually diminishing their capacities, notwithstanding the nearly rigorous constancy of their temperature, in order that they should not cease exactly to embrace their internal masses, the temperature of which diminishes sensibly. They must therefore depart, in a slight and progressive manner, from the spheroidal figure proper to them, and corresponding to a minimum of capacity; and the gradually-increasing tendency to revert to that figure, whether it acts alone, or whether it combines with other internal causes of change which the planets may contain, may, with great probability, completely account for the ridges and protuberances which have been formed at intervals on the external crust of the earth, and probably also of all the other planets."

have been several distinct systems of disturbance, each marked by a given direction. When more recently describing some lines of this kind which he considers referable to certain systems, succeeding each other in the order of geological time, and all of relatively ancient geological date, M. Élie de Beaumont takes occasion to remark, after alluding to the systems of small arcs of great circles, that, "the fundamental problem presented by a like system of small arcs observed on the surface of the globe, where they are marked by the crests of mountains or by the outcrop of beds, consists in determining the great circle of comparison, to one of the elements of which each of the small arcs observed is parallel."* Thus while

---

* "The small arcs determined by observation," continues M. Élie de Beaumont (Bulletin de la Soc. Géologique de France, t. 1846-7), "may be generally considered as being themselves infinitely small secants, or tangents to so many small circles resulting from the intersection of the surface of the sphere with planes parallel to the great circle of comparison, forming the equator of the whole system. Each of these small circles is a parallel with respect to the equator of the system; it has the same poles as it, and these poles are the two points where all the great circles perpendicular to the small arcs, constituting the system of parallel traces determined by observation, intersect.

"The problem arising from such a system of parallel traces observed on the surface of the globe consists in determining these two poles, or, which amounts to the same thing, its equator, *i.e.*, the great circle of comparison to which each of the small arcs observed may be considered as parallel. This determination," observes M. Élie de Beaumont, "would be easy, and might be made after two, or at least a few observations, if the condition of parallelism were rigorously satisfied: since, however, this in general is but approximately accomplished, the determination of the great circle of comparison can only follow from the means of numerous observations, well combined with each other; and thus, while the observations are not very multiplied or spread over a wide space, we can only advance towards this determination by successive approximations."

As it would be quite impossible to present a correct view of the different systems of disturbance, without the needful tables and calculations on which he has founded them, and which would be here out of place; and as it would moreover be extremely difficult satisfactorily to abridge the very condensed statements of M. Élie de Beaumont, we would refer the geological observer to his memoir in the "Dictionnaire Universelle d'Histoire Naturelle," t. xii. p. 167, and to his more extended and recent general work on this subject, entitled "Notice sur les Systèmes de Montagnes," Paris, 1852, in which he has treated the subject still more at large, and up to the present time, and where all his views respecting the great disturbances on the earth's surface, produced at distinct geological times, will be found.

In a note "Sur la Corrélation des Directions des differents Systèmes de Montagnes" (Comptes Rendus, 9 Septembre, 1850), M. Élie de Beaumont calls attention to the present known directions of mountains, and their adjustment to a *pentagonal network* formed by the intersection of fifteen great circles of the sphere. For the mode of investigation on which this view is founded, our limits compel us to refer to the memoir itself. M. Élie de Beaumont concludes his note by remarking that "the fifteen circles which divide the surface of the sphere into twelve regular pentagons possess the property of the *minimum contour* of the system of lines of most easy crushing (*plus facile écrasement*). If the ridging of the earth's crust were simultaneously produced, these fifteen circles would, perhaps, be alone traced; but as the production of the different systems of mountains has been successive, the octahedral, dodecahedral, and others, have probably been the forms necessarily intermediate in passing from one to the other of the fundamental circles."

estimating the directions of disturbance at different geological times, with reference to the views of M. Élie de Beaumont, the observer would have to bear in mind the great circles of comparison to which the directions of any ranges of mountains, or masses of disturbed beds are to be referred.

In investigations of this kind, the geologist has to consider not only any exertion of force tending to disrupt portions of the earth's surface, acting generally or partially, but also the kind of resistance offered, one which may be materially modified by any variable thickness of the solid matter acted upon, and by variations in the coherence of portions of that matter. As in all movements of this order, differences in the lines of least resistance to some given force, independently of those in that force itself, would produce very marked differences in the ranges of disturbed rocks, especially on the minor scale, in researches of this kind it may not be easy always to estimate very correctly the value of a so-called minor scale. If an observer, aware of the general geological structure of the British Islands, and of a few thousand square miles of the adjoining portion of the continent of Europe, duly weighing the probability of the mode of occurrence of the different rocks to a depth not extending to even more than three or four miles, suppose this mass of variably-accumulated matter to be ridged, squeezed, and contorted by a forec acting in some given direction, so as to produce a lofty chain of mountains like the Alps or the Himalaya, he would expect that very material minor modifications are not unlikely to be produced in the direction of the various parts, and even that these might extend and interfere with the direction of the range itself. If the great masses of igneous rocks, such as the granites of various parts of the area mentioned, are to be inferred as, so to speak, anchored somewhat firmly beneath, a crush, acting upon them and the detrital accumulations by which they may be surrounded superficially, or be covered by to various depths, would be expected to be marked by an arrangement of the mineral matter in accordance with its different coherence, form, and thickness.

As during the progress of geological time so much of the earth's surface, formed of either igneous products or strewed over with detrital, or chemically-deposited matter of various kinds, as also with the remains of animal and vegetable life, has been covered by more modern accumulations of the like kind, even now, over wide-spread areas, concealing them, it becomes no easy task for the geologist to picture to himself the surface conditions of our planet at given periods, so that the disturbed and undisturbed portions may be duly estimated. This becomes the more difficult as his investigations extend to the earlier periods, since not only may so much of the then surfaces of the earth be now buried beneath more modern accumulations, but even the ridging of such surfaces, constituting mountains, may have been obliterated by that action of the sea and atmospheric influences to which the term *denudation* has been applied. Looking at these sources of the removal of mineral matter, and for the moment inferring all other conditions to be equal, the older a range of mountains the less should we expect the remains of it; and conversely, the more modern the range the more should we expect to find it unaltered in its form and general character. Here at once the differences in the other conditions present themselves. Contemporaneously-produced ranges of mountains, and even portions of them, may have been acted upon very variously. One range, or part of it, in some given area, might remain as when thrust into the atmosphere, modified only by the influences to which it has been therein exposed, while in another area, or part of one, the land may have been depressed beneath and raised above the sea level, even several times, with the attendant consequences of either new coverings or the removal of mineral matter thence arising.

Fortunately in Europe and America large tracts are found, where the beds of the older fossiliferous rocks still occupy positions not very different from those of their accumulation, and wide-spread areas have changed their relative levels, as regards that of the sea, so in mass, that these old sea-bottoms became large portions of dry land, without the folding and crushing of their component beds. Other considerable areas of like kinds may probably be detected when extended, and as yet little explored regions become better known. Be that as it may, these old undisturbed portions of the world's surface become important, from pointing out those parts of it which have escaped the ridging, squeezing, and contortions to be found in many other localities. If we could obtain such somewhat widely dispersed, they would aid considerably in separating the undisturbed from the disturbed portions of the earth's crust, so far as regards the squeezing or contortion of them, though not, as is obvious, those which may have been lifted and let down bodily in a horizontal or nearly horizontal manner. When we find, as in the great north and south range of the Ural mountains, these same accumulations squeezed and disturbed as a whole, and in a marked line, and the relative date of the disturbance can be approximately inferred, as has been done by Sir

Roderick Murchison and his colleagues, Count Keyserling and M. de Verneuil,* the geologist obtains a knowledge, not only of the time up to which these portions of the earth's surface may have remained without such disturbance, but also of the direction of the line or lines along which it was effected.

The mountain ranges of the world, occurring in so many parts of its surface, seem all marked by evidence of the squeezing and contortion of the different accumulations disturbed, as far as researches have yet extended. While some show igneous matter to have risen up in somewhat considerable abundance, and apparently when these disturbances were effected, it is not discovered so commonly in others. This may merely depend, all other things being equal, upon the amount of mineral matter of another character which has been removed, or upon that matter having been so adjusted as to conceal the igneous rocks. Much caution is therefore needed when an observer may be engaged in this kind of inquiry. Thus, in some granitic ranges, such, for example, as those above noticed in South-western England and South-eastern Ireland, we may only have the remains of former chains of mountains.

To obtain very close approximations in ranges of mountains, to the amount of folding, contortion, or fracture of the various rocks acted upon in the manner mentioned, sections should be formed proportionally representing these circumstances. Usually, however, no great exactitude is attempted, so that sections of mountain districts merely afford very general views on the subject. Even these, nevertheless, are sufficient to show the great lateral pressure to which the whole, abstracting any igneous rocks (apparently introduced during the time or times of such disturbing action), has been commonly subjected. The observer often finds, when following some given series of beds, presenting characters sufficiently marked for the purpose, and properly weighing the evidence as to gaps, due to the openings from fractures, that if such contorted beds could be again laid out flat, as when deposited,

Fig. 259.

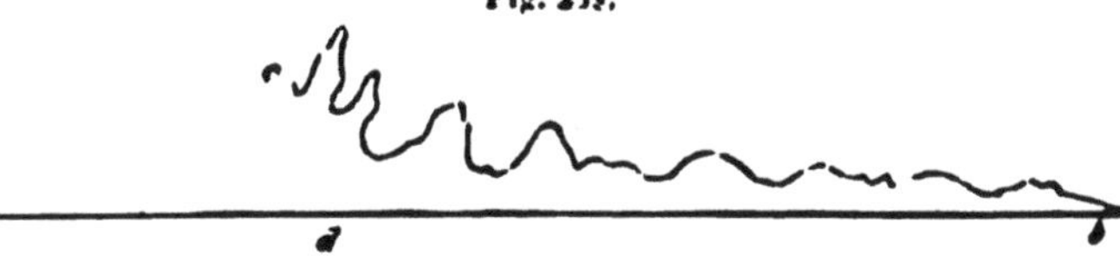

they would have to be spread over a greater superficial area than they now occupy. Thus, if in the annexed section (fig. 259),

* "Geology of European Russia and the Ural Mountains."

the curved and contorted line represents the foldings and contortions of a given series of beds, *c b*, on the flank of some mountain chain, such as the Alps, and, allowing for fractures and portions removed, if that line be reduced to a straight one, *a b*, it will be evident, that a lateral extension to the amount of the distance *a d*, will be required for the return of these beds to their original position, supposing, for illustration, the point *b* to have remained firm. In like manner, if instead of one flank only of a range of mountains, thus exhibiting a folding and contortion of its beds, both flanks do so, and a section across the whole range shows these to be of the kind represented beneath (fig. 260), then the

Fig. 260

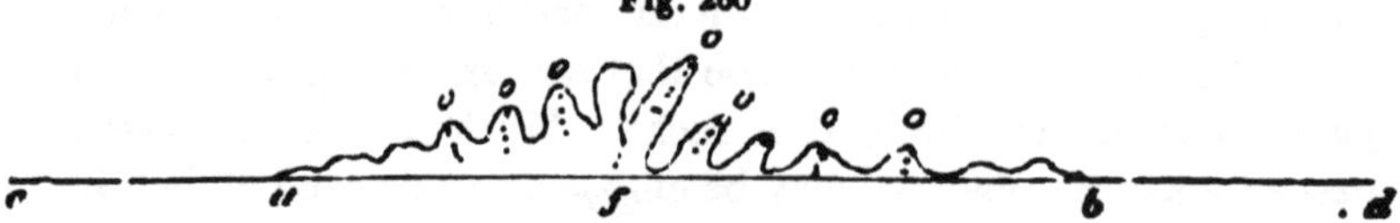

line *c d*, would represent the distance required for the flattening of the folded and contorted beds, instead of that of *a b*, giving the distance now occupied by them. If the points *a* and *b*, be inferred to have remained relatively firm, as respects distances outside them, then there has been a diminution in the distance between them, equal to $c\,a + b\,d$, the beds previously occupying the distance *c d*, being so folded and bent as not to extend beyond *a b*. Hence, also, a motion from *c* to *a*, and from *d* to *b* is inferred, and supposing the substances of these beds sufficiently yielding, this might be accomplished without a break at *f*. Considering breaks to have been formed at the chief bends, as at *o*, *o*, *o*, *o*, the distances for the relative movement *c a*, and *b d*, may be somewhat altered, fractures of the kind represented beneath, in fig. 261,

Fig. 261.

being to be taken into account, *c* being a line of fracture along which the beds *a* are considered to have slid to *b*.

A diminution of the area previously occupied by these folded and contorted beds having been thus effected, the observer has to see whether on the one side of a mountain chain or the other, or on both, there may be any evidence in favour of lateral pressure acting from without inwards, or if there may appear any in favour of a great fissure or fissures, in the ranges of the mountains themselves, against the sides of which the rocks moved, and had adjusted themselves according to the action of gravity, and the lateral thrust upon yielding materials outwards. The usual impression left, by even the general sections given of ranges of mountains, such, for example, as those of the Alps, is, that there has been an elevation of their component rocks in the direction of these main ranges, and that they have adjusted themselves laterally to meet the force of gravity acting vertically upon the upraised mass.* Inferring the needful pressure, it would be expected, that molten matter beneath the masses moved, would be ready to enter amid any openings effected, as far as that pressure permitted, this intruded matter tending to brace much of the fractured beds together, upon cooling. An intrusion of such molten rocks, might, therefore, be among the consequences of the action producing the elevation of the mountain range, and be more or less important, according to circumstances.

Geologists are indebted to the Professors Rogers for observations on an extensive district in North America, one of about 195,000 square miles, which have led them to point out an arrangement of the bends and foldings of disturbed rocks in accordance with the

* Respecting the folding of beds by vertical and lateral pressure, Sir James Hall, as long since as 1813 (Transactions of the Royal Society of Edinburgh, vol. vii., p. 86), showed that this could easily be imitated artificially by taking various pieces of cloth, placing them horizontally on some table, c (fig. 262), pressing them downwards by a

Fig. 262.

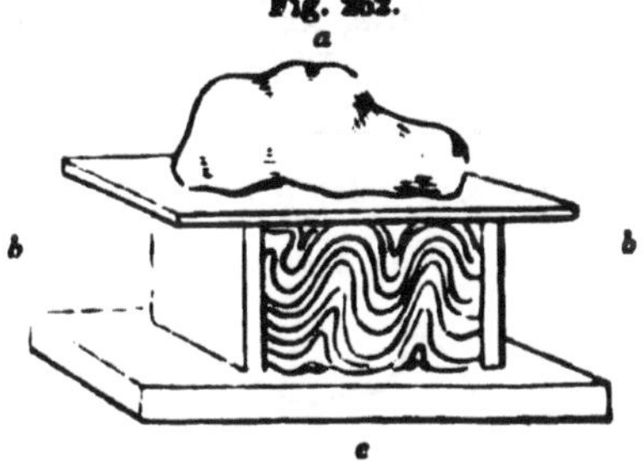

weight, a, acting parallel to the plane of the table beneath, and by applying force laterally, b, b. In experiments of this kind, it is not, however, necessary to have the top weight, a, since if the cloth be in proper quantity, its gravity alone will be sufficient to produce the contortions, and a more exact resemblance to nature be obtained. By moving only one side, or both, as thought desirable, a very interesting illustration of the contortions of beds may thus be easily seen.

distance from the application of force. A careful examination of the Appalachian zone, as they term that region, showed that it is marked by five great belts, which, when crossed from south-east to the north-west, exhibit the greater flexures in the first belt, or that on the south-east of the Blue Ridge or Green Mountain Chain. The component beds of the belt are doubled into enormous, closely compressed alternate folds, dipping almost exclusively to the south-east at angles varying from 45° to 70°. In the third belt, the beds are less compressed, the northern side of each anticlinal curve* approaching nearly to verticality. In the fourth belt, that of the central Appalachians of Pennsylvania, Virginia, and Tenessee, the convex and concave flexures progressively expand, the steepness of the north-west side of each anticlinal gradually diminishing. In the fifth belt, that of the coal region of the Alleghany and Cumberland Mountains, the curves dilate, and subside into broad symmetrical undulations with gentle dips. The folds and undulations of the beds occur in groups, the several axes being very nearly parallel and similar in the character of flexures, many of the larger anticlinals having a length of 80 or 100 miles.† With respect to dislocations of these beds, two systems are noticed, one of short fractures nearly perpendicular to the direction of the anticlinals, the other ranging with them, and often of considerable amount. The longitudinal dislocations (and some in Virginia have a length exceeding 100 miles) are inferred to be broken flexures, the fracture almost invariably occurring on the north-western or inverted sides of the anticlinals, and having a moderately steep south-eastern dip. Some of these great fractures have thrown the portions of once continuous beds not less than 8,000 feet asunder, measured perpendicularly to the surfaces of the strata. After an examination of the disturbed rocks of the Alps, Jura, and of the district of the more ancient fossiliferous rocks of the Rhine, Professor H. Rogers con-

* In a vertical section of rocks, of which the following line *b*, *c* (fig. 263) repre-

Fig. 263.

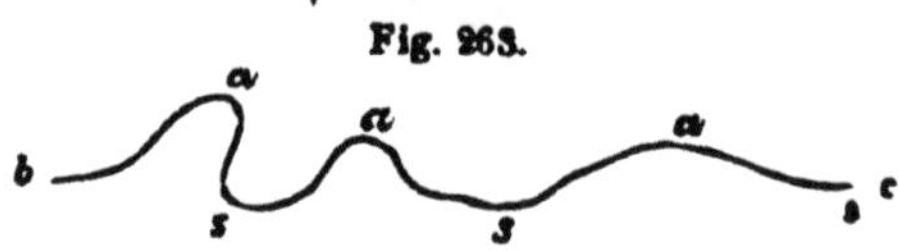

sents the bends from pressure, *a*, *a*, *a*, would be the anticlinal, and *s*, *s*, *s*, the synclinal curves.

† As regards the distances of the contiguous great folds, they are stated to be less than one mile in the south-eastern belt, in the central belt between one and two miles, and in the north-western belt the flexures have an amplitude of from five to ten miles.

siders that in these localities also the like flexures and plications are observable.*

To produce a system of flexures and plications, such as that described by the Professors Rogers as occurring in North America, would not only seem to require great lateral pressure, but also a somewhat uniform and general yielding of the various beds moved during the whole time that the needful action was prolonged. Had there been large volumes of intermingled or deep-seated masses of igneous rocks, offering different resistances to the force employed, much modifications in the resulting flexures and plications would be expected, the softer and less-consolidated beds being even occasionally squeezed over the large masses of the hard igneous rocks. Thus it might happen that when such igneous rocks were in abundance, many masses being deep-seated, the results of an application of force along an extended line of action would be so modified as to offer considerable difficulty in tracing the various flexures and plications to such line. In all cases, as well that of the great Appalachian zone, as in the masses piled up to more marked heights, such as in the Alps and Himalaya, a shortening of the space previously occupied by the component beds appears required (fig. 260, p. 642).

Whether the observer be engaged upon the examination of flexures or plications amid ranges of mountains or less highly-elevated portions of country, it is very desirable not only that he should duly appreciate the amount of the folding and bending of the accumulations disturbed, but also the real outline of the districts. Without a proper reference to this outline, the most exaggerated views may be entertained of the importance of heights and depressions, especially of mountainous regions, relatively to their dis-

* Upon examining the Devonian rocks of the Rhine, Prof. H. Rogers inferred, that the entire region composed of these and the carboniferous series exhibits the effects of the laws of flexure and plication found in the Appalachians, and he points to a section from south-east to north-west, either through the Taunus to Westphalia, or by the Rhine from Bingen to Remagen, or from the Hundsruck to the coal region of Liége, as showing an almost universal south-eastern dip, resulting from the close oblique folds with steep or inverted dips to the north-west of each large anticlinal. He further remarks, that on approaching the northern side of the district the flexures become progressively more open, and that the inequality in the dip of the sides of the anticlinals diminishes, so that in this case also the force would appear to have been applied on the south-east. In the Jura, the Professor considers the anticlinals to have one side of the arch more incurved than the other, but not inverted, and that while the ridges are higher next the great plain of Switzerland, all the individual flexures are steepest towards the Alps. In the Alps, he infers the axis-planes to dip inwards from both flanks towards the central portion, so that the masses are folded in opposite directions; the plications of the Bernese Oberland dipping south, those of the chain of the St. Gothard and the Simplon towards the north.

being the same for heights and distances. In this section *j* represents the Jura; *g*, the lake of Geneva; *v*, the Voirons; *m*, the Mole; *a*, the Aiguille de Varens; *b*, the Breven; MB, the Mont Blanc, and *c*, the Crament.

In certain regions where the diminution of an area, once occupied by a given series of beds, spread out horizontally, is effected by flexures and plications, these deposits even crumpled in various planes, and where also larger flexures may still be traced amid the complicated adjustment of particular portions, considerable masses of igneous rocks, often granitic, may be detected. The chances of some of these masses having risen in a molten state when they could move upwards from the required pressure upon them, has been already noticed. While the exposure of certain of them may have resulted from the removal of mineral matter by denudation (p. 574), others again appear more to have occupied some space against or between folded and contorted beds, into which they could freely enter in a viscous or pasty state. The Professors Rogers have pointed out that the mere injection of liquid and molten matter could scarcely produce the effects observed in the disturbed beds adjoining them, when such matter is considered to form a portion of a general mass of the same kind beneath. Whatever the cause of such juxtapositions of masses of igneous matter, they have to be properly considered, and it is always desirable to compare the space occupied by them with that lost by the folding and plication of the beds disturbed, so that the resemblance or difference may be apparent.† Under any hypothesis, the sliding of no inconsiderable portion of

* Reduced from the section by the Author, inserted in "Sections and Views illustrative of Geological Phenomena." 1830.

† With regard to large and wide-spread masses of granite amid disturbed detrital beds, it may be desirable to bear in mind, that, like volcanic matter of the present

mineral matter on the earth's surface seems required, and duly to appreciate its amount, it becomes needful to bear in mind the probable proportion of the original depth of the strata moved to the breadth of the surface acted upon. If, for example (some of the faults observed where plications in the Appalachian zone have snapped, being, according to the Professors Rogers, 8,000 feet), we take two miles of thickness for the beds moved, 150 miles for their present breadth, measured across their range, and allow one fifth more for their breadth in their prior extended form, the proportion of the thickness to the breadth of the mass disturbed, and more or less slid over some fitting surface beneath, would be about 1 : 90.

Of flexures and plications of beds, the fossiliferous rocks of Europe in many localities afford excellent examples, and of various geological dates. In the British Islands there are abundant opportunities for their study, as well on the minor as the larger scale. Some of those in Wales and parts of Ireland are well worthy of attention, not only for the folding of igneous products of various kinds amid the ordinary detrital deposits with which they are associated, but also for the apparent adjustment of more yielding to more resisting rocks to each other when exposed to lateral pressure. Some of the contortions of the coal measures of South Wales are of a very illustrative kind. As an example, the following section near Tenby (fig. 265) may be noticed, as the lowest

Fig. 265.

part of the rocks shown near that town are tilted over, so as to have the false appearance of having been deposited after those which they really support; the mountain limestone series, *a*, appearing to repose at Tenby, *r*, (from the part of the curve there visible,) upon the coal measures; *d d*, being the level of the sea, Certain lower beds of this limestone series are brought up, by a bend of the strata, at *b*. *c*, *c*, *c*, represent various shales and sandstones of the coal measures. There are dislocations, or faults, at *f* s, and *w v* is Waterwinch, on the northward of Tenby. A still more considerable apparent inversion, from the same reason, is to be seen

---

time, these may themselves be reheated in part after consolidation in their higher portions, and after the first uplifting, when fissures formed in the prior deposits were even filled with the then molten rock, so that, pressure continuing, these resoftened portions could be squeezed up like the beds of the prior-formed deposits, still further thrusting the latter on one side.

on the shores of part of Milford Haven (Langum Ferry), at a few miles westward from Tenby, where old red sandstone rests inclined on mountain or carboniferous limestone, and this again upon the coal measures.*

In movements of this kind, even disturbances in the arrangement of the component parts of the beds themselves would be expected according to their relative positions, and that of such component parts. Thus, with an interstratification of sand and mud, slightly, if at all, consolidated, if a squeezing lateral motion be applied to these beds collectively, they would yield relatively to their respective resistances. Of this class the minor contortion of the component parts of some sandstones interstratified with shale beds, of the older fossiliferous rocks at Bewly Bay, Waterford Harbour, as shown in the accompanying section (fig. 266), may be taken as an example. The minor portion of the sandstone beds, *a, a, a, a*, are there seen contorted, as in disturbed masses of rock on the large scale, while the shale *b, b, b*, (formerly mud) has slid and adjusted itself in a less marked manner, though its particles may have been also moved.

Fig. 266.

The sliding of more consolidated over less hard beds the observer will often find well shown, as also the marks of friction produced upon the adjustment of such consolidated beds as could move upon or against each other, the striation being often beautifully exhibited.† Pressure movements of this kind may be well seen in Pembrokeshire among the coal measures, some coal beds having so given way before the general force, that their component parts have been squeezed, in the manner represented (fig. 267), into the outer portion of the flexures, *a, a*, while the roofs and bases of the coal beds are brought into contact between them.

Fig. 267.

---

* With respect to such inversions, as they are sufficiently common amid series of beds bearing the same geological names, their occurrence in a sequence of accumulations is merely the same thing made to appear somewhat more important from different names being assigned to different parts of the accumulations moved.

† In the so-much-visited Alum Bay, in the Isle of Wight, where various tertiary beds are turned up vertically, the squeezing of parts of the clays against each other is well exhibited. This is particularly well seen in the white pipe-clay bed, containing fossil plants.

CHAPTER XXXIV.

FAULTS. — PRODUCTION AND DIRECTION OF FISSURES THROUGH ROCKS.— EVIDENCE OF THE RELATIVE DATES OF FISSURES.—FALLACIOUS APPEARANCE FROM A SINGLE MOVEMENT SHIFTING VARIOUS FISSURES.—FISSURES SPLIT AT THEIR ENDS.—LINES OF LEAST RESISTANCE TO FISSURES.—RANGE OF MINERAL VEINS AND COMMON FAULTS IN SOUTH-WESTERN ENGLAND. — RANGE OF FAULTS NEAR SWANSEA. — INCLINATION OF FAULTS.—PARTS OF DEPOSITS PRESERVED BY FAULTS. — COMPLICATED FAULTS.

NOT only has the geologist to direct his attention to the fractures effected by the snapping of plications, when the rocks acted upon have been incapable of further flexure, as a mass or in part; but also to numerous lines of fracture, sometimes of considerable length, which traverse beds and masses of rocks, where violent squeezing into great plications and flexures has not occurred. For such lines of fracture, the mining term *fault* has now been adopted.* Sometimes, when even of considerable length, they are accompanied by very minor dislocations, the sides of the fractures nearly corresponding; at others, the fracture has resulted in a separation of the beds, perpendicular to their surfaces, of several thousand feet, and yet the fracture not he on the bend of a plication. Being of importance in mining districts, and mineral veins being commonly the filling up of spaces consequent on them, the range of these fractures becomes better known in such districts than they would otherwise be; at the same time, however, in numerous other districts, where beds of marked and dissimilar mineral structure occur, they may be readily traced.†

The range of these fractures and the relative time of their pro-

* A term derived from the miners, chiefly those working coal, who, when these dislocations are met with, often find themselves *at fault*, the amount of the dislocation produced not being always clear. They are also known as *troubles* by the miners.

† The geologist will find faults traced with great care in many of the maps of the Geological Survey of the United Kingdom, as, for example, in Sheets 36, 37, 41, 42, 55, 56, 61, 74, and 79 of the Great Britain series.

duction have of late occupied much attention. Their mode of occurrence has especially engaged the attention of Mr. Hopkins, who has investigated the conditions under which directions would be taken by fissures, either formed at the same time, or at periods subsequently to each other, seeing if the anticlinal lines and other disturbances and dislocations of rocks may not be referable to some "widely-diffused action of some simple cause, general in its nature with respect to every part of the globe, and general in its action, at least with respect to the whole of each district, throughout which the phenomena are observed to approximate, without interruption, to the same geometric laws."* Mr. Hopkins commences, as to the action of an elevating force, with as simple an hypothesis as he conceives the subject will admit. "I assume this force," he observes, "to act under portions of the earth's crust of considerable extent at any assignable depth, either with uniform intensity at every point, or in some cases with a somewhat greater intensity at particular points; as, for instance, at points along the line of maximum elevation of an elevated range, or at other points where the actual phenomena seem to indicate a more than ordinary energy of this subterranean action. I suppose this elevatory force, whatever may be its origin, to act upon the lower surface of the uplifted mass, through the medium of some fluid which may be conceived to be an elastic vapour, or in other cases a mass of matter in a state of fusion from heat."†

---

* Hopkins, Researches in Physical Geology; Transactions of the Cambridge Philosophical Society, vol. vi., part i.

† "The first effect of our elevatory force," continues Mr. Hopkins, "will, of course, be to raise the mass under which it acts, and to place it in a state of *extension*, and, consequently, of *tension*. The increase of intensity in the elevatory force might be so rapid as to give it the character of an impulsive force, in which case it would be impossible to calculate the dislocating effects of it." He, therefore, always assumes "this intensity, and that of the consequent tensions to increase *continuously*, till the tension becomes sufficient to rupture the mass, thus producing fissures and dislocations," the nature and position of which are his first objects of investigation. "These will," he proceeds, "depend partly on the elevatory force, and partly on the resistance opposed to its action by the cohesive power of the mass. Our hypotheses respecting the constitution of the elevated mass are by no means restricted to that of perfect homogeneity; on the contrary, it will be seen that its cohesive power may vary in general, according to any continuous law, and, moreover, that this power, in descending along any vertical line, may vary according to any discontinuous law, so that the truth of our general results will be independent, for example, of any want of cohesion between contiguous horizontal beds of a stratified portion of the mass. Vertical, or nearly vertical, planes, however, along which the cohesion is much less than in the mass immediately on either side of them, may produce considerable modifications in the phenomena resulting from the action of an elevatory force. The existence of joints, for instance, or planes of cleavage in the elevated mass, supposing the regularly-jointed or slaty structure to prevail in it previous to its elevation, might affect in a most important degree the character of these phenomena."

After investigating the action of the elevatory force supposed upon a thin lamina, and the direction of the fissures according to various conditions, parallel upon the single application of that force, Mr. Hopkins in applying his researches* to a mass of three dimensions, deduces, among other important conclusions, that, "if the mass be subjected to two systems of parallel tensions of which the directions are perpendicular to each other, two systems of parallel fissures may be produced, of which the directions will be perpendicular to each other." "No two systems of parallel fissures," he infers, "could be thus formed, of which the directions should not be perpendicular to each other." "If the fissures in either of these systems be near to each other, they could not have been formed by such tensions as we have been considering, in succession. They must have been formed simultaneously in each system. One system, however, might be formed at any time subsequently to the other." The modifications produced by different conditions are pointed out, and Mr. Hopkins remarks upon the sense in which the term parallelism, in these investigations, should be regarded, He observes that, "if the size of the mass be comparatively small, and its boundary irregular, this property would altogether cease to characterise the phenomena."†

Reflecting upon the modes of accumulation, as well of igneous as of aqueous deposits, and upon their variable admixture in different localities and at different times, the observer will be led to infer that homogeneity of structure in considerable masses of the mineral matter distributed over the earth's surface would not very frequently be found. Bearing this in mind, as also that in the active volcanic districts of the world there is evidence of the varied intensity of igneous action somewhat irregularly distributed beneath a certain

* As it is out of place in a work of this kind to enter sufficiently into the investigations of Mr. Hopkins, further than to show their general bearing, we would refer for the mode of investigation, and the manner in which the varied results are progressively developed, to the Memoirs themselves, as given in the Cambridge Philosophical Transactions, where the observer will find the subject fully treated.

† Mr. Hopkins remarks, that "if we suppose the superficies of our elevated mass to be of finite length, and to be bounded, for instance, by a line approximating to the form of an elongated ellipse, the direction of the fissures in the transverse system, as we approach towards either extremity of the elevated range, will gradually change from perpendicularity with the major axis (the axis of elevation) till they become parallel to it at the extremities of the ellipse, always preserving their approximate coincidence with the directions of the lines of greatest inclination of the general surface of the mass. The fissures of the other system will be approximately perpendicular to these lines. In this case, then, the two systems will be no longer characterised by any constant relations which their directions bear to that of the axis of elevation, and, therefore, the terms longitudinal and transverse will cease to designate them so correctly as in other cases."

amount of the earth's crust, interferences with fractures of the regular kind above mentioned will probably suggest themselves. Nevertheless, it is highly desirable that he should endeavour to classify the fractures found so commonly in various parts of the world with reference to views on the large scale, so that he may look beyond the details of some given locality, and endeavour to arrive at general conclusions as to the cause of any faults and disturbances of deposits in it by following out their directions, differences of date, and such other circumstances as the conditions under which they are presented to his attention may permit.

The directions of fractures, if even merely those without that movement of either of their sides which should cause them to be faults, having been carefully noted, the relative geological dates of their production may not always be so easy to ascertain. It is found that, in certain districts, we may have several of different geological dates, and yet the whole be uncovered by any deposits of which the relative time of acccumulation may be ascertained, so that the probable date of the whole or some of these faults and fissures may remain uncertain. Unfortunately, this uncertainty too often prevails. At the same time, careful observation will sometimes enable the geologist to obtain somewhat fair evidence of the relative dates of these fractures, and from such evidence probable inferences as to those of others may be occasionally drawn. For example, there is evidence of north and south fractures having traversed the old red sandstone, mountain limestone, and coal measures of Somersetshire, anterior to the accumulation of the new red sandstone series of that district, and posterior to the bending and contortion of the former rocks, the faults traversing these contortions even at right angles, and the older rocks having been worn down after the fractures, the lowest beds of the new red sandstone series of that country reposing tranquilly upon the faulted and abraded older rocks. We may refer, in further illustration of this circumstance, to the geological map of the Mendip Hills, previously given (fig. 167, p. 478), where faults, *r*, *r*, *r*, *r*, somewhat parallel to each other, and having a north and south direction, cut through old red sandstone (1), carboniferous limestone (2), and coal measures (3), so that, from an irregular curve of these beds having been traversed, scarcely any horizontal movement in the present denuded exposure of this part of the Mendip Hills is seen on the north, while there appears a considerable shift on the south. These faults are observed, as far as the surface is concerned, to stop at the lias (6) and new red sandstone (5) on the north, and the only one traced completely across to terminate at the inferior oolite (7) on the south. This apparent and superficial termination of the faults, arises from their having been formed anterior to the deposits of the inferior oolite, lias, and new red sandstone. The chief fault is well known to traverse the coal mines beneath a continuation of these rocks, on its range northward, and is ascertained to be covered over horizontally by them all N. W. from Radstock. Thus, in this case, the date of these faults would be after the disturbance, and the flexure of the coal measures in that district, and anterior to the accumulation of the new red sandstone series (including its dolomitic conglomerate) in the same district. Hence other faults in the vicinity having the same range might be inferred to have been contemporaneously produced with them, the more especially as at Wick Rocks, five miles from Bath, there is also evidence of faults traversing the coal measures, these having been subsequently and quietly covered by beds of the new red sandstone series. That all the faults traversing any denuded or uncovered portion of the older rocks of the same district, were of the same relative date, is shown not to be probable by finding some traversing the higher deposits themselves, both on the north and south of the Mendip Hills, the chief of these taking an east and west direction, so that, fortunately, in this limited district, an observer may learn the value of caution, as to the relative dates of faults.*

As to the exposure of faults, and inferences as to the dislocation of one series by others, much caution is also often needed. For example, it does not follow, as in the subjoined plan (fig. 268),

Fig. 268.

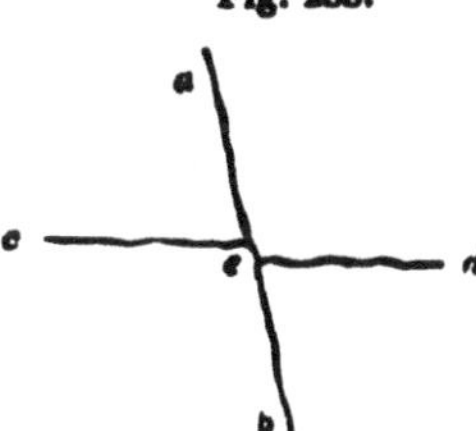

that the fissure *a b*, is posterior to another, *c d*, and has shifted it at *e*, because the one line is continuous and the other not, since

* As regards these subsequent faults, which have commonly an east and west direction, they are seen to have traversed deposits up to the chalk inclusive. A very considerable fault of the latter kind (*see* Sheets 18 and 19 of the Geological Survey of Great Britain) brings chalk into contact with the bed known as the Kimmeridge Clay, one of the oolitic series, at Mere, Wilts. Thus, in this district, there is evidence of an east and west disturbance between the deposit of the coal measures and that of the new red sandstone series, and of another posterior to the deposit and consolidation of the chalk.

such fractures, under fitting conditions, may have been contemporaneous portions of some far larger dislocation, of which these are only minor parts, with adjustments due to minor conditions. Such apparent shifting of one fissure by another is of the same kind as those small complicated fractures close to, or forming parts of, the fissures or faults themselves, and of which the following (fig. 269) is an example, from St. Agnes, Cornwall; small contemporaneous fractures in slate having been filled by peroxide of tin, and so that an apparent heave or shift took place at *h h*. When such appearances present themselves, it is needful to ascertain that any mineral matter, filling a fissure *c d* (fig. 268), has been dislocated and traversed by the fissure *a b*.

Fig. 269.

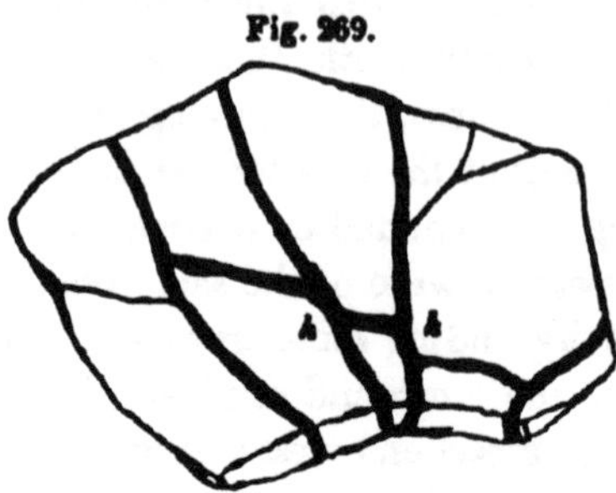

Evidence of the kind of dislocation mentioned is often to be found, so that no doubt remains of one fissure or set of fissures having been first formed, and also altogether or partially filled, prior to the production of another or others. Mining districts often present abundant opportunities for investigations of this kind. As an example, we may notice a well-known district near Redruth, Cornwall, where, as represented beneath (fig. 270), granite, *g*, slates, *s*, elvan dykes, *e*, *e*, *e*, and lodes or mineral

Fig. 270.

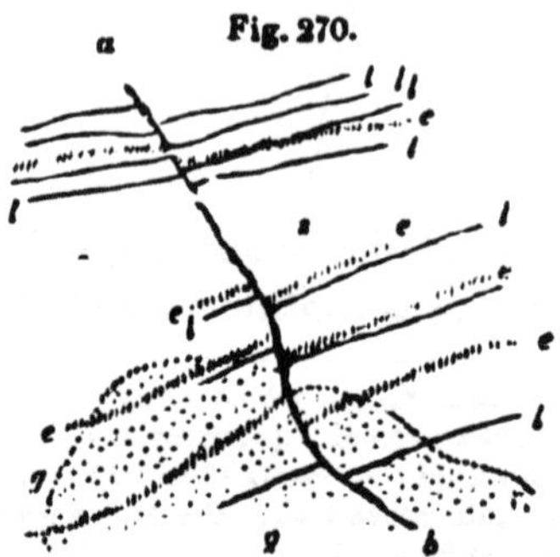

veins, *l*, *l*, *l*, are all cut through and dislocated by a fault *a b*, one of the great *cross courses*, as they are termed, of that country, having northerly and southerly ranges. This plan is also useful in showing the range of the fissures, *e*, *e*, filled with the granitic matter (elvan) introduced after the production of the granitic masses, *g* (p. 565), and the coincidence in range of parts of the fissures, *l*, *l*, of the country, containing copper and tin ores, and subsequently formed, since they traverse these elvans in the vertical section downwards.

With respect to sections in any planes, the horizontal, for example, in countries complicated by the occurrence of different rocks, variably situated as respects each other, or by fissures ranging differently and filled more or less with mineral substances of various kinds, even by mineral matter which has been raised in them in a molten state, some care is needed, so that an observer may properly appreciate the relative position of the parts of the general solid rock broken, shifted, and, as it were, rubbed down to some given plane. Let, for illustration, the following section (fig. 271) represent one of such a district as that of Cornwall, *a b*,

Fig. 271.

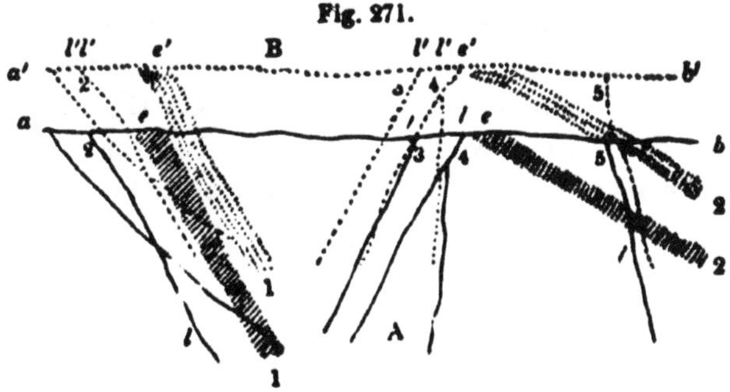

being the surface of the country, *e e*, elvan dykes, and *l*, *l*, *l*, lodes or mineral veins. Let this country be now dislocated in a plane perpendicular to the section, so that *a' b'* on the one side be lifted vertically above *a b* on the other. It will be seen that, on the level *a b*, though the amount of vertical elevation has been common to all the lodes and elvans, these now occupy, on the surface *a b*, very different distances from each other, according to the portions of their various dips or underlies intersected on that surface after the movement mentioned. This will be still further illustrated by the subjoined plan (fig. 272), supposed to be taken on the level *a b*, all above it, after the fault was effected, being considered as removed by denudation, as is commonly the case. As the letters and figures correspond on both the section and plan, it will be found that, while the lodes, *l* 1, *l* 2, and the elvan *e* 1, are shifted to the right, on the side of the dislocation marked B, the lodes *l* 3 and *l* 4 are shifted to the left; and that, in the latter

Fig. 272.

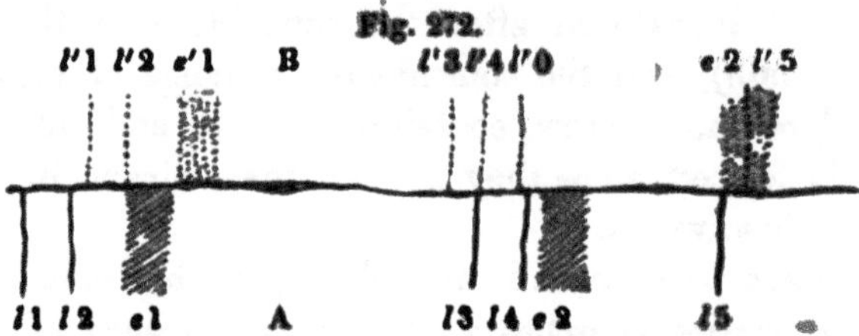

part of the section and plan, a lode or branch from a lode *l'* 0, appears on the side B, which was not at the surface on the side A, so that three lodes appear on the side B as continuations of the two lodes visible at the surface, on the side A. The elvan *e* 2, which was close to the lode *l* 4, on the side A, is apparently removed far from it on the side B, and moreover contains the lode *l* 5 in the latter case, one which was far removed from it, on the surface, on the side A.*

The evidence of a succession of fissures is often extremely interesting. While some clearly dislocate and shift the whole of a mass of rocks, with any prior-formed fissures included in them, others appear as mere fissures, with their walls slightly if at all moved from their former relative positions as continuous portions of the same mass of rocks. In the annexed plan (fig. 273), one of

Fig. 273.

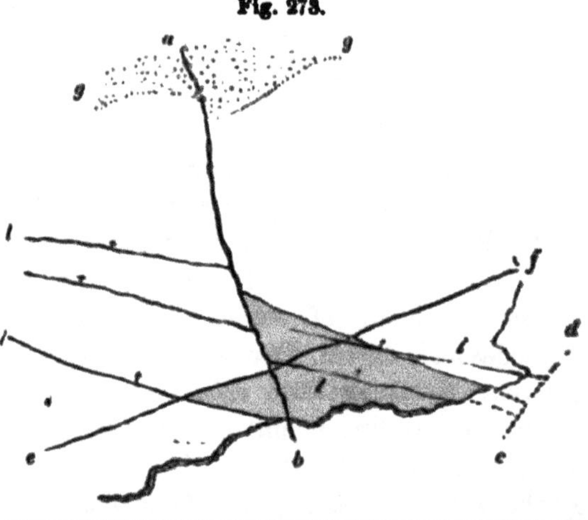

* Figures of this kind serve to illustrate the apparently contradictory facts sometimes observable on the sides of dislocations, denuded down to a common level, where elvans, or other dykes, and faults, or mineral veins, dip at various angles in opposite directions. In the illustration given in the text, the motion has been supposed vertical. As such movements are frequently otherwise, when it is desired to see how, by the

the mineral veins of the Charlestown, Pembroke, and Crinnis mines, St. Austell district, Cornwall, it will be seen that the granite boundary, *g g*, as well as the lodes *l l*, are shifted by the fault or cross course *a b*; (the same circumstance attending the fault, *c d*, though not shown on plan;) while another, and subsequent fissure, *e f*, traverses the whole without shifting it.

Fissures are often found to split at their ends after no very considerable course, when regarded in their horizontal range. Of mineral veins so divided at their extremities, when viewed horizontally, the following plan (fig. 274) of the Wheal Fortune range

Fig. 274.

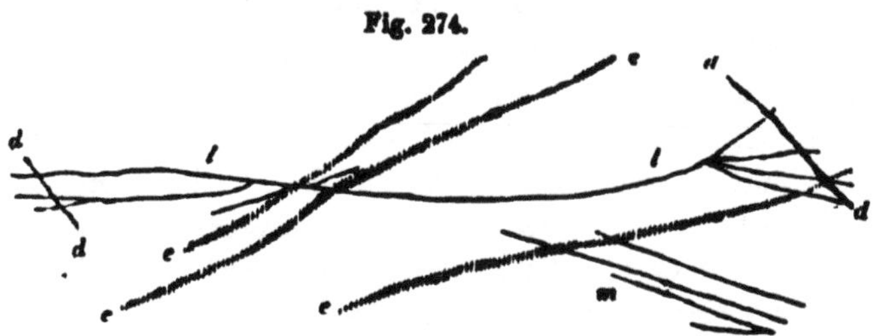

of mines, Breague district, Cornwall, may be taken as a good example. The main lode is there seen to be split on both the east and west after a range, as a marked fissure, for about a mile and a quarter (the plan is on a scale of one inch to the mile). The lodes, *m*, are those of Wheal Friendship mine, and, if prolonged, would also fall into the main vein of the Wheal Fortune mines. These various lodes traverse elvan dykes, *e*, *e*, or *courses*, as they are termed in Cornwall, and are cut by faults or cross courses, *d*, *d*, subsequently produced. It should be remarked, with reference to beds or other arrangements of rocks of variable toughness, traversed by fissures, that occasionally some care is needed not to be misled by minor appearances, for the fissures taking lines of least resistance may so run against or along harder beds, or dykes of mineral matter, as to lead to false impressions. Thus, in the annexed section or plan (it is immaterial which it may be considered), a fissure being opened from *d* towards *e*, and encountering an elvan

use of such sections, explanations of apparently complicated phenomena may be afforded, it becomes necessary not only to have the sections strictly accurate and proportional in all their details, but also to make the movement correspond with that found among the rocks themselves. If an observer will paint on two pieces of flat glass, a variety of sections of this kind, the same on both pieces, so that when held together they appear as one, and slide the glasses on their flat surfaces, a variety of interesting circumstances will be made apparent as to the consequences of fault movements in different directions; the surfaces of ground being supposed, as in nature, to be denuded down to some common levels.

dyke *a b*, might have resistances to the force employed so adjusted that it only traversed the latter at *c*, passing up the wall of

Fig. 275.

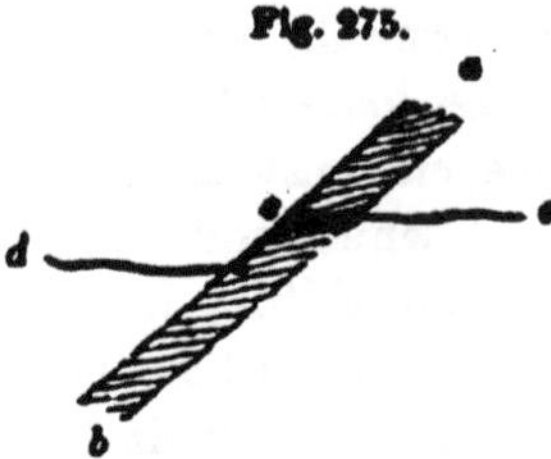

the elvan dyke for some distance, thence taking its course onwards to the right in a parallel line *c e*. It might be inferred, and in somewhat similar cases has been inferred, that the elvan filled a fissure, *a b*, produced subsequently to that noticed, *d c e*, the opening against the elvan on the side *c* being very slight, even forming a mere slide along the old plane of the fissure *a b*. The reverse would, in such a case, be the fact. Circumstances of a similar kind have sometimes occurred as respects the intermixture of an igneous rock (locally known as *toadstone*) (p. 559), and the limestone associated with it in Derbyshire, as will be hereafter noticed. Caution, therefore, on this head, is occasionally more needed, than at first sight might appear probable.

With respect to arrangements of the parts of a faulted country, and it is important to bear in mind how very extensively faults often prevail in otherwise undisturbed districts, their occurrence on the surface of land is sometimes such as to remind the geologist of inlaid marble work, curiously fitted together, and, as it were, polished down to some given plane. The observer will find a good example of a piece of natural inlaid work of this kind in Pembrokeshire, where the coal measures of Nolton and Wood appear as if inlaid among faults on the north, east, and south.*

* See Maps of the Geological Survey of Great Britain, Sheet 40. On the south a considerable fault throws the coal measures against lower Silurian rocks, on the north another brings them in contact with Cambrian rocks; both one and the other class of deposits being at the same time overlapped by them, so that it becomes needful to have a clear view of the amount of the overlaps as well as of the mode of occurrence due alone to the faults. The following section (fig. 276), north of Newgale

Fig. 276.

Sands, N, will show the manner in which the coal measures, *b*, *b*, are brought into contact with purple and grey sandstones, of the Cambrian series, *a*, *a*, by the fault, *f h*; *c* is a dyke of igneous rock filling a fissure traversing the latter beds.

As to the smoothing off of countries traversed by faults, these often considerable, so many regions present evidence of it that probably there are few portions of the earth's surface, even when offering scarcely any bending or contortion, which are not more or less cracked and broken in some form. It has been seen (p. 425) that in the earthquakes of the present day fissures are frequent, and there is every reason to suppose that such have occurred at all geological times. Whether faults arise from minor adjustments of the earth's crust, (the bending, contortion, and squeezing of various accumulations being regarded as more considerable consequences of those adjustments,) or from other causes, together with the greater plications and flexures, they show a broken and dislocated condition of that crust which it requires the geologist most carefully to bear in mind, when endeavouring to trace the facts he may observe in connexion with deposits, and their subsequent movements, to their sources.

As illustrative of the modes of occurrence of fissures and faults in mining districts, which usually afford, as above remarked, such good opportunities for their study, the following plans may be found useful. The first plan (fig. 277) represents a general view

Fig. 277.

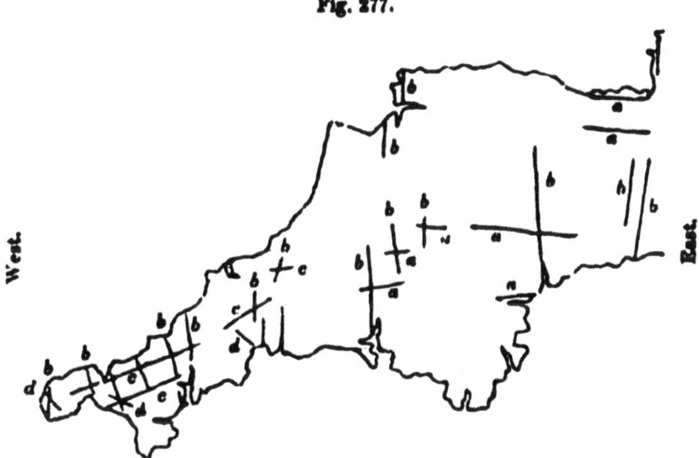

of the fissures, whether coming under the heads of mineral veins or ordinary fissures and faults, in Cornwall, Devon, and West Somerset. On the east, there is a tendency of nearly north and south faults to traverse others running east and west, while, on the west, fissures usually ranging about N.N.W. and S.S.E., cross others which take a course from W.S.W. to E.N.E., or from E.S.E. to W.N.W. It will be observed that towards the great metalliferous district of Cornwall, the lines *c c c* take a direction somewhat parallel to the general range of land, which is that also of the granitic masses of the district. Other lines, *d d d*, are observed as of importance in three situations (St. Austell, Marazion, and St. Just districts). The fissures and faults, *c c c* and *d d d*, contain the chief of the tin and copper ores of the district, while in the cross courses *b b b*, those of lead* and iron and some others are commonly found. The tin and copper veins or lodes, *a a*, near Tavistock, have a more east and west direction, the cross courses traversing them, *b b*, having a somewhat marked north and south range. The lines, both east and west, *a a a*, and north and south, *b b b*, on the east side of the plan, come under the heads of common faults; one, however, of the east and west lines, *a*, near Exeter, being connected with parallel fractures holding manganese.†

With respect to the relative geological age of these fissures, there is evidence that those having an easterly and westerly direction on the west, *c c c*, and *d d d*, were formed anterior to those traversing them in a northerly and southerly direction, since the former are not only shifted by the latter, but their contents are also broken through by them. The east and west fissures, *a a*, near Exeter, on the west of Dartmoor, were produced after the deposit and consolidation of the new red sandstone of that district, since that series has been dislocated by them. Fissures with the same direction, near Watchet, Somerset, *a a*, on the north-east corner of the plan, were produced after the deposit and consolidation of the lias. How far these latter may be contemporaneous with those containing tin and copper ores on either side of Dartmoor, and having the same direction, may not be clear, though they might be supposed to be so. Be this as it may, north and south faults have dislocated the chalk with other prior-formed deposits (of the oolitic series) near Lyme Regis, Chard, and Membury (*b b* on the south-

* The lead of Cornwall and Devon is not confined to the north and south fissures, though in certain districts it occurs in a somewhat marked manner in them.

† The following section illustrates these faults, one of which, *f*, can be traced for 10 miles from Poltimore, on the east, to Venny Tedborn, near Posbury Hill, on the

Fig. 278.

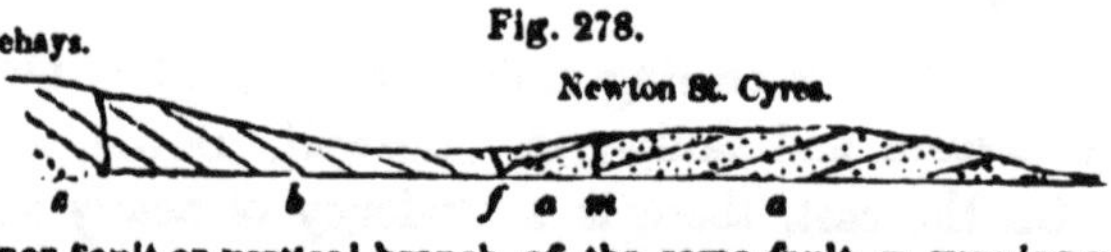

Shutehays. Newton St. Cyres.

a *b* *f* *a* *m* *a*

west; a minor fault or vertical branch of the same fault, *m*, running parallel to it, and having afforded a large quantity of valuable oxide of manganese (at Huxham, Upton Pyne, and Newton St. Cyres). *a*, *a*, are beds of the new red sandstone series of the district, brought into contact with the coal measure sandstones and shales, *b*, *b*, of the same country, in which there is another, and apparently parallel fissure, *l*, containing sulphuret of lead.

east of the plan). Taking these last in connexion with the north and south faults of the Mendip Hill district, near at hand eastward (fig. 167, p. 478), there have been fissures formed in the same general directions, north and south, at two distinct periods in this part of south-western England, one anterior to the deposit of the new red sandstone, and posterior to the flexures and plications of the coal measures, and the other after the deposit and consolidation of the chalk. The movements of different dates in east and west directions have already been noticed (p. 653).*

The following plan (fig 279) of part of Glamorganshire exhibits

Fig. 279.

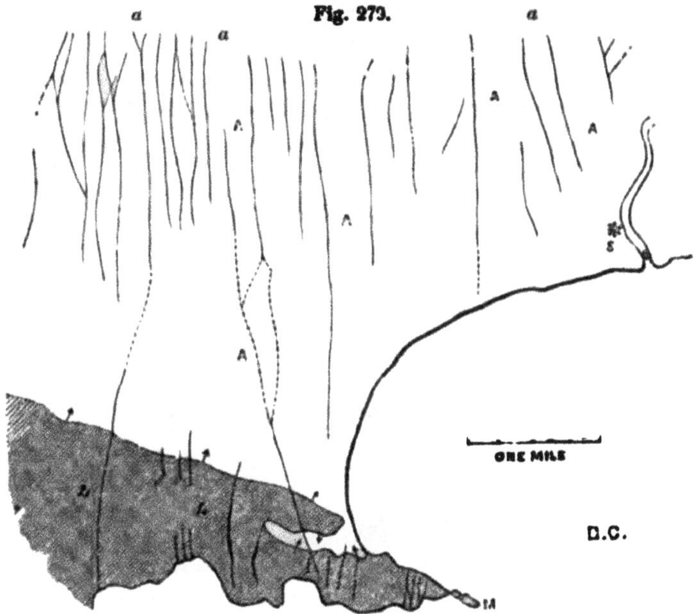

numerous parallel fractures traversing both mountain or carboniferous limestone and coal measures near Swansea; the working of the coal measures affording the needful evidence of many faults which are not so easily traced in an accumulation of such a gene-

* We would refer for more ample detail on the mode of occurrence of the faults and lodes, or mineral veins, of Cornwall, Devon, West Somerset, and a part of Dorsetshire, to the Author's Report on the Geology of that district, 1839.

As regards the range of east and west faults in neighbouring parts of England, it may be desirable to call the attention of the observer to the considerable fractures having that direction near Bridport and Weymouth (see Maps of the Geological Survey, Sheets 17, 18, where they have been most carefully laid down by Mr. H. W. Bristow), traversing a variety of beds up to the chalk inclusive; in the latter case, therefore, formed during some portion of the supracretaceous or tertiary period. In the Isle of Wight a great contortion, having an east and west direction, is seen to have occurred after the supracretaceous or tertiary rocks of that district had been accumulated.

have been of either of the dates previously noticed. It may not be improbable, however, that they were formed after the deposit of the lias, since somewhat more eastward, towards Cardiff, in the same general district, parallel faults dislocate the various accumulations up to that deposit inclusive.*

With respect to the manner in which portions of fractured masses are brought into contact in vertical sections by faults, the following sketch will serve to illustrate that of a simple kind, when

Fig. 280.

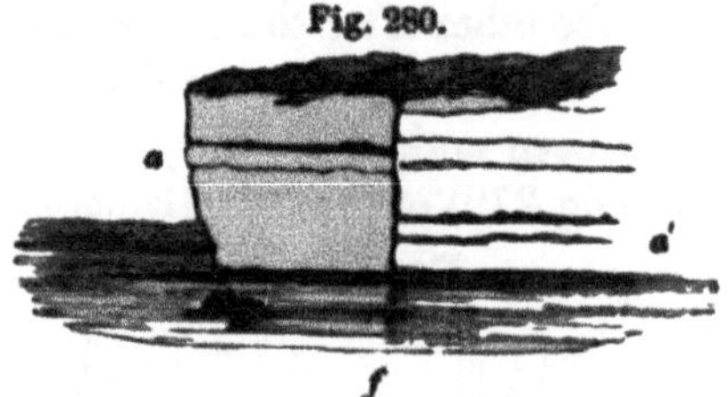

the amount of difference in the relative levels of the dislocated and once-continuous beds has been small, and the fissure nearly vertical, part of the bed *a* on the one side of the fault *f*, being separated from the portion *a'*, on the other. Faults are, as may be readily inferred, of all inclinations as regards the horizon, being sometimes sloping as beneath (fig. 281), so that to measure the amount of

Fig. 281.

* The observer is referred to various maps of the Geological Survey of the United Kingdom for numerous examples of faults traversing different rocks. Great care has been taken to have them properly examined and laid down, so that they may eventually constitute a body of evidence, of an accurate kind, for a due consideration of the various dislocations which the rocks, in the area of the British Islands, may have suffered during the lapse of the geological time of which such rocks may be the records.

geological dislocation produced by one at *f f*, the distance *b d*, extending vertically from the plane of the same bed *a* (supposed horizontal) on the one side, and *c* on the other, has to be ascertained.

In some districts faults are observed so to have occurred that several portions of country have been dropped down in one direction, prolonging the surface appearance of some rocks beyond that which would otherwise have happened after the various denudations to which they might have been exposed; portions being thus preserved which would otherwise have been swept away. The following section (fig. 282) may be taken in illustration of this subject, as

Fig. 282.

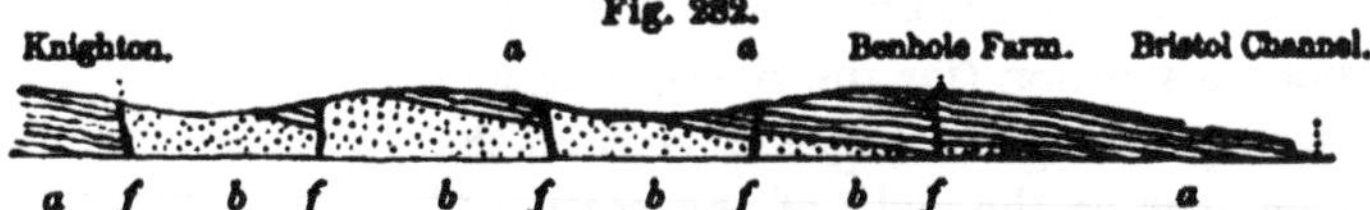

also of the vertical mode of occurrence of the faults near Watchet (*a a*, north-east corner of the plan, fig. 276), previously noticed. The deposits dislocated are lias *a*, and new red marl and sandstone, *b*; and it will be seen that parts of the lias have been preserved from denudation by being, as it were, dropped down by five faults, *f*, *f*, *f*, *f*, *f* (parallel to each other), into five sheltered depressions, succeeding each other in a southward direction. In this manner, valuable coal, in some coal districts, has been preserved from that removal by geological causes which it would otherwise have suffered. The amount of accumulations thus preserved, or the reverse, by systems of faults, is a subject which should engage the attention of the geologist as one of importance in investigations of this kind. The amount of various rocks so circumstanced is often very considerable.

As might be expected, lines of faults frequently exhibit minor complication and even disturbance, showing a certain amount of lateral pressure during the adjustment of their sides after the action of the force producing the original fracture. The following section, easily seen,* of a fault on the coast of Glamorganshire, west of

Fig. 283.

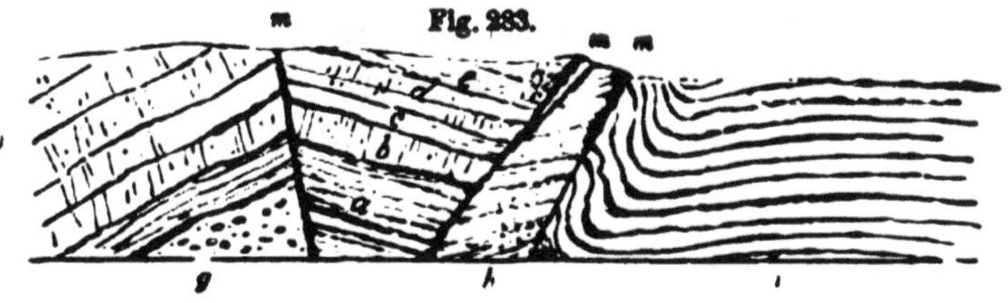

* Respecting illustrations of the various geologica. 'enomena noticed, the Author has endeavoured in this work as much as possible to select such ocalities as may be easily visited.

Lavernock Point, will illustrate minor com
and a bending of certain of the beds acted
minor parts of the same dislocation which h
lomitic limestone and marl *a*; varieties o
b, *c*, *d*, *e*, and *f*; dolomitic conglomerate, *g*
red sandstone series); and lias *l*. The bed
those on the left. While the fractures ha
former deposits, the edges of the lias have be
certain amount of lateral pressure. In some
of a portion of the beds acted upon, occasio
pect that, after the fracture, there has been
an upraised position (for the time), producin
even for upturning the edges of beds on the
as shown at *m* on the right of the section (fi
side relatively lowered is found raised at th
fault, the consequent friction turning up th
rock conformably with the movement.

As a vertical section may only give the s
the parts of rocks fractured and faulted, it is
server should search for the direction of any
ing pressure of the rocks on one side against
order to discover that in which the move
effected. This investigation will sometimes
though the general plane of a fault may dip
the movement has not always corresponde
these friction-marks bear evidence of the
pressure, more especially in those cases whe
in its plane, amount to several thousand feet,
moved against each other, and once so far a
jammed together. The contents of dislocati
common faults or mineral veins, often presen
of these friction-marks, parts of the walls
grating against each other in their moveme
cavities in which various mineral substanc
taking the form of the surfaces against which
effected.

CHAPTER XXXV.

FILLING OF FISSURES AND OTHER CAVITIES WITH MINERAL MATTER.—SULPHURETS OF LEAD, COPPER, ETC., REPLACING SHELLS.—FILLING OF MINOR FISSURES.—SOLUBILITY AND DEPOSIT OF MINERAL MATTER IN FISSURES.—SOLUBILITY OF SULPHATE OF BARYTA.—DEPOSITS FROM SOLUTIONS IN FISSURES.—EFFECTS PRODUCED IN HEATED FISSURES BENEATH SEAS.—MANY SIMILAR SUBSTANCES FOUND IN MINERAL SPRINGS AND VEINS.—FREQUENT OCCURRENCE OF SULPHUR, ARSENIC, ETC., WITH CERTAIN METALS IN MINERAL VEINS.—ACTION AND REACTION OF SUBSTANCES UPON EACH OTHER IN FISSURES AND CAVITIES.—CHARACTER OF METALLIFEROUS VEINS AMID ASSOCIATED DISSIMILAR ROCKS.—CONDITION OF MINERAL VEINS TRAVERSING ELVAN DYKES IN CORNWALL.—INFLUENCE OF THE DIFFERENT ROCKS TRAVERSED ON THE MINERAL CONTENTS OF FISSURES.—MODE OF OCCURRENCE OF LEAD ORES AMID THE LIMESTONES AND IGNEOUS ROCKS OF DERBYSHIRE.—'FLATS' OF LEAD ORE IN LIMESTONE DISTRICTS.—METALLIFEROUS DEPOSITS IN THE JOINTS OF ROCKS.—RELATIVE DIFFERENT DATES OF MINERAL VEINS.

THE filling of fissures and other cavities with mineral matter may, to a certain extent, be considered as in part connected with the changes and modifications of rocks above mentioned; since from the filling of minor cavities and fissures, such as occur in or traverse small portions of an accumulation, whether of igneous or aqueous origin, much change or modification may arise in the containing rocks. The filling of cavities, such as those previously noticed in vesicular lava and molten matter of all geological times, converting a highly porous and often originally light substance into a very solid rock, effects a marked change of structure. The infiltrations of the mineral substances into the cavities, in these cases, become important in the consideration of those which have filled various fissures and dislocations, as well as cavities of far greater size, since they seem to point to the solution of some substances, or of the elementary matter composing them, and to the power of such solutions to traverse the pores of rocks, even of those which are considered very solid and compact, in a manner which, at first sight, might not be expected. Let the observer, for example, study certain of the nodules of the impure carbonate of iron, known as *clay ironstones*, in many of the localities where they are obtained from the coal measures of the British Islands, opportunities for which are abundant in South Wales, Monmouthshire, Staffordshire, Derbyshire, and elsewhere. While in many of these nodules, the cracks, when they present themselves, as they often do, in the manner mentioned previously (fig. 227, p. 597), only contain more pure carbonate of iron or are entirely empty, at others they are incrusted or filled with such substances as copper pyrites, and the sulphurets of lead, zinc, nickel, and iron, with the occasional occurrence of other minerals of a different class. In such cases the observer can have little doubt that the component parts of these substances have come, by infiltration from without, into the cracks of the nodules of impure carbonate of iron, through their exterior pores, and through those and the laminæ of the surrounding argillaceous shales. He is therefore prepared to infer that these bodies, or their component parts were in a soluble state when they entered the cavities formed by the cracks in the nodules.

When he examines the minerals which have, under certain conditions, replaced organic remains in various rocks, the geologist may still further be prepared to regard the matter of these and other compound substances as having been introduced in solution into cavities left by the decomposition and disappearance of mollusc shells, or other organic bodies. Copper pyrites has been found to replace the shells of *spirifera*, at Doddington, Somersetshire*—sulphuret of lead various cavities left by the shells of molluscs in the lias near Merthyr Mawr, Glamorganshire†—and sulphate of baryta portions of corals in the mountain limestone of Cromford, Derbyshire.‡ Sulphuret of iron very frequently occupies the places of mollusc shells in many rocks, especially those which are argillaceous, even insinuating itself amid the matter of fossil bones, such as those of saurians in the lias, and other deposits. Silica, as

* In this locality there was a vein of copper. The ores raised were principally green and blue carbonates, and were first obtained in the new red sandstone conglomerate of the locality above a vein in the Devonian rocks beneath. Horner, Trans. Geol. Soc. London, vol. iii., pp. 352 and 363.

† The sulphuret of lead is much disseminated in this part of South Wales, and often in cavities. It occurs in the cracks of fossil wood in the lias near Dunraven Castle, in the same manner that the sulphuret of iron is often seen in coal beds, and in fossil wood in numerous clays of different geological dates.

‡ This fact is interesting in connexion with the considerable quantity of sulphate of baryta found in the lead veins and other cavities of that part of Derbyshire.

might be expected also, occupies the cavities left by shells, of which the chalcedonic replacements of the various shells of the greensand series at Blackdown, Devon and Somerset, are beautiful examples. Even the carbonate of lime of many fossil mollusc shells does not always appear to be that of the original, but to have been infiltrated into cavities left upon the disappearance of the matter of the actual shell, the particles of the carbonate of lime not being adjusted in the manner they usually are in shells of the same class by living animals, but as they would be upon simple infiltration and crystallization in any cavity. Again, in the crystals of felspars decomposed in the body of a rock, the original substance of the crystals removed, and replaced by peroxide of tin, even part of the original felspar crystal sometimes remaining, while the rest of its form is replaced by the peroxide of tin, as in an elvan at St. Agnes, Cornwall, the observer has another example of the inflow of mineral matter in solution into cavities and through the pores of the rock in which such cavities may be situated. In fact, looking at the subject generally, the various cavities in the rocks composing the crust of the earth, have a tendency to be filled by mineral matter, the component parts of which find their way to them in solution.

Passing from these cavities to those produced by cracks, these of minor size, and confined either to one, two, or some small number, of beds of sedimentary deposits, or some very limited volume of an igneous accumulation, it would be expected that, as a whole, the matter infiltrated into such cracks would chiefly partake of the mineral character of the rocks so broken, so that the substances principally filling the cracks in limestones would be calcareous, while those amid siliceous rocks would be quartzose, as is usually the fact. From the prevalence, however, of particular conditions, quartz veins are occasionally found in limestones, and calcareous matter among the siliceous rocks. This usually occurs when the limestone beds form a very subordinate portion of a sandstone or argillaceous accumulation, chiefly composed of silicates, or when calcareous deposits predominate among those of other kinds; as, for example, is the case with the igneous rocks of Derbyshire, where the vesicles and minor veins of the latter are often filled with calcareous spar.*

* The filling of cavities and small fissures in igneous rocks by carbonate of lime is not unfrequent, even when calcareous rocks do not constitute any very large proportion of a general mass of mixed accumulations. Thus at Trecarrell Bridge, between Launceston and Tavistock, the highly-vesicular rock of that locality, contempora-

Proceeding to examine the filling of ca larger dimensions, and such as, not confined rocks, can be traced for considerable distanc which are unknown, an observer will have mind the incrustations of the sides of such stances, which, passing amid the pores or sma the minor scale, are ready to fill up or inc senting themselves, no matter of what kind also to consider the kind of substances, and upon each other, which may be derived from sources. Viewing a considerable fissure, in what vertically traversing various beds of d the following section (fig. 284), *a* to *f*, each a

Fig. 284.

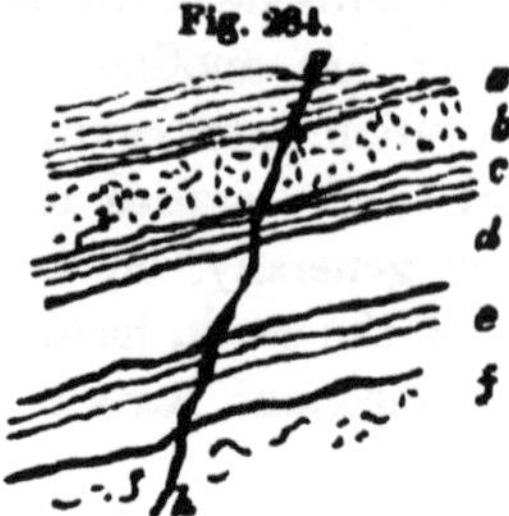

or variously combined matter in solution, a tion, at first, to solutions, the geologist ha problem presented to his attention than th mineral matter through the pores of rocks fissures in them. He has to regard not only tions and decompositions effected by a mixtu duced into the fissure, but also the motion liquid in it, according to temperature. The one through which waters rise to the surface it, thus discharging large volumes of wat matter in solution of various amount and ki merely rise to such a height in the fissure as it, and the portions of rocks adjacent, amid th of which it may also enter. According to te

neously formed with the Devonian rocks amid which it the infiltration of carbonate of lime from adjacent calcar or importance. The ready solubility of the carbonate o carbonic acid is present, has occasioned the passage of th calcareous beds into the vesicles of the igneous and jux there, when unless decomposed and again removed, it w other substances passing in solution through the pores of have to bear in mind the solubility or deposit of the matter generally in the liquid, permitting some of it to remain in solution while other parts were deposited, coating the walls of the fissures.

The observer will thus have to consider the probability of certain of the fissures extending to depths where the temperature may become very elevated, even to those depths where water, notwithstanding the great pressure, might be converted into steam, and numerous substances be vaporized. There is much to be accomplished with respect to our knowledge of the effects which would be produced under the conditions supposed. We may expect water to exist under pressure as such up to very high temperatures, and its power of dissolving various substances in that state to be so increased, that many viewed as insoluble at those temperatures at which experiments have been undertaken would become readily soluble.

The experiments of M. Gustav Bischoff on this subject are highly valuable. Impressed with the importance of the agency of steam in volcanic productions, and viewing the connexion of such agency and many substances found in mineral veins, he found that when galena (sulphuret of lead) was gently heated in a porcelain or glass tube, and steam driven over it, that sulphuretted hydrogen and sulphurous acid were evolved, and the ore reduced, and that if the lead thus obtained were wetted with distilled water, it was covered by the carbonate of lead. He remarks that some substances not known to us as evaporating at any temperature, are carried off by steam, as, for example, silica. Artificial sulphuret of silver was found to be very readily decomposed by steam, and more easily so at a moderate heat. At a temperature under the melting point of zinc, this was soon effected, and the silver effloresced in such forms as to induce M. Gustav Bischoff to regard the moss-like and filamentous occurrence of native silver in veins as very probably the result of the decomposition of the sulphurets. With respect to sulphate of baryta, usually termed insoluble, and yet so frequent in the veins of some districts, and in a manner to leave little doubt that it has been deposited from a solution, he found, by experiment, that when heated water, containing carbonate of soda or potash, came into contact with it (even when the temperature was not much elevated, and the water was only slightly charged with those substances), a partial decomposition took place, and that when the temperature was again lowered, a readjustment was effected, sulphate of baryta being again produced, and the carbonic acid, with which it was previously united, returning to the soda or potash. In this manner, M. Bischoff remarks, baryta may be separated from sulphuric acid in the lower part of a vein, where it could be exposed to the needful heat in waters containing the carbonates of soda or potash, and be removed to a cooler part of the vein, and be there deposited, again united with sulphuric acid.* Such decompositions and recompositions are evidently most important in explanation of the often complex contents of veins.

When we know that certain fissures in the earth's surface result from dislocations so great that beds of rock, once continuous, are thrown even several thousand feet distant from each other in the planes of the fissures, and vertically to the stratification, the depth to which some of these fissures must extend can scarcely have been otherwise than sufficiently considerable to afford conditions of an important kind, as respects the heating of water in them, and the consequent solubility of various substances not readily acted upon by water at more moderate temperatures, even to the solution of some forming parts of the rocks fissured.

The experiments of Professor Forchhammer have shown, though potash felspar, one so frequent among granites and felspar porphyries, may be exposed to the action of boiling water, under the ordinary pressure of the atmosphere, without obtaining the potash from it, that when that pressure is considerably increased, and the temperature augmented, this substance is obtained in solution.

As regards fissures, and heat at their greater depths sufficiently considerable to convert water into steam, even under great pressure, it may occur to the observer that, after these fissures were produced, many solutions percolating through the pores, or amid the beds and joints of the rocks broken through, would endeavour to deliver themselves into them. Where they entered any water in the cleft and various solutions in it, they would mix with them, obeying the same movements from differences of temperature, and acting upon them, or being acted upon, according to circumstances. Where the fissure was only filled by heated vapours, percolations into it at those depths might have a tendency to be vaporised also, and if any of them contained matters then rendered inso[illegible], it might be inferred that they were left incrusting the sides of the fissures, in the same manner that stalactitic incrustations of carbonate of lime cover the sides of caves and fissures when the water is evaporated, and the carbonic acid, rendering the carbonate of lime soluble, is removed.

* Poggendorf's Annalen, vol. lx.

Having considered the fissures with reference to waters dispersed amid rocks, and finding their way into them, as would happen when they rose to the surface of dry land, or were opened out only to situations where they did not reach any considerable superincumbent volumes of water, the geologist should direct his attention to the conditions which would obtain when these fissures rose to the bottom of the sea, either wholly, or so that the sea waters could readily rush into the clefts formed partly through dry land and partly under the sea. Looking at the present distribution of land and sea on the surface of the earth, many long and important fissures would be expected to occur beneath the sea. If *a b*, in the annexed section (fig. 285), be the level of the sea; *a c* and *b d*, depths of water; *e e*, rocks, such as argillaceous slates, resting upon or raised up by granite, *f f*; and *h*, a fissure traversing the whole, and opening to the sea water above, the latter would rush into the cleft or clefts at the prolongation of the fissure to the sea bottom, descending as far as any temperature in the cleft would permit. It may be assumed, for illustration, that, whatever may have been the effects of the first communication between considerable depths and the surface of the bottom of the sea, a time would come when the sea water could enter the fissure, unless any outflow of waters reaching it from the rocks traversed could prevent it. In certain situations, obstructing conditions of this kind might exist, the fissures answering the purpose of artesian wells to large tracts of country.

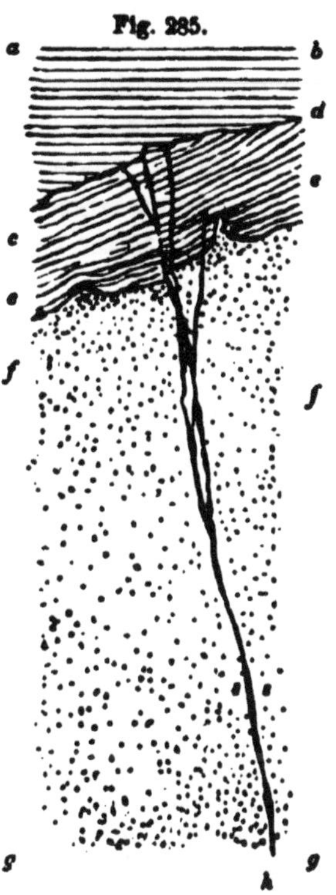

Fig. 285.

Taking, however, the conditions to be such as to permit the entrance of the sea water, and that at [illegible] depth, such as *s s*, the water was converted into steam, notwithstanding any pressure there might there be, the saline solutions, chloride of sodium constituting the important portion of them (p. 109), would be left to be dealt with according to the temperature existing at *s s*, and any vapours rising from beneath, *h*, where a still higher temperature might prevail. The production of chlorides of a volatile kind, such as those of copper, and others, might thence take place to a considerabl being again changed into other combinatio the fissures.*

When fissures are regarded as of deptl extend to such elevated temperatures, th fail to turn to the evidence respecting fis heated gaseous substances discharged fro quakes, whether these may traverse volc biting activity, or show no immediate co to regard the emanations which take pl themselves, since from such sources of com interior and exterior parts of the earth, evi as to the substances vaporized by heat l upwards. Neither should he neglect the as most of them are termed, *mineral* sprin only to be the condensation of vapours an tions of fissures, when the temperature becc With respect to the vapours and gaseous su from volcanos and found in mineral waters

* M. Élie de Beaumont remarks (Note sur les Éma lifères) that "iron as a chloride, often changing int oligiste), is among the most abundant of the subs emanations. Oxidulated iron is commonly dissemina volcanos, and it cannot be doubted that it exists in th ranean cavities. Iron in the form of an oxide or chl deposited in the fissures which volcanic emanations t surface." M. Élie de Beaumont also points out copp tions, and it may be observed, that chloride of copper

† Note sur les Émanations Volcaniques et Métallif de France, 2nd série, t. iv., p. 1249, 1847), wherein, ur important information will be found bearing on this su

Adverting to the various hypotheses which have be filling up of mineral veins, M. Élie de Beaumont rema butes ordinary mineral veins to emanations in the fo waters, enables us to comprehend the varied facts obs cially the development of those chemical affinities whi influencing the manner in which the metals are asso usually associated have much in common between the altogether analogous. Nickel and cobalt, so often f each other in their properties, and the same with iro and arsenic, the properties of which are so analogous, are frequently associated. Silver and lead have much quently united in veins. It is rare to find silver una scarcely happens except when the silver occurs native silver which most differ from the corresponding cond rare to find lead which is not argentiferous, the most the sulphuret, the properties of which are very analogo Lead and zinc, the sulphurets of which possess an associated together in the form of *galena* and *blende*; an in the great family of metals occurring in the stannifer tantalium, &c."

has pointed out, that the substances contained alike in them and in mineral veins, may be taken as 19, viz., *potassium*, *sodium*, *calcium*, *aluminum*, *manganese*, *iron*, *cobalt*, *lead*, *copper*, *hydrogen*, *silicon*, *carbon*, *boron*, *arsenic*, *nitrogen*, *selenium*, *sulphur*, *oxygen*, and *chlorine*. The substances found in mineral waters and veins, and not hitherto noticed in volcanic emanations, he notices as *lithium*, *barium*, *strontian*, *magnesium*, *phosphorus*, *iodine*, *bromine*, and *fluorine*.*

When the observer thus directs his attention to the consequences which may arise from the production of fissures, extending to portions of the earth where high temperature may be inferred, he should bear in mind that, of the substances occupying the interior of the earth, beyond such slight depths as the reasoning respecting the thicknesses of various rock deposits renders probable, nothing is known, except that their density, as a whole, must be much greater than that of the rocks at the surface, since, according to Laplace, the mean density of the earth is 1·55, while that of its solid surface is only 1. The substances chiefly forming the solid surface of the earth are oxides, those which are not of that character are very limited; and it is not a little interesting to find the latter, to a great extent, in the fissures under consideration, or so disposed as readily to have entered the cavities of deposits after their accumulation, there forming combinations other than oxides. As respects the frequent occurrence of certain of the metals with sulphur, arsenic, and other substances, which have been termed, with reference to their presence in veins, *mineralizers*, such frequent combinations, under conditions that may often be inferred as those which governed their original deposit in mineral veins, secondary actions having effected subsequent modifications and changes, are highly interesting. M. Élie de Beaumont has remarked, when treating of an initial volatilization of the metallic substances found in veins, that this hypothesis agrees with the fact that the metals, properly so called, are found in them much less frequently combined with oxygen than with sulphur, selenium, arsenic, phosphorus, antimony, tellurium, chlorine, iodine, and bromine. "These substances," he observes, "are not only in general volatile, as well as bismuth, which often accompanies them, but they have likewise the property of rendering many of those with which they combine also volatile. It is difficult to believe, that this property

* With respect to the substances contained in mineral waters, M. Élie de Beaumont mentions that he has taken them from the works of many chemists, and especially from those of MM. Berzelius, Bischoff, and Kopp.

has not acted a part in the filling of t expect to find in the contents of fissures cating with them, or disseminated amid may be inferred to have presented the re duction of mineral matter from them, som elsewhere, and under forms and combin kind, as well as those with which we may b forming the component parts of rocks gen may be sometimes discovered under new logist would expect also to find numero which he might refer to the reactions of c and to the readjustment of their compone governing conditions of the time.

With reference to slow secondary ele feeble currents, M. Becquerel pointed out, that various compounds are produced whi usual kind of experimental investigations, presented to each other in a nascent stat to such productions.† He observed tha termed insoluble, became crystallized, bec being slow, the chemical action was slo ponent molecules had time to arrange the laws governing crystallization, an advanta chemical forces have more intensity. various minerals by means of these secon oxides of copper and zinc, the sulphurets o iron, &c.‡ The action of bodies upon ea

* "These bodies," continues M. Élie de Beaumont, found among volcanic emanations, and also in ther in the veins contributes to corroborate the relations between these veins, volcanic emanations, and minera

† Traité Expérimental de l'Electricité et du Magnéti there remarked (t. iii., p. 295), that "it could not, for with apparently feeble electrical forces, strong affinit to decompose bodies and produce new combinations action of more or less energetic currents should alway ever, as the electrical effects which take place in chem it became clear that the same end might be obtain effects. It can be readily understood, that when any solution which reacts on one of the elements of this c tion, the moment they are brought into play by the o then, being in a nascent state, in the most favourable of the electric current produced by the couple."

‡ The observer will find much to interest him, bea veins in those experiments in which M. Becquerel e of a U, with clay moistened at the bottom, thus sep which solutions were placed to be acted upon, wires

experiments of M. Becquerel, so that after the production, and even crystallization, of some substances, they were again decomposed by the new action then set up among them, appears to have an important bearing upon the filling, and modifications of the contents of fissures and cavities.* He concluded, from his experiments, "that to obtain an insoluble crystallized substance by electro-chemical reactions, it is sufficient to make it combine with another which is soluble, and afterwards operate by means of very slow decomposition."†

In 1830, Mr. Robert Were Fox commenced a series of experiments in the mines of Cornwall, to ascertain the electro-magnetic properties of the mineral veins of that metalliferous district.‡ In 1837, he treated the connexion of electricity and mineral veins more at length, chiefly referring to the veins in Cornwall,§ observing, with respect to the present condition of mineral veins, that he found, "by an examination of water taken from different mines, and from various parts of the same mine, that different varieties of saline solutions now exist in neighbouring strata." In many instances the proportion of foreign matter in the water was very small, whilst in others it was very considerable; "but I have not," he adds, "yet tried any mine-water, that would not produce very decided electrical action, when the native sulphuret of copper, or of copper and iron (copper pyrites) were plunged into it, and the voltaic circuit was completed. The very superior conducting power of the saline water in the fissures, in relation to the merely moistened rocks, would always tend to supersede the transfer of electricity more or less through the latter. The contact of large

---

voltaic circuit; as also in those in which he placed substances in a tube, afterwards hermetically sealed, so that they formed voltaic circuits in the tube itself, the substances acting upon each other.

* M. Becquerel remarks, after describing some substances obtained by his experiments, "that all the chemical actions which lead to these compounds could only have arisen from certain electrical influences possessing little energy; for if we operate with apparatus the action of which is too strong, all the elements are isolated, and no combination is possible."

† Traité de l'Electricité, t. iii., p. 298. It is remarked, respecting truncations of the crystals of certain double chlorides obtained in some of the experiments, that, in the beginning, the crystals are perfectly formed; "but that when the apparatus has been in action for a long time, truncations of the angles are gradually produced; whence it seems to follow, that when the particles of the crystallizing substance are less abundant, the force which determines the regular grouping of them has no longer sufficient energy to complete the crystal."

‡ Philosophical Transactions, 1830.

§ "Observations on Mineral Veins;" Report of the Royal Polytechnic Society of Cornwall for 1836; Falmouth, 1837.

surfaces of rock, clay, &c., with water,
contents from them, must also have been
electrical excitement, and it should not be
culation of water would be liable to very frequ
in consequence of obstruction in the fissu
enlargement, so that the contents, as well
the water, would be subject to many modifi

The contents of fissures and cavities thro
not long have engaged attention before it
those districts where the ores of the usef
there is not unfrequently a marked associa
one or more of them being often of igneous
is far from being constant; at the same tir

---

* Robert Were Fox: Report of the Cornwall Polyte
mouth, 1837, p 110.

Adopting the view of M. Ampère, that the direction o
to the circulation of currents of electricity from east to
considers, that "if fissures happened to have opposite l
equally filled with water charged with saline matter, t
determined, in preference, through such of them as nea
netic east and west points at the time." The consequen
decomposition of the saline substances, and the determ
to the electro-negative, and the acids to the electro-posi
he remarks, "this process at first may have been, the d
cause it to become more and more energetic. The met
would, likewise, react on each other, and give rise to ne
ments till they arrived at a state of comparative equil
be very much the case with the lodes (mineral veins) a
which are capable of conducting electricity very nearly
the electrical scale, being more electro-negative than sili
so as platina; indeed, the grey oxide of manganese an
negative in a still higher degree. Arsenical pyrites, ir
hold rather a high place in the scale, and are electro-ne
copper and galena, but especially to the sulphuret or vi
produce a very decided action on the galvanometer whe
cuit with copper or iron pyrites."—p. 113.

M. Becquerel considers (Traité de l'Electricité, t. v.,
depth in the earth a multitude of electric currents ex
tions, the general result of which would produce an ac
He infers, that these are produced by the permanent
numerous fissures through which sea waters percolate
earths and alkalis, or to metallic chlorides, causing the
tricity, and the steam or other vapours positive elect
electricity, he considers, would be carried into the atmo
and the other would tend to combine with the negativ
passing through all the conducting bodies which est
between the metals or their chlorides in the solid, liqui
filled the fissures. Hence, he observes, a number of pa
circulate in the interior of the globe, producing electro-
we cannot appreciate the whole extent, but which c
numerous compounds.

long engaged the attention of miners, and in some mining countries, much importance has been attached to its practical bearings.* In the same countries also long experience has shown the miner that the ores he seeks are more likely to be discovered amid or against certain rocks than others, though the fissures in which they are found traverse several different kinds or modifications of rocks. It is very desirable that an observer should collect all facts of this kind, however ill-arranged they may sometimes be by those from whom he may derive them, and however needful their proper classification, from personal research, subsequently. At the contact of certain granites with other, and for the most part, sedimentary rocks, and especially where there may have been some modification or alteration of the latter from the intrusion of the former, fissures traversing them are often found productive of the ores of the useful metals, sufficiently abundant to be worked, provided the districts generally are metalliferous. In other words, such conditions, in a metalliferous district, are not uncommonly those under which the ores are the most abundant. In the mining districts of Cornwall and Devon the fissures through the junctions, or the vicinity of the junctions of the granite and schistose rocks, in those localities which may be termed metalliferous,† have been found to produce much ore, often not in the least quantity when they also traverse dykes, or *channels* as they are locally termed, of the porphyries and granitic rocks known as *elvans* (p. 565). Those irregular accumulations of ore usually termed *bunches* are often found at the junction of granite and the schistose rocks. In illustration also of the occurrence of similar accumulations of either tin or copper ores, in the same mining country, when a fissure traversing schistose and

---

* This somewhat common association of igneous rocks has also long since engaged the attention of geologists. Professor Necker adduced abundant evidence on this head in 1832 and 1833 (Proceedings of the Geological Society of London, March, 1832, vol. i., p. 392, and Jameson's Edinburgh Philosophical Journal, 1836). He thence inferred the filling of metalliferous veins by means of sublimation.

The observer will find the connexion of igneous rocks and mineral veins treated by M. Élie de Beaumont with precision and, at the same time, with ample detail, in his "Note sur les Émanations Volcaniques et Metallifères.—Bulletin de la Société Géologique de France, 1847, 2nde serie, t. iv.

† In illustration of the different distribution of chiefly metalliferous districts into which some areas, not unproductive of the useful metals, are sometimes naturally divided, it may be useful to mention, that Cornwall and Western Devon may be separated into six chief metalliferous districts. 1. That of Tavistock (including Dartmoor, and the mining country of Callington and Linkinghorne); 2, that of St. Austell (including the granitic mass of Hensbarrow, and its schistose skirts); 3, the St. Agnes district; 4, that of Gwennap, Redruth, and Camborne; 5, that of Breague, Marazion, and Gwinear; and 6, the district of St. Just and St. Ives, comprising the granitic country between these two places.

porphyritic dykes (elvans), passes through t
section (fig. 286), across the lode at Wheal

Fig. 286.

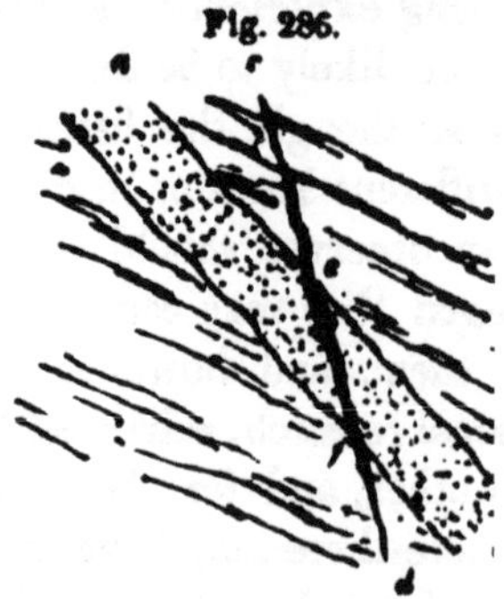

be useful. The elvan dyke, *a b*, is about 3(
direction about N.E. and S.W., and dipping a
northerly. The lode *c d*, dipping at an ang
traverses the elvan, *a b*, obliquely in its d
the fissure traversed the upper and adjoini
no great amount of ore was obtained, but u
it became more rich, and while passing thro
was found to be so abundant as to afford
After quitting the elvan at *f*, in its descent,
beneath, on the south, the lode became poor

The connexion of bunches of tin and
where these traverse elvan dykes, viewing
is well known practically to the Cornish min
as regards the abundant and profitable co
veins in that metalliferous land, can be c
many places.‡ An observer may sometime
a lode is split up into strings upon its entra
it may also be stated that it is thence in
however, when the facts are well investigat
cases that the ore itself continues sufficiently
even more abundant, though so divided i
amid fractured and highly-separated portion

* Those engaged in this mine reaped a profit, it is
time.

† The width of the lode was from six to nine feet i
increased in the latter to twenty-five feet, and decrease
feet.

‡ If the observer will direct his attention to the Geolog
he will find numerous examples of the intersection of el
The percentage of cases is considerable in which these i
by bunches of ore in fair quantities.

to be so profitably worked as previously. If elvans have been divided into joints, as often seems to have been the case, before the formation of the fissure traversing it and the adjoining rocks, it would probably happen that upon passing through them from these adjoining and less divided rocks, such joints would be the courses through which the force producing the general fissure would act, multiplying the parts of the general fracture in the elvan, so that when filled subsequently by mineral matter, the vein should appear split up into strings where the elvan occurred. If, as in the annexed section (fig. 287), a country composed of slate, *a b*, be

Fig. 287.

traversed by an elvan dyke, *c d*, having a jointed structure, and a fracture, *e f*, be made across the whole, it would be expected that where the fissure was effected across the jointed elvan dyke, the solids formed in the latter by the joints would be much dislocated, so that when the complicated fracture was subsequently filled by mineral matter, viewing such contents and their course alone, as is commonly the custom in mining countries, the vein would be considered as split into strings at *i g*.

The mineralogical modification of the various rocks in metalliferous districts, very commonly bearing the same names, is also a subject of no slight anxiety on the part of the miner, since, from experience in such districts, he finds that when it presents certain characters his chances of success as to the occurrence of the ores he seeks, are considerably increased. Thus, in Cornwall or Devon, he usually prefers a granite or elvan which is, to a certain extent, decomposed. The particular character of the various kinds of the schistose rocks and the harder beds associated with them, is also carefully noted, and from experience, some kinds, when forming the walls of the fissures, are known to carry more ore than the others, while some again are regarded as unfavourable.* In dis-

* As we have elsewhere stated (Report on the Geology of Cornwall, &c., 1839), in Gwennap (Cornwall) the more experienced miners seem to prefer those argillaceous beds which accompany the red or variegated slates of the district, and which have a fine grain and a blue-grey colour. Respecting the value of the red beds themselves, opinions somewhat differ. Mr. Carne states, that when the copper lodes in Gwennap

tricts where the rocks are more generally bedded, excellent opp tunities may often be obtained for studying the modification of contents of a metalliferous fissure according to the variation of

---

intersect the red beds they become unproductive, an immediate change ... when they pass beyond them into another slate. In most lodes the ... favourite kind of rock or *country*, so that the whole tendency of their ... to show that particular mineral structures, other circumstances being ... more favourable to the occurrence of the ores sought than others. The ... at Fowey Consols mines would seem to afford a good example of ore ... particular set of beds. The slate in this productive mine dips away from the ... of St. Blasey, on which it rests, towards the east, so that, as the lode has a ... east and west direction, the beds traversed by it on the lower part of the mine ... east rise towards the western end, and it is found that the bunches of ore accomp... this dip, coinciding with certain beds, viewing the subject on the large scale. The mode in which the *gossan* and other marks of the usually higher parts of a copper l... in Cornwall dip to the eastward in this mine is very interesting; gossan, with its ... common accompaniment of native copper, green carbonate of copper and ... sulphuret, descending, above the bunches of copper pyrites, to the depth of ab... 600 feet from the surface, with the dip of the beds, on the eastern part of the m... It is often very difficult to convey by words the differences in a rock which the pr... tised eye readily seizes as distinctive in these cases.

Regarding the changes in the metallic contents of the Cornish mineral veins acc... ing to the character of the adjoining rock, Mr. Carne observes, that "in Godolp... the lodes were rich where the killas (argillaceous slate) was of a bluish-white colo... but poor where it was black. In Poldice and Huel Fortune, the lodes in the kil... continued productive until they entered a stratum of blue hard killas, which cut ... the riches. In Huel Squire, the copper lodes were very productive when in the ... light-blue killas; but a stratum of hard black killas underlying (dipping) rapidly ... one lode at the depth of 44 fathoms, and the other at 120 fathoms, under the adit, ... at these levels both the lodes became poor. At Penstruthal copper mine the lode ... been tried unsuccessfully at various times in parts where the granite was hard, ... trial being made where that rock was soft, it became one of the most profitable mi... in Cornwall."—Trans. Geol. Soc. Cornwall, vol iii., p. 81; 1827.

M. Fournet has remarked on this subject that, commonly in Upper Hungary, ... largest copper lodes are found in fine clay slates; that in Saxony the silver ... occur in gneiss; and that in the Hartz certain ores are intimately connected w... grauwacke. The veins of Kongsberg, Norway, are sterile in mica slate, and bec... very productive in beds known by the name of *Faalbander*. At Andreasberg, Ha... the veins which pass from argillaceous slate into flinty slate lose their riches in ... latter rock.

M. Fournet gives, from the information of M. Voltz, the following remarka... example of the contents of a mineral vein varying according to the character of ... rocks on its sides:—The Wenzal vein at Furstenburg runs nearly vertically from ... to S., across many beds of gneiss, about 60 feet thick, dipping east. Each of th... beds forms a distinct variety of rock. The first is very micaceous; the second pas... into argillaceous slate; the third is hornblendic, and scarcely any mica can be detec... in the fourth. The vein is shifted in the depth to the westward by several cr... courses; and it was between two of these cross courses, distant from each other ab... 240 feet, that it contained those riches for which it has become so celebrated. In ... first bed of gneiss the vein merely formed a nearly imperceptible string of clay; ... the second it suddenly acquired a thickness of from 12 to 18 inches, and was co... posed of sulphate of baryta, antimonial silver, red silver, and argentiferous g... copper. The antimonial silver was always found in large masses. In the third ... the thickness of the vein is preserved, and the sulphate of baryta is continued in ... but the silver ores disappear, and a little sulphuret of lead is the only ore found. ... the fourth bed the silver ores become as abundant as in the second, but they gradu... disappear in depth and are replaced by selenite (sulphate of lime), a little sulphu... of lead, and some traces of pure sulphur.

rocks forming its walls. In Derbyshire, for example, where the same fissure not only passes through the mountain limestone, often with its associated igneous rocks (p. 558), but also across the surrounding, and higher accumulations of shales and sandstones, the lead ore, sulphuret of lead (that chiefly found in the Derbyshire veins), keeps generally, though not altogether to the limestone series, and appears most prevalent in the upper part of it. The igneous rocks, commonly compounds of felspar and hornblende, sometimes dense and hard, at others originally vesicular, though the vesicles may be now filled by infiltrated matter, are considered unfavourable for these ores of lead. Indeed, at one time, the opinion of the Derbyshire miners was, that the vein did not traverse the *toadstones* (p. 559), or *blackstones*, as these igneous rocks are locally termed, so unproductive are they.* It is now, however, well known that the veins, the true fissures, those locally termed *rakes*, pass through these rocks as well as the limestones, the ores being commonly absent where these igneous rocks constitute the walls of the vein, its contents in those situations being composed of other mineral substances.† Among the limestone beds themselves, some are considered more favourable, as walls to the vein, than others, and certain of them, in which much carbonate of magnesia occurs, are disliked and looked upon as somewhat unfavourable. Though the veins are known to be often continued into certain shales, not unfrequently black and containing much carbonaceous matter, above the limestones, and though these shales have occasionally *borne*, as the term is, a fair amount of ores, looking at the district generally, this is the exception, and it is a still greater exception when the sandstones surmounting these shales contain any appreciable amount

---

* With reference to this rock, which appears to be chiefly a compound of felspar and hornblende, with oxide of iron, and thus unfavourable generally as the wall of fissures for the lead ore in Derbyshire, it may not be out of place to remark that the greenstone of Devon and Cornwall, commonly of much the same composition, may be considered, as a whole, unfavourable to the ores of tin, copper, and lead. The mode of occurrence of these greenstones, as to proximity to granite, intermixture with elvan dykes, and the intersection of cross courses, is the same as that of the slates with which they are accompanied; the fissures, moreover, traversing them have the directions and are of the same kinds as those bearing ores elsewhere. Though certain mines at St. Just might be considered as exceptions, this is more apparent than real, abundant ores rarely being detected in the greenstone itself, which, from the dip of the beds near St. Just, often appears to occupy more of the mass of rocks there found than is really the fact.

† In the cases where a fair proportion of galena has been found in fissures through the *toadstones*, it has usually happened that the vein traversing the limestones above or beneath, and sometimes both, contained much ore; it thus appearing as if a superabundance of the ore found its way amid the toadstone, the effects due to the limestone being sufficiently powerful for the purpose.

of ores, though a fissure may have tra rocks, arranged as beds, and have been sp lar kind at the same time. Taken as a the mountain limestone series in Derbys liferous, and in it certain beds appear occurrence of the ores of lead than other equally well whether the sulphuret of lead versing all the rocks, or in the joints and limestone series. The metalliferous dep the irregular cavities so frequent in many different parts of the world, but extend to between the beds themselves, and arise removal of clays which were once interpo beds, or from the original small spaces be enlarged by the same causes as those whicl irregular and greater cavities.

Many of the small metalliferous veins stone are but joints (p. 624) in that rocl as to receive a deposit, which, when suff sulphuret of lead, the miner will follow i finding these above a bed of *toadstone* beneath that igneous rock, with no conn the impression seems in a great measure veins did not traverse the *toadstone*. Th wherein sulphuret of lead has been disco When they rise through the beds they and when interposed between them, *flat* the cavities in limestone districts of th evident that these distinctions are not alv that irregular cavities rising upwards may from them running amid the beds themse the cavities, and real dislocations traverse fully examined, *leaders*, as they are term such situations, so that dislocations havi munication was formed between them cavities, and thus any solutions or gaseo the dislocations would enter into them. C worked for lead ore seems to have been few years since considerable quantities w of lead encrusted, as well overhead as o

* Many, though not all, of the strings of ore whi *skrins*, are in joints.

fluor spar and sulphate of baryta, two very common veinstone minerals in certain parts of Derbyshire.

Let, in the annexed section (fig. 288) *a a'* represent a part of the

Fig. 288.

limestone series of Derbyshire, and *b* an interposed bed of *toadstone*, formed after the beds of limestone *a'*, and prior to the deposit of those at *a*, and *i*, *k*, *m* be fissures traversing all the rocks; *h*, *h*, *h*, *h*, being ordinary joints in the limestone which do not traverse the toadstone; *p*, *p*, irregular cavities in the limestone, and *f f*, the common interstices between the beds, enlarged by the removal of parts of the adjacent limestone in the usual manner; then a variety of spaces, differently communicating with each other, may be all filled with mineral matter contemporaneously derived from the same supply, and be all known by different terms among the miners. The fissures *g i*, *d k*, and *c m*, would be the channels (*rakes*) through which the various mineral substances introduced from beneath could pass into the irregular cavities *p p* (*pipes*), the enlarged spaces between the beds *f*, *f* (*flat work*), and into the joints *h*, *h*, *h*, *h* (*skrins*); all these varieties of open spaces occasionally intermingled, according as they locally occurred. The main fissures would be considered as passing through all the rocks, while the joints, or at least a large proportion of them, might terminate at the toadstone.*

Of the occurrence of the ores of lead in spaces between beds which were open when they and the other contents of such cavities were accumulated, that at Fawnog, two miles west from Mold, Flintshire, may be selected as an instructive example. From the information of Mr. Warington Smyth, it appears that after some unprofitable search for lead in shallow workings between the carboniferous limestone and its covering of the arenaceous rocks known

* The joint fissures through the toadstone appear to be very few when compared with those in the limestone. Although to render the complicated mode of occurrence somewhat more clear, the joints alone are noticed in the section (fig. 288), it should be stated that in some parts of Derbyshire, and independently of the joints, the fractures, when a main dislocation was effected, seem to have been more extensive in the limestone than in the toadstone, as might be expected from the application of the same force to bodies so different in tenacity, so that the same crack is more ramified in the one than in the other, and the minor fractures appear to terminate at the toadstone.

as *millstone grit*, it was discovered that o
tributed in a *flat*, or streak of ore, between
being elongated on an E.N.E. direction, th
vailing fissures, containing lead, in the s
following this "flat" downwards, on the
thousand tons of very excellent sulphuret o
a few years. Subsequently, another mining
still further upon the dip of the beds, an
deposit in a continuation of the same plan
grit and carboniferous limestone. From t
of ore, several thousand tons were also raise
ground being thus proved for half a mile in
several shafts, a very good illustration is af
occurrence of a metalliferous deposit betw
of rock, and probably also after their depo
the lead ore being simply in a cavity par
dissimilar beds, instead of in a vertical fissu

As regards the deposits of mineral mat
the useful metals, in joints of rocks (and the
is sometimes little else than the latter), the
traversing alike granite, the veins from it,
through which it has been protruded, may
Michael's Mount, Cornwall. The joints,
insular position of St. Michael's Mount, and
of the sea and atmosphere, give the granite
being regularly divided into vertical beds, r

* Warington Smyth, MSS., who further adds, that
semi-crystalline and grey, abounding in stems of encri
by "swallow-holes," or water-channels running in vari
which are smooth except where projecting fossils ar
resistance to the power which removed their once-cor
(millstone grit) is generally a sand, partially calcare
encrinites), about 18 feet thick, surmounted by a bed o
deep; this succeeded by various sandstone and conglon
are occasional lenticular masses of limestone. The *flat*
coloured argillaceous matter, from 15 inches to 8 feet in
portion averaging 14 inches thick, but in some places
feet, and consisting entirely of sulphuret of lead. Seve
with the same general direction. The third "flat"
often surmounted by 6 or 8 inches of compact carl
observes Mr. Warington Smyth, "of strings of ore fron
traced for 30 feet downwards amid the limestone, and
beds above, is sufficient to show that the introduction
been effected subsequently to the deposit of the mill
sandy character of the surmounting bed of rock, this w
the decomposition of the millstone grit above, while the
or from subsequent decomposition by the passage of w
careous particles.

and W. 10° S. A change in the structure of the granite is clearly perceptible towards the joints, and in them are found quartz, mica, topaz, apatite (phosphate of lime), peroxide of tin, wolfram (tungstate of iron), tin pyrites (sulphuret of tin and copper),* schorl, and occasionally other minerals. These are but mineral veins of a particular kind, and on the small scale. As to the peroxide of tin, it is one of the common ores in the fissure veins of the vicinity; and as to wolfram, more of it accompanies the tin ores in certain parts of Cornwall than is convenient for the miner. Quartz is the most abundant mineral in the open spaces, sometimes crystallized, at others filling the joint wholly to its sides. Where these joints traverse the granite veins, and the adjoining (altered) schistose rocks, they sometimes also present interesting examples of differences in their contents, according to the kind of rock forming their walls. The subjoined plan (fig. 289) is one of part of St. Michael's

Fig. 289.

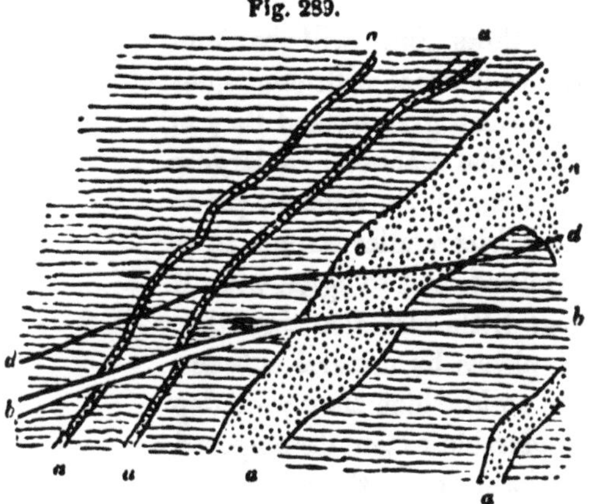

Mount (N.E. side), wherein the granite veins, *a*, *a*, *a*, are seen to traverse the altered slate rocks (which are shaded), a small included portion of the latter being seen in the largest granite vein at *c*. A joint *b*, *b*, traverses both the granite veins and the schistose rock, and *d d* is a parallel joint less wide. The latter is filled with mica where it crosses the slate, but contains also quartz, and is even occasionally altogether cemented together by that mineral where it traverses the granite veins.

---

* The variety of tin pyrites which we thence obtained also contained zinc. The following is an analysis of specimens of this mineral from St. Michael's Mount:—

Tin	31·618
Copper.	23·549
Zinc	10·113
Iron	4·790
Sulphur	29·929

The long-celebrated Carglaze tin mine, wall, also shows joints filled with mineral oxide of tin. Many of these have been granite in which they occur being soft fro granite being also white, these joint vei schorl and peroxide of tin, mingled with appearance, as represented in the annexed

Fig. 290.

large portion of these lines will be foun adjoining slates, as is usual with joint lines Devonian and Cornish granite, and they ar lines, also in the usual manner. A large p country on the north of St. Austell, partic Hensborough, exhibits similar strings, in wl of tin are intermixed, and so agree with li that they appear little else also than the such divisional planes by mineral substanc than the granite amid which they were dep become to a considerable extent decomposed. schorl in these joint veins, the observer w many of the highly schorlaceous portions Cornish granites, at their boundaries toward ing bosses of them, that it may often be fou near the joints, and, with quartz, sometime space between their walls, both minerals derived from the adjacent granite.

Not only are certain minerals, including metals, found in a fissure more frequently lated near, particular rocks or modificatior the manner above noticed, but also in som

* The works upon these small joint veins and upon t the country, the channels, it may have been, through w tents of the former have been derived, are very exten the tin ore having been of excellent quality, and the gra worked, from its state of decomposition.

ores than one occur in sufficient abundance to be profitably worked, so that the ground is well explored, fissures in given directions are observed to contain certain of these minerals more than others. Even as regards these also, there would appear to have frequently been conditions under which minerals, chiefly found in fissures taking given directions, were accumulated more in some parts of the same fissure than in others. The mining districts of Devon and Cornwall* may be studied with advantage in this respect, though similar facts are well known in other mining countries in different parts of the world.

Referring back to one of those districts (fig. 216, p. 566), it is chiefly in the fissures, *v, v, v*, having an easterly and westerly direction, that the tin and copper ores are obtained in profitable abundance, while those ranging northerly and southerly, *d, d, d*, often contain the ores of lead, iron, and some others. There are exceptions, but, as a whole, this distribution of ores is somewhat marked. Upon careful investigation it has been found that the north and south dislocations have been formed subsequently to those having an easterly and westerly direction, the proof being (p. 654), that the contents of the latter have been broken through, as well as the rocks forming their walls, and that new matter has been accumulated in the new fissures. The observer has, therefore, in such cases, not only to bear in mind the direction of the fissures, but also the difference in time when each of the two sets may have been produced, so that if at one time the conditions for the formation of the ores of tin and copper prevailed, and those of other ores at another, the opportunities for the production of various ores in all the fissures were not contemporaneous, but different. This circumstance has to be fully regarded, as well as any influences, causing the deposit of certain substances, which the direction of a fissure itself might occasion.†

---

* The observer will find a considerable mass of important information respecting these mines, in Mr. Henwood's work on the Metalliferous Deposits of Cornwall and Devon; with Appendices on Subterranean Temperature, the Electricity of Rocks and Veins, the Quantities of Water in the Cornish Mines, and Mining Statistics, forming, vol. v. of the "Transactions of the Royal Geological Society of Cornwall," Penzance, 1843.

† The study of the different fissures in Cornwall, some containing ores, others not, induced Mr. Carne, in 1822 (Trans. Geol. Society of Cornwall, vol. ii.), to class them under eight divisions, on the principle that the fissures of one epoch had a given direction, and were only cut through by those of subsequent times. By *east and west lodes*, Mr. Carne says that he means "metalliferous veins whose direction is not more than 30° from those points; by *contra-lodes*, metalliferous veins whose direction is from 30° to 60° from east and west; and by *cross courses*, veins whose direction is not more than 30° or 40° from north and south."

CHAPTER XXX\

MODIFICATIONS OF THE CONTENTS OF MINERAL \ RANGE.—MODIFICATION OF THE UPPER PART ATMOSPHERIC INFLUENCES.—SULPHURETS OF INTO CARBONATES.—REPLACEMENT OF ONE BY ANOTHER IN VEINS.—COATING OF THE WA OF MINERAL MATTER.—FISSURES COATED BY SEVERAL SUCCESSIVE MOVEMENTS THROUGH ING OF THE SIDES OF FISSURES ON MINERAL THEM AT INTERVALS.—FRACTURES THROUGI OF FISSURES.—MODIFICATIONS OF THE CON: CROSSING OF VEINS.—EFFECTS ON THE CONT OR CROSSING AT SMALL ANGLES.

TAKING certain minerals for study, and e useful metals, the observer will often find manner in which they may be distribute raneously in the same fissure. Certain co sometimes suggest themselves, if not as tl among the conditions which may have a ores of one metal more abundant than range of parts of the same mineral vein. does not so well accord with the facts obse of the great Crinnis lode, running from Island, Cornwall, into the granite, may b of the same fissure being cupriferous ami chiefly stanniferous towards the granite. tribution of tin and copper ores in Cornw mines in that country are well known to l upon their *backs*, as the upper parts of termed in some mining districts, and to ha the copper ore was attained beneath, such as worth raising at that time.* Some of these cases would not appear, as will be hereafter seen, to justify the view that the tin ores occurred to the exclusion of those of copper, but they nevertheless seem to show that tin ores were present in the higher parts of these lodes, and were scarce, if not absolutely absent, beneath.†

Points of this kind in connexion with other ores are also well known, and require similar attention, as, for instance, the frequent presence of copper pyrites on the backs of many of the lead veins in Cardiganshire (Goginan, Cwm Sebon, and others). In veins of mixed ores of different metals, where some of each are found disseminated through them, the relative abundance of the ores is sometimes found most materially modified at different depths, and this occasionally even to a certain extent irrespective of the kinds of rock forming the walls of the veins, though this influence requires always to be steadily borne in mind. Thus with some ores of zinc, lead, and copper, as, for example, in the well-known Ecton mine, Staffordshire, the sulphuret of zinc was found most abundant in the depth, the sulphuret of copper occupied a central position, and sulphuret of lead was found in the higher parts.‡ In the Spital vein at Schemnitz, according to Mr. Warington Smyth, where the sulphurets of silver and lead are raised, though the latter is argentiferous beneath, the ores towards the higher portions of the vein are chiefly sulphurets and other ores of silver, in which either lead is scarce or absent.

In this kind of investigation it is very requisite that attention should be directed, not only to the kinds of rock which may be traversed by the veins, as above noticed, but also to the decomposition and changes which may have taken place in a fissure, or

* Mr. Carne (Copper Mining of Cornwall; Trans. Geol. Soc. of Cornwall, vol. iii. p. 37) notices Wheal Damsel and Wheal Spinster copper mines as instances where the upper parts of the veins were taken away for the tin they contained. "The granite walls of the lode are still visible," he remarks, "at the surface and to the depth of three or four fathoms, having a space of about 4 feet between them. It is probable, that if the rubbish were taken away the space would be found to extend to the depth, perhaps, of 10 fathoms (60 feet), or as deep as the ancient miners could go without being obstructed by water. From this space the fine gossan of the copper lode was wholly taken away, and the tin ore extracted from it."

† At Dolcoath mine, Camborne, one which has been long in work, it has lately been found that tin occurred in profitable quantities in the depth after the vein had been worked chiefly for copper, the higher portions having formerly furnished tin as the principal ore. The ores of copper and tin are sometimes more mixed in Cornish mines than the distinctive names of "copper lode" or "tin lode" would lead those not familiar with those mines to suppose.

‡ In this mine, particular beds of limestone were found so much more favourable for ores than others, that they were always followed by the miners; and as the beds of that part of Staffordshire (adjoining Derbyshire) are much contorted, the workings have a remarkable appearance in consequence.

other cavity, after some original condition
itself of no slight importance. There
which appear clearly to show material c
from a previous state of mineral veins,
surface influences are often most marked.
pyrites is abundant, for instance, the ch
often extend to depths more considerabl
expected. This ore, essentially a combi
and iron, when exposed to surface influen
sulphuric acid being apparently produced
upon the sulphur, this acid then attacking
ore under the conditions in which it is p
that sulphate of copper is removed, and
left, forming eventually, from a continuan
a hydrated peroxide of iron, in which otl
entangled in the ore, may still remain. To
of iron the term *gossan* has been applied
and in them are sometimes found dissen
some other metals, those which were ming
of copper pyrites. Modifications of this ki
a certain amount of geological time dur
effected, but also sufficient proximity to
It will be evident, if geological changes sho
a large portion of the present surface in
numerous bunches of ore in the veins sho
exposed to atmospheric influences, that, t
of change and modification the contents
them may have been exposed, such newl
have to undergo the alteration noticed in th
and that the copper in solution, as a sulp
some situation where it would remain as s
change. The attention of the observer
directed to the conditions of a vein c
bunches of copper pyrites, before he will fi
copper, but also the oxides and carbonates

---

* The German miners term this decomposed or
chapeau de fer, expressive names showing their highe
The former say—

> " Es ist nie ein Gang so gut,
> Der trägt nicht einen eisernen H

† Several of the gossans in Cornwall have been f
this metal has not always been obtained in quantities su

frequent between them, as well as in situations where it may readily have happened that a solution of sulphate of copper, derived from decomposed copper pyrites above, could find its way. He may often, moreover, see the metallic copper in chinks and other situations, reminding him of the deposit of copper by the electrotype process from solutions of the same kind. He thus has to consider the vein and its walls with reference to their electrical conditions. If metallic copper were thrown down in fitting situations, he would probably have no difficulty in inferring that the oxides resulted from the action of the oxygen contained in the surface waters, as part of the common air disseminated in them, and that the carbonates were formed by the subsequent action of carbonic acid also contained in the same waters.* Bearing in mind the mode of occurrence of the mixed ores of tin and copper of Cornwall, and that some of the mines were formerly worked for tin, with a prevalence of copper ore beneath, it may readily have happened that the stanniferous parts of such veins may once have been more rich in copper pyrites, and that the latter had been removed, in the manner above mentioned, and the tin chiefly left in the gossan, so as to render it principally stanniferous.

Changes and alterations of a similar kind are, as might be expected, found at the higher parts of veins containing other metals, especially of those the ores of which seem more or less easily acted upon by atmospheric influences. Thus on the backs of veins containing sulphuret of lead, the carbonates of that metal may be often found. M. Hausman has suggested that the change has been effected by the conversion of the sulphur of the sulphuret of lead into sulphuric acid, which, combining with calcareous matter, set free carbonic acid, that, in its turn, combined with the lead, forming the carbonate. In those cases where the veins of sulphuret of lead are found amid limestone, and the carbonates in

* The condition of many small Roman copper coins found a few years since near Aberystwyth, Cardiganshire, illustrated these changes with more than the usual evidence, though the common *patina* or *ærugo* upon ancient copper coins also shows the same thing. The pot in which the coins were buried was found not far from the surface, and the coins themselves had been exposed to the action of atmospheric influences; the waters, containing common air and carbonic acid finding their way to them, had produced the red oxide of copper on the surface of some of the coins, beautifully crystallized, while on others the further change into the carbonate had been so effected as to present the usual mammillated character of malachite, sections of it showing the common variations in colour of that mineral. Illustrative specimens of these coins are now in the collections of the Museum of Practical Geology. Many ancient bronzes, when long exposed to the needful conditions, as has happened with most of those discovered at Nineveh, well illustrate the changes of the copper in them from the metallic state, through the red oxide, to the carbonates.

their higher parts are sufficiently common,
have been thus effected.

In illustration of the conversion of the
into a carbonate, it may be useful to re
ground refuse of the old workings* of the
of which may reach back to about 1700 y
turned over for the ores it may contain,
phuret of lead are found wholly changed
the larger pieces are thickly coated with th
miners observe that in the places where th
the alteration is most marked and conside
tration of this kind of alteration is to be fo
pieces of sulphuret of lead are distributed
fragments of limestone, in a few localities
for example, at a mine named the Green L
where pieces of lead ore appearing to h
some neighbouring vein and distributed i
bably at the tertiary period, are found. Th
of lead are sometimes wholly, at others pa
crystalline carbonate of lead. Similar illus
the phosphate are also to be well observed
fragments of sulphuret of lead having been
verted into the phosphate of lead,§ a subst
in the higher parts of lead veins in other s
depth of 150 feet and more, as, for examp
mine, and at other places in the vicinity of

Calamine seems frequently little else tha
similar manner, the sulphuret being conv
of zinc, the sulphur having disappeared,
oxygen and carbon. The conditions unde
tion of calamine is so often found, especial
would appear to render this view extrem
goch, Flintshire, it is seen now forming,
decomposition of the sulphuret of zinc i

* Among the ancient pigs of lead found in Derbysh
at Cromford Moor, with the inscription, in raised letter
AUG. MEI. LVI. According to Mr. Pegge (Archæ
have been cast about the year 130.

† This refuse, left in the workings, is locally know

‡ It should be noticed, with reference to this water
bicarbonate of lime in solution.

§ Illustrative specimens, exhibiting these modificat
in the collections of the Museum of Practical Geology

brought in solution, like bicarbonate of lime into limestone caves, and deposited in the workings of the mine.*

Independently of these modifications and changes in the higher portions of mineral veins, there are others to be found occasionally in all parts of them, showing that the substances thrown down in them, or against the walls of the fissures have been again removed, their places either vacant, or replaced by other substances filling the cavities which were thus left. A large proportion of the pseudomorphous crystals of different minerals has been thus produced—at least those of them which have, as it were, filled moulds prepared for them in a vein, by the removal of some first-formed substances, coated by others prior to such removal. Of this kind of change the observer may often study examples in the fissures and cavities of mining districts, as also in the half-decomposition and partial removal of various mineral bodies. Vein quartz is sometimes found as if it had been partially attacked by solvents and left in a highly-porous state, evidently from a loss of a portion of its substance, and not from the removal of more soluble substances which may have been once included in it, a circumstance to be carefully investigated, since the latter has sometimes been clearly the case.

In some veins the changes of conditions for the deposit and removal of mineral matter are highly interesting. Some circumstances observed in the Virtuous Lady Mine, near Tavistock, Devon, a few years since, may serve to illustrate these changes. The fissure in which this vein occurred was very irregular, and sometimes the cavity between the walls extended to many feet. Upon the walls there had been first deposited, (1) a mixture of quartz and copper pyrites, the latter often crystallized, the original tetrahedrons so elongated as to form rough prisms with pyramidal summits; (2) upon these, cubic crystals, often of considerable size, were accumulated, and were probably of fluor spar (fluoride of calcium); (3) an incrustation of carbonate of iron then completely covered the whole; and, (4) the substance forming the cubes being dissolved and removed, and so as not to injure the carbonate of iron, (5) cavities were left in which silica and the sulphuret of

---

* The composition of calamine is very variable. According to the analyses of M. Berthier. the carbonate of zinc varies in its ores from 30 to 90 per cent.; the other substances in the ore being carbonates of iron, manganese, lead, and lime. It often appears as if the carbonate of zinc had been deposited from solution, and was mixed with other substances thrown down at the same time, these commonly carbonates, the whole not unfrequently deposited in Veins amid limestones.

copper and iron (copper pyrites) entered a
these latter minerals not, however, entir
Thus, after the formation of the fissure, th
at least five changes of condition in that
these facts were observed, during one of
calcium, previously deposited in a crystalli
while its coating of carbonate of iron rem
very desirable that modifications and chan
be carefully investigated, especially with re
of the rocks in which the fissures and other
and to the probabilities of any new fractu
introducing new solutions or gaseous matte
any substances previously accumulated in
the observer have to study those more co
crystallizations where one substance is de
by another, and is removed and replaced
the manner above noticed, but also the ren
certain minerals, as if molecule by molecul
nal crystal remaining unchanged. With re
of the carbonate for the sulphuret of lead, a
though fragments would serve to show th
form of the sulphuret of lead being still
latter have been found completely replac
lead, as for example, in Derbyshire, at a
Matlock. Copper pyrites has been found
where, replacing carbonate of iron, the form
latter completely preserved; and numerou
known, where there seems no reason to s
morphous minerals have been the result o
latter removed, so that no trace of them ha
doubt this is a circumstance to be regarded
at all times in investigations of this kind.†

---

* M. Becquerel, while noticing these changes, ment
trating their production from those analogous to ceme
an electrical origin. M. Darcet left a plate of steel d
the Mint, at Paris, in contact, by means of one of its
of silver, which reached it very slowly from a fissure in
half of this steel plate was entirely changed into very
mass without the least trace of iron. The volume of
the same with that of the plate of steel.

† With reference to these changes, it may often
certain solutions flowing over the surface of land fr
metalliferous matter in some of these solutions was
combined with some ordinary sedimentary deposits w
some localities deposits of this kind, forming bands

Examining the manner in which the substances have been arranged in the fissures, or other cavities, after the action regulating the deposit of certain bodies against their walls, more upon some rocks than others, and the direction of the fissures themselves are considered; the most simple mode of occurrence presented is that where a single substance may be found in them. This substance may either have been derived from the adjoining rocks, by solutions formed in and percolating through them, such as is shown by quartz veins amid siliceous rocks, and calcareous veins among limestones, or be derived, such as the peroxide of tin, the sulphurets of lead, copper, and antimony, and other ores from other sources, strings even of metallic silver and larger breadths of metallic copper presenting themselves. In the first case it may be apparent from successive coatings of crystals, each pointing inwards, and from both sides of the fissure, that the filling has been a work of time, during which the conditions only for the deposit of the single substance prevailed in the part of the fissure seen. The following section (fig. 291) will illustrate these successive

Fig. 291.

deposits, *a b* being a line of fissure, cutting through any class of rocks, *d d*; successive coatings of a single substance, *c c c c*, filling it up towards the centre, where, for still further illustration, occasional cavities may be supposed to remain. The probability of the single substance so found being more or less directly or indirectly derived from the rocks traversed, will have to be weighed with

assisting the geologist in the relative dates for the filling of fissures with ores in a district. For example, at the Hook Point, county Wexford, and also nearer the town of Wexford, the upper part of the old red sandstone series contains carbonate of copper mingled with vegetable remains, the bands with this interspersed carbonate of copper several times repeated. It might thence be inferred that this carbonate had been derived from veins of copper ores traversing the adjacent Silurian rocks, so that these veins were formed anterior to the deposit of the old red sandstone of that part of Ireland. In certain mining districts solutions of sulphate of copper are sufficiently strong to produce marked effects among any ordinary deposits to which they could flow uninterruptedly for a long lapse of time. The cupriferous slates of Mansfield are well-known examples of detrital accumulations containing disseminated copper, usually copper pyrites, profitably worked. And here again organic matter is often mingled with the ore, reminding us of the mode of occurrence of certain iron pyrites.

reference to the conditions of the locali
country, such as the mining districts of
found the successive coatings composed of
be led to infer that this substance was der
beds, so abundant around, and in which the
while, if composed of sulphate of baryta
parts of that district, he might be induce
of supply. Should the single substance in
be fluoride of calcium (fluor spar), while
abundant supply of calcium was at hand t
he might question the limestone having
latter substance in the quantities found in

Single substances are not only found t
fissures together, but also cementing fragm
or have been introduced into them, and so
show that such fragments may have fallen
matter while it was being deposited, sin
isolated, and suspended amid the cement
thus, not only bind those together which l
another previous to its introduction, but al
are occasionally well separated from any ot
not have received support from them. Th
of mineral veins is to be seen in some parts
and may often be as well studied in fissur
the useful metals do not occur, and as they
sections, such as in cliffs on the sea-coasts
ing substance of fragments may be any

* In Cornwall, where fragmentary lodes are not unc
of Cornwall, &c., p. 323), that of the Relistian mine
rounded pebbles of slate and quartz (the latter from th
formed parts of some vein) cemented by peroxide of
chief mass of this conglomerate was about 12 feet l
thickness, and was found in the lode (stanniferous) al
Scattered pebbles of the same kind were discovered be
Badger lode, one near Relistian, pebbles of granite w
slate and quartz; and Mr. Carne mentions pebbles hav
of Ding Dong Mine, near Penzance, and in the lode of
The rounded state of these fragments renders them
they had been introduced in that condition into the fis
that part of the fissure, at least, the deposit of the tin
In other respects the mode of occurrence is the sam
cemented by similar substances M. Fournet (Étude
notices pebbles of gneiss as found in the vein at Joa
fathoms (1,152 feet). With respect to the fragments o
as might be anticipated, they are usually more commo
inclined than when it becomes vertical, though th
in the latter.

fissures,—as well the ores of lead, copper, tin, and the useful metals generally, as others. Where the fragments are those of rocks, which may be easily removed by decomposition from the action of solutions or gaseous matter in the veins, as for example, those of limestone by carbonic acid in water, cavities are then left, into which other substances, those found in other parts of the vein, may be introduced, in the same manner as the cavities left by the crystals above noticed have been filled by any other substances introduced into them.*

The coatings of different kinds of mineral matter on the sides of fissures, as if from successive deposits of dissimilar substances from solutions in them, seem to have been first clearly pointed out by Werner, in 1791.† Among the examples adduced, he particularly noticed the vein at Segen-Gottes, Gersdorf, "which, reckoning from the middle (composed of two layers of calcareous spar, in which small druses here and there occur), thirteen beds of different minerals are arranged in the same order on each side of the vein, these are fluor spar, heavy spar, galena, &c. In the southern vein, Gregorius, the two layers which adhere to the sides of the vein are composed of crystallized quartz; next to these, on each side, is a layer of sulphuret of zinc, mixed with sulphuret of iron; this is followed by sulphuret of lead, carbonate of iron, sulphuret of lead, carbonate of silver, red silver ore, and sulphuret of silver. The central part, which, of course, is most recently formed, is of calcareous spar."‡

That this regularity should always be found, even in a fissure which has probably remained in its first state, as regards any subsequent movement of sides, could scarcely be expected, more especially in cases where the walls opposite to each other may be formed of different rocks (when the fissure is a fault), or where, from any other cause, one side may have received deposits which the other did not. One main cause of difference often appears to arise from

* In the cases of fragmentary veins in limestone districts, interesting removals of the ordinary calcareous matter, with the preservation of organic remains in a fragment, may occasionally be seen. A good example of this fact occurred in a large vein of peroxide of iron (hematite) in the carboniferous limestone near Wrington, Somersetshire, where, from a somewhat considerable fragment containing a fossil coral, one common in beds of the fissure above, the ordinary calcareous matter had been removed, probably by the action of water containing carbonic acid derived from the atmosphere, and the parts so removed were replaced by the peroxide of iron of the vein, so that the piece of fossil coral, preserving all the angular character of the original fragment, appeared in the midst of the vein of iron ore.

† "Theory of Mineral Veins," published at Freiberg, in 1791.

‡ Ibid, chap. iv., sec. 52.

that of the rocks on each side, a deposit, fi
tinuing to receive additions to it in prefe
circumstance previously noticed as to d
(p. 374).

The geologist, when studying the conter
districts, where so many small strings of
found in chinks and cavities, more or les
main veins, or even in them, will have his
evidences of many main fissures having
once, while the new cracks thus produced l
traversed any mineral deposits which ma
formed in the fissure, in its first state, but s
Werner seems to have been aware of thi
remarked that "we meet with distinct
formed in the direction of those of older
substance or by the side of them), formi
individual substance."* Very material ch
of veins may thence arise, both as to the de
of substances previously in the fissure, and
mineral matter.† Good evidence of such m
of the first contents of a vein being travers
ing as much a part of the walls of the s
other portion of the mineral mass broken
found in the mode of occurrence of the c
where these still retain a certain parallelis
in the following section (fig. 292), part of

Fig. 292.

* "Theory of Mineral Veins," chap. v., sec. 55.

† When describing the successive openings of th
Fournet (Études sur les Dépôts Métallifères) points
mineral substances, or modifications of such substan
of the six successive dislocations which he was en
marked by the introduction of pebbles and sand, a co
which still covers the country in many places, and
lavas of the extinct volcanos of Louchadière and Pran

Binner Downs, Cornwall, the central deposit of quartz crystals, pointing inwards to *e*, is only one of four other similar arrangements of parts, *b*, *d*, *g*, and *h*. To effect this crystallization, the increase of the crystals having taken place inwards, five different openings, at different times, have been effected, so that the needful walls for the commencement of crystallization in each case could be afforded. Without additional evidence, such as the openings in other parts of the vein, or amid the adjoining rocks, might present, it would be difficult to determine which of these openings might have preceded the others. Commencing on the left, *a*, the section gives a layer of copper pyrites and sulphuret of zinc (blende) upon the wall of the lode in that direction; to this succeeds quartz crystals pointing inwards, *b*; indurated argillaceous matter, *c*, and another collection of quartz crystals pointing inwards at *d*; then a set of crystals, *e*, commencing on either side upon a layer of crystallized copper pyrites and zinc blende, *f f*, first lining the opening in which this arrangement is seen. Still proceeding from left to right, quartz crystals have been deposited upon another cavity, leaving an open space at *g* (*vug* of Cornish miners), these succeeded by layers of quartz crystals, pointing inwards at *h*, resting against the wall of the vein in that direction. Examples of this kind, with considerable modifications, could be readily multiplied to a great extent.

With reference to the occurrence of lines of different substances, when they do not resemble each other on both sides of a vein, though it may be suspected that there has been more than one movement in the fissure containing them, after its first production, the evidence is often by no means so clear. As, for example, in the subjoined section (fig. 293) of part of a lode at Godolphin Bridge,

Fig. 293.

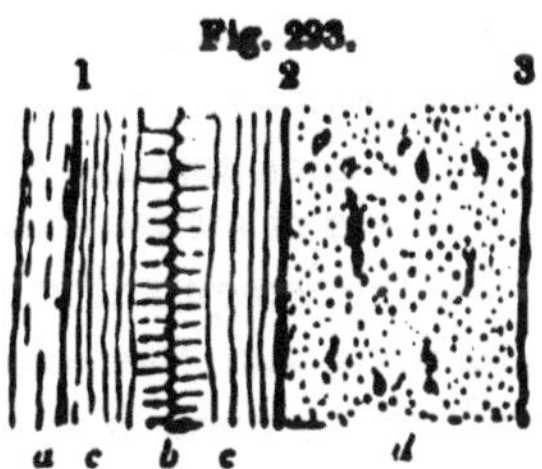

Cornwall, where *a* represents a layer of quartz resting against the wall of the vein in that direction, *b* quartz crystals pointing inwards, and based on agatiform bands of silica on either side, *c c*, and *d* is a thick layer of copper pyrites, mingled with some quartz. At *b* there appears decisive evidence that when the silica there found was deposited, the walls of the opening extended from 1 to 2; in the absence of any arrangement of parts showing that the co pyrites, with its quartz, had 2 for one of its walls, it could no proved that there was a distinct opening between 2 and 3, in w it was accumulated. It may have happened that the copper pyr with some intermingled quartz, was deposited against the wal the vein on the right, while quartz only was accumulated on wall to the left.

With respect to the movements producing these parallel arra ments of parts, without much, if any, evidence of the prev mineral accumulations in the veins having been broken or distur it will soon be found, while studying those fissures which are simple cracks, but faults (p. 649), that this may be produced the mere slipping of the uneven sides of the fractures, with cer intervals of repose between each movement, so that mineral ma might be deposited in the cavities during each period of repose. *a b* in the annexed diagram (fig. 294), represent a line of fiss

Fig. 294.

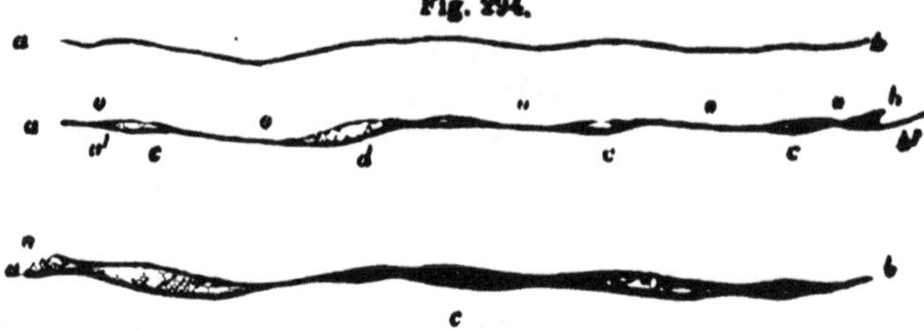

and a movement be effected so that one side, *a' b'*, be slid in manner by which the parts *o o o o* touch, supporting the press on either side of the fissure, cavities will be formed at *c c c*, and *d*, the latter somewhat more extended than the first. If movement has taken place in a contrary direction, to the left (as the third line), to the same amount as previously to the right, openings, *c c*, would be more considerable, so far as respects illustration given, and the points of contact less numerous.*

In a fracture across rocks, the irregularities of the fissures bei usually in a variety of directions, and the points of contact in co sequence variously situated, the general inequality of the walls veins has to be regarded.

When movements have been considerable, a polish and striation

* If a piece of paper or card be cut in this, or any other manner, representing section of an irregular fissure, as such fractures more or less are, and the pieces slid on a table, so that parts of the cut paper touch each other, this illustration w readily be seen.

the sides have often been effected, the striation corresponding with the direction of the movement; evidence of importance when that direction may not be clearly seen by the bedding or other mode of occurrence of the rocks fractured. In fissures where there has been more than one movement they are also valuable, especially when the evidence of crystallization having taken place at separate times in different cavities being absent, the contents of the veins themselves exhibit this striation and polish in given directions.* In illustration of movements not marked by considerable dislocations, yet where successive new apertures or cavities have been formed, filled either in the manner above noticed, crystallization towards the interior spaces affording evidence on this head, or where simple polish and striation may appear, perhaps the following diagram (fig. 295) may be useful. If *a b* be a fissure, and the portions of

Fig. 295.

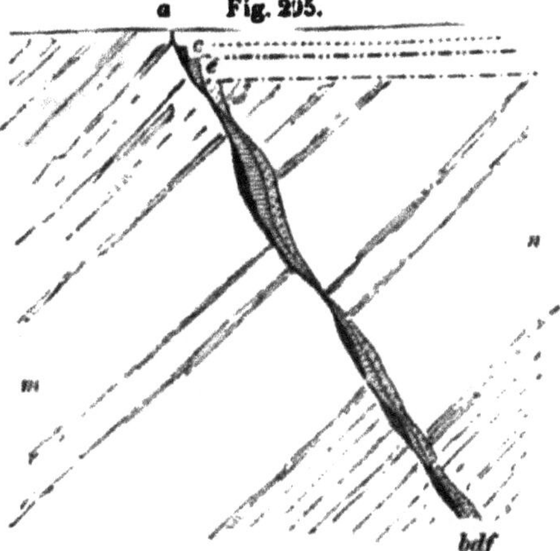

separated rock, *m n*, slide against each other, so that the beds at *n* are lowered to *c* along the line of the fissure, touching in sufficient points for support, until a fresh set of cavities be formed, represented by the dark portion, *b*, and these cavities be filled by mineral matter, a further movement in the same direction would cause friction on these contents which would thus constitute an uneven surface towards the side *n*. Assuming these contents somewhat solid (they may be even more so than the sides of the old fracture), this second movement might extend, for illustration, to *e*, sufficient points of support existing as before, so that the cavities *d* were produced. A third movement taking place, after

* When either the striation or polish has been effected in accumulations of ores themselves, or coatings of ores have covered them over, taking as it were a cast of them in their uninjured state, these polished and striated surfaces are commonly termed *slickensides*.

these second cavities were filled, similar s
a third set of cavities, *f*, would be filled.
ments or movements of the masses of rock,
seen in some localities to have been very fi

When the fissures are more complicat
different cavities have been subsequently fi
of mineral matter into those formed by
intervals, of the rocks on one side, but fr
substances variously accumulated in them,
them, extending to the adjacent rocks on e
the walls into fragments, and the whole is ag
matter newly introduced, the study of the
requires no little caution. In such cases
substances traversed will usually lead to the
ticularly when combined with any eviden
from fractures in the adjoining rocks. If
(fig. 296) *r r* represent the rocks on eith

Fig. 296.

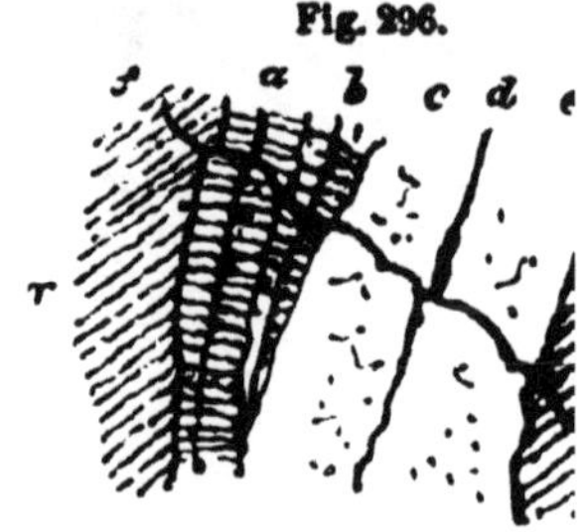

arrangement of quartz or other crystals,
separate cavity now occupied by them;
mode of occurrence of any crystallized min
any other substance separated from anothe
and striation (*slickensides*), *d*, and the w
string of some other body, *f, g*, such as
independently of the evidence of separat
by the two sets of crystallized substances, *a*
and striated surface, *d*, between the sub
movement would be shown by the string
more especially if, as in the figure, a kin
various substances, including the walls
divided, has been effected. In this mann
of the substances first introduced into fis
taken place, while mineral matter, new t
introduced.

We have hitherto supposed the successive movements to have been effected in planes either corresponding with those of the original fractures, or not far removed from them. As has been shown, veins traverse each other at considerable and even right angles, one series being formed subsequently to another, as proved by the contents of one vein forming parts of the walls of the other (p. 654). In such cases the study of any change which have taken place in the veins cut through is one of much interest in itself, and is still more so when combined with any modifications found in the contents of the traversing vein. It sometimes occurs that the mineral matter in the traversed vein is much modified, so that, independently of any removal of its parts into the new cavity, it might appear as beneath (fig. 297), where the vein *a b* being cut at a considerable angle by another, *c d*, a movement of certain substances from *b* to *e*, and a removal of similar substances from *f* to *a*, is supposed to have occurred, so that if the vein *c d* were again removed, the parts broken through would not correspond to the same extent that they did prior to the second fracture.

Fig. 297.

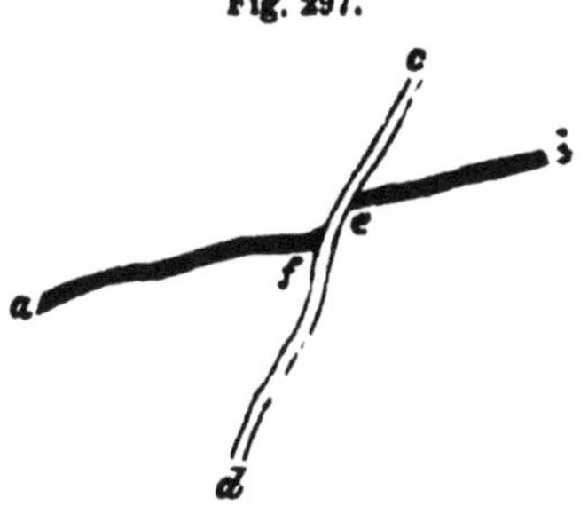

When the mineral matter is apparently moved on the one side, and on the other ore is sought by the miner, this becomes at once known; but it will reward the attention of an observer, not only to study the veins with reference to ore, but also to all the substances so traversed. It is also important to direct his attention to the deposits of any substance or larger bodies of them where the walls may be formed of the contents of a first-formed fissure, since these then occupy the position of any dissimilar rocks (p. 679), and may be found favourable or unfavourable to the deposit of some given substances, whether ores of the useful metals or otherwise. In some mining countries certain mineral substances are found more abundantly where one vein crosses another, or within moderate distances from the intersection, than elsewhere in such veins.

When engaged in this investigation, it becomes needful well to consider the fact, so commonly observed in mining districts of various parts of the world, that where two or more veins come together at a moderate angle, the ores of the useful metals sought in them usually occur more abundantly than elsewhere. This may be far from constant, nevertheless the percentage of instances in which this happens is so considerable as commonly to have arr the attention of the miners in numerous districts. The cir stance has been remarked as well when the fissures cut each somewhat vertically, as beneath (fig. 298), representing a ve

Fig. 298.

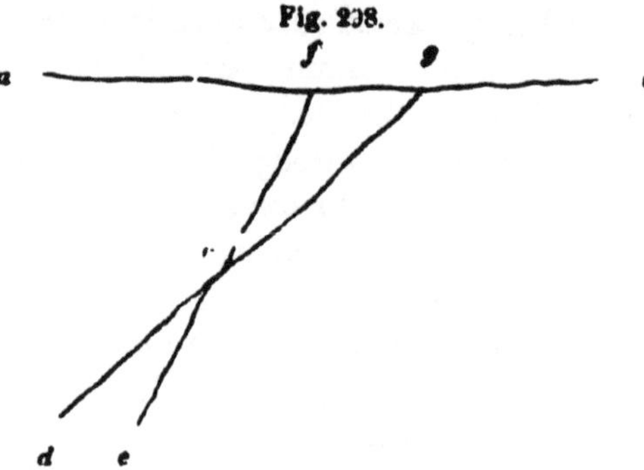

a (North), b, (South) surface of the country; f, e, north lode; g, d, iron shaft l c, their intersection, where the riches of the mine were improved, a b, slate trav by the lodes.

section of the lodes at East Wheal Crinnis Mine, Cornwall, as wi they traverse each other somewhat horizontally, when a resembling this section would also suffice for numerous examp Even a large *horse** has been observed to afford more ore at sharp ends of the fragment than in other adjacent portions. miners often give the significant name of *leaders* to the conten small fissures ranging out from a main fissure, it frequen though by no means constantly, occurring that a bunch of o found where they unite.

When the geologist regards the infiltration of solutions cavities of various kinds, the deposits of different mineral stances in such cavities, whether fissures or otherwise, the select as it were, of certain rocks by them, and the various modificat and changes which the arrangement of the different kinds of m ral matter found in such situations have sustained, he will prob be led to consider the general subject as one of no slight scien interest, while, at the same time, the investigations into which will have to enter also possess no little importance from their di bearing on the discovery and extraction of many substances so portant to the progress of mankind.

---

* This term is applied in many of our mining districts to large fragment mineral veins, portions of the adjoining rocks, which, from a complication of fracture forming the main fissure, become isolated and jammed in between its sid

CHAPTER XXXVII.

PARTIAL REMOVAL OR DENUDATION OF ROCKS.—GREAT DENUDATION ARISING FROM THE ACTION OF BREAKERS.—ANCIENT EXPOSURE OF THE AREA OF THE BRITISH ISLANDS TO THE ATLANTIC.—CARE REQUIRED RESPECTING THE AMOUNT OF DENUDATION OF OVERLAPPING ROCKS.—ISLAND MASSES OF DEPOSITS LEFT BY DENUDATION.—DISTRICTS OF BENT AND PLICATED BEDS WORN DOWN BY DENUDATION.—NEW SLICES OF LAND NOW BEING CUT AWAY BY BREAKER ACTION.—AMOUNT OF MATTER REMOVED BY DENUDATION IN PARTS OF ENGLAND AND WALES.—NEEDFUL ATTENTION TO THE GREATER GEOLOGICAL PROBLEMS.

ALTHOUGH mention has often been previously made of the partial removal of the mineral accumulations of various geological times, and this has been referred to the action of breakers on coasts, to the decomposition of different rocks by atmospheric influences, and to their erosion and subsequent transport by running waters, a few words on geological denudation, regarded by itself, may be usefully added. Viewed with reference to the causes of the partial removal of rock accumulations, there are few subjects connected not only with the present, but also many prior conditions of the earth's surface in different regions, which more impress the geologist with the great lapse of time required in explanation of the facts observed. Making all due allowance for the abrasion and transport of an immense mass of mineral matter by means of atmospheric influences, the kind of smoothing or levelling of great areas so often found, and the isolation of portions of various dimensions, outstanding like islands from a kind of main coast composed of like deposits, of which they only constitute detached parts, so forcibly remind him of the action of breakers on land, that there seems much difficulty in avoiding the inference that to this source of abrasion geological denudation is chiefly due.

As an intermixture of land and water is needed in explanation of the production and distribution of detrital matter at all geological times, and as tides, or the absence of them, would be produced by the action of the same causes as at present during the geological time, so far the general causes for the abrasion of land, at the of waters, would have been always the same as now. Winds, ever, being, with the exception of such movements as th earthquakes, the cause of the waves which break on coasts, to the geologist has to look for the continued disturbance of where its surface so cuts the land as to permit the abrasi quired. As the winds of the present day are arranged genera their courses by the action of the sun on the earth, and the ment of the latter around its axis, the observer has to co whether the former has ever been less or more intense than it is (assuming the rate of the latter, for better illustration, to remained the same), or whether counteracting or modifyin fluences, such as a temperature sufficiently high of the whol face of our planet to prevent the present differences of heat a from the sun, have had an appreciable effect in that direction. has thus to see if there be evidence of breaker action at diff and especially at early geological times. The facts noticed beaches adjoining land in the midst of the accumulations of so a date as that of the Silurian series (p. 475) may satisfy him breakers were in action at that time as now on sea-c Evidence of the like kind being seen amid detrital depos various geological periods, the geologist may feel assured that the earliest times when the remains of marine life were entomb the present day, breaker action was in force.

Being satisfied that this action has been an important geolo agent for so long a lapse of time, the observer may be ind to inquire from such evidence as may present itself, if particular exposures to great oceans can be traced. Assu winds to produce the ordinary waves breaking on shore, and the present causes of the general arrangement or modificatio land and sea have continued far back into geological time, greater the ocean exposure of given coasts to prevalent and st winds, the greater, all other conditions being equal, would be abrasion experienced. Hence it becomes important to study d dation in connexion with evidence as to the distribution of and sea at different geological times, and especially as to the d tion in which great portions of the ocean of any of those times have been situated. We have been often struck, while exami the geological structure of different portions of the British Isl with the length of time during which the various modificatio coasts attending the elevation or depression of the area at diffe periods, above and beneath the level of the sea, may have been exposed to, or sheltered from, such an ocean as the Atlantic on the westward. The position of the conglomerates of the new red sandstone time, arranged against and around the older rocks of the Mendip Hills, and portions of the adjacent country (fig. 167, p. 478), remind us, while on the spot, very forcibly of an exposure to a considerable range of water on the westward. Respecting surfaces of land left by various denudations at more recent times, the impression as to a great exposure to western waters is still greater, inland cliffs presenting themselves precisely where they would be expected, and even the shelter derived from protruding land being found. In illustration of this circumstance, an observer regarding the present escarpments of the oolitic series, including the lias, from their comparative shelter on the eastward of the high land of Wales to the English Channel, cannot fail to be struck, making all due allowance for subsequent changes arising from atmospheric influences, with the mode of occurrence of the old cliffs, and of their character, according to exposure or shelter, where the Mendip Hills, and other elevations of the same character, would modify the force of the great Atlantic waves breaking on such a range of coast.

The removal of portions of the accumulations of various periods, extending back in time to those of the Cambrian series in our islands, is something very considerable, and to be measured by thousands of cubic miles of mineral matter. In Ireland, the abrasion of the Silurian rocks, anterior to the conglomerates of the old red sandstone, is well marked. Large surfaces were evidently there shorn down by denudation previous to, or at that time, the granitic protrusions then existing in parts of that land being also cut partially away, and having from that period to the present been exposed to abrasion and loss of volume whenever brought within the range of the breakers, or raised up within the influences of the atmosphere. In regarding, therefore, such masses of mineral matter, the geologist has before him the results of many partial removals of rock accumulations, and sees the remains only of many modifications of surfaces, wrought by denudations during a long lapse of time.

In estimating any amount of matter which may have been removed, either from igneous rocks or from deposits formed by the aid of water, much care is sometimes needed, so that a correct estimate may be formed of the probable prior arrangement of the parts of it. This is especially necessary when a series, or parts of a series, of deposits may extend over or conceal another, or portions of itself, otherwise, by supposing an extension of accumulatio directions where they never existed, far more mineral matter r be inferred to have been removed than has been the case. example, the old red sandstone of Herefordshire, South Wales parts of southern Ireland, so occurs, that, viewed as a whole higher portion gradually overlapped or passed over a older and subjacent beds in a southerly and westerly directi relative depression of land and sea bottom in that direction ha probably been the cause of this overlap.* In such a case probable range and limit of the conditions for the deposit of higher portion of the old red sandstone have to be carefully timated, so that the removal of matter, never deposited, may be inferred. It fortunately happens that, in southern Pembr shire, the extension of the old red sandstone, at the time of deposit of the carboniferous limestone of the same district, ca clearly seen at Slebech, the latter there overlapping the sandst as, in like manner, in the same county, the extent of the carb ferous limestone, at the time of the accumulation of the lower of the coal measures, can also be well observed at the overlap of latter over the former, near Haverfordwest. The following sec (fig. 299) may serve to illustrate the need of caution on this

Fig. 299.

ject. If *a b* be a series of consecutive detrital deposits, for under conditions which allowed them completely to cover other, so far as the area represented in the section is concern and *c*, *d*, *e*, be other accumulations, formed after a movement the direction of *b*, had lowered that end of the deposits *a b*, so *c*, *d*, *e* were successively formed in planes parallel to each ot but at an angle with that of *a b*, the successive extensions of *c* and *e*, in the direction *a*, being governed by causes such as a lin coast, and its gradual depression beneath a sea level in that di tion, the rocks *c*, *d*, and *e*, would gradually spread beyond, overlap each other towards *a*. If now, from subsequent brea action, atmospheric influences, and final elevation of the mas now found, a surface should be exposed corresponding with

* Part of this overlap will be seen by reference to Sheets 38, 40, of the Geolo Survey of Great Britain, and also to the small General Map of South Wales, inse in the "Memoirs of the Geological Survey," vol. i.

line $f\,g$, a section of the following kind (fig. 300) would be obtained, one very deceptive, without care and attention to the general geo-

Fig. 300.

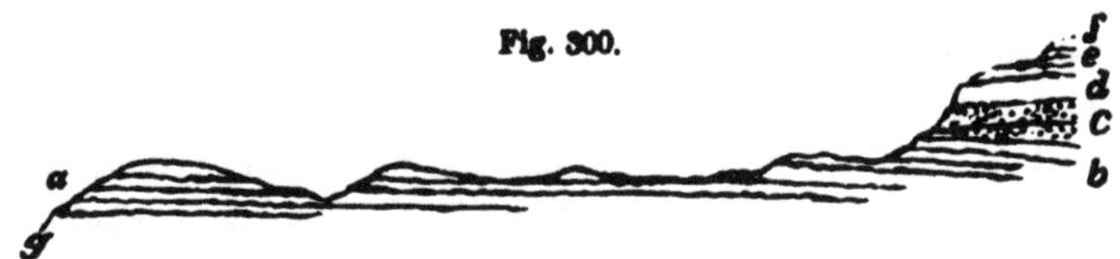

logical structure of the country as to the prior extension of the beds *c, d, e,* towards *a*, inasmuch as a vast extent of country composed of them might be assumed as shorn down in that direction, and a mass of mineral removed, which never existed.*

It thus becomes essential, when endeavouring to estimate any mineral matter removed by denudation, carefully to weigh the probability of the conditions under which the accumulations, supposed to be removed, may have extended. We may here again refer with advantage to the small area represented in the map of the Mendip Hills and adjoining country (fig. 167, p. 478,) for illustration on this head. Upon the portion of country where the new red marl and sandstone (5) occupies the surface, several isolated patches of lias (6) may be seen distributed. These patches are only the remains of beds which once continuously covered the former deposits, and joined the main mass of the lias seen on the southern portion of the map. The mode of accumulation of this lias, and its modification where adjoining the dry land of the time, have been previously mentioned (p. 481). It is sufficient here to remark, that while this rock was thus modified near the Mendip Hills, it stretched away with common characters in other directions, patches of it being found in various other localities scattered over the prior and subjacent rocks, such being the remains of a wide area once occupied by this deposit. As with the lias, so also with the accumulations which immediately succeeded it—those of the inferior oolite (7, fig. 167), detached patches of which are also seen isolated upon prior-formed rocks, such patches once portions of a continuous deposit in a sea. One small islet of inferior oolite will be observed towards the left corner of the map (fig. 167). Other patches will, however, be seen at Dundry and Brent Knoll, by reference to more extended geological maps. By careful examination it is found that the mode of occurrence of their component

* Those who have examined the escarpments of the coal measures, carboniferous limestone, and old red sandstone, on the northern boundary of the great coal districts of Monmouthshire and South Wales, will at once see the application of these sections.

beds is such as to point towards their termination, as a depo the time, at no great distance westward, so that when conside the mass of mineral matter removed by denudation in that direc with reference to the general structure of the district, due al ance has to be made for this probable termination of these westward. As illustrative of denudation towards the termin of accumulations, the following section (fig. 301) may be useful

Fig. 301.

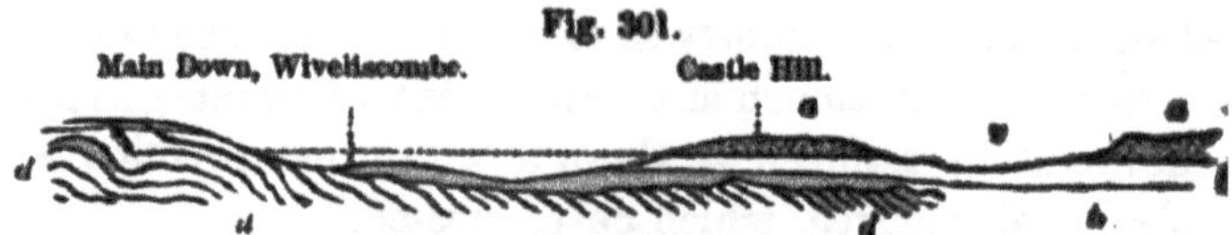

is one of the country near Wiveliscombe, Somersetshire, d d b rocks of the Devonian series, disturbed and denuded anterio the accumulation of the sandstones, *b*, of the new red sands series of that district, and of a conglomerate, *a a*, compose rounded portions of the adjacent disturbed rocks, *d d*, cemente calcareo-magnesian matter. In this case, the former contin portion of the conglomerate *a a*, and of part of the sandst beneath it, *b*, have been removed by denudation, not only the minor valley, *v*, but also from the larger surface hollowed between Main Down and Castle Hill. The conglomerates, have thus been cut off, as it were, from the source of their peb the country towards Main Down, the portion left being only remains of a once continuous accumulation of them, probably a a line of coast existing at that time at a short distance westwar

Notwithstanding these modifications required in our estima the amount of accumulations removed in some districts, there c remains so great a gap in the connexion which evidently existed between portions of deposits in others, that, collectiv the removed portions can only be satisfactorily measured by a siderable amount of cubic miles of missing mineral matter. smoothing off of the surfaces of even the most disturbed depo portions of these deposits now variously consolidated, and appare possessing that character when so worn down, is often to be fo The coasts of northern Devonshire, where the variously-consolid beds of the lower coal measures are so much bent and plicated, afford the geologist excellent oportunities for observation. It the order represented opposite (fig. 302), where *a*, *b*, being the line, cliffs, *c*, are exposed, showing numerous flexures, the conti tions of which can be readily supplied by the dotted lines above present surface of the land, *d*, and beneath the level of the sea,

The importance of cliffs, either inland or on the sea-coast, in properly appreciating the matter removed from a district presenting

Fig. 302.

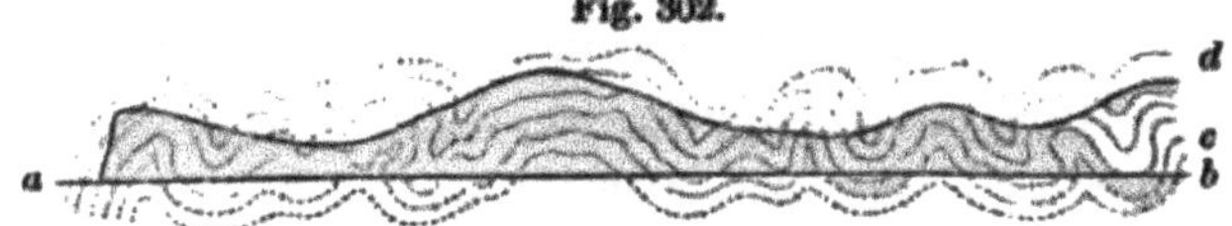

no very great differences of surface elevation and depression (mere common undulations of country), is very considerable. Without them the observer might often be uninstructed as to the real amount of flexure and plication planed down and concealed beneath the ordinary smooth surfaces exposed. Of this the following (fig. 303) is an example. It is a sketch, from Trenance Point, of

Fig. 303.

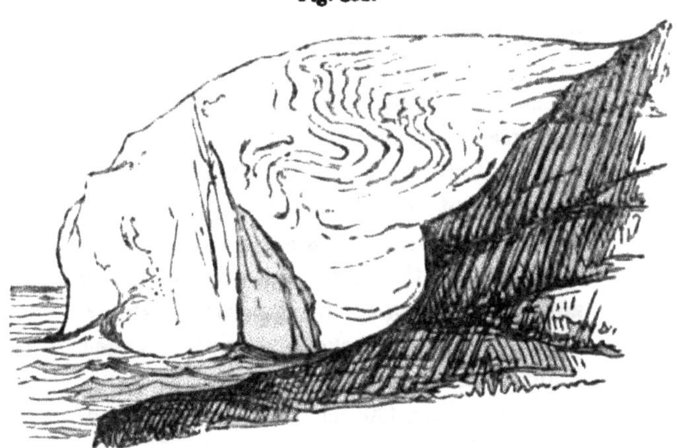

High Cove, on the north coast of Cornwall, exhibiting a section of hard sandstones of the older rocks of that country worn down, probably, by the same kind of heavy Atlantic breakers which are now cutting off a further slice of these ancient accumulations down to a newlevel.*

By consulting the geological maps of various lands, when such maps have been carefully prepared, abundant evidence will frequently appear of the surface denudation to which very extended areas have been exposed. Taken alone, without carefully constructed sections, and a due regard to the circumstances above noticed, no very accurate estimate can usually be formed.

* With regard to the present action of breakers thus, as it were, removing new slices of deposits, whether contorted, or in their original planes of accumulation, it not only effects this, but also, when the conditions for the elevation and depression of masses of land have been favourable, will cut away the accumulations of former times so as to restore old worn surfaces. Of this examples may be seen on the shores of

We may, indeed, infer, when we find near Bideford, Devon,* a part of the cretaceous series, possessing all the essenti portion when last seen in mass to the east Hills, forty-two miles distant, that the wh was once covered with the sands and che of the cretaceous series in that part of Eng by taking a given area and the thickness number of cubic miles of the deposit rem be approximately estimated. When, ho with a bent or contorted set of beds in only by very carefully following these that any fair approximation to the matter mass can be made. The various bends a only be properly ascertained, due allowan tures at the extreme parts of flexures (p. 6 of faults that have occurred, and which flexures and plications, should be carefully

As a locality, easily visited, and having sections not far distant from each other, t regarded as by no means ill-calculated to denudation; the more especially as the p action, according to the exposure of the co resistances of the rocks composing that coas From a somewhat sharp contortion, rang direction, from Culver Cliff to the Need

---

the Bristol Channel in several places (Aust Passag following section (fig. 304), near Portishead, exhibit

Fig. 304.

upturned prior to the accumulation of dolomitic co stone, *b*, the part towards *e* being now swept by break the beds, *b*, to the cliff, *d*; a portion being, for the pr of an old surface becomes again exposed at the level o of time.

As to denudation generally, which has removed l now exposed on the surface of land, the observer will preceding figures, viz., figs. 7, 10, 11, 15, 47, 57, 79, 80, 169, 171, 174, 177, 181, 209, 212, 213, 217, 218.

* See Geological Survey of Great Britain, sheet No.

beds, resting on chalk, are thrown into a vertical position (well seen at Alum and Whitecliff Bays), the chalk itself near them highly inclined, even nearly vertical, but dipping at minor angles on the southward. All these tertiary beds are cut off by denudation on the south of the east and west contortion, there being little doubt that a considerable portion, at least, of them once ranged over the chalk in that direction. On the south of the highland of chalk, left from its better resistance to denudation, along this sharp contortion, the mass of rocks is curved into a low arch, and the chalk itself and inferior beds (upper greensand, gault, and in two situations, east and west sides of the island, the lower greensand) have been removed by denudation, portions of the chalk remaining upon the high ground of St. Catherine's, Rew, and Boniface Downs, on the south of the island. The observer, placed on Ashley or Arreton Down, southward from Ryde, may have an excellent view, on the south-west, of the break into the arch, made by denudation, between the chalk of Chillerton Down and St. Catherine's Down, a long ridge of upper greensand (harder there than on the north) stretching out from the latter. The volume of mineral matter removed by denudation, including part of the tertiary rocks, in the lower country, between the high grounds extending from Culver Cliff to the Needles and the heights of St. Catherine's, Rew, and Boniface Downs, must, in some places, be from 2,000 to 3,000 feet thick.

Availing himself of the observations made on the Geological Survey, Professor Ramsay has pointed out the vast mass of matter which has been removed by denudation from bent and contorted portions of South Wales and the adjacent English counties, the amount of which may to a certain extent be estimated.* Taking the evidence obtained respecting the thicknesses of the various beds of Silurian rocks and their curvature, in the Woolhope district (fig. 305), he infers that, if the amount of mineral matter now removed were restored, so as to complete the section of the manner in which the rocks probably occurred at the time they were bent, it would give them the great additional height of about 3,500 feet. Employing the same kind of evidence for the old red sandstone, carboniferous limestone, and coal measures of the Mendip Hills district, the Professor considers that an additional mass of accumulations, 4,000 feet in thickness, would be required above the Mendip Hills to supply the place of beds removed by denudation,

* "Memoirs of the Geological Survey of Great Britain," vol. i., p. 297.

Fig. 305.

a a, Old red Sandstone; *b b*, Upper Ludlow Rocks; *c c*, Aymestry Limestone; *d d*, Lower Ludlow Rocks: *f f*, Wenlock Shales; *g g*, Woolhope Limestone; *h*, Caradoc Sandstone.

Fig. 306.

and another of 5,000 feet, for the denudation of the same rocks on the north of Bristol.* The annexed section (fig. 306) is one of those given in illustration of this subject, and will serve to exhibit the disturbed beds of carboniferous limestone, old red sandstone, and Silurian rocks, shorn down by denudation in the slightly elevated district of Southern Pembrokeshire, as also, with every allowance for numerous fractures and gaps in the higher part carried off, the great mass of matter thus removed.

Viewing such considerable denudations, quite as readily seen in many other, and distant regions, as in the minor districts noticed, and combining them with the elevations and depressions of land that have taken place at all geological times, sometimes slowly moving large surfaces above and beneath the level of the sea, at others squeezing and folding various mineral accumulations into great ranges of mountains, the geologist will be at no loss for evidence, not only that the surface of the earth has long continued in an unquiet state, but also that the same amount of mineral matter may have been repeatedly employed in part, or as a whole, in the production of deposits spread over various areas for the time being, these deposits either fossiliferous or without organic exuviæ, as the conditions for the preservation of the remains of the animal and vegetable life of different times, may or may not have prevailed. As considerations of this kind constitute a part of those which lead to the most extended views, by the aid of which we endeavour to trace back the past conditions of our planet, they, and the class to which they belong, tending, as they do, to keep attention alive to the greater problems, while the detail necessary for their solution is collected, cannot be too frequently present to the mind of the geological observer.

---

* Referring to the reduction of the Horizontal Section of the Geological Survey, Sheet 17, extending from Glastonbury Tor, across the Mendip Hills, by Clifton, Bristol, to the flat land at the Severn, and to the sketch for filling up denudation, pl. 4, fig. 4, in the "Memoirs of the Geological Survey," vol. I.

APPENDIX.

Geological Maps and Sections—Though an observer may be supposed usually to have access to the best maps of any country he may examine geologically, and, in general, to find such maps containing the information which is desirable, as well respecting the natural physical features of the country, as the artificial modifications of, or arrangements on, its surface, so that he can always ascertain his exact position, and possess the power of recording any circumstances considered sufficiently important in their true relative places on such maps; it, nevertheless, sometimes happens that the maps of a district are inaccurate, or do not contain those things which are needed. A geologist will not long have endeavoured to record his observations upon maps, before he will ascertain that many a beautiful engraving may be worthless, while some coarse, unpromising plan may be most valuable. In case of need, therefore, it becomes important for an observer to be so far skilled in the construction of maps, as not only to be able to correct one which may be imperfect in an efficient manner, but also to make such a sketch of ground as may suffice for his purpose. A knowledge of the kind of surveying commonly termed *military drawing*, will be found most advantageous for his progress. He will scarcely accomplish much on this head by the aid of books alone, and therefore should study it in the field. If possessing a good eye for form, he will by no means find the acquisition of this knowledge difficult, while he will soon perceive that it affords him great additional power in satisfactorily recording his observations.

Even in many a map where the lines representing rivers, coasts, and other natural features, are exceedingly accurate, as also those showing the roads, canals, villages, and other artificial arrangements, are equally so, it too often happens that the relief of the ground, the true forms of the inequalities of surface of the hills and mountains, is either not given, or so inaccurately that it would have been better if no attempts had been made to represent it.* Now, the true relief of the surface of a district is often of the greatest

* The method, too often adopted, of representing the lines of water-shed as those of the highest grounds in many regions, cannot be too much deprecated, leading, as it often does, to the most imperfect views as to the real inequalities of surface in them, and as to the action of those geological causes which have produced such inequalities.

value to the observer. It is only necessary fo
his observations upon maps with and without
this relief, fully to appreciate the difference. A
to sketch in the forms required becomes of no
may be accomplished by the improved prismat
in districts where rocks containing much protox
as many of those of igneous origin in which be
which will divert the magnetic needle), and by
of a spirit level, by which close approximation
For approximations to heights within a certain
will be found useful, especially in regions wh
and sympesometer may not be easily carried, and
affecting such instruments, are not considerabl
instruments, and a fair knowledge of military
make many a sketch of a country, the geologic
otherwise have been imperfectly represented.

Supposing the possession of a proper map, eit
structed during the progress of his work, it is ne
follow up the rocks he may be investigating
opportunities permit, so that all connected wit
may be carefully ascertained, and all the points
fully entered upon his map. Without cauti
errors may readily be committed, inaccurate inf
ing the mode of occurrence of accumulation
frequently distant points, which more detailed
vented, and this more especially in districts of
rocks. As to the boundaries of the different
be thought desirable to insert on geological m
upon the scale of the map on the one hand, a
relative value on the other. Whatever the s
great distinctions considered important should b
the detail should be so represented as not to imp
It is better to select such portions of a map
required for the illustration of any particular de
comprehensive views for the sake of its int
needed in particular localities.

As with the objects to be represented on g
colouring employed upon them. Comprehensi

* With reference to the sketching of ground, the m
mountainous country by lines, approximating to thos
by fig. 180, p. 493, will be found very serviceable
especially those where relative altitudes are importan
of beds can only be measured by the heights they o
siderable to be ascertained by instruments constructe
(clinometers). The contour lines, or those of equal le
Maps in England and Ireland, and so valuable for n
representation of this method.

sacrificed to attempts to introduce detail only important locally, and which can be best shown by the enlargement of such portions of a map. The employment of given colours to represent certain divisions of the geological series has been considered very desirable, so that the eye becoming accustomed to them may, as it were, currently read off maps thus coloured. This is certainly important, and might be accomplished, to a considerable extent, in the general maps, national maps for example, of different countries.

Much may be, and has been effected as to the information to be afforded in geological maps, by a mixture of signs and colours; the latter representing some accumulation, or series of accumulations; and the former certain modifications of it considered important. In this manner, for example, igneous rocks may be represented by some given tint or colour; and the variations in their mineral structure, so far as regards the surface of the land, by various signs. The like with those divisions in the sedimentary deposits of different geological dates to which names have been given, various signs also readily show their mineral structure in different parts of their surface exposure. Among the signs employed amid the stratified rocks, it is very needful to have a sufficient number representing their modes of occurrence as to the position of their beds, showing when these are horizontal, inclined at any particular angle, or contorted; when the latter, the kind of contortion, and the like. The following signs (fig. 307) have been found useful for this purpose. The point of the arrow,

Fig. 307.

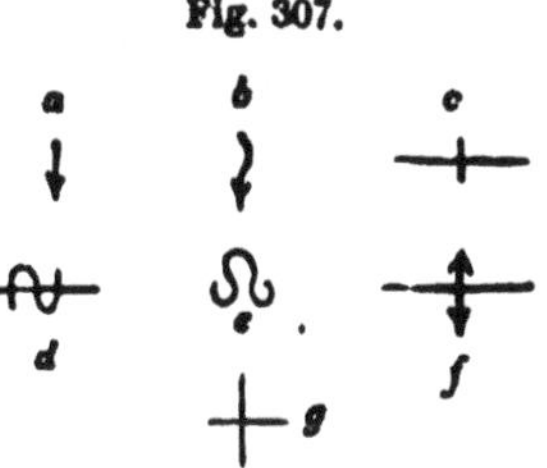

a, shows the dip or inclination of the beds as respects the horizon; and it is desirable to place on one side of this sign the amount of the dip, such as 5°, 15°, 23°, as it may happen to be. The sign *b* is intended to point out that, while the general inclination, or dip of the beds may be in the direction corresponding with that of the arrow, they undulate on the minor scale; *c*, shows that the strata are vertical, their range, or *strike*, as it is often termed, being in the direction of the longest line. Beds much plicated on the minor scale, while they have a general range, are shown by *d*, the straight line pointing out the general range. An anticlinal ridge is represented by the sign *f*, the two arrow-heads showing the direction of the dip on either side, and the cross line that of the range of this form of beds; *e* is intended to indicate the occurrence of beds so contorted and folded in various planes, that no definite dip or range of them can be inferred in the locality where this sign may be entered upon the map. The cross *g*,

represents a horizontal arrangement of the be
any other system of signs considered effective,
ponent beds of stratified rocks is so exhibited
with evidence enabling him to take a compreh
subject. By combining any system of this k
tribution of organic remains, in the fossilife
advances his general views; so that with
mineral structure, distribution of organic re
which the beds may have sustained since th
becomes a record of his observations, but also
collection of facts from which he may deduce
that otherwise might not so readily be attained

Geological maps conveying information only
of rocks, vertical sections of the country, e
natural or artificial exposures of the various ac
abundant and satisfactory information, collec
line of section or within safe distances from i
view of the manner in which the various rocks
much stress cannot be laid on the importance
strictly proportional, so that they should, as n
representations of nature. The distortions an
arise from a want of attention to this point ar
the outlines of countries are usually so exag
depressions, that the real forms of the surfa
appreciation of them, which would lead to j
portance, and of the conditions which have
by most imperfect ideas as to the real relief
may be mountainous. True surface sections o
needed to afford the geologist a correct view
depressions, so that he may insert his observ
deceive him in his endeavours to trace the
general disturbance, to ascertain the value of a
and to restore the various component parts t
tions in such regions.

Though with known altitudes at sufficien
given line of proposed section, the various dist
known, much may be accomplished by a
sketching in the intermediate ground; and th
command in somewhat rapid excursions:* th
work, when time suffices, is the only real me
sought. All lines of section are thus run
Great Britain, and the results thence obtain

* For example, the section from the Jura to, a
given (fig. 264, p. 646), was obtained, from known l
line, by barometric measurements of intermediate
ground on the spot.

that few, once experiencing the advantages so derived, would probably be disposed to abandon this method of observation. In certain districts, such as those where that important product coal is obtained, exact sections of surface are as indispensable as the exact relative positions of the beds themselves with reference to them, so that the true positions of the coal-beds may appear. With his level or his theodolite, an observer feels that confidence in his labours which he might not otherwise possess. Having the surface right, he can enter the dips, and other modes of occurrence of the rocks found, in their real relative situations on his section, and thus possess a collective miniature representation of the needful circumstances, such as no other less correct method will insure, be his powers of generalization what they may.

As in many lines of section, all the various accumulations cannot be so traversed, as to have all deposits cut at right angles by them, care is required to represent only the relative thickness of such deposits where the section passes; that is, the lines of separation of beds, or collections of them, as given in the section, should correspond exactly with those which would appear if the rocks supposed to be vertically cut through, were really so; and the beds on one side of the cut being removed, the face of the other was exposed, as if on a cliff. By turning to fig. 257 (p. 635), it will be found that a line of section parallel to the cliff represented would even give the beds there shown as horizontal, while they really dipped considerably at right angles to it, as seen in the sketch, fig. 258. It is easy in such cases to notice the true amount and direction of the beds on the section, and thus make the real value of the lines on such section clear. By giving more dip than such lines represent, a greater thickness is shown than really exists, and the total amount of mineral matter which the surface of the ground and the line of section should exhibit, is misrepresented.

In addition to these vertical and proportional sections, it sometimes becomes necessary to enlarge a part, so far as regards a column rising vertically to the plane of accumulations. In like manner, also, this should be proportional, and on a scale sufficient to render the object sought by the enlargement clear. The scale of such sections adopted by the Geological Survey is that of 40 feet to the inch; and it has been found one amply sufficient for very considerable detail, as may be seen by reference to the vertical sections of the coal measures, those which can be used for mining purposes (sections, Nos. 1 to 11 and 16 to 18).

Vertical sections, deposits represented as piled one above the other horizontally, whatever may be their real inclinations in different localities, may also be usefully employed for comparing distant accumulations with each other, especially as regards their thickness. As, for example, the following section (fig. 308), serves to show the different thicknesses and modifications of the cretaceous and oolitic groups as developed in southern and northern England. In these sections (the same letters being employed to represent equivalent deposits in both), the cretaceous series in Wilts and

Somerset is divided into chalk, a; upper green sand, [illegible]

Fig. 393.

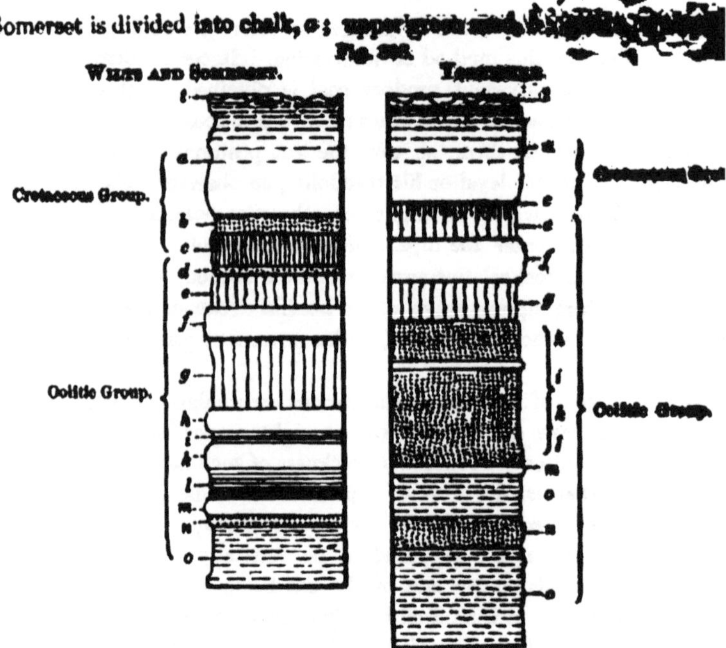

named, c; and d, lower green sand; while in the same series in Yorks b and d are supposed to be absent. As regards the oolitic grou represents the Kimmeridge clay; f, coral rag and its calcareous grits Oxford clay and the Kelloway rock in its lower part; h, Cornbrash Forest Marble; i, Bradford clay; k, great oolite; l, Fuller's earth inferior oolite; n, marlstone; o, lias. The superficial gravels, &c., al the chalk, are represented by t. As respects the divisions e, f, and g, two sections do not much vary, while a considerable difference is seen in beds h, i, k, and l, as found developed in southern and northern Engl

There has apparently been a modification of the conditions under w these equivalent portions of the oolitic series were deposited in the localities, so that while, in the south, marine remains point merel deposits beneath the waters of a sea, shales and sandstones on the n contain the remains of terrestrial fossil plants (p. 516) so occurring that only the close proximity of land has to be inferred, but also the existen marshy land itself, supporting a growth of certain plants (*Equiset* entombed as they stood. The different character of the lias on the n and on the south will also appear, this deposit being not only thicke the north, but also there exhibiting a certain depth of upper lias marls, continued to Wiltshire, though it can be seen gradually fining off south into Gloucestershire. In this manner, it will be obvious that much u evidence may be embodied; so that, by the combined aid of the maps, the sections of various kinds, a sound and comprehensive view of different rock accumulations of a country may be obtained.

Page 103. *Great Salt Lake of North America.*—The recent researches of Captain Stansbury, of the United States' Topographical Engineers (Expedition to the Great Salt Lake of Utah, 1852), have made us acquainted, from actual survey, with many important circumstances connected with the region in which the Great Salt Lake of North America is found. A considerable area, on the western watershed of that continent, is separated from the general drainage. In it a chief depression finally receives the waters of the country, forming a lake from which the evaporation is such that no surplus waters escape out of that area, as a whole, into the general watershed to the westward. A minor depression first receives the drainage of part of the general area. The waters of this lake, named Utah, are fresh, while those of the other, the Great Salt Lake, are highly impregnated with saline matter. The evaporation over the area draining into the Lake Utah (30 miles long and 10 miles in breadth,) is such, compared with the supply of water, that a river, named the Jordan, by the Mormon settlers on its banks, flows from it and falls, after a course of about 50 miles, into the Great Salt Lake. Notwithstanding this supply of fresh water, as also from other streams, the chief of which enters it on the N.E., rounding the Wahsatch Mountains, on the east, the waters of the Great Salt Lake are so entirely saline as to form a strong brine, one that in 100 parts, by weight, of the water, contains more than 22 parts of soluble salts, common salt (chloride of sodium) by itself constituting 20 parts. The exact amount of soluble substances, as determined by Dr. L. D. Gale, is as follows:—

Solid contents in 100 parts of the water = 22·422 parts.

Specific gravity of the water =	1·17
Chloride of sodium . .	20·196
Sulphate of soda . .	1·834
Chloride of magnesium .	0·252
Chloride of calcium . .	a trace.

The Great Salt Lake is situated, according to Captain Stansbury's maps, between about 40° 40′ and 41° 42′ north latitude, and between about 112° to 113° 10′ west longitude. The western side of the lake is bounded by a level plain country, little elevated above its waters, so little indeed that the pressure of the wind is sufficient considerably to change the coastline in that direction, according to its duration, the difference "amounting, in many cases, to miles in width" (p. 186). In dry weather the plain, abounding in soft ground, is covered with a coating of salt. Captain Stansbury mentions that the minute crystals of salt "glisten brilliantly in the sunlight, and present the appearance of a large piece of water so perfectly, that it is difficult, at times, for one to persuade himself that he is not standing on the shore of the lake itself" (p. 119). The Salt Lake, exclusive of offsets, is 291 miles in circumference, the Wahsatch Mountains rising above it on the eastward; several islands rise above its waters, which

would seem generally shallow, 36 feet between
Islands, being apparently the most considerable c
Salt Lake is estimated at about 4,300 feet above

Upon the slope of one of the ridges connect
Salt Lake, Captain Stansbury states that "th
watermarks were counted, which had evidently
by the lake, and must have been the result o
some time at each level." The highest of the
feet above the present level of the lake. He in
period a vast inland sea extended over this region

We would here seem to have an isolated, and
exposed, like the region of the Caspian, to cond
to adjust itself to the supply of the fresh water i
or rivers. This supply not being equal to th
contents gradually formed an increasing relati
in the chief depression, until they became that
Great Salt Lake. Whether this adjustment
probably uncertain; but even assuming that it
things remained the same, there is a disturbing
the amount of detritus now thrown by the stre
All would appear to add something, but the
bringing down an immense quantity of sedi
freshet. The exact difference of elevation betw
Salt Lakes does not appear to have been ascert
this level is not considerable, and that formerl
whole, there is no difficulty in seeing that La
water than covered its evaporation (the lake is s
ground), would gradually become fresh from the
saline solutions, while the lower lake became as
the reverse action. Thus a fresh-water and a ve
resulted from the imprisonment of parts of th
upon, under modified conditions, in a given s
now obtaining a supply of water beyond its gene

Page 368. *Volcanic Rocks of Iceland.*—Profes
after personally studying the volcanic rocks of Ice
gating their chemical relation in the laboratory, ha
of great value and importance, not only respecting
also as regards the volcanic products ejected in
Processes which have taken place during the Form
of Iceland; Poggendorff's 'Annalen,' 1851, No. 6
new series, vol. i, p. 33, 1852.) After alluc
character of the mixed silicates ejected in a sta
the separation of the component matter into dif
according to the physical conditions to which

divides the Iceland volcanic rocks into two groups, which he names *normal-trachytic* and *normal-pyroxenic*. He understands by the former the trachytic rocks richest in silica, and by the latter the basaltic and doloritic rocks poorest in silica. These constitute extreme members of a general mass, and graduate into each other. "The first distinguishing characteristic of the normal-trachytic rocks is, that they represent almost exactly a mixture of bisilicates of alumina and alkali, in which lime, magnesia, and protoxide of iron are either wholly wanting or are present only in insignificant quantities." * * * * * "The normal-pyroxenic rocks, which as basic silicates of alumina and protoxide of iron, in combination with lime, magnesia, potash and soda, form the most extreme members towards the opposite end of the series, present a similar correspondence in their average composition." * * * * "As the proportion of the oxygen of the silicic acid to that of the bases is here, with slight variations, as 3 : 1·998, all these rocks may be regarded as a constant mixture of bibasic silicates, when we consider only their entire mass, without reference to the fact of their constituents being grouped into minerals of definite composition." "The quantity of silica almost always bears a constant proportion to the lime and magnesia, while the relation between the quantities of alumina and protoxide of iron is subject to considerable variations." Thus there is a mass of matter having a tendency to arrange itself into two distinct sub-masses, each distinguished by a definite composition, the whole governed in its lithological character by the variable manner in which a mixture was effected, and by the physical conditions to which the molten mass may have been exposed.

The Professor gives the following as the composition of the normal-trachytic (1) and the normal-pyroxenic rocks (2,) observing that from it the mean relation between the oxygen of the acid and that of the bases can be calculated, being for the trachytic rock as 3 : 0·596, and for the pyroxenic as 3 : 1·998 :—

	(1)	(2)
Silica	76·67 .	48·47
Alumina and protoxide of iron .	14·23 .	30·16
Lime	1·44 .	11·87
Magnesia	0·28 .	6·89
Potash	3·20 .	0·65
Soda	4·18 .	1·96
	100·00	100·00

Selecting silica, as the most convenient constituent of these rocks for calculation, because it can be well determined, and is the least variable in them, in order to estimate the mixture of these normal rocks which may exist in any compound of them, Professor Bunsen remarks that, "If we characterise by S the per centage of silica in a mixed rock, by *s* the per-

centage of silica in the normal-trachytic, and by
in the normal-pyroxenic rocks, then

$$\frac{s - S}{S - \sigma} = a$$

in which equation a represents the quantity
mass which must be mixed with one part of
order to give the composition of the mixe
"All the other constituents of the mixed roc
means of a. For if the weight of the separat
the normal-pyroxenic rock-mass is taken as p_0
manner the weight of the separate constituen
the normal-trachytic rock-mass as $t_0\ t_1 \ldots t_n$,
constituents may be ascertained by means of th

$$1 = \frac{(a p_0 + t_0)}{(a + 1)} + \frac{(a p_1 + t_1)}{(a + 1)} + \ldots$$

The abstract given by the Professor abounc
should be consulted by the observer anxiou
interesting class of inquiries. The memoir
relations existing between non-metamorphic r
above given has been taken); and 2, Genetic
phic rocks, the latter being subdivided into
Zeolitic formations; and 3, Formation of roc
morphism. In the first subdivision of the sec
tuff is pointed out as a mixture of hydrated
latter belonging exclusively to the pyroxenic
cementing the fragmentary rock, being regard
nation of two silicates, one represented by th
Aq, and the other by $3Al^2O^3$, SiO^3 +
Professor considers as appearing to combine
which he proposes the formula $3RO^2$ $2Si$(
He adds, that the palagonitic substance occurs
the pyroxenic volcanic rocks, mentioning be
accumulations of Germany and France, the I
Azores, Canaries, Cape de Verd Islands, a
Under the head of *Zeolitic Formations*, their i
palagonitic and pyroxenic classes is pointed
Formation of Rocks by Pneumatolytic Metamorp
from the action of volcanic gases and vapours
been treated of.

INDEX.

LONDON: PRINTED BY W. CLOWES AND SONS, STAMFORD

OF

NEW WORKS IN GENERAL LITERATUR

PUBLISHED BY

LONGMAN, BROWN, GREEN, AND LONGMANS,

39, PATERNOSTER ROW, LONDON.

CLASSIFIED INDEX.

Pages.

Natural History in general.

1-Volume Encyclopædias and Dictionaries.

Religious & Moral Works.

Poetry and the Drama.

Political Economy and Statistics.

The Sciences in General and Mathematics.

Rural Sports ...

Veterinary ...

Voyages and ...

Works of Fi...

ALPHABETICAL CATALOGUE

OF

NEW WORKS AND NEW EDITION

PUBLISHED BY

Messrs. LONGMAN, BROWN, GREEN, and LONGMANS,

PATERNOSTER ROW, LONDON.

Miss Acton's Modern Cookery-Book.—Modern Cookery in all its Branches, reduced to a System of Easy Practice. For the use of Private Families. In a Series of Receipts, all of which have been strictly tested, and are given with the most minute exactness. By ELIZA ACTON. New Edition; with various Additions, Plates and Woodcuts. Fcp. 8vo. price 7s. 6d.

Aikin.—Select Works of the British Poets, from Ben Jonson to Beattie. With Biographical and Critical Prefaces by Dr. AIKIN. New Edition, with Supplement by LUCY AIKIN; consisting of additional Selections from more recent Poets. 8vo. price 18s.

Arnold.—Poems. By Matthew Arnold. Second Edition. Fcp. 8vo. price 5s. 6d.

Arnold.—Oakfield; or, Fellowship in the East. By W. D. ARNOLD, Lieutenant 58th Regiment, Bengal Native Infantry. The Second Edition, revised. 2 vols. post 8vo. price 21s.

Atkinson (G.)—Sheriff-Law; or, a Practical Treatise on the Office of Sheriff, Undersheriff, Bailiffs, &c.: Their Duties at the Election of Members of Parliament and Coroners, Assizes, and Sessions of the Peace; Writs of Trial; Writs of Inquiry; Compensation Notices; Interpleader; Writs; Warrants; Returns; Bills of Sale; Bonds of Indemnity, &c. By GEORGE ATKINSON, Serjeant-at-Law. Third Edition, revised. 8vo. price 10s. 6d.

Atkinson (G.)—The Shipping Laws of the British Empire: Consisting of *Park on Marine Assurance*, and *Abbott on Shipping*. Edited by GEORGE ATKINSON, Serjeant-at-Law. 8vo. price 10s. 6d.

Atkinson (W.)—The Church: An Ex[amin]ation of the Meaning contained in [the] Bible; shewing the Ancient, Conti[nued] and Prevailing Error of Man, the Sub[stitu]tion of Worship for Religion; and she[wing] that the Principles of all Right Indiv[idual] Action and of General Government, o[r the] Government of all Nations, are compris[ed in] Revealed Religion. By WILLIAM ATKI[NSON]. 2 vols. 8vo. price 30s.

Austin.—Germany from 1760 to 1[814]; Or, Sketches of German Life from the [Decay] of the Empire to the Expulsion of [the] French. By Mrs. AUSTIN. Post [8vo.] price 12s.

Joanna Baillie's Dramatic and Poet[ical] Works, complete in One Volume: [com]prising the Plays of the Passions, Misc[ella]neous Dramas, Metrical Legends, Fug[itive] Pieces (several now first published), [and] Ahalya Baee. Second Edition, inclu[ding] a new Life of Joanna Baillie; with [a Por]trait, and a View of Bothwell Manse. Sq[uare] crown 8vo. price 21s. cloth; or 42s. b[ound] in morocco.

Baker.—The Rifle and the Hound [in] Ceylon. By S. W. BAKER, Esq. [With] several Illustrations printed in Colours, [and] Engravings on Wood. 8vo. price 14s.

Balfour.—Sketches of English Literat[ure] from the Fourteenth to the Present Cent[ury]. By CLARA LUCAS BALFOUR. Fcp. 8vo

Barter.—Homer's Iliad, transla[ted] almost literally into the Spenserian Sta[nza], with Notes. By W. G. T. BARTER. [8vo.] price 18s.

Banfield.—The Statistical Companion for 1854: Exhibiting the most Interesting Facts in Moral and Intellectual, Vital, Economical, and Political Statistics, at Home and Abroad. Corrected to the Present Time; and including the Census of the British Population taken in 1851. Compiled from Official and other Authentic Sources, by T. C. BANFIELD, Esq. Fcp. 8vo. price 6s.

Baylden's Art of Valuing Rents and Tillages, and Tenant's Right of Entering and Quitting Farms, explained by several Specimens of Valuations; with Remarks on the Cultivation pursued on Soils in different Situations. Adapted to the Use of Landlords, Land-Agents, Appraisers, Farmers, and Tenants. New Edition; corrected and revised by JOHN DONALDSON. 8vo. 10s. 6d.

Berkeley.—Reminiscences of a Huntsman. By the Honourable GRANTLEY F. BERKELEY. With Four Etchings by John Leech (one coloured). 8vo. price 14s.

Black's Practical Treatise on Brewing, Based on Chemical and Economical Principles: With Formulæ for Public Brewers, and Instructions for Private Families. New Edition, with Additions. 8vo. price 10s. 6d.

Blaine's Encyclopædia of Rural Sports; Or, a complete Account, Historical, Practical, and Descriptive, of Hunting, Shooting, Fishing, Racing, and other Field Sports and Athletic Amusements of the present day. A new and thoroughly revised Edition: The Hunting, Racing, and all relative to Horses and Horsemanship, revised by HARRY HIEOVER; Shooting and Fishing by EPHEMERA; and Coursing by Mr. A. GRAHAM. With upwards of 600 Wood Engravings. 8vo. price 50s. half-bound.

Blair's Chronological and Historical Tables, from the Creation to the present time: With Additions and Corrections from the most authentic Writers; including the Computation of St. Paul, as connecting the Period from the Exode to the Temple. Under the revision of Sir HENRY ELLIS, K.H. New Edition, with Corrections. Imperial 8vo. price 31s. 6d. half-morocco.

Bloomfield.—The Greek Testament: With copious English Notes, Critical, Philological, and Explanatory. Especially formed for the use of advanced Students and Candidates for Holy Orders. By the Rev. S. T. BLOOMFIELD, D.D., F.S.A. New Edition. 2 vols. 8vo. with Map, price £2.

Dr. Bloomfield's Additional Annotations on the above. 8vo. price 15s.

The Maternal Management of
ren in Health and Disease. By
ULL, M.D., Member of the Royal
ge of Physicians; formerly Physician-
cheur to the Finsbury Midwifery
ution. New Edition. Fcp. 8vo.
5s.

Hints to Mothers, for the Ma-
ent of their Health during the Period
gnancy and in the Lying-in Room:
an Exposure of Popular Errors in
xion with those subjects, &c.; and
upon Nursing. By T. BULL, M.D.
Edition. Fcp. price 5s.

.—Christianity and Mankind,
Beginnings and Prospects. By
STIAN CHARLES JOSIAS BUNSEN, D.D.,
., D.Ph. Being a New Edition, cor-
, remodelled, and extended, of *Hip-
s and his Age*. 7 vols. 8vo. price
.

This Second Edition of the *Hippolytus*
posed of three distinct works, which
be had separately, as follows:—

Historical Section.

Hippolytus and his Age; or, the Beginnings and Prospects of Christianity. 2 vols. 8vo. price £1. 10s.

. Hippolytus and the Teachers of the
Apostolical Age;

. The Life of the Christians of the
Apostolical Age.

Philological Section.

Outline of the Philosophy of Universal History applied to Language
and Religion. 2 vols. 8vo. price
£1. 13s.

Philosophical Section.

Analecta Ante-Nicaena. 3 vols. 8vo.
price £2. 2s.

. Reliquiae Literariae;

. Reliquiae Canonicae;

. Reliquiae Liturgicae: Cum Appen-
dicibus ad Tria Analectorum
Volumina.

.—Egypt's Place in Universal
: An Historical Investigation, in
ooks. By C. C. J. BUNSEN, D.D.
, D.Ph. Translated from the Ger-
by C. H. COTTRELL, Esq. M.A.
. with many Illustrations. 8vo.
8s.

The Second Volume is preparing for
ation.

—The History of Scotland, from
volution to the Extinction of the last
te Insurrection (1689—1748). By
HILL BURTON, Author of *The Life of
Hume*, &c. 2 vols. 8vo. price 26s.

Bishop Butler's General Atlas of Modern
and Ancient Geography; comprising Fifty-
two full-coloured Maps; with complete Indices. New Edition, nearly all re-engraved,
enlarged, and greatly improved; with Corrections from the most authentic sources in
both the Ancient and Modern Maps, many
of which are entirely new. Edited by the
Author's Son. Royal 4to. 24s. half-bound.

Separately { The Modern Atlas of 28 full-coloured Maps. Rl. 8vo. 12s.
The Ancient Atlas of 24 full-coloured Maps. Rl. 8vo. 12s. }

Bishop Butler's Sketch of Modern and
Ancient Geography. New Edition, carefully revised, with such Alterations introduced as continually progressive Discoveries
and the latest Information have rendered
necessary. 8vo. price 9s.

The Cabinet Gazetteer: A Popular Exposition of all the Countries of the World;
their Government, Population, Revenues,
Commerce, and Industries; Agricultural,
Manufactured, and Mineral Products; Religion, Laws, Manners, and Social State:
With brief Notices of their History and Antiquities. From the latest Authorities. By
the Author of *The Cabinet Lawyer*. Fcp. 8vo.
price 10s. 6d. cloth; or 13s. calf lettered.

The Cabinet Lawyer: A Popular Digest
of the Laws of England, Civil and Criminal;
with a Dictionary of Law Terms, Maxims,
Statutes, and Judicial Antiquities; Correct
Tables of Assessed Taxes, Stamp Duties,
Excise Licenses, and Post-Horse Duties;
Post-Office Regulations, and Prison Disci-
pline. 16th Edition, comprising the Public
Acts of the Session 1853. Fcp. 8vo. 10s. 6d.

Caird.—English Agriculture in 1850 and
1851; Its Condition and Prospects. By
JAMES CAIRD, Esq., of Baldoon, Agricultural
Commissioner of *The Times*. The Second
Edition. 8vo. price 14s.

Calvert.—The Wife's Manual; or,
Prayers, Thoughts, and Songs on Several
Occasions of a Matron's Life. By the Rev.
WILLIAM CALVERT, Rector of St. Antholin,
and one of the Minor Canons of St. Paul's.
Printed by C. Whittingham; and orna-
mented from Designs by the Author in the
style of *Queen Elizabeth's Prayer Book*.
Crown 8vo. price 10s. 6d.

Carlisle (Lord).—A Diary in Turkish and
Greek Waters. By the Right Hon. the
Earl of CARLISLE. Post 8vo.
[Nearly ready.

Catlow.—Popular Conchology; or, the Shell Cabinet arranged according to the Modern System: With a detailed Account of the Animals; and a complete Descriptive List of the Families and Genera of Recent and Fossil Shells. By AGNES CATLOW. Second Edition, much improved; with 405 Woodcut Illustrations. Post 8vo. price 14s.

Cecil.—The Stud Farm; or, Hints on Breeding Horses for the Turf, the Chase, and the Road. Addressed to Breeders of Race Horses and Hunters, Landed Proprietors, and especially to Tenant Farmers. By CECIL. Fcp. 8vo. with Frontispiece, 5s.

Cecil's Records of the Chase, and Memoirs of Celebrated Sportsmen; Illustrating some of the Usages of Olden Times and comparing them with prevailing Customs: Together with an Introduction to most of the Fashionable Hunting Countries; and Comments. With Two Plates by B. Herring. Fcp. 8vo. price 7s. 6d. half-bound.

Cecil's Stable Practice; or, Hints on Training for the Turf, the Chase, and the Road; with Observations on Racing and Hunting, Wasting, Race Riding, and Handicapping: Addressed to owners of Racers, Hunters, and other Horses, and to all who are concerned in Racing, Steeple Chasing, and Fox Hunting. Fcp. 8vo. with Plate, price 5s. half-bound.

Chalybaeus's Historical Survey of Modern Speculative Philosophy, from Kant to Hegel: Designed as an Introduction to the Opinions of the Recent Schools. Translated from the German by ALFRED TULK. Post 8vo. price 8s. 6d.

Captain Chesterton's Autobiography.— Peace, War, and Adventure: Being an Autobiographical Memoir of George Laval Chesterton, formerly of the Field-Train Department of the Royal Artillery, subsequently a Captain in the Army of Columbia, and at present Governor of the House of Correction at Cold Bath Fields. 2 vols. post 8vo. price 16s.

Chevreul on Colour.—The Principles of Harmony and Contrast of Colours, and their Applications to the Arts: Including Painting, Interior Decoration, Tapestries, Carpets, Mosaics, Coloured Glazing, Paper-Staining, Calico Printing, Letterpress Printing, Map Colouring, Dress, Landscape and Flower Gardening, &c. By M. E. CHEVREUL, Membre de l'Institut de France, etc. Translated from the French by CHARLES MARTEL; and illustrated with Diagrams, &c. Crown 8vo. price 12s. 6d.

Clinton.—The Autobio Literary Journal of the la Clinton, Esq., M.A., Aut *Hellenici*, the *Fasti Romani*, the Rev. C. J. Fynes-C Rector of Cromwell, Notts.

Conversations on Botany. improved; with 22 Plates. 7s. 6d.; or with the Plates c

Conybeare and Howson.— Epistles of Saint Paul: complete Biography of the a Translation of his Epist Chronological Order. By t CONYBEARE, M.A., late Fel College, Cambridge; and t HOWSON, M.A., Principal of Institution, Liverpool. With on Steel and 100 Woodcut price £2. 8s.

Copland.—A Dictionary Medicine: Comprising Gen the Nature and Treatment Morbid Structures, and the pecially incidental to Climat to the different Epochs of Lif rous approved Formulae of recommended. By JAMES C Consulting Physician to Qu Lying-in Hospital, &c. Vol price £3; and Parts X. to X

The Children's Own Sund JULIA CORNER, Author of *the History of Europe*. Wit tions. Square fcp. 8vo. pri

Cresy.—An Encyclopædia neering, Historical, Theoretic By EDWARD CRESY, F.S. trated by upwards of 3, explanatory of the Princip and Constructions which c direction of the Civil E price £3. 13s. 6d.

The Cricket-Field; or, the History of the Game of Cr Author of *Principles of Sc* Second Edition, greatly im Plates and Woodcuts. Fcp. half-bound.

Lady Cust's Invalid's Bo valid's Own Book: A Colle from various Books and var By the Honourable LADY C price 3s. 6d.

The Domestic Liturgy and Family
ain, in Two Parts: The First Part Church Services adapted for Domestic ith Prayers for every day of the week, dexclusively from the Book of Common ; Part II. comprising an appropriate s for every Sunday in the year. By w. THOMAS DALE, M.A., Canon Resi-y of St. Paul's. Second Edition. o. price 21s. cloth; 31s. 6d. calf; 10s. morocco.

ly { THE FAMILY CHAPLAIN, 12s.
THE DOMESTIC LITURGY, 10s.6d.

he.—The Geological Observer.
HENRY T. DELABECHE, F.R.S., or-General of the Geological Survey of ited Kingdom. New Edition; with ous Woodcuts. 8vo. price 18s.

he.—Report on the Geology of
all, Devon, and West Somerset. By NRY T. DELABECHE, F.R.S., Director-l of the Geological Survey. With Woodcuts, and 12 Plates. 8vo. 4s.

ve.—A Treatise on Electricity,
ory and Practice. By A. DE LA RIVE, or in the Academy of Geneva. In Volumes, with numerous Wood En-gs. Vol. I. 8vo. price 18s.

ne. By the Author of "Letters
Unknown Friends," &c. Second n, enlarged. 18mo. price 2s. 6d.

e.—Materials for a History of Oil
g. By Sir CHARLES LOCK EASTLAKE, , F.S.A., President of the Royal ny. 8vo. price 16s.

lipse of Faith; or, a Visit to a
us Sceptic. Fifth *and cheaper* Edition. vo. price 5s.

ce of The Eclipse of Faith, by
hor: Being a Rejoinder to Professor n's *Reply:* Including a full Exami-of that Writer's Criticism on the ter of Christ; and a Chapter on the and Pretensions of Modern Deism. *Edition*, revised. Post 8vo. 5s. 6d.

glishman's Greek Concordance of
w Testament: Being an Attempt at a Connexion between the Greek and glish Texts; including a Concordance Proper Names, with Indexes, Greek-and English-Greek. New Edition, new Index. Royal 8vo. price 42s.

The Englishman's Hebrew and Chaldee
Concordance of the Old Testament: Being an Attempt at a Verbal Connection between the Original and the English Translations; with Indexes, a List of the Proper Names and their occurrences, &c. 2 vols. royal 8vo. £3. 13s. 6d.; large paper, £4. 14s. 6d.

Ephemera.—A Handbook of Angling;
Teaching Fly-fishing, Trolling, Bottom-fishing, Salmon fishing; with the Natural History of River Fish, and the best modes of Catching them. By EPHEMERA. Third *and cheaper* Edition, corrected and improved; with Woodcuts. Fcp. 8vo. 5s.

Ephemera.—The Book of the Salmon:
Comprising the Theory, Principles, and Practice of Fly-fishing for Salmon; Lists of good Salmon Flies for every good River in the Empire; the Natural History of the Salmon, all its known Habits described, and the best way of artificially Breeding it explained. With numerous coloured Engravings. By EPHEMERA; assisted by ANDREW YOUNG. Fcp. 8vo. with coloured Plates, price 14s.

W. Erskine, Esq.—History of India
under Báber and Humáyun, the First Two Sovereigns of the House of Taimur. By WILLIAM ERSKINE, Esq., Editor of *Memoirs of the Emperor Báber*. 8vo. price 32s.

Faraday (Professor).—The Subject-
Matter of Six Lectures on the Non-Metallic Elements, delivered before the Members of the Royal Institution in 1852, by Professor FARADAY, D.C.L., F.R.S., &c. Arranged by permission from the Lecturer's Notes by J. SCOFFERN, M.B. Fcp. 8vo. price 5s. 6d.

Norway in 1848 and 1849: Contain-
ing Rambles among the Fjelds and Fjords of the Central and Western Districts; and including Remarks on its Political, Military, Ecclesiastical, and Social Organisation. By THOMAS FORESTER, Esq.; and Lieutenant M. S. BIDDULPH, Royal Artillery. With Map, Woodcuts, and Plates. 8vo. 18s.

Francis.—Annals, Anecdotes, and
Legends: A Chronicle of Life Assurance. By JOHN FRANCIS, Author of *The History of the Bank of England*, &c. Post 8vo. 8s. 6d.

Fullom.—The Marvels of Science and
their Testimony to Holy Writ: A Popular System of the Sciences. By S. W. FULLOM, Esq. The *Eighth* and cheaper *Edition;* with numerous Illustrations. Post 8vo. price 5s.

ho Poetical Works of Oliver Goldsmith. Edited by Bolton Corney, Esq. Illustrated by Wood Engravings, from Designs by Members of the Etching Club. Square crown 8vo. cloth, 21s.; morocco, £1. 16s.

osse.—A Naturalist's Sojourn in Jamaica. By P. H. Gosse, Esq. With Plates. Post 8vo. price 14s.

r. W. R. Greg's Contributions to The Edinburgh Review.—Essays on Political and Social Science. Contributed chiefly to the *Edinburgh Review*. By William R. Greg. 2 vols. 8vo. price 24s.

urney.—Historical Sketches; illustrating some Memorable Events and Epochs, from A.D. 1,400 to A.D. 1,546. By the Rev. John Hampden Gurney, M.A., Rector of St. Mary's, Marylebone. Fcp. 8vo. 7s. 6d.

wilt.—An Encyclopædia of Architecture, Historical, Theoretical, and Practical. By Joseph Gwilt. Illustrated with more than One Thousand Engravings on Wood, from Designs by J. S. Gwilt. Third Edition (1854). 8vo. price 42s.

idney Hall's General Large Library Atlas of Fifty-three Maps (size, 20 in. by 16 in.), with the Divisions and Boundaries carefully coloured; and an Alphabetical Index of all the Names contained in the Maps. New Edition, corrected from the best and most recent Authorities; with the Railways laid down and many entirely new Maps. Colombier 4to. price £5. 5s. half-russia.

amilton.—Discussions in Philosophy and Literature, Education and University Reform. Chiefly from the *Edinburgh Review*; corrected, vindicated, enlarged, in Notes and Appendices. By Sir William Hamilton, Bart. Second Edition, with Additions. 8vo. price 21s.

are (Archdeacon).—The Life of Luther, in Forty-eight Historical Engravings. By Gustav König. With Explanations by Archdeacon Hare. Square crown 8vo. [*In the press.*

arrison.—The Light of the Forge; or, Counsels drawn from the Sick-Bed of E. M. By the Rev. William Harrison, M.A., Rector of Birch, Essex, and Domestic Chaplain to H.R.H. the Duchess of Cambridge. With 2 Woodcuts. Fcp. 8vo. price 5s.

Harry Hieover.—Th By Harry Hieover. One representing *The R* *The Wrong Sort.* Fcp. 8

Harry Hieover.— ship. By Harry Hie —One representing other, *Going like* half-bound.

Harry Hieover.—The St Purposes and Practical to the Choice of a Horse for show. By Harry Plates—One representing *for most purposes*; the other *sort for any purpose.* Fcp bound.

Harry Hieover.—The Stud; or, Practical Hint ment of the Stable. By Second Edition; with Port on his favourite Horse *Ha* price 5s. half-bound.

Harry Hieover.—Stable Talk; or, Spectacles for Y By Harry Hieover. Ne 8vo. with Portrait, price 2

Haydon.—The Life of B Haydon, Historical Paint biography and Journals. piled by Tom Taylor, M Temple, Esq.; late Fellow Cambridge; and late Profe Language and Literature lege, London. Second E tions and an Index. 3 vo 31s. 6d.

Haydn's Book of Digniti Rolls of the Official Person Empire, Civil, Ecclesiastic tary, Naval, and Municipal Periods to the Present chiefly from the Record Offices. Together with Europe, from the founda spective States; the Peera Great Britain, and nume Being a New Edition, im nued, of Beatson's Polit Joseph Haydn, Compiler *of Dates*, and other Work half-bound.

n Herschel.—Outlines of Astro-
. By Sir John F. W. Herschel,
&c. New Edition; with Plates and
Engravings. 8vo. price 18s.

Travels in Siberia. By S. S. Hill,
Author of *Travels on the Shores of*
altic. With a large coloured Map of
pean and Asiatic Russia. 2 vols. post
rice 24s.

on Etiquette and the Usages of
y: With a Glance at Bad Habits.
Edition, revised (with Additions) by a
of Rank. Fcp. 8vo. price Half-a-Crown.

-Prize Essay on the History and
gement of Literary, Scientific, and
anics' Institutions, and especially how
ey may be developed and combined so
promote the Moral Well-being and
try of the Country. By James Hole,
Secretary of the Yorkshire Union of
anics' Institutes. 8vo. price 5s.

Holland's Memoirs.—Memoirs of
Whig Party during my Time. By
y Richard Lord Holland. Edited
Son, Henry Edward Lord Holland.
I. and II. post 8vo. price 9s. 6d. each.

Holland's Foreign Reminiscences.
d by his Son, Henry Edward Lord
and. Second Edition; with Fac-
. Post 8vo. price 10s. 6d.

d.—Chapters on Mental Physio-
By Sir Henry Holland, Bart.,
., Physician-Extraordinary to the
; and Physician in Ordinary to His
Highness Prince Albert. Founded
on Chapters contained in *Medical*
and Reflections by the same Author.
rice 10s. 6d.

-The Last Days of Our Lord's
try: A Course of Lectures on the
pal Events of Passion Week. By
er Farquhar Hook, D.D., Chaplain
dinary to the Queen. New Edition.
vo. price 6s.

and Arnott.—The British Flora;
ising the Phaenogamous or Flowering
, and the Ferns. The Sixth Edition,
Additions and Corrections; and nu-
Figures illustrative of the Umbelli-
Plants, the Composite Plants, the
s, and the Ferns. By Sir W. J.
er, F.R.A. and L.S., &c., and G. A.
er-Arnott, LL.D., F.L.S. 12mo.
2 Plates, price 14s.; with the Plates
d, price 21s.

Hooker.—Kew Gardens; or, a Popular
Guide to the Royal Botanic Gardens o
Kew. By Sir William Jackson Hooker,
K.H., D.C.L., F.R.A., and L.S., &c. &c.
Director. New Edition; with numerous
Wood Engravings. 16mo. price Sixpence.

Horne.—An Introduction to the Critical
Study and Knowledge of the Holy Scriptures.
By Thomas Hartwell Horne,
B.D. of St. John's College, Cambridge; Prebendary
of St. Paul's. New Edition, revised
and corrected; with numerous Maps and
Facsimiles of Biblical Manuscripts. 5 vols
8vo. price 63s.

Horne.—A Compendious Introduction to
the Study of the Bible. By Thomas
Hartwell Horne, B.D., of St. John's
College, Cambridge. Being an Analysis o
his *Introduction to the Critical Study and*
Knowledge of the Holy Scriptures. New
Edition, corrected and enlarged; with Maps
and other Engravings. 12mo. price 9s.

Howitt (A. M.)—An Art-Student in
Munich. By Anna Mary Howitt. 2
vols. post 8vo. price 14s.

Howitt.—The Children's Year. By Mary
Howitt. With Four Illustrations, engraved
by John Absolon, from Original Designs by
Anna Mary Howitt. Square 16mo. price 5s.

William Howitt's Boy's Country Book;
Being the Real Life of a Country Boy,
written by himself; exhibiting all the Amusements,
Pleasures, and Pursuits of Children
in the Country. New Edition; with 40
Woodcuts. Fcp. 8vo. price 6s.

Howitt.—The Rural Life of England.
By William Howitt. New Edition, corrected
and revised; with Woodcuts by
Bewick and Williams: Uniform with *Visits*
to Remarkable Places. Medium 8vo. 21s.

Howitt.—Visits to Remarkable Places;
Old Halls, Battle-Fields, and Scenes illustrative
of Striking Passages in English History
and Poetry. By William Howitt. New
Edition, with 40 Woodcuts. Medium 8vo.
price 21s.

Second Series, chiefly in the
Counties of Northumberland and Durham,
with a Stroll along the Border. With upwards
of 40 Woodcuts. Medium 8vo. 21s.

Hudson.—Plain Directions for Making Wills in Conformity with the Law: with a clear Exposition of the Law relating to the distribution of Personal Estate in the case of Intestacy, two Forms of Wills, and much useful information. By J. C. HUDSON, Esq., late of the Legacy Duty Office, London. New and enlarged Edition; including the provisions of the Wills Act Amendment Act of 1852 (introduced by Lord St. Leonard's). Fcp. 8vo. price 2s. 6d.

Hudson. — The Executor's Guide. By J. C. HUDSON, Esq. New and enlarged Edition; with the Addition of Directions for paying Succession Duties on Real Property under Wills and Intestacies, and a Table for finding the Values of Annuities and the Amount of Legacy and Succession Duty thereon. Fcp. 8vo. price 6s.

Hulbert.—The Gospel Revealed to Job; or, Patriarchal Faith illustrated in Thirty Lectures on the principal Passages of the Book of Job: With Explanatory, Illustrative, and Critical Notes. By the Rev. C. A. HULBERT, M.A. 8vo. price 12s.

Humbley.—Journal of a Cavalry Officer: Including the memorable Sikh Campaign of 1845–6. By W. W. W. HUMBLEY, M.A. Trinity College, Cambridge; Fellow of the Cambridge Philosophical Society; Captain, 9th Queen's Royal Lancers. With Plans and Map. Royal 8vo. price 21s.

Humboldt's Aspects of Nature. Translated, with the Author's authority, by Mrs. SABINE. New Edition. 16mo. price 6s.: or in 2 vols. 3s. 6d. each, cloth; 2s. 6d. each, sewed.

Humboldt's Cosmos. Translated, with the Author's authority, by Mrs. SABINE. Vols. I. and II. 16mo. Half-a-Crown each, sewed; 3s. 6d. each, cloth: or in post 8vo. 12s. 6d. each, cloth. Vol. III. post 8vo. 12s. 6d. cloth: or in 16mo. Part I. 2s. 6d. sewed, 3s. 6d. cloth; and Part II. 3s. sewed, 4s. cloth.

Humphreys.—Sentiments and Similes of Shakspeare: A Classified Selection of Similes, Definitions, Descriptions, and other remarkable Passages in Shakspeare's Plays and Poems. With an elaborately illuminated border in the characteristic style of the Elizabethan Period, massive carved covers, and other Embellishments, designed and executed by H. N. HUMPHREYS. Square post 8vo. price 21s.

Hunt.—Researches on Lig Chemical Relations; embrac a Consideration of all the Photograph By ROBERT HUNT, F.R.S., Physics in the Metropolitan Science. Second Edition, th vised; with extensive Additi and Woodcuts. 8vo. price 10s

Jameson. — A Commonplace Thoughts, Memories, and Fanc and Selected. Part I. Ethics an Part II. Literature and Art. JAMESON. With Etchings and gravings. Square crown 8vo. [

Mrs. Jameson's Legends of and Martyrs. Forming the F *Sacred and Legendary Art.* Seco with numerous Woodcuts, and by the Author. Square crown 8

Mrs. Jameson's Legends of the Orders, as represented in the Forming the Second Series of *Legendary Art.* Second Editio and enlarged; with 11 Etchi Author, and 88 Woodcuts. Sq 8vo. price 28s.

Mrs. Jameson's Legends of the as represented in the Fine Arts. the Third Series of *Sacred and Art.* With 55 Drawings by the A 152 Wood Engravings. Square price 28s.

Lord Jeffrey's Contributions Edinburgh Review. A New Ed plete in One Volume, with a P graved by Henry Robinson, and View of Craigcrook engraved by Square crown 8vo. 21s. cloth; or

*** Also a LIBRARY EDIT vols. 8vo. price 42s.

Bishop Jeremy Taylor's Entire With Life by Bishop HEBER. R corrected by the Rev. CHARLES PA Fellow of Oriel College, Oxfor complete in Ten Volumes 8vo. pri Guinea each.

Jesse. — Russia and the W Captain JESSE (late Unattached) of Murray's *Handbook for Russia*, a Plan of the Town and Harbour tapool, shewing the Batteries proaches. Crown 8vo. price 2s. 6

Dr. Latham on Diseases of the Heart. Lectures on Subjects connected with Clinical Medicine: Diseases of the Heart. By P. M. LATHAM, M.D., Physician Extraordinary to the Queen. New Edition. 2 vols. 12mo. price 16s.

Mrs. R. Lee's Elements of Natural History; or, First Principles of Zoology: Comprising the Principles of Classification, interspersed with amusing and instructive Accounts of the most remarkable Animals. New Edition, enlarged, with numerous additional Woodcuts. Fcp. 8vo. price 7s. 6d.

L. E. L.—The Poetical Works of Letitia Elizabeth Landon; comprising the *Improvisatrice*, the *Venetian Bracelet*, the *Golden Violet*, the *Troubadour*, and Poetical Remains. New Edition; with 2 Vignettes by R. Doyle. 2 vols. 16mo. 10s. cloth; morocco, 21s.

Letters on Happiness, addressed to a Friend. By the Author of *Letters to My Unknown Friends*, &c. Fcp. 8vo. price 6s.

Letters to my Unknown Friends. By a LADY, Author of *Letters on Happiness*. Fourth *and cheaper* Edition. Fcp. 8vo. price 5s.

Lindley.—The Theory of Horticulture; Or, an Attempt to explain the principal Operations of Gardening upon Physiological Principles. By John Lindley, Ph.D. F.R.S. New Edition, revised and improved; with Wood Engravings. 8vo. [*In the press.*

Dr. John Lindley's Introduction to Botany. New Edition, with Corrections and copious Additions. 2 vols. 8vo. with Six Plates and numerous Woodcuts, price 24s.

Linwood.—Anthologia Oxoniensis, sive Florilegium e lusibus poeticis diversorum Oxoniensium Græcis et Latinis decerptum. Curante GULIELMO LINWOOD, M.A. Ædis Christi Alummo. 8vo. price 14s.

Dr. Little on Deformities.—On the Nature and Treatment of Deformities of the Human Frame. By W. J. LITTLE, M.D., Physician to the London Hospital, Founder of the Royal Orthopædic Hospital, &c. With 160 Woodcuts and Diagrams. 8vo. price 15s.

tton.—The Church of Christ, in its dea, Attributes, and Ministry: With a articular Reference to the Controversy on he Subject between Romanists and Proestants. By the Rev. EDWARD ARTHUR ITTON, M.A., Vice-Principal of St. Edmund all, Oxford. 8vo. price 16s.

ch.—A Practical Legal Guide for ailors and Merchants during War: Comrising Blockade, Captors, Cartel, Colours, Contraband, Droits of Admiralty, Flag hare, Freight, Head Money, Joint Capture, Neutrals and Neutral Territory, Prizes, Recapture of Property of Ally, Rescue, Right of Visit and Search, Salvage, Derelict, Trading with the Enemy, Orders in Council, &c., Prize Act, Proclamation as to Colours. With Appendices containing the Orders in Council and other Official Documents elating to the Present War. By WILLIAM ADAM LOCH, of the Hon. Society of Linoln's Inn. 8vo. price 9s. 6d.

rimer's (C.) Letters to a Young Master Mariner on some Subjects connected with his calling. New Edition. Fcp. 8vo. 5s. 6d.

udon's Self-Instruction for Young Gardeners, Foresters, Bailiffs, Land Stewards, nd Farmers; in Arithmetic, Book-keeping, Geometry, Mensuration, Practical Trigonometry, Mechanics, Land-Surveying, Levelling, Planning and Mapping, Architectural Drawng, and Isometrical Projection and Perspective: With Examples shewing their applications to Horticulture and Agricultural Purposes; a Memoir, Portrait, and Woodcuts. 8vo. price 7s. 6d.

udon's Encyclopædia of Gardening; omprising the Theory and Practice of Horiculture, Floriculture, Arboriculture, and Landscape Gardening: Including all the latest improvements; a General History of Gardening in all Countries; a Statistical View of its Present State; and Suggestions for its Future Progress in the British Isles. With many hundred Woodcuts. New Edition, corrected and improved by Mrs. LOUDON. 8vo. price 50s.

udon's Encyclopædia of Trees and Shrubs; or, the *Arboretum et Fruticetum ritannicum* abridged: Containing the Hardy rees and Shrubs of Great Britain, Native nd Foreign, Scientifically and Popularly Described; with their Propagation, Culture, nd Uses in the Arts; and with Engravings f nearly all the Species. Adapted for the se of Nurserymen, Gardeners, and Foresters. With about 2,000 Woodcuts. 8vo. price 50s.

London's Encyclopædia of Agriculture comprising the Theory and Practice of th Valuation, Transfer, Laying-out, Improve ment, and Management of Landed Property and of the Cultivation and Economy of th Animal and Vegetable Productions of Agri culture; Including all the latest Improv ments, a general History of Agriculture i all Countries, a Statistical View of its presen State, and Suggestions for its future progres in the British Isles. New Edition; wit 1,100 Woodcuts. 8vo. price 50s.

London's Encyclopædia of Plants, in cluding all which are now found in, o have been introduced into, Great Britain Giving their Natural History, accompanie by such descriptions, engraved figures, an elementary details, as may enable a beginner who is a mere English reader, to discove the name of every Plant which he may fin in flower, and acquire all the informatio respecting it which is useful and interesting New Edition, corrected throughout an brought down to the year 1855, by Mrs LOUDON and GEORGE DON, Esq., F.L.S. &c. 8vo. [*In the Spring*.

London's Encyclopædia of Cottage Farm, and Villa Architecture and Furniture containing numerous Designs, from the Vill to the Cottage and the Farm, including Fa Houses, Farmeries, and other Agricultu Buildings; Country Inns, Public Houses and Parochial Schools; with the requisit Fittings-up, Fixtures, and Furniture, an appropriate Offices, Gardens, and Garde Scenery: Each Design accompanied b Analytical and Critical Remarks. Ne Edition, edited by Mrs. LOUDON; with mor than 2,000 Woodcuts. 8vo. price 63s.

London's Hortus Britannicus; or, Cata logue of all the Plants indigenous to, culti vated in, or introduced into Britain. A entirely New Edition, corrected throughout With a Supplement, including all the Nev Plants, and a New General Index to th whole Work. Edited by Mrs. LOUDON assisted by W. H. BAXTER and DAVI WOOSTER. 8vo. price 31s. 6d.—The SUP PLEMENT separately, price 14s.

Mrs. London's Amateur Gardener' Calendar: Being a Monthly Guide as t what should be avoided as well as wha should be done, in a Garden in each Month with plain Rules *how to do* what is requisite Directions for Laying Out and Plantin Kitchen and Flower Gardens, Pleasure Grounds, and Shrubberies: and a shor Account, in each Month, of the Quadrupeds Birds, and Insects then most injurious t Gardens. 16mo. with Woodcuts, price 7s.6d

Mrs. Loudon's Lady's Country Companion; or, How to enjoy a Country Life Rationally. Fourth Edition; with Plates and Wood Engravings. Fcp. 8vo. price 5s.

Low.—A Treatise on the Domesticated Animals of the British Islands: Comprehending the Natural and Economical History of Species and Varieties; the Description of the Properties of external Form; and Observations on the Principles and Practice of Breeding. By D. Low, Esq., F.R.S.E. With Wood Engravings. 8vo. price 25s.

Low.—Elements of Practical Agriculture; comprehending the Cultivation of Plants, the Husbandry of the Domestic Animals, and the Economy of the Farm. By D Low, Esq. F.R.S.E. New Edition; with 200 Woodcuts. 8vo. price 21s.

Macaulay.—Speeches of the Right Hon. T. B. Macaulay, M.P. Corrected by Himself. 8vo. price 12s.

Macaulay.—The History of England from the Accession of James II. By Thomas Babington Macaulay. New Edition. Vols. I. and II. 8vo. price 32s.

Mr. Macaulay's Critical and Historical Essays contributed to The Edinburgh Review. Four Editions, as follows:—

1. Library Edition (the *Seventh*), in 3 vols. 8vo. price 36s.
2. Complete in One Volume, with Portrait and Vignette. Square crown 8vo. price 21s. cloth; or 30s. calf.
3. A New Edition, in 3 vols. fcp. 8vo. price 21s.
4. People's Edition, in 2 vols. crown 8vo. price 8s. cloth.

Macaulay.—Lays of Ancient Rome, with Ivry and the Armada. By Thomas Babington Macaulay. New Edition. 16mo. price 4s. 6d. cloth; or 10s. 6d. bound in morocco.

Mr. Macaulay's Lays of Ancient Rome. With numerous Illustrations, Original and from the Antique, drawn on Wood by George Scarf, Jun., and engraved by Samuel Williams. New Edition. Fcp. 4to. price 21s. boards; or 42s. bound in morocco.

Macdonald.—Villa Verocchio; or, the Youth of Leonardo da Vinci: A Tale. By the late Diana Louisa Macdonald. F 8 rice 6s.

Mac
E
E
R
an
an
R
th
Al
G
F.
in
pc

Sir
la
E
E
8v

Sir

et's Conversations on Chemis- ich the Elements of that Science iarly explained and illustrated by ats. New Edition, enlarged and , 2 vols. fcp. 8vo. price 14s.

et's Conversations on Natural y, in which the Elements of ce are familiarly explained. New enlarged and corrected; with 23 Fcp. 8vo. price 10s. 6d.

et's Conversations on Political , in which the Elements of that are familiarly explained. New Fcp. 8vo. price 7s. 6d.

cet's Conversations on Vege- siology; comprehending the Ele- Botany, with their Application ulture. New Edition; with 4 Fcp. 8vo. price 9s.

cet's Conversations on Land ter. New Edition, revised and .; with a coloured Map, shewing parative Altitude of Mountains. . price 5s. 6d.

u.—Church History in England: Sketch of the History of the Church nd from the Earliest Times to the of the Reformation. By the Rev. MARTINEAU, M.A. late Fellow of ollege, Cambridge. 12mo. price 6s.

s Biographical Treasury; con- of Memoirs, Sketches, and brief of above 12,000 Eminent Persons of s and Nations, from the Earliest History; forming a new and com- ctionary of Universal Biography. hth Edition, revised throughout, ght down to the close of the year cp. 8vo. 10s. cloth; bound in roan, f lettered, 12s. 6d.

s Historical Treasury; com- General Introductory Outline of l History, Ancient and Modern, ries of separate Histories of every Nation that exists; their Rise, , and Present Condition, the Moral l Character of their respective in- , their Religion, Manners and Cus- &c. New Edition; revised through- brought down to the Present Time. , 10s. cloth; roan, 12s.; calf, 12s. 6d.

Maunder's Scientific and Literary ury: A new and popular Encyclop Science and the Belles-Lettres; in all Branches of Science, and every connected with Literature and Art. Edition. Fcp. 8vo. price 10s. cloth; in roan, 12s.; calf lettered, 12s. 6d.

Maunder's Treasury of Natural Hi Or, a Popular Dictionary of An Nature: In which the Zoological Cha istics that distinguish the different C Genera, and Species, are combined variety of interesting Information illus of the Habits, Instincts, and Genera nomy of the Animal Kingdom. Wi Woodcuts. New Edition. Fcp. 8vo 10s. cloth; roan, 12s.; calf, 12s. 6d.

Maunder's Treasury of Knowledge Library of Reference. Comprising a lish Dictionary and Grammar, an Un Gazetteer, a Classical Dictionary, a C logy, a Law Dictionary, a Synopsis Peerage, numerous useful Tables, &c Twentieth Edition, carefully revise corrected throughout: With some Add Fcp. 8vo. price 10s. cloth; bound i 12s.; calf lettered, 12s. 6d.

Merivale.—A History of the Ro under the Empire. By the Rev. Ch MERIVALE, B.D., late Fellow of St. College, Cambridge. Vols. I. and I price 28s.; and Vol. III. price 14s.

Merivale.—The Fall of the Roma public: A Short History of the Las tury of the Commonwealth. By th CHARLES MERIVALE, B.D., late Fel St. John's College, Cambridge. price 7s. 6d.

Merivale.—An Account of the Lif Letters of Cicero. Translated fro German of Abeken; and edited by th CHARLES MERIVALE, B.D. 12mo. 9

Milner.—The Baltic; Its Gates, S and Cities: With a Notice of the Sea, &c. By the Rev. T. MILNER, F.R.G.S. Post 8vo. [*Just re*

Milner's History of the Church of C With Additions by the late Rev. MILNER, D.D., F.R.S. A New E revised, with additional Notes by th T. GRANTHAM, B.D. 4 vols. 8vo. pri

Montgomery.—Memoirs of the Lif Writings of James Montgomery: Inc Selections from his Correspondence Conversations. By JOHN HOLLAN JAMES EVERETT. [*In the p*

James Montgomery's Poetical Works: Collective Edition; with the Author's Autobiographical Prefaces, complete in One Volume; with Portrait and Vignette. Square crown 8vo. price 10s. 6d. cloth; morocco, 21s.—Or, in 4 vols. fcp. 8vo. with Portrait, and 7 other Plates, price 20s. cloth; morocco, 36s.

James Montgomery's Original Hymns for Public, Social, and Private Devotion. 18mo. price 5s. 6d.

Moore.—Man and his Motives. By GEORGE MOORE, M.D., Member of the Royal College of Physicians. *Third* and cheaper *Edition.* Fcp. 8vo. price 6s.

Moore.—The Power of the Soul over the Body, considered in relation to Health and Morals. By GEORGE MOORE, M.D., Member of the Royal College of Physicians. *Fifth* and cheaper *Edition.* Fcp. 8vo. price 6s.

Moore.—The Use of the Body in relation to the Mind. By GEORGE MOORE, M.D., Member of the Royal College of Physicians. *Third* and cheaper *Edition.* Fcp. 8vo. 6s.

Moore.—Health, Disease, and Remedy, familiarly and practically considered in a few of their relations to the Blood. By GEORGE MOORE, M.D., Post 8vo. 7s. 6d.

Moore.—Memoirs, Journal, and Correspondence of Thomas Moore. Edited by the Right Hon. LORD JOHN RUSSELL, M.P. With Portraits and Vignette Illustrations. Vols. I. to VI. post 8vo. price 10s. 6d. each.

*** Vols. VII. and VIII., completing the work, are *nearly ready.*

Thomas Moore's Poetical Works. Containing the Author's recent Introduction and Notes. Complete in One Volume; with a Portrait, and a View of Sloperton Cottage. Medium 8vo. price 21s. cloth; morocco, 42s. Or in 10 vols. fcp. 8vo. with Portrait, and 19 Plates, price 35s.

Moore.—Songs, Ballads, and Sacred Songs. By THOMAS MOORE, Author of *Lalla Rookh,* &c. First collected Edition, with Vignette by R. Doyle. 16mo. price 5s. cloth; 12s. 6d. bound in morocco.

Moore's Irish Melodies. New Edition, with the Autobiographical Preface from the Collective Edition of Mr. Moore's Poetical Works, and a Vignette Title by D. Maclise, R.A. 16mo. price 5s. cloth; 12s. 6d. bound in morocco.

Moore's Irish Melodies. Illust[rated by] D. Maclise, R.A. New and cheaper [Edition,] with 161 Designs, and the who[le of the] Letterpress engraved on Steel, [by F. P.] Becker. Super-royal 8vo. pric[e] boards; bound in morocco, £2. 1[...]

The Original Edition of th[...] in imperial 8vo. price 63s. boards; [morocco,] £4. 14s. 6d.; proofs, £6. 6s. [...] *still be had.*

Moore's Lalla Rookh: An [Oriental] Romance. New Edition, with t[he Auto]biographical Preface from the [Collective] Edition of Mr. Moore's Poetical W[orks, and] a Vignette Title by D. Maclise, R.[A. 16mo.] price 5s. cloth; 12s. 6d. bound in [morocco.]

Moore's Lalla Rookh: An [Oriental] Romance. With 13 highly-finis[hed Steel] Plates from Designs by Corbould, [Meadows,] and Stephanoff, engraved under [the super]intendence of the late Charles Hea[th. New] Edition. Square crown 8vo. [price 15s.] cloth; morocco, 28s.

*** A few copies of the Origin[al Edition,] in royal 8vo. price One Guinea, st[ill remain.]

Morton's Manual of Pharmacy [for the] Student of Veterinary Medicine; [contain]ing the Substances employed at t[he Royal] Veterinary College, with an attem[pt at their] Classification; and the Pharma[copoeia of] that Institution. *Fifth Edition* (12[mo.]) 8vo. price 10s.

Moseley.—The Mechanical Prin[ciples of] Engineering and Architecture. [By the Rev.] H. MOSELEY, M.A., F.R.S., Pr[ofessor of] Natural Philosophy and Astro[nomy in] King's College, London. 8vo. p[rice ...]

Mure.—A Critical History of t[he Lan]guage and Literature of Ancie[nt Greece.] By WILLIAM MURE, M.P. of [Caldwell.] Vols. I. to III. 8vo. price 36s.—[Vol. IV.] price 15s.

"We hail with great satisfactio[n the con]tinuation of a work so eminently c[alculated to] promote the knowledge of Greek clas[sical litera]ture, and to increase the taste for it [among] educated classes. Singularly felicitou[s in] the salient points in the character o[f the diffe]rent tribes of the ancient Greeks, [and of the] different branches and periods of thei[r literature,] and in conveying to his readers,—ev[en those] unable to follow the train of his discu[ssion in the] originals,—clear and vivid ideas on t[hese points,] Mr. Mure has in the present volume [treated in a] most effective manner the general [history of] Greek literature during the Attic p[eriod] from the usurpation of the suprem[acy of] Athens by Pisistratus, A. 560 B.C. to t[he death of] Alexander the Great, 323 B.C." Jo[...]

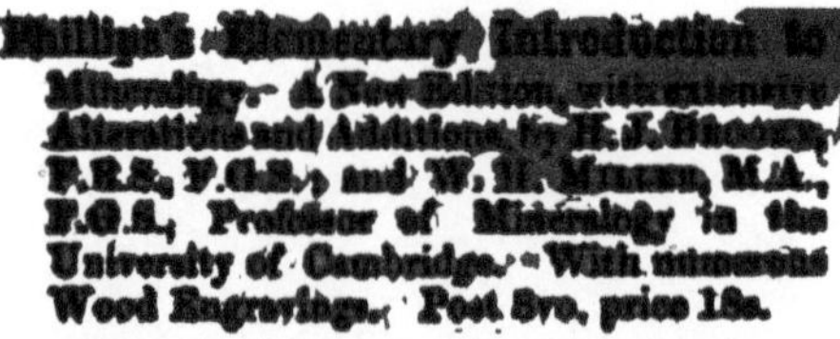

Phillips's Elementary Introduction to Mineralogy. A New Edition, with extensive Alterations and Additions, by H. J. Brooke, F.R.S., F.G.S.; and W. H. Miller, M.A., F.G.S., Professor of Mineralogy in the University of Cambridge. With numerous Wood Engravings. Post 8vo. price 18s.

Phillips.—Figures and Descriptions of the Palæozoic Fossils of Cornwall, Devon, and West Somerset; observed in the course of the Ordnance Geological Survey of that District. By John Phillips, F.R.S. F.G.S. &c. 8vo. with 60 Plates, price 9s.

Captain Portlock's Report on the Geology of the County of Londonderry, and of Parts of Tyrone and Fermanagh, examined and described under the Authority of the Master-General and Board of Ordnance. 8vo. with 48 Plates, price 24s.

Power's Sketches in New Zealand, with Pen and Pencil. From a Journal kept in that Country, from July 1846 to June 1848. With Plates and Woodcuts. Post 8vo. 12s.

Psychological Inquiries, in a Series of Essays intended to illustrate the Influence of the Physical Organisation on the Mental Faculties. Fcp. 8vo. price 5s.

Pulman's Vade-mecum of Fly-Fishing for Trout; being a complete Practical Treatise on that Branch of the Art of Angling; with plain and copious Instructions for the Manufacture of Artificial Flies. Third Edition, with Woodcuts. Fcp. 8vo. price 6s.

Pycroft's Course of English Reading, adapted to every Taste and Capacity: With Literary Anecdotes. New and cheaper Edition. Fcp. 8vo. price 5s.

Dr. Reece's Medical Guide; for the Use of the Clergy, Heads of Families, Schools, and Junior Medical Practitioners: Comprising a complete Modern Dispensatory, and a Practical Treatise on the distinguishing Symptoms, Causes, Prevention, Cure and Palliation of the Diseases incident to the Human Frame. With the latest Discoveries in the different departments of the Healing Art, Materia Medica, &c. Seventeenth Edition, corrected and enlarged by the Author's Son, Dr. H. Reece, M.R.C.S. &c. 8vo. price 12s.

...on's Debater: A Series of complete ...ssays, Outlines of Debates, and Questions ... Discussion; with ample References ... the best Sources of Information on ... particular Topic. New Edition. Fcp. ... price 6s.

...rs of Rachel Lady Russell. A New ...tion, including several unpublished Let... together with those edited by Miss ...ry. With Portraits, Vignettes, and ...simile. 2 vols. post 8vo. price 15s.

...Life of William Lord Russell. By ... Right Hon. Lord John Russell, M.P. ... Fourth Edition, complete in One ...ume; with a Portrait engraved on Steel ... Bellin, from the original by Sir Peter ... at Woburn Abbey. Post 8vo. 10s. 6d.

...ohn (the Hon. F.) — Rambles in ...ch of Sport, in Germany, France, Italy, ... Russia. By the Honourable Ferdinand ...John. With Four coloured Plates. ... 8vo. price 9s. 6d.

...n (H.) — The Indian Archipelago; ... History and Present State. By Horace ...John, Author of *The British Conquests in* ... &c. 2 vols. post 8vo. price 21s.

...n (J. A.) — There and Back Again ...rch of Beauty. By James Augustus ...John, Author of *Isis, an Egyptian Pil...ge*, &c. 2 vols. post 8vo. price 21s.

... John's Work on Egypt. ... An Egyptian Pilgrimage. By James ...gustus St John. 2 vols. post 8vo. 21s.

...ints our Example. By the Author ...*ters to My Unknown Friends*, &c. Fcp. ... price 7s.

...tz. — History of Greece, from the ...iest Times to the Taking of Corinth by ... Romans, B.C. 146, mainly based upon ...op Thirlwall's History of Greece. By ... Leonhard Schmitz, F.R.S.E., Rector ...he High School of Edinburgh. New ...ion. 12mo. price 7s. 6d.

... Eighteenpence.

The Sermon on the Mount. Print... Silver; with Picture Subjects, nu... Landscape and Illustrative Vignett... Illuminated Borders in Gold and C... designed by M. Lepelle du Bois-G... Square 18mo. price in ornamental ... One Guinea; or 31s. 6d. bound in m...

Self-Denial the Preparation for E... By the Author of *Letters to my Un... Friends*, &c. Fcp. 8vo. price 2s. 6d.

Sharp's New British Gazetteer, or ... graphical Dictionary of the British ... and Narrow Seas: Comprising conci... scriptions of about Sixty Thousand ... Seats, Natural Features, and Objects o... founded on the best Authorities; fu... ticulars of the Boundaries, Registered ... tors, &c. of the Parliamentary Bor... with a reference under every name ... Sheet of the Ordnance Survey, as far ... pleted; and an Appendix, contai... General View of the Resources of the ... Kingdom, a Short Chronology, a... Abstract of certain Results of the Ce... 1851. 2 vols. 8vo. price £2. 16s.

Sewell. — Amy Herbert. By a ... Edited by the Rev. William Sewel... Fellow and Tutor of Exeter College, ... New Edition. Fcp. 8vo. price 6s.

Sewell. — The Earl's Daughter. ... Author of *Amy Herbert*. Edited by t... W. Sewell, B.D. 2 vols. fcp. 8vo. ...

Sewell. — Gertrude: A Tale. B... Author of *Amy Herbert*. Edited by t... W. Sewell, B.D. New Edition. ... 8vo. price 6s.

Sewell. — Laneton Parsonage: A T... Children, on the Practical Use of a ... of the Church Catechism. By the ... of *Amy Herbert*. Edited by the R... Sewell, B.D. New Edition. 3 vo... 8vo. price 16s.

Sewell. — Margaret Percival. By the Author of *Amy Herbert*. Edited by the Rev. W. Sewell, B.D. New Edition. 2 vols. fcp. 8vo. price 12s.

By the same Author,

Katharine Ashton. New Edition. 2 vols. fcp. 8vo. price 12s.

The Experience of Life. New Edition. Fcp. 8vo. price 7s. 6d.

Readings for a Month preparatory to Confirmation: Compiled from the Works of Writers of the Early and of the English Church. Fcp. 8vo. price 5s. 6d.

Readings for Every Day in Lent: Compiled from the Writings of Bishop Jeremy Taylor. Fcp. 8vo. price 5s.

The Family Shakspeare; in which nothing is *added* to the Original Text, but those words and expressions are *omitted* which cannot with propriety be read aloud. By T. Bowdler, Esq. F.R.S. New Edition, in Volumes for the Pocket; with 36 Wood Engravings, from Designs by Smirke, Howard, and other Artists. 6 vols. fcp. 8vo. price 30s.

*** Also a Library Edition, in One Volume, medium 8vo. price 21s.

Short Whist; Its Rise, Progress, and Laws: With Observations to make any one a Whist Player. Containing also the Laws of Piquet, Cassino, Ecarté, Cribbage, Backgammon. By Major A * * * * *. New Edition; to which are added, Precepts for Tyros, by Mrs. B * * * *. Fcp. 8vo. 3s.

Sinclair. — The Journey of Life. By Catherine Sinclair, Author of *The Business of Life* (2 vols. fcp. 8vo. price 10s.) New Edition, corrected and enlarged. Fcp. 8vo. price 5s.

Sir Roger de Coverley. From The Spectator. With Notes and Illustrations, by W. Henry Wills; and Twelve fine Wood Engravings, by John Thompson, from Designs by Frederick Tayler. Crown 8vo. price 15s. boards; or 27s. bound in morocco.—A Cheap Edition, without Woodcuts, in 16mo. price One Shilling.

hey's Commonplace Books. Containing—1. Choice Passages: With Collections for the History of Manners and Literature in England; 2. Special Collections on various Historical and Theological Subjects; 3. Analytical Readings in various branches of Literature; and 4. Original Memoranda, Literary and Miscellaneous. Edited by the Rev. J. W. Warter, B.D. vols. square crown 8vo. price £3. 18s.

Each Commonplace Book, complete in itself, may be had separately as follows:—

First Series—CHOICE PASSAGES, &c. 18s.

Second Series—SPECIAL COLLECTIONS. 18s.

Third Series—ANALYTICAL READINGS. 21s.

Fourth Series—ORIGINAL MEMORANDA, &c. 21s.

Southey's The Doctor &c. Complete in One Volume. Edited by the Rev. J. W. Warter, B.D. With Portrait, Vignette, Bust, and coloured Plate. New Edition. Square crown 8vo. price 21s.

Robert Southey's Complete Poetical Works; containing all the Author's last Introductions and Notes. Complete in One Volume, with Portrait and Vignette. Medium 8vo. price 21s. cloth; 42s. bound in morocco. Or in 10 vols. fcp. 8vo. with Portrait and Plates, price 35s.

Select Works of the British Poets; from Chaucer to Lovelace, inclusive. With Biographical Sketches by the late Robert Southey. Medium 8vo. price 30s.

Stephen.—Lectures on the History of France. By the Right Hon. Sir James Stephen, K.C.B. LL.D. Professor of Modern History in the University of Cambridge. Second Edition. 2 vols. 8vo. price 24s.

Stephen.—Essays in Ecclesiastical Biography; from The Edinburgh Review. By the Right Hon. Sir James Stephen, K.C.B. LL.D. Third Edition. 2 vols. 8vo. 24s.

Stonehenge.—The Greyhound: Being a Treatise on the Art of Breeding, Rearing, and Training Greyhounds for Public Running; their Diseases and Treatment: Containing also, Rules for the Management of Coursing Meetings, and for the Decision of Courses. By Stonehenge. With numerous Portraits of Greyhounds, &c. engraved on Wood, and a Frontispiece engraved on Steel. Square crown 8vo. price 21s.

Stow.—The Training System, the Moral Training School, and the Normal Seminary for preparing School-Trainers and Governesses. By David Stow, Esq., Honorary Secretary to the Glasgow Normal Free Seminary. Tenth Edition; with Plates and Woodcuts. Post 8vo. price 6s.

Dr. Sutherland's Journal of a Voyage in Baffin's Bay and Barrow's Straits, in the Years 1850 and 1851, performed by H.M. Ships *Lady Franklin* and *Sophia*, under the command of Mr. William Penny, in search of the missing Crews of H.M. Ships *Erebus* and *Terror*. With Charts and Illustrations. 2 vols. post 8vo. price 27s.

Tate.—On the Strength of Materials; Containing various original and useful Formulæ, specially applied to Tubular Bridges, Wrought Iron and Cast Iron Beams, &c. By Thomas Tate, F.R.A.S. 8vo. price 5s. 6d.

Taylor.—Loyola: And Jesuitism in its Rudiments. By Isaac Taylor. Post 8vo. with Medallion, price 10s. 6d.

Taylor.—Wesley and Methodism. By Isaac Taylor. Post 8vo. with a Portrait, price 10s. 6d.

Theologia Germanica: Which setteth forth many fair lineaments of Divine Truth, and saith very lofty and lovely things touching a Perfect Life. Translated by Susanna Winkworth. With a Preface by the Rev. Charles Kingsley; and a Letter by Chevalier Bunsen. Fcp. 8vo. price 5s.

Thirlwall.—The History of Greece. By the Right Rev. the Lord Bishop of St. David's (the Rev. Connop Thirlwall). An improved Library Edition; with Maps. 8 vols. 8vo. price £4. 16s.

*** Also, an Edition in 8 vols. fcp. 8vo. with Vignette Titles, price 28s.

Thomson (The Rev. W.)—An Outline of the Laws of Thought: Being a Treatise on Pure and Applied Logic. By the Rev. W. Thomson, M.A. Fellow and Tutor of Queen's College, Oxford. Third Edition, enlarged. Fcp. 8vo. price 7s. 6d.

Thomson's Tables of Interest, at Three, Four, Four-and-a-Half, and Five per Cent., from One Pound to Ten Thousand, and from 1 to 365 Days, in a regular progression of single Days; with Interest at all the above Rates, from One to Twelve Months, and from One to Ten Years. Also, numerous other Tables of Exchanges, Time, and Discounts. New Edition. 12mo. price 8s.

Thomson's Seasons. Edited by Bolton Corney, Esq. Illustrated with Seventy-seven fine Wood Engravings from Designs by Members of the Etching Club. Square crown 8vo. price 21s. cloth; or, 36s. bound in morocco.

Thornton.—Zohrab; or, a Midsummer Day's Dream: And other Poems. By William Thomas Thornton. Fcp. 8vo. price 2s. 6d.

The Thumb Bible; or, Verbum Sempiternum. By J. Taylor. Being an Epitome of the Old and New Testaments in English Verse. Reprinted from the Edition of 1693; bound and clasped. 64mo. 1s. 6d.

Todd (Charles).—A Series of Tables of the Area and Circumference of Circles; the Solidity and Superficies of Spheres; the Area and Length of the Diagonal of Squares; and the Specific Gravity of Bodies, &c.: To which is added, an Explanation of the Author's Method of Calculating these Tables. Intended as a Facility to Engineers, Surveyors, Architects, Mechanics, and Artisans in general. By Charles Todd, Engineer. The Second Edition, improved and extended. Post 8vo. price 6s.

Townsend.—The Lives of Twelve Eminent Judges of the Last and of the Present Century. By W. C. Townsend, Esq., M.A., Q.C. 2 vols. 8vo. price 28s.

Townsend.—Modern State Trials revised and illustrated with Essays and Notes. By W. C. Townsend, Esq. M.A. Q.C. 2 vols. 8vo. price 30s.

Sharon Turner's Sacred History of the World, attempted to be Philosophically considered, in a Series of Letters to a Son. New Edition, edited by the Author's Son, the Rev. S. Turner. 3 vols. post 8vo. price 31s. 6d.

Sharon
during
Reign
Access
revise
8vo. p

Sharon
Sense
Norm
revised
8vo. p

Dr. Tur
Fresh-
A New
by Jos
and 12

Dr. Ure
tures,
sition
Fourth
rected
compr
provem
Time
Most
written
added.
vols. 8

Watert
chiefly
With
Views
Editio

Sep
Vol. I

Alaric V
other
Line E
work
Engra
boards
Impre

Webste
Domes
jects a
House
Domes
ing, V
scripti
with t
Servan
1,000

render them worthy of preservation.

List of the VOLUMES *already published.*

I.	MACAULAY's ESSAYS on WARREN HASTINGS and LORD CLIVE......	2/6
II.	———— ESSAYS on PITT and CHATHAM, RANKE and GLADSTONE,	2/6
III.	LAING's RESIDENCE in NORWAY	2/6
IV.	PFEIFFER's VOYAGE ROUND the WORLD..........................	2/6
V.	EOTHEN, TRACES of TRAVEL from the EAST	2/6
VI.	MACAULAY's ESSAYS on ADDISON, WALPOLE, and LORD BACON	2/6
VII.	HUC's TRAVELS in TARTARY, &c..............................	2/6
VIII.	THOMAS HOLCROFT's MEMOIRS..................................	2/6
IX.	WERNE's AFRICAN WANDERINGS................................	2/6
X.	MRS. JAMESON's SKETCHES in CANADA	2/6
XI.	JERRMANN's PICTURES from ST. PETERSBURG....................	2/6
XII.	THE REV. G. R. GLEIG's LEIPSIC CAMPAIGN	2/6
XIII.	HUGHES's AUSTRALIAN COLONIES	2/6
XIV.	SIR EDWARD SEAWARD's NARRATIVE	2/6
XV.	ALEXANDRE DUMAS' MEMOIRS of a MAITRE-D'ARMES	2/6
XVI.	OUR COAL FIELDS and OUR COAL PITS..........................	2/6
XVII.	M'CULLOCH's LONDON and GIRONIÈRE's PHILIPPINES.............	2/6
XVIII.	SIR ROGER DE COVERLEY and SOUTHEY's LOVE STORY	2/6
XIX.	LORD CARLISLE's LECTURES and ADDRESSES; and JEFFREY's ESSAYS on SWIFT and RICHARDSON .... }	2/6
XX.	HOPE's BIBLE in BRITTANY and CHASE in BRITTANY	2/6
XXI.	THE ELECTRIC TELEGRAPH and NATURAL HISTORY of CREATION,	2/6
XXII.	MEMOIR of DUKE of WELLINGTON and LIFE of MARSHAL TURENNE,	2/6
XXIII.	RANKE's FERDINAND and MAXIMILIAN & TURKEY and CHRISTENDOM	2/6
XXIV.	BARROW's CONTINENTAL TOUR and FERGUSON's SWISS MEN and SWISS MOUNTAINS .. }	2/6
XXV.	SOUVESTRE's ATTIC PHILOSOPHER in PARIS, and WORKING MAN'S CONFESSIONS .... }	2/6
XXVI.	MACAULAY's ESSAYS on LORD BYRON and the COMIC DRAMATISTS and his SPEECHES on PARLIAMENTARY REFORM (1831-32).... }	2/6
XXVII.	SHIRLEY BROOKS's RUSSIANS of the SOUTH; and DR. KEMP'S INDICATIONS of INSTINCT .... }	2/6
XXVIII.	LANMAN'S ADVENTURES in the WILDS of NORTH AMERICA.......	2/6
XXIX.	DE CUSTINE's RUSSIA, abridged.............................	2/6
XXX.	SELECTIONS from SYDNEY SMITH's WRITINGS, Vol. I.	2/6
XXXI.	BODENSTEDT and WAGNER's SCHAMYL; and M'CULLOCH's RUSSIA and TURKEY .... }	2/6
XXXII.	LAING's NOTES of a TRAVELLER, First Series	2/6
XXXIII.	DURRIEU'S MOROCCO; and an ESSAY on MORMONISM	2/6

worthy of preservation.

List of the PARTS already published:—

1. Mr. Macaulay's Essay on Warren Hastings.
2. " " Lord Clive.
3. London in the years 1850 and 1851. By J. R. M'Culloch, Esq.
4. Sir Roger De Coverley. From the Spectator.
5. Mr. Macaulay's Two Essays on William Pitt and the Earl of Chatham.
6–7. Laing's Residence in Norway.
8. Mr. Macaulay's Essays on Ranke and Gladstone.
9–10. Ida Pfeiffer's Lady's Voyage round the World.
11–12. Eothen, or Traces of Travel from the East.
13. Mr. Macaulay's Essays on Addison and Walpole.
14–15. Huc's Travels in Tartary, Thibet, and China.
16–17. Thomas Holcroft's Memoirs.
18. The Earl of Carlisle's Lectures and Addresses.
19–20. Werne's African Wanderings.
21–22. Mrs. Jameson's Sketches in Canada.
23. Brittany and the Bible. By I. Hope.
24. The Natural History of Creation. By Dr. L. Kemp.
25. Mr. Macaulay's Essay on Lord Bacon.
26. The Electric Telegraph, etc. By Dr. G. Wilson.
27–28. Jerrmann's Pictures from St. Petersburg.
29–30. The Rev. G. R. Gleig's Leipsic Campaign.
31. Memoir of the Duke of Wellington.
32–33. The Australian Colonies. By William Hughes, F.R.G.S.
34–35. Sir Edward Seaward's Narrative Abridged.
36. Lord Jeffrey's Essays on Swift and Richardson.
37. Ranke's Ferdinand I. and Maximilian II.
38–39. Memoirs of a Maître d'Armes. By Alexandre Dumas.
40. Byron and the Comic Dramatists. By T. B. Macaulay.
41. Marshal Turenne. By the Rev. T. O. Cockayne, M.A.
42–43. Our Coal Fields and Coal Pits.
44. Barrow's Tour on the Continent in MDCCCLII.
45. Swiss Men and Swiss Mountains. By R. Ferguson.
46. Gironière's Philippine Islands.
47. Turkey and Christendom.
48. Confessions of a Working Man. By Emile Souvestre.
49. The Chase in Brittany. By I. Hope.
50. The Love Story from Southey's Doctor.
51. An Attic Philosopher in Paris. By E. Souvestre.
52. Mr. Macaulay's Speeches on Parliamentary Reform.
53. The Russians of the South. By Shirley Brooks.
54. Indications of Instinct. By Dr. Lindley Kemp.
55–56. Lanman's Adventures in the Wilds of North America.
57–59. De Custine's Russia.
60. Durrieu's Morocco.
61–62. Selections from Sydney Smith's Writings.
63. Schamyl, the Chieftain of the Caucasus.
64. Russia and Turkey. By J. R. M'Culloch, Esq.
65–66. Laing's Notes of a Traveller, First Series.
– – Mormonism.

Geography, &

Lady Willoug
Printed, ornam
of the period
New Edition
8vo. price 8
morocco, 18s

Wilmot's Ab
Commentarie
tended for th
comprised in
to his Daugh
and brought
Sir John E
18mo. price

Yonge.—A N
Containing
Writers of
Yonge, B.A

Yonge's New
every Word
authority.
of Eton, We
Charterhous
College, Lon
Third Editi
rected. Po

Youatt.—Th
With a Trea
with numer
Designs by
Longman a
dered.) 8vo

Youatt.—Th
A New Editi
from Design

Zumpt's Gr
guage. Tr
use of Engli
F.R.S.E.:
Corrections
The Third
with an Ind

WILSON AND OGILVY, SKINNER STREET, SNOW H

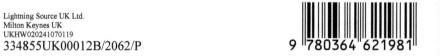

Lightning Source UK Ltd.
Milton Keynes UK
UKHW020241070119
334855UK00012B/2062/P

9 780364 621981